Statistics FROM DATA TO DECISIO..

Second Edition

Vice President and Executive Publisher	Laurie Rosatone
Acquisitions Editor	Joanna Dingle
Project Editor	Ellen Keohane
Editorial Assistant	Beth Pearson
Marketing Manager	Sarah Davis
Production Manager	Dorothy Sinclair
Senior Production Editor	Sandra Dumas
Creative Director	Harry Nolan
Art Director	Jeof Vita
Senior Designer	Kevin Murphy
Senior Media Editor	Melissa Edwards
Media Assistant	Lisa Sabitini
Photo Department Manager	Hilary Newman
Photo Editor	Sarah Wilkin
Cover Photo	Bruno Morandi/Getty Images
Cover Design	Michael Boland
Production Management Services	MPS Limited, A MacMillan Company

This book was typeset in 10/12 Janson at MPS, and printed and bound by Courier/Kendallville. The cover was printed by Courier/Kendallville.

The paper in this book was manufactured by a mill whose forest management programs include sustained yield-harvesting of its timberlands. Sustained yield harvesting principles ensure that the number of trees cut each year does not exceed the amount of new growth.

This book is printed on acid-free paper.

ISBN 13 978-0470-65097-4 (Preview Edition)
ISBN 13 978-0470-45851-8
ISBN 13 978-0470-55994-9

Printed in the United States of America.
10 9 8 7 6 5 4 3 2 1

Statistics FROM DATA TO DECISION

Ann E. Watkins

Richard L. Scheaffer

George W. Cobb

Second Edition

WILEY

John Wiley & Sons, Inc.

About the Authors

Ann E. Watkins is Professor of Mathematics at California State University, Northridge (CSUN). She received her doctorate in education (with an emphasis on statistics) from the University of California, Los Angeles. She is a former president of the Mathematical Association of America (MAA) and a fellow of the American Statistical Association (ASA). Professor Watkins has served as coeditor of *College Mathematics Journal*, as a member of the board of editors of the *American Mathematical Monthly* and the *Journal of Statistics Education*, and as chair of the Advanced Placement Statistics Development Committee. She was selected as the 1994–1995 CSUN Outstanding Professor and won the 1997 CSUN Award for the Advancement of Teaching Effectiveness. Besides the many journal articles that she has written, she is the coauthor of books based on work produced by the Activity-Based Statistics Project (coauthor, Richard Scheaffer) and the American Statistical Association's Quantitative Literacy Project.

Richard L. Scheaffer, Professor Emeritus of Statistics at the University of Florida, received his doctorate in statistics from Florida State University. Professor Scheaffer's research interests are in the areas of sampling and applied probability, especially in their applications to industrial processes. He has coauthored five college-level textbooks. Over the years, much of his effort has been directed toward statistics education at the elementary, secondary, and college levels. He was one of the developers of the American Statistical Association's Quantitative Literacy Project and an author of the *ASA's Guidelines for Assessment and Instruction in Statistics Education (GAISE)* report. Professor Scheaffer directed the task force that developed the Advanced Placement Statistics Program and served as its first chief faculty consultant. He is fellow and past president of the American Statistical Association, from which he received a Founders Award.

George W. Cobb is the Robert L. Rooke Professor of Statistics at Mt. Holyoke College, where he served a three-year term as dean of studies. He received his doctorate in statistics from Harvard University and is an expert in statistics education. He chaired the joint committee on undergraduate statistics of the Mathematical Association of America and the American Statistical Association and is a fellow of the American Statistical Association. He led the Statistical Thinking and Teaching Statistics (STATS) project of the Mathematical Association of America, which helped professors of mathematics learn to teach statistics. He co-authored the ASA's Guidelines for Assessment and Instruction in Statistics Education (GAISE) report, and recently completed a three-year term as vice-president of the American Statistical Association. Over the past two decades, Professor Cobb has frequently served as an expert witness in lawsuits involving alleged employment discrimination.

Acknowledgments

This book is a product of what we have learned from the statisticians and instructors who have been actively involved in helping the introductory statistics course evolve into one that emphasizes student engagement with statistical concepts while incorporating modern statistical practices. This book is written in the spirit of the recommendations from the Mathematical Association of America's STATS project and Focus Group on Statistics, and the American Statistical Association's *Guidelines for Assessment and Instruction in Statistics Education (GAISE)*. We hope that it adequately reflects the wisdom and experience of those with whom we have worked and who have inspired and taught us.

We would like to thank the following reviewers of the second edition of this book, whose comments and suggestions helped us improve the text:

Vittorio Addona, Macalester College

Nazanin Azarnia, Santa Fe College

Narayanaswamy Balakrishnan, McMaster University

Chad Birger, University of Sioux Falls

Thomas R. Boucher, Plymouth State University

Ryan Charest, James Madison University

Smiley Cheng, University of Manitoba

Patti Costello, Eastern Kentucky University

Yixin Fang, Georgia State University

Diane Fisher, University of Louisiana at Lafayette

Robert Gould, University of California, Los Angeles

K. L. D. Gunawardena, University of Wisconsin–Oshkosh

Sat N. Gupta, University of North Carolina at Greensboro

James E. Helmreich, Marist College

Rasul A. Khan, Cleveland State University

Patrick Lang, Idaho State University

Carl Lee, Central Michigan University

Ginny Powell, Georgia Perimeter College, Decatur Campus

Julia Soulakova, University of Nebraska–Lincoln

Robert Stack, Chadron State College

Gail Tudor, Husson University

Diane Van Deusen, Napa Valley College

Timothy M. Walker, Ohio Dominican University

John Wilkins, California State University, Dominguez Hills

Also thanks to Paul Lorczak, Jada Hill, Georgia Mederer, Jon Booze, Susan Herring, Sandra Zirkes, and Ann Ostberg for their work ensuring the accuracy of this textbook and to developmental editor Donald Gecewicz for inspiring us through his intellectual curiosity and insight into what makes a textbook a true medium for learning. The thoroughly professional staff at Wiley has been wonderful to work with. Finally, we would like to thank Carol Noble and her staff, Md. Furqan, Gagan Gusai, Kumar Anuranjan, Neelam Kaul, Dharamraj, Satinder Singh Baveja, and Kailash Sharma, for their skilled production of the text.

Ann Watkins
Dick Scheaffer
George Cobb

Contents

Preface

A Note to Students from the Authors

Whether you talk about income, the prices of goods and services, sports, health, politics, or the weather, data enter your conversations every day. In fact, in this age of information technology, data come at you so rapidly that you can catch only a glimpse of the masses of numbers. Further, you frequently hear statistical terms in the media:

- The *margin of error for this poll* is 3%.
- The *experiment* showed that people lost more weight with the new diet than with the old, but the *difference* wasn't *significant*.
- The *median* starting salary for college graduates in this field remains $40,000 but is more *variable* than for last year's graduates.
- Income and education are *correlated*.

The only way to cope intelligently with this quantitative world and make informed decisions is to gain an understanding of the basic concepts of statistics by working with real data.

What's in *Statistics: From Data to Decision*

We wrote this book for students taking an introductory statistics course that covers all of the standard topics for that course, taught from a modern perspective. Beginning in Chapter 1, you'll be immersed in real problems that can be solved only with statistical methods. You will learn to do the following:

- explore, summarize, and display data
- design surveys and experiments
- use probability to understand random behavior, how randomness is used in designing statistical studies, and why many statistical techniques actually work
- make inferences about the qualities and characteristics of populations by looking at samples taken from those populations
- make inferences about the causes and effects of treatments from designed experiments

We want to underscore the usefulness of statistics for your career as well as in your daily life. For that reason, we sought to enliven the chapters with recent, relevant examples. We chose examples from a wide array of interests, as well as from the many academic majors that require the course, particularly the social, life, and biological sciences. We especially wanted to include many examples (and problems) that show how statistics is used in research.

How *Statistics*: *From Data to Decision* Is Different

The features of this book grow out of the vigorous changes that have been reshaping the practice and teaching of statistics over the last quarter century. All these features help to show you how to go from data to decisions about bigger questions that spring up as you look at data.

Statistical work can rely more on technology than it could a generation ago. Computers and graphing calculators have automated the graphical exploration of data.

They have made the practice of statistics more visual. Statistical techniques are also changing as technology allows statisticians (and you) to shift the emphasis from following recipes for calculations to interpreting the output of the technology. Thus, you can pay more attention to statistical concepts and less to doing the computations. Your instructor has selected this book for you because he or she:

- wants you to learn this modern, data-analytic approach to statistics
- encourages you to be an active participant in the classroom
- wants you to see real data (because, if you have only pretend data, you can only pretend to analyze it)
- believes that statistical analysis must be tailored to the data
- uses graphing calculators or statistical software for data analysis

Throughout this textbook, you will see many graphical displays, lots of real data, activities that support each major topic, output from statistical software, and questions for you to discuss with other students in your class. Also, we have written many engaging practice problems (coded with a P) so that you can be sure you understand the basics before you move on. Work all of the practice problems for the topics that your instructor assigns. These practice problems will lead you into the exercises (coded with an E) that involve practice with the basics in a context that involves more thought-provoking work with real data from many different fields, from agriculture to zoology. Answers to the practice problems and the odd-numbered exercises can be found at the back of the book.

In this textbook, the visuals (both the displays and the photographs) are meant to be more than just pictures. We have taken great care to make sure that the displays are helpful visualizations of the data. We have chosen photographs, including many of actual experiments in various fields, that will help you understand the situation you are analyzing in a way that words cannot.

What You Should Know Before You Start

Because you will be using this book in an introductory statistics course, you aren't required to know anything yet about statistics. You may find that your perseverance in trying to understand what you read will contribute more to your success in statistics than your skill with algebra. However, basic topics from algebra, such as the equation for a line and understanding of slope, will come up throughout the book. Be prepared to review those as you go along, if necessary.

Ann Watkins
Dick Scheaffer
George Cobb

To the Instructor

Statistics: From Data to Decision teaches a modern approach that allows for more emphasis on statistical concepts and data analysis than on following recipes for calculations. The text is intended to be used as a facilitator of dialog and as a reference. We encourage students to be thoughtful as they read. We begin early on—in Chapter 1 in fact—setting up the basics of statistical inference. We want to engage the students in thinking statistically as soon as they can. Doing so makes the discipline of statistics more meaningful. It also gives the students a hint of where the course is heading (inference) and why (statistical inference is a powerful way of investigating the world).

One way to get students to think statistically is to get them to understand how to act statistically: An activity therefore supports each major topic, allowing students to experience the major concepts of statistical thinking before going on to confirm them through further analysis (often involving simulation) and practice with data. *Statistics: From Data to Decision* also gives the design and analysis of both experiments and surveys, which are the basic information-gathering activities of statistics—an honest treatment—rather than forcing them into simplistic models.

New to the Second Edition

We carefully considered reviewer feedback as we worked on the revision of this book. In particular, we decided to change the name of the text to highlight two emphases of the introductory statistics course: data and decision making. An informed citizen understands where data come from and has a feel for data and for how to assess real-life data. Also, familiarity with statistics as a discipline and a mind-set leads to critical thinking and sound decision making.

For the second edition, we significantly reorganized the text. In particular, we made the following changes:

- Chapter 3 (Relationships Between Two Quantitative Variables) was streamlined to focus more clearly on exploring correlation and regression.

- We wanted to bring the most important rules of probability to the foreground, while strengthening coverage of two central ideas, conditional probability and independence. For these reasons, Chapter 5 (Probability Models) was also reorganized.

- We wanted to streamline coverage and to focus on the discrete distributions in Chapter 6 (Probability Distributions). (The continuous distributions are introduced somewhat earlier, in Section 2.4.) To give the discrete distributions their due, Chapter 6 was heavily rewritten and reorganized, highlighting the big ideas of these probability distributions—especially the binomial setting and binomial distribution.

- We rethought our ideas about how statistical inference should be ordered. Chapter 8 (Inference for a Proportion) now covers inference on a single proportion, whereas the comparisons of two proportions were separated into Chapter 9 (Comparing Two Populations: Inference for the Difference of Two Proportions). Likewise, the tests on means were divided into Chapter 10 (Inference for Means) on a single mean and Chapter 11 (Comparing Two Populations: Inference for the Difference of Two Means) on the differences between two means. This new lineup met with approval from reviewers, and we think that it is a more effective and flexible way to approach statistical inference.

- New Chapter 14 (One-Way Analysis of Variance) covers one-way ANOVA. Section 14.1 offers a new look at why we study ANOVA. We consider one-way ANOVA a staple of the course (14.2).
- New Chapter 15 (Multiple Regression) presents an introduction to multiple regression.
- Finally, Chapter 16 (*Martin vs. Westvaco* Revisited) takes another look at our opening case study from Chapter 1 and offers analysis through nonparametric tests, which some instructors cover as an alternate way of testing data.

In addition to reorganizing topic coverage, we also have added new examples throughout the book, selecting them from a wider array of scenarios, in line with the interests of students and the many majors that require the course. We have also included more examples and exercises that show how statistics is used for research to impress upon students the usefulness of statistics for their future careers as well as in the social, life, and physical sciences.

Hallmark Features of This Text

Chapter Opener
The openers are designed to introduce the topics covered in the chapter and pique student interest. Each chapter begins with a look at a statistics case study, which is explored in greater detail later in the chapter.

Exercise Sets
We have created two distinctive categories of problems: The P sets are *practice problems* for reinforcing newly acquired concepts and techniques. The E sets are *exercises* intended to challenge the student with a call to engage in higher-order thinking, but also to review the basics. Throughout the book, we revised the Practice problems and Exercise sets by updating data and changing scenarios.

We encourage instructors to assign the Practice (P) problems before progressing to the Exercises (E) because the Practice problems start at a simpler level than the Exercise sets and cover all of the important ideas of the chapter. Because we believe that students need the practice and will benefit from working through the entire Practice set, we provide brief answers to all P problems at the back of the book.

In contrast, the review Exercises are meant to be more challenging. At the end of the Exercises are more difficult, concept-extending, or more mathematical problems. These were placed there to challenge more advanced students. Because the Exercise sets are intended to be used as homework, only odd answers are provided at the end of the book.

The Concept Tests, which contain the C exercises, appear at the end of each chapter. In general, there are about ten Concept exercises per chapter, most of them multiple-choice. The last one or two are investigations, which require more sustained student analysis than the exercises. The answers to these exercises are not provided in the book.

Discussion Questions
The Discussion Questions, a trademark of the text, help students frame major concepts as well as lead them to new ideas and realizations about what they have just learned. Discussion Questions can be used during lectures by the instructor, in smaller sections by teaching assistants, or in small group work to spark thinking about the topics being taught.

Statistics in Action
The Statistics in Action feature encourages students to be active participants in the classroom through activities and dialog. It supports major topics, allowing students to experience the salient concepts of statistical thinking before going on to confirm them

through further analysis and practice with data. This feature gives tasks and activities that make statistics more hands-on. If the Discussion Questions help students who are willing to participate in class, the Statistics in Action feature is good for those who learn by doing. Statistics in Action features vary from taking one's pulse seated or standing, to the classic penny-spinning experiment.

Pedagogical Use of Color in Displays and Graphs

In addition to the main text, various problem sets, discussion questions, and Statistics in Action activities, the book includes displays and photos rich with information. In this edition, colored displays help students identify visual information. We have used the following code for graphs.

- For data, we used light blue, with dot plots in a darker blue.

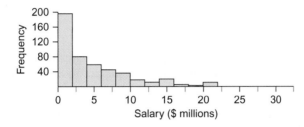

- To highlight sampling distributions, a concept that can cause students difficulties, we are using light green with dot plots in a darker green.

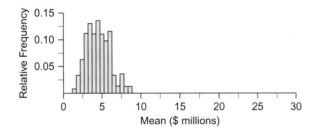

- Because of their importance as a bridge to statistical inference, we highlight probability distributions in a reddish brown with dot plots in a darker red-brown.

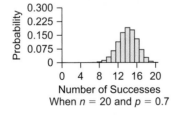

Summary

Each section and chapter ends with a summary of the important topics presented. We find that students like to have one place to go to for a brief recap of the main topics of the chapter. They can rely on our detailed reviews to help them reinforce concepts after reading the chapter or to help them recall the main ideas of the chapter as they prepare for exams. Each chapter summary is also an excellent way of preparing for the end-of-chapter Practice problems (P) and Exercise (E) sets, the Concept Test, or an investigations extended analysis.

Data Analysis

This text emphasizes data analysis. As a result, examples and problems in the text are based on real data. Realistic models are provided for the honest and thorough design and analysis of experiments and surveys at an introductory level. Students learn and practice concepts and computations through the study of interesting surveys and experiments, using data and photos provided by the researchers whenever possible.

Technology

We recognize that many calculators and software packages are used across the country. Likewise, some schools rely on TI calculators, whereas other schools are less concerned with calculators in the classroom. Despite the many devices and programs, students still have to learn to interpret statistical displays and various kinds of output. For these reasons, we show output throughout. Further, because of the emphasis on a variety of technologies such as graphing calculators and software, this text features many graphical displays and computer printouts.

WileyPLUS

This online teaching and learning environment integrates the *entire digital textbook* with the most effective instructor and student resources to fit every learning style. With *WileyPLUS*:

- Students achieve concept mastery in a rich, structured environment that's available 24/7.
- Instructors personalize and manage their course more effectively with assessment, assignments, grade tracking, and more.

WileyPLUS can complement your current textbook or replace the printed text altogether.

For Students

Personalize *the learning experience*. Different learning styles, different levels of proficiency, different levels of preparation—each of your students is unique. *WileyPLUS* empowers them to take advantage of their individual strengths:

- Students receive timely access to resources that address their demonstrated needs, and they get immediate feedback and remediation when needed.
- Integrated, multimedia resources—including audio and visual exhibits, demonstration problems, and much more—provide multiple study paths to fit each student's learning preferences and encourage more active learning.
- *WileyPLUS* includes many opportunities for self-assessment linked to the relevant portions of the text. Students can take control of their own learning and practice until they master the material.

For Instructors

Personalize *the teaching experience*. *WileyPLUS* empowers you with the tools and resources you need to make your teaching even more effective:

- You can customize your classroom presentation with a wealth of resources and functionality from PowerPoint slides to a database of rich visuals. You can even add your own materials to your *WileyPLUS* course.
- With *WileyPLUS*, you can identify those students who are falling behind and intervene accordingly, without having to wait for them to come to office hours.
- *WileyPLUS* simplifies and automates such tasks as student performance assessment, making assignments, scoring student work, keeping grades, and more.

Supplements

Web Site: www.wiley.com/college/watkins

The web site for this text provides additional resources for instructors and students:

- Computerized Test Bank
- Instructor's Solutions Manual
- PowerPoint Slides
- Data Sets
- Technology Resource Manuals

  These manuals provide step-by-step instructions, screen captures, and examples for using technology in the introductory statistics course. Also provided are exercise tables, indicating which exercises from the text best lend themselves to the use of the package presented.

  - TI Graphing Calculator Manual
  - Minitab Manual
  - Excel Manual

Supplements

The following supplements are available to accompany this text:

- *Instructor's Solutions Manual* (978-0-470-53059-7): This manual contains solutions to all of the exercises in the text.
- *Student Solutions Manual* (978-0-470-53060-3): This manual contains solutions to all Practice (P) problems and odd-numbered Exercises.
- *Printed Test Bank*: The test bank contains a number of problems for each chapter.
- *Computerized Test Bank:* All the questions in the *Printed Test Bank* are available electronically and can be obtained from the publisher.

Chapter 1

Statistical Reasoning: Investigating a Claim of Discrimination

Were older workers discriminated against during a company's downsizing? When an older worker thought that he had been unfairly laid off, his lawyers called on a statistician to help them evaluate the claim.

In the year Robert Martin turned 54, the Westvaco Corporation decided to downsize. They laid off more than half of the engineering department, including Robert Martin. Later that year, he sued Westvaco, claiming he had been laid off because of his age. A major piece of Martin's case was based on a statistical analysis of the ages of the Westvaco employees.

In the two sections of this chapter, you will get a chance to try your hand at two very different kinds of statistical work, exploration and inference. **Exploration** is an informal, open-ended examination of data. Your goal in the first section will be to uncover and summarize patterns in data from Westvaco that bear on the *Martin* case. You will try to formulate and answer basic questions such as, "Were those who were laid off older on average than those who weren't laid off?" You can use any tools—graphs, averages, and so on—that you think might be useful.

Inference, which you'll use in the second section, is quite different from exploration in that it follows strict rules and focuses on judging whether the patterns you found are the sort you might expect from a company that does not discriminate on the basis of age, or whether further investigation into possible age discrimination is needed.

You may wonder what possible bearing an age discrimination case has on your life. After all, if you are a typical college student, you are not "old." This case is not about age; it is about discrimination. It is about an employee fighting back against what he sees as unfair treatment. You are not going to be discriminated against for being too old anytime soon, but you may have trouble getting a job or be turned down when you try to rent an apartment or get a car loan. If you are denied a loan, it could be that you aren't qualified. If you are fired from a job, maybe you weren't very good at it. If you were laid off, maybe you were just unlucky.

On the other hand, maybe it was because someone thought you were the "wrong" gender, or your skin was the "wrong" color, or you looked too young. *How do you know?* While the use of statistics alone cannot *prove* discrimination, statistics can provide evidence by detecting patterns that are consistent with the practice of discrimination.

In this chapter, you will learn the basic ideas of

▶ exploring data—uncovering and summarizing patterns

▶ making inferences from data—deciding whether an observed feature of the data could reasonably be attributed to chance alone

These ideas will remain key components of the statistical concepts you'll develop and study throughout this course.

1.1 ▶ Discrimination in the Workplace: Data Exploration

Robert Martin was one of 50 people working in the engineering department of Westvaco's envelope division. One spring, Westvaco's management went through five rounds of planning for a reduction in their work force. In Round 1, they eliminated 11 positions, and they eliminated 9 more in Round 2. By the time the layoffs ended, after all five rounds, only 22 of the 50 workers had kept their jobs. The average age in the department had fallen from 48 to 46.

After Martin, age 54, was laid off, he sued Westvaco for age discrimination. Display 1.1 shows the data provided by Westvaco to Martin's lawyers. The statistical analysis in the lawsuit used all 50 employees in the engineering department of the envelope division, with separate analyses for salaried and hourly workers. Each row in Display 1.1

Row	Job Title	Pay (hourly or salaried)	Seniority (yr)	Round Laid Off (6 = retained)	Age (yr)
1	Engineering Clerk	H	1.5	6	25
2	Engineering Tech II	H	12.4	6	38
3	Engineering Tech II	H	25.5	6	56
4	Secretary to Engin Manag	H	24.3	6	48
5	Engineering Tech II	H	16.3	1	53
6	Engineering Tech II	H	30.8	1	55
7	Engineering Tech II	H	27.9	1	59
8	Parts Crib Attendant	H	1.2	1	22
9	Engineering Tech II	H	13.8	2	55
10	Engineering Tech II	H	39.1	2	64
11	Technical Secretary	H	17.2	2	55
12	Engineering Tech II	H	28.8	3	55
13	Engineering Tech II	H	14.2	4	33
14	Engineering Tech II	H	13.7	4	35
15	Customer Serv Engineer	S	24.3	6	61
16	Customer Serv Engr Assoc	S	2.7	6	29
17	Design Engineer	S	23.3	6	48
18	Design Engineer	S	16.6	6	54
19	Design Engineer	S	12.9	6	55
20	Design Engineer	S	23.8	6	60
21	Engineering Assistant	S	4.5	6	31
22	Engineering Associate	S	5.8	6	34
23	Engineering Manager	S	27.2	6	59
24	Machine Designer	S	0.8	6	32
25	Packaging Engineer	S	7.2	6	53
26	Prod Spec-Printing	S	16.2	6	47
27	Proj Engineer-Elec	S	19.8	6	48
28	Project Engineer	S	17.3	6	42
29	Project Engineer	S	26.8	6	48
30	Project Engineer	S	9.4	6	57
31	Supv Engineering Serv	S	18.6	6	37
32	Supv Machine Shop	S	26.8	6	54
33	Chemist	S	36.8	1	69
34	Design Engineer	S	3.1	1	53
35	Engineering Associate	S	5.3	1	30
36	Machine Designer	S	5.8	1	52
37	Machine Parts Cont-Supv	S	37.4	1	63
38	Prod Specialist	S	47.2	1	64
39	Project Engineer	S	31.3	1	66
40	Chemist	S	38.2	2	61
41	Design Engineer	S	1.7	2	31
42	Electrical Engineer	S	4.8	2	42
43	Machine Designer	S	22.1	2	56
44	Machine Parts Cont Coor	S	23.2	2	54
45	VH Prod Specialist	S	35.3	2	56
46	Printing Coordinator	S	29.0	3	50
47	Prod Dev Engineer	S	5.2	3	32
48	Prod Specialist	S	36.0	4	59
49	VH Prod Specialist	S	28.8	4	49
50	Engineering Associate	S	1.7	5	23

Go to www.wiley.com/college/ Watkins to download data sets.

Display 1.1 The data in *Martin v. Westvaco*. [Source: *Martin v. Envelope Division of Westvaco Corp.*, CA No. 92-03121-MAP, 850 Fed. Supp. 83 (1994).]

Variables provide information about cases.

Variability is what statistics is all about.

A distribution is a record of variability.

corresponds to one worker, and each column corresponds to a characteristic of the worker: job title, whether hourly or salaried, number of years the employee worked for Westvaco (seniority), and age on his or her birthday the year layoffs began. The next-to-last column (Round) tells how the worker fared in the downsizing: 1 means chosen for layoff in Round 1 of planning for the reduction in force, 2 means chosen in Round 2, and so on for Rounds 3, 4, and 5; 6 means the employee was not chosen for layoff.

The subjects (or objects) of statistical examination often are called **cases.** In the rows in Display 1.1, the cases are individual Westvaco employees. Their characteristics, in the columns, are the **variables.** If you pick a row and read across, you find information about a single case. For example, Robert Martin, in Row 44, was salaried, had worked for Westvaco for 23.2 years, was chosen for layoff in Round 2, and was 54 years old the year layoffs began. Although reading across might seem the natural way to read the table, in statistics you will often find it useful to pick a column and read down. This gives you information about a single variable as you range through all the cases. For example, pick Age, read down the column, and notice the **variability** in the ages. It is variability like this—the fact that individuals differ—that can make it a challenge to see patterns in data and figure out what they mean.

Imagine: If there had been no variability—if all the workers had been of just two ages, say, 30 and 50, and Westvaco had laid off all the 50-year-olds and kept all the 30-year-olds—the conclusion would be obvious and there would be no need for statistics. But real life is more subtle than that. The ages of the laid-off workers varied, as did the ages of the workers retained. *Statistical methods were designed to cope with such variability.* In fact, you might define statistics as the science of learning from data in the presence of variability.

Although the bare fact that the ages vary is easy to see in the data table, the pattern of those ages is not so easy to see. This pattern—what the values are and how often each occurs—is their **distribution.** In order to see that pattern, a graph is better than a table. The **dot plot** in Display 1.2 shows the distribution of the ages of the 36 salaried employees who worked in the engineering department just before the layoffs began.

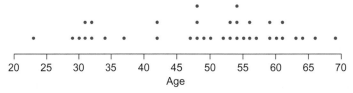

Display 1.2 Ages of the salaried workers. Each dot represents a worker; the age is shown by the position of the dot along the scale below it.

Display 1.2 provides some useful information about the variability in the ages, but by itself doesn't tell anything about possible age discrimination in the layoffs. For that, you need to distinguish between those salaried workers who lost their jobs and those who didn't. The dot plot in Display 1.3, which shows those laid off and those retained, provides weak evidence for Martin's case. Those laid off generally were older than those who kept their jobs, but the pattern isn't striking.

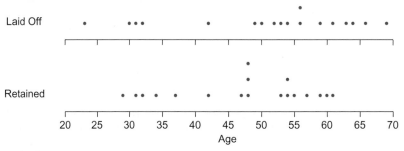

Display 1.3 Salaried workers: ages of those laid off and those retained.

Display 1.3 shows that most salaried workers who were laid off were age 50 or older. However, this alone doesn't support Martin's case because most of the workers were age 50 or older to begin with.

One way to proceed is to make a summary table. The table shown here classifies the 36 salaried workers according to age and whether they were laid off or retained. (Using 50 as the dividing age between "younger" and "older" is a somewhat arbitrary, but reasonable, decision that could be made by Martin's lawyer. You will check another dividing age in the first exercise.)

	Laid Off	Retained	Total
Under 50	6	10	16
50 or Older	12	8	20
Total	18	18	36

> The ages of those laid off must be compared to the ages of those retained.

There are more workers 50 or older than there are workers under 50. Thus, to decide whether Martin has a case, you must compare the *proportion* of salaried workers under 50 who were laid off with the *proportion* of those 50 or older who were laid off. Of the younger workers, 6 out of 16, or 0.375, were laid off. Of the older workers, 12 out of 20, or 0.60, were laid off. These proportions are quite different, an argument in favor of Martin.

Looking at the layoffs of the salaried workers round by round provides further evidence in favor of Martin. The dot plots in Display 1.4 show the ages of the salaried workers laid off and retained by round. These new dot plots use different symbols for laid-off workers and for retained workers. For example, in the top dot plot, the open circles represent the salaried workers whose jobs did not survive Round 1. In this round, the four oldest workers were laid off, but only one worker under age 50 was laid off. In the second round, one of the two oldest employees was laid off. But then this pattern stopped. Again, the evidence favors Martin but is far from conclusive.

> Compare the proportions laid off in each age group rather than the numbers laid off.

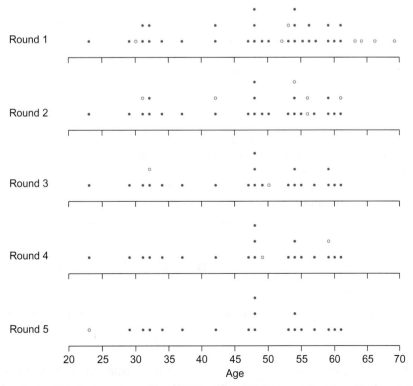

Display 1.4 Salaried workers: ages of those laid off (open circles) and those retained (solid dots) in each round.

You might feel as if the analysis so far ignores important facts, such as worker qualifications. That's true. However, the first step is to decide whether, based on the data in Display 1.1, older workers were more likely to be laid off. If not, Martin's case fails. If so, it is then up to Westvaco to justify its actions.

DISCUSSION
Exploring the *Martin v. Westvaco* Data

D1. Suppose you were the judge in the *Martin v. Westvaco* case. How would you use the information in Display 1.1 to decide whether Westvaco tended to lay off older workers in disproportionate numbers (for whatever reason)?

D2. Display 1.5 is like Display 1.3 except that it gives data for the 14 hourly workers. Compare the plots for the hourly and salaried workers. Which provides stronger evidence in support of Martin's claim of age discrimination?

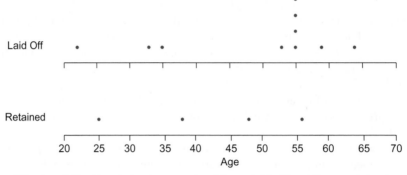

Display 1.5 Hourly workers: ages of those laid off and those retained.

D3. Whenever you think you have a message from data, you should be careful not to jump to conclusions. The patterns in the Westvaco data might be "real"—they reflect age discrimination on the part of management. On the other hand, the patterns might be the result of chance—management wasn't discriminating on the basis of age but simply by chance happened to lay off a larger percentage of older workers. What's your opinion about the Westvaco data: Do the patterns seem "real"—too strong to be explained by chance?

D4. The analysis up to this point ignores important facts such as worker qualifications. Suppose Martin makes a convincing case that older workers were more likely to be laid off. It is then up to Westvaco to justify its actions. List several specific reasons Westvaco might give that would justify laying off a disproportionate number of older workers.

Summary 1.1: Discrimination in the Workplace—Data Exploration

Data exploration, or exploratory analysis, is a purposeful investigation to find patterns in data, using tools such as tables and graphs to display those patterns and statistical concepts such as distributions and averages to summarize them.

- A table display, with cases listed in rows and variables in columns, helps you look at how variables differ from case to case.

- The distribution of a variable tells you the set of values that the variable takes on, together with how often each value occurs.

- A dot plot, which shows the values of a variable along a number line, provides you with a visual display of the distribution of the variable and gives you a sense of how large or small the values are, which values occur most often, how spread out the values are, and whether any values appear to be unusually large or small.

Statistics involves coping with variability, so you have to understand the causes of that variability before you can draw informed conclusions. All the features of data exploration that you have investigated here will be important when you move on to the inference phase of a statistical investigation.

Practice

Practice problems help you master basic concepts and computations. Throughout this textbook, you should work *all* the practice problems for each topic you want to learn. The answers to all practice problems are given in the back of the book.

Exploring the *Martin v. Westvaco* Data

P1. Construct a dot plot similar to Display 1.3, comparing the ages of hourly workers who lost their jobs during Rounds 1–3 to the ages of hourly workers who still had their jobs at the end of Round 3. How do the ages differ?

P2. This summary table classifies the hourly workers according to age and whether they were laid off or retained. Choose the best conclusion to draw from this table.

	Laid Off	Retained	Total
Under 50	3	3	6
50 or Older	7	1	8
Total	10	4	14

A. The table supports a claim of age discrimination because most of the people who were laid off were 50 or older.

B. The table supports a claim of age discrimination because a larger percentage of the people age 50 or older were laid off than people under 50.

C. The table supports a claim of age discrimination because a larger percentage of the laid-off people were age 50 or older than were under 50.

D. The table does not support a claim of age discrimination because the number of people under age 50 who were laid off is equal to the number of people age 50 or older who were laid off.

E. The table does not support a claim of age discrimination because a larger percentage of people were 50 or older to begin with.

P3. Display 1.6 shows layoffs and retentions by round for hourly workers. (There is no plot for Round 5 because no hourly workers were chosen for layoff in that round.) Compare the pattern for the hourly workers with the pattern for the salaried workers in Display 1.4 on page 5. For which group does the evidence more strongly favor Martin's case?

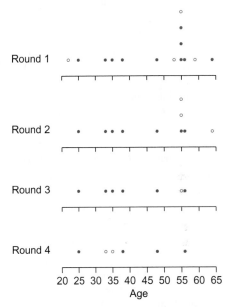

Display 1.6 Hourly workers: ages of those laid off (open circles) and those retained (solid dots) in each round.

Exercises

Exercises are mixed in difficulty—some are like the more routine practice problems while others require original thought and understanding of several concepts. Also, unlike practice problems, exercises are not necessarily in the order that the concepts were introduced in the section. Each odd exercise and the even exercise following it are somewhat alike, should you want more practice. Answers to the odd-numbered exercises are given in the back of the book.

E1. Refer to the table in P2.

a. What proportion of hourly workers under age 50 were laid off? Were not laid off?

b. What proportion of laid-off hourly workers were under age 50? Were age 50 or older?

c. What two proportions should you compute and compare in order to decide whether older hourly workers were disproportionately laid off? Make these computations and give your conclusion.

d. Compare this table with the table for the salaried workers on page 5. For which group does the evidence more strongly favor Martin's case?

E2. This summary table classifies salaried workers as to whether they were laid off and their age, this time using 40 as the cutoff between younger and older workers.

	Laid Off	Retained	Total
Under 40	4	5	9
40 or Older	14	13	27
Total	18	18	36

a. What proportion of workers age 40 or older were laid off? What proportion of laid-off workers were age 40 or older?

b. What proportion of workers under age 40 were laid off? What proportion were not laid off?

c. What two proportions should you compute and compare to decide whether older workers were disproportionately laid off? Make these computations and give your conclusion.

d. Compare this table with the table for the salaried workers on page 5, where 50 was the age cutoff. If you were Martin's lawyer, would you present a table using 40 or 50 as the cutoff?

E3. Explore whether the age distributions are similar for the 4 hourly and the 18 salaried workers who kept their jobs.

a. Show the two age distributions on a pair of dot plots that have the same scale. How do these distributions differ?

b. Do your dot plots in part a support a claim that Westvaco was more inclined to keep older workers if they were salaried rather than hourly?

E4. Explore whether hourly workers at Westvaco were more likely than salaried workers to lose their jobs.

a. Start by constructing a summary table to display the relevant data.

b. Compute two proportions that will allow you to make this comparison.

c. What do you conclude from comparing the proportions?

E5. Refer to Display 1.1 on page 3.

a. Create a summary table whose five cases are Round 1 through Round 5 and whose three variables are *total number of employees laid off in that round*, *number of employees laid off in that round who were 40 or older*, and *percentage laid off in that round who were 40 or older*.

b. Describe any patterns you find in the table and what you think they might mean.

E6. "Last hired, first fired" is shorthand for "When you have to downsize, start by laying off the newest person, then the person hired next before that, and work back in reverse order of seniority." (The person who's been working longest will be the last to be laid off.)

a. How was seniority related to the decisions about layoffs in the engineering department at Westvaco?

b. What explanation(s) can you suggest for any patterns you find?

E7. Many tables in the media are arranged with cases as rows and variables as columns. For Displays 1.7 and 1.8 in parts a and b, identify the cases and the variables. Then compute the values missing from each table.

a. Display 1.7 lists the five Major League baseball players with the highest number of lifetime stolen bases. The rule for deciding when a base was stolen was changed in 1898. Before that time, if the batter hit a single and a runner advanced two bases, the runner got credit for a stolen base. Thus, the only players included here are those who played under the modern rule.

b. When the stock market takes a precipitous fall, invariably the current situation is compared with the stock market crash of 1929, which led to the Great Depression of the 1930s. Display 1.8 gives some information about how much the value of various stocks changed on "Black Tuesday" of 1929.

Wall Street, October 1929.

Player	Games	At Bats	Hits	Home Runs	Stolen Bases	Batting Average
R. Henderson	3081	10961	3055	297	1406	0.279
L. Brock	2616	10332	3023	149	938	—?—
B. Hamilton	1591	—?—	2159	40	912	0.344
T. Cobb	3035	11429	4191	117	892	0.367
T. Raines	2502	8872	—?—	170	808	0.294

Display 1.7 Number of bases stolen in a single season by the top five Major League baseball players.
[Source: *mlb.mlb.com.*]

Stock	Closing Price on 10/28	Closing Price on 10/29	Change	Percentage Change	Volume
Chrysler	40	$33\frac{1}{2}$	$-6\frac{1}{2}$	-16.25	269,100
Coca-Cola	137	$128\frac{3}{8}$	–?–	–?–	14,100
Eastman Kodak	$181\frac{1}{8}$	–?–	$-11\frac{1}{8}$	-6.14	27,800
General Electric	250	222	-28	-11.20	136,300
General Motors	–?–	40	$-7\frac{1}{2}$	–?–	971,300
Procter & Gamble	$77\frac{3}{4}$	$66\frac{1}{2}$	$-11\frac{1}{4}$	-14.47	13,800
US Steel	186	174	–?–	–?–	307,300

Display 1.8 New York Stock Exchange activity for October 29, 1929.
[Source: *marketplace.publicradio.org*.]

E8. Suppose you are studying the effects of poverty and plan to construct a data set whose cases are the villages in Bolivia. Name some variables that you might study.

1.2 ▶ Discrimination in the Workplace: Inference

Overall, the exploratory work on the Westvaco data set in Section 1.1 shows that a larger proportion of older workers than younger workers were laid off and were laid off earlier. One of the main arguments in the court case was about what those patterns mean: Can you infer from the patterns that Westvaco has some explaining to do, or are they the sort of patterns that tend to happen even in the absence of discrimination?

A comprehensive analysis of *Martin v. Westvaco* will have to wait for its reappearance in Chapter 16, when you'll be more familiar with the concepts and tools of statistics. For now, you can get a pretty good idea of how the analysis goes by working with a subset of the data.

Picking Workers at Random

The ages of the ten hourly workers involved in Round 2 of the layoffs, arranged from youngest to oldest, were 25, 33, 35, 38, 48, 55, 55, 55, 56, and 64. The three workers who were laid off were ages 55, 55, and 64. Display 1.9 shows these data on a dot plot.

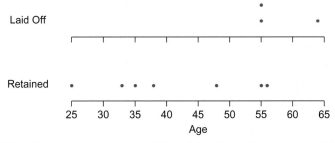

Display 1.9 Hourly workers: ages of those laid off and those retained in Round 2.

To simplify the statistical analysis to come, it helps to "condense" the data into a single number, called a **summary statistic.** One possible summary statistic is the **average,** or **mean,** age of the three workers who lost their jobs:

> Use a summary statistic to "condense" the data.

$$\frac{55 + 55 + 64}{3} = 58 \text{ years}$$

Knowing what to make of the data requires balancing two points of view. On one hand, the pattern in the data is pretty striking. Of the five workers under age 50, all kept their jobs. Of the five who were 55 or older, only two kept their jobs. On the other hand, the number of workers involved is small: only three out of ten. Should you take seriously a pattern involving so few cases? Imagine two people taking sides in an argument that was at the center of the statistical part of the *Martin* case.

Martin: Look at the pattern in the data. All three of the workers laid off were much older than the average age of all workers. That's evidence of age discrimination.

Westvaco: Not so fast! You're looking at only ten workers total, and only three positions were eliminated. Just one small change and the picture would be entirely different. For example, suppose it had been the 25-year-old instead of the 64-year-old who was laid off. Switch the 25 and the 64, and you get a totally different set of averages. (Ages in red are those selected for layoff.)

Actual data: 25   33   35   38   48   55   55   55   56   64
Altered data: 25   33   35   38   48   55   55   55   56   64

See! Make just one small change, and the average age of the three who were laid off is *lower* than the average age of the others.

Average Age	Laid Off	Retained
Actual Data	58.0	41.4
Altered Data	45.0	47.0

Martin: Not so fast yourself! Of all the possible changes, you picked the one most favorable to your side. If you'd switched one of the 55-year-olds who got laid off with the 55-year-old who kept his or her job, the averages wouldn't change at all. Why not compare what actually happened with the other possibilities?

Westvaco: What do you mean?

Martin: Start with the ten workers, and pick three at random. Do this over and over, to see what typically happens, and compare the actual data with the results. Then we'll find out how likely it is that the average age of those laid off would be 58 or greater.

> In random sampling, all possible samples of a given size are equally likely to be the sample selected.

The dialog between Martin and Westvaco describes one age-neutral method for choosing which workers to lay off: Pick three workers completely at random, with all sets of three having the same chance to be chosen.

DISCUSSION
Picking Workers at Random

D5. If you pick three of the ten ages at random, do you think you are likely to get an average age of 58 or greater?

D6. If the probability of getting an average age of 58 or greater turns out to be small, does this favor Martin or Westvaco?

STATISTICS IN ACTION 1.1 ▶ By Chance or Design?

This activity shows you how to use simulation to estimate the probability of getting an average age of 58 years or greater if you choose three workers at random.

What You'll Need: paper or 3 × 5 cards, a box or other container

1. *Create a model of a chance process.* Write each of the ten ages on identical pieces of paper or 3 × 5 cards, and put the ten cards in a box. Mix them thoroughly, draw out three (the ones to be laid off), and record the ages.

2. *Compute a summary statistic.* Compute the average of the three numbers in your sample to one decimal place.

3. *Repeat the process.* Repeat steps 1 and 2 nine more times.

4. *Display the distribution.* Pool your results with the rest of your class and display the average ages on a dot plot.

5. *Estimate the probability.* Count the number of times your class computed an average age of 58 years or greater. Estimate the probability that simply by chance the average age of those chosen would be 58 years or greater.

6. *Interpret your results.* What do you conclude from your class's estimate in step 5?

In the dialog, Martin suggested selecting three workers at random to lay off, computing their average age, and repeating this many times. This procedure, called a **simulation,** sets up a **model** of a chance process (drawing three ages at random) that copies, or simulates, a real situation (selecting three employees at random to lay off). Such a simulation allows you to decide how likely it is to get an average age as high as 58, like that of the actual layoffs, from a completely age-blind process.

Shown here are the first 4 of 200 repetitions from such a simulation. (The ages in red are those selected for layoff.) The average ages of the workers selected for layoff—42.7, 48.0, 42.7, and 37.0—are highlighted by the red dots in the distribution of 200 repetitions in Display 1.10.

> The simulation tells what kind of data to expect if workers are selected at random for layoff.

										Average Age
25	33	35	38	48	55	55	55	56	64	42.7
25	33	35	38	48	55	55	55	56	64	48.0
25	33	35	38	48	55	55	55	56	64	42.7
25	33	35	38	48	55	55	55	56	64	37.0

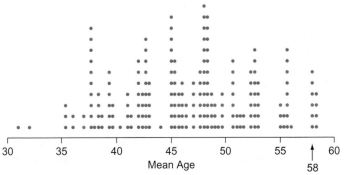

Display 1.10   Results of 200 repetitions: the distribution of the average age of the three workers chosen for layoff by chance alone.

Typically, when deciding whether it is reasonable to attribute a result to chance alone, the cutoff is 5%, or 0.05. (In discrimination lawsuits, the cutoff may be 2.5%.)

Out of 200 repetitions, only 10, or 5%, gave an average age of 58 or greater. So it is not at all likely that simply by chance you'd pick workers as old as the three Westvaco picked. Did the company discriminate? There's no way to tell from the numbers alone—Westvaco might have a good explanation. On the other hand, if this simulation had told you that an average of 58 or greater is easy to get by chance alone, then the data would provide no evidence of discrimination and Westvaco wouldn't need to explain.

The Logic of Inference

To better understand how this logic applies to *Martin v. Westvaco*, imagine a realistic argument between the advocates for each side.

> **Martin:** Look at the pattern in the data. All three of the workers laid off were much older than average.
>
> **Westvaco:** So what? You could get a result like that just by chance. If chance alone can account for the pattern, there's no reason to ask us for any other explanation.
>
> **Martin:** Of course you *could* get this result by chance. The question is whether it's easy or hard to do so.
>
> **Westvaco:** Well . . . I'll agree that it's really hard to get an average age as high as 58 simply by chance, but that by itself still doesn't prove discrimination.
>
> **Martin:** No, but I think it leaves you with some explaining to do!

The statistician hired by Bob Martin was able to demonstrate that if employees had been selected at random for layoff, the probability of getting an average age as large or even larger than Westvaco got is only 0.05. This probability was considered small enough to warrant asking for an explanation from Westvaco but not small enough to present in court as clear evidence of discrimination.

Even though the probability is small, the statistics can't tell you whether age discrimination was the cause. All the statistics can tell you is that older workers were disproportionately laid off and this cannot reasonably be attributed to chance alone. Westvaco might have had a perfectly reasonable and legal explanation for why such a large proportion of older workers happened to have been laid off. But we will never know. There was an out-of-court settlement before the case went to trial.

DISCUSSION
The Logic of Inference

D7. Why must you estimate the probability of getting an average age of 58 *or greater* rather than the probability of getting an average age of 58?

D8. *How unlikely is "too unlikely"?* The probability estimated in Display 1.10 on page 11 is in fact exactly equal to 0.05. In a typical court case, a probability of 0.025 or less is required to serve as evidence of discrimination.

    a. Did the Round 2 layoffs of hourly workers in the *Martin* case meet the court requirement?

    b. If the probability in the *Martin* case had been 0.01 instead of 0.05, how would that have changed your conclusions? 0.10 instead of 0.05?

D9. A friend wants to bet with you on the outcome of a coin flip. The coin looks fair, but you decide to do a little checking. You flip the coin, and it lands heads. You flip again—also heads. A third flip—heads. Flip—heads. Flip—heads. You continue and the coin lands heads 19 times in 20 flips.

    a. Explain why the evidence—19 heads in 20 flips—makes it hard to believe the coin is fair.

    b. Design and carry out a simulation to estimate how unusual this result would be if the coin were fair.

Summary 1.2: Discrimination in the Workplace—Inference

Inference is a statistical procedure that involves deciding whether an event can reasonably be attributed to chance or whether you should look for—and perhaps investigate—some other explanation. In the *Martin* case, you used inference to determine that the relatively high average age of the laid-off hourly employees in Round 2 could not reasonably be attributed to chance alone.

The steps in statistical inference include:

- First, set up a *model* of a process in which chance is the only factor influencing the outcomes.

- Then, create the *distribution* of the possible summary statistics. You can do this with simulation, as in Display 1.10, or sometimes you can create the distribution using mathematical and statistical theory.

- *Locate* the actual summary statistic on the distribution of summary statistics to determine how likely the actual result or one even more extreme would be if chance was the only factor influencing which outcome you got.

- Finally, reach—or infer—a *conclusion*. If the probability of getting a summary statistic as extreme as or more extreme than that from your actual data is small, conclude that chance isn't a reasonable explanation. If the probability isn't small, conclude that you can reasonably attribute the result to chance alone. Typically, the cutoff point for a "small" probability is 5%, or 0.05. (In discrimination cases, the cutoff point may be 2.5%.)

The logic you've just seen is basic to all statistical inference, but it's not easy to understand. In fact, it took mathematicians centuries to come up with the ideas. It wasn't until the 1920s that a brilliant British biological statistician and mathematician, R. A. Fisher, realized that results of agricultural experiments could be analyzed using a process similar to that used for the Westvaco case. Fisher could then decide whether the differences observed in, say, wheat growth under two different fertilizers could reasonably be attributed to chance alone or should be attributed to the fertilizer used. Since then, statisticians building on the work of Fisher and his contemporaries have applied randomization methods, and approximations to them, to almost all human endeavors, especially medical experiments and social science research.

Practice

Picking Workers at Random

P4. Suppose three workers were laid off from a set of ten whose ages were the same as those of the hourly workers in Round 2 in the *Martin* case. This time, however, the ages of those laid off were 48, 55, and 55.

25 33 35 38 48 55 55 55 56 64

    a. Use the dot plot in Display 1.10 on page 11 to estimate the probability of getting an average age as large as or larger than that of those laid off in this situation.

    b. What would your conclusion be if Westvaco had laid off workers of these three ages?

P5. At the beginning of Round 1, there were 14 hourly workers. Their ages were 22, 25, 33, 35, 38, 48, 53, 55, 55, 55, 55, 56, 59, and 64. After the layoffs were complete, the ages of those left were 25, 38, 48, and 56. In this problem, you will use these data to create a simulation similar to the one whose results are summarized in Display 1.10 on page 11.

a. What is the average age of the ten workers laid off?

b. Describe a simulation for finding the distribution of the average age of ten workers selected at random to be laid off.

c. The results of 200 repetitions from a simulation are shown in Display 1.11. Suppose 10 workers are picked at random for layoff from the 14 hourly workers. Make a rough estimate of the probability of getting, just by chance, the same or larger average age as that of the workers who actually were laid off (from part a).

d. Does this ihanalysis provide evidence in Martin's favor?

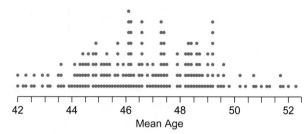

Display 1.11   Results of 200 repetitions.

Exercises

E9. In this exercise, you will revisit the idea of the simulation whose results are summarized in Display 1.10 on page 11. This time you will consider all 14 hourly workers and will use a different summary statistic, the number of hourly workers laid off who were 40 or older. The ages listed here are those of the hourly workers, with the ages of those laid off in red. Note that, of the ten hourly workers laid off by Westvaco, seven were age 40 or older.

22   25   33   35   38   48   53
55   55   55   55   56   59   64

a. Write the 14 ages on 14 slips of paper and draw 10 at random to be chosen for layoff. How many of the 10 are age 40 or older?

b. The dot plot in Display 1.12 shows the results of 50 repetitions of this simulation. Estimate the probability that, by chance, seven or more of the ten hourly workers who were laid off would be age 40 or older.

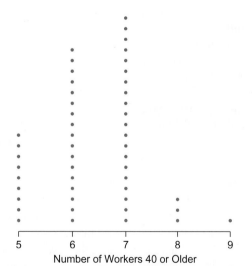

Display 1.12   Results of 50 repetitions: the distribution of the number of workers age 40 or older from ten randomly selected workers.

c. Do you conclude that the proportion of laid-off workers age 40 or older could reasonably be due to chance alone, or should Westvaco be asked for an explanation?

E10. The ages of the ten hourly workers left after Round 1 are given here. The ages of the four workers laid off in Rounds 2 and 3 are shown in red. Their average age is 57.25.

25   33   35   38   48   55   55   55   56   64

a. Describe how to simulate the chance of getting an average age of 57.25 or more if four workers are selected from these ten at random to be laid off.

b. Perform your simulation once and compute the average age of the four hourly workers laid off.

c. The dot plot in Display 1.13 shows the results of 200 repetitions of this simulation. What is your estimate of the probability of getting an average age as great as or greater than Westvaco did if four workers are picked at random for layoff in Rounds 2 and 3 from the ten hourly workers remaining after Round 1?

d. What is your conclusion?

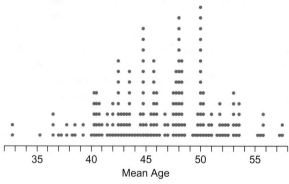

Display 1.13   Results of 200 repetitions: mean ages of three randomly selected workers.

E11. Mrs. Garcia was not happy when she found that her baker had raised the price of a loaf of bread—and she let him know it. However, she did buy her usual three loaves of bread. They seemed a little light, so she asked that they be weighed and that the other eight loaves the baker could have given her also be weighed. The other eight loaves weighed 14, 15, 15, 16, 16, 17, 17, and 18 ounces. The three loaves Mrs. Garcia was given weighed 14, 15, and 16 ounces. The baker claimed that he picked the three loaves at random.

In Display 1.14, each dot represents the average weight of a random sample of three loaves.

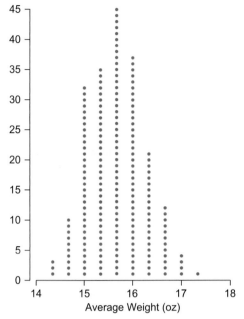

Display 1.14 Results of 200 repetitions: average weight of three randomly selected loaves.

Which of these conclusions should Mrs. Garcia draw?

A. Because the probability of getting an average weight as low as or lower than that of Mrs. Garcia's three loaves is small, Mrs. Garcia should not be suspicious that the baker deliberately gave her lighter loaves.

B. Because the probability of getting an average weight as low as or lower than that of Mrs. Garcia's three loaves is small, Mrs. Garcia should be suspicious that the baker deliberately gave her lighter loaves.

C. Because the probability of getting an average weight as low as or lower than that of Mrs. Garcia's three loaves is fairly large, Mrs. Garcia should not be suspicious that the baker deliberately gave her lighter loaves.

D. Because the probability of getting an average weight as low as or lower than that of Mrs. Garcia's three loaves is fairly large, Mrs. Garcia should be suspicious that the baker deliberately gave her lighter loaves.

E12. *Statistical Reasoning in Oz.* You have spent some time in Oz. You think the date back in Kansas is July 4, but you can't be sure because days might not have the same length in Oz as on Earth. A friendly tornado puts you

and your dog Toto down in Kansas. However, you see snow falling (data). Which of these inferences should you make?

A. If this is Kansas, it is very unlikely to be snowing on July 4. Therefore, this probably isn't Kansas.

B. If it is July 4, it is very unlikely to be snowing in Kansas. Therefore, this probably isn't July 4.

C. If it is snowing in Kansas on July 4, it is time to go back to Oz.

D. If this is Kansas and it is July 4, it probably isn't really snowing.

E13. For some situations, instead of using simulation, it is possible to find exact probabilities by counting equally likely outcomes. Suppose only two out of the ten hourly workers had been laid off in Round 2 and that those two workers were ages 55 and 64, with an average age of 59.5. It is straightforward, though tedious, to list all 45 possible pairs of workers who might have been chosen. Here's the beginning of a systematic listing. The first nine outcomes include the 25-year-old and one other. The next eight outcomes include the 33-year-old and one other, but not the 25-year-old because the pair {25, 33} was already counted.

Count	Pair Chosen (red indicates laid off)										Average Age
1	25	33	35	38	48	55	55	55	56	64	29.0
2	25	33	35	38	48	55	55	55	56	64	30.0
3	25	33	35	38	48	55	55	55	56	64	31.5
⋮											
9	25	33	35	38	48	55	55	55	56	64	44.5
10	25	33	35	38	48	55	55	55	56	64	34.0
11	25	33	35	38	48	55	55	55	56	64	35.5
⋮											

Using counting methods you may have seen in a mathematics class, the number of ways to pick two out of ten workers to lay off is

$$_{10}C_2 = \binom{10}{2} = 45$$

a. How many of the 45 pairs give an average age of 59.5 or greater? (List them.)

b. If the pair is chosen completely at random, then all possibilities are equally likely and the probability of getting an average age of 59.5 or greater equals the number of pairs with an average of 59.5 or more divided by the total number of possible pairs. What is the probability?

c. Does the evidence of age discrimination meet the 2.5% standard used by the courts?

E14. In this exercise, you will follow the same steps as in E13 to find the probability of getting an average age of 58 or greater when drawing three hourly workers at random in Round 2. The number of ways to pick three different workers from ten to lay off is

$$_{10}C_3 = \binom{10}{3} = 120$$

a. List the ways that give an average age of 58 or greater.

b. Compute the probability of getting an average age of 58 or greater when three workers are selected for layoff at random.

c. How does this probability compare to the results of the simulation shown in Display 1.10 on page 11? Why do the two probabilities differ (if they do)?

Chapter Summary

In this chapter, you explored the data from an actual case of alleged age discrimination, looking for evidence you considered relevant. You then saw how to use statistical reasoning to test the strength of the evidence: Are the patterns in the data solid enough to support Martin's claim of age discrimination, or are they the sort that you would expect to occur even if there was no discrimination? Along the way you made a substantial start at learning many of the most important statistical terms and concepts: distribution, cases and variables, summary statistic, simulation, and how to determine whether the result from the real-life situation can reasonably be attributed to chance alone or whether an explanation is called for.

You have practiced both thinking like a statistician and reporting your results like a statistician. Throughout this textbook, you will be asked to justify your answers in the real-world context. This includes stating assumptions, giving appropriate plots and computations, and writing a conclusion in context.

The last chapter of this book includes a final look at the *Martin* case.

Review Exercises

E15. An instructor had two statistics classes, and students could enroll in either the earlier class or the later class. The final grades in the courses are given here.

Earlier class:

99	95	69	91	79	67	64	54	68
47	53	86	100	95	45	41	59	66

Later class:

84	68	94	77	88	75	88	91	83
61	97	75	37	82	62	49	43	93

a. Display these data on dot plots so that you can compare the two classes.

b. What conclusion can you draw from the dot plots? Could the difference between the two classes reasonably be attributed to chance, or do you think the instructor should look for an explanation?

E16. Refer to the data in E15.

a. Make a table that divides the course grades into "fail" (less than 60) and "pass" (60 or more) for the two classes.

b. What proportion of the students in the earlier class passed? What proportion of students who passed were in the earlier class? What proportion of students passed overall?

c. Suppose that you want to decide whether a disproportionate number of "passing" students enrolled in the earlier class. What two proportions should you compute and compare? Make these computations and give your conclusion.

E17. This table classifies the Westvaco workers by whether they were laid off and whether they were under age 50 or were age 50 or older.

	Laid Off	Retained	Total
Under 50	9	13	22
50 or Older	19	9	28
Total	28	22	50

Choose the best conclusion to draw from this table.

A. The table supports a claim of age discrimination because most of the people who were laid off were 50 or older.

B. The table supports a claim of age discrimination because a larger percentage of the people age 50 or older were laid off than people under 50.

C. The table supports a claim of age discrimination because a larger percentage of the laid-off people were age 50 or older than were under 50.

D. The table does not support a claim of age discrimination because the number of people under age 50 who were laid off is equal to the number of people age 50 or older who were retained.

E. The table does not support a claim of age discrimination because a larger percentage of people were 50 or older to begin with.

E18. Display 1.15 contains information about the planets in our solar system. The radius of each planet is given in miles, and the temperature is the average at the surface. What are the cases? What are the variables?

Planet	Radius (mi)	Moons	Temperature
Mercury	1,516	0	332°F
Venus	3,760	0	67°F
Earth	3,963	1	59°F
Mars	2,111	2	−82°F
Jupiter	44,423	63	−163°F
Saturn	37,449	56	−218°F
Uranus	15,882	27	−323°F
Neptune	15,388	13	−330°F

Display 1.15 Data about planets in our solar system.
[Source: *solarsystem.nasa.gov.*]

E19. On page 5 you studied the summary table of salaried workers classified according to age and to whether they were laid off or retained, using 50 as the dividing age. That table is shown again here.

	Laid Off	Retained	Total
Under 50	6	10	16
50 or Older	12	8	20
Total	18	18	36

The proportion of those under age 50 who were laid off (6 out of 16, or 0.375) is smaller than the proportion of those age 50 or older who were laid off (12 out of 20, or 0.60). The key question, however, is, "Is the actual 12 versus 6 split of those laid off consistent with selecting workers at random for layoff?"

a. Which of these demonstrations would help Martin's case?

   I. Showing that it's not unusual to get 12 or more older workers if 18 workers are selected at random for layoff

   II. Showing that it's pretty unusual to get 12 or more older workers if 18 workers are selected at random for layoff

b. To investigate this situation using simulation, follow these steps once and record your results.

   i. Make 36 identical white cards. Label 20 cards with an O for "50 or older" and 16 cards with a U for "under 50."

   ii. Mix the cards in a bag and select the 18 to be laid off at random.

   iii. Count the number of O's among the 18 selected.

c. Display 1.16 shows the results of a computer simulation of 200 repetitions conducted according to the steps given in part b. Where would your result from part b be placed on this dot plot?

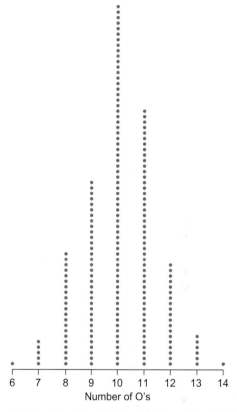

Display 1.16 Results of 200 repetitions: the distribution of the number of O's in 18 randomly selected cards.

d. From this simulation, a total of 26 out of 200 repetitions gave counts of "50 or older" that were 12 or more. What percentage of the repetitions is this? Is this percentage small enough to cast serious doubt on a claim that those laid off were chosen by chance?

E20. The Eastbanko Company had fifteen workers before laying off five of them. The ages of the fifteen workers were 22, 23, 25, 31, 34, 36, 37, 40, 41, 43, 44, 50, 55, 55, and 60, with the ages of the five laid-off workers in red.

The dot plot in Display 1.17 gives the results of a simulation with 600 repetitions for the average age of five of these workers chosen for layoff at random. Each dot represents the average of five ages, rounded down to a whole number.

a. Estimate the probability that if workers were selected by chance alone for layoff, the average age of those laid off would be as large as or larger than the average age of those in the actual layoffs.

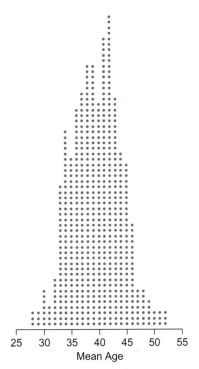

Display 1.17 Results of 600 repetitions: the distribution of the average age of five randomly chosen workers.

   b. If the 60-year-old sues for age discrimination, would Eastbanko have some explaining to do?

E21. In E9, you conducted a simulation to estimate the probability that if 10 of the 14 hourly workers are laid off, then, just by chance, 7 or more would be age 40 or older. The ages of the 14 hourly workers were 22, 25, 33, 35, 38, 48, 53, 55, 55, 55, 55, 56, 59, and 64.

   a. How many ways can 10 workers be selected from 14 workers for layoff?

   b. If there are a total of 10 layoffs, what numbers of older workers would it have been possible to lay off?

   c. Using your calculator, find the number of ways that you can lay off

      i. seven older workers and three younger workers

      ii. eight older workers and two younger workers

      iii. nine older workers and one younger worker

   d. What is the probability that you will get 7 or more workers age 40 or older if you select 10 of the 14 workers completely at random for layoff?

E22. Refer to your reasoning in E14, where you computed the probability that the three workers laid off in Round 2 would have an average age of 58 or greater. Describe how your reasoning and conclusions would be different if the workers' ages were 25, 33, 35, 38, 48, 55, 55, 55, 55, and 55, and the three workers chosen for layoff were all age 55. Is the evidence stronger or weaker for Martin in this situation than in E14?

E23. The Society for the Preservation of Wild Gnus held a raffle last week and sold 50 tickets. The two lucky participants whose tickets were drawn received all-day passes to the Wild Gnu Park in Florida. But there was a near riot when the winners were announced—both winning tickets belonged to society president Filbert Newman's cousins. After some intense questioning by angry ticket holders, it was determined that only 4 of the 50 tickets belonged to Newman's cousins and the other 46 tickets belonged to people who were not part of his family. Newman's final comment to the press was "Hey kids, I guess we were just lucky. Deal with it."

One member of the Gnu Society was taking a statistics class and decided to deal with it by simulating the drawing. He put 50 tickets in a bowl; 4 of the tickets were marked "C" for "cousin" and 46 were marked "N" for "not a cousin." The statistics student drew two tickets at random and kept track of the number of cousins picked. After doing this 1000 times, the student found that 844 draws resulted in two N's, 149 in one N and one C, and only 7 in two C's.

   a. Use the results of the simulation to estimate the probability that, in a fair drawing, both winning tickets would be held by Newman's cousins.

   b. Using the probability you estimated in part a, write a short paragraph that the statistics student can send to other members of the Gnu Society.

   c. Is it possible that Newman's cousins won the prizes by chance alone? Explain.

   d. Using reasoning like that in E13 and E14, compute the exact probability that, in a fair drawing, both winning tickets would be held by Newman's cousins.

C1. This plot shows the ages of the part-time and full-time students who receive financial aid at a small college. Which of the following is a conclusion about students at this college that cannot be drawn from the plot alone?

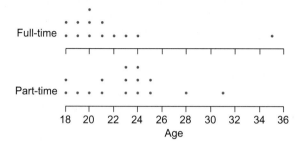

Ⓐ Part-time students who receive financial aid tend to be older than full-time students who receive financial aid.

Ⓑ A larger proportion of part-time students than full-time students receive financial aid.

Ⓒ The oldest student receiving financial aid is a full-time student.

Ⓓ No student under age 18 receives financial aid.

Ⓔ More part-time students than full-time students receive financial aid.

C2. This table classifies hourly and salaried workers as to whether they were laid off. Do the data support a claim that hourly workers are being treated unfairly?

	Laid Off	Not Laid Off	Total
Hourly	31	31	62
Salaried	24	14	38
Total	55	45	100

Ⓐ Yes, because most people laid off were hourly workers.

Ⓑ Yes, because a bigger proportion of hourly workers were laid off than salaried workers.

Ⓒ No, because half of hourly workers were laid off and half were not.

Ⓓ No, because more than half of the workers were hourly and less than half salaried.

Ⓔ No, because half of hourly workers were laid off, but more than half of salaried workers were laid off.

C3. This table shows the number of male and female applicants who applied and were either admitted to or rejected from a graduate program. What proportion of admitted applicants were female?

	Admitted	Rejected	Total
Male	17	33	50
Female	8	12	20
Total	25	45	70

Ⓐ $\frac{20}{70}$ Ⓑ $\frac{8}{25}$ Ⓒ $\frac{8}{20}$ Ⓓ $\frac{8}{70}$ Ⓔ $\frac{25}{70}$

C4. For the data in C3, in order to determine if there is evidence to continue investigating whether the graduate admissions process discriminates against females, a study takes a random sample of 25 out of the 70 applicants to be the "admitted" group. The proportion of females in the sample was computed. This process was repeated for a total of 50 random samples and the results are graphed below. What is the best conclusion to draw from this simulation?

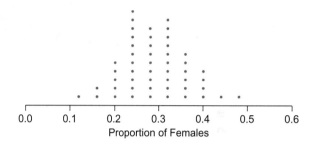

Ⓐ The actual proportion of females among those admitted is very near the center of this distribution, so there is no evidence of discrimination.

Ⓑ The actual proportion of females among those admitted is very near the center of this distribution, so there is strong evidence of discrimination in favor of female applicants.

Ⓒ The actual proportion of females among those admitted is very near the center of this distribution, so there is strong evidence of discrimination against female applicants.

Ⓓ The actual proportion of females among those admitted is quite a bit above the center of this distribution, so there is strong evidence of discrimination in favor of female applicants.

Ⓔ The actual proportion of females among those admitted is quite a bit below the center of this distribution, so there is strong evidence of discrimination against female applicants.

Investigations

C5. People with asthma often use an inhaler to help open up their lungs and breathing passages. A pharmaceutical company has come up with a new compound to put in the inhaler that, they believe, will open up the lungs of the user even more than the standard compound tends to do. Ten volunteers with asthma are randomly split into two groups: one group uses the new compound B and the other uses the standard compound A. The measurements listed in Display 1.18 are the increase in lung capacity (in liters) 1 hour after the use of the inhaler.

Compound A	Compound B
1.03	1.11
0.45	1.01
0.32	0.44
0.64	1.41
1.29	1.04

Display 1.18 Increase in lung capacity, in liters, 1hr after use of an inhaler containing a compound.

a. From simply studying the data in the table, do you think compound B does better than compound A in increasing lung capacity?

b. Construct dot plots for compounds A and B. Does it now appear that compound B tends to give larger measurements than compound A?

c. Find the average increase in lung capacity for compound A and for compound B. When you compare these means, does it look to you as if compound B is better than compound A at opening up the lungs?

C6. Refer to C5. Your task now is to see whether the observed difference in the means of each treatment group reasonably could be attributed to chance alone.

a. Place the ten measurements on separate slips of paper and mix them in a bag. Select five at random to play the role of the A treatment group; the other five play the role of the B treatment group. This time you will use as your summary statistic the difference between the means of each treatment group. Calculate this difference, mean *(compound B)* – mean *(compound A)*, for your sample.

b. The dot plot in Display 1.19 shows the results of 50 repetitions of this simulation. Compute the difference between the means for the actual data. Locate this difference on the dot plot. How many simulated differences exceed the actual difference? What proportion?

c. In light of this simulation, do you think it is reasonable to attribute the actual difference to chance alone? Explain.

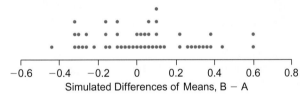

Display 1.19 Results of 50 repetitions: the distribution of the difference between the means of two randomly selected groups of five values.

Chapter 2
Exploring Distributions of Data

These women are using minimal resources while paddling themselves along Chaa Creek in a Central American jungle. But their ecological footprint is much larger if they took a plane to get there. Which of the hundreds of countries of the world have residents with the largest ecological footprints? Statistics gives you the tools to visualize and describe such large sets of data.

This chapter is about statistical plots and numerical summaries, which often are thought to be boring and impersonal. Yet the dot plot in Display 2.1 evokes a strong emotional reaction in most people. Each dot represents a country in Europe, Central Asia, or the Americas. The red dot represents the United States. A country's ecological footprint is the number of hectares used per person for food, clothing, timber, fishing, waste absorption, and infrastructure such as roads and housing. (A hectare is about 2.5 acres.) The world average is 2.7 hectares per person, but the U.S. ecological footprint is 3.5 times that. The fact that people in the United States have a very large ecological footprint, even compared to other developed countries, stands out clearly in the dot plot.

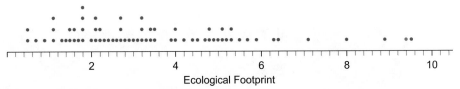

Display 2.1   Ecological footprint of countries in three regions of the world.
[Source: World Wildlife Federation Living Planet Report 2008. *http://assets. panda.org/downloads/living_planet_report_2008.pdf.*]

In this chapter, you begin a systematic study of distributions like the one above by learning how to

▷ make and interpret different kinds of plots

▷ describe the shapes of distributions

▷ choose and compute a measure of center

▷ choose and compute a measure of spread

▷ work with the normal distribution

2.1 ▷ Visualizing Distributions: Shape, Center, and Spread

"Raw" data—a long list of values—are hard to make sense of. Suppose, for example, that you are thinking of applying to the law school at the University of Texas and wonder how your college GPA of 3.59 compares with those of the students in that program. If all you have are raw data—a list of the GPAs of the over 400 students who were admitted last year—it would take a lot of time and effort to make sense of the numbers.

Fortunately, the law school web site gives you a summary: The middle half of the students had college GPAs between 3.42 and 3.82, with half having a GPA above 3.62 and half below. Now you know that your GPA of 3.59, though in the bottom half, is not far from the center value of 3.62 and is above the bottom quarter. [Source: *www. utexas.edu/law/depts/admissions/application/quickfacts.html.*]

Notice that the summary on the web site gives two different kinds of information: the *center*, 3.62, and the *spread* of the middle half of the GPAs, from 3.42 to 3.82. Often center and spread will be all you need, especially if the *shape* of the distribution is one of a few standard shapes described in this section.

Shapes of Distributions

To help build your visual intuition about how shape and summary statistics are related, this section introduces four important shapes and shows you how to estimate some summary statistics from a plot. In Section 2.2, you will learn how to compute summary statistics using formulas.

Uniform (Rectangular) Distributions

Calculators and computer software generate random numbers between 0 and 1 in such a way that the next number is equally likely to fall in any subinterval, no matter what numbers have been generated in the past. In other words, the next number generated is just as likely to be above 0.5 as it is below 0.5, or it is just as likely to be in the interval 0.2 to 0.4 as it is in the interval 0.7 to 0.9. Display 2.2 shows a dot plot of 1000 random numbers. This may look a bit ragged, but about 100 of the numbers are in each of the subintervals (0, 0.1), (0.1, 0.2), and so on.

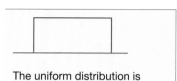

The uniform distribution is rectangular.

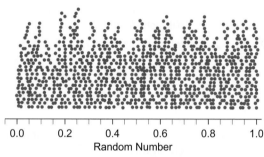

Display 2.2   Dot plot of 1000 random numbers between 0 and 1.

The 1000 random numbers shown in Display 2.2 are a sample from the infinite population of random digits that, conceptually, is available from any random number generator. The model for the population is called a **uniform distribution**, or sometimes a **rectangular distribution**, for obvious reasons.

Here are two possible ways to describe the distribution of the sample of random digits:

- The distribution is approximately uniform with values ranging from 0 to 1.

- The distribution is approximately uniform with a center at 0.5 and spreads out across an interval 0.5 unit long to either side of the center.

DISCUSSION
Uniform Distributions

D1. What variables would you expect to be approximately uniformly distributed?

D2. What variables would you expect to be very nonuniformly distributed?

Normal Distributions

Quite often, measurements tend to pile up around a central value and then become less frequent away from the center, with values far from the center occurring very rarely. SAT mathematics scores for any given year, for example, pile up around 500, with fewer scores around 600 or 400, and very few close to 800 or 200. Distributions of the heights of males or females in the general population have this same characteristic shape—many people of medium height with a few quite short or quite tall.

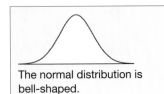

The normal distribution is bell-shaped.

Measurements of objects produced in a controlled environment such as a manufacturing process often display a similarly shaped distribution. For example, Display 2.3 shows the distribution of measurements of the diameter of tennis balls. Variables that produce a distribution with this bell shape are said to be **normally distributed,** and the distribution can be approximated by a **normal curve** drawn over the tops of the dots. A normal curve is symmetric—the right side is the mirror image of the left side—with a single peak at the line of symmetry. The normal distribution serves as a model for tennis ball diameters. Normal distributions serve as good models for many other distributions as well.

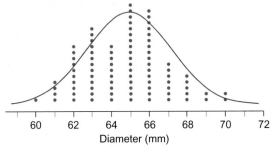

Display 2.3   Dot plot of diameters of tennis balls.

STATISTICS IN ACTION 2.1 ▶ Measuring Diameters

What shaped distribution can you expect when different people measure the length or weight of the same object? In this activity, you'll measure a tennis ball with a ruler, but the results you get will reflect what happens even when very precise instruments are used under carefully controlled conditions.

What You'll Need: a tennis ball, a ruler with a millimeter scale

1. Pass a tennis ball (or a few identical tennis balls) and a ruler (or identical rulers) around the class.

2. When the ball gets to you, measure the diameter as carefully as you can, to the nearest millimeter.

3. If you combine your measurement with those of the rest of your class, what shape do you think the distribution will have? Make a dot plot of the measurements to check your prediction.

4. *Shape.* What is the approximate shape of the plot? Are there clusters and gaps or unusual values (outliers) in the data?

5. *Center and spread.* Choose two numbers that seem reasonable for completing this sentence: "Our typical diameter measurement is about —?—, give or take about —?—."

6. Discuss some possible reasons for the variability in the measurements. How could the variability be reduced? Can the variability be eliminated entirely?

Just as it is common for repeated measurements of the *same* object to be normally distributed, it is common for the measurements of similar but *different* objects to be normally distributed. The "living" plot in Display 2.4 shows the heights of 175 male students at Connecticut State Agricultural College (now the University of Connecticut) back in 1914. Each man is standing behind the card that gives his height. The heights pile up in the center with a few heights far away on both the low and high sides, in a somewhat symmetric arrangement. You easily can picture a normal curve, drawn over the top, that would serve as a model of men's heights.

| 4:10 | 4:11 | 5:0 | 5:1 | 5:2 | 5:3 | 5:4 | 5:5 | 5:6 | 5:7 | 5:8 | 5:9 | 5:10 | 5:11 | 6:0 | 6:1 | 6:2 |

Display 2.4   Living plot showing the height of 175 male students. [Source: A. F. Blakeslee, "Corn and Men," *Journal of Heredity*, Vol. 5 (1914), pp. 511–518.]

You should use the **mean** (or *average*) to describe the center of a normal distribution. The mean falls below the high point. For example, the curve on the left in Display 2.5 represents a normal distribution with mean 20.

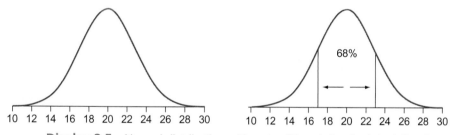

Display 2.5   Normal distribution with mean 20 and standard deviation 3.

Use the **standard deviation** to describe the spread of a normal distribution. About 68% of the values in a normal distribution are within one standard deviation of the mean. So to estimate the standard deviation, measure how far you must go to either side of the mean in order to enclose the middle 68% (roughly two-thirds) of the distribution. As you can see from the plot on the right in Display 2.5, the values 17 and 23 enclose the middle 68% of the distribution. Thus the standard deviation is 3; you have to go 3 units to either side of the mean of 20 (20 ± 3) to enclose 68% of the values. (We sometimes use SD as an abbreviation for standard deviation.)

> Use the mean and standard deviation to describe the center and spread of a normal distribution.

Averages of Random Samples

Example 2.1

Display 2.6 shows the distribution of average ages computed from 200 sets of five workers chosen at random from the ten hourly workers in Round 2 of the Westvaco case discussed in Chapter 1. Notice that, apart from the bumpiness, the shape is roughly normal. Estimate the mean and standard deviation.

Solution
The curve in the display has center at 47, and the middle 68% of the dots fall roughly between 43 and 51. Thus, the estimated mean is 47, and the estimated standard deviation is 4. A typical random sample of five workers has average age 47 years, give or take about 4 years.

> For a random sample, individuals are selected by chance.

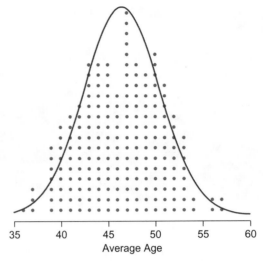

Display 2.6 Distribution of average age for groups of five workers drawn at random.

You now have seen the three most common ways that normal distributions arise in practice:

- through variation in repeated measurement of the same object (diameter of a tennis ball)
- through natural variation in populations (heights of male college students)
- through variation in averages computed from random samples (average of workers' ages)

All three scenarios are common, which makes the normal distribution especially important. In fact, the normal distribution is the most important distribution you will encounter in statistics and will be used widely in the remainder of this course.

DISCUSSION
Normal Distributions

D3. Estimate the mean and standard deviation of the following distributions visually and then write a statement summarizing the distribution.

   a. the tennis ball measurements in Display 2.3

   b. the weights of the sample of 100 different pennies shown in Display 2.7

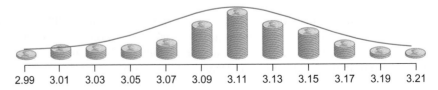

Display 2.7 Weights of pennies (gm). [Source: W. J. Youden, *Experimentation and Measurement* (U.S. Department of Commerce, 1984), p. 108.]

Skewed Distributions

Skewed distributions show bunching at one end and a long tail stretching out in the other direction. The direction of the tail tells whether the distribution is skewed right (tail stretches right, toward the high values) or skewed left (tail stretches left, toward the low values).

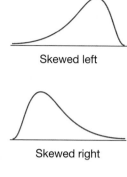

Skewed left

Skewed right

The dot plot in Display 2.8 shows the weights, in pounds, of a sample of 143 wild bears. It is skewed right (toward the higher values) because the tail of the distribution stretches out in that direction. If someone shouts "Abnormal bear loose!" you should run for cover—that bear is likely to be big!

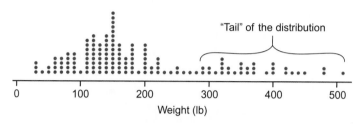

Display 2.8 Weights of bears of all ages in pounds. [Source: Minitab® Statistical Software data set.]

Often the bunching in a skewed distribution happens because values "bump up against a wall"—either a minimum that values can't go below, such as 0 for measurements and counts, or a maximum that values can't go above, such as 100 for percentages. For example, the distribution in Display 2.9 shows the grade point averages of students taking an introductory statistics course at the University of Florida. The skew is to the left—an unusual GPA would be one that is low compared to most GPAs of students in the class. The maximum grade point average is 4.0, for all As, and the distribution is bunched at the high end near this wall. (A GPA of 0.0 wouldn't be called a wall, even though GPAs can't go below 0.0, because the values aren't bunched up against it.)

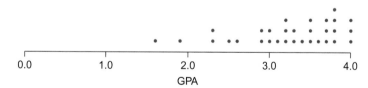

Display 2.9 Grade point averages of 31 statistics students.

Because the values in the tail have a strong influence on the mean, typically you should use the **median** to describe the center of a skewed distribution. To estimate the median from a dot plot, locate the value that divides the dots into two halves, with equal numbers of dots on either side.

Use the lower and upper quartiles to indicate spread. The **lower quartile** is the value that divides the lower half of the distribution into two halves. The **upper quartile** is the value that divides the upper half of the distribution into two halves. These three values—lower quartile, median, and upper quartile—divide the distribution into quarters. The following example will show you how to estimate them.

Use the median along with the lower and upper quartiles to describe the center and spread of a skewed distribution.

Example 2.2 | Median and Quartiles for Bear Weights

Divide the bears' weights in Display 2.10 into four groups of equal size, and estimate the median and quartiles. Write a short summary of this distribution.

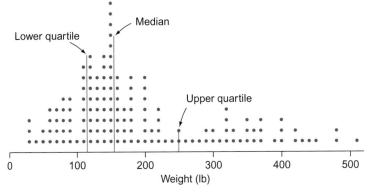

Display 2.10 Estimating center and spread for the weights of bears.

Solution

There are 143 dots in Display 2.8, so there are 71 or 72 dots in each half and 35 or 36 dots in each quarter. The weight of 155 lb divides the dots in half. The values that divide the two halves in half are roughly 115 lb and 250 lb. Thus, the middle 50% of the bear weights are between about 115 lb and 250 lb, with half above about 155 lb and half below.

DISCUSSION
Skewed Distributions

D4. Decide whether each distribution described will be skewed. Is there a wall that leads to bunching near it and a long tail stretching out away from it? If so, describe the wall.

    a. the sizes of islands in the Caribbean

    b. the average per capita incomes for the countries of the United Nations

    c. the lengths of pants legs cut and sewn to be 32 in. long

    d. the times for 300 university students of introductory psychology to complete a 1-hour timed exam

    e. the lengths of reigns of Japanese emperors

D5. Which would you expect to be the more common direction of skew, right or left? Why?

Outliers, Gaps, and Clusters

An unusual value, or **outlier**, is a value that stands apart from the bulk of the data. Outliers always deserve special attention. Sometimes they are mistakes (a typing mistake, a measuring mistake). Sometimes they are just atypical (a really big bear). Sometimes unusual features of the distribution are the key to an important discovery.

In the late 1800s, John William Strutt, third Baron Rayleigh, was studying the density of nitrogen using samples from the air outside his laboratory (from which known impurities were removed) and samples produced by a chemical procedure in his lab. He saw a pattern in the results that you can observe in the plot of his data in Display 2.11.

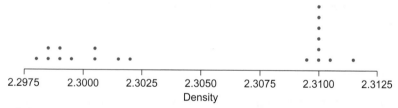

Display 2.11   Lord Rayleigh's densities of nitrogen. [Source: *Proceedings of the Royal Society*, 55 (1894).]

Lord Rayleigh saw two clusters separated by a gap. (There is no formal definition of a **gap** or a **cluster**; you have to use your best judgment about them. For example, some people call a single outlier a cluster of one; others don't. You also could argue that the value at the extreme right is an outlier, perhaps because of a faulty measurement.)

When Rayleigh checked the clusters, it turned out that the ten values to the left had all come from the chemically produced samples and the nine to the right had all come from the atmospheric samples. What did this great scientist conclude? The air samples on the right might be denser because of something in them besides nitrogen. This hypothesis led him to discover inert gases in the atmosphere.

Discoveries like this demonstrate why you should always plot your data.

Many data distributions are not simply mound-shaped, symmetric, or skewed, or have clusters as pronounced in those in the Rayleigh data. They may well be combinations of the above. Splitting such data into meaningful groups may produce distributions that have both simpler statistical properties and clearer practical interpretations.

For example, Display 2.12 shows the life expectancies of females from countries on two continents, Europe and Africa. The shape of this plot is difficult to describe succinctly. The left part looks skewed to the left, and the right part appears to have a couple of peaks of about equal height. These continents differ greatly in their socio-economic conditions. Life expectancies reflect these conditions, so you might guess that most of the African countries are in the left group and most of the European countries are in the right group.

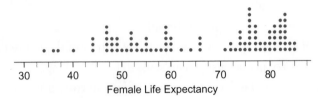

Display 2.12 Life expectancy of females by country on two continents.
[Source: Population Reference Bureau, World Population Data Sheet, 2008.]

A separate plot for each of the two continents (Display 2.13) shows a clearer picture. The life expectancies in African countries are not skewed to the left, but are spread far to the right as well, because some countries in Africa have life expectancies that rival those of Europe. The life expectancies in European countries are skewed to the left, as might be expected for a continent with a high standard of living.

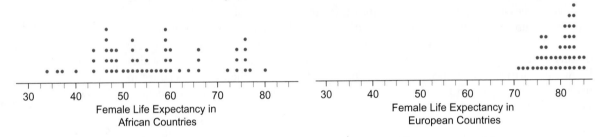

Display 2.13 Life expectancy of females in African and European countries.

DISCUSSION
Outliers, Gaps, and Clusters

D6. For which of the following distributions would you expect splitting into groups to be advantageous? Explain your reasoning.

a. heights of all the students in your school

b. gas mileages of all the cars in the school parking lots

c. ages of all the people in Sudan who died last year

Graphical Displays of Distributions

To see the shape, center, and spread of a distribution, you need a suitable plot. In this section, you'll learn more about three kinds of plots for quantitative variables (dot plot, histogram, and stemplot). When you look at a plot of quantitative data, you should attempt to answer these four questions:

- Where did this set of data come from?

- What are the cases and the variables?

- What are the shape, center, and spread of this distribution? Does the distribution have any unusual characteristics such as clusters, gaps, or outliers?

- What are possible interpretations or explanations of any patterns in the distribution?

The table in Display 2.14 gives various measurements or characteristics (called **variables**) of a sample of mammals. In this table, each mammal is a **case**, or object of study. These data will be used for some of the examples in this section.

Research about mammals is progressing so rapidly that values in this table may soon be out of date. For example, as zoologists continue to improve the housing and diet of mammals in captivity, the maximum longevity of some species increases almost yearly. For up-to-date information, consult the Animal Ageing & Longevity Database at http://genomics.senescence.info/species/. This database is useful for comparative biology studies, for ecological and conservation studies, and as a reference for students and zoologists.

Go to www.wiley.com/college/ Watkins to download data sets.

Mammal	Gestation Period (days)	Average Longevity (yr)	Maximum Longevity (yr)	Speed (mph)	Wild (1 = yes; 0 = no)	Predator (1 = yes; 0 = no)
Baboon	187	20	45	*	1	1
Bear, grizzly	225	25	50	30	1	1
Beaver	105	5	50	*	1	0
Bison	285	15	40	*	1	0
Camel	406	12	50	*	1	0
Cat	63	12	28	30	0	1
Cheetah	*	*	14	70	1	1
Chimpanzee	230	20	53	*	1	0
Chipmunk	31	6	8	*	1	0
Cow	284	15	30	*	0	0
Deer	201	8	20	30	1	0
Dog	61	12	20	39	0	1
Donkey	365	12	47	40	0	0
Elephant	660	35	70	25	1	0
Elk	250	15	27	45	1	0
Fox	52	7	14	42	1	1
Giraffe	425	10	34	32	1	0
Goat	151	8	18	*	0	0
Gorilla	258	20	54	*	1	0
Guinea pig	68	4	8	*	0	0
Hippopotamus	238	41	54	20	1	0
Horse	330	20	50	48	0	0
Kangaroo	36	7	24	40	1	0
Leopard	98	12	23	*	1	1
Lion	100	15	30	50	1	1
Monkey	166	15	37	*	1	0
Moose	240	12	27	*	1	0
Mouse	21	3	4	*	1	0
Opossum	13	1	5	*	1	1
Pig	112	10	27	11	0	0
Puma	90	12	20	*	1	1
Rabbit	31	5	13	35	0	0
Rhinoceros	450	15	45	*	1	0
Sea lion	350	12	30	*	1	1
Sheep	154	12	20	*	0	0
Squirrel	44	10	23	12	1	0
Tiger	105	16	26	*	1	1
Wolf	63	5	13	*	1	1
Zebra	365	15	50	40	1	0

(Asterisks [*] mark missing values.)

Display 2.14 Facts about a sample of mammals. [Source: *World Almanac and Book of Facts 2001*, p. 237].

Dot Plots

As the name suggests, dot plots such as those in Display 2.13 show individual cases as dots (or other plotting symbols, such as X). When you read a dot plot, keep in mind that different statistical software packages make dot plots in different ways. For example, sometimes one dot represents two or more cases, and sometimes values have been rounded. With a small data set, different rounding rules can give different shapes.

Dot plots tend to work best when

- you have a relatively small number of values to plot
- you want to see individual values, at least approximately
- you want to see the shape of the distribution
- you have one group, or a small number of groups, you want to compare
- you are making the plot by hand

Histograms

Much of the raggedness in the dot plot of the 1000 random numbers in Display 2.2 on page 23 can be smoothed out if the plot is changed from a dot plot to a **histogram** like the one in Display 2.15. The vertical axis gives the number of cases (called **frequency** or **count**) that are represented by each bar. You can think of a histogram as a dot plot with bars drawn around the dots and the dots erased. This makes the height of the bar a visual substitute for the number of dots.

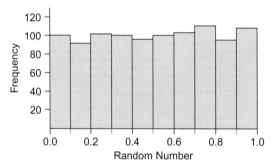

Display 2.15 Histogram of 1000 random numbers between 0 and 1.

To make a histogram, you divide the number line into subintervals, called **bins,** and construct a bar over each bin that has a height equal to the number of cases in that bin. In Display 2.15, each bin is of length 0.1.

Most calculators and statistical software place a value that falls at the dividing line between two bars into the bar on the right. For example, in Display 2.15, the bar going from 0.2 to 0.3 represents the 101 random numbers for which

$$0.2 \le random\ number < 0.3.$$

When data sets are large, you typically will find it more informative to change the vertical axis to relative frequency, measuring the proportion of the values in each bin. Display 2.15 was changed to the **relative frequency histogram** in Display 2.16 by dividing each frequency by 1000.

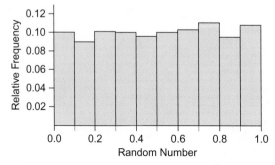

Display 2.16 Relative frequency histogram of 1000 random numbers.

Relative Frequency of Life Expectancies	**Example 2.3**

Display 2.17 shows the relative frequency distribution of life expectancies for 223 countries around the world. How many countries have a life expectancy of at least 75 but less than 80 years? Give the proportion of countries that have a life expectancy of 70 years or more.

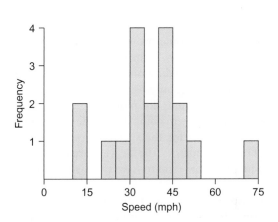

Elderly Tibetan couple.

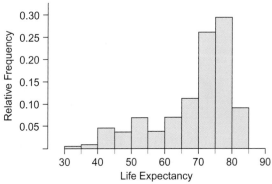

Display 2.17   Life expectancies of people by country. [Source: *www.cia.gov/library/publications/the-world-factbook/ rankorder/2102rank.html.*]

Solution

The bar including 75 years and up to 80 years has a relative frequency of about 0.29, so the number of countries with a life expectancy of at least 75 years but less than 80 years is about 0.29 · 223, or approximately 65.

The proportion of countries with life expectancy of 70 years or greater is the sum of the heights of the three bars to the right of 70—about 0.26 + 0.29 + 0.08, or 0.63.

Changing the width of the bars in your histogram can sometimes change your impression of the shape of the distribution, especially when there are few values. The two histograms in Display 2.18 show the distribution of the speeds of a sample of mammals (from the table in Display 2.14 on page 31). The shapes appear somewhat different as the two peaks disappear when the histogram has wider bars. There is no "right" answer to the question of which bar width is best, just as there is no rule that tells a photographer when to use a zoom lens for a close-up.

 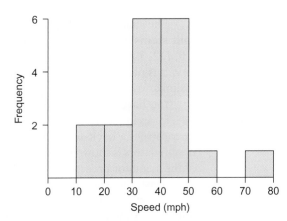

Display 2.18   Speeds of mammals using two different bar widths.

Histograms tend to work best when

- you have a large number of values to plot
- you don't need to see individual values exactly
- you want to see the general shape of the distribution
- you have only one distribution, or a small number of distributions, you want to compare
- you can use a calculator or computer to make the plot for you

---

DISCUSSION
Histograms

D7. Does using relative frequencies change the shape of a histogram? What information is lost and gained by using a relative frequency histogram rather than a frequency histogram?

D8. Refer to Display 2.18. In what sense does a histogram with narrow bars give you more information than a histogram with wider bars? In light of your answer, why don't we always make histograms with very narrow bars?

Stemplots

The plot in Display 2.19 is a **stem-and-leaf plot,** or **stemplot,** of the mammal speeds. It shows the key features of the distribution and preserves all the original numbers. The numbers on the left, called the **stems,** are the tens digits of the speeds. The numbers on the right, called the **leaves,** are the ones digits of the speeds. The leaf for the dog's speed of 39 mph is printed in bold. If you turn your book 90° counterclockwise, you will see that a stemplot looks something like a dot plot or histogram. Again, you can see the shape, center, and spread of the distribution.

```
1 | 1 2
2 | 0 5
3 | 0 0 0 2 5 9
4 | 0 0 0 2 5 8
5 | 0
6 |
7 | 0
```
3 | 9 represents 39 mph

Display 2.19   Stemplot of mammal speeds.

The stemplot in Display 2.20 displays the same speeds as Display 2.19, but with **split stems:** Each stem from the original plot has become two stems. If the ones digit is 0, 1, 2, 3, or 4, it is placed on the first line for that stem. If the ones digit is 5, 6, 7, 8, or 9, it is placed on the second line for that stem. Spreading out the stems in this way is similar to changing the width of the bars in a histogram.

You have compared two distributions by examining dot plots on the same scale (see, for example, Display 1.9 on page 9). Another way to compare two distributions is to construct a **back-to-back stemplot.** Such a plot for the speeds of predators and nonpredators is shown in Display 2.21. (For nonpredators, the "leaves" are to the right of the stem; for predators, to the left.) The predators tend to have the faster speeds— or, at least, there are no slow predators!

The stemplot of mammal speeds in Display 2.22 was made by Minitab® Statistical Software. Although different in format from the handmade plot in Display 2.20, it has the same basic structure. In the first two lines, N = 18 means that 18 cases were plotted;

```
1 │ 1 2
  •
2 │ 0
  • │ 5
3 │ 0 0 0 2
  • │ 5 9
4 │ 0 0 0 2
  • │ 5 8
5 │ 0
  •
6 │
  •
7 │ 0
```

3 │ 9 represents 39 mph

Display 2.20 Stemplot of mammal speeds, using split stems.

```
Predator          │ │ Nonpredator
                  │1│ 1 2
                  │•│
                  │2│ 0
                  │•│ 5
             0 0  │3│ 0 2
               9  │•│ 5
               2  │4│ 0 0 0
                  │•│ 5 8
               0  │5│
                  │•│
                  │6│
                  │•│
               0  │7│
```

3 │ 9 represents 39 mph

Display 2.21 Back-to-back stemplot of mammal speeds for predators and
nonpredators.

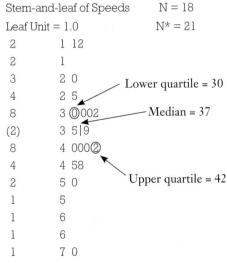

Stem-and-leaf of Speeds   N = 18
Leaf Unit = 1.0   N* = 21

```
  2      1  12
  2      1
  3      2  0            ─ Lower quartile = 30
  4      2  5
  8      3  ⓪002         ── Median = 37
 (2)     3  5│9
  8      4  000②
  4      4  58
  2      5  0            ── Upper quartile = 42
  1      5
  1      6
  1      6
  1      7  0
```

Display 2.22 Stemplot of mammal speeds with quartiles and median (vertical line).

$N^* = 21$ means that there were 21 cases in the original data set for which speeds were missing; and Leaf Unit = 1.0 means that the ones digits were graphed as the leaves. The numbers in the left column keep track of the cumulative number of cases, counting in from the extremes. The 2 on the left in the first line means that there are two cases on that stem. If you skip down three lines, the 4 on the left means that there are a total of four cases on the first four stems (speeds of 11, 12, 20, and 25). The left column makes it easy to identify the median (37) and to count in from either end to find the lower quartile (30) and upper quartile (42).

Usually, only two digits are plotted on a stemplot, one digit for the stem and one digit for the leaf. If the values contain more than two digits, the values may be truncated (the extra digits simply cut off) or rounded. For example, if the speeds had been given to the nearest tenth, 32.6 mph could be either truncated to 32 mph or rounded to 33 mph. As with the other types of plots, the rules for making stemplots are flexible. Do what seems to work best to reveal the important features of the data.

Stemplots are useful when

- you have a relatively small number of values to plot
- you would like to see individual values exactly, or, when the values contain more than two digits, you would like to see approximate individual values
- you want to see the shape of the distribution clearly
- you have two groups you want to compare

DISCUSSION
Stemplots

D9. What information is given by the numbers in the bottom half of the far left column of the plot in Display 2.22? What does the 2 in parentheses indicate?

D10. How might you construct a stemplot of the gestation periods for the mammals listed in Display 2.14 on page 31? Construct the stemplot and describe the shape of the distribution.

Bar Charts for Categorical Data

The variables plotted so far in this section have been **quantitative** (numerical) measurements of something—diameter of a tennis ball, weight of a penny, height of a student, and speed of an animal. The other type of variable characterizes an item as being in a certain category: a mammal is a predator or a nonpredator; a shirt is small, medium. or large; a vote on a ballot measure is yes or no; and an animal is male or female. Such **categorical data** are typically displayed using a **bar chart (bar graph)** of the frequencies.

The bar chart in Display 2.23 shows the frequency of mammals that fall into the categories "wild" and "domesticated," coded 1 and 0, respectively. You easily can see

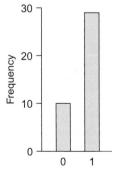

Display 2.23 Bar chart showing frequency of domesticated (0) and wild (1) mammals.

that there were about three times as many domesticated mammals as wild mammals in this sample. Note that the bars are separated so that there is no suggestion that the variable can take on a value of, say, 0.5.

Bar charts, like histograms, can be scaled in terms of frequencies or relative frequencies. Display 2.24 shows the proportion (relative frequency) of the female labor force age 25 and older in the United States who fall into various educational categories. The educational categories have a natural ordering from least education to most and are coded 1 through 9:

1. less than 9th grade
2. 9th to 12th grade, no diploma
3. high school graduate (includes equivalency)
4. some college, no degree
5. associate's degree
6. bachelor's degree
7. graduate or professional degree

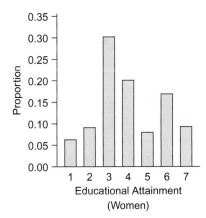

Display 2.24   The female labor force age 25 and older by educational attainment. [Source: U.S. Census Bureau, *March 2007 Current Population Survey, www.census.gov.*]

Because the ordering of categories in a bar chart is often arbitrary and the names of the categories need not be numbers, it makes little sense to talk about center and spread. But often it does make sense to talk about the **modal category**—the category with the highest frequency. More women in the labor force fall into the "high school graduate" category than into any other category.

Bar charts can be segmented to display two categorical variables on the same plot. In Display 2.25, the bars representing predators and nonpredators each are segmented into wild and domestic categories. The **segmented bar chart** makes it clear, for example, that there are more wild animals among the nonpredators than among the

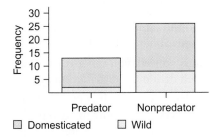

Display 2.25   A segmented bar chart of predator and nonpredator mammals.

predators, but the proportion of predators that are wild is larger than the proportion of nonpredators that are wild.

The analysis of categorical data is covered in Chapter 12. The analysis of quantitative data is the main theme of most of the remainder of the book.

DISCUSSION
Bar Charts for Categorical Data

D11. In the bar chart in Display 2.23 (domesticated/wild), would it matter if the order of the bars were reversed?

D12. Construct a segmented bar chart in which the wild and domesticated bars are segmented by the predator versus nonpredator categories. Explain how it relates to the chart of Display 2.25.

D13. The Gallup Poll of March 14, 2007 asked 1010 randomly sampled adults across the United States, "Which comes closest to your view about what government policy should be toward illegal immigrants currently residing in the United States?" The table in Display 2.26 gives the responses, by political party preference (in percent).

	Republicans	Independents	Democrats
Deport all	29	26	18
Allow to work in the U.S. for a limited time	20	12	13
Remain in the U.S. to become citizens	50	60	66
No opinion	1	2	3

Display 2.26 Response by political party preference (%).

a. Explain what the 29 in the Republicans column represents.

b. Construct a bar chart of the results for Democrats. Does the order of the bars matter?

c. Can you construct a bar chart of the type explained in this section for the data in the "Remain in the U.S. to become citizens" category? Why or why not? If not, what further information would you need to construct such a bar chart?

Summary 2.1: Visualizing Distributions— Shape, Center, and Spread

Distributions have different shapes, and different shapes call for different summaries.

- If your distribution is uniform (rectangular), it's often enough simply to tell the range of the set of values.

- If your distribution is approximately normal, you can give a good summary with the mean and the standard deviation. The mean lies at the center of the distribution, and the standard deviation is the distance on either side of the mean that encloses about 68% of the cases.

- If your distribution is skewed, you can give the lower quartile, median, and upper quartile, which divide the distribution into fourths.

- If your distribution has two peaks, it isn't useful to report a single center. One reasonable summary is to report the two peaks. However, it is even more useful if you can find another variable that divides your set of cases into two groups centered at the two peaks.

When a variable is quantitative, you can use a dot plot, stemplot (or stem-and-leaf plot), or histogram to display the distribution of values. From each, you can see the shape, center, and spread. However, the amount of detail varies, and you should choose a plot that fits both your data set and your reason for analyzing it.

- A dot plot is best used with a small number of values and shows roughly where each value lies on a number line.

- A stemplot is best used with a small number of values. Sometimes the actual values can be read from the plot.

- A histogram shows frequencies on the vertical axis and is most appropriate for large data sets. A relative frequency histogram shows relative frequencies on the vertical axis. To compute a relative frequency, divide the frequency by the total number of values in the data set.

When a variable is categorical, a bar chart is the best way to display the distribution. Unless the bars are rather uniform in height, the modal category is often of interest.

Practice

Practice problems help you master basic concepts and computation. Throughout this textbook, you should work *all* the practice problems for each topic you want to learn. The answers to all practice problems are given in the back of the book.

Shapes of Distributions

P1. The dot plot in Display 2.27 shows the distribution of the ages of 172 pennies in a sample collected by a statistics class.

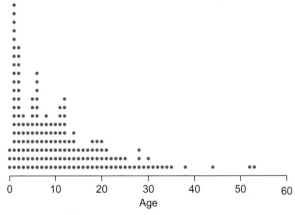

Display 2.27 Ages of 172 pennies (yr).

a. Where did this data set come from? What are the cases and the variables?

b. What are the shape, center, and spread of this distribution?

c. Does the distribution have any unusual characteristics? What are possible interpretations or explanations of the patterns you see in the distribution? That is, why does the distribution have the shape it does?

P2. The plots of variables w, x, and y in Display 2.28 show random samples, each a size of 200, from three different populations.

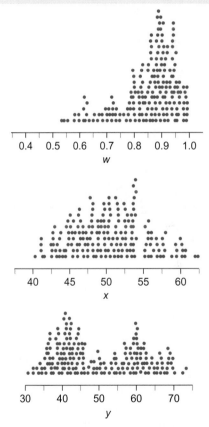

Display 2.28 Samples from three different populations.

a. Describe the shape of each distribution.

b. Which distribution, in your opinion, can best be modeled by a normal distribution?

c. Approximate the mean and standard deviation of the distribution that you chose in part b.

P3. For each of the following normal distributions, estimate the mean and standard deviation visually, and use your estimates to write a verbal summary of the form

"A typical SAT math score is roughly [mean], give or take [standard deviation] or so."

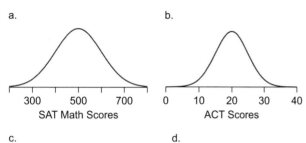

a.

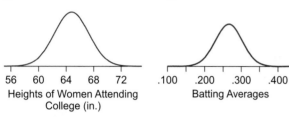

b.

SAT Math Scores

ACT Scores

c.

d.

Heights of Women Attending College (in.)

Batting Averages

P4. The dot plot in Display 2.29 gives the ages of people who died at rock concerts over a 12-year period, most of them crushed by the crowd. Estimate the median and quartiles of the distribution. Then write a verbal summary of the distribution.

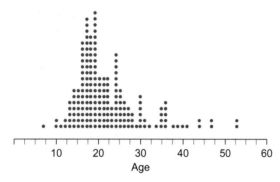

Display 2.29 Ages of people who died at rock concerts. [Source: Crowd Management Strategies, *www.crowdsafe.com/thewall.html.*]

Graphical Displays of Distributions

P5. Refer to the gestation periods of the mammals listed in Display 2.14 on page 31.

a. Make a dot plot of these gestation periods.

b. Write a sentence summarizing the shape, center, and spread of this distribution.

c. What kinds of mammals have longer gestation periods?

P6. The histogram in Display 2.30 gives the ages of a sample of 1000 people.

a. Describe the shape, center, and spread of this distribution.

b. Convert the histogram into a relative frequency histogram.

c. About what proportion of the people are age 50 or older?

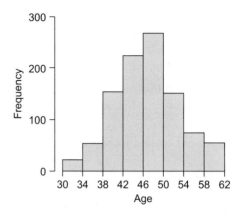

Display 2.30 Ages of 1000 people.

P7. Refer to the relative frequency histogram of life expectancy in countries around the world in Display 2.17 on page 33.

a. Estimate the proportion of countries with a life expectancy of less than 50 years.

b. Estimate the *number* of countries with a life expectancy of less than 50 years.

c. Describe the shape, center, and spread of this distribution.

P8. Refer to the table in Display 2.14 on page 31.

a. Make a back-to-back stemplot of the average longevities and maximum longevities.

b. Describe how the distributions differ in terms of shape, center, and spread.

c. Why do the differences occur?

P9. Using the technology available to you, make histograms of the average longevity and maximum longevity data in Display 2.14 on page 31, using bar widths of 4, 8, and 16 years. Comment on the main features of the shapes of these distributions. Which bar width appears to display these features best?

Bar Charts for Categorical Data

P10. The plot in Display 2.31 gives the number of deaths in the United States per month in 2007, with January coded as 1, February as 2, and so on. Does the number of deaths appear to be uniformly distributed over the months? Give a verbal summary of the way deaths are distributed over the months of the year.

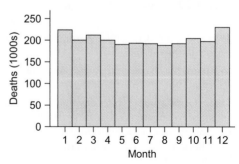

Display 2.31 Deaths, in thousands, per month, 2007. [Source: *www.cdc.gov/nchs/data/nvsr/nvsr57/ nvsr57_06.html.*]

P11. Suppose you collect this information for each student in your class: age, hair color, number of siblings, gender, and miles he or she lives from school. What are the cases? What are the variables? Classify each variable as quantitative or categorical.

P12. The plot in Display 2.32 shows the last digit of the Social Security numbers of the students in a statistics class. Describe this distribution.

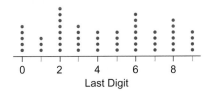

Display 2.32   Last digit of a sample of Social Security numbers.

P13. Display 2.33, which gives the educational attainment of the male labor force, is the counterpart of Display 2.24 on page 37.

   a.   What are the cases, and what is the variable?

   b.   Describe the distribution you see here.

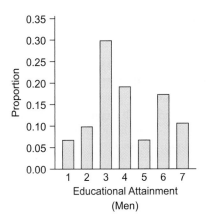

Display 2.33   The male labor force age 25 years and older by educational attainment. [Source: U.S. Census Bureau, March 2007 Current Population Survey, *www.census.gov*.]

   c.   How does the distribution of female education compare to the distribution of male education?

   d.   Why is it better to look at relative frequency bar charts rather than frequency bar charts to make this comparison?

Exercises

Exercises are mixed in difficulty—some are like the more routine practice problems while others require original thought and understanding of several concepts. Also unlike practice problems, exercises are not necessarily in the order that the concepts were introduced in the section. Each odd exercise and the even exercise following it are somewhat alike, should you want more practice. Answers to the odd numbered exercises are given in the back of the book.

E1. Using your knowledge of the variables and what you think the shape of the distribution might be, match each variable in this list with the appropriate histogram in Display 2.34.

   i.   scores on a fairly easy examination in statistics

   ii.   heights of a group of mothers and their 12-year-old daughters

   iii.   numbers of medals won by medal-winning countries in the 2008 Summer Olympics

   iv.   weights of grown hens in a barnyard

E2. The distribution in Display 2.35 shows measurements of the strength in pounds of 22s yarn (22s refers to a standard unit for measuring yarn strength). What is the basic shape of this distribution? What feature makes it uncharacteristic of distributions with that shape?

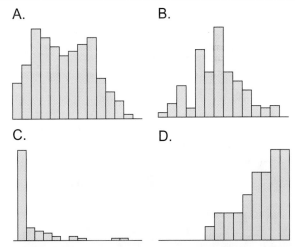

Display 2.34   Four histograms with different shapes.

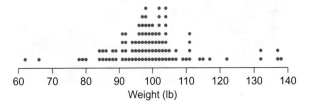

Display 2.35   Strength of yarn. [Source: Data and Story Library at Carnegie-Mellon University, *lib.stat.cmu.edu*.]

E3. The dot plot in Display 2.36 gives the ages of the officers who attained the rank of colonel in the Royal Netherlands Air Force.

    a. What are the cases? Describe the variables.

    b. Describe this distribution in terms of shape, center, and spread.

    c. What kind of wall might there be that causes the shape of the distribution? Generate as many possibilities as you can.

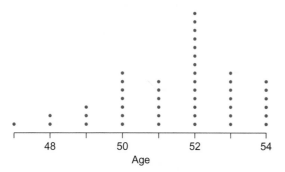

Display 2.36 Ages of colonels. Each dot represents two points. [Source: Data and Story Library at Carnegie-Mellon University, *lib.stat.cmu.edu.*]

E4. The dot plot in Display 2.37 shows the distribution of the number of inches of rainfall in Los Angeles for the seasons 1899–1900 through 1999–2000.

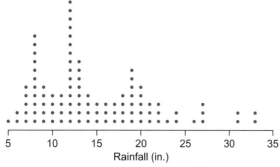

Display 2.37 Los Angeles rainfall during the twentieth century. [Source: National Weather Service.]

Los Angeles has wet winters and dry summers.

    a. What are the cases? Describe the variables.

    b. Describe this distribution in terms of shape, center, and spread.

    c. What kind of wall might there be that causes the shape of the distribution? Generate as many possibilities as you can.

E5. Describe each distribution as clustered, skewed right, skewed left, approximately normal, or roughly uniform.

    a. ages of all people who died last year in the United States

    b. ages of all people who got their first driver's license in your state last year

    c. SAT scores for all students in your state taking the test this year

    d. selling prices of all cars sold by General Motors this year

E6. Describe each distribution as clustered, skewed right, skewed left, approximately normal, or roughly uniform.

    a. the incomes of the world's 100 richest people

    b. the birthrates of Africa and Europe

    c. the heights of soccer players on the last Women's World Cup championship team

    d. the last two digits of telephone numbers in the town where you live

    e. the length of time students used to complete a chapter test, out of a 50-minute class period

Women's World Cup soccer.

E7. Sketch these distributions.

    a. a uniform distribution that shows the sort of data you would get from rolling a fair die 6000 times

    b. a roughly normal distribution with mean 15 and standard deviation 5

    c. a distribution that is skewed left, with half its values above 20 and half below, and with the middle 50% of its values between 10 and 25

    d. a distribution that is skewed right, with the middle 50% of its values between 100 and 1000 and with half the values above 200 and half below

    e. a normal distribution with mean 0 and standard deviation 1 (You will study this *standard normal distribution* in Section 2.4.)

E8. The U.S. Environmental Protection Agency keeps a list of hazardous waste sites for each of the 57 states and territories. The number of sites per state or territory ranges from 1 to 141. The middle 50% of the values lie between 12 and 34. Half of the values are above 18 and half are below. Sketch what the distribution might look like. [Source: Superfund Site Information, U.S. Environmental Protection Agency, 2009, *www.epa.gov/superfund/sites.*]

E9. Display 2.38 shows the distribution of the heights of U.S. males between the ages of 18 and 24. The heights are rounded to the nearest inch.

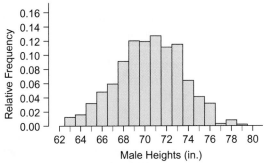

Display 2.38   Heights of males, ages 18 to 24. [Source: U.S. Census Bureau, *Statistical Abstract of the United States, 2009,* Table 201.]

a. Draw a smooth curve to approximate the histogram.

b. Without doing any computing, estimate the mean and standard deviation.

c. Estimate the proportion of men ages 18 to 24 who are 74 in. tall or less.

d. Estimate the proportion of heights that fall below 68 in.

e. Why should you say that the distribution of heights is "approximately" normal rather than simply saying that it is normally distributed?

E10. The histogram in Display 2.39 shows the distribution of SAT I math scores.

a. Without doing any computing, estimate the mean and standard deviation.

b. Roughly what percentage of the SAT I math scores would you estimate are within one standard deviation of the mean?

c. For SAT I critical reading scores, the shape was similar, but the mean was 10 points lower and the standard deviation was 2 points smaller. Draw a

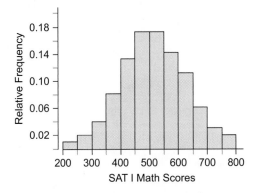

Display 2.39   Relative frequency histogram of SAT I math scores, 2004–2005. [Source: College Board Online, *www.collegeboard.org.*]

smooth curve to show the distribution of SAT I critical reading scores.

E11. The table in Display 2.41 provides the area, the population, and population density of the U.S. states. The histogram in Display 2.40 shows the areas of the states. It does not include Alaska because Alaska is so large compared to the other states that it doesn't fit on the plot.

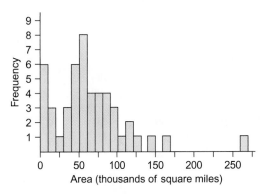

Display 2.40   Area of the U.S. states, excluding Alaska.

a. The distribution has two peaks. What simple geographic factor could help explain this?

b. Split the states into two groups according to that factor and use the technology available to you to make a plot of areas for each group. Do two peaks appear in each?

State	Area (sq mi)	Population (thousands)	Density (people/ sq mi)
Alabama	52,419	4,662	88.9
Alaska	663,267	686	1.0
Arizona	113,998	6,500	57.0
Arkansas	53,179	2,855	53.7
California	163,696	36,757	224.5
Colorado	104,094	4,939	47.4
Connecticut	5,543	3,501	631.6
Delaware	2,489	873	350.7
Florida	65,755	18,328	278.7
Georgia	59,425	9,686	163.0
Hawaii	10,931	1,288	117.8
Idaho	83,570	1,524	18.2
Illinois	57,914	12,902	222.8
Indiana	36,418	6,377	175.1
Iowa	56,272	3,003	53.4
Kansas	82,277	2,802	34.1
Kentucky	40,409	4,269	105.6
Louisiana	51,840	4,411	85.1
Maine	35,385	1,316	37.2
Maryland	12,407	5,634	454.1

Display 2.41   *(Continued on next page)*

State	Area (sq mi)	Population (thousands)	Density (people/ sq mi)
Massachusetts	10,555	6,498	615.6
Michigan	96,716	10,003	103.4
Minnesota	86,939	5,220	60.0
Mississippi	48,430	2,939	60.7
Missouri	69,704	5,912	84.8
Montana	147,042	967	6.6
Nebraska	77,354	1,783	23.0
Nevada	110,561	2,600	23.5
New Hampshire	9,350	1,316	140.7
New Jersey	8,721	8,683	995.6
New Mexico	121,590	1,984	16.3
New York	54,556	19,490	357.2
North Carolina	53,819	9,222	171.4
North Dakota	70,700	641	9.1
Ohio	44,825	11,486	256.2
Oklahoma	69,898	3,642	52.1
Oregon	96,381	3,790	38.5
Pennsylvania	46,055	12,448	270.3
Rhode Island	1,545	1,051	680.3
South Carolina	32,020	4,480	139.9
South Dakota	77,117	804	10.4
Tennessee	42,143	6,215	147.5
Texas	268,581	24,327	90.6
Utah	84,899	2,736	32.2
Vermont	9,614	621	64.6
Virginia	42,774	7,769	181.6
Washington	71,300	6,549	91.9
West Virginia	24,230	1,814	74.9
Wisconsin	65,498	5,628	85.9
Wyoming	97,814	533	5.4

Display 2.41 Area and population of U.S. states. [Source: U.S. Census, *State and Metropolitan Handbook, 2009.*]

E12. Refer to the table in Display 2.41 of E11. The histogram in Display 2.42 shows the population densities of all 50 states.
   a. Show how the population density for Vermont was computed.

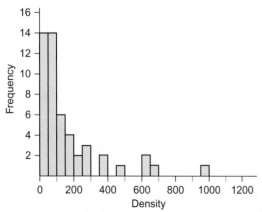

Display 2.42 Population density (people per square mile) of the U.S. states.

b. Which states are outliers?

c. Is Alaska an outlier for this variable? How can you tell from the table? From the plot?

E13. How do countries compare with respect to the value of the goods they produce? Display 2.43 shows gross domestic product (GDP) per capita, a measure of the total value of all goods and services produced divided by the number of people in a country, and the average number of people per room in housing units, a measure of crowdedness, for a selection of countries in Asia, Europe, and North America.

A dot plot of the per capita GDP by country is shown in Display 2.44.

a. How would you describe this distribution?

b. Which two countries have the highest per capita GDP? Do they appear to be outliers?

Country	Per Capita GDP (U.S. $)	Average Number of People per Room
Austria	44,652	0.7
Azerbaijan	3,691	2.1
Belgium	43,470	0.6
Bulgaria	5,178	1.0
Canada	43,368	0.5
China	2,604	1.1
Croatia	11,256	1.2
Cyprus	27,465	0.6
Czech Republic	16,881	1.0
Finland	46,371	0.8
France	40,090	0.7
Germany	40,162	0.5
Hungary	13,777	0.8
India	976	2.7
Iraq	2,404	1.5
Israel	23,383	1.2
Japan	34,225	0.8
Korea, Republic of	19,841	1.1
Kuwait	38,574	1.7
Netherlands	46,669	0.7
Norway	82,465	0.6
Pakistan	996	3.0
Poland	11,008	1.0
Portugal	20,990	0.7
Romania	7,523	1.3
Serbia-Montenegro	5,383	1.2
Slovakia	13,702	1.2
Sri Lanka	1,676	2.2
Sweden	49,873	0.5
Switzerland	56,579	0.6
Syria	1,883	2.0
Turkey	6,511	1.3
United Kingdom	46,549	0.5
United States	46,047	0.5

Display 2.43 Per capita GDP and crowdedness for a selection of countries. [Source: United Nations, *unstats.un.org.*]

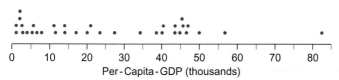

Display 2.44   Dot plot of per capita GDP.

c. A gap appears near the middle of the distribution. Which of the two clusters formed by this gap contains mostly Western European and North American countries? In what part of the world are most of the countries in the other cluster?

d. Is it surprising to find clusters and gaps in data that measure an aspect of the economies of the countries?

E14. The dot plot in Display 2.45 allows you to compare the countries listed in Display 2.43 in terms of the crowdedness of their residents.

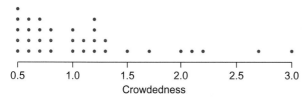

Display 2.45   Dot plot of crowdedness.

a. Describe this distribution in terms of shape, center, and spread.

b. Which countries appear to be outliers? Are they the same as the countries that appeared to be outliers for the per capita GDP data?

c. Where on the dot plot is the cluster that contains mostly Western European and North American countries?

E15. This diagram shows a uniform distribution on [0, 2], the interval from 0 through 2.

a. What value divides the distribution in half, with half the numbers below that value and half above?

b. What values divide the distribution into quarters?

c. What values enclose the middle 50% of the distribution?

d. What percentage of the values lie between 0.4 and 0.7?

e. What values enclose the middle 95% of the distribution?

E16. Match each plot in Display 2.46 with its median and quartiles (the set of values that divide the area under the curve into fourths).

a. 15, 50, 85

b. 50, 71, 87

c. 63, 79, 91

d. 35, 50, 65

e. 25, 50, 75

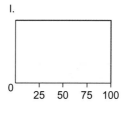

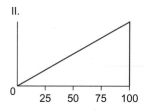

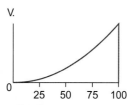

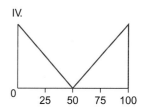

Display 2.46   Five distributions with different shapes.

E17. In this section, you looked at various characteristics of mammals.

a. Would you predict that wild mammals or domesticated mammals generally have greater longevity?

b. Using the data in Display 2.14 on page 31, make a back-to-back stemplot to compare the average longevities.

c. Write a short summary comparing the two distributions.

E18. The plots in Display 2.47 (on the next page) show a form of back-to-back histogram called a *population pyramid*. Describe how the distribution of ages in the United States differs from the distribution of ages in Mexico.

E19. Using the Westvaco data in Display 1.1 on page 3, make a bar chart showing the number of workers laid off in each round. In addition to a bar showing layoffs for each of the five rounds, include a bar showing the number of workers not laid off. Then make a relative frequency bar chart. Describe any patterns you see.

E20. In the listing of the Westvaco data in Display 1.1 on page 3, which variables are quantitative? Which are categorical?

E21. Examine the grouped bar chart in Display 2.48 (on the next page), which summarizes some of the information from Display 2.14 on page 31.

a. For each of the first three bars, describe what the height represents.

b. How can you tell from this bar chart whether a predator from the list in Display 2.14 is more likely to be wild or domesticated?

c. How can you tell from this bar chart whether a nonpredator or a predator is more likely to be wild?

E22. Make a grouped bar chart similar to that in E21 for the hourly and salaried Westvaco workers (see Display 1.1 on page 3), with bars showing the frequencies of *laid off* and *not laid off* for the two categories of workers.

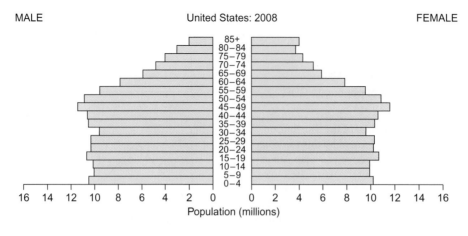

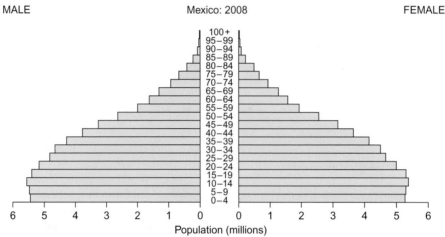

Display 2.47 Population pyramids for the United States and Mexico, 2008. [Source: U.S. Census Bureau, International Data Base, *www.census.gov/ipc/www/idb/pyramids.html*.]

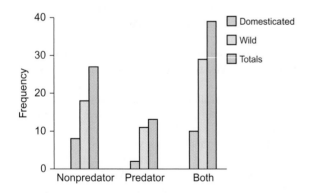

Display 2.48 Bar chart for nonpredators and predators, showing frequency of wild and domesticated mammals.

E23. When using published sources of data, you should try to understand exactly how they were collected. Otherwise, you may misinterpret them. For example, consider the mammal speeds in Display 2.14 on page 31.

   a. Count the number of mammals that have speeds ending in 0 or 5.

   b. How many speeds would you expect to end in 0 or 5 just by chance?

   c. What are some possible explanations for the fact that your answers in parts a and b are so different?

E24. Look through newspapers and magazines to find an example of a graph that is either misleading or difficult to interpret. Redraw the graph to make it clear.

2.2 ▶ Summarizing Center and Spread

Distributions of data typically are described by giving their shape, center, and spread. Summary statistics that locate the center of a distribution include the mean (average) and median. Summary statistics that measure the spread of a distribution include the range, distance between the quartiles, and standard deviation. So far you have relied on visual methods for estimating these summary statistics. In this section, you will learn how to compute their exact values.

Measures of Center: Mean and Median

The two most commonly used **measures of center** for quantitative data are the mean and the median.

Computing the Mean, x̄

The **mean**, $\bar{x}$, is the same number that many people call the "average." To compute the mean, sum all the values of the variable x and divide by the number of values, n:

$$\bar{x} = \frac{\Sigma x}{n}$$

(The symbol Σ, for sum, means to add up all the values of x.)

The mean is the balance point of a distribution. To estimate the mean visually on a dot plot or histogram, find where you would have to place a finger below the horizontal axis in order to balance the distribution, as if it were a tray of blocks (see Display 2.49).

The mean is the balance point.

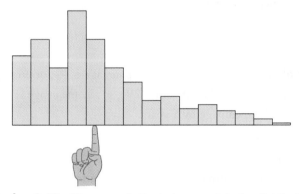

Display 2.49   The mean is the balance point of a distribution.

Finding the Median

The **median** is the number that divides the values into halves. To find it, first list the values in order. If there are an odd number of values, select the middle one. If there are n values and n is odd, you will find the median at position $\frac{n+1}{2}$. If n is even, the median is the average of the two values on either side of position $\frac{n+1}{2}$.

The median is the halfway point.

To estimate the median visually on a histogram, find the point that divides the total area of the bars into two equal parts, as shown in Display 2.50.

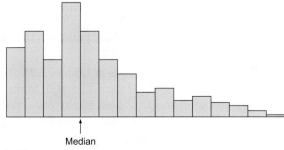

Median

Display 2.50   The median divides the distribution into two equal areas.

| Example 2.4 | **Effect of Round 2 Layoffs on Measures of Center** |

Ten workers were involved in the second round of layoffs at the Westvaco Corporation. Three workers were laid off, aged 55, 55, and 64. Seven workers were retained, aged 25, 33, 35, 38, 48, 55, and 56. The two dot plots in Display 2.51 show the distributions of hourly workers before and after the second round of layoffs. What was the effect of Round 2 on the mean age? On the median age?

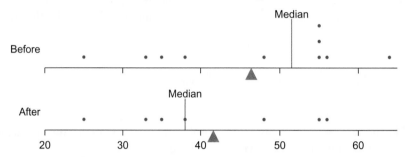

Display 2.51   Ages of Westvaco hourly workers before and after Round 2, showing the means(▲) and medians.

Solution

Means

Before: The sum of the ten ages is 464, so the mean age is $\frac{464}{10}$, or 46.4 years.

After: There are seven ages and their sum is 290, so the mean age is $\frac{290}{7}$, or 41.4 years.

The layoffs reduced the mean age by 5 years.

Medians

Before: Because there are ten ages, $n = 10$, so $\frac{(n + 1)}{2} = \frac{(10 + 1)}{2}$ or 5.5, and the median is halfway between the fifth ordered value, 48, and the sixth ordered value, 55. The median is $\frac{(48 + 55)}{2}$, or 51.5 years.

After: There are seven ages, so $\frac{(n + 1)}{2} = \frac{(7 + 1)}{2}$ or 4. The median is the fourth ordered value, or 38 years.

The layoffs reduced the median age by 13.5 years.

DISCUSSION
Measures of Center: Mean and Median

D14. Find the mean and median of each ordered list, and contrast their behavior.

    a. 1, 2, 3   **b.** 1, 2, 6

    c. 1, 2, 9   **d.** 1, 2, 297

D15. As you saw in D14, typically an outlier affects the mean more than the median.

    a. Use the fact that the median is the halfway point and the mean is the balance point to explain why this is true.

    b. For the distributions of mammal speeds in Display 2.21 on page 35 the means are 43.5 mph for predators and 31.5 mph for nonpredators. The medians are 40.5 mph and 33.5 mph, respectively. What about the distributions causes the means to be farther apart than the medians?

    c. What about the shapes of the plots in Display 2.51 explains why the means change so much less than the medians?

Measuring Spread Around the Median: Quartiles and Interquartile Range

You can locate the median of a distribution by dividing your data into a lower and upper half. You can use the same idea to measure spread: Find the values that divide each half in half. These two values, the lower quartile, Q_1, and the upper quartile, Q_3, together with the median, divide your data into four quarters. The distance between the upper and lower quartiles, called the interquartile range, or *IQR*, is a measure of spread.

Use the *IQR* as the measure of spread when the median is the measure of center.

$$IQR = Q_3 - Q_1$$

San Francisco, California, and Springfield, Missouri, have about the same median temperature over the year. In San Francisco, half the months of the year have a normal temperature above 56.5°F, half below. In Springfield, half the months have a normal temperature above 57°F, half below. If you judge by these medians, the difference hardly matters. But if you visit San Francisco, you had better take a jacket, no matter what month you go. If you visit Springfield, take your shorts and a T-shirt in the summer and a heavy coat in the winter.

The difference in temperatures between the two cities is not in their centers but in their variability. In San Francisco, the middle 50% of normal monthly temperatures lie in a narrow 9° interval between 52.5°F and 61.5°F, whereas in Springfield the middle 50% of normal monthly temperatures range over a 31° interval, varying from 40.5°F to 71.5°F. In other words, the *IQR* is 9°F for San Francisco and 31°F for Springfield.

Finding the Quartiles and *IQR*

If you have an even number of cases, finding the quartiles is straightforward: Order your observations, divide them into a lower and upper half, and then divide each half in half. If you have an odd number of cases, the idea is the same, but there's a question of what to do with the middle value when you form the upper and lower halves.

There is no one standard answer. Different statistical software packages use different procedures that can give slightly different values for the quartiles. Which procedure is used matters little with large data sets. In this book, the procedure is to omit the middle value when you form the two halves.

Finding the Quartiles and Interquartile Range for Workers' Ages	Example 2.5

Refer to Example 2.4. Find the quartiles and *IQR* for the ages of the hourly workers at Westvaco before and after Round 2 of the layoffs.

Solution

Before: There are ten ages: 25, 33, 35, 38, 48, 55, 55, 55, 56, 64. Because ten is even, the median is halfway between the two middle values, 48 and 55, so it is 51.5. The lower half of the data is made up of the first five ordered values, and the median of five values is located at the third value, so Q_1 is 35. The upper half of the data is the set of the five largest values, and the median of these is again the third value, so Q_3 is 55. The *IQR* is $55 - 35$, or 20.

25   33   35   38   48   55   55   55   56   64

Q_1 M Q_3

After: After the three workers are laid off in Round 2, there are seven ages: 25, 33, 35, 38, 48, 55, 56. Because *n* is odd, the median is the middle value, 38. To find the quartiles, ignore this one number. The lower half of the data is made up of the three ordered values to the left of position 4. The median of these is the second value, so Q_1 is 33. The upper half of the data is the set of the three ordered values to the right of position 4, and the median of these is again the second value, so Q_3 is 55. The *IQR* is 55 − 33, or 22.

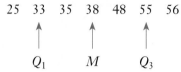

Unexpectedly, the *IQR* is slightly larger after three older ages are removed. This illustrates why you should be cautious using the median and quartiles with small data sets and when there are gaps in the distribution.

DISCUSSION
Finding the Quartiles and *IQR*

D16. The following quote is from the mystery *The List of Adrian Messenger*, by Philip MacDonald (Garden City, NY: Doubleday, 1959, p. 188). Detective Firth asks Detective Seymour if eyewitness accounts have provided a description of the murderer:

"Descriptions?" he said. "You must've collected quite a few. How did they boil down?"

"To a no-good norm, sir." Seymour shrugged wearily. "They varied so much, the average was useless."

Explain what Detective Seymour means.

Five-Number Summaries, Outliers, and Boxplots

The visual, verbal, and numerical summaries you've seen so far tell you about the middle of a distribution but not about the extremes. If you include the minimum and maximum values along with the median and quartiles, you get the five-number summary.

> ### The Five-Number Summary
>
> The five-number summary for a set of values includes:
>
> **Minimum:** the smallest value
>
> **Lower** or **first quartile, Q_1:** the median of the lower half of the ordered set of values
>
> **Median** or **second quartile:** the value that divides the ordered set of values into halves
>
> **Upper** or **third quartile, Q_3:** the median of the upper half of the ordered set of values
>
> **Maximum:** the largest value

The difference of the maximum and the minimum is called the **range.**

Display 2.52 shows the five-number summary for the speeds of the mammals listed in Display 2.14.

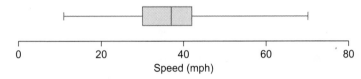

1	① 2		min	11
2	0 5		Q1	30
3	⓪ 0 0 2 5 │ 9		median	37
4	0 0 0 ② 5 8		Q3	42
5	0		max	70
6				
7	⓪			

Display 2.52 Five-number summary for the mammal speeds.

Display 2.53 shows a basic boxplot of the mammal speeds. A **basic boxplot** (or **box-and-whiskers plot**) is a graphical display of the five-number summary. The "box" extends from Q_1 to Q_3, with a line at the median. The "whiskers" run from the quartiles to the extreme values.

Display 2.53 Basic boxplot of mammal speeds.

The maximum speed of 70 mph for the cheetah is 20 mph from the next fastest mammal (the lion) and 28 mph from the nearest quartile. It is handy to have a version of the boxplot that shows isolated cases—outliers—such as the cheetah. Informally, outliers are any values that stand apart from the rest. You can use the following guideline to identify values that may qualify as outliers.

Guideline for Identifying Possible Outliers

Quantifying outliers in terms of the *IQR*.

A value may be an **outlier** if it is more than 1.5 times the *IQR* from the nearest quartile.

Note that "more than 1.5 times the *IQR* from the nearest quartile" is another way of saying "either greater than $Q_3 + 1.5 \cdot IQR$ or less than $Q_1 - 1.5 \cdot IQR$."

Outliers in the Mammal Speeds

Example 2.6

Use the $1.5 \cdot IQR$ guideline to identify outliers and the largest and smallest nonoutliers among the mammal speeds.

Solution
From Display 2.52, $Q_1 = 30$ and $Q_3 = 42$, so the *IQR* is $42 - 30$ or 12, and $1.5 \cdot IQR$ equals 18.

At the low end:

$$Q_1 - 1.5 \cdot IQR = 30 - 18 = 12$$

The pig, at 11 mph, is an outlier.

The cheetah, at 70 mph, is an outlier. The gazelle is not.

The squirrel, at 12 mph, is the smallest nonoutlier.

At the high end:

$$Q_3 + 1.5 \cdot IQR = 42 + 18 = 60$$

The cheetah, at 70 mph, is an outlier.
The lion, at 50 mph, is the largest nonoutlier.

The boxplot shown in Display 2.54 is like the basic boxplot except that the whiskers extend only as for as the largest and smallest nonoutliers (sometimes called adjacent values) and any outliers appear as individual dots, asteristes, or other symbols.

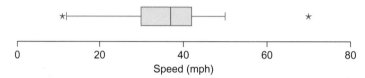

Display 2.54 Boxplot of mammal speeds with the outliers shown—the pig and the cheetah.

Boxplots are particularly useful for comparing several distributions.

| Example 2.7 | **Using Boxplots to Compare Strategies for Remembering People's Names** |

Two psychologists randomly divided 139 students into three groups. Group 1 learned the names of the others using the "name game," where the first student states his or her full name, the second student states his or her full name and that of the first student, and so on. Group 2 learned the names using the name game with the addition that each student stated a favorite activity. In Group 3, each student learned the names by going around and introducing him- or herself to each of the other students. One year later, the students were sent photos of the others in the group and asked to give the names of as many other students as he or she could remember. The variable recorded was the percentage of names recalled by each student. The results are shown in the boxplots in Display 2.55. [Source: Peter E. Morris and Catherine O. Fritz, "The Name Game: Using Retrieval Practice to Improve the Learning of Names," *Journal of Experimental Psychology: Applied*, Vol. 6, 2 (June 2000), pp. 124–129. Data from William Mendenhall and Terry L. Sincich, *Second Course in Statistics, Regression Analysis*, 6th ed. (Upper Saddle River, NJ: Prentice Hall, 2003), p. 265.]

   a. Compare the three distributions.

   b. Which method of recalling names would you choose as most effective and why?

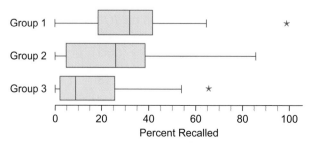

Display 2.55 Comparison of three methods of recalling names.

Solution

a. The shape of the distribution of Group 1 is roughly symmetric except for the outlier, while those of Groups 2 and 3 are skewed right. You can see that Group 3 is strongly skewed right because the left whisker is shorter than the left side of the box, which is shorter than the right side of the box, which is shorter than the right whisker. The distribution for Group 3 is bumping up against the wall of 0% where a person would recall no names at all. Each of Group 1 and Group 3 has an outlier on the high side. One person in Group 1 remembered all but one of the names! Because of the outliers in two of the groups, it is best to use the median as the measure of center. The median for Group 1 is higher than that for Group 2 and much higher than that for Group 3. The medians are about 32%, 26%, and 9%. The percentages for Group 2 have the largest spread.

b. When there are around 40 or 50 names to learn, the name game appears to be the most effective method of learning people's names. Having each person add a fact about him- or herself appears to change the results very little, especially with respect to the center of the distribution. Having people introduce themselves in pairs appears to be the least effective method.

An outlier may lie away from the general pattern for many reasons. Sometimes, an outlier results from mistakes in reading or recording measurements, so looking for outliers is a good way to check for such errors. Often, however, outliers are natural occurrences in distributions, as in the typically skewed distributions of salaries of employees of large firms. (There are generally a few very high salaries, which is to be expected.) Occasionally, the identification of outliers leads to insights into the process from which the measurements came, as in the following examples:

- In 1901, a dentist in Colorado Springs, Colorado, by the name of Fred McKay observed that children coming to him had unusually healthy teeth. After some research, he found the cause to be the presence of fluoride in the local water supply, leading to the common practice of fluoridation of water and adding fluoride to toothpaste.

- A British meteorologist, Sir Gilbert Walker, observed that the extreme values of sea surface pressure differences between Darwin, Australia, and Tahiti helped him to predict El Niño. (See *www.it.usyd.edu.au/~chawla/dmkd/dmkd-cfp.html*.)

- In analyzing reasons for the financial meltdown of 2008–2009, many economists were talking about the fact that banks and individual investors should have paid more attention to the likelihood of "tail events" (unlikely, but surely possible, disastrous financial scenarios).

Boxplots are useful when you are plotting a single quantitative variable and

| When are boxplots most useful? |

- you want a rough idea of the shape of one distribution
- you want to compare the shapes, centers, and spreads of two or more distributions
- you don't need to see individual values, even approximately
- you don't need to see more than the five-number summary but would like outliers to be clearly indicated

DISCUSSION
Five-Number Summaries, Outliers, and Boxplots

D17. Test your ability to interpret boxplots by answering these questions.

a. Approximately what percentage of the values in a data set lie within the box? Within the lower whisker, if there are no outliers? Within the upper whisker, if there are no outliers?

b. How would a boxplot look for a data set that is skewed right? Skewed left? Symmetric?

c. How can you estimate the *IQR* directly from a boxplot? How can you estimate the range?

d. Is it possible for a boxplot to be missing a whisker? If so, give an example. If not, explain why not.

e. Contrast the information you can learn from a boxplot with what you can learn from a histogram. List the advantages and disadvantages of each type of plot.

Measuring Spread Around the Mean: The Standard Deviation

Researchers in the sciences, social sciences, business, and medicine typically use the standard deviation as the measure of spread (variability). For example, in finance, to measure the volatility of a stock, analysts compute the standard deviation of the daily closing prices of the stock over, say, the past 30 days. A lower standard deviation means a more stable price and so less risk to the investor, while a larger standard deviation means more variability from day to day in the prices and so indicates a more risky investment. The upper curve in Display 2.56 gives the daily prices of the S&P 500 stock index over 2007 and 2008. Each standard deviation plotted in the lower curve was computed from the prices of the previous 30 days. Notice that during 2007, when the price was relatively stable, the standard deviation was low. During October 2008, when the price dropped precipitously, the standard deviation increased dramatically.

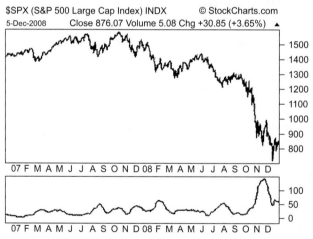

Display 2.56   S&P 500 stock index, 2007–2008. [Source: *http://stockcharts.com/*.]

STATISTICS IN ACTION 2.2 ► Comparing Hand Spans: How Far Are You from the Mean?

There are various ways you can measure the spread of a distribution around its mean. This activity gives you a chance to create a measure of your own.

What You'll Need: a ruler

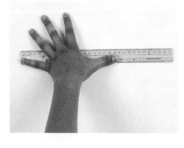

1. Spread your hand on a ruler and measure your hand span (the distance from the tip of your thumb to the tip of your little finger when you spread your fingers) to the nearest half centimeter.

2. Find the mean hand span for your group.

3. Make a dot plot of the results for your group. Write names or initials above the dots to identify the cases. Mark the mean with a wedge (▲) below the number line.

4. Give two sources of variability in the measurements. That is, give two reasons why all the measurements aren't the same.

5. How far is your hand span from the mean hand span of your group? How far from the mean are the hand spans of the others in your group?

6. Make a plot of differences from the mean. Again label the dots with names or initials. What is the mean of these differences? Tell how to get the second plot from the first without computing any differences.

7. Using the idea of differences from the mean, invent at least two measures that give a "typical" distance from the mean.

8. Compare your measures with those of the other groups in your class. Discuss the advantages and disadvantages of each group's method.

The first step in understanding the standard deviation is to learn to compute deviations from the mean. The difference between a value and the mean of its distribution, $x - \bar{x}$, is called the **deviation from the mean.** Because the mean is the balance point of the distribution, the set of deviations from the mean will always sum to zero.

Deviations from the Mean, $x - \bar{x}$

Deviations from the mean sum to zero:

$$\Sigma(x - \bar{x}) = 0$$

Pain Threshold and Hair Color: Deviations from the Mean

Example 2.8

Psychologists have conducted many studies to determine the characteristics of people who have a high threshold for pain. One of the first, at the University of Melbourne, compared the pain tolerance of people of various hair colors. The results are shown in the following table. The higher the score, the higher the person's pain threshold. [Source: *www.statsci.org/data/oz/blonds.html.*]

Light Blond	Dark Blond	Light Brunette	Dark Brunette
62	63	42	32
60	57	50	39
71	52	41	51
55	41	37	30
48	43		35

Compute the mean score for the dark brunettes and the deviations from the mean. Which score is farthest from the mean? Verify that the sum of the deviations from the mean is 0.

Solution

The scores 32, 39, 51, 30, and 35 have mean 37.4. The deviations from the mean are

$$32 - 37.4 = -5.4$$
$$39 - 37.4 = 1.6$$
$$51 - 37.4 = 13.6$$
$$30 - 37.4 = -7.4$$
$$35 - 37.4 = -2.4$$

The sum of the deviations is $(-5.4) + 1.6 + 13.6 + (-7.4) + (-2.4)$, which equals 0. The score of 51 is farthest from the mean.

This dot plot shows how the deviations are balanced around zero:

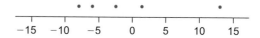

How can you use the deviations from the mean to get a measure of spread? You can't simply find the average of the deviations, because you will get 0 every time. You could find the average of the *absolute values* of the deviations. That gives a perfectly reasonable measure of spread, but it does not turn out to be very easy to use or very useful. Think of how hard it is to deal with an equation that has sums of absolute values in it, for example, $y = |x - 1| + |x - 2| + |x - 3|$. On the other hand, if you *square* the deviations, which also gets rid of the negative signs, you get a sum of squares. Such a sum is always quadratic no matter how many terms there are, for example,

$$y = (x - 1)^2 + (x - 2)^2 + (x - 3)^2 = 3x^2 - 12x + 14.$$

| Use the standard deviation as the measure of spread when the mean is the measure of center.

The measure of spread that incorporates the square of the deviations is the standard deviation, abbreviated s, that you met in Section 2.1. Because sums of squares are easy to work with mathematically and because the standard deviation is the natural partner with the normal distribution and the mean, the standard deviation offers important advantages that other measures of spread don't have. You will learn more about these advantages in Chapter 7. The formula for the standard deviation, s, is given in the box.

| The square root changes the units back to those of the original measurements.

Formula for the Standard Deviation, s

$$s = \sqrt{\frac{\Sigma(x - \bar{x})^2}{n - 1}}$$

Here, $\bar{x}$ is the mean and n is the number of values of x. The square of the standard deviation, s^2, is called the **variance**.

It might seem more natural to divide by n to get the average of the squared deviations. In fact, two versions of the standard deviation formula are used: One divides by the sample size, n; the other divides by $n - 1$. Dividing by $n - 1$ gives a slightly larger value. This is useful because otherwise the standard deviation computed from a sample would tend to be smaller than the standard deviation of the population from which the sample came. (You will learn more about this in Chapter 7.) In practice, dividing by $n - 1$ is almost always used for real data even if they aren't a sample from a larger population.

Pain Threshold and Hair Color: The Standard Deviation	Example 2.9

Refer to Example 2.8 on page 55. Compute the standard deviation of the pain threshold scores of the dark brunettes.

Solution
Using a table is a good way to organize the steps in computing the standard deviation. First find the mean score, $\bar{x}$, which is 37.4. Then subtract it from each observed score, x, to get the deviations, $(x - \bar{x})$. Square each deviation to get $(x - \bar{x})^2$.

Score, x	Mean, $\bar{x}$	Deviation, $(x - \bar{x})$	Squared Deviation, $(x - \bar{x})^2$
32	37.4	−5.4	29.16
39	37.4	1.6	2.56
51	37.4	13.6	184.96
30	37.4	−7.4	54.76
35	37.4	−2.4	5.76
	Sum	0	277.2

To compute the standard deviation, sum the squared deviations, divide the sum by $n - 1$, and take the square root.

$$s = \sqrt{\frac{\Sigma(x - \bar{x})^2}{n - 1}} = \sqrt{\frac{277.2}{5 - 1}} \approx 8.32$$

You can think of the standard deviation of 8.32 as representing a typical deviation from the mean.

Notice how much the score of 51, which had a large deviation from the mean, contributed to the size of the standard deviation.

> Think of the standard deviation as a typical deviation from the mean.

David Brooks, a columnist for the *New York Times*, once advised college students to "Take statistics. Sorry, but you'll find later in life that it's handy to know what a standard deviation is" (*New York Times*, March 2, 2006). Why did he mention the standard deviation in particular? Because it is the most widely used measure of variability in data, and variability is often not recognized as central to understanding and interpreting data correctly. You score 120 on an aptitude test for a job where the average score of all test takers is about 100. Is your score good or not so good? If, in addition, you know that the standard deviation is 22, then your score is just a typical deviation away for the mean and not all that exciting because many others will have similar scores. If, on the other hand, you know that scores on this test have a standard deviation of 10, then you also know that you have done very well, indeed.

As another example, suppose that the price of a stock you carefully selected gained 14% over the last year while the average percentage gain was 10%. Is that something to get excited about? If the standard deviation of percentage gains over the year was 2%, then you can pat yourself on the back; the 14% increase was a very large gain relative to other stocks. If, on the other hand, the standard deviation of percentage gains over the year was 5%, then your selection wasn't anything special because the 14% gain was less than a typical deviation from average.

Your calculator or computer will compute all common summary statistics. For example, here are the summary statistics, as computed by the statistical software Minitab, for the pain threshold scores of each of the hair color groups and by a TI-84

Plus calculator for the dark brunettes. The SE Mean (standard error of the mean) is the subject of the second section of Chapter 7.

Descriptive Statistics: Light Blond, Dark Blond, Light Brunette, Dark Brunette

Variable	N	N*	Mean	SE Mean	StDev	Minimum	Q1	Median	Q3	Maximum
Light Blond	5	0	59.20	3.81	8.53	48.00	51.50	60.00	66.50	71.00
Dark Blond	5	0	51.20	4.15	9.28	41.00	42.00	52.00	60.00	63.00
Light Brunette	4	0	42.50	2.72	5.45	37.00	38.00	41.50	48.00	50.00
Dark Brunette	5	0	37.40	3.72	8.32	30.00	31.00	35.00	45.00	51.00

```
1-Var Stats L1



```

```
1-Var Stats
x̄=37.4
Σx=187
Σx²=7271
Sx=8.324662155
σx=7.445804188
↓n=5
```

```
1-Var Stats
↑n=5
minX=30
Q1=31
Med=35
Q3=45
maxX=51
```

DISCUSSION
The Standard Deviation

D18. The dot plots give the pain threshold scores for people of various hair colors (a higher score indicates a higher threshold).

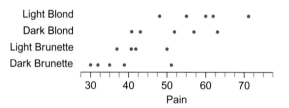

a. Which color has the largest standard deviation? Which has the smallest? Refer to the output of the summary statistics for the pain threshold and hair color example above to check your answer.

b. Refer to the means in the output of the summary statistics. What general statement can you make about hair color and pain threshold?

c. What concerns should you have about this study?

D19. The standard deviation, if you look at it the right way, is a generalization of the usual formula for the distance between two points, (x_1, y_1) and (x_2, y_2), which is

$$\sqrt{(x_2 - x_1)^2 + (y_2 - y_1)^2}$$

How does the formula for the standard deviation remind you of the formula for the distance between two points?

D20. What effect does dividing by $n - 1$ rather than by n have on the standard deviation? Does which one you divide by matter more with a large number of values or with a small number of values?

Summaries from a Frequency Table

To find the mean of the numbers 5, 5, 5, 5, 5, 5, 8, 8, and 8, you could sum them and divide their sum by how many numbers there are. However, you could get the same answer faster by taking advantage of the repetitions:

$$\bar{x} = \frac{5 \cdot 6 + 8 \cdot 3}{6 + 3}$$

$$= \frac{30 + 24}{9} = \frac{54}{9} = 6$$

You can use formulas to find the mean and standard deviation of values in a frequency table, like the one in Example 2.10.

Formulas for the Mean and Standard Deviation of Values in a Frequency Table

If each value x occurs with frequency f, the mean of a frequency table is given by

$$\bar{x} = \frac{\sum x \cdot f}{n}$$

The standard deviation is given by

$$s = \sqrt{\frac{\sum(x - \bar{x})^2 \cdot f}{n - 1}}$$

where n is the sum of the frequencies, or $n = \Sigma f$.

Aircraft Struck by Reptiles

Example 2.10

The following table gives the number of aircraft struck by reptiles (alligators, turtles, and iguanas) in the United States in each of the years 1990 through 2007. About half of these strikes occurred in Florida and only one resulted in substantial damage to the aircraft. Organize the number of strikes into a frequency table and compute the mean and standard deviation of the number of reptile strikes per year. [Source: U.S. Department of Transportation, *Wildlife Strikes to Civil Aircraft in the United States, 1990–2007*, June 2008, Table 1.]

Year	Reptile Strikes	Year	Reptile Strikes
1990	0	1999	1
1991	0	2000	3
1992	1	2001	8
1993	0	2002	15
1994	1	2003	5
1995	8	2004	6
1996	3	2005	7
1997	14	2006	9
1998	7	2007	7

Solution

In three years, there were 0 strikes, in three years there was 1 strike, and so on. The resulting frequency table in Display 2.57 shows a way to organize the steps in computing the mean.

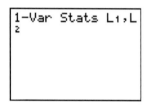

Number of Strikes, x	Frequency, f	$x \cdot f$
0	3	0
1	3	3
3	2	6
5	1	5
6	1	6
7	3	21
8	2	16
9	1	9
14	1	14
15	1	15
Sum	18 years	95 strikes

$$\bar{x} = \frac{\sum x \cdot f}{n} = \frac{95}{18} \approx 5.28$$

Display 2.57 Steps in computing the mean from a frequency table.

Display 2.58 gives an extended version of the table, organizing the steps in computing both the mean and the standard deviation.

```
1-Var Stats
x̄=5.277777778
Σx=95
Σx²=859
Sx=4.586496235
σx=4.457273152
↓n=18
```

Number of Strikes, x	Frequency, f	$x \cdot f$	$x - \bar{x}$	$(x - \bar{x})^2$	$(x - \bar{x})^2 \cdot f$
0	3	0	−5.28	27.88	83.64
1	3	3	−4.28	18.32	54.96
3	2	6	−2.28	5.20	10.40
5	1	5	−0.28	0.08	0.08
6	1	6	0.72	0.52	0.52
7	3	21	1.72	2.96	8.88
8	2	16	2.72	7.40	14.80
9	1	9	3.72	13.84	13.84
14	1	14	8.72	76.04	76.04
15	1	15	9.72	94.48	94.48
Sum	18 years	95 strikes			357.61

$$s = \sqrt{\frac{\sum (x - \bar{x})^2 \cdot f}{n - 1}} = \sqrt{\frac{357.61}{18 - 1}} \approx 4.59$$

Display 2.58 Steps in computing the standard deviation from a frequency table.

In a typical year, the United States had 5.28 aircraft that were struck by reptiles, give or take 4.59 strikes.

DISCUSSION
Summaries from a Frequency Table

D21. Explain why the formula for the standard deviation in the box on page 59 gives the same result as the formula on page 56.

Summary 2.2: Summarizing Center and Spread

Boxplots, rules for outliers, and stemplots were the brainchild of one of the most extraordinary statisticians of the 20th century—John Tukey. Although a brilliant mathematician himself, Professor Tukey came to see the practice of statistics as more closely aligned with science. Tukey's methods and the technological revolution of the last half of the 20th century made data analysis possible and brought data analysis to the center of modern statistical practice.

Your first step in any data analysis should always be to look at a plot of your data, because the shape of the distribution will help you determine what summary measures to use for center and spread.

- To describe the center of a distribution, the two most common summaries are the median and the mean. The median, or halfway point, of a set of ordered values is either the middle value (if n is odd) or halfway between the two middle values (if n is even). The mean, or balance point, is the sum of the values divided by the number of values.

- To measure spread around the median, use the interquartile range, or *IQR*, which is the width of the middle 50% of the values and equals the distance from the lower quartile to the upper quartile. The quartiles are the medians of the lower half and upper half of the ordered list of values.

- To measure spread around the mean, use the standard deviation. To compute the standard deviation for a data set of size n, first find the deviations from the mean, then square them, sum the squared deviations, divide by $n - 1$, and take the square root.

A boxplot is a useful way to compare the general shape, center, and spread of two or more distributions with a large number of values. A boxplot also shows outliers. Examine any value that is more than 1.5 times the *IQR* from the nearest quartile to see if it should be considered an outlier.

Practice

Measures of Center: Mean and Median

P14. Find the mean and median of these ordered lists.

    a. 1 2 3 4   **b.** 1 2 3 4 5

    c. 1 2 3 4 5 6   **d.** 1 2 3 4 5 … 97 98

    e. 1 2 3 4 5 … 97 98 99

P15. Five 3rd graders, all about 4 ft tall, are standing together when their teacher, who is 6 ft tall, joins the group. What is the new mean height? The new median height?

P16. The stemplots in Display 2.59 show the life expectancies (in years) for females in the countries of Africa and Europe. The means are 56.5 years for Africa and 79.6 years for Europe.

    a. Find the median life expectancy for each set of countries.

    b. Is the mean or the median smaller for each distribution? Why is this so?

```
         Africa                    Europe
N = 56   Leaf Unit = 1.0    N = 43   Leaf Unit = 1.0
 3 |4                        7 |1
 3 |6 7                      7 |2 3
 4 |0 4 4 4                  7 |4 5 5
 4 |7 7 7 7 7 8 8 8 9 9 9    7 |6 6 6 7 7 7 7
 5 |0 1 2 2 2 2 3 3 4        7 |8 8 9 9
 5 |5 5 5 6 7 8 9 9 9 9 9    8 |0 0 0 1 1 1 1 1
 6 |0 0 0 2 4                8 |2 2 2 2 2 2 3 3 3 3 3 3 3
 6 |6 6 6                    8 |4 4 5 5
 7 |2 4 4 4
 7 |6 6 6 6 7
 8 |0
```

Display 2.59 Female life expectancy in Africa and Europe. [Source: Population Reference Bureau, *World Population Data Sheet*, 2008.]

Measuring Spread Around the Median: Quartiles and *IQR*

P17. Find the quartiles and *IQR* for these ordered lists.

    a. 1 2 3 4 5 6   **b.** 1 2 3 4 5 6 7

    c. 1 2 3 4 5 6 7 8   **d.** 1 2 3 4 5 6 7 8 9

P18. Display 2.60 shows a back-to-back stemplot of the average longevity of predators and nonpredators.

```
      Predators  |   | Nonpredators
              1  | 0 | 3 4
            7 5  | • | 5 5 6 7 8 8
      2 2 2 2 2  | 1 | 0 0 0 2 2 2 2
            6 5  | • | 5 5 5 5 5 5
            0    | 2 | 0 0 0
            5    | • |
                 | 3 |
                 | • | 5
                 | 4 | 1
                   1 | 5 stands for 15 years
```

Display 2.60 Average longevities of predators and nonpredators.

    a. By counting on the plot, find the median and quartiles for each group of mammals.

    b. Write a pair of sentences summarizing and comparing the shape, center, and spread of the two distributions.

Five-Number Summaries, Outliers, and Boxplots

P19. Display 2.61 shows parallel boxplots of the percentages of people living in urban areas for the countries of Europe and the countries of Africa. Describe the distributions as well as the key differences in shape, center, and spread between the two continents. The outlier is Liechtenstein.

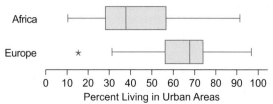

Display 2.61 Percentage of residents living in urban areas. [Source: Population Reference Bureau, *World Population Data Sheet*, 2008.]

P20. a. Display 2.62 shows side-by-side boxplots of average longevity for wild and domesticated mammals. Why does the boxplot for domesticated animals have no median line? Compare the two distributions.

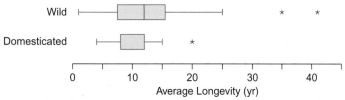

Display 2.62 Comparison of average longevity.

b. Use the medians and quartiles for the speeds of the domesticated and wild mammals given below and the data in Display 2.14 on page 31 to construct side-by-side basic boxplots of the speeds of wild and domesticated mammals. (Don't show outliers in these plots.)

	Q1	Median	Q3
Domesticated	30	37	40
Wild	27.5	36	43.5

P21. The stemplot of average mammal longevities is shown in Display 2.63.

```
0 | 1 3 4              1 | 5 stands for 15 years
· | 5 5 5 6 7 7 8 8
1 | 0 0 0 2 2 2 2 2 2 2 2 2
· | 5 5 5 5 5 5 6
2 | 0 0 0 0
· | 5
3 |
· | 5
4 | 1
```

Display 2.63 Average longevity (in years) of 38 mammals.

a. Use the stemplot to find the five-number summary.
b. Find the *IQR*.
c. Compute $Q_1 - 1.5 \cdot IQR$. Identify any outliers (give the animal name and longevity) at the low end.

d. Identify any outlier at the high end and the largest nonoutlier.
e. Draw a boxplot.

The Standard Deviation

P22. Verify that the sum of the deviations from the mean is 0 for the numbers 1, 2, 4, 6, and 9. Find the standard deviation.

P23. Without computing, match each list of numbers in the left column with its standard deviation in the right column. Check any answers you aren't sure of by computing.

a.	1 1 1 1	**i.**	0
b.	1 2 2	**ii.**	0.058
c.	1 2 3 4 5	**iii.**	0.577
d.	10 20 20	**iv.**	1.581
e.	0.1 0.2 0.2	**v.**	3.162
f.	0 2 4 6 8	**vi.**	3.606
g.	0 0 0 0 5 6 6 8 8	**vii.**	5.774

Summaries from a Frequency Table

P24. How did family sizes change over the years from 1967 to 2007? Display 2.64 shows the data on family size for two representative sets of 100 families each, one from 1967 and one from 2007. (The U.S. Census Bureau defines a family as two or more people related by birth, marriage, or adoption living together.)

Family Size	1967	2007
2	33	45
3	21	22
4	19	20
5	13	8
6	7	3
7	5	1
8	2	1

Display 2.64 Family sizes for representative samples of 100 families in two different years. [Source: U.S. Census Bureau.]

a. Sketch a histogram of family size for each year and comment on how the distribution changed from 1967 to 2007.
b. Use the formulas for the mean and standard deviation of values in a frequency table to compute the mean and standard deviation of the family sizes for each year. Compare.

P25. Refer to Display 2.64 and your work in P24.
a. Find the median family size for each year and compare.
b. Find the quartiles for each year and compute the interquartile range.
c. Make parallel boxplots of the family sizes for the two years.
d. Write a brief summary statement (as for a local newspaper) on how family size has changed over the 40 years.

Exercises

E25. The mean of a set of seven values is 25. Six of the values are 24, 47, 34, 10, 22, and 28. What is the 7th value?

E26. The sum of a set of values is 84, and the mean is 6. How many values are there?

E27. Here are the medians and quartiles for the speeds of the domesticated and wild mammals:

	Q_1	Median	Q_3
Domesticated	30	37	40
Wild	27.5	36	43.5

   a. Use the information in Display 2.14 on page 31 to verify these values, and then use them to summarize and compare the two distributions.

b. Why might the speeds of domesticated mammals be less spread out than the speeds of wild mammals?

E28. Make a back-to-back stemplot comparing the ages of those retained and those laid off among the salaried workers in the engineering department at Westvaco (see Display 1.1 on page 3). Find the medians and quartiles, and use them to write a verbal comparison of the two distributions.

E29. Four histograms and four boxplots appear in Display 2.65. Each shows 1000 values.

   a. Describe the shape of the distribution in each histogram.

   b. Match each histogram to its boxplot.

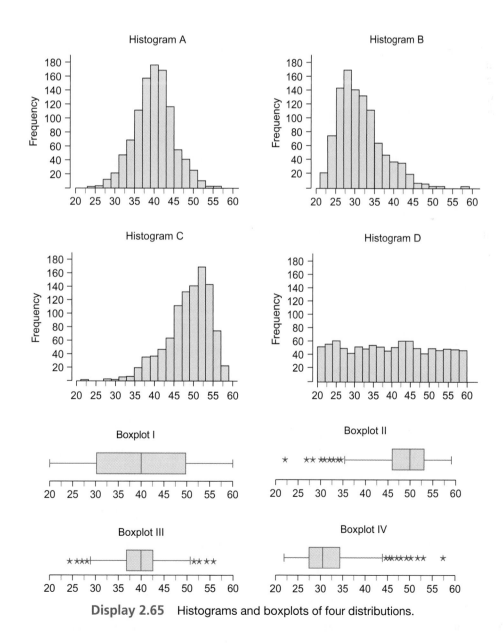

Display 2.65 Histograms and boxplots of four distributions.

E30. Three histograms and three boxplots appear in Display 2.66. Which boxplot displays the same information as

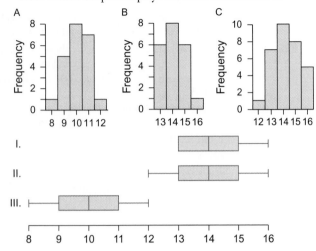

Display 2.66 Match the histograms with their boxplots.

a. histogram A?

b. histogram B?

c. histogram C?

E31. *Ecological footprint.* According to the World Wildlife Foundation, "a country's Ecological Footprint is the sum of all the cropland, grazing land, forest and fishing grounds required to produce the food, fiber and timber it consumes, to absorb the wastes emitted when it uses energy, and to provide space for its infrastructure." The total footprint, measured in hectares (ha), for the world's population first exceeded the world supply in the 1980s and now exceeds that supply by about 30%.

a. Make side-by-side boxplots showing the ecological footprint for the countries in the Americas (A), the European Union (EU), and Central Asia (CE) given in Display 2.67.

Country	Region	Population (millions)	Eco Footprint (ha per person)
Argentina	A	38.7	2.5
Bolivia	A	9.2	2.1
Brazil	A	186.4	2.4
Chile	A	16.3	3.0
Colombia	A	45.6	1.8
Costa Rica	A	4.3	2.3
Cuba	A	11.3	1.8
Dominican Republic	A	8.9	1.5
Ecuador	A	13.2	2.2
El Salvador	A	6.9	1.6
Guatemala	A	12.6	1.5
Haiti	A	8.5	0.5
Honduras	A	7.2	1.8
Jamaica	A	2.7	1.1
Mexico	A	107.0	3.4
Nicaragua	A	5.5	2.0
Panama	A	3.2	3.2
Paraguay	A	6.2	3.2
Peru	A	28.0	1.6
Trinidad and Tobago	A	1.3	2.1
Uruguay	A	3.5	5.5
Venezuela	A	26.7	2.8
Canada	A	32.3	7.1
United States	A	298.2	9.4
Austria	EU	8.2	5.0
Belgium	EU	10.4	5.1
Bulgaria	EU	7.7	2.7
Czech Republic	EU	10.2	5.3
Denmark	EU	5.4	8.0
Estonia	EU	1.3	6.4
Finland	EU	5.2	5.2
France	EU	60.5	4.9
Germany	EU	82.7	4.2
Greece	EU	11.1	5.9
Hungary	EU	10.1	3.5

Country	Region	Population (millions)	Eco Footprint (ha per person)
Ireland	EU	4.1	6.3
Italy	EU	58.1	4.8
Latvia	EU	2.3	3.5
Lithuania	EU	3.4	3.2
Netherlands	EU	16.3	4.0
Poland	EU	38.5	4.0
Portugal	EU	10.5	4.4
Romania	EU	21.7	2.9
Slovakia	EU	5.4	3.3
Slovenia	EU	2.0	4.5
Spain	EU	43.1	5.7
Sweden	EU	9.0	5.1
United Kingdom	EU	59.9	5.3
Afghanistan	CE	29.9	0.5
Armenia	CE	3.0	1.4
Azerbaijan	CE	8.4	2.2
Georgia	CE	4.5	1.1
Iran	CE	69.5	2.7
Iraq	CE	28.8	1.3
Israel	CE	6.7	4.8
Jordan	CE	5.7	1.7
Kazakhstan	CE	14.8	3.4
Kuwait	CE	2.7	8.9
Kyrgyzstan	CE	5.3	1.1
Lebanon	CE	3.6	3.1
Oman	CE	2.6	4.7
Saudi Arabia	CE	24.6	2.6
Syria	CE	1.9	2.1
Tajikistan	CE	6.5	0.7
Turkey	CE	73.2	2.7
Turkmenistan	CE	4.8	3.9
United Arab Emirates	CE	4.5	9.5
Uzbekistan	CE	26.6	1.8
Yemen	CE	21.0	0.9

Display 2.67 Population and ecological footprints in three regions of the world. [Source: *WWF Living Planet Report*, 2008.]

b. Describe how the three regions of the world compare with regard to their ecological footprints.

c. Which countries are the outliers in each region?

d. By what percentage would the United States have to reduce its footprint to be close to the upper quartile of that for the Americas?

E32. The U.S. Supreme Court instituted a temporary ban on capital punishment between 1967 and 1976. Between 1977 and 2008, 34 states (among the 36 that have a death penalty) used it to carry out 1133 executions. The five states that executed the most prisoners were Texas (423), Virginia (102), Oklahoma (88), Missouri (66), and Florida (66). The remaining 29 states carried out these numbers of executions: 43, 43, 40, 38, 28, 27, 27, 23, 19, 14, 13, 12, 12, 10, 6, 5, 4, 4, 3, 3, 3, 3, 2, 1, 1, 1, 1, 1, 1. For all 50 states, what was the mean number of executions per state? The median number? What were the quartiles? Draw a boxplot, showing any outliers, of the number of executions for all 50 states. [Source: *Death Penalty Information Center Fact Sheet*, December 2008.]

E33. The test scores of 40 students in a first-period class were used to construct the first boxplot in Display 2.68, and the test scores of 40 students in a second-period class were used to construct the second. Can the third plot be a boxplot of the combined scores of the 80 students in the two classes? Why or why not?

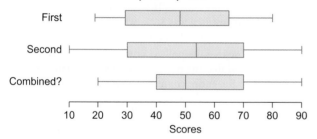

Display 2.68 Boxplots of three sets of test scores.

E34. The boxplots in Displays 2.69 and 2.70 show the average longevity of mammals, from Display 2.14.

a. Using only the basic boxplot in Display 2.69, show that there must be at least one outlier in the set of average longevity.

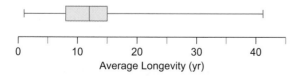

Display 2.69 Boxplot of average longevity of mammals.

b. How many outliers are there in the boxplot of average longevity in Display 2.70?

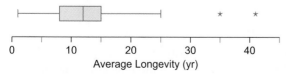

Display 2.70 Boxplot of average longevity of mammals, showing outliers.

c. How many outliers are shown in Display 2.62 on page 62? How can that be, considering the boxplot in Display 2.70?

E35. No computing should be necessary to answer these questions.

a. The mean of each of these sets of values is 20, and the range is 40. Which set has the largest standard deviation? Which has the smallest?

I. 0 10 20 30 40

II. 0 0 20 40 40

III. 0 19 20 21 40

b. Two of these sets of values have a standard deviation of about 5. Which two?

I. 5 5 5 5 5 5

II. 10 10 10 20 20 20

III. 6 8 10 12 14 16 18 20 22

IV. 5 10 15 20 25 30 35 40 45

E36. The standard deviation of the first set of values listed here is about 32. What is the standard deviation of the second set? Explain. (No computing should be necessary.)

16 23 34 56 78 92 93

20 27 38 60 82 96 97

E37. Consider the set of the heights of all female National Collegiate Athletic Association (NCAA) athletes and the set of the heights of all female NCAA basketball players. Which distribution will have the larger mean? Which will have the larger standard deviation? Explain.

E38. Consider the set of values 15, 8, 25, 32, 14, 8, 25, and 2. You can replace any one value with a number from 1 to 10. How would you make this replacement

a. to make the standard deviation as large as possible?

b. to make the standard deviation as small as possible?

c. to create an outlier, if possible?

E39. Another measure of center that sometimes is used is the *midrange*. To find the midrange, compute the mean of the largest value and the smallest value.

The statistics in the following computer output summarize the number of viewers of prime-time television shows (in millions) for the week of the last new *Seinfeld* episode.

Variable	N	Mean	Median	TrMean	StDev	SeMean
Viewers	101	11.187	10.150	9.831	9.896	0.985

Variable	Min	Max	Q1	Q3
Viewers	2.320	76.260	5.150	12.855

a. Using these summary statistics alone, compute the midrange both with and without the value representing the *Seinfeld* episode. (*Seinfeld* had the largest number of viewers and *Seinfeld Clips*, with 58.53 million viewers, the second largest.) Is the midrange affected much by outliers? Explain.

b. Compute the mean of the ratings without the *Seinfeld* episode, using only the summary statistics in the computer output.

E40. In computer output like that in E39, TrMean is the **trimmed mean.** It typically is computed by removing the largest 5% of values and the smallest 5% of values from the data set and then computing the mean of the remaining middle 90% of values. (The percentage that is cut off at each end can vary depending on the software.)

a. Find the trimmed mean of the maximum longevities in Display 2.14 on page 31.

b. Is the trimmed mean affected much by outliers?

E41. This table shows the weights of the pennies in Display 2.7 on page 26.

Weight (gm)	Frequency
2.99	1
3.01	4
3.03	4
3.05	4
3.07	7
3.09	17
3.11	24
3.13	17
3.15	13
3.17	6
3.19	2
3.21	1

a. Find the mean weight of the pennies.

b. Find the standard deviation.

c. Does the standard deviation appear to represent a typical deviation from the mean?

E42. Suppose you have five pennies, six nickels, four dimes, and five quarters.

a. Sketch a dot plot of the values of the 20 coins, and use it to estimate the mean.

b. Compute the mean using the formula for the mean of values in a frequency table.

c. Estimate the standard deviation from your plot: Is it closest to 0, 5, 10, 15, or 20?

d. Compute the standard deviation using the formula for the standard deviation of values in a frequency table.

E43. On the first test of the semester, the scores of the first-hour class of 30 students had a mean of 75 and a median of 70. The scores of the second-hour class of 22 students had a mean of 70 and a median of 68.

a. To the nearest tenth, what is the mean test score of all 52 students? If you cannot calculate the mean of the two classes combined, explain why.

b. What is the median test score of all 52 students? If you cannot find the median of the two classes combined, explain why.

E44. The National Council on Public Polls rebuked the press for its coverage of a Gallup poll of Islamic countries. According to the Council:

News stories based on the Gallup poll reported results in the aggregate without regard to the population of the countries they represent. Kuwait, with fewer than 2 million Muslims, was treated the same as Indonesia, which has over 200 million Muslims. The "aggregate" quoted in the media was actually the average for the countries surveyed regardless of the size of their populations.

The percentage of people in Kuwait who thought the September 11, 2001 terrorist attacks were morally justified was 36%, while the percentage in Indonesia was 4%. [*Source: www.ncpp.org.*]

a. Suppose that the poll covered only these two countries and that the people surveyed were representative of the entire country. What percentage of all the people in these two countries thought that the terrorist attacks were morally justified?

b. What percentage would have been reported by the press?

E45. Prove that $\Sigma(x - \bar{x}) = 0$. Here, the sum is over all values of the variable x.

E46. Prove that if it is the case that $\Sigma(x - C) = 0$ for some constant C, then $C = \bar{x}$. Here, the sum is over all values of the variable x.

2.3 ▶ Working with Summary Statistics

Summary statistics are very useful but only when they are used with good judgment. Summaries can oversimplify and can mislead, which means it is important to know when to use summaries and which summaries to use. This section will teach you how to tell which summary statistic to use, how changing units of measurement and the presence of outliers affect your summary statistics, and how to interpret percentiles.

Which Summary Statistic?

Plot first. Then choose a summary statistic.

Which summary statistics should you use to describe a distribution? Should you use mean and standard deviation? Median and quartiles? Something else? The right choice can depend on the shape of your distribution, so you should always start with a plot. For normal distributions, the mean and standard deviation are nearly always the most suitable. For skewed distributions, the median and quartiles are often the most useful, in part because they have a simple interpretation based on dividing a data set into fourths.

Sometimes, however, the mean and standard deviation will be the right choice even if you have a skewed distribution. For example, if you have a representative sample of house prices for a town and you want to use your sample to estimate the total value of all the town's houses, the mean is what you want, not the median. In Chapter 7, you'll see why the mean and standard deviation are the most useful choices when doing statistical inference.

Choosing the right summary statistics is something you will get better at as you build your intuition about the properties of these statistics and how they behave in various situations.

DISCUSSION
Which Summary Statistic?

D22. Explain how to determine the total amount of property taxes for a city if you know the number of properties, the mean dollar value of all properties, and the tax rate. In what sense is knowing the mean equivalent to knowing the total?

D23. When a measure of center for the income of a community's residents is given, that number is usually the median. Why do you think that is the case?

The Effects of Changing Units

You are studying world temperature change and have recorded temperatures and computed summary statistics in degrees Fahrenheit. For publication of your report outside of the United States, you have been asked to convert your summary statistics to degrees Celsius. Must you convert each individual measurement from Fahrenheit to Celsius and then compute new summary statistics? In this section, you will learn an easier way to adjust your summary statistics.

The table and dot plot in Display 2.71 show the lowest recorded temperatures for seven capital cities of the world.

City	Country	Temperature (°F)
Addis Ababa	Ethiopia	26
Algiers	Algeria	25
Bangkok	Thailand	50
Madrid	Spain	10
Nairobi	Kenya	36
Brasilia	Brazil	36
Warsaw	Poland	−29

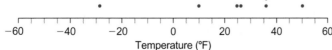

Display 2.71   Record low temperatures for seven capitals. [Source: Extreme temperatures around the world, *www.mherrera.org/temp.htm*, 2009.]

When you convert from Fahrenheit to Celsius, you use the formula

$$C = \frac{5}{9}(F - 32)$$

This formula amounts to first subtracting 32 from each temperature and then multiplying by 5/9.

What happens to the shape, center, and spread of the distribution of temperatures when you begin by subtracting 32? To find out, subtract 32 from each value, and plot the new values. Display 2.72 shows that the center of the distribution is 32 degrees lower, moving the mean from about 22 degrees to −10, but the spread and shape are unchanged.

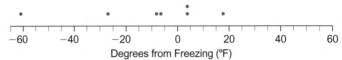

Display 2.72   Dot plot of the number of degrees Fahrenheit above or below freezing for record low temperatures for the seven capitals.

Adding (or subtracting) a constant to each value in a set of data doesn't change the spread or the shape of a distribution, but slides the entire distribution a distance equivalent to the constant. Thus, the transformation of subtracting 32 from each temperature amounts to a *recentering* of the distribution.

Now, what happens to the shape, center, and spread of this distribution when you multiply each recentered value by 5/9? Display 2.73 shows that the shape has not changed, but the mean has changed from −10 degrees to about $\frac{5}{9}(-10)$, or −5.7, and the spread is 5/9ths as large as before.

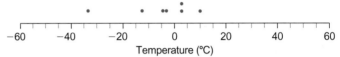

Display 2.73   Dot plot of the record low temperatures for seven capitals in degrees Celsius.

Multiplying or dividing each value in a set of data by a positive constant doesn't change the basic shape of the distribution. The mean and the spread both are multiplied by that number. Thus, this transformation amounts to a *rescaling* of the distribution.

Recentering and Rescaling a Data Set

Recentering a data set—adding the same number *c* to all the values in the set—doesn't change the shape or spread but slides the entire distribution by the amount *c*, adding *c* to the median and the mean.

Rescaling a data set—multiplying all the values in the set by the same positive number *d*—doesn't change the basic shape but stretches or shrinks the distribution, multiplying the spread (*IQR* or standard deviation) by *d* and multiplying the center (median or mean) by *d*.

DISCUSSION
The Effects of Changing Units

D24. Suppose a U.S. dollar is worth 14.5 Mexican pesos.

    a. A set of prices, in U.S. dollars, has mean $20 and standard deviation $5. Find the mean and standard deviation of the prices expressed in pesos.

    b. Another set of prices, in Mexican pesos, has a median of 145.0 pesos and quartiles of 72.5 pesos and 29 pesos. Find the median and quartiles of the same prices expressed in U.S. dollars.

D25. The median of the temperatures in Display 2.71 on page 67 is 26°F. What is the median of the temperatures in Celsius? The standard deviation of the temperatures is 25.6°F. What is the standard deviation of the temperatures in Celsius?

D26. You are attempting to replicate a classic study in sociology that found that, on average, it takes people longer to vacate a parking spot when another car is waiting than when no car is waiting. You measure the time, in seconds, with a stopwatch for large samples of cars in the two situations. When you have finished your report, you decide that it would be better to give your summary statistics in minutes rather than in seconds. How would you transform the mean, median, standard deviation, and interquartile range so that the units are minutes rather than seconds? How would you transform the variance? [Source: R. Barry Ruback and Daniel Juieng, "Territorial Defense in Parking Lots: Retaliation Against Waiting Drivers," *Journal of Applied Social Psychology*, Vol. 27, 9 (1997), pp. 821–834.]

The Influence of Outliers

A summary statistic is **resistant to outliers** if the summary statistic is not changed very much when an outlier is removed from the set of data. If the summary statistic tends to be affected by the removal of outliers, it is **sensitive to outliers**.

> ### Sensitive and Resistant Summary Statistics
>
> In general, the mean is more sensitive (less resistant) to outliers than are the median and quartiles.
>
> The standard deviation is more sensitive to outliers than is the interquartile range.

Display 2.74 shows a dot plot of the number of viewers of the 68 television series finales that had the largest number of viewers. The three shows with the largest numbers of viewers—*M*A*S*H*, *Cheers*, and *Seinfeld*—are outliers.

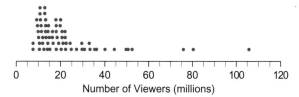

Display 2.74 Number of viewers (in millions) of the most watched series finales in television history. [Source: *http://en.wikipedia.org/wiki/List_of_most_watched_television_broadcasts*. Their sources: Reuters, Variety, Nielsen Media Research, ratings data from *USA Today* weekly ratings charts.]

The Minitab output in Display 2.75 gives the summary statistics for all 68 shows. The output in Display 2.76 gives the summary statistics when the three outliers are removed from the set of shows. Compare this output with the one in Display 2.75 and notice which summary statistics are most sensitive to the outliers.

Descriptive Statistics: Viewers

Variable	N	N*	Mean	SE Mean	StDev	Minimum	Q1	Median	Q3	Maximum
Viewers	68	0	23.04	2.14	17.63	7.40	12.95	18.00	25.05	105.90

Display 2.75 Summary statistics for the number of viewers (in millions) of the most watched series finales in television history.

Descriptive Statistics: No Outliers

Variable	N	N*	Mean	SE Mean	StDev	Minimum	Q1	Median	Q3	Maximum
No Outliers	65	0	20.07	1.32	10.65	7.40	12.60	17.50	23.40	52.50

Display 2.76 Summary statistics for the number of viewers (in millions) of the most watched series finales in television history, without the top three shows.

DISCUSSION
The Influence of Outliers

D27. Are these measures of center for the number of television viewers affected much by the three outliers? (Refer to Displays 2.74–2.76.) Explain.

  a. mean   b. median

D28. Are these measures of spread for the number of television viewers affected much by the three outliers? Explain why or why not.

  a. range   b. standard deviation   c. interquartile range

Percentiles and Cumulative Relative Frequency Plots

Percentiles measure position within a data set. The first quartile, Q_1, of a distribution is the 25th percentile—the value that separates the lowest 25% of the ordered values from the rest. The median is the 50th percentile, and Q_3 is the 75th percentile. You can define other percentiles in the same way. The 10th percentile, for example, is the value that separates the lowest 10% of ordered values in a distribution from the rest. In general, a value is at the kth **percentile** if k% of all values are less than or equal to it.

If you decide to get an MBA, it is likely that you will take the GMAT (Graduate Management Admission Test). When you get your official score report, you will find that the scores of hundreds of thousands of students are summarized in a table like that in Display 2.77. The plot beside the table is sometimes called a **cumulative percentage**

Total GMAT Score	Percentile
760	99
750	98
700	90
650	80
600	66
550	51
500	36
450	23
400	13
350	7
300	3
250	1

Display 2.77 Cumulative relative frequency plot of GMAT scores and percentiles, 2008. [Source: GMAT Sample Score Report, *www.mba.com.*]

plot or a **cumulative relative frequency plot.** The table and plot show that, for example, 90% of the students received a score of 700 or lower and 10% received a score between 700 and 650.

DISCUSSION
Percentiles and Cumulative Relative Frequency Plots

D29. Display 2.78 shows the percentile scores that correspond to the life expectancies for the 43 countries of Europe.

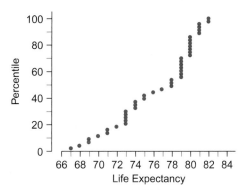

Display 2.78 Cumulative relative frequency (percentile) plot of mean life expectancy, in years, for countries in Europe. [Source: Population Reference Bureau, *World Population Data Sheet*, 2008.]

   a. Latvia has a mean life expectancy of 72 years. What is the approximate percentile for Latvia? Ireland has a percentile of about 60. What is the mean life expectancy in Ireland?

   b. How many countries have a mean life expectancy of 73 years?

   c. Discuss the shape of this plot and how it differs from the plot of percentiles from the GMAT scores in Display 2.77. Do you think a plot of the actual life expectancies for European countries would be approximately normal in shape? Sketch a shape for the data distribution.

Summary 2.3: Working with Summary Statistics

Knowing which summary statistic to use depends on what use you have for that summary statistic.

If a summary statistic doesn't change much whether you include or exclude outliers from your data set, it is said to be resistant to outliers.

- The median and quartiles are resistant to outliers.

- The mean and standard deviation are sensitive to outliers.

Recentering a data set by adding the same number to each value (or subtracting the same number from each value) doesn't change the shape or spread of the distribution, but to find the new mean and median, add the number to the original mean and median. Converting to another unit of measurement by multiplying each value by the same positive number doesn't change the basic shape, but to find the new summary statistics of center (mean, median) and spread (interquartile range and standard deviation), multiply the original ones by the number.

The percentile of a value tells you what percentage of all values lie at or below the given value. The 30th percentile, for example, is the value that separates the distribution into the lowest 30% of values and the highest 70% of values.

Practice

Which Summary Statistic?

P26. A community in Nevada has 9751 houses, with a median house price of $320,000 and a mean price of $392,059.

    a. Why is the mean larger than the median?

    b. The property tax rate is about 1.15%. What total amount of taxes will be assessed on these houses?

    c. What is the average amount of taxes per house?

P27. A news release at www.polk.com stated that the median age of cars being driven in 2007 was 9.2 years, tying the record age set in 2006. The median was 8.3 years in 2000 and 7.7 years in 1995.

    a. Why were medians used in this news story?

    b. What reasons might there be for the increase in the median age of cars? (The median age in 1970 was only 4.9 years!)

The Effects of Changing Units

P28. The mean height of a class of 15 children is 48 in., the median is 45 in., the standard deviation is 2.4. in., and the interquartile range is 3 in. Find the mean, standard deviation, median, and interquartile range if

    a. you convert each height to feet

    b. each child grows 2 in.

    c. each child grows 4 in. and you convert the heights to feet

P29. The standard deviation of the set of numbers {1, 2, 3} is equal to 1. Without doing any computing, give the standard deviation of each of the following sets of numbers.

    a. 11   12   13

    b. 10   20   30

    c. 105   110   115

    d. 800   900   1000

The Influence of Outliers

P30. The histogram and boxplot in Display 2.79 and the summary statistics in Display 2.80, show the record low temperatures for the 50 states.

    a. Hawaii has a lowest recorded temperature of 12°F. The boxplot shows Hawaii as an outlier. Verify that this is justified.

    b. Suppose you exclude Hawaii from the data set. Copy the table in Display 2.80, substituting the value (or your best estimate if you don't have enough information to compute the value) of each summary statistic with Hawaii excluded.

Percentiles and Cumulative Relative Frequency Plots

P31. Refer to Display 2.77 on page 70.

    a. Use the plot to estimate the percentile for a GMAT score of 425.

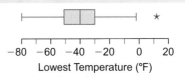

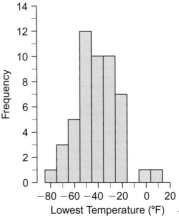

Display 2.79 Record low temperatures for the 50 states. [Source: National Climatic Data Center, 2007, *www.ncdc.noaa.gov.*]

Summary of Lowest Temperature

Count	50
Mean	−40.3
Median	−40
StDev	17.7
Min	−80
Max	12
Range	92
Q_1	−51
Q_3	−30

Display 2.80 Summary statistics for lowest temperatures for the 50 states.

    b. What two values enclose the middle 90% of GMAT scores? The middle 95%?

    c. Use the table to estimate the score that falls at the 40th percentile.

P32. Give the proportion of cases that lie between the 5th and 95th percentiles of a distribution. What percentiles enclose the middle 95% of the cases in a distribution?

P33. Estimate the quartiles and the median of the GMAT scores in Display 2.77 on page 70, and then use these values to draw a boxplot of the distribution. What is the *IQR?*

Exercises

E47. Discuss whether you would use the mean or the median to measure the center of each set of data and why you prefer the one you chose.

    a. the prices of single-family homes in your neighborhood

    b. the yield of corn (bushels per acre) for a sample of farms in Iowa

    c. the survival time, following a diagnosis of terminal cancer, for a sample of cancer patients

E48. *Mean versus median.*

    a. You are tracing your family tree and would like to go back to the year 1700. To estimate how many generations back you will have to trace, would you need to know the median length of a generation or the mean length of a generation?

    b. If a car trip takes 3 hours, do you need to know the mean speed or the median speed to find the total distance?

    c. Suppose all trees in a forest are right circular cylinders with radius 3 ft. The heights vary, but the mean height is 45 ft, the median is 43 ft, the *IQR* is 3 ft, and the standard deviation is 3.5 ft. From this information, can you compute the total volume of wood in all the trees?

E49. The histogram in Display 2.81 shows record high temperatures for the 50 states.

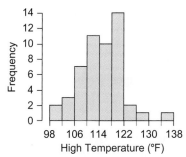

Display 2.81   Record high temperatures for the 50 U.S. states. [Source: National Climatic Data Center, 2007, *www.ncdc.noaa.gov.*]

    a. Suppose each temperature is converted from degrees Fahrenheit, *F*, to degrees Celsius, *C*, using the formula

$$C = \frac{5}{9}(F - 32)$$

    If you make a histogram of the temperatures in degrees Celsius, how will it differ from the one in Display 2.81?

    b. The summary statistics in Display 2.82 are for record high temperatures in degrees Fahrenheit. Make a similar table for the temperatures in degrees Celsius.

    c. Are there any outliers in the data in °C?

Variable	N	Mean	Median	StDev
HighTemp	50	114.10	114.00	6.69

Variable	Min	Max	Q1	Q3
HighTemp	100.00	134.00	110.00	118.00

Display 2.82   Summary statistics for record high temperatures for the 50 U.S. states.

E50. Tell how you could use recentering and rescaling to simplify the computation of the mean and standard deviation for this list of numbers:

    5478.1   5478.3   5478.3   5478.9   5478.4   5478.2

E51. Suppose a constant *c* is added to each value in a set of data, x_1, x_2, x_3, x_4, and x_5. Prove that the mean increases by *c* by comparing the formula for the mean of the original data to the formula for the mean of the recentered data.

E52. Suppose a constant *c* is added to each value in a set of data, x_1, x_2, x_3, x_4, and x_5. Prove that the standard deviation is unchanged by comparing the formula for the standard deviation of the original data to the formula for the standard deviation of the recentered data.

E53. The cumulative relative frequency plot in Display 2.83 shows the amount of change carried by a group of 200 students. For example, about 80% of the students had $0.75 or less in coins.

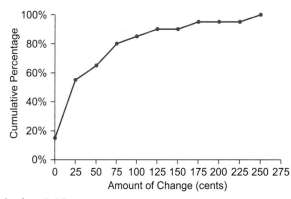

Display 2.83   Cumulative percentage plot of amount of change.

    a. From this plot, estimate the median amount of change.

    b. Estimate the quartiles and the interquartile range.

    c. Is the original set of amounts of change skewed right, skewed left, or symmetric?

E54. Use Display 2.83 to make a boxplot of the amounts of change carried by the students.

E55. Did you ever wonder how speed limits on roadways are determined? Most government jurisdictions set speed limits using the same method, described on the website of the Michigan State Police.

> Speed studies are taken during times that represent normal free-flow traffic. Since modified speed limits are the maximum allowable speeds, roadway conditions must be close to ideal. The primary basis for establishing a proper, realistic speed limit is the nationally recognized method of using the 85th percentile speed. This is the speed at or below which 85% of the traffic moves. [Source: *www.michigam.gov*.]

The 85th percentile speed typically is rounded down to the nearest 5 mph. The table and histogram in Display 2.84 give the measurements of the speeds of 1000 cars on a stretch of road in Mellowville with no curviness or other additional factors. At what speed would the speed limit be set if the guidelines described were followed?

E56. Refer to the distribution of speeds in E55. Make a cumulative relative frequency plot of these speeds.

E57. The cumulative relative frequency plot in Display 2.85 gives the ages of the CEOs (Chief Executive Officers) of the 500 largest U.S. companies. Does A, B, or C give its median and quartiles? Using the diagram, explain why your choice is correct.

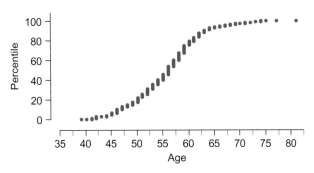

Display 2.85 Cumulative relative frequency plot of CEO ages. [Source: *www.forbes.com*]

A. Q_1 51; median 56; Q_3 60
B. Q_1 50; median 60; Q_3 70
C. Q_1 25; median 50; Q_3 75

E58. Match each histogram in Display 2.86 with its cumulative relative frequency plot.

Speed (mph)	Count
25	2
26	31
27	92
28	149
29	178
30	156
31	157
32	99
33	74
34	31
35	16
36	13
37	1
38	1
Total	1000

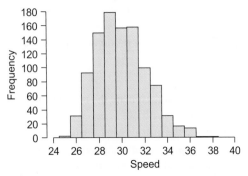

Display 2.84 Speed of 1000 cars in Mellowville.

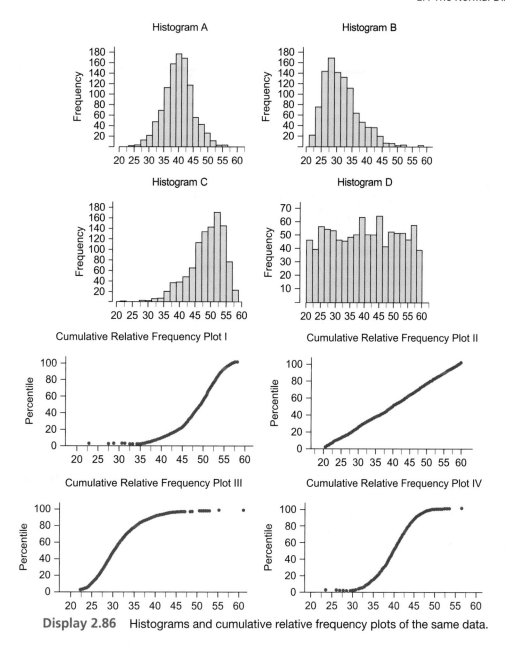

Display 2.86 Histograms and cumulative relative frequency plots of the same data.

2.4 ▶ The Normal Distribution

You have seen several reasons why the normal distribution is so important:

- It tells you how variability in repeated measurements often behaves (diameters of tennis balls).
- It tells you how variability in populations often behaves (weights of pennies, SAT scores).
- It tells you how means (and some other summary statistics) computed from random samples behave (the Westvaco case).

In this section, you will learn that if you know that a distribution is normal (shape), then the mean (center) and standard deviation (spread) tell you everything else about the distribution. The reason is that, whereas skewed distributions come in many different shapes, there is only one normal shape. It's true that one normal distribution might appear tall and thin while another looks short and fat. However, the horizontal axis of the tall, thin distribution can be stretched out so that it looks exactly the same as the short, fat one.

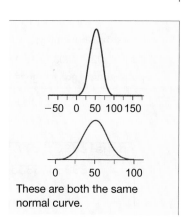

These are both the same normal curve.

The Standard Normal Distribution

Because all normal distributions have the same basic shape, you can use recentering and rescaling to change any normal distribution to the one with mean 0 and standard deviation 1.

The Standard Normal Distribution

The normal distribution with mean 0 and standard deviation 1 is called the **standard normal distribution**. In this distribution, the variable along the horizontal axis is called a **z-score**.

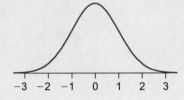

The standard normal distribution is symmetric, with total area under the curve equal to 1, or 100%. To find the proportion of the area to the left of the corresponding z-score, you can use the z-table, your calculator, or your computer.

The next two examples show you how to use the z-table, Table A on pages 759–760.

Example 2.11	Finding the Percentage When You Know the z-Score

Find the percentage of the values in a standard normal distribution that are less than $z = 1.23$, represented by the shaded area in Display 2.87. Find the percentage of the values that are greater than $z = 1.23$.

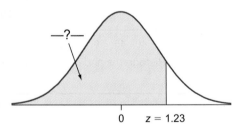

Display 2.87 The percentage of values less than $z = 1.23$.

Solution

	Tail Probability, p		
z	.02	.03	.04
1.2	.8888	**.8907**	.8925

Think of 1.23 as $1.2 + 0.03$. In Table A on pages 759–760, find the row labeled 1.2 and the column headed .03. Where this row and column intersect, you find the number .8907. That means that 89.07% of standard normal scores are less than 1.23.

The total area under the curve is 1, so the proportion of values greater than $z = 1.23$ is $1 - .8907$, or .1093, which is 10.93%.

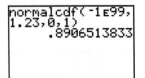

A graphing calculator or statistical software will give you greater accuracy in finding the proportion of values that lie between two specified values in a standard normal distribution. For example, you can find the proportion of values that are less than 1.23 in a standard normal distribution using the command on the screen to the left.

| **Finding the z-Score When You Know the Percentage** | **Example 2.12** |

Find the z-score that falls at the 75th percentile of the standard normal distribution, that is, the z-score that divides the smallest 75% of the values from the rest.

Solution
First make a sketch of the situation, as in Display 2.88.

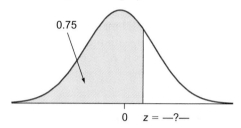

Display 2.88   The z-score that corresponds to the 75th percentile.

Look for .75 in the body of Table A. No value in the table is exactly equal to .75. The closest value is .7486. The value .7486 sits at the intersection of the row labeled 0.6 and the column headed .07, so the corresponding z-score is roughly 0.6 + .07, or 0.67.

Tail Probability, *p*

z	.06	.07	.08
0.6	.7454	**.7486**	.7517

You can use a calculator to find the 75th percentile of a standard normal distribution.

```
invNorm(.75,0,1)
          .6744897495
```

DISCUSSION
The Standard Normal Distribution

D30. For the standard normal distribution,

    a. what is the median?

    b. what is the lower quartile?

    c. what z-score falls at the 95th percentile?

    d. what is the *IQR*?

Standard Units: How Many Standard Deviations Is It from Here to the Mean?

Converting to standard units, or **standardizing,** is the two-step process of recentering and rescaling that turns any normal distribution into the standard normal distribution.

First you recenter all the values of the normal distribution by subtracting the mean from each. This gives you a distribution with mean 0. Then you rescale by dividing all the values by the standard deviation. This gives you a distribution with standard deviation 1. You now have a standard normal distribution. You can also think of the two-step process of standardizing as answering two questions: How far above or below the mean is my score? How many standard deviations is that?

| **Formula for a z-Score** |

The standard units or z-score is the number of standard deviations that a given value, *x*, lies above or below the mean.

    How far and which way to the mean?

$$x - mean$$

How many standard deviations, *SD*, that?

$$z = \frac{x - mean}{SD}$$

| Example 2.13 | **Computing a z-Score** |

In a recent year, the distribution of SAT I math scores for the incoming class at the University of Georgia was roughly normal, with mean 610 and standard deviation 69. What is the z-score for an incoming University of Georgia student who got 560 on the math SAT I?

Solution
A score of 560 is 50 points below the mean of 610. This is $\frac{-50}{69}$, or 0.725, standard deviations below the mean. Alternatively, using the formula,

$$z = \frac{x - mean}{SD} = \frac{560 - 610}{69} \approx -0.725$$

the student's z-score is −0.725.

Solving for x

To unstandardize, think in reverse. Alternatively, you can solve the z-score formula for *x* and get

$$x = mean + z \cdot SD$$

| Example 2.14 | **Finding the Value When You Know the z-Score** |

What was a University of Georgia student's SAT I math score if his or her score was 1.6 standard deviations above the mean?

Solution
The score that is 1.6 standard deviations above the mean is

$$x = mean + z \cdot SD = 610 + 1.6(69) \approx 720$$

| Example 2.15 | **Using z-Scores to Make a Comparison** |

In the United States, heart disease and cancer are the two leading causes of death. If you look at the death rate per 100,000 residents by state, the distributions for the two diseases are roughly normal, provided you leave out Alaska and Utah, which are outliers because of their unusually young populations. The means and standard deviations for all 50 states are given here.

	Mean	SD
Heart disease	219	46
Cancer	194	30

Alaska had 94 deaths per 100,000 from heart disease and 110 from cancer. Explain which death rate is more extreme compared to the other states. [Source: *National Vital Statistics Reports*, Vol. 56, 10 (April 24, 2008).]

Solution

$$z_{\text{heart}} = \frac{94 - 219}{46} = -2.72$$

$$z_{\text{cancer}} = \frac{110 - 194}{30} = -2.80$$

Alaska's death rate for heart disease is 2.72 standard deviations below the mean. The death rate for cancer is 2.80 standard deviations below the mean. These rates are about equally extreme, but the death rate for cancer is slightly more extreme.

DISCUSSION
Standard Units

D31. Standardizing is a process that is similar to more familiar computations.

  a. You're driving at 60 mph on the interstate and are now passing the marker for mile 200, and your exit is at mile 80. How many hours from your exit are you?

  b. What two arithmetic operations did you do to get the answer in part a? Which operation corresponds to recentering? Which corresponds to rescaling?

Solving the Unknown Percentage Problem and the Unknown Value Problem

Now you know all you need to know to analyze situations involving two related problems concerning a normal distribution: finding a percentage when you know the value, and finding the value when you know the percentage.

Percentage of Males Taller Than 74 Inches	**Example 2.16**

For groups of similar individuals, heights often are approximately normal in their distribution. For example, the heights of 20- to 29-year-old males in the United States are approximately normal, with mean 70.4 in. and standard deviation 3.0 in. What percentage of these males are more than 74 in. tall? [Source: U.S. Census Bureau, *Statistical Abstract of the United States.*]

Solution
First make a sketch of the situation, as in Display 2.89. Draw a normal shape above a horizontal axis. Place the mean in the middle on the axis. Then mark and label the points that are two standard deviations either side of the mean, 64.4 and 76.4, so that about 95% of the values lie between them. Next, mark and label the points that are one and three standard deviations either side of the mean (67.4 and 73.4, and 61.4 and 79.4). Finally, estimate the location of the given value of x and mark it on the axis.

Display 2.89 The percentage of heights greater than 74 in.

Standardize:

$$z = \frac{x - mean}{SD} = \frac{74 - 70.4}{3.0} = 1.20$$

Look up the proportion: The area to the left of the z-score of 1.20 is 0.8849, so the proportion of males taller than 74 in. is 1 – 0.8849, or 0.1151 or 11.51%.

| Example 2.17 | **Percentage of Males Between 72 and 74 Inches Tall** |

The heights of 20- to 29-year-old males in the United States are approximately normal, with mean 70.4 in. and standard deviation 3.0 in. What percentage of these males are between 72 and 74 in. tall?

Solution
First make a sketch, as in Display 2.90.

Display 2.90 The percentage of male heights between 72 and 74 in.

Standardize: From the previous example, a height of 74 in. has a z-score of 1.20. For a height of 72 in.,

$$z = \frac{x - mean}{SD} = \frac{72 - 70.4}{3.0} \approx 0.53$$

Look up the proportion: The area to the left of a z-score of 1.20 is 0.8849. The area to the left of a z-score of 0.53 is 0.7019. The area you want is the area between these two z-scores, which is 0.8849 – 0.7019, or 0.1830. So the percentage of 20- to 29-year-old males in the United States who are between 72 and 74 in. tall is about 18.30%.

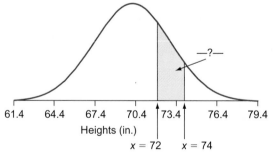

You can also use a graphing calculator or computer to find this value.

75th Percentile of Female Heights | Example 2.18

The heights of 20- to 29-year-old females in the United States are approximately normal, with mean 65.1 in. and standard deviation 2.6 in. What height separates the shortest 75% from the tallest 25%?

Solution
First make a sketch, as in Display 2.91.

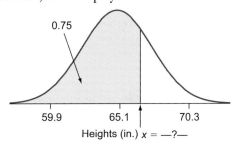

Display 2.91   The 75th percentile in height for women ages 20 to 29.

Look up the z-score: If the proportion P is 0.75, then from Table A, you find that z is approximately 0.67.
Unstandardize:

$$x = mean + z \cdot SD \approx 65.1 + 0.67(2.6) = 66.842 \text{ in.}$$

Solving the Two Types of Problems

For an unknown percentage problem:
   First standardize by converting the given value to a z-score:

$$z = \frac{x - mean}{SD}$$

   Then look up the percentage.
For an unknown value problem, reverse the process:
   First look up the z-score corresponding to the given percentage.
   Then unstandardize:

$$x = mean + z \cdot SD$$

DISCUSSION
Solving the Unknown Percentage Problem and the Unknown Value Problem

D32. The cars in Clunkerville have a mean age of 12 years and a standard deviation of 8 years. What percentage of the cars are more than 4 years old? (Warning: This is a trick question.)

Central Intervals for Normal Distributions

You learned in Section 2.1 that if a distribution is roughly normal, about 68% of the values lie within one standard deviation of the mean. It is also the case that about 95%

of the values lie within about two standard deviations of the mean and almost all (99.7%) values lie within three standard deviations of the mean. These facts are often cited together and referred to as the **Empirical Rule** or the **68-95-99.7 Rule**. They are so commonly used in statistics that it is helpful to memorize them, along with the other fact that 90% of the values in a normal distribution lie within 1.645 standard deviations of the mean. These facts are presented in the following box.

Central Intervals for Normal Distributions

68% of the values lie within 1 standard deviation of the mean.

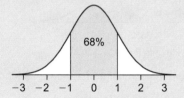

90% of the values lie within 1.645 standard deviations of the mean.

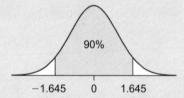

95% of the values lie within 1.96 (or about 2) standard deviations of the mean.

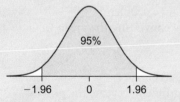

99.7% (or almost all) of the values lie within 3 standard deviations of the mean.

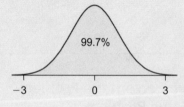

| Example 2.19 | **Middle 90% of Death Rates from Cancer** |

According to the table on page 78, the state death rates from cancer per 100,000 residents are approximately normally distributed with mean 194 and standard deviation 30. The middle 90% of death rates are between what two numbers?

Solution

The middle 90% of values in this distribution lie within 1.645 standard deviations of the mean, 194. That is, about 90% of the values lie in the interval $194 \pm 1.645(30)$, or between about 145 and 243.

DISCUSSION
Central Intervals for Normal Distributions

D33. Use Table A on pages 759 to verify that 99.7% of the values in a normal distribution lie within three standard deviations of the mean.

Summary 2.4: The Normal Distribution

The standard normal distribution has mean 0 and standard deviation 1. All normal distributions can be converted to the standard normal distribution by converting to standard units:

- First, recenter by subtracting the mean.
- Then rescale by dividing by the standard deviation:

$$z = \frac{x - mean}{SD}$$

Standard units z tell how far a value x is from the mean, measured in standard deviations. If you know z, you can find x by using the formula $x = mean + z \cdot SD$.

If your population is approximately normal, you can compute z and then use Table A or technology to find the corresponding proportion. Be sure to make a sketch so that you know whether to use the proportion in the table or to subtract that proportion from 1.

For any normal distribution,

- 68% of the values lie within 1 standard deviation of the mean
- 90% of the values lie within 1.645 standard deviations of the mean
- 95% of the values lie within 1.96 (or about 2) standard deviations of the mean
- 99.7% (or almost all) of the values lie within 3 standard deviations of the mean

---

Practice

The Standard Normal Distribution

P34. Find the percentage of values below each given z-score in a standard normal distribution.

    a. -2.23 **b.** -1.67 **c.** -0.40 **d.** 0.80

P35. Find the z-score that has the given percentage of values below it in a standard normal distribution.

    a. 32%   **b.** 41%   **c.** 87%   **d.** 94%

P36. Give the percentage of the values in a standard normal distribution that fall between

    a. -1.46 and 1.46.

    b. -3 and 3.

P37. For a standard normal distribution, what interval contains

    a. the middle 90% of z-scores?

    b. the middle 95% of z-scores?

Standard Units

P38. Refer to the table in the example on page 78.

    a. California had 180 deaths from heart disease and 151 deaths from cancer per 100,000 residents. Which rate is more extreme compared to other states? Why?

    b. Florida had 260 deaths from heart disease and 228 deaths from cancer per 100,000 residents. Which rate is more extreme?

    c. Colorado had an unusually low rate of heart disease, 135 deaths per 100,000 residents; Georgia had an unusually low rate of cancer, 158 deaths per 100,000 residents. Which is more extreme?

Solving the Unknown Percentage Problem and the Unknown Value Problem

P39. The heights of 20- to 29-year-old males in the United States are approximately normal, with mean 70.4 in. and standard deviation 3.0 in. The heights of 20- to 29-year-old females in the United States are approximately normal, with mean 65.1 in. and standard deviation 2.6 in.

 a. Estimate the percentage of U.S. males between 20 and 29 who are 6 ft tall or taller.

 b. How tall does a U.S. woman between 20 and 29 have to be in order to be at the 35th percentile of heights?

Central Intervals for Normal Distributions

P40. Refer to the table in Example 2.15 on page 78.

 a. The middle 90% of the states' death rates from heart disease fall between what two numbers?

 b. The middle 68% of death rates from heart disease fall between what two numbers?

P41. Refer to the information in P39. Which of the following heights are outside the middle 95% of the distribution? Which are outside the middle 99%?

 a. a male who is 79 in. tall

 b. a female who is 68 in. tall

 c. a male who is 65 in. tall

 d. a female who is 70 in. tall

Exercises

E59. What percentage of values in a standard normal distribution fall

 a. below a z-score of 1.00? 2.53?

 b. below a z-score of -1.00? -2.53?

 c. above a z-score of -1.5?

 d. between z-scores of -1 and 1?

E60. On the same set of axes, draw two normal curves with mean 50, one having standard deviation 5 and the other having standard deviation 10.

E61. *Standardizing.* Convert each of these values to standard units, z. (Do not use a calculator. These are meant to be done in your head.)

 a. $x = 12$, mean 10, standard deviation 1

 b. $x = 12$, mean 10, standard deviation 2

 c. $x = 12$, mean 9, standard deviation 2

 d. $x = 12$, mean 9, standard deviation 1

 e. $x = 7$, mean 10, standard deviation 3

 f. $x = 5$, mean 10, standard deviation 2

E62. *Unstandardizing.* Find the value of x that was converted to the given z-score.

 a. $z = 2$, mean 20, standard deviation 5

 b. $z = -1$, mean 25, standard deviation 3

 c. $z = -1.5$, mean 100, standard deviation 10

 d. $z = 2.5$, mean -10, standard deviation 0.2

E63. SAT I critical reading scores are scaled so that they are approximately normal, with mean about 505 and standard deviation about 111.

 a. Find the probability that a randomly selected student has an SAT I critical reading score

  i. between 400 and 600

  ii. over 700

  iii. below 450

 b. What SAT I critical reading scores fall in the middle 95% of the distribution?

E64. SAT I math scores are scaled so that they are approximately normal, with mean about 511 and standard deviation about 112. A college wants to send letters to students scoring in the top 20% on the exam. What

SAT I math score should the college use as the dividing line between those who get letters and those who do not?

E65. *Height limitations for flight attendants.* To work as a flight attendant for United Airlines, you must be between 5 ft 2 in. and 6 ft tall. [Source: *www.ual.com.*] The heights of 20- to 29-year-old males in the United States have mean 70.4 in. and standard deviation 3.0 in. The heights of 20- to 29-year-old females in the United States have mean 65.1 in. and standard deviation 2.6 in. Both distributions are approximately normal. Give the percentage of men this age who meet United's height limitation. Give the percentage of women this age who meet the height limitation.

E66. *Where is the next generation of male professional basketball players coming from?*

 a. The heights of 20- to 29-year-old males in the United States are approximately normal, with mean 70.4 in. and standard deviation 3.0 in. Use this information to approximate the percentage of men in the United States between the ages of 20 and 29 who are as tall as or taller than each basketball player listed here. Then using the fact that there are about 19 million men between the ages of 20 and 29 in the United States, estimate how many are as tall as or taller than each player.

  i. Shawn Marion, 6 ft 7 in.

  ii. Allen Iverson, 6 ft 0 in.

  iii. Shaquille O'Neal, 7 ft 1 in.

 b. Distributions of real data that are approximately normal tend to have heavier "tails" than the ideal normal curve. Does this mean your estimates in part a are too small, too big, or just right?

E67. *Puzzle problems.* Problems that involve computations with the normal distribution have four quantities: mean, standard deviation, value x, and proportion P below value x. Any three of these values are enough to determine the fourth. Think of each row in this table as a little puzzle, and find the missing value in each case. This isn't the sort of thing you are likely to run into in practice, but solving the puzzles can help you become more skilled at working with the normal distribution.

Mean	SD	x	Proportion
3	1	2	—a—
10	2	—b—	0.18
—c—	3	6	0.09
10	—d—	12	0.60

E68. *More puzzle problems.* In each row of this table, assume the distribution is normal. Knowing any two of the mean, standard deviation, Q_1, and Q_3 is enough to determine the other two. Complete the table.

Mean	SD	Q_1	Q_3
10	5	—a—	—b—
—c—	—d—	120	180
—e—	10	100	—f—
10	—g—	—h—	11

E69. ACT scores are approximately normally distributed, with mean 21 and standard deviation 5. Without using your calculator, roughly what percentage of scores are between 16 and 26? Between 11 and 31? Above 26? Below 26? Below 11? Above 11? [Source: *www.actstudent.org/pdf/norms.pdf*, 2009.]

E70. A group of subjects tested a certain brand of foam earplug. The number of decibels (dB) that noise was reduced for these subjects was approximately normally distributed, with mean 30 dB and standard deviation 3.6 dB. The middle 95% of noise reductions were between what two values?

E71. The heights of 20- to 29-year-old males in the United States are approximately normal, with mean 70.4 in. and standard deviation 3.0 in.

  a. If you select a U.S. male between ages 20 and 29 at random, what is the approximate probability that he is less than 68 in. tall?

  b. There are roughly 19 million 20- to 29-year-old males in the United States. About how many are between 67 and 68 in. tall?

  c. Find the height of 20- to 29-year-old males that falls at the 90th percentile.

E72. If the measurements of height are transformed from inches into feet, will that change the shape of the distribution in E71? Describe the distribution of male heights in terms of feet rather than inches.

E73. *The British monarchy.* Over the 1200 years of the British monarchy, the average reign of kings and queens has lasted 18.5 years, with a standard deviation of 15.4 years.

  a. What can you say about the shape of the distribution based on the information given?

  b. Suppose you made the mistake of assuming a normal distribution. What fraction of the reigns would you estimate lasted a negative number of years?

  c. Use your work in part b to suggest a rough rule for using the mean and standard deviation of a set of positive values to check whether it is possible that a distribution might be approximately normal.

E74. *NCAA scores.* The histogram in Display 2.92 was constructed from the total of the scores of both teams in all NCAA basketball play-off games over a 57-year period.

  a. Approximate the mean of this distribution.

  b. Approximate the standard deviation of this distribution.

  c. Between what two values do the middle 95% of total points scored lie?

  d. Suppose you chose a game at random from next year's NCAA play-offs. What is the approximate probability that the total points scored in this game will exceed 150? 190? Do you see any potential weaknesses in your approximations?

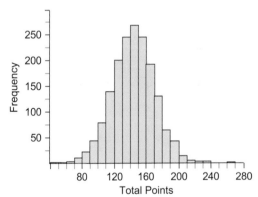

Display 2.92   Total points scored in NCAA play-off games [Source: *www.ncas.com.*]

Chapter Summary ◂

Distributions come in various shapes, and the appropriate summary statistics (for center and spread) usually depend on the shape, so you should always start with a plot of your data.

Common symmetric shapes include the uniform (rectangular) distribution and the normal distribution. There are also various skewed distributions. Distributions with two clusters often result from mixing cases of two kinds.

Dot plots, stemplots, and histograms show distributions graphically and let you estimate center and spread visually from the plot.

For approximately normal distributions, you ordinarily use the mean (balance point) and standard deviation as the measure of center and spread. If you know the mean and standard deviation of a normal distribution, you can use z-scores and Table A or technology to find the percentage of values in any interval.

The mean and standard deviation are not resistant—their values are sensitive to outliers. For a description of a skewed distribution, you should consider using the median (halfway point) and quartiles (medians of the lower and upper halves of the data) as summary statistics.

Later on, when you make inferences about the entire population from a sample taken from that population, the sample mean and standard deviation will be the most useful summary statistics, even if the population is skewed.

Review Exercises

E75. The map in Display 2.93, from the U.S. National Weather Service, gives the number of tornadoes by state, plus the District of Columbia.

a. Make a stemplot of the number of tornadoes.

b. Write the five-number summary.

c. Identify any outliers.

d. Draw a boxplot.

e. Compare the information in your stemplot with the information in your boxplot. Which plot is more informative?

f. Describe the shape, center, and spread of the distribution of the number of tornadoes.

Display 2.93 The number of tornadoes per state in a recent year. [Source: *www.ncdc.noss.com.*]

E76. Pecos Historical National Park, 25 miles east of Santa Fe, New Mexico, contains 629 sites of interest to archeologists and anthropologists. Researchers classified each site as to the time it was occupied, its elevation, and whether the site was used for habitation by large groups, for seasonal use by small groups (often near some natural resource), or for special use on a limited basis. The set of boxplots, for sites occupied between 1450 and 1575, was printed in their report. [Source: Genevieve N. Head and Janet D. Orcutt, eds., *From Folsom to Fogelson: The Cultural Resources Inventory Survey of Pecos National Historical Park*, Intermountain Cultural Resources Management Professional Paper No. 66, Anthropology

Projects Cultural Resources Management, Intermountain Region, National Park Service, Department of the Interior, 2002, *www.nps.gov/history/history/online_books/pecos/cris/chap6.htm.*]

a. Estimate the median elevation for the special use sites. Use this median in a sentence or two that makes it clear what the median represents in this context.

b. Why are there no whiskers on the habitation sites boxplot?

c. Compare the elevation for the three types of site.

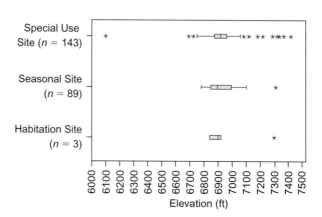

E77. A university reports that the middle 50% of the SAT I math scores of its students were between 585 and 670, with half the scores above 605 and half below.

a. What SAT I math scores would be considered outliers for that university?

b. What can you say about the shape of this distribution?

E78. These statistics summarize a set of television ratings from a week without any special programming. Are there any outliers among the 113 ratings?

Variable	N	Mean	Median	StDev
Ratings	113	6.867	6.900	3.490

Variable	Min	Max	Q1	Q3
Ratings	1.400	20.700	4.550	8.250

E79. The boxplots in Display 2.94 show the life expectancies for the countries of Africa, Europe, and the Middle East. The table shows a few of the summary statistics for each of the three data sets.

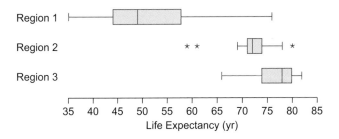

Group	Mean	Median	StdDev
A	76.44	78	4.15
B	72	72	5.24
C	52.20	49	11.04

Display 2.94 Life expectancies for the countries of Africa, Europe, and the Middle East, [Source: Population Reference Bureau, *World Population Data Sheet*, 2005.]

a. From your knowledge of the world, match the box-plots to the correct region.

b. Match the summary statistics (for Groups A–C) to the correct boxplot (for Regions 1–3).

E80. Pecos Historical National Park, 25 miles east of Santa Fe, New Mexico, contains 251 "ceramic assemblages." These are specific sites that have more than five datable pottery shards. For each assemblage, archeologists determined the (mean) date of the shards at that site. The following stem-and-leaf plot was printed in their report. Each stem represents a 20-year period. For example, the years 1740–1759 are on the stem 174. (See also E76.)

Pottery shards in New Mexico.

a. What is the earliest (mean) date for any assemblage? The latest date?

b. Describe the shape of the distribution. Do there appear to be any outliers?

```
176 | 3 3 3 3
174 | 4 6 5
172 |
170 | 0 5 5 5 5 5 0 8
168 | 5 0 9
166 | 1 3 4 4 4 4 8
164 | 0 1 2 6 6 9 3 5
162 | 3 4 4 7 0 0 0 0 1 3 4 8 9
160 | 3 3 5 5 8 8 8 8 8 8 8 8 2 3 3
158 | 2 9 4 9
156 | 2 4 6 7
154 | 3 5 5
152 | 2 4 4 9 9 3 4 9 9
150 | 3 4 5 6 8 1 2 2 5 5 6 6 8 9
148 | 2 3 8 9 5 5 5 9
146 | 0 0 0 2 4 6 7 8 8 9 9 0 0 2 6 6
144 | 1 2 3 4 5 6 7 8 8 9 9 2 4 8 8 8 8 8 9
142 | 1 4 5 6 6 7 7 7 8 8 9 0 1 2 2 3 4 4 5 9
140 | 0 3 3 5 6 9 3 4 5 6 8 8 9
138 | 0 2 2 2 5 9 0 0 0 3 3 4 4 6 7 8
136 | 0 4 4 6 6 7 8 9 9 0 0 0 1 2 3 5 6 6 8 9
134 | 1 5 6 7 9 0 0 0 1 2 4 5 9
132 | 0 8 1 6 6 9 9
130 | 1 4 4 6 7 9 1 5 5 5
128 | 0 1 1 3 6 7 8 8 8 1 9
126 | 3 3 8 8
```

174 | 465 represents 1744, 1746, and 1755

c. The report says, "Several suggestive natural breaks indicate relatively fewer sites between A.D. 1320 and 1330, during the A.D. 1500s, and after A.D. 1710." Do you agree?

d. Convert the stem-and-leaf plot to a histogram with a 40-year interval in each bin, beginning with the year 1260. Does your impression of the shape of the distribution change?

E81. A distribution is symmetric with approximately equal mean and median. Is it necessarily the case that about 68% of the values are within one standard deviation of the mean? If yes, explain why. If not, give an example.

E82. Display 2.95 (on the next page) shows two sets of graphs. The first set shows smoothed histograms I–IV for four distributions. The second set shows the corresponding cumulative relative frequency plots, in scrambled order A–D. Match each plot in the first set with its counterpart in the second set.

E.83 Pedestrians account for about 11% of deaths from motor vehicle crashes. Alcohol use often is involved, either by the pedestrian or by the driver. Typically, the pedestrian is taking a risk at the time of the accident. Not surprisingly, about two-thirds of the deaths occur in cities. The number of pedestrian deaths for 41 cities is given in Display 2.96.

Distributions

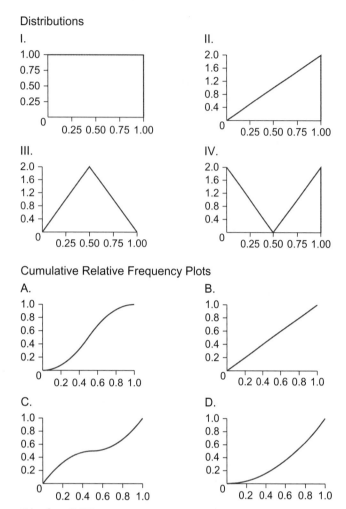

Cumulative Relative Frequency Plots

Display 2.95 Four distributions with different shapes and their cumulative relative frequency plots.

a. What is the median number of deaths? Write a sentence explaining the meaning of this median.

b. Is any city an outlier in terms of the number of deaths? If so, what is the city, and what are some possible explanations?

c. Make a plot of the data that you think will show the distribution in a useful way. Describe why you chose that plot and what information it gives you about the pedestrian deaths.

d. The population of New York is about 8,214,000 and that of Miami is about 404,000. Compute the pedestrian fatality rate per 100,000 people for each city. Why might giving the death rate be more meaningful than giving the number of deaths?

E84. The boxplots in Display 2.97 give the percentage of 5th-grade-age children who are still in school in various regions. Each case is a country. The four regions marked 1, 2, 3, and 4 are Sub-Saharan Africa, Asia, Europe, and South/Central America, not necessarily in that order.

a. Which region do you think corresponds to which number?

b. Is the distribution of values for any region skewed left? Skewed right? Symmetrical?

Display 2.98 shows dot plots of the same data.

City	Annual Pedestrian Fatalities
Atlanta	6
Baltimore	16
Boston	7
Charlotte, NC	15
Chicago	48
Cincinnati	5
Cleveland	8
Columbus, OH	9
Dallas	30
Denver	14
Detroit	28
Fort Lauderdale	15
Honolulu	12
Houston	45
Indianapolis	10
Kansas City	9
Los Angeles	99
Miami	27
Milwaukee	14
Minneapolis	1
Newark, NJ	10
New Orleans	0
New York	157
Norfolk, VA	4
Orlando, FL	13
Philadelphia	36
Phoenix	59
Pittsburgh	5
Portland, OR	8
Riverside, CA	6
Rochester, NY	3
Sacramento, CA	9
Salt Lake City	10
San Antonio	28
San Diego	22
San Francisco	19
San Jose, CA	22
Seattle	8
St. Louis	8
Tampa	13
Washington, DC	17

Display 2.96 Annual pedestrian deaths. [Source: U.S. Department of Transportation, *National Pedestrian Crash Report*, June 2008, Table A-7, *www-nrd.nhtsa.dot.gov/Pubs/810968.PDF*.]

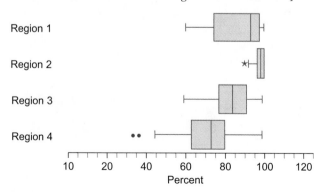

Display 2.97 Boxplots of the percentage of 5th-grade-age children still in school in countries of the world, by region. [Source: UNESCO.]

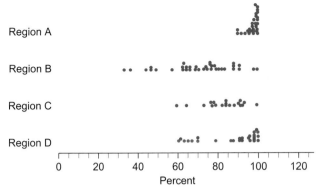

Display 2.98 Dot plots of the percentage of 5th-grade-age children still in school in countries of the world, by region.

c. Match each dot plot to the corresponding boxplot.

d. In what ways do the boxplots and dot plots give different impressions? Why does this happen? Which type of plot gives a better impression of the distributions?

E85. The first AP Statistics Exam was given in 1997. The distribution of scores received by the 7667 students who took the exam is given in Display 2.99. Compute the mean and standard deviation of the scores.

Score	Number of Students
5	1205
4	1696
3	1873
2	1513
1	1380

Display 2.99 Scores on the first AP Statistics Exam. [Source: The College Board.]

E86. For the countries of Europe, many average life expectancies are approximately the same, as you can see from the stemplot in Display 2.59 on page 61. Use the formulas for the summary statistics of values in a frequency table to compute the mean and standard deviation of the life expectancies for the countries of Europe.

E87. Construct a set of values in which all values are larger than 0, but one standard deviation below the mean is less than 0.

E88. Without computing, what can you say about the standard deviation of this set of values: 4, 4, 4, 4, 4, 4, 4, 4?

E89. In this exercise, you will compare how dividing by n versus $n - 1$ affects the standard deviation for various values of n. So that you don't have to compute the sum of the squared deviations each time, assume that this sum is 400.

a. Compare the standard deviation that would result from

   i. dividing by 10 versus dividing by 9

   ii. dividing by 100 versus dividing by 99

   iii. dividing by 1000 versus dividing by 999

b. Does the decision to use n or $n - 1$ in the formula for the standard deviation matter very much if the sample size is large?

E90. If two sets of test scores aren't normally distributed, it's possible to have a larger z-score on Test II than on Test I yet be in a lower percentile on Test II than on Test I.

The computations in this exercise will illustrate this point.

a. On Test I, a class got these scores: 11, 12, 13, 14, 15, 16, 17, 18, 19, 20. Compute the z-score and the percentile for the student who got a score of 19.

b. On Test II, the class got these scores: 1, 1, 1, 1, 1, 1, 1, 18, 19, 20. Compute the z-score and the percentile for the student who got a score of 18.

c. Make a case that the student who got a score of 19 on Test I did better, relative to the rest of the class, than the student who got a score of 18 on Test II. Then make a case that the student who got a score of 18 on Test II did better, relative to the rest of the class, than the student who got a score of 19 on Test I.

E91. The income of the residents varies dramatically from state to state. Summary statistics for three different years for the per capita income (mean income per person across the state) for the 50 states are listed in Display 2.100.

a. Explain the meaning of $6573 for the minimum in 1980.

b. Are any states outliers for any year?

c. In 2007, the per capita income in Alabama was $32,404 and in 1980 it was $7,465. Did the per capita income change much in relation to the other states? Explain your reasoning.

d. Histograms for the incomes for the three years follow. Describe, in general, the changes in shape, center, and spread across the years.

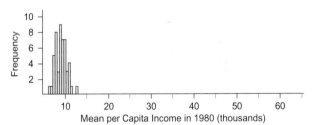

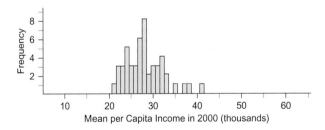

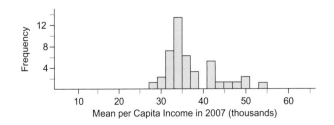

e. For which of the three years would it be most appropriate to use a z-score to describe the position of the per capita income for Alabama? Explain your reasoning.

E92. In this exercise, you will find percentiles of the GMAT distribution using Display 2.77 on page 70 or the fact that the scores have a normal distribution with mean 525.2 and standard deviation 120.2.

a. Estimate the percentile for a GMAT score of 425 using the cumulative relative frequency plot. Then find the percentile for a score of 425 using a z-score. Are the two values close?

b. Estimate the GMAT score that falls at the 40th percentile, using the table in Display 2.77. Then find the 40th percentile using a z-score. Are the two values close?

c. Estimate the median from the cumulative relative frequency plot. Is this value close to the median you would get by assuming a normal distribution of scores?

d. Estimate the quartiles and the interquartile range using the plot. Find the quartiles and interquartile range assuming a normal distribution of scores.

E93. For 17-year-olds in the United States, blood cholesterol levels in milligrams per deciliter have an approximately normal distribution with mean 176 mg/dL and standard deviation 30 mg/dL. The middle 90% of the cholesterol levels are between what two values?

E94. Display 2.101 shows the distribution of batting averages for all 187 American League baseball players who batted 100 times or more in a recent season. (A batting "average" is the fraction of times that a player hits safely—that is, the hit results in a player advancing to a base—usually reported to three decimal places.)

a. Do the batting averages appear to be approximately normally distributed?

b. Approximate the mean and standard deviation of the batting averages from the histogram.

c. Use your mean and standard deviation from part b to compute an estimate of the percentage of players who batted .300 or over.

d. Now use the histogram to estimate the percentage of players who batted .300 or over. Compare to your estimate from part c.

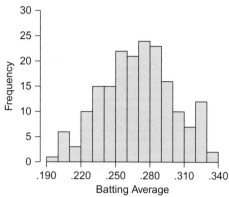

Display 2.101 American League batting averages. [Source: *www.cbssports.com.*]

E95. How good are batters in the National League? Display 2.102 shows the distribution of batting averages for all 223 National Leaguers who batted 100 times or more in a recent season.

a. Approximate the mean and standard deviation of the batting averages from the histogram.

b. Compare the distributions of batting averages for the two leagues. (See E94 for the American League.) What are the main differences between the two distributions?

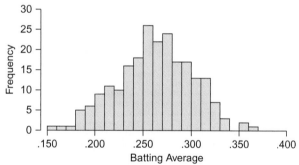

Display 2.102 National League batting averages. [Source: CBS, *www.cbssports.com.*]

c. A batter hitting .300 in the National League is traded to a team in the American League. What batting average could be expected of him in his new league if he maintains about the same position in the distribution relative to the other players?

	Year	Mean	SD	Minimum	Lower Quartile	Median	Upper Quartile	Maximum
1	1980	9,144.18	1,295.61	6,573	7,957	9,226.5	10,103	13,007
2	2000	27,972.7	4,322.7	20,900	24,708	27,645	31,012	40,702
3	2007	37,172.6	5,516.57	28,845	33,457	35,566.5	40,480	54,117

Display 2.100 Per capita income for the 50 states, in dollars, unadjusted for inflation. [Source: U.S. Department of Commerce, Bureau of Economic Analysis, *www.infoplease.com/ipa/A0104652.html.*]

C1. These summary statistics are for the distribution of the populations of the major cities in Brazil.

Variable	N	Mean	Median	StDev	SeMean
Population	209	409,783	204,400	867,140	59,981

Variable	Min	Max	Q1	Q3
Population	103,100	10,057,700	138,050	351,250

Which of the following best describes the shape of this distribution?

Ⓐ skewed right without outliers

Ⓑ skewed right with at least one outlier

Ⓒ roughly normal, without outliers

Ⓓ skewed left without outliers

Ⓔ skewed left with at least one outlier

C2. Which of these lists contains only summary statistics that are sensitive to outliers?

Ⓐ mean and median

Ⓑ standard deviation, *IQR*, and range

Ⓒ mean and standard deviation

Ⓓ median and *IQR*

Ⓔ five-number summary

C3. This stem-and-leaf plot shows the ages of CEOs of 60 corporations with annual sales between $5 million and $350 million. Which of the following is not a correct statement about this distribution?

Ⓐ The distribution is skewed left (towards smaller numbers).

Ⓑ The oldest of the 60 CEOs is 74 years old.

Ⓒ The distribution has no outliers.

Ⓓ The range of the distribution is 42.

Ⓔ The median of the distribution is 50.

3	2 3
3	6 7 8
4	0 1 3 3 4 4
4	5 5 5 5 6 6 6 7 7 7 8 8 8 8 9
5	0 0 0 0 0 0 1 1 2 3 3 3
5	5 5 5 6 6 6 6 6 7 7 8 8 9
6	0 1 1 1 2 2 3
6	9 9
7	0 4

C4. A traveler visits Europe and stays 30 days in 30 different hotels, paying each day with her credit card. The hotels charge a mean price of 50 euros, with a standard deviation of 10 euros. When the charges appear on her credit card statement in the United States, she finds that her bank charged her $1.20 per euro, plus a $5 fee for each transaction. What are the mean and standard deviation of the 30 daily hotel charges in dollars, including the fee?

Ⓐ mean $50, standard deviation $17

Ⓑ mean $60, standard deviation $12

Ⓒ mean $60, standard deviation $17

Ⓓ mean $65, standard deviation $12

Ⓔ mean $65, standard deviation $17

C5. The scores on a nationally administered test are approximately normally distributed with mean 47.3 and standard deviation 17.3. Approximately what must a student have scored to be in the 95th percentile nationally?

Ⓐ 55 **Ⓑ** 61 **Ⓒ** 73 **Ⓓ** 76 **Ⓔ** 81

C6. A particular brand of cereal boxes is labeled "16 oz." This dot plot shows the actual weights of 100 randomly selected boxes. Which of the following is the best estimate of the standard deviation of the these weights?

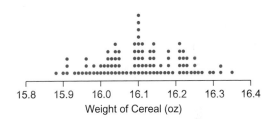

Weight of Cereal (oz)

Ⓐ 0.04 oz **Ⓑ** 0.1 oz

Ⓒ 0.2 oz **Ⓓ** 0.4 oz

Ⓔ between 16.0 and 16.2 oz

C7. The distribution of the number of points earned by the thousands of contestants in the Game of Pig World Championship has mean 20 and standard deviation 6. What percentage of the contestants earned more than 26 points?

Ⓐ less than 1% **Ⓑ** 16%

Ⓒ 32% **Ⓓ** 84%

Ⓔ This percentage cannot be determined from the information given.

C8. Anya scored 70 on a statistics test for which the mean was 60 and the standard deviation was 10. She also scored 60 on a chemistry test for which the mean was 50 and the standard deviation was 5. If the scores for both tests were approximately normally distributed, which best describes how Anya did relative to her classmates?

Ⓐ Anya did better on the statistics test than she did on the chemistry test because she scored 10 points higher on the statistics test than on the chemistry test.

Ⓑ Anya did equally well relative to her classmates on each test, because she scored 10 points above the mean on each.

Ⓒ Anya did better on the chemistry test than she did on the statistics test because she scored two standard deviations above the mean on the chemistry test and only one standard deviation above the mean on the statistics test.

D It's impossible to tell without knowing the number of points possible on each test.

E It's impossible to tell without knowing the number of students in each class.

Investigation

C9. A game invented by three college students involves giving the name of an actor and then trying to connect that actor with actor Kevin Bacon, counting the number of steps needed. For example, Sarah Jessica Parker has a "Bacon number" of 1 because she appeared in the same movie as Kevin Bacon, *Footloose* (1984). Angelina Jolie has a Bacon number of 2. She has never appeared in a movie with Kevin Bacon; however, she was in *Sledge* (2005) with Jim Cody Williams who was in *Rails & Ties* (2007) with Kevin Bacon. This table gives the number of links required to connect each of the 982,586 actors in the Internet Movie Database to Kevin Bacon. [Source: The Oracle of Bacon, *oracleofbacon.org*]

Bacon Number	Number of Actors
0	1
1	2,156
2	209,477
3	620,848
4	140,656
5	8,573
6	767
7	102
8	6

a. How many people have appeared in a movie with Kevin Bacon?

b. Who is the person with Bacon number 0?
   It has been questioned whether Kevin Bacon was the best choice for the "center of the Hollywood universe." A possible challenger is Will Smith.

Will Smith Number	Number of Actors
0	1
1	3,865
2	257,603
3	607,200
4	106,210
5	7,062
6	608
7	33
8	4

c. Do you think Kevin Bacon or Will Smith better deserves the title "Hollywood center"? Make your case using statistical evidence (as always).

d. (For movie fans.) What is Bacon's Smith number? What is Smith's Bacon number?

Chapter 3
Relationships Between Two Quantitative Variables

What variables predict whether a college has a high graduation rate? Scatterplots, correlation, and regression are the basic tools used to describe relationships between two quantitative variables.

In this chapter, you'll learn how to explore and summarize the relationship between two quantitative variables. The data set on mammals in Display 2.14 on page 31 raises many questions of this sort: Do mammals with longer average longevity also have longer gestation period? Is there a relationship between speed and longevity?

The approach to describing distributions in Chapter 2 boiled down to finding shape, center, and spread. For distributions that are approximately normal, two numerical summaries—the mean for center, the standard deviation for spread—tell you basically all you need to know. With paired quantitative variables, you can see the shape of the distribution by making a scatterplot. For scatterplots with points that lie in an oval cloud, it turns out once again that two summaries tell you pretty much all you need to know: the regression line and the correlation. The regression line tells about center: What is the equation of the line that best fits the cloud of points? The correlation tells about spread: How spread out are the points around the line?

In this chapter, you will learn to

▶ describe the pattern in a scatterplot, and decide what its *shape* tells you about the relationship between the two variables

▶ find a regression line through the *center* of a cloud of points to summarize the relationship

▶ use the correlation as a measure of how *spread* out the points are from this line

▶ use diagnostic tools to check for information the summaries don't tell you, and decide what to do with that information

3.1 ▶ Scatterplots

A **scatterplot** shows the relationship between two quantitative variables.

Chapter 1 contained an exploration of the relationship between the age of an employee at the Westvaco Corporation and whether the employee was laid off when the company downsized. When trying to help Robert Martin establish a case that he was discriminated against on the basis of his age, the statistician employed by Martin's lawyer found that there is more to see. In the scatterplot in Display 3.1, for

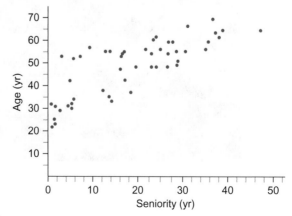

Display 3.1   *Age* versus *seniority* for the 50 employees in Westvaco Corporation's engineering department.

example, each employee is represented by a dot that shows the age of the employee when layoffs began plotted against the number of years the employee had worked for Westvaco.

In this scatterplot, you can see a moderate positive association: Employees with more seniority generally were the older employees, and employees with less seniority generally were younger. This trend is fairly linear. You can visualize a summary line going through the center of the points from lower left to upper right. As you move to the right along this line, the points cluster more closely around the line.

DISCUSSION
Interpreting Scatterplots

D1. You will examine Display 3.1 more closely in these questions.

    a. Why should the two variables plotted in Display 3.1 show a positive association?

    b. Why do all of the points in Display 3.1 lie above a diagonal line running from the lower left to the upper right?

    c. Is this sentence a reasonable interpretation of Display 3.1? "As time passed, Westvaco tended to hire younger and younger people."

Describing the Pattern in a Scatterplot

For the distribution of a single quantitative variable, "shape, center, and spread" is a useful summary. For **bivariate** (two-variable) quantitative data, the summary becomes "shape, trend, and strength."

Shape, trend, and strength.

You might find it helpful to follow this set of steps as you practice describing scatterplots.

1. Identify the **variables** and **cases**. On a scatterplot, each point represents a case, with the x-coordinate equal to the value of one variable and the y-coordinate equal to the value of the other variable. You should describe the scale (units of measurement) and range of each variable.

2. Describe the overall **shape** of the relationship, paying attention to
   - **linearity:** Is the pattern linear (scattered about a line) or curved?
   - **clusters:** Is there just one cluster, or is there more than one?
   - **outliers:** Are there any striking exceptions to the overall pattern?

3. Describe the **trend.** If as x gets larger y tends to get larger, there is a **positive trend.** (The cloud of points tends to slope upward as you go from left to right.) If as x gets larger y tends to get smaller, there is a **negative trend.** (The cloud of points tends to slope downward as you go from left to right.)

4. Describe the **strength** of the relationship. If the points cluster closely around an imaginary line or curve, the association is **strong.** If the points are scattered farther from the line, the association is **weak.**

   If, as in Display 3.1, the points tend to fan out at one end (a tendency called heteroscedasticity), the relationship **varies in strength.** If not, it has **constant strength.**

5. Does the pattern **generalize** to other cases, or is the relationship an instance of "what you see is all there is"?

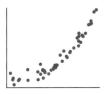
Curved and strong

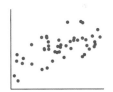
Linear and moderate

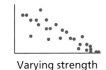
Varying strength

6. Are there plausible **explanations** for the pattern? Is it reasonable to conclude that one variable causes the other? Is there a third or **lurking variable** that might be causing both?

Example 3.1	**Dormitory Populations**

A plot of variable *A* against (or versus) variable *B* shows *A* on the *y*-axis and *B* on the *x*-axis.

A student investigating the effect of surrounding urban areas on resident college students began by creating the scatterplot in Display 3.2. The plot shows, for the 50 states in the United States, the number of people living in cities versus the number of people living in college dormitories. Describe the pattern in the plot. What can the student conclude about the relationship between the two variables?

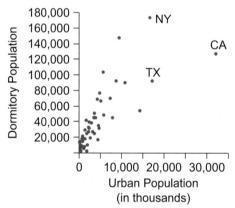

Display 3.2 Number of people living in college dormitories versus number of people living in cities for the 50 states in the United States. [Source: U.S. Census Bureau, 2000 Census of Population and Housing.]

Solution

1. *Variables and cases.* The scatterplot shows *dormitory population* plotted against *urban population*, in thousands, for the 50 U.S. states. Dormitory population ranges from near 0 to a high of more than 174,000 in New York. The urban population ranges from near 0 to about 17 million in Texas and New York and 32 million in California.

2. *Shape.* While most states follow a linear trend, the three states with the largest urban population suggest curvature in the plot because, for those states, the number of people living in dormitories is proportionately lower than in the smaller states. California can be considered an outlier with respect to its urban population, which is much larger than that of other states. It is also an outlier with respect to the overall pattern, because it lies far below the generally linear trend.

3. *Trend.* The trend is positive—states with larger urban populations tend to have larger dormitory populations, and states with smaller urban populations tend to have smaller dormitory populations.

4. *Strength.* The relationship varies in strength. For the states with the smallest urban populations, the points cluster rather closely around a line. For the states with the largest urban populations, the points are scattered farther from the line. Overall, the strength of the relationship is moderate.

5. *Generalization.* The 50 states aren't a sample from a larger population of cases, so the relationship here does not generalize to other cases. Because both variables tend to change rather slowly, however, we can expect the relationship in Display 3.2 to be similar to that of other years.

6. *Explanation.* It was tempting for the student to attribute the positive relationship to the idea that students from states with large urban populations prefer to live in dorms. The main reason for the positive relationship, however, is not nearly so interesting: Both variables are related to a state's population. The more people in a state, the more people live in dormitories and the more people live in cities. (There's a moral here: Interpreting association can be tricky, in part because the two variables you see in a plot often will be related to some lurking variable that you don't see.)

DISCUSSION
Describing the Pattern in a Scatterplot

D2. Display 3.3 is derived from the data for Display 3.2 by converting the variables to the *proportion* of a state's population living in college dormitories (given as the number living in dorms per 1000 state residents) and the proportion of the state's population living in cities.

    a. Follow steps 1–6 in the previous example to describe what you see in this new plot.

    b. When you go from totals (Display 3.2) to proportions of total population (Display 3.3), the relationship changes from positive to negative and becomes weaker. Give an explanation for the differences in these two plots.

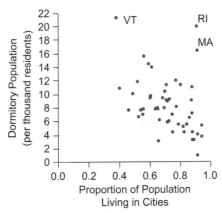

Display 3.3   The proportion of people living in college dormitories versus the proportion of people living in cities for the 50 U.S. states.

Summary 3.1: Scatterplots

A scatterplot shows the relationship between two quantitative variables. Each case is a point, with the *x*-coordinate equal to the value of one variable and the *y*-coordinate equal to the value of the other variable.

    In describing a scatterplot, be sure to cover all of the following:

- cases and variables (What exactly does each point represent?)
- shape (linear or curved, clusters, outliers)
- trend (positive, negative, or none)
- strength (strong, moderate, or weak; constant or varying)
- generalization (Does the pattern generalize?)
- explanation (Is there an explanation for the pattern?)

Practice

Describing the Pattern in a Scatterplot

P1. For each of the scatterplots in Display 3.4, give the trend (positive or negative), strength (strong, moderate, or weak), and shape (linear or curved). Which plots show varying strength?

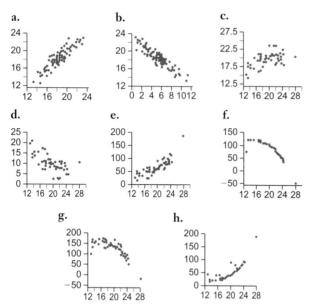

Display 3.4 Eight scatterplots with various distributions.

P2. *Growing kids.* This table gives median heights of boys at ages 2, 3, 4, 5, 6, and 7 years.

Age (yr)	Height (in.)	Age (yr)	Height (in.)
2	35.8	5	44.2
3	39.1	6	46.8
4	41.4	7	49.6

a. *Scatterplot.* Plot *height* versus *age*; that is, put height on the *y*-axis and age on the *x*-axis.

b. *Shape, trend, and strength.* Describe the shape, trend, and strength of the relationship.

c. *Generalization.* Would you expect these data to allow you to make good predictions of the median height of 8-year-olds? Of 50-year-olds?

P3. The scatterplot in Display 3.5 shows the cross-sectional areas (in square inches at chest height) of oak trees plotted against the ages of the trees (in years).

a. *Shape, trend, and strength.* Describe the shape, trend, and strength of the relationship.

b. *Generalization.* All of these trees are on the same ranch. Do you think this same pattern would continue to hold if you had measurements from oak trees taken at a nearby ranch.

c. *Explanation.* Provide a realistic explanation for the pattern seen here. Why are there no points in the upper left of the display?

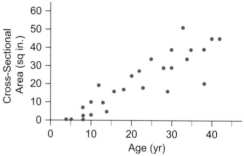

Display 3.5 *Cross-sectional area* versus *age* of oak trees.

Exercises

E1. For each set of cases and variables, tell whether you expect the relationship to be (i) positive or negative and (ii) strong, moderate, or weak.

Cases	Variable 1	Variable 2
a. Hens' eggs	Length	Width
b. High school seniors	SAT I math score	SAT I critical reading score
c. Trees	Age	Number of rings
d. People	Age	Body flexibility
e. U.S. states	Population	Number of representatives in Congress
f. Countries of the United Nations	Land area	Population
g. Olympic games	Year	Winning time in the women's 100-meter race

E2. Match each set of cases and variables (A–D) with the short summary (I–IV) of its scatterplot.

Cases	Variable 1	Variable 2
A. Earth	Year	Human population
B. Toddlers	Age (12 to 36 mo)	Number of times fell today
C. U.S. states	Population	Number of doctors
D. Cars in the United States	Weight	Gas mileage

 I. strong negative relationship, somewhat curved

 II. strong, curved positive relationship

 III. moderate, roughly linear, positive relationship

 IV. weak negative relationship

E3. *Late planes and lost bags.* A great way to cap off a long day of travel is to have your plane arrive late and then find that the airline has lost your luggage. As Display 3.6 shows, some airlines handle baggage better than others.

  a. Which airline has the worst record for mishandled baggage? For being on time?

  b. Where on the plot would you find the airline with the best on-time record and the best mishandled-baggage rate? Considering both variables together, which airlines are best overall?

  c. Determine whether this statement is true or false, and explain your answer: American had a mishandled-baggage rate that was more than twice the rate of Southwest.

  d. Is there a positive or a negative relationship between the on-time percentage and the rate of mishandled baggage? Is it strong or weak?

  e. Would you expect the relationship in this plot to be roughly the same for data from 10 years ago? For next year?

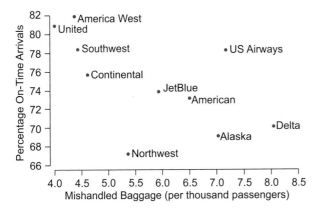

Display 3.6 *On-time arrivals* versus *mishandled baggage.* [Source: U.S. Department of Transportation, *Air Travel Consumer Report*, October 2005.]

E4. *SAT I math scores.* In 2008, the average SAT I math score across the United States was 515. North Dakota students averaged 604, Illinois students averaged 601, and students from the nearby state of Iowa did even better, averaging 612. Why do states from the Midwest do so well? It is

easy to jump to a false conclusion, but the scatterplot in Display 3.7 can help you find a reasonable explanation.

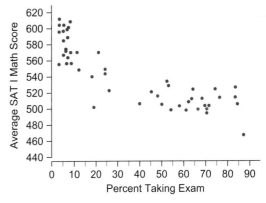

Display 3.7 Average SAT I math scores by state versus the percentage of high school graduates who took the exam. [Source: College Board, *www.collegeboard.com.*]

  a. Estimate the percentage of students in Iowa and in Illinois who took the SAT I. Maine had the highest percentage of students who took the SAT I. Estimate that percentage and the average SAT I math score for students in that state.

  b. Describe the shape of the plot. Do you see any clusters? Are there any outliers? Is the relationship linear or curved? Is the overall trend positive or negative? What is the strength of the relationship?

  c. Would a plot of the distribution of the percentage of students taking the SAT I divide into clusters? Explain how the scatterplot shows this. Would the distribution of SAT I math scores divide into clusters?

  d. The cases used in this plot are the 50 U.S. states in 2008. Would you expect the pattern to generalize to other years? Why or why not?

  e. Suggest an explanation for the trend. (*Hint:* The SAT is administered from Princeton, New Jersey. An alternative exam, the ACT, is administered from Iowa. Many colleges and universities in the Midwest either prefer the ACT or at least accept it in place of the SAT, whereas colleges in the eastern states tend to prefer the SAT.) Is there anything in the data that you can use to help you decide whether your explanation is correct?

E5. Each of the 51 cases plotted on the scatterplots in Display 3.8 is a top-rated university. The *y*-coordinate of a point tells the graduation rate, and the *x*-coordinate tells the value of some other quantitative variable—the percentage of alumni who gave that year, the student/faculty ratio, the 75th percentile of the SAT scores (math plus critical reading) for a recent entering class, and the percentage of incoming students who ranked in the top 10% of their high school graduating class.

  a. Compare the *shapes* of the four plots.

   i. Which plots show a linear shape? Which show a curved shape?

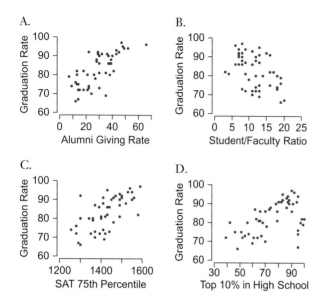

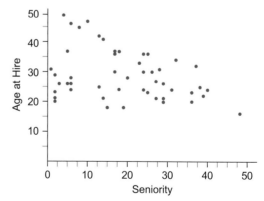

Display 3.8 Scatterplots showing the relationship between graduation rate and four other variables for 51 top-rated universities. [Source: *U.S. News and World Report*, 2000.]

    ii.   Which plots show just one cluster? Which show more than one?

    iii.   Which plots have outliers?

  b. Compare the *trends* of the relationships: Which plots show a positive trend? A negative trend? No trend?

  c. Compare the *strengths* of the relationships: Which variables give more precise predictions of the graduation rate? Which variable is almost useless for predicting graduation rate?

  d. *Generalization.* The cases in these plots are the 51 universities that happened to come out at the top of one particular rating scheme. Do you think the complete set of all U.S. universities would show pretty much the same relationships? Why or why not?

  e. *Explanation.* Consider the two variables with the strongest relationship to graduation rates. Offer an explanation for the strength of these particular relationships.

E6. *Hat size.* What does hat size really measure? A group of students investigated this question by collecting a sample of hats. They recorded the size of the hat and then measured the circumference, the major axis (the length across the opening in the long direction), and the minor axis. (See Display 3.9) Is hat size most closely related to circumference, major axis, or minor axis? Answer this question by making appropriate plots and describing the patterns in those plots.

E7. To help determine whether the Westvaco Corporation may have discriminated by age in laying off employees, you could investigate whether it might have discriminated in hiring. Look at Display 3.10, where the age at hire is plotted against how many years ago the person was hired.

  a. Does this plot provide evidence that Westvaco did or did not discriminate by age in hiring? Explain.

Hat Size	Circumference	Major Axis	Minor Axis
6.625	20.00	7.00	5.75
6.750	20.75	7.25	6.00
6.875	20.50	7.50	6.00
6.875	20.75	7.25	6.00
6.875	20.75	7.50	6.00
6.875	21.50	7.25	6.25
7.000	21.25	7.50	6.00
7.000	21.00	7.50	6.00
7.000	21.00	7.50	6.25
7.000	21.75	7.50	6.25
7.125	21.50	7.75	6.25
7.125	21.75	7.75	6.50
7.125	21.50	7.75	6.25
7.125	22.25	7.75	6.25
7.250	22.00	7.75	6.25
7.250	22.50	7.75	6.50
7.375	22.25	7.75	6.50
7.375	22.25	8.00	6.50
7.375	22.50	8.00	6.50
7.375	22.75	8.00	6.50
7.375	23.00	8.00	6.50
7.500	22.75	8.00	6.50
7.500	22.50	8.00	6.50
7.625	23.00	8.25	6.50
7.625	23.00	8.25	6.50
7.625	23.25	8.25	6.75

Display 3.9 Hat sizes, as decimals, with circumference and axes in inches. [Source: Roger Johnson, Carleton College, data from student project.]

Display 3.10 *Age at hire* versus *seniority* for the 50 employees in Westvaco Corporation's engineering department.

  b. Display 3.11 shows the age of the Westvaco employees plotted against the number of years of seniority. Does this scatterplot suggest a reason why older employees tended to be laid off more frequently?

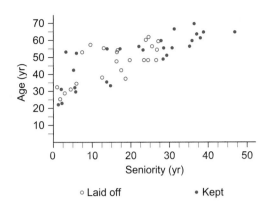

Display 3.11 *Age at hire* versus *seniority* for the 50 employees in Westvaco Corporation's engineering department. Open circles represent employees laid off, and solid circles represent employees kept.

E8. *NASCAR.* When turns on a racetrack are banked, cars can go faster through them. See the photo. The table in Display 3.14 gives information collected over a representative set of races at 21 NASCAR racetracks: the angle at which the turns are banked, the circumference of the raceway, the top speed in the qualifying trials among all cars that qualified for the racers (TQS), and the top average speed in the set of races (TRAS).

  a. Describe the pattern in the plot in Display 3.12, which shows TRAS versus the banking angle of the track.

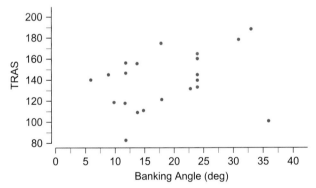

Display 3.12 Plot of TRAS versus banking angle of the raceway.

  b. Display 3.13 shows a plot of TRAS versus the banking with the long tracks indicated by a circle and the short tracks indicated by a square (with "long" being defined as 2 miles or more). Describe the pattern in the plot, taking the length of the track into consideration.

  c. Which track is an outlier in the scatterplots? Suggest a possible reason for this.

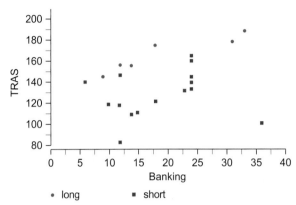

Display 3.13 Plot of TRAS versus banking angle of the raceway, with length of track.

E9. *School expenditures.* You are an employee of your state department of education and have been assigned to investigate how to improve high school graduation rates. Specifically, the governor would like to know what variables are good predictors of graduation rates. Exploring data provided by the states may result in the discovery of some patterns that could shed light on the question. You collect the information in the table and scatterplots of Display 3.15, which give the reported high school graduation rates (percentage of 9th graders who actually graduate), the per-pupil expenditure (PPE) in dollars, and the student/teacher ratio (number of students per teacher) for each state.

  a. Use the scatterplots in Display 3.15 to investigate possible associations between graduation rates and the explanatory variables *PPE* and *student-teacher ratio*. Which explanatory variable appears to do the better job of predicting graduation rates? Does either do a good job of predicting graduation rates?

  b. Considering graduation rates versus PPE, which state appears to be getting the best bargain in graduation rates for funds expended?

  c. Do you see different patterns for the different regions of the United States with regard to graduation rates as related to student-teacher ratios?

  d. Your state is Delaware. Write a description for the governor of the situation in Delaware, taking all three variables into account. Include your recommendation.

E10. *Passenger aircraft.* Airplanes vary in their size, speed, average flight length, and cost of operation. You can probably guess that larger planes use more fuel per hour and cost more to operate than smaller planes, but the shapes of

Speedway	Banking Angle (deg)	Circumference (mi)	TQS (mph)	TRAS (mph)
Atlanta Motor Speedway	24	1.54	197.478	163.633
Bristol Motor Speedway	36	0.53	126.37	101.074
California Speedway	14	2	187.432	155.012
Chicagoland Speedway	18	1.5	183.717	121.200
Darlington Raceway	24	1.37	173.797	139.958
Daytona International Speedway	31	2.5	210.364	177.602
Dover International Speedway	24	1	159.964	132.719
Homestead-Miami Speedway	6	1.5	156.44	140.335
Indianapolis Motor Speedway	12	2.5	181.072	155.912
Kansas Speedway	15	1.5	176.449	110.576
Las Vegas Motor Speedway	12	1.5	172.563	146.530
Lowes Motor Speedway	24	1.5	186.034	160.306
Martinsville Speedway	12	0.53	95.371	82.223
Michigan International Speedway	18	2	191.149	173.997
New Hampshire International Speedway	12	1.06	132.089	117.134
North Carolina Speedway	23	1.02	158.035	131.103
Phoenix International Raceway	10	1	134.718	118.132
Pocono Raceway	9	2.5	172.391	144.892
Richmond International Raceway	14	0.75	126.499	108.707
Talladega Superspeedway	33	2.66	212.809	188.354
Texas Motor Speedway	24	1.5	192.137	144.276

Display 3.14 NASCAR track data. [Source: *www.NASCAR.com* or *www.stat.ufl.edu/~winner/*. Go to *www.wiley.com/college/ Watkins* to download data sets.]

State	Region	Graduation Rate	PPE	Student/Teacher Ratio
Alabama	S	66.2	8,004	14.9
Alaska	W	66.5	11,726	16.6
Arizona	W	70.5	5,791	23.6
Arkansas	S	80.4	9,586	14.1
California	W	69.2	8,823	21
Colorado	W	75.5	9,555	17
Connecticut	E	80.9	12,936	13.4
Delaware	E	76.3	12,770	15.7
Florida	S	63.6	8,305	16.4
Georgia	S	62.4	9,176	14.8
Hawaii	W	75.5	10,696	16.2
Idaho	W	80.5	7,475	18.1
Illinois	M	79.7	10,477	16
Indiana	M	73.3	9,696	17.1
Iowa	M	86.9	8,325	13.7
Kansas	M	77.6	9,613	14.4
Kentucky	S	77.2	9,031	15.9
Louisiana	S	59.5	8,602	14.8
Maine	E	76.3	12,223	12.2
Maryland	S	79.9	10,497	14.9
Massachusetts	E	79.5	13,407	13.2
Michigan	M	72.2	10,818	16.6
Minnesota	M	86.2	10,361	16.3
Mississippi	S	63.5	7,503	15.4
Missouri	M	81.0	8,518	13.9

State	Region	Graduation Rate	PPE	Student/Teacher Ratio
Montana	W	81.9	9,620	14.1
Nebraska	M	87.0	8,534	13.7
Nevada	W	55.8	7,200	19
New Hampshire	E	81.1	11,110	13.3
New Jersey	E	84.8	14,149	12.6
New Mexico	W	67.3	9,580	14.9
New York	E	67.4	14,568	12.4
North Carolina	S	71.8	8,163	15
North Dakota	W	82.1	8,458	12.9
Ohio	M	79.2	11,316	15.4
Oklahoma	M	77.8	7,449	15.3
Oregon	W	73.0	9,846	19
Pennsylvania	E	82.2	11,521	14.9
Rhode Island	E	77.8	11,693	10.5
South Carolina	S	60.6	8,971	14.8
South Dakota	W	84.5	8,334	13.6
Tennessee	S	70.6	7,453	15.7
Texas	S	72.5	8,078	15
Utah	W	78.6	5,815	22.3
Vermont	E	82.3	14,836	10.8
Virginia	S	74.5	10,011	13.2
Washington	W	72.9	8,563	19.3
West Virginia	S	76.9	10,103	14.2
Wisconsin	M	87.5	10,672	14.7
Wyoming	W	76.1	12,484	12.7

E = Eastern state; M = Midwestern state; S = Southern state; W = Western state.

[Source: National Center for Education Statistics; *U.S. Statistical Abstract 2009*.]

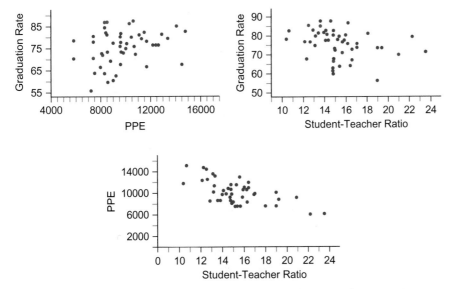

Display 3.15 High school graduation rates, per-pupil, expenditures, and student-teacher ratio by state. [Source: *http://nces.ed.gov/* and U.S. Statistical Abstract 2009.]

the relationships are less obvious. Display 3.16 lists data on the 33 most commonly used passenger airplanes in the United States. The variables are the number of seats, average cargo payload in tons, airborne speed in miles per hour, flight length in miles, fuel consumption in gallons per hour, and operating cost per hour in dollars.

a. *Cost per hour*
   i. Make scatterplots with *cost per hour* on the y-axis to explore this variable's dependence on the other variables. Report your most interesting findings. Here are examples of some questions you could investigate: For which

Aircraft	Number of Seats	Cargo (tons)	Speed (mph)	Flight Length (mi)	Fuel (gal/h)	Cost ($/h)
B747-200/300	370	16.6	520	3148	3625	9153
B747-400	367	8.06	534	3960	3411	8443
L-1011	325	0.04	494	2023	1981	8042
DC-10	286	24.87	497	1637	2405	7374
B767-400	265	6.26	495	1682	1711	3124
B-777	263	9.43	525	3515	2165	5105
A330	261	11.12	509	3559	1407	3076
MD-11	261	45.07	515	2485	2473	7695
A300-600	235	19.12	460	947	1638	6518
B757-300	235	0.3	472	1309	985	2345
B767-300ER	207	7.89	497	2122	1579	4217
B757-200	181	1.41	464	1175	1045	3312
B767-200ER	175	3.72	487	1987	1404	3873
A321	169	0.44	454	1094	673	1347
B737-800/900	151	0.37	454	1035	770	2248
MD-90	150	0.25	446	886	825	2716
B727-200	148	6.46	430	644	1289	4075
A320	146	0.31	454	1065	767	2359
B737-400	141	0.25	409	646	703	2595
MD-80	134	0.19	432	791	953	2718
B737-700LR	132	0.28	441	879	740	1692
B737-300/700	132	0.22	403	542	723	2388
A319	122	0.27	442	904	666	1913
B737-100/200	119	0.11	396	465	824	2377
B717-200	112	0.22	339	175	573	3355
B737-500	110	0.19	407	576	756	2347
DC-9	101	0.15	387	496	826	2071
F-100	87	0.05	398	587	662	2303
B737-200C	55	2.75	387	313	924	3421
ERJ-145	50	0	360	343	280	1142
CRJ-145	49	0.01	397	486	369	1433
ERJ-135	37	0	357	382	267	969
SD340B	33	0	230	202	84	644

Display 3.16 Data on passenger aircraft. [Source: Air Transport Association of America, 2005, *www.air-transport.org.*]

variable is the relationship to the cost per hour strongest? Is there any one airplane whose cost per hour, in relation to other variables, makes it an outlier?

ii. Do your results mean that larger planes are less efficient? Choose your own variable, and plot it against other variables to judge the relative efficiency of the larger planes.

b. *Flight length*

i. Make scatterplots with *length of flight* on the *x*-axis to explore this variable's relationship to the other variables. Report your most interesting findings. Here is an example of a question you could investigate: Which variable, *cargo* or *number of seats*, shows a stronger relationship to *flight length*? Propose a reasonable explanation for why this should be so.

  ii. Do planes with a longer flight length tend to use less fuel per mile than planes with a shorter flight length?

c. *Speed, seats,* and *cargo*

  i. Make scatterplots to explore the relationships between the variables *speed, seats,* and *cargo.* Report your most interesting findings. Here are some examples of questions you could

investigate: For which variable, *cargo* or *number of seats,* is the relationship to *speed* more obviously curved? Explain why that should be the case. Which plane is unusually slow for the amount of cargo it carries? Which plane is unusually slow for the number of seats it has?

  ii. The plot of *cargo* against *seats* has two parts: a flat stretch on the left and a fan on the right. Explain, in the language of airplanes, seats, and cargo, what each of the two patterns tells you.

3.2 ▶ Regression: Getting a Line on the Pattern

In this section, you will learn how to model a linear relationship between two quantitative variables by finding the equation of the line that best summarizes the relationship. In the next section you will learn to measure the variability from that line. You've worked with linear equations before, so the following review will be brief.

Lines as Summaries

Linear equations have the important property that for any two points (x_1, y_1) and (x_2, y_2) on the line, the ratio

$$\frac{\text{rise}}{\text{run}} = \frac{\text{change in } y}{\text{change in } x} = \frac{y_2 - y_1}{x_2 - x_1}$$

is a constant. This ratio is the **slope** of the line. The rise and run are illustrated in Display 3.17. The slope is the ratio of the two legs of the right triangle. This ratio is the same for any two points on the line because the triangles formed are all similar.

The slope tells you how much y changes when x changes by one unit.

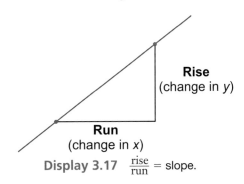

Rise
(change in *y*)

Run
(change in *x*)

Display 3.17 $\frac{\text{rise}}{\text{run}}$ = slope.

Stopping on a Dime?

Example 3.2

In an emergency, the typical driver requires about 0.75 second to get his or her foot onto the brake. The distance that the car travels during this reaction time is called the reaction distance. The equation that expresses this relationship is $y = 1.1x$, where y is reaction distance in feet and x is speed in miles per hour. The table and plot in Display 3.18 show the typical reaction distance for cars traveling at various speeds.

  a. What is the slope of the line? Interpret the slope in the context of the situation.

  b. What is the y-intercept of the line? Interpret the y-intercept in the context of the situation.

Speed (mph), x	Reaction Distance (ft), y = 1.1x
20	22
30	33
40	44
50	55
60	66
70	77

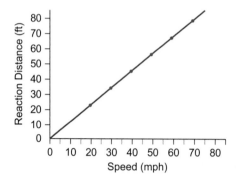

Display 3.18 Reaction distance at various speeds.

c. How would the equation change if it took a typical driver longer than 0.75 second to react?

Solution

a. You can find the slope from the equation of the line or by constructing a ratio using information in the table:

$$\frac{33 - 22}{30 - 20} = \frac{44 - 33}{40 - 30} = \cdots = \frac{77 - 66}{70 - 60} = \frac{11}{10} = 1.1$$

The slope of the line is 1.1. This means that for every 1-mph increase in speed, the reaction distance is an additional 1.1 feet. Alternatively, you could say that the reaction distance increases by 11 feet for every increase of 10 mph in the speed.

b. The y-intercept is 0, which makes sense. If the car has zero speed, the reaction distance should be 0.

c. If the reaction time were longer, the reaction distance would be more than 1.1 feet for every 1-mph increase in speed. Thus, the slope of the line would increase but the y-intercept would remain 0.

Values of the dependent variable observed in the data are denoted by y, and values computed from the regression line are denoted by $\hat{y}$.

$\hat{y}$ is read "y-hat."

In the previous example, the equation came first, so all of the points in the table fell exactly on a line. However, points that represent real data do not all lie exactly on a line. Thus, you will be using technology to give you a summary linear equation, $\hat{y} = b_0 + b_1 x$, called the *regression equation* or *least squares regression equation*. The notation $\hat{y}$ is used rather than y to emphasize that the equation is a summary of the points, much like $\bar{x}$, the sample mean, is a summary of the values of x.

Statisticians use the form $\hat{y} = b_0 + b_1 x$ because it generalizes easily to the case where there is more than one independent variable: $\hat{y} = b_0 + b_1 x_1 + b_2 x_2 + \cdots + b_n x_n$. This generalization is the subject of Chapter 5 on multiple regression.

The following two examples show you how to interpret the slope and intercept of a regression equation that models real data.

Example 3.3

The Federal Minimum Wage

In an effort to keep the wages of hourly workers up to a level that allows some possibility of making a decent living, the United States government establishes a minimum hourly wage. The scatterplot of Display 3.19 shows the minimum wage (in dollars) for every 5 years from 1960 through 2010. The line fitted to the points on the plot is the least squares regression line. Its equation is $\hat{y} = -222.1 + 0.11364x$, where x is the year and $\hat{y}$ is the predicted minimum wage.

   a. What is the slope of the line? What does the slope tell you?
   b. What is the y-intercept? What does it tell you?

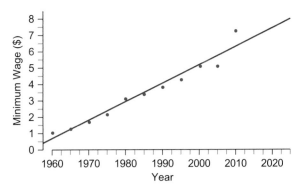

Display 3.19   Minimum wage at 5-year intervals, 1960–2010.

Solution

   a. The slope is 0.11364. This means that the minimum wage has tended to increase by 11.364¢ per year over this 50-year time period. Note that you shouldn't say that the minimum wage went up *exactly* 11.364¢ per year. It didn't. Say that it *tended* to go up this amount or went up *about* this amount each year. Or say that the regression equation *predicts* an increase in the minimum wage of 11.364¢ per year.

   b. The y-intercept is -222.1, which would mean that in the year 0, the minimum wage is predicted to have been $-\$222.10$. This makes no sense. Often the y-intercept won't have a sensible interpretation because 0 is too far from the values of x in the data.

Interpreting the slope.

Be sure to qualify your interpretations of the slope with "the regression equation predicts that."

Example 3.4

Doctoral Degrees in the Sciences, Social Sciences, and Engineering

One way to measure how close women are to achieving equality is to compare the number of women to the number of men who get doctoral degrees in the sciences. Display 3.20 gives the number of U.S. citizens and permanent residents who earned a doctorate in the sciences, social sciences, or engineering in 2005. Men earned more doctorates than women in all fields but psychology and the social sciences. The line on the plot is the least squares regression line, $\hat{y} = 61 + 0.741x$, where x is the number of males and $\hat{y}$ is the predicated number of females who earned a doctorate in a field.

a. What is the slope of the line? What does the slope tell you?

b. What is the y-intercept? What does it tell you?

Field of Study	Total	Male	Female
Agricultural sciences	557	333	224
Biological sciences	4395	2227	2168
Computer sciences	473	368	105
Earth, atmospheric, and ocean sciences	442	278	164
Mathematics and statistics	541	390	151
Physical sciences	1900	1338	562
Psychology	2891	901	1990
Social sciences	2540	1244	1296
Engineering	2283	1805	478

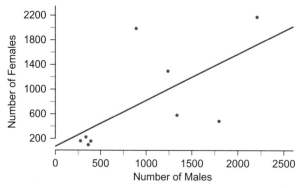

Display 3.20 Number of science and engineering doctorates awarded to U.S. citizens and permanent residents, by field and sex. [Source: National Science Foundation, Division of Science Resources Statistics, *Survey of Earned Doctorates, 1998–2005*; last updated October 2007, *www.nfs.gov/statistics/wmpd/figf-1.htm*.]

Solution

a. The slope is 0.741. This means that if one field has 1 more male graduate than another field, the equation predicts that it would have 0.741 more female graduates. Alternatively, if one field has 100 more male graduates than another field, the equation predicts that it would have 74.1 more female graduates.

b. The y-intercept is 61, which means that if 0 males earn a doctorate in a field, the predicted number of women who would earn a doctorate is 61. As is typical, this interpretation of the y-intercept isn't very useful.

STATISTICS IN ACTION 3.1 ► Pinching Pages

How thick is a single sheet of your book? One sheet alone is too thin to give an accurate answer if you measure directly with a ruler, but you could measure the thickness of 50 sheets together, then divide by 50. This method would give you an estimate of the thickness but no information about the precision of your measurement. The approach in this activity lets you judge precision as well as thickness.

What You'll Need: a ruler with a millimeter scale, a copy of your textbook

1. Pinch together the front cover and first 50 sheets of your book. Then measure and record the thickness of the front cover and 50 sheets to the nearest millimeter.

2. Repeat for the cover plus 100, 150, 200, and 250 sheets.

3. Plot your data on a scatterplot, with number of sheets on the horizontal scale and thickness on the vertical scale.

Pinch 50 *sheets* of paper, not up to page 50.

4. Does the plot look linear? Should it? Discuss why or why not, and make your measurements again if necessary. Place a straight line on the plot that fits best through the points.

5. Estimate the slope and *y*-intercept of your line. What does the *y*-intercept tell you? What does the slope tell you? What is your estimate of the thickness of a sheet?

6. Use the information in your graph to discuss the precision of your estimate in step 5.

7. How would your line have changed if you hadn't included the front cover?

DISCUSSION
Lines as Summaries

D3. The Consumer Price Index (CPI) is a measure of the change over time in the prices paid by urban consumers for a selected group of goods and services (called a "market basket") thought to be typical of urban households. The CPI is a commonly used measure of price increases and is often used to adjust salaries, rents, and other monetary values for inflation. The CPI is, itself, a statistical estimate based on a number of large-scale surveys conducted by agencies of the federal government. Display 3.21 shows the CPI for every 5 years from 1970 to 2005. The equation of the least squares regression line drawn on the plot is $\hat{y} = -27102 + 13.812x$, where x is the year and $\hat{y}$ is the predicted CPI.

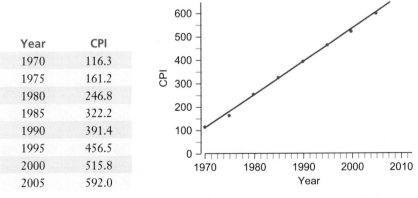

Year	CPI
1970	116.3
1975	161.2
1980	246.8
1985	322.2
1990	391.4
1995	456.5
2000	515.8
2005	592.0

Display 3.21 CPI at 5-year intervals, 1970–2005. [Source: U.S. Bureau of Labor Statistics, *www.bls.gov/cpi*.]

a. What is the slope of the line? What does the slope tell you?

b. What is the *y*-intercept? Interpret it. Does the interpretation seem reasonable in this setting?

Using Lines for Prediction

Lines are summaries and can be used to predict.

There are two main reasons that you would want to fit a line to bivariate data:

- to find a summary, or model, that describes the relationship between the two variables

- to use the line to predict the value of y when you know the value of x. In cases where it makes sense to do this, the variable on the x-axis is called the **predictor** or **explanatory variable,** and the variable on the y-axis is called the **response variable.**

$\hat{y}$ may be called the "predicted" value or the "fitted" value.

| Example 3.5 | Modeling and Predicting the Minimum Wage |

In Example 3.3, on page 107, the equation

$$\hat{y} = -222.1 + 0.11364x$$

models the rise in the minimum wage for the years 1960 through 2010. Using this equation as a model enabled you to make general statements about the minimum wage throughout these years: "Roughly, the minimum wage went up 11.364¢ per year."

You also might want to use this line to predict the minimum wage in some other year. For example, use the equation, $\hat{y} = -222.1 + 0.11364x$, to predict the minimum wage in 2003 and in 1950.

Solution
The predicted minimum wage for 2003 is

$$\hat{y} = -222.1 + 0.11364x = -222.1 + 0.11364(2003) = 5.52, \text{ or } \$5.52$$

Assuming the linear trend continues backward, the predicted minimum wage for 1950 is

$$\hat{y} = -222.1 + 0.111364x = y = -222.1 + 0.11364(1950) = -0.50, \text{ or } -\$0.50$$

The predicted minimum wage for 2003 of $5.52 is fairly close to the actual minimum wage of $5.15 an hour. But the actual minimum wage in 1950 was 75¢ an hour, not some negative number! As you can see, making the assumption that the linear trend continues can be a very risky business. This type of prediction is called **extrapolation**—making a prediction when the value of x falls outside the range of the actual data. **Interpolation**—making a prediction when the value of x falls inside the range of the data, as does 2003—is safer.

Extrapolation is risky.

When you know the value of x and use a line to predict the corresponding value of y, you know that your prediction for y probably won't be exact, but you hope that the error will be small. The **prediction error** is the difference between the observed value of y and the **predicted value** of y (denoted $\hat{y}$). You usually don't know what that error is, or else you wouldn't need to use the line to predict the value of y. You do, however, know the errors for the points used to construct the line. These errors are called residuals.

| Residuals |

The residual is the signed vertical distance of the point from the line.

For a point used to construct the regression equation, the **residual** is the difference between the value of y observed and the value of y predicted by the regression equation:

residual = observed value of y − predicted value of y = y − ŷ

The geometric interpretation of the residual is shown in Display 3.22. A residual is the signed vertical distance from an observed data point to the regression line. The sign of the residual is positive if the point is above the line and negative if the point is below the line

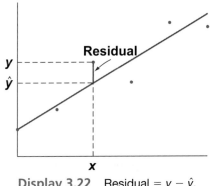

Display 3.22   Residual = $y - \hat{y}$.

<table>
<tr><td>**Doctoral Degrees in the Sciences, Social Sciences, and Engineering: Computing Residuals**</td><td>**Example 3.6**</td></tr>
</table>

Refer to Example 3.4, on page 107. Find the two points that have the largest residuals (in absolute value). What fields of study do these points represent? Compute and then interpret the residuals for these fields of study.

Solution

The two points that have the largest residuals in absolute value are circled in Display 3.23, where the residuals are drawn in as vertical lines.

The two fields are psychology and engineering. There were 1990 women who got doctorates in psychology. Using the equation of the fitted line, the predicted number is

$$\hat{y} = 61 + 0.741x = 61 + 0.741(901) = 728.641$$

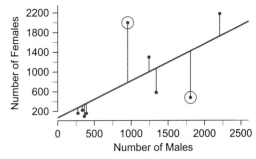

Display 3.23   Residuals for science and engineering doctorates awarded to U.S. citizens and permanent residents, by field and sex.

To find the residual, subtract the predicted value from the observed value:

$$y - \hat{y} = 1990 - 728.641 = 1261.359$$

This means that about 1261 more women got doctorates in psychology than would be predicted by knowing the number of men who got doctorates in psychology.

There were 478 women who got doctorates in engineering. Using the equation of the fitted line, the predicted number is

$$\hat{y} = 61 + 0.741x = 61 + 0.741(1805) = 1398.505$$

Technology will calculate all residuals at once.	The residual is $y - \hat{y} = 478 - 1398.505 = -920.505$. This means that about 921 fewer women got doctorates in engineering than would be predicted by knowing the number of men who got doctorates in engineering.

DISCUSSION
Using Lines for Prediction

D4. Refer to D3 on page 109.

    a. Use the equation to predict the Consumer Price Index (CPI) in 1960. The CPI in 1960 actually was 88.7. What is the error in your prediction?

    b. For which year is the residual largest in absolute value? Compute this residual. What was happening about that time that might affect the CPI in this way?

D5. Test how well you understand residuals.

    a. If a residual is large and negative, where is the point located with respect to the line? Draw a diagram to illustrate. What does it mean if the residual is 0?

    b. If someone said that he or she had fit a line to a set of points and all the residuals were positive, what would you say?

D6. What do you think of the arithmetic and the reasoning in this passage from Mark Twain's *Life on the Mississippi* (1883)?

> In the space of one hundred and seventy-six years the Lower Mississippi has shortened itself two hundred and forty-two miles. That is an average of a trifle over one mile and a third per year. Therefore, any calm person, who is not blind or idiotic, can see that in the Old Oölitic Silurian Period, just a million years ago next November, the Lower Mississippi River was upwards of one million three hundred thousand miles long, and stuck out over the Gulf of Mexico like a fishing rod.
>
> And by the same token any person can see that seven hundred and forty-two years from now the Lower Mississippi will be only a mile and three quarters long, and Cairo and New Orleans will have joined their streets together, and be plodding comfortably along under a single mayor and a mutual board of aldermen. There is something fascinating about science. One gets such wholesale returns of conjecture out of such a trifling investment of fact.

> Given that the Mississippi/Missouri river system was about 3710 miles long in the year 2000, write an equation that Twain would say gives the length of the river in terms of the year.

Least Squares Regression Lines

The general approach to fitting lines to data is called the **method of least squares.** The method was invented about 200 years ago by Carl Friedrich Gauss (1777–1855), Adrien-Marie Legendre (1752–1833), and Robert Adrain (1775–1843), who were working independently of one another in Germany, France, and the United States, respectively.

The **least squares regression line,** also called the **least squares line** or **regression line,** for a set of data points (x, y) is the line for which the **sum of squared errors** (residuals), or SSE, is as small as possible.

$$\text{SSE} = \sum(\textit{residual})^2$$
$$= \sum(y - \hat{y})^2$$

| **Regression Equation for Predicting Passenger Jet Cost** | **Example 3.7** |

This table shows *cost per hour* versus *number* of *seats* for three models of passenger jets from the data in Display 3.16 on page 104. (Some of the values have been rounded.)

Aircraft	Number of Seats	Cost ($/h)
ERJ-145	50	1100
DC-9	100	2100
MD-90	150	2700
Mean	100	1967

One of these two equations gives the least squares regression line for predicting *cost* from *number of seats*. Which one is it?

$$\hat{y} = 367 + 16x$$
$$\hat{y} = 300 + 16x$$

Solution
The least squares regression line minimizes the sum of the squared errors, SSE, so the equation with the smaller SSE must be the equation of the regression line.
For the equation $\hat{y} = 367 + 16x$:

x	y	$\hat{y}$	$y - \hat{y}$	$(y - \hat{y})^2$
50	1100	1167	−67	4,489
100	2100	1967	133	17,689
150	2700	2767	−67	4,489
				SSE = 26,667

For the equation $\hat{y} = 300 + 16x$:

x	y	$\hat{y}$	$y - \hat{y}$	$(y - \hat{y})^2$
50	1100	1100	0	0
100	2100	1900	200	40,000
150	2700	2700	0	0
				SSE = 40,000

The first equation has the smaller SSE, so it must be the equation of the least squares regression line. Note that for this line, except for rounding error, the sum of the residuals, $\Sigma(y - \hat{y})$, is equal to 0. This is always the case for the least squares regression line, but it can be true for other lines, too.

In addition to making the sum of the squared errors as small as possible, the least squares regression line has some other properties, given in the following box.

Properties of the Least Squares Regression Line

The fact that the sum of squared errors, or SSE, is as small as possible means that, for the least squares regression line, these properties also hold:

- The sum (and mean) of the residuals is 0.
- The point of averages, $(\bar{x}, \bar{y})$, lies on the regression line.

- The standard deviation of the residuals is smaller than for any other line.
- The line has slope b_1, where

$$b_1 = \frac{\Sigma(x - \bar{x})(y - \bar{y})}{\Sigma(x - \bar{x})^2}$$

There are some appealing mathematical relationships among these properties, which, taken together, show that the line through the point of averages $(\bar{x}, \bar{y})$ having slope b_1 does, in fact, minimize the sum of the squared errors. This gives you a way to find the equation of the least squares line: $\hat{y} = b_0 + b_1 x$. To find the y-intercept, b_0, use the fact that the point $(\bar{x}, \bar{y})$ lies on the regression line and so satisfies the equation $\bar{y} = b_0 + b_1 \bar{x}$. Solve this equation for b_0 to get $b_0 = \bar{y}\, b_1 \bar{x}$.

Example 3.8	**Least Squares Line for the Passenger Jets**

Find the equation of the least squares line for the passenger jets data given in the previous example.

Solution
Finding the equation requires three main steps: Find the point of averages $(\bar{x}, \bar{y})$, find the slope b_1, and use the point and slope to find the y-intercept b_0.

A convenient way to organize the computations is to work from a table.

Aircraft	Seats, x	Cost ($/h), y
ERJ-145	50	1100
DC-9	100	2100
MD-90	150	2700
Sum	300	5900
Mean	100	1966.$\overline{6}$

Point of averages: The point of averages $(\bar{x}, \bar{y})$ is $(100, 1966.\overline{6})$, and the least squares regression line passes through this point.

Slope: To compute the slope, first create two new columns for deviations from the mean, one for $x - \bar{x}$ and the other for $y - \bar{y}$:

Aircraft	Seats, x	Cost ($/h), y	$x - \bar{x}$	$y - \bar{y}$
ERJ-145	50	1100	−50	−866.$\bar{6}$
DC-9	100	2100	0	133.$\bar{3}$
MD-90	150	2700	50	733.$\bar{3}$
Sum	300	5900	0	0
Mean	100	1966.$\bar{6}$	0	0

Now create two more columns, one for $(x - \bar{x}) \cdot (y - \bar{y})$ and the other for $(x - \bar{x})^2$:

Aircraft	Seats, x	Cost ($/h), y	$x - \bar{x}$	$y - \bar{y}$	$(x - \bar{x}) \cdot (y - \bar{y})$	$(x - \bar{x})^2$
ERJ-145	50	1,100	−50	−866.$\bar{6}$	43,333.$\bar{3}$	2500
DC-9	100	2,100	0	133.$\bar{3}$	0	0
MD-90	150	2,700	50	733.$\bar{3}$	36,666.$\bar{6}$	2500
Sum	300	5,900	0	0	80,000	5000
Mean	100	19,66.$\bar{6}$	0	0		

The ratio of the sums of the last two columns gives the slope

$$b_1 = \frac{\Sigma(x - \bar{x})(y - \bar{y})}{\Sigma(x - \bar{x})^2} = \frac{80,000}{5,000} = 16$$

y-intercept: Now that you have a point on the line, $(100, 1966.\bar{6})$, and the slope, 16, you can find the y-intercept from the equation

$$b_0 = \bar{y} - b_1\bar{x}$$

$$= 1966.\bar{6} - 16(100)$$

$$= 366.\bar{6}$$

This agrees with what you found in the previous example. That is, the equation of the least squares regression line (with rounded y-intercept) is

$$\hat{y} = 367 + 16x$$

DISCUSSION
Least Squares Regression Lines

D7. You might have wondered why statisticians don't fit a regression line by minimizing the sum of the absolute values of the residuals, $\Sigma|y - \hat{y}|$, rather than the sum of the squares of the residuals. Here you will learn one reason why.

a. Plot the points in the table and the line $y = 1 + x$. Explain why this is the line that best fits these points. Compute the sum of the absolute values of the residuals.

x	y
0	0
0	2
2	2
2	4

b. Draw another line that passes between the two points at $x = 0$ and also passes between the two points at $x = 2$. Compute the sum of the absolute values of the residuals for this line and compare it to your sum from part a.

c. Draw yet another line that passes between the two points at $x = 0$ and also passes between the two points at $x = 2$. Find the sum of the absolute values of the residuals for this line and compare it to your sums from parts a and b.

d. Draw a line that does *not* pass between the two points at $x = 0$. Find the sum of the absolute values of the residuals for this line and compare it to your sums from parts a and b.

e. Now find the least squares regression line and compute the sum of the squared residuals. Compute the sum of the squared residuals for your lines in parts b, c, and d. What can you conclude?

f. Find the standard deviation of the residuals for the least squares regression line and for the lines in parts b, c, and d. What can you conclude?

Reading Computer Output

When you are working with real data, the best way to get the least squares line is by computer or calculator. Display 3.24 shows typical computer output for the minimum wage data in Display 3.19 on page 107.

Regression Analysis: Minimum Wage Versus Year

The regression equation is
Minimum Wage = −222 + 0.114 Year

Predictor	Coef	SE Coef	T	P
Constant	−222.11	16.41	−13.54	0.000
Year	0.113636	0.008266	13.75	0.000

S = 0.433450   R-Sq = 95.5%   R-Sq(adj) = 94.9%

Analysis of Variance

Source	DF	SS	MS	F	P
Regression	1	35.511	35.511	189.01	0.000
Residual error	9	1.691	0.188		
Total	10	37.202			

Display 3.24   Minitab output giving the equation of the least squares line for the minimum wage data.

You can ignore most of the output for now. You will learn how to interpret it in Chapter 13. For the time being, focus on the first two columns of the first table. The *y*-intercept is the coefficient in the row labeled "Constant" and is –222.11. The slope is the coefficient of the predictor variable Year and is 0.113636. The SSE for the regression line is found in the "Residual error" row and is 1.691.

Display 3.25 shows the steps for getting a regression equation using a calculator.

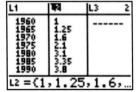  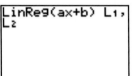 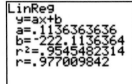

Display 3.25   TI calculator steps for regression equation for the minimum wage data.

DISCUSSION
Reading Computer Output

D8. *Doctors' incomes.* Display 3.26 shows the average net income y of family practitioners versus year x, with computer output.

a. What is the equation of the least squares line? Estimate the SSE from the scatterplot in Display 3.26, and then find it in the computer output.

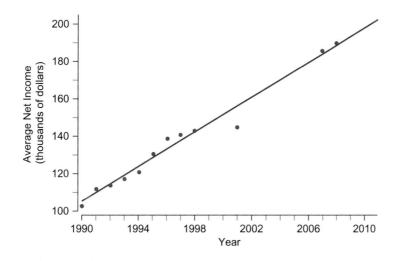

Regression Analysis: Income Versus Year

The regression equation is
Income $= -9125 + 4.64$ Year

Predictor	Coef	SE Coef	T	P
Constant	−9124.7	467.7	−19.51	0.000
Year	4.6381	0.2342	19.80	0.000

S = 4.55204   R-Sq = 97.5%   R-Sq(adj) = 97.3%

Analysis of Variance

Source	DF	SS	MS	F	P
Regression	1	8124.5	8124.5	392.09	0.000
Residual error	10	207.2	20.7		
Total	11	8331.7			

Display 3.26 Scatterplot of average net income (thousands of dollars) of doctors board-certified in family practice, 1990–2008, and Minitab output for the regression. [Source: U.S. Census Bureau, *Statistical Abstract of the United States, 2004–2005*. 2007 and 2008 Medical Group Compensation and Financial Survey, *www.cms.hhs.gov/AcuteInpatientPPS/Downloads/AMGA_2007%20Report.pdf*, *www.cms.hhs.gov/AcuteInpatientPPS/Downloads/AMGA_08_data.pdf*.]

b. The StatPlus software output for this regression is shown in Display 3.27. How is it different from the Minitab output?

Linear Regression

Regression Statistics

R^2	0.9875
R^2	0.9751
Adjusted R^2	0.9726
Standard error	4.5520
Total number of cases	12.0000

Income = −9124.7370 + 4.6381 * Year

ANOVA

	DF	SS	MS	F	p-level
Regression	1.0000	8124.4560	8124.4560	392.0862	0.0000
Residual	10.0000	207.2109	20.7211		
Total	11.0000	8331.6670			

	Coefficients	Standard Error	t-Stat	P-level
Intercept	−9124.7370	467.7303	−19.5085	0.0000
Year	4.6381	0.2342	19.8012	0.0000

Display 3.27 StatPlus output for the regression of family practitioners' income versus year.

Summary 3.2: Regression: Getting a Line on the Pattern

For many quantitative relationships, it makes sense to use one variable, x, called the predictor or explanatory variable, to predict values of the other variable, y, called the predicted or response variable. When the data are roughly linear, you can use a fitted line, called the least squares regression line, as a summary or model that describes the relationship between the two variables. You might also use it to predict the value of an unknown value y when you know the value of x.

Interpolation—using a fitted relationship to predict a response value when the predictor value falls *within* the range of the data—generally is much more trustworthy than extrapolation—predicting response values based on the assumption that a fitted relationship applies *outside* the range of the observed data.

Each residual from a fitted line measures the vertical distance from a data point to the line:

$$residual = (observed\ value) - (predicted\ value) = y - \hat{y}$$

The least squares regression line for a set of pairs (x, y) is the line for which the sum of squared errors, or SSE, is as small as possible. For this line, these properties hold:

- The sum (and mean) of the residuals is 0.
- The line contains the point of averages, $(\bar{x}, \bar{y})$.
- The variation in the residuals is as small as possible.
- The line has slope b_1, where

$$b_1 = \frac{\Sigma(x - \bar{x})(y - \bar{y})}{\Sigma(x - \bar{x})^2}$$

To find the equation, $\hat{y} = b_0 + b_1 x$, of the regression line,

- compute $\bar{x}$ and $\bar{y}$
- find the slope using the formula for b_1
- compute the y-intercept: $b_0 = \bar{y} - b_1\bar{x}$

Remember to use a hat, $\hat{y}$, to indicate a predicted value of y.

Practice

Lines as Summaries

P4. Display 3.28 shows the weight of Zach's pink eraser, in grams, plotted against the number of days into the academic year. The equation of the regression line, drawn on the plot, is $\hat{y} = 91.48 - 0.01x$.

a. What is the slope of the line? Interpret the slope in the context of the situation.

b. What is the y-intercept? Interpret the y-intercept in the context of the situation. Does this make sense?

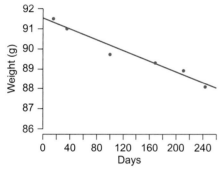

Display 3.28 Weight of pink eraser. [Source: Zach's Eraser, *CMC ComMuniCator*, Vol. 28 (June 2004), p. 28.]

P5. Display 3.29 shows the hand width of 383 students plotted against hand length. The line drawn on the plot is the least squares line. Its equation is $\hat{y} = 1.69 + 0.80x$.

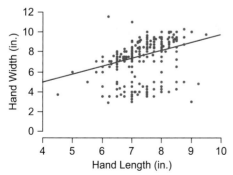

Display 3.29 Hand width and hand length, for 383 students.

a. What is the slope of the line?

b. What does the slope tell you?

c. What is the y-intercept? Interpret it in the context of the situation. Does the interpretation make sense in this situation?

d. Students were instructed to measure their "hand width" with their fingers spread apart as far as possible. The scatterplot shows a smaller cloud of points below the main one. How do you think these students measured? What would happen to the regression line if those points were removed?

Using Lines for Prediction

P6. Sophia measured the thickness of the first 50, 100, 150, 200, and 250 sheets of paper in her textbook, including the front cover. Her data are given in Display 3.30. The equation of the least squares regression line is $\hat{y} = 2.7 + 0.072x$, where x is the number of sheets of paper and $\hat{y}$ is the predicted thickness.

a. What is the slope of this line? Interpret the slope.

b. What is the y-intercept? Interpret it. Does the interpretation seem reasonable in this setting?

c. Predict the thickness of 175 sheets of paper and front cover.

d. Which number of sheets has the biggest residual? Is the residual positive or negative?

e. Compute the residuals and verify that their sum is 0.

Number of Sheets	Thickness (mm)
50	6.0
100	11.0
150	12.5
200	17.0
250	21.0

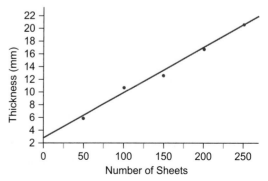

Display 3.30 Measured thicknesses of cover and sheets of textbook paper.

Least Squares Regression Line

P7. Three different kinds of pizza have the following amounts of fat and calories in a 5-oz slice: 9 g and 305 cal; 11 g and 309 cal; and 13 g and 316 cal.

   a. Plot the three points with *fat* as the predictor variable and *calories* as the response.

   b. Compute the equation of the least squares regression line by following the steps in Example 3.8 on page 114.

P8. Display 3.31 shows the birth weight and adult weight of the six species in the family camelidae (camels).

   a. Use your calculator or software to find the equation of the regression equation for predicting adult weight from birth weight.

   b. Compu0te and interpret the residual for the llama.

   c. Verify that the sum of the residuals is 0.

   d. Find the sum of the squared residuals (SSE).

   e. Verify that the regression line goes through the point of averages, $(\bar{x}, \bar{y})$.

Common Name	Birth Weight (kg)	Adult Weight (kg)
Bactrian camel	36	475
Dromedary	37	434
Llama	11	140
Guanaco	11.5	100
Alpaca	7.21	62
Vicuña	5.74	50

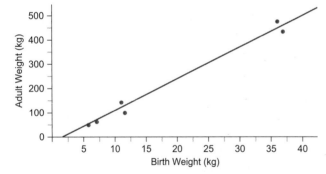

Display 3.31 Table and scatterplot of adult weight versus birth weight for the six species of camel. [Source: AnAge, the Animal Ageing and Longevity Database, *http://genomics. senescence.info/species/*.]

Reading Computer Output

P9. The JMP-IN computer output in Display 3.32 is for the pizza data in P7. Does it give the same results that you computed by hand? Where in the output is the SSE found?

P10. Display 3.33 shows the scatterplot of the oak tree data from P3 on page 98 with the least squares regression line added, along with the regression output from Minitab.

   a. Find the slope of the line on the output. Give an interpretation of this slope in context.

   b. Find the *y*-intercept of the line on the output. Does it have a practical interpretation in this setting?

Linear Fit

Calories = 279.75 + 2.75 Fat

Summary of Fit

RSquare	0.975806
RSquare Adj	0.951613
Root Mean Square Error	1.224745
Mean of Response	310
Observations (or Sum Wgts)	3

Analysis of Variance

Source	DF	Sum of Squares	Mean Square	F Ratio
Model	1	60.500000	60.5000	40.3333
Error	1	1.500000	1.5000	Prob>F
C. Total	2	62.000000		0.0994

Parameter Estimates

Term	Estimate	Std Error	t Ratio	Prob>|t|
Intercept	279.75	4.81534	58.10	0.0110
Fat	2.75	0.433013	6.35	0.0994

Display 3.32 JMP-IN computer output for pizza.

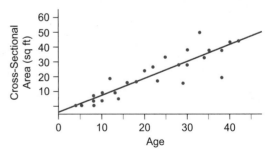

The regression equation is
Area = −3.70 + 1.13 Age

Predictor	Coef	Stdev	t-Ratio	P
Constant	−3.697	2.919	−1.27	0.217
Age	1.1308	0.1173	9.64	0.000

S = 7.120   R-squ = 78.8%   R-sq(adj) = 77.9%

Display 3.33 Scatterplot with regression line and Minitab output for oak tree data.

Exercises

E11. Display 3.34 shows *cost in dollars per hour* versus *number of seats* for three aircraft models. Five lines, labeled A–E, are shown on the plot. Their equations, listed below, are labeled I–V.

  a. Match each line (A–E) with its equation (I–V).

  I. $cost = -290 + 15.8 \ seats$

  II. $cost = 400 + 15.8 \ seats$

  III. $cost = 1000 + 15.8 \ seats$

  IV. $cost = -370 + 25 \ seats$

  V. $cost = 900 + 10 \ seats$

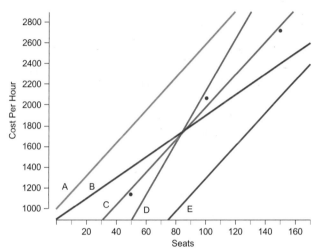

Display 3.34   Cost in dollars per hour versus number of seats for three aircraft models.

  b. Match each line (A–E) with the appropriate verbal description (I–V).

  I.   This line overestimates cost.

  II. This line underestimates cost.

  III. This line overestimates cost for the smallest plane and underestimates cost for the largest plane.

  IV. This line underestimates cost for the smallest plane and overestimates cost for the largest plane.

  V.   On balance, this line gives a better fit than the other lines.

E12. Examine the scatterplot in Display 3.35.

  a. Which two kinds of pizza in Display 3.35 have the fewest calories? Which two have the least fat? Which region of the graph has the pizzas with the most fat?

  b. Display 3.36 shows the data again, with five possible summary lines. Match each equation (I–V) with the appropriate line (A–E).

  I. $calories = 70 + 15 \ fat$

  II. $calories = -10 + 25 \ fat$

  III. $calories = 150 + 15 \ fat$

  IV. $calories = 110 + 15 \ fat$

  V. $calories = 170 + 10 \ fat$

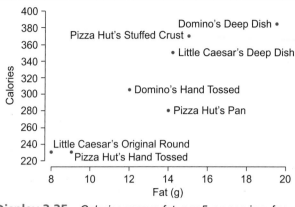

Display 3.35   *Calories* versus *fat*, per 5-oz serving, for seven kinds of pizza. [Source: *Consumer Reports*, July 2003.]

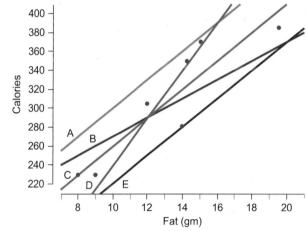

Display 3.36   Five possible fitted lines for the pizza data.

  c. Consider the possible summary lines in Display 3.36.

  i.   Which line gives predicted values for calorie content that are too high? How can you tell this from the plot?

  ii. Which line tends to give predicted calorie values that are too low?

  iii. Which line tends to overestimate calorie content for lower-fat pizzas and underestimate calorie content for higher-fat pizzas?

  iv. Which line has the opposite problem, under-estimating calorie content when fat content is lower and overestimating calorie content when fat content is higher?

  v.   Which line fits the data best overall?

E13. *Heights of boys.* The scatterplot in Display 3.37 shows the median height, in inches, for boys ages 2 through 14 years. The equation of the regression lines is $\hat{y} = 31.60 + 2.47x$.

  a. What is the slope of the line? Explain the meaning of the slope with respect to boys and their median height.

b. What is the *y*-intercept? Interpret it. Does the interpretation make sense in this context?

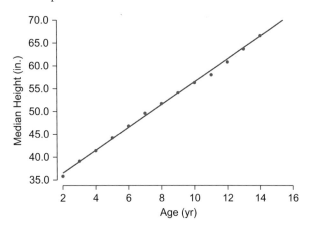

Display 3.37 *Median height* versus *age* for boys. [Source: National Health and Nutrition Examination Survey (NHANES), 2002, *www.cdc.gov*.]

E14. *Pizza again.* Display 3.38 shows the calorie and fat content of 5 oz of various kinds of pizza. The equation of the regression line is $\hat{y} = 112 + 14.9x$.

Pizza	Fat (g)	Calories
Pizza Hut's Hand Tossed	9	230
Domino's Deep Dish	19.5	385
Pizza Hut's Pan	14	280
Domino's Hand Tossed	12	305
Little Caesar's Original Round	8	230
Little Caesar's Deep Dish	14.2	350
Pizza Hut's Stuffed Crust	15	370

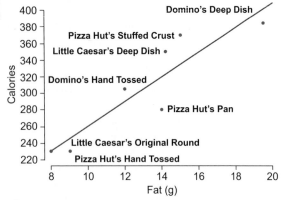

Display 3.38 Calories and fat content per 5-oz serving, for seven kinds of pizza. [Source: *Consumer Reports*, January 2002.]

a. Use the line on the scatterplot to predict the calorie content of a pizza with 10.5 g of fat, then use the line to predict the calorie content of a pizza with 15 g of fat.

b. What is the slope of the line? Interpret the slope in the context of this situation.

c. There are 9 calories in a gram of fat. How is the slope related to this number?

E15. Arsenic is a potent poison sometimes found in groundwater. Long-term exposure to arsenic in drinking water can cause cancer. How much arsenic a person has absorbed can be measured from a toenail clipping. The plot in Display 3.39 shows the arsenic concentrations in the toenails of 21 people who used water from their private wells plotted against the arsenic concentration in their well water. Both measurements are in parts per million.

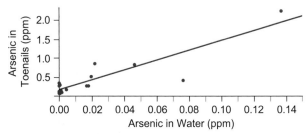

Display 3.39 Arsenic concentrations. [Source: M. R. Karagas et al., "Toenail Samples as an Indicator of Drinking water Arsenic Exposure," *Cancer Epidemiology, Biomarkers and Prevention.* Vol. 5 (1996), pp. 849–852.]

a. What is the predictor variable, and what is the response variable?

b. Describe the relationship.

c. Estimate the residual for the person with the highest concentration of arsenic in the well water.

d. Find the person on the plot with the residual that is the largest in absolute value. What was the concentration of arsenic in that person's toenails?

e. The World Health Organization has set a standard that the concentration of arsenic in drinking water should be less than 0.01mg/L. (1 mg/L = 1 ppm.) Is this standard exceeded in any of these wells? [Source: *www.who.int*.]

E16. The scatterplot in Display 3.40 shows *operating cost* (in dollars per hour) versus *fuel consumption* (in gallons per hour) for a sample of commercial aircraft.

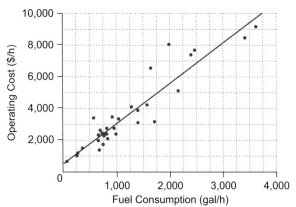

Display 3.40 *Operating cost* versus *fuel consumption* for commercial aircraft.

a. Which is the explanatory variable and which is the response variable?

b. The slope of the regression line is approximately 2.5. Interpret this slope in the context of this situation.

c. The *y*-intercept is 470. Does this value have a reasonable interpretation in this situation?0

d. Write the equation of the regression line and use it to predict the cost per hour for a plane that consumes fuel at the rate of 1500 gal/h.

e. Estimate the residual for the aircraft with the largest fuel consumption. Is this aircraft more or less cost efficient than predicted from its fuel consumption?

f. Find the aircraft with the largest residual. What is the cost per hour for that aircraft?

E17. *More NASCAR.* The plot in Display 3.41 shows top qualifying speed (TQS) plotted against banking angle for 21 NASCAR racetracks, with the regression line.

a. The five largest residuals in absolute value are 64.13, 63.64, 35.14, 26.92, and 26.42. Refer to the table of NASCAR data of E8 on page 104. Match these residuals with the speedway they represent.

b. What does the residual for Bristol Motor Speedway tell you about how that racetrack fits in with the other racetracks so far as these two variables are concerned?

c. Does Talladega Superspeedway have a positive or negative residual? How about Richmond International Speedway? How can you tell by looking at the plot?

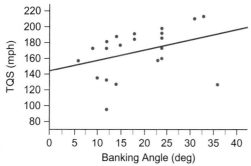

Display 3.41 Top qualifying speed plotted against banking angle for 21 NASCAR racetracks.

E18. *More pizza.* Refer to the pizza data in E14 on page 122.

a. The least squares residuals for the pizza data are, in order from smallest to largest, −40.58, −17.66, −15.95, −1.03, 14.28, 26.44, and 34.50. Match each residual with its pizza.

b. What does the residual for Pizza Hut's Pan pizza tell you about the pizza's number of calories versus fat content?

c. For Pizza Hut's Hand Tossed and Domino's Deep Dish, are the residuals positive or negative? How can you tell this from the scatterplot in Display 3.38?

E19. The level of air pollution is indicated by a measure called the air quality index (AQI). An AQI greater than 100 means the air quality is unhealthy for sensitive groups such as children. The table and plot in Display 3.42 show the number of days in Detroit that

Years Since 2000	Number of Days AQI > 100
1	31
2	28
3	19

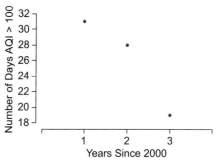

Display 3.42 Air quality index for 2001–2003. [Source: U.S. Environmental Protection Agency, *www.epa.gov.*]

the AQI was greater than 100 for the years 2001, 2002, and 2003.

a. Compute the equation of the least squares line by following the steps in Example 3.8 on page 114.

b. Interpret the slope in the context of this situation.

c. Which year has the residual that is the largest in absolute value? What is this residual?

d. Compute the SSE for this line.

e. Verify that the sum of the residuals is 0.

f. Find the SSE for the line that has the same slope as the least squares line but passes through the point for 2002. Is this SSE larger or smaller than the SSE for the least squares line? According to the least squares approach, which line fits better?

g. Find the slope of the line that passes through the points for 2001 and 2003. Then find the fitted value for 2002. Finally, find all three residuals and the value of the SSE for this line.

h. The least squares line doesn't pass through any of the points, and yet judging by the SSE that line fits better than the one in part g. Do you agree that the least squares line fits better than the lines in parts f and g? Explain why or why not.

E20. *Even more pizza.* Refer again to the table and scatterplot in Display 3.38 on page 122.

a. Compute the equation of the least squares regression line for using *fat* to predict *calories* by following the steps in Example 3.8 on page 114.

b. Which of these values do you think is the SSE for this regression? Explain your answer.

$$0 \qquad 29.3 \qquad 861.4 \qquad 4307$$

E21. *Heights of girls.* Display 3.43 gives the median height in inches for girls ages 2–14.

a. Practice using technology by making a scatterplot, finding the equation of the least squares line for *median height* versus *age*, and graphing the equation on the plot.

b. Judging from the plot, is the residual for 11-year-olds positive or negative? Compute this residual to check your answer.

c. Verify that the line contains the points of averages, $(\bar{x}, \bar{y})$.

d. How does the regression line for girls compare to the line for boys in E13?

Age (yr)	Median Height (in.)
2	35.1
3	38.7
4	41.3
5	44.1
6	46.5
7	48.6
8	51.7
9	53.7
10	56.1
11	59.5
12	61.2
13	62.9
14	63.6

Display 3.43 Median height for girls ages 2–14.

e. Calculate the residual for girls ages 2, 8, and 14. What does this suggest about the pattern of growth beyond what is summarized in the equation of the regression line?

E22. *Body mass and blood pressure.* According to a large national health study, the percentage of adults with hypertension (high blood pressure) increases with body mass index (BMI). Hypertension is defined as mean systolic blood pressure of at least 140 and mean diastolic blood pressure of at least 90. The BMI is given by the formula

$$\text{BMI} = \frac{\text{weight (kg)}}{[\text{height (m)}]^2}$$

The percentages of males and females with hypertension for various BMIs are given in Display 3.44.

BMI	Percentage of Males with Hypertension	Percentage of Females with Hypertension
24	14.9	15.2
26	22.1	27.7
28	27.0	32.7
32	41.9	37.8

Display 3.44 Percentages of males and females with hypertension by BMI. [Source: NHANES III.]

a. Practice using technology by plotting the percentages of males with hypertension (*y*) as a function of BMI (*x*). Describe the shape of the plot.

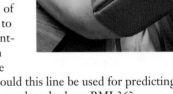

b. Use your technology to find the equation of the least squares regression line for males. Explain what the slope represents in the context of these data.

c. Verify that the line contains the point of averages, $(\bar{x}, \bar{y})$.

d. Use the equation of the line in part b to predict the percentage of males with BMI 30 who have hypertension. Should this line be used for predicting the percentage for males who have BMI 36?

e. Plot the percentages of females who have hypertension as a function of BMI. Describe the shape of the plot. Should a straight line be used to summarize this relationship?

E23. Display 3.45 shows the scatterplot of the oak tree data from P3 on page 98, with the least squares regression line added.

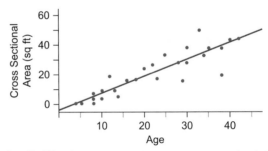

Display 3.45 *Cross-sectional area* versus *age* of oak trees.

a. Which is the explanatory variable and which is the response variable?

b. Explain how you can see from the graph that an increase in age of 10 years corresponds to an increase of about 10 square feet in the cross-sectional area.

c. When you use the regression line to predict the cross-sectional area, would you expect a larger error in prediction for a 10-year-old tree or a 30-year-old tree?

d. Use the plot to estimate the residual for the oldest tree. Interpret this residual.

e. For which tree is the residual largest in absolute value? Describe how this tree compares to others about the same age.

E24. If you attend a university where class sizes tend to be small, are you more likely to give to your alumni fund after you graduate than if you graduate from a university with large classes? Display 3.46 shows a scatterplot of a sample of 40 universities. Each university appears as a point. The vertical coordinate, *y*, tells the percentage of alumni who gave money. Each *x*-coordinate tells the student/faculty ratio (number of students per faculty member). The equation of the fitted line is approximately $\hat{y} = 55 - 2x$.

a. Which is the explanatory variable and which is the response variable?

b. Explain how you can see from the graph that an increase of five students per faculty member corresponds to a decrease of about 10 percentage points in the giving rate. Explain how you can see this from the equation of the fitted line.

c. Does the y-intercept have a useful interpretation in this situation?

d. Use the regression line to predict the giving rate for a university with a student/faculty ratio of 12. When you use the regression line to predict the giving rate, would you expect a rather large error or a relatively small error in your prediction?

e. Use the plot to estimate the residual for the university with the highest student/faculty ratio and for the university with the highest giving rate.

f. The university with the lowest student/faculty ratio, 6 to 1, had a giving rate of 32%. Use the equation of the fitted line to find the residual for that university.

g. Suppose the Alumni Association at Piranha State University boasts a giving rate of 80%. Without knowing the student/faculty ratio at PSU, can you tell whether the prediction error will be positive or negative?

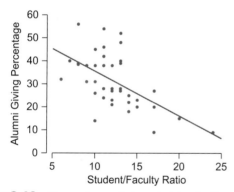

Display 3.46 Percentage of alumni giving to the alumni fund versus the student/faculty ratio for 40 highly rated U.S. universities.

E25. *Height versus age.* Display 3.47 shows standard computer output for the *median height* versus *age* data of E13.

Predictor	Coef	Stdev	t-ratio	p
Constant	31.5989	0.3063	103.17	0.000
Age	2.47418	0.03468	71.34	0.000

$s = 0.4679$ R-sq = 99.8%   R-sq(adj) = 99.8%

Analysis of Variance

SOURCE	DF	SS	MS	F	p
Regression	1	1114.1	1114.1	5089.73	0.000
Error	11	2.4	0.2		
Total	12	1116.5			

Display 3.47 Computer output of *median height* versus *age* data.

a. Write the equation of the regression line.

b. What is the SSE for this least squares line? Does its value seem reasonable given the scatterplot in Display 3.37 on page 122.

E26. Part of an output for the percentage of alumni who give to their colleges versus the student/faculty ratio is shown in Display 3.48. (These are the data in the scatterplot shown in Display 3.46.)

a. What equation is given in the printout for the least squares regression line?

b. Examine the table of unusual observations. What is the student/faculty ratio at the college with the largest residual (in absolute value)? Find this college in Display 3.46.

Predictor	Coef	Stdev	t-ratio	p
Constant	54.979	5.477	10.04	0.000
S/F Ratio	−1.9455	0.4354	−4.47	0.000

$s = 9.704$ R-sq = 34.4%   R-sq (adj) = 32.7%

Analysis of Variance

SOURCE	DF	SS	MS	F	P
Regression	1	1880.2	1880.2	19.97	0.000
Error	38	3578.5	94.2		
Total	39	5458.7			

Unusual Observations

Obs.	S/F Ratio	% Giving	Fit	Residual	St. Resid
9	10.0	14.00	35.52	−21.52	−2.26
14	11.0	54.00	33.58	20.42	2.13
30	13.0	52.00	29.69	22.31	2.33
39	20.0	15.00	16.07	−1.07	−0.12
40	24.0	9.00	8.29	0.71	0.09

Display 3.48 Computer output: regression analysis of *percentage giving to alumni fund* versus *student/faculty ratio.*

c. Verify that the fit ($\hat{y}$) and the value of the largest residual were computed correctly.

d. Locate the SSE on the output. Why is this value so large?

E27. *Sum of residuals.* In this exercise, your will show that the sum of the residuals is equal to 0 if and only if the regression line passes through the point of averages, $(\bar{x}, \bar{y})$.

a. Write the equation of the horizontal line through $(\bar{x}, \bar{y})$. Show that for a horizontal line the sum of the residuals will be 0 if and only if the line passes through the point of averages.

b. Write the equation of the line through $(\bar{x}, \bar{y})$ with slope b_1. Show that no matter what the slope of the line is, the sum of the residuals will be 0 if and only if the line passes through the point of averages.

c. Why isn't it good enough to define the regression line as the line that makes the sum of the residuals equal 0?

E28. *More about slope.*

    a. You and three friends, one right after the other, each buy the same kind of gas at the same pump. Then you make a scatterplot of your data, with one point per person, plotting the number of gallons on the *x*-axis and the total price paid on the *y*-axis. Will all four points lie on the same line? Explain.

    b. You and the same three friends each drive 80 mi but at different average speeds. Afterward, you plot your data twice, first as a set of four points with coordinates average speed, *x*, and elapsed time, *y*, and then as a set of points with coordinates average speed, *x*, and *y** defined as $\frac{1}{elapsed\ time}$. Which plot will give a straight line? Explain your reasoning. Will the other plot be a curve opening up, a curve opening down, or neither?

E29. The data set in Display 3.49 is the pizza data of E14 augmented by other brands of cheese pizza typically sold in supermarkets.

    a. Plot *calories* versus *fat*. Does there appear to be a linear association between *calories* and *fat*? If so, fit a least squares line to the data, and interpret the slope of the line.

    b. Plot *fat* versus *cost*. Does there appear to be a linear association between *cost* and *fat*? If so, fit a least squares line to the data, and interpret the slope of the line.

    c. Plot *calories* versus *cost*. Does there appear to be a linear association between *cost* and *calories*?

    d. Write a summary of your findings.

E30. *Poverty.* What variables are most closely associated with poverty? Display 3.50 provides information on population characteristics of the 50 U.S. states plus the District of Columbia. Each variable is measured as a percentage of the state's population, as described here:

    Percentage living in metropolitan areas

    Percentage white

    Percentage of adults who have graduated from high school

    Percentage of families with incomes below the poverty line

    Percentage of families headed by a single parent

Construct scatterplots to determine which variables are most strongly associated with poverty.

Write a letter to your representative in Congress about poverty in America, relying only on what you find in these data. Point out the variables that appear to be most strongly associated with poverty and those that appear to have little or no association with poverty.

Pizza	Calories	Fat (g)	Cost ($)
Domino's Deep Dish	385	19.5	1.87
Pizza Hut's Stuffed Crust	370	15	1.83
Pizza Hut's Pan	280	14	1.83
Domino's Hand Tossed	305	12	1.67
Pizza Hut's Hand Tossed	230	9	1.63
Little Caesar's Deep Dish	350	14.2	1.06
Little Caesar's Original Round	230	8	0.81
Freshchetta Bakes & Rises 4-Cheese	364	15	0.98
Freshchetta Bakes & Rises Sauce Stuffed Crust 4-Cheese	334	11	1.23
DiGiorno Rising Crust Four Cheese	332	12	0.94
Amy's Organic Crust & Tomatoes Cheese	341	14	1.92
Safeway Select Verdi Quattro Formaggio Self-Rising Crust	307	9	0.84
Tony's Super Rise Crust Four-Cheese	335	12	0.96
Kroger Self-Rising Crust Four-Cheese	292	9	0.80
Tombstone Stuffed Crust Cheese	364	18	0.96
Red Baron Classic 4-Cheese	384	20	0.91
Boboli Original	333	12	0.89
Tombstone Original Extra Cheese	328	14	0.94
Reggio's Chicago Style Cheese	367	13	1.02
Jack's Original Cheese	325	13	0.92
Celeste Pizza for One Cheese	346	17	1.17
McCain Ellio's Cheese	299	9	0.54
Michelina's Zap 'ems That'za Pizza!	394	19	1.28
Totino's The Original Crisp Crust Party Cheese	322	14	0.67

Display 3.49 Food values and cost per 5-oz serving of pizza. [Source: *Consumer Reports*, January 2002.]

State	Metropolitan Residence	White	Graduates	Poverty	Single Parent
Alabama	55.4	71.3	79.9	14.6	14.2
Alaska	65.6	70.8	90.6	8.3	10.8
Arizona	88.2	87.7	83.8	13.3	11.1
Arkansas	52.5	81.0	80.9	18.0	12.1
California	94.4	77.5	81.1	12.8	12.6
Colorado	84.5	90.2	88.7	9.4	9.6
Connecticut	87.7	85.4	87.5	7.8	12.1
Delaware	80.1	76.3	88.7	8.1	13.1
District of Columbia	100.0	36.2	86.0	16.8	18.9
Florida	89.3	80.6	84.7	12.1	12.0
Georgia	71.6	67.5	85.1	12.1	14.5
Hawaii	91.5	25.9	88.5	10.6	12.4
Idaho	66.4	95.5	88.2	11.8	8.7
Illinois	87.8	79.5	85.9	11.2	12.3
Indiana	70.8	88.9	86.4	8.7	11.1
Iowa	61.1	94.9	89.7	8.3	8.6
Kansas	71.4	89.3	88.6	9.4	9.3
Kentucky	55.8	90.3	82.8	13.1	11.8
Louisiana	72.6	64.2	79.8	17.0	16.6
Maine	40.2	97.1	86.6	11.3	9.5
Maryland	86.1	65.6	87.6	7.3	14.1
Massachusetts	91.4	87.2	87.1	9.6	11.9
Michigan	74.7	81.5	87.6	10.3	12.5
Minnesota	70.9	90.2	91.6	6.5	8.9
Mississippi	48.8	61.2	81.2	17.6	17.3
Missouri	69.4	85.3	88.3	9.6	11.6
Montana	54.1	90.9	90.1	13.7	8.9
Nebraska	69.8	92.1	90.8	9.5	9.1
Nevada	91.5	84.1	85.6	8.3	11.1
New Hampshire	59.3	96.3	92.1	5.6	9.1
New Jersey	94.4	77.3	86.2	7.8	12.6
New Mexico	75.0	84.9	81.7	17.8	13.2
New York	87.5	73.6	84.2	14.0	14.7
North Carolina	60.2	74.1	81.4	13.1	12.5
North Dakota	55.9	92.5	89.7	11.9	7.8
Ohio	77.4	85.4	87.2	10.1	12.1
Oklahoma	65.3	78.4	85.7	14.7	11.4
Oregon	78.7	90.8	86.9	11.2	9.8
Pennsylvania	77.1	86.4	86.0	9.2	11.6
Rhode Island	90.9	89.2	81.0	10.3	12.9
South Carolina	60.5	67.7	80.8	13.5	14.8
South Dakota	51.9	88.8	88.7	10.2	9.0
Tennessee	63.6	80.8	81.0	14.2	12.9
Texas	82.5	83.6	77.2	15.3	12.7
Utah	88.2	93.6	89.4	9.3	9.4
Vermont	38.2	96.9	88.9	9.9	9.3
Virginia	73.0	73.9	87.8	8.7	11.9
Washington	82.0	85.5	89.1	10.8	9.9
West Virginia	46.1	95.0	78.7	16.0	10.7
Wisconsin	68.3	90.1	88.6	8.6	9.6
Wyoming	65.1	94.7	90.9	9.5	8.7

Display 3.50 Characteristics of state populations, as percentage of populations. [Source: U.S. Census Bureau, *www.census.gov.*]

3.3 ▶ Correlation: The Strength of a Linear Trend

Some of the linear relationships you've seen in this chapter have been extremely strong, with points packed tightly around the regression line. Other linear relationships have been quite weak—although there was a general linear trend, a lot of variation was present in the values of *y* associated with a given value of *x*. Still other linear relationships have been in between.

In this section, you'll learn how to measure the strength of a linear relationship numerically by using the correlation coefficient, r (which, from this point on, will be referred to simply as the **correlation**), where $-1 \leq r \leq 1$. Just as in the last section, you'll start by working intuitively and visually and then move on to a computational approach.

> The correlation measures the amount of variation from the regression line.

Estimating the Correlation

Examine the scatterplots and their correlations in Display 3.51. To get a rough idea of the size of a correlation, it is helpful to sketch an ellipse around the cloud of points in the scatterplot. If the ellipse has points scattered throughout and the points appear to follow a linear trend, then the correlation is a reasonable measure of the strength of the association. If the ellipse slants upward as you go from left to right, the correlation is **positive.** If the ellipse slants downward as you go from left to right, the correlation is **negative.** If the ellipse is fat, the correlation is *weak* and the absolute value of r is close to 0. If the ellipse is skinny, the correlation is *strong* and the absolute value of r is close to 1.

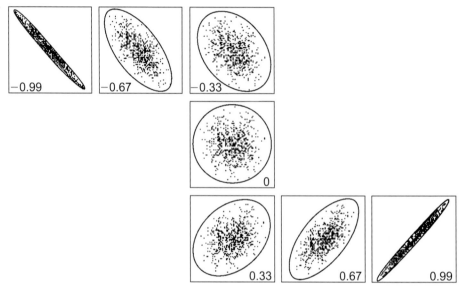

Display 3.51   Scatterplots with ellipses and their correlations. [Source: George Cobb, *Electronic Companion to Statistics* (Cogito Learning Media, Inc., 1997), p. 114.]

DISCUSSION
Estimating the Correlation

D9.   Match each of the scatterplots with its correlation, choosing from the values 0.783, 0.908, and 0.999.

a.

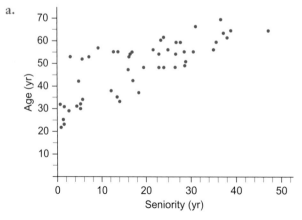

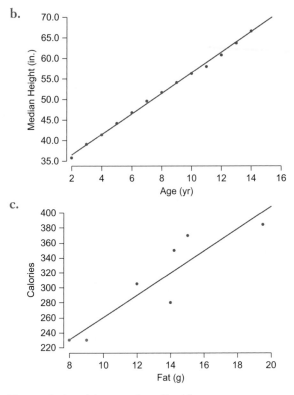

D10. Four relationships are described here.

    I. For a random sample of students from the senior class, x represents the day of the month of the person's birthday and y represents the cost of the person's most recent haircut.

    II. For a random collection of U.S. coins, x represents the diameter and y represents the circumference.

    III. For a random sample of bags of white socks, x represents the number of socks and y represents the price per bag.

    IV. For a random sample of bags of white socks, x represents the number of socks and y represents the price per sock.

  a. Which of these relationships have a positive correlation, and which a negative correlation? Which has the strongest relationship? The weakest?

  b. For each of the four relationships, discuss the connection between your ability to precisely predict a value of y for a given value of x and the size of the correlations.

A Formula for the Correlation, *r*

A formula for the correlation, r, follows. It looks impressive, but the basic idea is simple—you convert x and y to standardized values (z-scores), and then find their average product (dividing by $n - 1$).

Computing the Correlation

$$r = \frac{1}{n-1}\sum\left(\frac{x-\bar{x}}{s_x}\right)\left(\frac{y-\bar{y}}{s_y}\right) = \frac{1}{n-1}\sum z_x \cdot z_y$$

In this formula, s_x is the standard deviation of the x's, and s_y is the standard deviation of the y's. Remember that the z-score tells you how many standard deviations the value lies above or below the mean.

> You can think of the correlations, r, as the average product of the z-scores.

| Example 3.9 | **Computing the Correlation Between Television Watching and GPA** |

Many professional and student research projects have documented a negative relationship between a student's GPA and the number of hours per week spent playing video games or the number of hours per week spent watching television. (For example, V. Anand, "A Study of Time Management: The Correlation Between Video Game Usage and Academic Performance Markers," *CyberPsychology & Behavior*, Vol. 10, 4 (2007), pp. 552–559.)

The data in Display 3.52 are from a psychology student's research project. Compute the correlation between a student's GPA and the number of hours spent watching television per week.

| Student | Original Units | | Standard Units (z-scores) | | Product |
	TV Watching (h), x	GPA, y	$z_x = \dfrac{x - \bar{x}}{s_x}$	$z_y = \dfrac{y - \bar{y}}{s_y}$	$z_x \cdot z_y = \left(\dfrac{x - \bar{x}}{s_x}\right)\left(\dfrac{y - \bar{y}}{s_y}\right)$
A	3	3.5	−0.95	1.04	−0.98
B	3	3.1	−0.95	0.00	0.00
C	4	2.9	−0.83	−0.52	0.43
D	5	3.4	−0.71	0.78	−0.55
E	9	3.8	−0.24	1.81	−0.43
F	10	2.8	−0.12	−0.78	0.09
G	12	2.8	0.12	−0.78	−0.09
H	15	3.2	0.47	0.26	0.12
I	20	3.0	1.07	−0.26	−0.28
J	29	2.5	2.13	−1.55	−3.32
Sum	110	31.0	0	0	−5.00942
Mean	11	3.1	0	0	
Standard deviation	8.433	0.386	1	1	

Display 3.52 Calculations for the correlation between *TV watching* and *GPA*.

Solution

If you must compute a correlation, r, by hand, it is easiest to organize your work in a table as in Display 3.52. First, compute the mean of the values of x and of the values of y and then compute their standard deviations, s_x and s_y.

$$\bar{x} = \frac{110}{10} = 11 \qquad \bar{y} = \frac{31}{10} = 3.1$$

$$s_x \approx 8.433 \qquad s_y \approx 0.386$$

Then convert each value of x and of y to standard units, or z-scores. The correlation, r, is the average product of the z-scores: the sum of last column divided by $n - 1$, which is

$$r = \frac{1}{n - 1}\sum z_x \cdot z_y$$

$$= \frac{1}{10 - 1}(-5.00942)$$

$$\approx -0.557$$

A way to visualize these computations is to look at the four quadrants formed on the scatterplot by dividing it vertically at the mean value of x and horizontally at the mean value for y. Such a plot is shown in Display 3.53. Points in Quadrant I, such as the point for Student H, have positive z-scores for both x and y, so their product contributes a positive amount to the calculation of the correlation. Points in Quadrant III, such as the point for Student C, have negative z-scores for both x and y, so their product contributes a positive amount to the calculation of the correlation. Points in Quadrants II or IV, such as for Student A and Student J, contribute negative amounts to the calculation for the correlation since one z-score is positive and the other is negative.

The points in Quadrants II and IV have negative products $z_x \cdot z_y$.

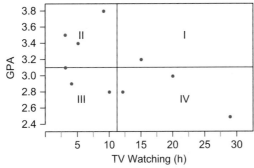

Display 3.53 Scatterplot divided into quadrants at $(\bar{x}, \bar{y})$.

Because r is the average of the products $z_x \cdot z_y$ and z-scores have no units, r has no units. In fact, r does not depend on the units of measurement in the original data. For example, if you measure arm spans and heights in inches and your friend measures then in centimeters, you will both compute the same value for the correlation between height and arm span.

The correlation, r, is a quantity without units.

DISCUSSION
A Formula for the Correlation, r

D11. Look at Display 3.52 showing the calculations for the correlation, r.

a. Confirm the calculations for Student A.

b. Which point makes the largest contribution to the correlation? Where is this point on the scatterplot?

c. Which point makes the smallest contribution to the correlation? Where is this point on the scatterplot?

D12. *Understanding r.*

a. Explain in your own words what the correlation measures.

b. Explain in your own words why r has no units.

c. When computing the correlation between two variables, does it matter which variable you select as y and which you select as x? Explain.

D13. Refer to Display 3.53.

a. What can you say about r if there are many points in Quadrants I and III and few in Quadrants II and IV?

b. What can you say about r if there are many points in Quadrants II and IV and few in Quadrants I and III?

c. What can you say about r if points are scattered randomly in all four quadrants?

STATISTICS IN ACTION 3.2 ▶ Was Leonardo Correct?

In this activity, you will learn more about correlation and you'll practice finding the value of *r* using your calculator.

What You'll Need: a measuring tape, yardstick, or meterstick

Leonardo da Vinci was a scientist and an artist who combined these skills to draft extensive instructions for other artists on how to proportion the human body in painting and sculpture. Three of Leonardo's rules were

- height is equal to the span of the outstretched arms
- kneeling height is three-fourths of the standing height
- the length of the hand is one-ninth of the height

1. Work with a partner to measure your height, kneeling height, arm span, and hand length. Combine your measurements with the rest of your class's.
2. Check Leonardo's three rules visually by plotting the data on three scatterplots.
3. For the plots that have a linear trend, use technology to find the equation of the regression line and the value of *r* (the correlation).
4. Interpret the slopes of the regression lines. Interpret the correlations.
5. Do the three relationships described by Leonardo appear to hold? Do they hold strongly?

Correlation and the Appropriateness of a Linear Model

It is tempting to believe that a high correlation (either positive or negative) is evidence that a linear model is appropriate for your data. Alas, the real world is not so simple. For example, Display 3.54 shows the number of blogs (Web-based periodic postings of a person's thoughts) for the first few years after 2003, when blogging's popularity took off. The growth is exponential, yet the (linear) correlation is very strong, $r = 0.91$. The points do cluster fairly tightly about the linear regression line, but that line is not the best model for the data. The plot on the right shows the points and the graph of the best-fitting exponential equation, $\hat{y} = 0.353 \cdot 1.140^x$.

Month	Months from March 2003	Number of Blogs (millions)
August 2003	5	0.6
January 2004	10	1.5
June 2004	15	2.7
November 2004	20	4.6
April 2005	25	9.3
September 2005	30	17.8

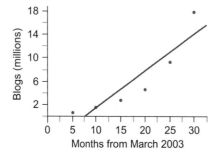
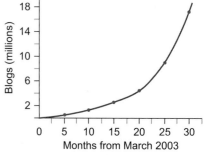

Display 3.54 The number of blogs growth with graph on an experimental equation, $y = 0.353 \cdot 1.140^x$ versus the number of months after March 2003. [Source: *State of the Blogosphere, October 2005, Part1: Blogosphere Growth*, posted by Dave Sifry, October 17, 2005, *Technorati News, www.technorati.com.* Table numbers estimated from graph.]

The quiz scores for 22 students in Display 3.55, on the other hand, have a correlation, r, of only 0.48. There is quite a bit of scatter, partly because the quizzes covered very different topics. Quiz 2 covered exponential growth, and Quiz 3 covered probability. In spite of the scatter, a line is the most appropriate model because there is no curvature in the pattern of data points.

> The moral: Always plot your data before computing summary statistics!

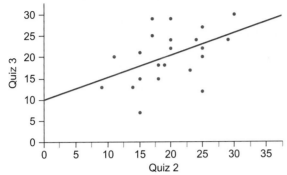

Display 3.55 Scores on two consecutive 30-point quizzes.

DISCUSSION
Correlation and the Appropriateness of a Linear Model

D14. When the correlation is small in absolute value, what does it mean for the prediction error? Why would anyone want to fit a line to data in a case in which the correlation is small, as in Display 3.55 (quiz scores)?

D15. Provide a real-life scenario involving two variables for each situation. Assume r is positive in each case.

    a. r is small and you do not want to fit a line.

    b. r is small and you do want to fit a line.

    c. r is large and you do not want to fit a line.

    d. r is large and you do want to fit a line.

D16. It's common in situations similar to the growth in blogs for the numbers to increase exponentially for the first few years. What do you think would happen if the table in Display 3.54 were continued to include numbers for months up to the current year?

The Relationship Between the Correlation and the Slope

By now you might have observed that the slope of the regression line, b_1, and the correlation, r, always have the same sign. But they have a more specific relationship.

Finding the Slope from the Correlation and the Standard Deviations

The slope of a least squares regression line, b_1, and the correlation, r, are related by the equation

$$b_1 = r \cdot \frac{s_y}{s_x}$$

where s_x is the standard deviation of the x's and s_y is the standard deviation of the y's. This means that if you standardize the data so that $s_x = 1$ and $s_y = 1$, then the slope of the regression line is equal to the correlation.

Example 3.10	**Critical Reading and Math SAT Scores**

In 2008, the mean critical reading score for all SAT I test takers was 502, with a standard deviation of 112. For math scores, the mean was 515, with a standard deviation of 116. The correlation between the two scores was not given but is known to be quite high. If you can estimate this correlation as, say, 0.7, you can find the equation of the regression line and use it to estimate the math score from a student's critical reading score. [Source: College Board, *2008 College-Bound Seniors: Total Group Profile Report*.]

Solution
The formula gives an estimate of the slope of

$$b_1 = r \cdot \frac{s_y}{s_x} = 0.7 \cdot \frac{116}{112} = 0.725$$

To find the y-intercept, use the fact that the point $(\bar{x}, \bar{y}) = (502, 515)$ is on the regression line:

$$\hat{y} = y\text{-intercept} + slope \cdot x$$
$$\hat{y} = b_0 + b_1 x$$
$$515 = b_0 + 0.725(502)$$
$$b_0 = 515 - 363.95 = 151.05$$

The equation is $\hat{y} = 0.725x + 151.05$.

DISCUSSION
The Relationship Between the Correlation and the Slope

D17. Find the equation of the regression line for predicting an SAT I critical reading score given the student's SAT I math score.

Correlation Does Not Imply Causation

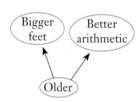

In a sample of elementary school students, there is a strong positive relationship between shoe size and scores on a standardized test of ability to do arithmetic. Does this mean that studying arithmetic makes your feet bigger? No. Shoe size and skill at arithmetic are related to each other because both increase as a child gets older. Age is an example of a *lurking variable*.

Beware the Lurking Variable

A **lurking variable** is a variable that you didn't include in your analysis but that might explain the relationship between the variables you did include. When variables x and y are correlated, it might be because both are consequences of a third variable, w, that is lurking in the background.

Even if you can't identify a lurking variable, you should be careful to avoid jumping to a conclusion about cause and effect when you observe a strong relationship. *The value of r does not tell you anything about* why *two variables are related*. The statement "Correlation does not imply causation" can help you remember this. To conclude that one thing causes another, you need data from a randomized experiment, as you'll learn in the next chapter.

Two variables might be highly correlated without one causing the other.

DISCUSSION
Correlation Does Not Imply Causation

D18. Display 3.53 on page 131 shows a negative association between the number of hours a student watches TV per week and his or her GPA. Discuss whether you think one of these variables might cause the other, or whether a lurking variable might account for both.

D19. A sample of 51 top-rated universities shows a very strong positive relationship between acceptance rate (percentage of applicants who are offered admission) and SAT scores (the 75th percentile for an entering class). Explain why these two variables have such a strong relationship. Does one "cause" the other? If not, how might you account for the strong relationship?

D20. People who argue about politics and public policy often point to relationships between quantitative variables and then offer a cause-and-effect explanation to support their points of view. For each of these relationships, first give a possible explanation by assuming a causal relationship and then give another possible explanation based on a lurking variable.

    a. Faculty positions in academic subjects with a higher percentage of male faculty tend to pay higher salaries. (For example, engineering and geology have high percentages of male faculty and high average salaries; journalism, music, and social work have much lower percentages of male faculty and much lower academic salaries.)

    b. States with larger reported numbers of hate groups tend to have more people on death row. (Here, also, tell how you could adjust for the lurking variable to uncover a more informative relationship.)

    c. States with higher reported rates of gun ownership tend to have lower reported rates of violent crime.

Interpreting r^2

You might have noticed that computer outputs for regression analysis, like that in Display 3.24 on page 116, give the value of *R*-squared, or r^2, rather than the value of *r*.

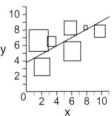

On page 116, you learned that the least squares regression line was chosen so as to minimize the sum of squared residuals, $SSE = \Sigma(y - \hat{y})^2$, where the residuals are the differences between the observed and predicted values of *y*. You can think of the SSE as a measure of the total variation in the residuals. If you don't take into account the relationship of *y* with *x*, the total variation in the *y*'s can be measured by the sum of the squared differences between the values of *y* and their mean:

$$SST = \Sigma(y - \bar{y})^2$$

The SST is the numerator of the sample variance, as defined on page XX.

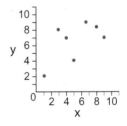

Now, if two variables are strongly correlated, then the residuals will tend to be small and the corresponding SSE will be small relative to the SST. That is, the ratio SSE/SST will be close to 0. If, on the other hand, the correlation is weak, the residuals will tend to be large and the ratio SSE/SST will be close to 1. (The SSE can never exceed the SST because the variation in the residuals can never exceed the variation in the *y*'s.)

That the ratio SSE/SST goes in the opposite direction to the correlation, *r*, is a bit disconcerting. To correct this little anomaly, a statistic r^2 is defined as

$$r^2 = 1 - \frac{SSE}{SST}$$

The value of r^2 is close to 1 for a strong correlation between the variables and close to 0 for a weak correlation. A more practical interpretation of this statistic can be seen by writing it as

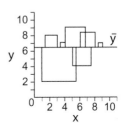

$$r^2 = \frac{SST - SSE}{SST}$$

This form shows that r^2 measures the proportion of the total variation of the y's that is taken away by the least squares regression line. (If, for example, SSE = 0, then the line fits the points perfectly; thus, $r^2 = 1$ and all of the variation in the y direction is taken away by the linear relationship with x.) The statistic r^2 is called the **coefficient of determination** and is said to measure "the proportion of the variation in the y's that is explained by x." The next example shows the calculations and interpretation in a practical setting.

Example 3.11 | Predicting Pizzas

Here again is the pizza data from E14 on page 122. Compute the SST, the SSE, and the value of r^2. Then interpret r^2.

Fat (g) x	Calories y
9.0	230
19.5	385
14.0	280
12.0	305
8.0	230
14.2	350
15.0	370

Solution

If you had to pick a single number as your predicted calorie content, you might choose the mean, 307.14 calories per serving. The SST is a measure of the error in the resulting predictions: SST = $\Sigma(y - \bar{y})^2$. The plot in Display 3.56 shows these "squared" errors geometrically.

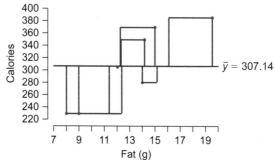

Display 3.56 Squared deviations around the mean of y.

The computation of the SST is shown in Display 3.57.

Fat (g) x	Calories y	Predicted $\bar{y}$	Error $(y - \bar{y})$	Squared Error $(y - \bar{y})^2$
9.0	230	307.14	−77.14	5951.02
19.5	385	307.14	77.86	6061.73
14.0	280	307.14	−27.14	736.74
12.0	305	307.14	−2.14	4.59
8.0	230	307.14	−77.14	5951.02
14.2	350	307.14	42.86	1836.73
15.0	370	307.14	62.86	3951.02
				SST = 24,492.86

Display 3.57 Computing the SST.

If you use the regression equation, $\widehat{calories} = 112 + 14.9\,fat$, to predict calories, the resulting errors are much smaller for most pizzas, as shown on the plot in Display 3.58.

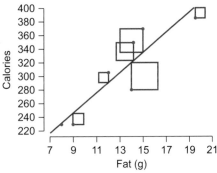

Display 3.58 Squared deviations around the least squares line.

The computation of the SSE is shown in Display 3.59.

Fat (g) x	Calories y	Predicted $\hat{y}$	Error $(y - \hat{y})$	Squared Error $(y - \hat{y})^2$
9.0	230	245.95	−15.95	254.42
19.5	385	402.66	−17.66	311.97
14.0	280	320.58	−40.58	1646.36
12.0	305	290.73	14.27	203.76
8.0	230	231.03	1.03	1.05
14.2	350	323.56	26.44	699.06
15.0	370	335.50	34.50	1190.23
				SSE = 4306.85

Display 3.59 Computing the SSE.

Using the formula,

$$r^2 = \frac{SST - SSE}{SST} = \frac{24{,}492.86 - 4{,}306.85}{24{,}492.86} = 0.8242$$

> The coefficient of determination tells the proportion of the total variation in y that can be explained by x.

About 82% of the variation in calories among these brands of pizza can be attributed to fat content.

DISCUSSION
Interpreting r^2

D21. The scatterplot in Display 3.60 shows IQ plotted against head circumference for a sample of 20 people. The mean IQ was 101, and the mean head circumference was 56.125 cm. The correlation is 0.138.

a. If you knew nothing about any possible relationship between head circumference and IQ, what IQ would you predict for a person with a head circumference of 54 cm?

b. The regression equation is $\widehat{IQ} = 0.997 \cdot head\ circumference + 45$. If one person has a head that is 1 cm larger in circumference than the head of a second person, by about how much does this equation predict that the IQ of the first person will differ from that of the second person?

c. What IQ does this equation predict for a person with a head circumference of 54 cm? How much faith do you have in this prediction?

d. How much of the variability in IQ is accounted for by the regression? Does the regression equation help you predict IQ in any practical sense?

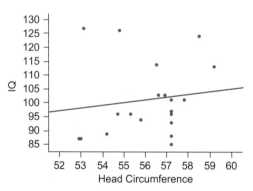

Display 3.60   [Source: M. J. Tramo, W. C. Loftus, R. L. Green, T. A. Stukel, J. B. Weaver, and M. S. Gazzaniga, "Brain Size, Head Size, and IQ in Monozygotic Twins," *Neurology*, Vol. 50 (1998), pp. 1246–1252.]

e. If more people were added to the plot, how do you think the regression equation and correlation would change?

D22. Use the formula for r^2 to explain why $-1 \le r \le 1$.

D23. In a study of the effect of temperature on household heating bills, an investigator said, "Our research shows that about 70% of the variability in the number of heating units used by a particular house over the years can be explained by outside temperature." Explain what the investigator might have meant by this statement.

Regression Toward the Mean

Display 3.61 shows a hypothetical data set with the height of younger adult sisters plotted against the height of their older sisters. There is a moderate positive association: $r = 0.337$. For both younger and older sisters, the mean height is 65 in. and the standard deviation is 2.5 in. The line drawn on the first plot, $y = x$, indicates the location of points representing the same height for both sisters. If you rotate your book and sight down the line, you can see that the points are scattered symmetrically about it.

In the second plot, look at the vertical strip for the older sisters with heights between 62 in. and 63 in. The X is at the mean height of the younger sisters with older sisters in this height range. It falls at about 64 in., not between 62 in. and 63 in. as you would expect. Looking at the vertical strip on the right, the mean height of younger sisters with older sisters between 68 in. and 69 in. is only about 66 in. If you were to use the line $y = x$ to predict the height of the younger sister, you would tend to predict a height that is too small if the older sister is shorter than average and a height that is too large if the older sister is taller than average.

The flatter line through the third scatterplot in Display 3.61 is the least squares regression line. Notice that this line gets as close as it can to the center of each vertical strip. Thus, the least squares line is sometimes called the **line of means.** The predicted value of y at a given value of x, using the regression line as the model, is the estimated mean of all responses that can be produced at that particular value of x.

The regression line is a line of means.

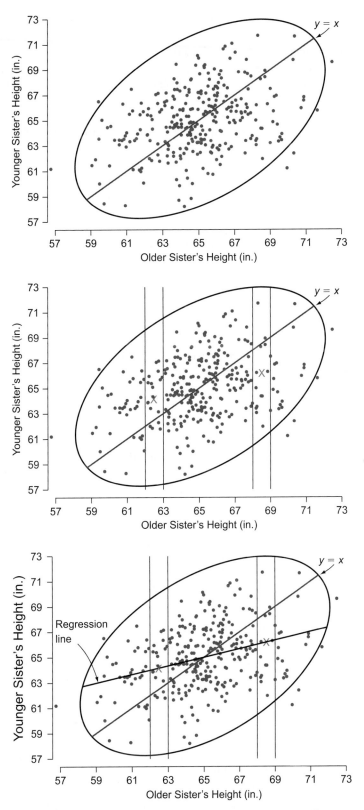

Display 3.61 Scatterplots showing the regression effect.

Notice that the regression line has a smaller slope than the major axis ($y = x$) of the ellipse. This means that the predicted values are closer to the mean than you might expect, which will always be the case for positively correlated data following a linear trend. The difference between these two lines is sometimes called the **regression effect.** If the correlation is near $+1$ or -1, the two lines will be nearly on top of each other and the regression effect will be minimal. For a moderate correlation such as that for the sisters' heights, the regression effect will be quite large.

The regression effect was first noticed by British scientist Francis Galton around 1877. Galton noticed that the largest sweet-pea seeds tended to produce daughter seeds that were large but smaller than their parent. The smallest sweet-pea seeds tended to produce daughter seeds that were also small but larger than their parent. There was, in Galton's words, a **regression toward the mean.** This is the origin of the term *regression line*. [Source: D. W. Forrest, *Francis Galton: The Life and Work of a Victorian Genius* (Taplinger, 1974).]

> "Regression toward the mean" is another term for the regression effect.

The regression effect is with us in everyday life whenever some element of chance is involved in a person's score. For example, athletes are said to experience a "sophomore slump." That is, athletes who have the best rookie seasons do not tend to be the same athletes who have the best second year. The top students on the second exam in your class probably did not do as well, relative to the rest of the class, on the first exam. The children of extremely tall or short parents do not tend to be as extreme in height as their parents. There does, indeed, seem to be a phenomenon at work that pulls us back toward the average. As Galton noticed, this prevents the spread in human height, for example, from increasing. Look for this effect as you work on regression analyses of data.

DISCUSSION
Regression Toward the Mean

D24. Why is the regression line sometimes called the "line of means"?

D25. The equation of the regression line for the scatterplot in Display 3.61 is $\hat{y} = 43.102 + 0.337x$. Interpret the slope of this line in the context of the situation and compare it to the interpretation of the slope of the line $y = x$.

Summary 3.3: Correlation: The Strength of a Linear Trend

In your study of normal distributions in Chapter 2, you used the mean to tell the center and then used the standard deviation as the overall measure of how much the values deviated from that center. For "well-behaved" quantitative relationships—that is, those whose scatterplots look elliptical—you use the regression line as the center and then measure the overall amount of variation from the line using the correlation, r. You can think of the correlation, r, as the average product of the z-scores.

$$r = \frac{1}{n-1}\sum\left(\frac{x - \bar{x}}{s_x}\right)\left(\frac{y - \bar{y}}{s_y}\right) = \frac{1}{n-1}\sum z_x \cdot z_y$$

Geometrically, the correlation measures how tightly packed the points of the scatterplot are about the regression line.

The correlation has no units and ranges from -1 to $+1$. It is unchanged if you interchange x and y or if you make a linear change of scale in x or y, such as from feet to inches or from pounds to kilograms.

In assessing correlation, begin by making a scatterplot and then follow these steps:

1. *Shape:* Is the plot linear, shaped roughly like an elliptical cloud, rather than curved, fan-shaped, or formed of separate clusters? If so, draw an ellipse to enclose the cloud of points. The data should be spread throughout the ellipse; otherwise, the pattern might not be linear or might have unusual

features that require special handling. You should not calculate the correlation for patterns that are not linear.

2. *Trend:* If your ellipse tilts upward to the right, the correlation is positive; if it tilts downward to the right, the correlation is negative. The relationship between the correlation and the slope, b_1, of the regression line is given by

$$b_1 = r \cdot \frac{s_y}{s_x}$$

3. *Strength:* If your ellipse is almost a circle or is horizontal, the relationship is weak and the correlation is near zero. If your ellipse is so thin that it looks like a line, the relationship is very strong and the correlation is near $+1$ or -1.

Correlation is not the same as causation. Two variables may be highly correlated without one having any causal relationship with the other. The value of r tells nothing about *why* x and y are related. In particular, a strong relationship between x and y might be due to a lurking variable.

You can interpret the value r^2 as the proportion of the total variation in y that can be accounted for by using x in the prediction model:

$$r^2 = \frac{\text{SST} - \text{SSE}}{\text{SST}}$$

The regression effect (or regression toward the mean) is the tendency of y-values to be closer to their mean than you might expect. That is, the regression line is flatter than the major axis of the ellipse surrounding the data.

Practice

Estimating the Correlation

P11. By comparing to the plots in Display 3.51 on page 128, match each of the five scatterplots in Display 3.62 with its correlation, choosing from -0.95, -0.5, 0, 0.5, and 0.95.

P12. The table in P8 (Display 3.31 on page 120) gives the birth weight and adult weight of species of camels.

   a. Guess a value for the correlation, r.

   b. Calculate r using technology.

A Formula for the Correlation, *r*

P13. For each set of pairs, (x, y), compute the correlation by hand, standardizing and finding the average product, as in Example 3.9 on page 130.

   a. $(-2, -1), (-1, 1), (0, 0), (1, 1), (2, 1)$

   b. $(-2, 2), (0, 2), (0, 3), (0, 4), (2, 4)$

P14. The table in P8 (Display 3.31 on page 120) gives the birth weight and adult weight of species of camels. In P12, you used technology to find the correlation, r. This time, make a table like that in Display 3.52 on page 130, and use the formula to find r. What do you notice about the products $z_x \cdot z_y$?

P15. The scatterplot in Display 3.63 is divided into quadrants by vertical and horizontal lines that pass through the point of averages, $(\bar{x}, \bar{y})$.

   a. Is the correlation positive or negative?

   b. Give the coordinates of the point that will contribute the most to the correlation, r.

   c. Consider the product

$$z_x \cdot z_y = \left(\frac{x - \bar{x}}{s_x}\right)\left(\frac{y - \bar{y}}{s_y}\right)$$

a.
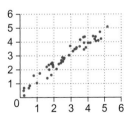

b.
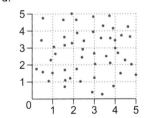

c.
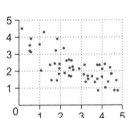

d.
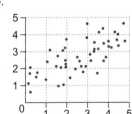

e.
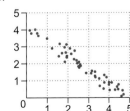

Display 3.62 Five scatterplots.

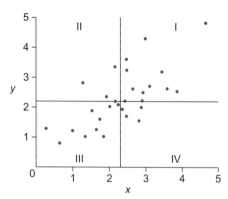

Display 3.63 Scatterplot divided into quadrants at the point of averages, $(\bar{x}, \bar{y})$.

Where are the points that have a positive product? How many of the 30 points have a positive product?

d. Where are the points that have a negative product? How many of the 30 points have a negative product?

Correlation and the Appropriateness of a Linear Model

P16. Both plots in Display 3.64 have a correlation of 0.26. For each plot, is fitting a regression line (as shown on the plot) an appropriate thing to do? Why or why not?

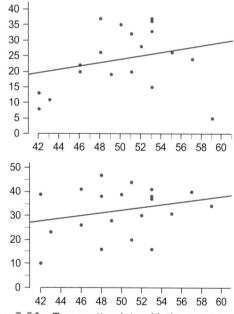

Display 3.64 Two scatterplots with the same correlation.

The Relationship Between the Correlation and the Slope

P17. Imagine a scatterplot of two sets of exam scores for students in a statistics class. The score for a student on Exam 1 is graphed on the x-axis, and his or her score on Exam 2 is graphed on the y-axis. The slope of the regression line is 0.368. The mean of the Exam 1 scores is 72.99, and the standard deviation is 12.37. The mean of the Exam 2 scores is 75.80, and the standard deviation is 7.00.

a. Find the correlation between the two sets of scores.

b. Find the equation of the regression line for predicting an Exam 2 score from an Exam 1 score. Predict

the Exam 2 score for a student who got a score of 80 on Exam 1.

c. Find the equation of the regression line for predicting an Exam 1 score from an Exam 2 score.

d. Sketch a scatterplot that could represent the situation described.

Correlation Does Not Imply Causation

P18. If you take a random sample of U.S. cities and measure the number of fast-food franchises in each city and the number of cases of stomach cancer per year in the city, you find a high correlation.

a. What is the lurking variable?

b. How would you adjust the data for the lurking variable to get a more meaningful comparison?

P19. If you take a random sample of public school students in grades K–12 and measure weekly allowance and size of vocabulary, you will find a strong relationship. Explain in terms of a lurking variable why you should not conclude that raising a student's allowance will tend to increase his or her vocabulary.

P20. For the countries of the United Nations, there is a strong negative relationship between the number of TV sets per thousand people and the birthrate. What would be a careless conclusion about cause and effect? What is the lurking variable?

Interpreting r^2

P21. Data on the association between high school graduation rates and the percentage of families living in poverty for the 50 U.S. states were presented in E30 on page 126. Display 3.65 contains the scatterplot and a standard computer output of the regression analysis.

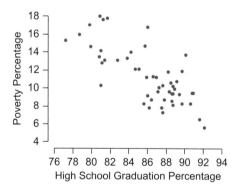

The regression equation is
Poverty = 64.8 − 0.621 HSG

Predictor	Coefficient	SD	t-Ratio	P
Constant	64.781	6.803	9.52	0.000
HSG	−0.62122	0.07902	−7.86	0.000

S = 2.082   R-sq = 55.8%   R-sq (adj) = 54.9%

Analysis of Variance

Source	DF	SS	MS	F	P
Regression	1	267.88	267.88	61.81	0.000
Error	49	212.37	4.33		
Total	50	480.25			

Display 3.65 Poverty rates versus high school graduation rates.

a. Under "Source," the "Total" variation is the SST, and the "Error" variation is the SSE. From this information, find r, the correlation.

b. Write an interpretation for r^2 in the context of these data.

c. What are the units for each of the values x, y, b_1, and r?

Regression Toward the Mean

P22. The plot in Display 3.66 shows the heights of older sisters plotted against the heights of their younger sisters. On a copy of this scatterplot, draw vertical lines to divide the points into six groups. Mark the approximate location of the mean of the y-values of each vertical strip. Sketch the regression line, $\hat{y} = 43 + 0.337x$. Note that the regression line comes as close as possible to the mean of each vertical strip. Now draw an ellipse around the data and connect the two ends of the ellipse. Is the regression line "flatter" than this line? Does this plot show the regression effect?

P23. Display 3.67 shows the first two exam scores for 29 college students enrolled in an introductory statistics course. Do you see any evidence of regression to the mean? If so, explain the nature of the evidence.

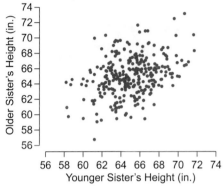

Display 3.66 The heights of older sisters versus the heights of their younger sisters.

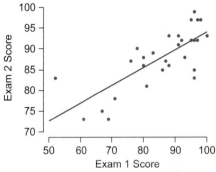

Display 3.67 Exam scores.

Exercises

E31. Each scatterplot in Display 3.68 was made on the same set of axes. Match each scatterplot with its correlation, Choosing from -0.06, 0.25, 0.40, 0.52, 0.66, 0.74, 0.85, and 0.90.

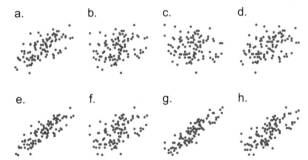

Display 3.68 Eight scatterplots with various correlations.

E32. Estimate the correlation between the variables in these scatterplots.

a. The proportion of the state population living in dorms versus the proportion living in cities in Display 3.43 on page 124.

b. The graduation rate versus the 75th percentile of SAT scores in E5 on pages 99–100.

c. The college graduation rate versus the percentage of students in the top 10% of their high school graduating class in E5 on pages 99–100.

E33. Eight artificial "data sets" are shown here. For each one, find the value of r, without computing if possible. Drawing a quick sketch might be helpful.

a.

x	y
-1	-1
0	
1	1

b.

x	y
-1	-1
0	1
1	1

c.

x	y
-1	0
1	0
1	-1

d.

x	y
-1	1
0	0
1	-1

e.

x	y
99	9
100	10
101	11

f.

x	y
15	30
20	40
25	20

g.

x	y
1003	80
1006	82
1009	81

h.

x	y
9.9	1000
10.0	2000
10.1	0

E34. Compute the correlation for the thickness of a piece of textbook paper data in P6 on page 119, which is reproduced below. Do this by hand, standardizing and finding the average product.

Sheets	Thickness
50	6.0
100	11.0
150	12.5
200	17.0
250	21.0

E35. The scatterplot in Display 3.69 shows part of the hat size data of E6 on page 100. The plot is divided into quadrants by vertical and horizontal lines that pass through the point of averages, $(\bar{x}, \bar{y})$.

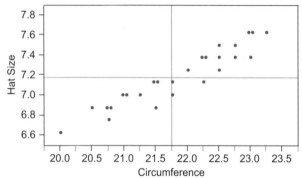

Display 3.69 Hat circumference versus hat size for 26 hats.

a. Estimate the value of the correlation.

b. Using the idea of standardized scores, explain why the correlation is positive.

c. Identify the point that contributes the most to the correlation. Explain why the contribution it makes is large.

d. Identify a point that contributes little to the correlation. Explain why the contribution it makes is small.

E36. The ellipses in Display 3.70 represent scatterplots that have a basic elliptical shape.

Display 3.70 Three pairs of elliptical scatterplots.

a. Match these conditions with the corresponding pair of ellipses.
  I. One s_y is larger than the other, the s_x's are equal, and the correlations are strong.
  II. One of the correlations is stronger than the other, the s_x's are equal, and the s_y's are equal.
  III. One s_x is larger than the other, the s_y's are equal, and the correlations are weak.

b. Draw a pair of elliptical scatterplots to illustrate each comparison.
  i. One s_y is larger than the other, the s_x's are equal, and the correlations are weak.
  ii. One s_x is larger than the other, the s_y's are equal, and the correlations are strong.

E37. Several biology students are working together to calculate the correlation for the relationship between air temperature and how fast a cricket chirps. They all use the same crickets and temperatures, but some measure temperature in degrees Celsius and others measure it in degrees Fahrenheit. Some measure chirps per second,

and others measure chirps per minute. Some use x for temperature and y for chirp rate, while others have it the other way around.

a. Will all the students get the same value for the slope of the least squares line? Explain why or why not.

b. Will they all get the same value for the correlation? Explain why or why not.

E38. For the sample of top-rated universities in E5 on page 99, the graduation rate has mean 82.7% and standard deviation 8.3%. The student/faculty ratio has mean 11.7 and standard deviation 4.3. The correlation is −0.5.

a. Find the equation of the least squares line for predicting graduation rate from student/faculty ratio.

b. Find the equation of the least squares line for predicting student/faculty ratio from graduation rate.

E39. These questions concern the relationship between the correlation, r, and the slope, b_1, of the regression line.

a. If y is more variable than x, will the slope of the least squares line be greater (in absolute value) than the correlation? Justify your answer.

b. For a list of pairs (x, y), $r = 0.8$, $b_1 = 1.6$, and the standard deviation of x and y are 25 and 50. (Not necessarily in that order.) Which is the standard deviation for x? Justify your answer.

c. Students in a statistics class estimated and then measured their head circumferences in inches. The actual circumferences had standard deviation 0.93, and the estimates had standard deviation 4.12. The equation of the least squares line for predicting estimated values from actual values was $\hat{y} = 11.97 + 0.36x$. What was the correlation?

d. What would be the slope of the least squares line for predicting actual head circumferences from the estimated values?

E40. *Lost final exam.* After teaching the same history course for about a hundred years, an instructor has found that the correlation, r, between the students' total number of points before the final examination and the number of points scored on their final examination is 0.8. The pre-final-exam point totals for all students in this year's course have mean 280 and standard deviation 30. The final exam points have mean 75 and standard deviation 8. The instructor's dog ate Julie's final exam, but the instructor knows her total number of points before the exam was 300. He decides to predict her final exam score from her pre-final-exam total. What value will he get?

E41. *Lurking variables.* For each scenario, state a careless conclusion assuming cause and effect, and then identify a possible lurking variable.

    a. For a large sample of different animal species, there is a strong positive correlation between average brain weight and average life span.

    b. Over the last 30 years, there has been a strong positive correlation between the average price of a cheeseburger and the average tuition at private liberal arts colleges.

    c. Over the last decade, there has been a strong positive correlation between the price of an average share of stock, as measured by the S&P 500, and the number of web sites on the Internet.

E42. Manufacturers of low-fat foods often increase the salt content in order to keep the flavor acceptable to consumers. For a sample of different kinds and brands of cheeses, *Consumer Reports* measured several variables, including calorie content, fat content, saturated fat content, and sodium content. Using these four variables, you can form six pairs of variables, so there are six different correlations. These correlations turned out to be either about 0.95 or about −0.5.

    a. List all six pairs of variables, and for each pair decide from the context whether the correlation is close to 0.95 or to −0.5.

    b. State a careless conclusion based on taking the negative correlation as evidence of cause and effect.

    c. Explain the negative correlation using the idea of a lurking variable.

E43. A study to determine how ice cream consumption depends on the outside temperature gave the results shown in Display 3.71.

    a. Use the values of SST and SSE in the regression analysis to compute r, the correlation for the relationship

Mean Temperature (°F)	Pints per Person
56	0.386
63	0.374
68	0.393
69	0.425
65	0.406
61	0.344
47	0.327
32	0.288
24	0.269
28	0.256
26	0.286
32	0.298
40	0.329
55	0.318
63	0.381
72	0.381
72	0.470
67	0.443
60	0.386
44	0.342
40	0.319
32	0.307
27	0.284
28	0.326
33	0.309
41	0.359
52	0.376
64	0.416
71	0.437

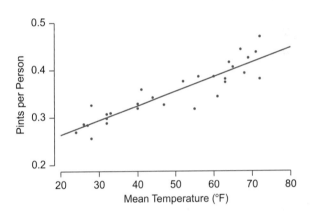

The regression equation is
pts/pers = 0.202 + 0.00306 Temp F

Predictor	Coef	Stdev	t-Ratio	P
Constant	0.20200	0.01452	13.91	0.000
Temp F	0.0030567	0.0002791	10.95	0.000

S = 0.02457   R-sq = 81.6%   R-sq(adj) = 80.9%

Analysis of variance

Source	DF	SS	MS	F	P
Regression	1	0.072436	0.072436	119.96	0.000
Error	27	0.016304	0.000604		
Total	28	0.088740			

Display 3.71 Data table, scatterplot, and regression analysis for the effects of outside temperature on ice cream consumption. [Source: Koteswara Rao Kadiyala, "Testing for the Independence of Regression Disturbances," *Econometrica*, Vol. 38 (1970), pp. 97–117.]

between the temperature in degrees Fahrenheit and the number of pints of ice cream consumed per person. Check your answer against R-sq in the analysis.

b. Compute the value of the residual that is largest in absolute value.

c. Is there a cause-and-effect relationship between the two variables?

d. What are the units for each of x, y, b_1, and r?

e. The letters MS stand for "mean square." How do you think the MS is computed?

E44. *Air quality.* In efforts to improve air quality, savings in cost and time might be possible if scientists knew more about how various contributors to poor air quality worked together. Two of the major components of poor air are carbon monoxide (CO) and ozone (O_3). The data of Display 3.72 show the maximum concentrations for these two compounds for a recent year in 15 major cities from across the United States. The carbon monoxide concentration is measured in parts per million (ppm) and ozone is also measured in parts per million. The display

also provides a typical computer analysis for these data, with O_3 as the explanatory variable.

a. Find the values of SST and SSE in the summary tables and use them to compute r, the correlation between CO and O_3 concentrations. Check your answer against the r^2 given in the analysis.

b. Compute the value of the residual that is largest in absolute value.

c. Do you think there is a cause-and-effect relationship between these two variables, or is it more plausible that other variables are causing these two variables to rise together? Explain your reasoning.

d. What are the units for each of x, y, b_1, and r?

e. How do you think the mean square for error is calculated? What does this measure?

f. Do you see anything in the pattern of these data to suggest that simple linear regression may not be the best way to conduct the analysis? Explain.

E45. The scatterplot in Display 3.73 shows part of the aircraft data of Display 3.16 on page 104. For these data, $r^2 = 0.83$. Should r^2 be used as a statistical measure for

City	CO	O_3
Atlanta	4.1	0.023
Boston	3.4	0.029
Chicago	8.3	0.032
Dallas	3.5	0.014
Denver	2.1	0.016
Detroit	4.4	0.024
Houston	4.2	0.021
Kansas City	1.8	0.017
Los Angeles	9.5	0.044
New York	9.3	0.038
Philadelphia	5.1	0.028
Pittsburgh	2.4	0.025
San Francisco	1.7	0.020
Seattle	2.4	0.021
Washington, D.C.	4.9	0.023

Linear Fit

CO = −2.452415 + 277.02991.03

Summary of Fit

RSquare	0.7548
RSquare Adj	0.735939
Root Mean Square Error	1.33979
Mean of Response	4.473333
Observations (or Sum Wgts)	15

Analysis of Variance

Source	DF	Sum of Squares	Mean Square	F Ratio
Model	1	71.833857	71.8339	40.0180
Error	13	23.335476	1.7950	Prob > F
C. Total	14	95.169333		<0.0001

Parameter Estimates

Term	Estimate	Std Error	t Ratio	Prob > Itl
Intercept	−2.452415	1.148163	−2.14	0.0523
O_3	277.02991	43.7924	6.33	<0.0001

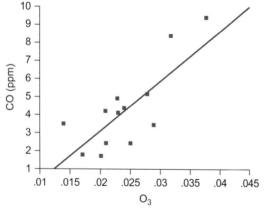

Display 3.72 Maximum ozone and carbon monoxide concentration for selected cities. [Source: *National Air Quality and Emissions Trends Report,* 2003.]

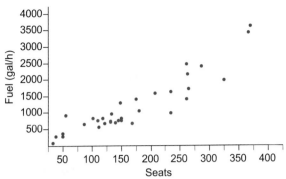

Display 3.73 Scatterplot of *number of seats* versus *fuel consumption* (gal/h) for passenger aircraft.

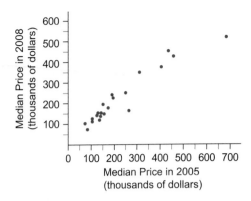

Display 3.74 Median sales prices of houses for sample metropolitan areas ($1000). [Source: National Association of Realtors.]

these data? If so, interpret this value of r^2 in the context of the data. If not, explain why not.

E46. *Housing slump.* How did housing prices change during a recessionary economic period? Display 3.74 provides the median sales price of existing single-family houses for a sample of metropolitan areas around the country for the years 2005 and 2008. Describe the pattern of the association in the scatterplot. Should $r^2 = 0.959$ be used as a statistical measure of the strength of the linear relationship between these two variables? If so, interpret its value in the context of these data. If not, explain why not.

E47. Suppose an instructor always praises students who score exceptionally well on a test and always scolds students who score exceptionally poorly. Use the

notion of regression toward the mean to explain why the results will tend to suggest the false conclusion that scolding leads to improvement whereas praise leads to slacking off.

E48. A few years ago, a school in New Jersey tested all its 4th graders to select students for a program for the gifted. Two years later, the students were retested, and the school was shocked to find that the scores of the gifted students had dropped, whereas the scores of the other students had remained, on average, the same. What is a likely explanation for this disappointing development?

3.4 ▶ Diagnostics: Looking for Features That the Summaries Miss

As you learned in Chapter 2, summaries simplify. They are useful because they omit detail in order to emphasize a few general features. This quality also makes summaries potentially misleading, because sometimes the detail that is ignored has an important message to convey. Knowing just the mean and the standard deviation of a distribution doesn't tell you if there are any outliers or whether the distribution is skewed. The same is true of the regression line and the correlation.

This section is about "diagnostics"—tools for looking beyond the summaries to see how well they describe the data and what features they leave out. The first part of this section deals with individual cases that stand apart from the overall pattern and with how these cases influence the regression line and the correlation. The second part shows you how to identify systematic patterns that involve many or all of the cases—the "shape" of the scatterplot.

Which Points Have the Influence?

Not all data points are created equal. You saw in the calculation of the correlation in Display 3.52 on page 130 that some points make large contributions and some small. Some make positive contributions and some negative. Your goal is to learn to recognize the points in a data set that might have an unusually large influence on where the regression line goes or on the size and sign of the correlation.

Just as among people, some data points have more influence than others.

STATISTICS IN ACTION 3.3 ▶ Near and Far

What You'll Need: an open area in which to step off distances

In this activity, you compare the actual distance to an object with what the distance appears to be.

1. Go to an open area, such as the hall or lawn of your campus, and pick a spot as your origin. Choose six objects at various distances from the origin. Five of the objects should be within 10 to 20 paces, and the other should be a long way away (at least 100 paces).

2. For each of the six objects, estimate the number of paces from the origin to the object. Record your estimates.

3. From your origin, walk to each of your objects and count the actual number of regular paces it takes you to get there. Record this number beside your estimate.

4. Plot your data on a scatterplot, with your estimated value on the *x*-axis and the actual value on the *y*-axis. Does the plot show a linear trend?

5. Determine the equation of the regression line, and calculate the correlation.

6. Delete the point for the object that is farthest away from the origin. Determine the equation of the regression line and calculate the correlation for the reduced data set.

7. Did the extreme point have any influence on the regression line? On the correlation? Explain.

In Chapter 2, you learned about outliers for distributions—values that are separated from the bulk of the data. Outliers are atypical cases, and they can exert more than their share of influence on the mean and standard deviation. For scatterplots, as you will soon see, working with two variables together means that there can be outliers of various kinds. Different kinds of outliers can have different types of influence on the least squares line and the correlation. Unfortunately, there is no rule you can use to identify outliers in bivariate data. Just look for points surrounded by white space.

Judging a Point's Influence

Points separated from the bulk of the data may be outliers and are potentially influential. To judge a point's influence, compare the regression equation and correlation computed first with and then without the point in question.

To see these ideas in action, turn to the data on mammal longevity in Display 2.14 on page 32 and think about how to summarize the relationship between maximum and average longevity.

Influential Mammals

Example 3.12

The average elephant lives 35 years. The oldest elephant on record lived 70 years. The average hippo lives 41 years—longer than the average elephant—but the record-holding hippo lived only 54 years. The oldest-known beaver lived 50 years, almost as long as the champion hippo, but the average beaver cashes in his wood chips after only 5 short years of making them. Other mammals, however, are more predictable. If you look at the entire sample, shown in Display 3.75, it turns out that the elephant (E), hippo (H), and beaver (B) are the oddballs of the bunch. For the rest, there's an almost linear relationship between average longevity and maximum longevity. The least squares line for the entire sample has the equation

$$\hat{M} = 10.53 + 1.58A$$

where $\hat{M}$, or "M-hat," stands for predicted maximum longevity and A stands for observed average longevity. For every increase of 1 year in average longevity, the model predicts a 1.58-year increase in maximum longevity. The correlation for the relationship between these two variables is 0.77. How much influence do the oddballs have on these summaries?

Points separated from the bulk of the data may have strong influence.

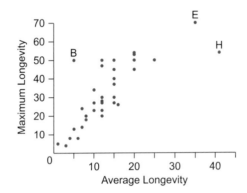

Display 3.75   *Maximum longevity* versus *average longevity* for selected mammals.

Solution
The hippo has the effect of pulling the right end of the regression line downward (like putting a heavy weight on one end of a seesaw), as you can see in Display 3.76. When the hippo is removed, that end of the regression line will "spring upward" and the slope will increase. Because one large residual has been removed and many of the remaining residuals have been reduced in size, the correlation will increase. The new slope is 1.96, and the new correlation is 0.80. The hippo has considerable influence on the slope and some influence on the correlation.

Now envision the scatterplot with just the elephant, E, missing. Because E is close to the straight line fit to the data, it produces a small residual. Thus, you would expect that removing E should not change the slope of the regression line much (not nearly as much as removing H did) and should reduce the correlation just a bit. In fact as the regression analysis on the next page shows, the correlation does decrease some, to 0.72 from 0.77. However, the new slope is 1.53. It turns out that removing the elephant gives the hippo even more influence, and the slope decreases.

Finally, envision the scatterplot with just the beaver, B, removed. B produces a large, positive residual close to the left end of the regression line. Thus, removing B should allow the left end of the line to drop, increasing the slope, and removing a large residual should increase the correlation. As the regression analysis on the next page shows, the new slope is 1.69 (an increase from 1.58), and the new correlation is 0.83 (an increase from 0.77). The beaver also has considerable influence on both slope and correlation.

Regression analysis with the beaver removed.

LinReg
y=ax+b
a=1.691644344
b=8.143991729
r²=.6862141995
r=.8283804678

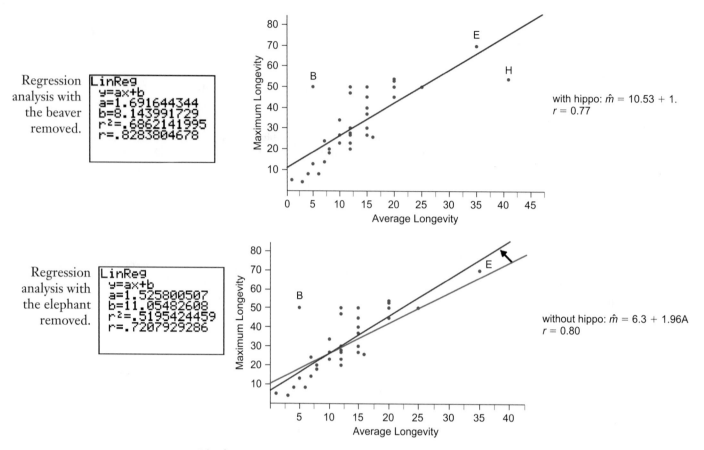

with hippo: $\hat{m} = 10.53 + 1.$
$r = 0.77$

Regression analysis with the elephant removed.

LinReg
y=ax+b
a=1.525800507
b=11.05482608
r²=.5195424459
r=.7207929286

without hippo: $\hat{m} = 6.3 + 1.96A$
$r = 0.80$

Display 3.76 Regression lines for *maximum longevity* versus *average longevity*, with (top) and without (bottom) the hippo.

With a little practice, you often can anticipate the influence of certain points in a scatterplot, as in the previous example, but it is difficult to state general rules. The best rule is the one given in the box on page 150: Fit the line with and without the questionable point and see what happens. Then report all the results, with appropriate explanations.

DISCUSSION
Why the Anscombe Data Sets Are Important

Display 3.77 shows four scatterplots. These plots, known as "the Anscombe data" after their inventor, are arguably the most famous set of scatterplots in all of statistics. The questions that follow invite you to figure out why statistics books refer to them so often. In the process, you'll learn more about what a summary doesn't tell you about a data set.

D26. For each plot in Display 3.77, first give a short verbal description of the pattern in the plot. Then

  a. either fit a line by eye and estimate its slope or tell why you think a line is not a good summary

  b. either estimate the correlation by eye or tell why you think a correlation is not an appropriate summary

D27. Display 3.78 shows computer output for one of the four Anscombe data set plots. Can you tell which one? If so, tell how you know. If not, explain why you can't tell.

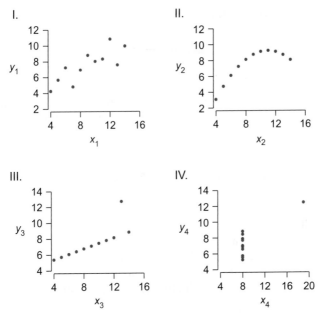

Display 3.77 Four regression data sets invented by Francis J. Anscombe. [Source: Francis J. Anscombe "Graphs in Statistical Analysis" *American Statistician*, Vol. 27 (1973), pp. 17–21.]

$R^2 = 66.6\%$ R^2 (adj) = 62.9%

S = 1.297 with $11 - 2 = 9$ degrees of freedom

Source	Sum of Squares	DF	Mean Square	F-Ratio
Regression	27.5000	1	27.5000	18.0
Residual	13.7763	9	1.53070	

Variable	Coefficient	SE of Coefficient	t-Ratio	Prob.
Constant	3.00091	1.125	2.67	0.0258
x	0.500000	0.1180	4.24	0.0022

Display 3.78 Regression analysis for one of the Anscombe data sets.

D28. Display 3.79 lists the ordered pairs for the Anscombe plots.

Data Set I		Data Set II		Data Set III		Data Set IV	
x_1	y_1	x_2	y_2	x_3	y_3	x_4	y_4
10	8.04	10	9.14	10	7.46	8	6.58
8	6.95	8	8.14	8	6.77	8	5.76
13	7.58	13	8.74	13	12.74	8	7.71
9	8.81	9	8.77	9	7.11	8	8.84
11	8.33	11	9.26	11	7.81	8	8.47
14	9.96	14	8.10	14	8.84	8	7.04
6	7.24	6	6.13	6	6.08	8	5.25
4	4.26	4	3.10	4	5.39	19	12.50
12	10.84	12	9.13	12	8.15	8	5.56
7	4.82	7	7.26	7	6.42	8	7.91
5	5.68	5	4.74	5	5.73	8	6.89

Display 3.79 Anscombe plot data values.

a. Which plot has a point that is highly influential both with respect to the slope of the regression line and with respect to the correlation?

b. Compared to the other points in the plot, does the influential point lie far from the least squares line or close to it?

c. How would the slope and correlation change if you were to remove this point? Discuss this first without actually performing the calculations. Then carry out the calculations to verify your conjectures.

Residual Plots: Putting Your Data Under a Microscope

As you can see from the Anscombe plots, there are many features of the shape of a scatterplot that you can't learn from the standard set of summary numbers. Only when the cloud of points is elliptical, as in Display 3.51 on page 152, does the least squares line, together with the correlation, give a good summary of the relationship described by the plot. If the cloud of points isn't elliptical, these summaries aren't appropriate. How can you decide?

Residual plots may uncover more detailed patterns.

A special kind of a scatterplot, called a residual plot, often can help you see more clearly what's going on. For some data sets, a residual plot can even show you patterns you might otherwise have overlooked completely. Statisticians use residual plots the way a doctor uses a microscope or an X-ray—to get a better look at less obvious aspects of a situation. (Plots you use in this way are called "diagnostic plots" because of the parallel with medical diagnosis.) Push the analogy just a little. You're the doctor, and data sets are your patients. Sets with elliptical clouds of points are the "healthy" ones. They don't need special attention.

Residual Plot

A **residual plot** is a scatterplot of residuals, $y - \hat{y}$, versus predictor values, x (or, sometimes, versus predicted values, $\hat{y}$).

Example 3.13	**Constructing a Residual Plot**

Return to Example 3.9 on page 130. Fit a regression line, calculate the residuals, and make a residual plot.

Solution
The graph of the regression equation, $\hat{y} = 3.38 - 0.026x$, is shown on the plot in Display 3.80. Visualize each residual—the difference between the observed value of y and the predicted value, $\hat{y}$—as a vertical segment on the scatterplot.

The calculations of the residuals are shown in Display 3.81. The point for Student E lies far above the line and so produces the largest positive residual. The point for Student C lies farthest below the line and so produces the negative residual that is largest in absolute value.

The residual plot, Display 3.82, is simply a scatterplot of the residuals versus the original predictor variable, *TV watching*. Note that 0 is at the middle of the residuals on the vertical scale.

The residual plot shows nearly random scatter, with no obvious trends. This is the ideal shape for a residual plot because it indicates that a straight line is a reasonable model for the trend in the original data.

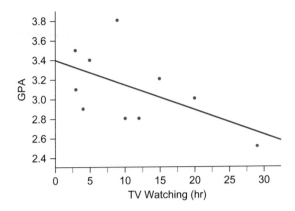

Display 3.80 Visualize the residuals.

Student	TV Watching (h), x	GPA, y	Predicted Value, $\hat{y}$	Residual, $y-\hat{y}$
A	3	3.5	3.30	0.20
B	3	3.1	3.30	−0.20
C	4	2.9	3.28	−0.38
D	5	3.4	3.25	0.15
E	9	3.8	3.15	0.65
F	10	2.8	3.13	−0.33
G	12	2.8	3.07	−0.27
H	15	3.2	3.00	0.20
I	20	3.0	2.87	0.13
J	29	2.5	2.64	−0.14

Display 3.81 Table showing residual calculations.

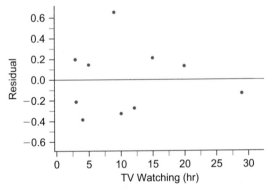

Display 3.82 Residual plot.

A careful data analyst always looks at a residual plot.

What to Look For in a Residual Plot

If the original cloud of points is elliptical, so that a line is an appropriate summary, the residual plot will look like a random scatter of points.

Use residual plots to check for systematic departures from constant slope (linear trend) and constant strength (same vertical spread). Look in particular for plots that are curved or fan-shaped. It's true that for data sets with only one predictor value (like those

Residual plots sometimes yield surprises.

in this chapter), you often can get a good idea of what the residual plot will look like by carefully inspecting the original scatterplot. Once in a while, however, you get a surprise.

| **Example 3.14** | **Interpreting a Residual Plot** |

E21 on page 123 introduced data on median height versus age for young girls. Display 3.83 shows the scatterplot of these data, with the regression line. The overall average growth rate for the 12-year period is the slope of the regression line.

The plot looks nearly linear, but is a line a suitable model?

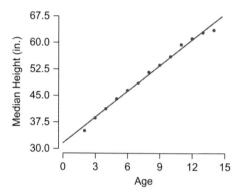

Display 3.83 *Median height* versus *age* for young girls.

Solution
The residual plot, shown in Display 3.84, quite dramatically reveals that the trend is not as linear as first imagined. The curvature in the residual plot mimics the curvature in the original scatterplot, which is harder to see. A line is not a good model for these data.

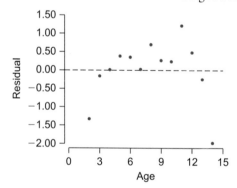

Display 3.84 Residual plot of *median height* versus *age* for young girls.

Residuals sometimes are plotted against the predicted values, $\hat{y}$.

Statistical software often plots residuals against the predicted values, $\hat{y}$, rather than against the predictor values, x. For simple linear regression, both plots have exactly the same shape as long as the slope of the regression line is positive. Plotting the residual against $\hat{y}$, turns out to be more helpful in multiple regression, when there is more than one predictor variable

DISCUSSION
Residual Plots

D29. In Display 3.82, identify which residual belongs to Student A and which to Student J.

D30. To see how residual plots magnify departures from the regression line, compare the four Anscombe plots in Display 3.77 on page 151 with the four corresponding residual plots in scrambled order in Display 3.85 below.

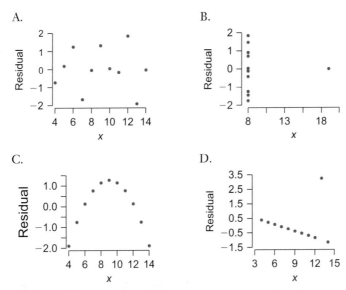

Display 3.85 Residual plots for the four Anscombe data sets.

a. Match each of the original scatterplots in Display 3.77 with its corresponding residual plot in Display 3.85.

b. Describe the overall difference between the original scatterplots and the residual plot. What do the scatterplots show that the residual plots don't? What do the residual plots show that the scatterplots don't?

D31. Statistical software typically makes residual plots by plotting residuals against the predicted values of y rather than against the values of x. For example, the two types of residual plots for the television watching and GPA example follow. What are the similarities and differences in the plots? Why does this occur? Does it matter when you are interpreting a residual plot whether the values of the predictor variable, x, or the predicted values, $\hat{y}$, are plotted on the horizontal axis?

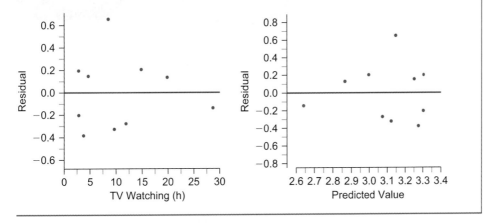

Summary 3.4: Diagnostics: Looking for Features That the Summaries Miss

For the simplest clouds of data points—elliptical in shape, with linear trend and no outliers—you can summarize all the main features of a scatterplot with just a few numbers, mainly the slope of the fitted line, y-intercept, and correlation. Not all plots are this simple, however, and a good statistician always does diagnostic checks for outliers and influential points and for departures from constant slope or constant strength.

- Points separated from the bulk of the data may be outliers and potentially influential.
- To judge a point's influence, fit a line to the data and compute a regression equation and a correlation first with and then without the point in question. If the change in the regression equation and correlation is meaningful in your situation, report both sets of summary statistics.

For some data sets, a residual plot can show patterns you might otherwise overlook. A residual plot is a scatterplot of residuals, $y - \hat{y}$, versus predictor values, x. A residual plot also can be constructed as a scatterplot of residuals, $y - \hat{y}$, versus fitted values, $\hat{y}$. Use residual plots to check for systematic departures from linearity and for constant variability in y across the values of x. If the data aren't linear, the residual plot doesn't look random. If the data have nonconstant variability, the residual plot is fan-shaped.

Practice

Which Points Have the Influence?

P24. The data in Display 3.86 show some interesting patterns in the relationship between domestic and international gross income from the ten movies with the highest domestic gross ticket sales.

    a. Construct a scatterplot suitable for predicting international sales from domestic sales. Describe the pattern in the data.

    b. Find the least squares line and the correlation for these data.

    c. Remove the most influential data point and recalculate the least squares line and correlation. Describe the influence of the removed point.

P25. A student estimated the number of paces to six objects and then counted the actual number of paces. The student's results are shown in Display 3.87.

    a. How well did the student do in estimating the number of paces?

    b. Which point appears to be most influential?

    c. Calculate the slope of the regression line and the correlation with and without this point. Describe the influence of this point.

Movie	Domestic ($ million)	International ($ million)
Titanic	601	1235
The Dark Knight	533	464
Star Wars	461	337
Shrek 2	436	444
E.T.: The Extra-Terrestrial	435	322
Star Wars: Episode I— The Phantom Menace	431	491
Pirates of the Caribbean: Dead Man's Chest	423	637
Spider-Man	404	418
Star Wars: Episode III— Revenge of the Sith	380	468
The Lord of the Rings: The Return of the King	377	752

Display 3.86 Ticket sales for the ten highest-grossing domestic (United States) movies of all time. [Source: Internet Movie Database, *www.imdb.com*, February 18, 2009.]

Estimate	Actual
3	3
21	34
25	48
28	63
40	146
180	350

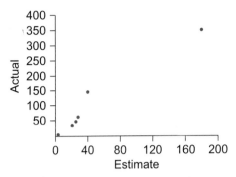

Display 3.87 Actual number of paces versus a student's estimate for six distances.

Residual Plots

P26. For the set of (x, y) pairs $(0, 0)$, $(0, 1)$, $(1, 1)$, and $(3, 2)$, the equation of the least squares line is $\hat{y} = 0.5 + 0.5x$.

a. Plot the data and graph the least squares line.

b. Next complete a table for the predicted values and residuals, like the table in Display 3.81 on page 153.

c. Using the values in your table, plot residuals versus predictor, x.

d. How does the residual plot differ from the scatterplot?

P27. How has the population density (people per square mile) of the United States changed over the years? Display 3.88 provides the densities for each census year from 1850 to 2000, along with the scatterplot and residual plot.

Year	Density	Year	Density
1850	7.9	1930	41.2
1860	10.6	1940	44.2
1870	13.4	1950	50.7
1880	16.9	1960	50.6
1890	21.2	1970	57.4
1900	25.6	1980	64.0
1910	31.0	1990	70.3
1920	35.6	2000	79.6

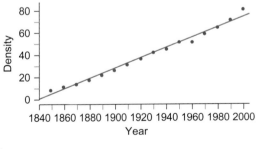

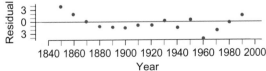

Display 3.88 Population density of the United States since 1850.

a. Looking at the scatterplot alone, describe the general pattern of population density over the years.

b. Describe the pattern in the deviations from this general pattern that is revealed by the residual plot.

P28. Display 3.89 shows four scatterplots (A–D) for the data from a sample commercial aircraft. Display 3.90 shows four corresponding residual plots (I–IV).

a. Match the residual plots to the scatterplots.

b. Using scatterplots A–D as examples, describe how you can identify each of these in a scatterplot from the residual plot.

   i. a curve with increasing slope

   ii. unequal variation in the responses

   iii. a curve with decreasing slope

   iv. two linear patterns with different slopes

c. For one of the plots, two line segments joined together seem to give a better fit than either a single line or a curve. Which plot is this? Is this pattern easier to see in the original scatterplot or in the residual plot?

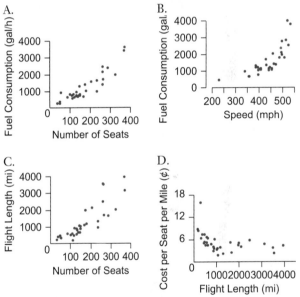

Display 3.89 Four scatterplots for the sample of commercial aircraft.

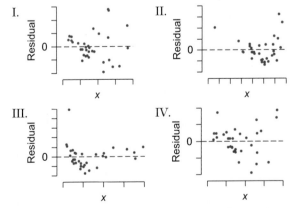

Display 3.90 Four residual plots corresponding to the scatterplots in Display 3.89.

Exercises

E49. *Extreme temperatures.* The data in Display 3.91 are the maximum and minimum temperatures ever recorded on each continent.

Continent	Maximum Temperature (°F)	Minimum Temperature (°F)
Africa	136	−11
Antarctica	59	−129
Asia	129	−90
Australia	128	−9
Europe	122	−67
North America	134	−81
Oceania	108	12
South America	120	−27

Display 3.91 Maximum and minimum recorded temperatures for the continents. [Source: National Climatic Data Center, 2005, *www.ncdc.noaa.gov.*]

a. Construct a scatterplot of the data suitable for predicting the minimum temperature from a given maximum temperature. Is a straight line a good model for these points? Explain.

b. Fit a least squares line to the points and calculate the correlation, even if you thought in part a that a straight line was not a good model.

c. Explain, in words and numbers, what influence Antarctica has on the slope of the regression line and on the correlation. How could a summary of these data be misleading if it were not accompanied by a plot?

E50. The data and plot in Display 3.92 are from E15 on page 122. They show the arsenic concentrations in the toenails of 21 people who used water from their private wells. Both measurements are in parts per million.

a. Which point do you think has the most influence on the slope and correlation? What would be the effect of removing this point? Perform the calculations to see if your intuition is correct.

b. Find a point that you think has almost no influence on the slope and correlation. Perform the calculations to see if your intuition is correct.

c. Find a point whose removal you think would make the correlation increase. Perform the calculations to see if your intuition is correct.

E51. *How effective is a disinfectant?* The data in Display 3.93 show (coded) bacteria colony counts on skin samples before and after a disinfectant is applied.

a. Plot the data, fit a regression line to them, and complete a copy of the table, filling in the predicted values and residuals.

b. Plot the residuals versus *x*, the count before the treatment. Comment on the pattern.

Arsenic in Water (ppm)	Arsenic in Toenails (ppm)
0.00087	0.119
0.00021	0.118
0	0.099
0.00115	0.118
0	0.277
0	0.358
0.00013	0.08
0.00069	0.158
0.00039	0.31
0	0.105
0	0.073
0.046	0.832
0.0194	0.517
0.137	2.252
0.0214	0.851
0.0175	0.269
0.0764	0.433
0	0.141
0.0165	0.275
0.00012	0.135
0.0041	0.175

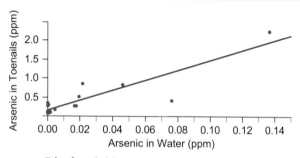

Display 3.92 Arsenic concentrations.

x	y	Predicted	Residual
11	6	—?—	—?—
8	0	—?—	—?—
5	2	—?—	—?—
14	8	—?—	—?—
19	11	—?—	—?—
6	4	—?—	—?—
10	13	—?—	—?—
6	1	—?—	—?—
11	8	—?—	—?—
3	0	—?—	—?—

Display 3.93 Coded bacteria colony counts before (*x*) and after (*y*) treatment. [Source: Snedecor and Cochran, *Statistical Methods* (Iowa State University Press, 1967), p. 422.]

c. Use the residual plot to determine for which skin sample the disinfectant was unusually effective and for which skin sample it was not very effective.

E52. *Textbook prices.* Display 3.94 compares prices at a college bookstore to those of a large online bookstore.

a. The equation of the regression line is $\widehat{online} = -3.57 + 1.03\ college$. Interpret this equation in terms of textbook prices.

Type of Text book	College Bookstore Price ($)	Online Bookstore Price ($)
Chemistry	93.40	94.18
Classic Fiction	9.95	7.96
English Anthology	46.70	48.75
Calculus	76.00	94.15
Biology	86.70	80.95
Statistics	7.95	6.36
Dictionary	24.00	16.80
Style Manual	12.70	10.66
Art History	66.00	45.50

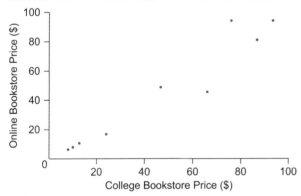

Display 3.94 Prices for a sample of textbooks at a college bookstore and an online bookstore.

b. Construct a residual plot. Interpret it and point out any interesting features.

c. In comparing the prices of the textbooks, you might be more interested in a different line: $y = x$. Draw this line on a copy of the scatterplot in Display 3.94. What does it mean if a point lies above this line? Below it? On it?

d. A boxplot of the differences *college price—online price* is shown in Display 3.95. Interpret this boxplot.

E53. *Pizzas, again.* Display 3.96 shows the pizza data from E12 on page 121, with its regression line.

a. Estimate the residuals from the graph, and use your estimates to sketch a rough version of a residual plot for this data set.

b. Which pizza has the largest positive residual? The largest negative residual? Are any of the residuals so extreme as to suggest that those pizzas should be regarded as exceptions?

Display 3.95 A boxplot of the differences between the college price and the online price for various textbooks.

c. Are any of the pizzas highly influential? If so, specify which one(s), and describe the effect on the slope of the fitted line and the correlation of removing the influential point or points from the analysis.

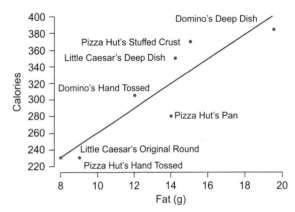

Display 3.96 *Calories* versus *fat*, per 5-oz serving, for seven kinds of pizza.

E54. *Aircraft.* Look again at Display 3.89 on page 157, which shows a scatterplot of *flight length* versus *number of seats*.

a. Does the slope of the pattern increase, decrease, or stay roughly constant as you move from left to right across the plot?

b. Focusing on the variation (spread) in flight length, *y*, for planes with roughly the same seating capacity, compare the spreads for planes with few seats, a moderate number of seats, and a large number of seats. As you move from left to right across the plot, how does the spread change, if at all?

c. Suppose a friend chose a plane from the sample at random and told you the approximate number of seats. Could you guess its flight length to within 500 mi if the number of seats was between 50 and 150? If it was between 200 and 300? Explain.

d. What is the relationship between your answer in part b and residual plot I in Display 3.90?

E55. Match each scatterplot (A–D) in Display 3.97 with its residual plot (I–IV) in Display 3.98. For which plots is a linear regression appropriate?

E56. Can either of the plots in Display 3.99 be a residual plot? Explain your reasoning.

E57. Display 3.100 gives the data set for three passenger jets 160, along with a scatterplot showing the least squares line. (Values have been rounded.)

a. Use the equation of the line to find predicted values and residuals to complete the table in Display 3.100.

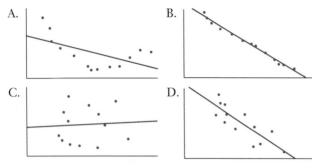

Display 3.97 Four scatterplots.

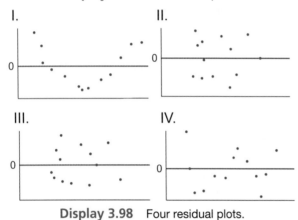

Display 3.98 Four residual plots.

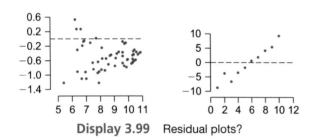

Display 3.99 Residual plots?

Aircraft	Seats	Cost	Predicted	Residual
ERJ-145	50	1100	—?—	—?—
DC-9	100	2100	—?—	—?—
MD-90	150	2700	—?—	—?—

Display 3.100 *Cost per hour* versus *number of seats* for three models of the passenger aircraft.

b. Use your numbers from part a to construct two residual plots, one with the predictor, x, on the horizontal axis and the other with the predicted value, $\hat{y}$, on the horizontal axis. How do the two plots differ?

E.58 Explain why a residual plot of (x, *residual*) and a plot of (*predicted value*, *residual*) have exactly the same shape if the slope of the regression line is positive. What changes if the slope is negative?

E59. *Can you recapture the scatterplot from the residual plot?* The residual plot in Display 3.101 was calculated from data showing the recommended weight (in pounds) for men at various heights over 64 in. The fitted weights ranged from 145 lb to 187 lb. Make a rough sketch of the scatterplot of these data.

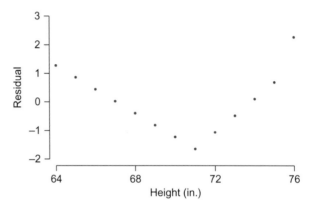

Display 3.101 Residuals of *recommended weight* versus *height* for men.

E60. The plot in Display 3.102 shows the residuals resulting from fitting a line to the data for female life expectancy (*life exp*) versus gross national product (*GNP*, in thousands of dollars per capita) for a sample of countries from around the world. The regression equation for the sample data was

$$\widehat{life\ exp} = 67.00 + 0.63\ GNP$$

Sketch the scatterplot of *life exp* versus *GNP*.

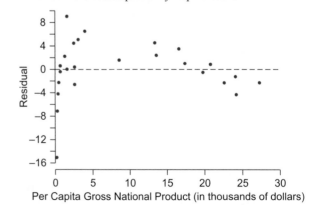

Display 3.102 Residuals of *female life expectancy* versus *gross national product*.

Chapter Summary

In Chapter 2, you worked with univariate data. In Chapter 3, you learned how to uncover information for bivariate (two-variable) data, using plots and numerical summaries of center and spread. In Chapter 2, the basic plot was the histogram. For histograms of distributions that are approximately normal, the fundamental measures of center and spread are the mean and the standard deviation. For bivariate data, the basic plot is the scatterplot. For scatterplots that have an elliptical shape, the fundamental summary measures are the regression line (which you can think of as the measure of center) and the correlation (which you can think of as the measure of spread).

For now, correlation and regression merely describe your data set. In Chapter 13, you will learn to use numerical summaries computed from a sample to make inferences about a larger population. Using diagnostic tools such as residual plots will come in especially handy because you won't be able to make valid inferences unless the points form an elliptical cloud.

Review Exercises

E61. *Leonardo's rules.* A class of 15 students recorded the measurements in Display 3.103.

Student	Height	Arm Span	Kneeling Height	Hand Length
1	170.5	168.0	126.0	18.0
2	170.0	172.0	129.5	18.0
3	107.0	101.0	79.5	10.0
4	159.0	161.0	116.0	16.0
5	166.0	166.0	122.0	18.0
6	175.0	174.0	125.0	19.5
7	158.0	153.5	116.0	16.0
8	95.5	95.0	71.5	10.0
9	132.5	129.0	95.0	11.5
10	165.0	169.0	124.0	17.0
11	179.0	175.0	131.0	20.0
12	149.0	154.0	109.5	15.5
13	143.0	142.0	111.5	16.0
14	158.0	156.5	119.0	17.5
15	161.0	164.0	121.0	16.5

Display 3.103 Measurements, in centimeters.

a. You may have heard the following three rules for artists who want to make body proportions look realistic.
  - Height is equal to the span of the outstretched arms.
  - Kneeling height is three-fourths of the standing height.
  - The length of the hand is one-ninth of the height.

  Using the data in Display 3.103, construct scatterplots and regression lines to help you evaluate these rules. Do the rules appear to hold?

b. Interpret the slopes of your regression lines.

c. If appropriate, find the value of *r* for each of the three relationships. Which correlation is strongest? Which is weakest?

E62. *Space shuttle* Challenger. On January 28, 1986, because two O-rings did not seal properly, space shuttle *Challenger* exploded and seven people died. The temperature predicted for the morning of the flight was between 26°F and 29°F. The engineers were concerned that the cold temperatures would cause the rubber O-rings to malfunction. On seven previous flights at least one of the 12 O-rings had shown some distress. The NASA officials and engineers who decided not to delay the flight had available to them data like those on the scatterplot in Display 3.104 before they made that decision.

Crew members on the space shuttle *Challenger*.

a. Why did it seem reasonable to launch despite the low temperature?

b. Display 3.104 contains information only about flights that had O-ring failures. Data for all flights were available on a table like the one in Display 3.105. Add the missing points to a copy of the scatterplot in Display 3.104. How do these data

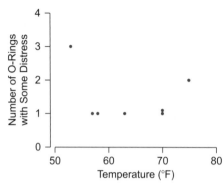

Display 3.104 Flights when at least one O-ring showed some distress.

affect any trend in the scatterplot? Would you have recommended launching the space shuttle if you had seen the complete plot? Why or why not?

Temperature (°F)	Number of O-Rings with Some Distress	Temperature (°F)	Number of O-Rings with Some Distress
53	3	69	0
57	1	70	0
58	1	70	0
63	1	72	0
70	1	73	0
70	1	75	0
75	2	76	0
66	0	76	0
67	0	78	0
67	0	79	0
67	0	80	0
68	0	81	0

Display 3.105 Space shuttle flights and O-ring distress. [Source: Siddhartha R. Dalal et al., "Risk Analysis of the Space Shuttle: Pre-Challenger Prediction of Failure," *Journal of the American Statistical Association*, Vol. 84 (1989), pp. 945–947.]

E63. *Exam scores.* Students' scores on two exams in a statistics course are given in Display 3.106 along with a scatterplot with regression line and a residual plot. The regression equation is $exam\ 2 = 51.0 + 0.430\ (exam\ 1)$, and the correlation, r, is 0.756.

a. Is there a point that is more influential than the other points on the slope of the regression line? How can you tell from the scatterplot? From the residual plot?

b. How will the slope change if the scores for this one influential point are removed from the data set? How will the correlation change? Calculate the slope and correlation for the revised data to check your estimate.

c. Construct a residual plot of the revised data. Does a linear model fit the data well?

Exam 1	Exam 2	Exam 1	Exam 2
80	88	96	99
52	83	78	90
87	87	93	88
95	92	92	92
67	75	91	93
71	78	96	92
97	97	69	73
96	85	76	87
88	93	91	91
100	93	98	97
88	86	83	89
86	85	96	83
81	81	95	97
61	73	80	86
97	92		

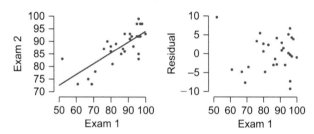

Display 3.106 Data for exam scores in a statistics class, with scatterplot and residual plot.

d. Refer to the scatterplot of *Exam 2* versus *Exam 1* in Display 3.106. Does this plot illustrate regression to the mean? Explain your reasoning.

E64. Suppose you have the Exam 1 and Exam 2 scores of all students enrolled in U.S. History.

a. The slope of the regression line for predicting the scores on Exam 2 from the scores on Exam 1 is 0.51. The standard deviation for Exam 1 scores is 11.6, and the standard deviation for Exam 2 scores is 7.0. Use only this information to find the correlation between Exam 1 scores and Exam 2 scores.

b. Suppose you know, in addition, that the means are 82.3 for Exam 1 and 87.8 for Exam 2. Find the equation of the least squares line for predicting Exam 2 scores from Exam 1 scores.

E65. You are given a list of six values, $-1.5, -0.5, 0, 0, 0.5,$ and 1.5, for x and the same list of six values for y. Note that the list has mean 0 and standard deviation 1.

a. Match each x-value with a y-value so that the resulting six pairs (x, y) have correlation 1.

b. Match the x- and y-values again so that the pairs have the largest possible correlation less than 1.

c. Match the values again, this time to get a correlation as close to 0 as possible.

d. Match the values a fourth time to get a correlation of -1.

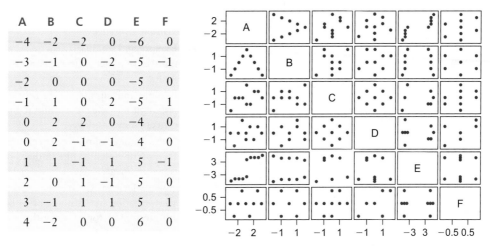

A	B	C	D	E	F
−4	−2	−2	0	−6	0
−3	−1	0	−2	−5	−1
−2	0	0	0	−5	0
−1	1	0	2	−5	1
0	2	2	0	−4	0
0	2	−1	−1	4	0
1	1	−1	1	5	−1
2	0	1	−1	5	0
3	−1	1	1	5	1
4	−2	0	0	6	0

Display 3.107   Data table for six variables and a "scatterplot matrix" of all 30 possible scatterplots for the variables.

E66. Display 3.107 lists the values of six variables, with a "scatterplot matrix" showing all 30 possible scatterplots for these variables. For example, the first scatterplot in the first row has variable B on the x-axis and variable A on the y-axis. The first scatterplot in the second row has variable A on the x-axis and variable B on the y-axis.

   a. For five pairs of variables the correlation is exactly 0, and for one other pair it is 0.02, or almost 0. Identify these six pairs of variables. What do they have in common?

   b. At the other extreme, one pair of variables has correlation 0.87; the next highest correlation is 0.58, and the third highest is 0.45. Identify these three pairs, and put them in order from strongest to weakest correlation.

   c. Of the remaining six pairs, four have correlations of about 0.25 (give or take a little) and two have correlations of about 0.1 (give or take a little). Which four pairs have correlations around 0.25?

   d. Choose several scatterplots that you think best illustrate the statement "The correlation measures direction and strength but not shape," and use them to show what you mean.

E67. Decide whether each statement is true or false, and then explain your decision.

   a. The correlation is to bivariate data what the standard deviation is to univariate data.

   b. The correlation measures direction and strength but not shape.

   c. If the correlation is near 0, knowing the value of one variable gives you a narrow interval of likely values for the other variable.

   d. No matter what data set you look at, the correlation coefficient, r, and least squares slope, b_1, will always have the same sign.

E68. Look at the scatterplot of average SAT I math scores versus the percentage of students taking the exam in Display 3.7 on page 99.

   a. Estimate the correlation.

   b. What possibly important features of the plot are lost if you give only the correlation and the equation of the least squares line?

   c. Sketch what you think the residual plot would look like if you fitted one line to all the points.

E69. The correlation between in-state tuition and out-of-state tuition, measured in dollars, for a sample of public universities is 0.80.

   a. Rewrite the sentence above so that someone who does not know statistics can understand it.

   b. Does the correlation change if you convert tuition costs to thousands of dollars and recompute the correlation?

   c. Does the slope of the least squares line change if you convert tuition costs to thousands of dollars and recompute the slope?

E70. Display 3.108 shows a scatterplot divided into quadrants by vertical and horizontal lines that pass through the point of averages, $(\bar{x}, \bar{y})$.

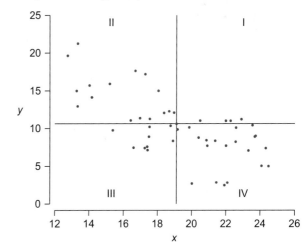

Display 3.108   A scatterplot divided into quadrants by lines passing through the means.

   a. For each of the four quadrants, give the sign of z_x (the standardized value of x), z_y (the standardized value of y), and their product $z_x \cdot z_y$.

   b. Which point(s) make the smallest contribution to the correlation? Explain why the contributions are small.

E71. Rank these summaries for three sets of bivariate data by the strength of the relationship, from weakest to strongest.

A. $\hat{y} = 90 + 100x$ $s_x = 5$ $s_y = 1000$

B. $\hat{y} = \frac{x}{3} - 12$ $s_x = 0.9$ $s_y = 1$

C. $\hat{y} = 1.05 + 0.01x$ $s_x = 0.05$ $s_y = 0.002$

E72. There's an extremely strong relationship between the price of books online and the price at your local bookstore.

  a. Does this mean the prices are almost the same?

  b. Explain why it is wrong to say that the price online "causes" the price at your local bookstore. Why is the relationship so strong if neither set of prices causes the other?

E73. Describe a set of cases and two variables for which you would expect to see regression toward the mean.

E74. *Life span.* In Chapter 2, you looked at the characteristics of mammals (given in Display 2.14 on page 32) one at a time. Now you can look at the relationship between two variables. For example, is longevity associated with gestation period?

  a. Construct a scatterplot of gestation period versus maximum longevity. Describe what you see, including an estimate of the correlation.

  b. Repeat part a with average longevity in place of maximum longevity. Does the average longevity or the maximum longevity give a better prediction of the gestation period?

  c. Does speed appear to be associated with average longevity?

E75. *Spending for police.* The data in Display 3.109 give the number of police officers, the total expenditures for police officers, the population, and the violent crime rate for a sample of states.

  a. Explore and summarize the relationship between the number of police officers and total expenditures for police.

  b. Explore and summarize the relationship between the population of the states and the number of police officers they employ.

  c. Is the number of police officers strongly related to the rate of violent crime in these states? Is the relationship linear?

E76. *Home prices.* Display 3.110 gives the selling prices and other information for a sample of 53 homes sold in Washington, D.C. for which both the location and the area of the home were reported.

  a. Construct a model to predict the selling price from the area for homes in the NW section of the city. Would you use the same model for homes in the NE section? Explain.

  b. Are there any potentially influential observations that might have a serious effect on the model? If so, what would happen to the slope of the regression equation and the correlation if you removed these observations from the analysis?

  c. Predict the selling price of a home in the NW section measuring 1000 square feet. Do the same for an home measuring 2000 square feet. Which prediction do you feel more confident about? Explain.

  d. Explain the effect of the number of bathrooms and of bedrooms on the selling price of the homes. Is either a better predictor than area? Is the pattern the same in all parts of the city?

State	Number of Police Officers (thousands)	Expenditures for Police (millions of dollars)	Population (millions)	Violent Crime Rate (number per 100,000 of state population)
California	96.9	7653	34.0	622
Colorado	12.0	753	4.3	334
Florida	55.2	3371	16.0	812
Illinois	44.1	2718	12.4	657
Iowa	7.3	346	2.9	266
Louisiana	16.1	635	4.5	681
Maine	3.1	118	1.3	110
Mississippi	8.6	337	2.8	361
New Jersey	33.4	1829	8.4	384
Tennessee	18.1	828	5.7	707
Texas	58.9	2866	20.9	545
Virginia	18.8	965	7.1	282
Washington	14.1	854	5.9	370

Display 3.109 Number of police officers and related variables. [Source: U.S. Census Bureau, *Statistical Abstract of the United States*, 2004–2005.]

Beds	Baths	Area (sq ft)	Location	Price	Beds	Baths	Area (sq ft)	Location	Price
5	2	1900	NE	506080	1	1	608	NW	340000
1	1	747	NW	380000	3	2	1464	NE	310000
2	1	735	NW	399000	1	1	904	NW	310000
6	5.5	4100	NW	1695000	3	2	2460	NE	210000
1	1	725	NW	491500	0	1	442	NW	195000
1	1	946	NW	397500	3	2	2800	NW	1300000
2	2.5	2000	NE	449000	1	1	717	NE	275000
1	1	888	NW	265000	2	2.5	2067	NW	800000
1	1	600	NW	220000	3	2	1300	NW	705000
2	2	1282	NW	835000	3	2	2460	NE	210000
2	2	850	NW	383000	3	2.5	1683	NW	850000
2	2.5	1200	NW	559000	2	2	1065	NW	510000
1	1	718	NW	383000	3	2	1464	NE	310000
3	2.5	1683	NW	850000	2	1	872	NW	365000
2	2.5	2067	NW	800000	1	2	844	SE	251501
3	2	1300	NW	705000	2	2	1200	NW	502500
2	2	1250	NW	620000	2	1	800	SE	385000
2	2	1105	NW	615000	2	1	1200	NW	533000
1	2	1350	NW	559000	1	1	904	NW	310000
2	1	1200	NW	533000	1	1	792	NE	350000
2	2	1157	NW	520000	1	1	610	NW	309000
2	2	1200	NW	502500	1	1	806	NW	132000
2	2	1030	NW	500000	1	1	690	NW	335000
3	2	1000	SE	411000	2	1.5	1235	NW	530000
1	1	718	NW	383000	3	4.5	3911	NW	1600000
2	1	872	NW	365000	0	1	585	NW	210000
1	1	792	NE	350000					

Display 3.110 Selling prices of 53 homes in Washington, D.C. [Source: *www.redfin.com/* December 2009.]

E77. Return to the data on schools given in E9, Display 3.15, page 103.

a. Find the regression line for predicting graduation rates from student-teacher ratios. Is there a strong linear relationship here?

b. Find the regression line for predicting graduation rates from per pupil expenditures (PPE). Is there a strong linear relationship here?

C1. This scatterplot shows the age in years of the oldest and youngest child in 116 households with two or more children age 18 or younger living with their parents. (Some points have been moved slightly to show that multiple households are at each coordinate.) Which of the following is *not* a reasonable interpretation of this scatterplot?

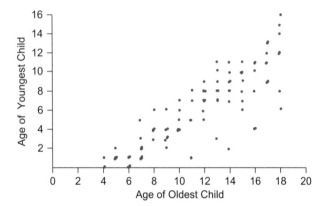

- **A** There aren't any points in the upper-left region because older children tend to move out to go to college or to get married.
- **B** The older the oldest child in a household, the older the youngest child tends to be.
- **C** There are no households represented here in which the only children are twins.
- **D** The variability in the age of the youngest child tends to increase with the age of the oldest child.
- **E** Few households have a range of more than 10 years in the ages of all of the children in the household.

C2. In a study of 190 nations, the least squares line for the relationship between birthrate (per thousand women per year) and female literacy rate (in percent) is $\widehat{birthrate} = -0.38\,literacy + 53.5$, with $r = 0.8$. Uganda has a birthrate of 47 and a female literacy rate of 60. What is the residual for Uganda?

- **A** −17.1
- **B** 29.3
- **C** 16.3
- **D** 64.1
- **E** 69.8

C3. In a linear regression of the heights of a group of trees versus their circumferences, the pattern of residuals is U-shaped. Which of the following must be true?

I. A nonlinear regression would be a better model.

II. For trees near the middle of the range of tree circumferences studied, the predicted tree height tends to be too tall.

III. r will be close to 0.

- **A** II only
- **B** III only
- **C** I and II
- **D** I and III
- **E** I, II, and III

C4. A recent study modeled the relationship between the number of employees in a company and the total number of sick days taken by the employees last year. This scatterplot shows these data for all companies with over 15 employees in a small city.

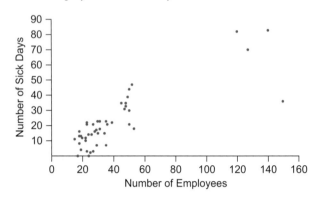

Which of the following is *not* a reasonable way to proceed with the analysis?

- **A** Remove the four outliers permanently from the data set.
- **B** Run the regression again without the four outliers to see how much the slope and correlation change.
- **C** Verify the values for the four outliers to make sure they are correct.
- **D** Convert the total number of sick days to the average number of sick days per employee.
- **E** Make a residual plot in order to judge the linearity of the points.

C5. Upon checking out of a large hospital, 2000 patients rated their satisfaction with their stay on a scale of 0–10, with 10 indicating complete satisfaction. The relationship between the satisfaction rating and the patient's length of stay (in days) was analyzed using linear regression. Here is part of the computer printout for this regression.

The regression equation is
Satisfaction = 4.10 + 0.231 Stay

Predictor	Coef	SE Coef	T	P
Constant	4.1018	0.3050	13.45	0.000
Stay	0.23110	0.07979	2.90	0.004

S = 1.50287   R-sq = 5.2%   R-sq (adj) = 4.6%

Which is a correct interpretation of this regression?

- **A** Patients who stay longer in the hospital tend to be more satisfied than patients with shorter stays, although this relationship is weak.
- **B** The correlation is ± 0.228. However, the sign on the correlation cannot be determined.
- **C** The value of R^2 indicates that the relationship between *satisfaction rating* and *length of stay* is weak but linear.
- **D** The y-intercept of the regression line indicates that no patients rated their satisfaction less than 4.

E The slope of the regression line indicates that as a patient stays longer in the hospital, his or her satisfaction tends to increase day by day.

C6. The ruler of Barbaria has seen a scatterplot of *average life span in years* plotted against *proportion of people with Internet access* for 100 countries. This scatterplot shows a strong positive but nonlinear association. So the ruler proposes a new plan to increase the life span of Barbaria's citizens: Connect them to the Internet! Which of these is the best advice the health minister of Barbaria might give to the ruler?

A The positive association means that in order to increase life span you should *limit* Internet access.

B Increased life span results in more people with Internet access, not the other way around.

C A lurking variable such as wealth of the country could be the cause of both longer life span and more people with Internet access.

D Look at other variables besides Internet access to see if you can find an even stronger association.

E Because the association isn't linear, you can't come to any conclusion about causation.

C7. The Barbarian Aptitude Test (BAT) gives each Barbarian two scores, one for pillaging and one for burning. The scores range from a low of 0 to a high of 50. The least squares equation for a large group of Barbarians who took the BAT is $\widehat{burning} = pillaging + 19$, with $r = 0.6$. Which is the best interpretation of the slope of this line?

A A Barbarian who studies harder and improves her pillaging score by 1 point on the next BAT will tend to increase her burning score by about 0.3 point as well.

B Barbarians tend to score about 0.3 point higher on burning than on pillaging.

C Barbarians score about 30% as many points on burning as on pillaging.

D The burning score is highly correlated with the pillaging score.

E A Barbarian who earned 1 more point on pillaging than another Barbarian tended to earn only 0.3 point more on burning.

C8. A least squares regression analysis using a rating of each Barbarian's personal cleanliness as the explanatory variable and the number of raids he or she has carried out as the response variable found a positive relationship with $R^2 = 0.81$. Which is not a correct interpretation of this information?

A The correlation between personal cleanliness and the number of raids is 0.9.

B There is a strong relationship between personal cleanliness and number of raids among Barbarians.

C A Barbarian who is more personally clean than another also tends to have made more raids.

D There is an 81% chance that the relationship between personal cleanliness and number of raids is linear.

E The variation in the residuals for the number of raids among Barbarians is about 19% of the variation in the original responses.

Investigations

C9. *Siri's equation.* Athletes and exercise scientists sometimes use the proportion of fat to overall body mass as one measure of fitness, but measuring the percentage of body fat directly poses a challenge. Fortunately, some good statistical detective work by W. E. Siri in the 1950s provided an alternative to direct measurement that is still in use today. Siri's method lets you estimate the percentage of body fat from body density, which you can measure directly by hydrostatic (underwater) weighing. In this exercise, you'll see how transformations and residual plots play a crucial role in finding Siri's model.

a. *A first model.* Use the data in Display 3.111.

Density	Percentage of Body Fat	Density	Percentage of Body Fat
1.053	19.94	1.001	44.33
1.053	20.04	1.052	20.74
1.055	19.32	1.014	38.32
1.056	18.56	1.076	9.91
1.048	22.08	1.063	15.81
1.040	25.81	1.023	34.02
1.030	30.78	1.046	23.15
1.064	15.33		

Display 3.111 The percentage of body fat and body density of 15 women. [Source: M. L. Pollock, University of Florida, 1956.]

   i. Plot *percentage of body fat* versus *body density*, and from your plot explain whether you think a line gives a poor fit, a moderately good fit, or an extremely good fit.

   ii. Write the equation of the least squares line.

   iii. Does the value of r^2 tend to confirm your opinion about how well the line fits?

   iv. Construct a residual plot and describe the pattern. Does the plot tend to confirm or raise questions about your opinion? Explain.

b. *A new model*

   i. Explain how knowing that fat is less dense than the rest of the body might have led Siri to plot the percentage of body fat against the reciprocal of density, 1/*density* . Construct this plot; fit a least squares line and compare its equation to the one Siri found:

$$\% \, body \, fat = -450 + 495\left(\frac{1}{density}\right)$$

   Next, plot residuals versus $\frac{1}{density}$.

What features of this plot confirm that the transformation has improved the linear fit?

ii. Find the correlation between the percentage of body fat and body density and the correlation between the percentage of body fat and the reciprocal of body density. Comment on using correlation as the only criterion for assessing the usefulness of a model.

iii. Suggest another model for the percentage of body fat and body density data that might work nearly as well as Siri's.

C10. In C9, you explored the relationship between the percentage of body mass that is fat and body density.

Woman being hydrostatically weighed as part of a body fat test.

Display 3.112 is an extension of the data in C9, including skinfold measurements and data for men.

a. Does Siri's model for relating percentage of body fat to body density hold for men as well as it did for women? That model was

$$\% \text{ body fat} = -450 + 495\left(\frac{1}{density}\right)$$

b. The variable *skinfold* is the sum of a number of skin-fold thicknesses taken at various places on the body. (The units are millimeters.) The skinfold measurements are used to predict body density. Find a good model for predicting density from skinfold measurements based on these data for women.

	Women			Men	
%Fat	Density	Skinfold (mm)	%Fat	Density	Skinfold (mm)
19.94	1.053	126.5	31.60	1.029	223.0
20.04	1.053	69.0	17.99	1.058	149.5
19.32	1.055	98.0	16.10	1.602	148.0
18.56	1.056	85.5	8.68	1.079	112.5
22.08	1.048	104.5	26.87	1.038	141.5
25.81	1.040	163.0	18.06	1.058	220.5
30.78	1.030	192.0	28.71	1.034	252.5
15.33	1.064	67.0	21.71	1.049	177.5
44.33	1.001	228.0	23.38	1.046	152.0
20.74	1.052	102.0	9.09	1.078	110.0
38.32	1.014	248.5	10.77	1.074	100.5
9.91	1.076	73.0	4.58	1.089	72.0
15.81	1.063	92.5	21.93	1.049	219.5
34.02	1.023	144.5	3.82	1.091	85.5
23.15	1.046	86.5			

Display 3.112 Percentage of body fat, body density, and skinfold for 15 women and 14 men. [Source: M. L. Pollock, University of Florida, 1956.]

Chapter 4
Sample Surveys and Experiments

What prompts a hamster to prepare for hibernation? A student designed an experiment to see whether the number of hours of light in a day affects the concentration of a key brain enzyme.

Most of what you've done in Chapters 2 and 3, as well as in the first part of Chapter 1, is part of data exploration—uncover, display, and describe patterns in data. Methods of exploration can help you look for patterns in just about any set of data, but they can't take you beyond the data in hand. *With exploration, what you see is all you get.* Often, that's not enough.

Methods of *inference* can take you beyond the data you actually have, but only if your numbers come from the right kind of process. If you want to use 100 likely voters to tell you about *all* likely voters, how you choose those 100 voters is crucial. The quality of your inference depends on the quality of your data. In other words, bad data lead to bad conclusions. This chapter tells you how to gather data through surveys and experiments in ways that make sound conclusions possible.

Here's a simple example.[1] When you taste a spoonful of chicken soup and decide it doesn't taste salty enough, that's exploratory analysis: You've found a pattern in your one spoonful of soup. If you generalize and conclude that the whole pot of soup needs salt, that's an inference. To know whether your inference is valid, you have to know how your one spoonful—the data—was taken from the pot. If a lot of salt is sitting on the bottom, soup from the surface won't be representative, and you'll end up with an incorrect inference. If you stir the soup thoroughly before you taste, your spoonful of data will more likely represent the whole pot. Sound methods for producing data are the statistician's way of making sure the soup gets stirred so that a single spoonful—the sample—can tell you about the whole pot. Instead of using a spoon, the statistician relies on a chance device to do the stirring and on probability theory to make the inference.

Soup tasting illustrates one kind of question you can answer using statistical methods: Can I generalize from a small sample (the spoonful) to a larger population (the whole pot of soup)? To use a sample for inference about a population, you must *randomize,* that is, use chance to determine who or what gets into your sample.

The other kind of question is about comparison and cause. For example, if people eat chicken soup when they catch a cold, will this cause the cold to go away more quickly? When designing an experiment to determine if a pattern in the data is due to cause and effect, you also must randomize; that is, you must use chance to determine which subjects get which treatments. To answer the question about chicken soup, you would use chance to decide which of your subjects eat chicken soup and which don't, and then compare the duration of their colds.

The first part of this chapter is about designing surveys. A well-designed survey enables you to make inferences about a population by looking at a sample from that population. The second part of the chapter introduces experiments. An experiment enables you to determine cause by comparing the effects of two or more treatments.

In this chapter, you will learn

▷ the reasons for using samples when conducting a survey

▷ how to design a survey by randomly selecting participants

▷ how surveys can go wrong (bias)

▷ how experiments can determine cause

▷ how to design a sound experiment by randomly assigning treatments to subjects

▷ how experiments can go wrong (confounding)

---

[1]The inspiration for this metaphor came from Gudmund Iversen, who teaches statistics at Swarthmore College.

4.1 ▶ Random Sampling: Playing It Safe by Taking Chances

"Inaugural boosts American hopes," declares *USA Today* on January 21, 2009. The article under the headline says the 62% of Americans feel more hopeful after the inauguration of President Obama. Where did this percentage come from? Were all the adults in America asked whether they were more hopeful? The article explains that the result came from a *USA Today*/Gallup poll of about 1000 adults conducted by phone, using both landlines and cell phones. This commonly used method of opinion polling is a good example of what statistics is largely about: getting good information from a sample and using that information to draw a conclusion about the larger group from which the sample was selected.

How are samples chosen? Suppose your class of 40 students is asked to choose a committee of three students to represent the class in an important college matter. What is the fairest way to make the choice? If asked this question, someone in the class is likely to say, "Put all the names in a hat, mix them up, and draw out three." That, in fact, is a very fair way to make the selection because every possible group of three students has the same chance of being selected—no favoritism is shown. As you will see in later chapters, this method also has very important probabilistic and statistical properties that are the basis of statistical inference.

Thus, the best-planned surveys leave a lot to chance. The key idea is to **randomize,** that is, let chance choose the sample. This might seem like a paradox at first, but it makes sense once you understand a basic fact about chance-like behavior: *Selecting your sample by chance is the only method guaranteed to be fair to the larger group to which the sample results will be generalized.* In selecting a committee from your class, people who care enough to volunteer might have a strong personal opinion in the matter under discussion, and not fairly represent the opinions of the class. The three students in the front row might be among the better students, again not truly representative of the class. Even relying on expert judgment (the instructor?) often doesn't work well. Over the long run, chance beats all these methods. In addition, chance is the basis for all the statistical inference procedures you will see in later chapters.

Population and Sample

In statistics, the set of people or objects that you want to know about is called the **population.** The individual elements of the population sometimes are called **units.** In everyday language, *population* often refers to the number of units in the set, as when you say, "In 1776, the population of the colonies was about 2.5 million." In statistics, *population* refers to the set itself (for example, the people of the colonies). The number of units is called the **population size.** Ordinarily, you don't get to record data on all the units in the population, so you use a sample. The **sample** is the set of units you do get to study. The **sample size** is the number of units in the sample. The special case where you collect data on the entire population is a **census.**

Nielsen Media Research takes a survey so that they can get an estimate of the proportion of all U.S. households that are tuned to a particular television program. The true proportion that Nielsen would get from a survey of *every* household is called a population **parameter.** Nielsen uses the proportion in the sample as an *estimate* of this parameter. Such an estimate from a sample is called a **statistic.**

Not all statistics are created equal. Some aren't very good estimators of the population parameter. For example, the maximum in the sample isn't a very good estimator of the maximum in the population; it is almost always too small. You will learn more about the properties of estimators in Chapter 7.

> A *statistic* is a numerical summary of the sample. A *parameter* is a numerical characteristic of a population.

DISCUSSION
Population and Sample

D1. Describe the population in each of the following situations. In which of these situations do you think a sample is used to collect data, and in which do you think a census is used?

    a. An automobile manufacturer inspects its new models.

    b. A cookie producer checks the number of chocolate chips per cookie.

    c. The U.S. president is determined by an election.

    d. Weekly movie attendance figures are released each Sunday.

    e. A Los Angeles study does in-depth interviews with teachers in order to find connections between nutrition and health.

Simple Random Samples

When each unit in the population has a fixed probability of ending up in the sample, the sample is called a **probability sample.** The simplest type of probability sample is a simple random sample, or SRS.

> ### Simple Random Sampling
>
> In **simple random sampling** (sometimes called just **random sampling**), all possible samples of a given size are equally likely to be the sample selected.

In a simple random sample of size n, each possible sample of size n has the same chance of being selected.

    The simplest way to let chance choose your simple random sample is to model the way the three students were selected randomly from your class: Put the names of all the individuals from your population into a gigantic hat, mix them thoroughly, and draw names out one at a time until you have enough for your sample. Although this method is exactly right in theory, in practice mixing thoroughly is almost impossible. So it's actually better to get a random sample using random numbers.

> ### Steps in Choosing a Simple Random Sample (SRS)
>
> 1. Start with a list of all N units in the population.
> 2. Number the units in the list from 1 to N.
> 3. Use a random digit table or generator to choose numbers from 1 to N, one at a time, until you have as many units as you need.

STATISTICS IN ACTION 4.1 ▶ Random Rectangles

This activity provides an opportunity for you to see if you can select representative sampling units as well as a random digit table can.

What You'll Need: Display 4.1 and a method of producing random digits

Your goal is to choose a sample of five rectangles from which to estimate the average area of the 100 rectangles in Display 4.1.

    1. Without studying the display of rectangles too carefully, quickly choose five that you think represent the population of rectangles on the page. This is your judgment sample.

Display 4.1 A population of rectangles.

2. Find the area of each rectangle in your sample of five and compute the sample mean, that is, the average area of the rectangles in your sample.

3. List your sample mean with those of the other students in the class. Construct a plot of the means for your class.

4. Describe the shape, center, and spread of the plot of sample means from the judgment samples.

5. Now generate five distinct random integers between 1 and 100. Find the rectangles whose numbers correspond to your random numbers. This is your random sample of five rectangles.

6. Repeat steps 2–4, this time using your random sample.

7. Discuss how the two distributions of sample means are similar and how they differ.

8. Which method of producing sample means do you think is better if the goal is to use the sample mean to estimate the population mean? (The mean area of all 100 rectangles is 7.4 square units.)

| Example 4.1 | Judgment Samples Versus Random Samples |

Students selected samples of five rectangles from the 100 rectangles in Display 4.1. The goal was to use the mean area of the five rectangles as an estimate of the mean area of all 100 rectangles. Each student used two different sampling methods:

- Used his or her best judgment to pick five representative rectangles.
- Selected five rectangles at random using a random digit table to pick five numbers from 1 to 100.

Display 4.2 shows the results. Describe these plots and their key differences. Given further information that the mean area of all 100 rectangles is 7.4, what can you conclude about the two methods of sampling?

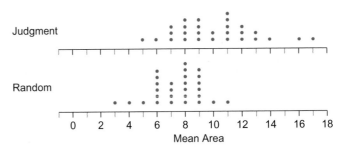

Display 4.2   Means of five rectangles for judgment and random samples for 28 students.

Solution
Both methods of sampling produced distributions that are somewhat mound-shaped, although two judgment samples resulted in unusually large estimates. The judgment samples tended to produce much larger estimates, centered around 10, while the estimates from the random samples were centered around 7. The estimates from the judgment samples also were more variable than the estimates from the random samples.

You can conclude that the method of random sampling is better than that of judgment sampling because the estimates from random sampling are centered near the population mean of 7.4 while the estimates from judgment sampling tend to be too high. Further, there is less variability in the estimate of the population mean from one sample to another when random sampling is used.

Although this one example does not prove that random sampling is always better than judgment sampling, the fact that similar patterns appear in almost every test of people's ability to select a representative sample provides strong evidence in favor of random sampling.

DISCUSSION
Simple Random Samples

D2. Use a table of random digits to take a simple random sample of six students from your statistics class. Did your sample turn out to be representative of your class with respect to gender? Hair color? Class year? Academic major?

D3. You are with a team of inspectors who have arrived at a large lake that serves as a city water reservoir.

    a. One of your tasks is to measure the water quality. Today, you have with you 20 test tubes that you can fill with water to take back to the lab. Describe the population. Describe how you might select the sample.

    b. You also have been asked to estimate the proportion of recreational boats on the lake that are properly equipped for water safety. Describe the population. Describe how you might select the sample.

    c. Compare the nature of the two populations and samples. Is a census a possibility in either case?

D4. *Readability.* You've decided you simply *must* know the proportion of letters in this textbook that are capitalized because it can indicate the complexity of the sentences on a page. Think of the printed characters in the book itself as your population. How could you get a simple random sample of 10,000 characters? What are the advantages and disadvantages of this sampling method?

Stratified Random Samples

Suppose you are planning a sample survey for an international outdoor clothing manufacturer to see if their image has suffered among their retailers due to negative press coverage regarding their use of sweatshop labor. The company headquarters furnishes you with a list of retail outlets throughout the world that sell their products. Because press coverage is largely national, it will be more informative to organize this list by country and take a random sample from each country. You'll be gathering data using each country's phone or postal system in the language of that country, so categorizing the retail outlets by country will also be more convenient.

    Classifying the retail outlets by country is called **stratification.** If you can divide your population into subgroups that do not overlap, the subgroups are called **strata.** If the sample you take within each subgroup is a simple random sample, your result is a **stratified random sample.** This is a second type of probability sample.

Steps in Choosing a Stratified Random Sample

    1. Divide the units of the population into nonoverlapping subgroups.

    2. Take a simple random sample from each subgroup.

Most national polls, like Gallup, for example, use some form of stratification in their sampling designs to make sure that they cover all geographic regions of the country and have good precision for the overall results.

Why stratify?

Why might you want to stratify a population? Here are the three main reasons:

- *Convenience* in selecting the sampling units is enhanced. It is easier to sample in smaller, more compact groups (countries) than in one large group spread out over a huge area (the world).

- *Coverage* of each stratum is assured. The company might want to have data from each country in which it sells products. A simple random sample from the population does not guarantee that this would happen.

- *Precision* in the estimate may be improved; that is, stratification tends to give estimates that are closer to the value for the entire population than does an SRS. This is the fundamental statistical reason for stratification.

STATISTICS IN ACTION 4.2 ▶ Sampling to Monitor Air Quality: How Stratification Can Result in a Better Estimate

What You'll Need: a copy of the blank grid shown in Display 4.3 and a method of producing random digits

A cement production plant is being built close to a residential neighborhood, but with strong air quality safeguards. Once the plant is in production, air quality will be monitored by sampling air from houses in the neighborhood on a regular basis. Suppose the 100 neighborhood houses are built on a square grid ten lots wide and ten lots deep, as depicted in the schematic drawing in Display 4.3. Your task is to get a sample of ten houses for the first air quality check.

1. Select ten houses from the neighborhood using:

    a. simple random sampling

    b. stratified random sampling with columns of the grid (avenues) as strata

    c. stratified random sampling with rows of the grid (streets) as strata

2. Which sampling plan do you think will produce the estimate of mean air quality across the neighborhood with the smallest variation?

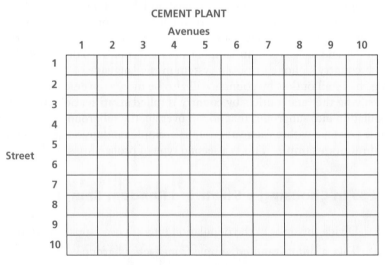

Display 4.3   Grid for sampling of houses.

You will measure air quality using the Pollution Standards Index (PSI). Suppose the PSI readings for the day your sampling plan is carried out in the field are as shown in Display 4.4. The mean PSI for all 100 houses is 97.7.

3. Identify the PSI for the houses in your simple random sample, in your stratified random sample using columns as strata, and in your stratified random sample using rows as strata. For each sample, calculate the mean PSI.

4. For each sampling plan, combine your sample mean with those of the rest of the class and make an appropriate plot of the sample means for each plan.

5. Comment on the shape, center, and spread of each of the plots. Which sampling plan appears to produce estimates with the smallest variability? The next smallest variability?

CEMENT PLANT

Avenues

	1	2	3	4	5	6	7	8	9	10
1	121	118	124	123	116	118	120	118	114	122
2	116	118	118	113	117	116	117	112	112	115
3	114	107	109	106	112	108	112	110	111	111
4	105	104	103	101	103	105	104	106	109	107
5	100	100	101	96	98	96	100	100	105	100
6	97	95	96	94	96	95	96	97	96	97
7	92	90	91	89	93	94	93	92	92	90
8	86	81	85	87	85	85	86	87	83	84
9	80	78	80	79	77	81	81	79	84	81
10	76	77	74	77	75	74	80	75	77	74

Street (rows 1–10)

Display 4.4 Grid of houses showing PSI measurements.

In Example 4.1 on page 174, you saw that random sampling can result in greater precision in the estimates than judgment sampling. The following example shows how stratification can further increase the precision of the estimate.

Air Quality

Example 4.2

Air quality is measured by the Pollutant Standards Index (PSI), which is a composite measure based on particulate matter (PM_{10}) that can penetrate into the lungs, sulfur dioxide (SO_2), carbon monoxide (CO), ozone (O_3), and nitrogen dioxide (NO_2). A PSI over 100 indicates air quality in the unhealthy range.

The grid in Display 4.4 shows PSI measurements for 100 houses in a neighborhood abutting a cement plant. To estimate the mean PSI for all 100 houses, samples of size 10 were selected by each of 30 students in a class. Each student used three different sampling plans: a simple random sample from the 100 houses, a column-sampling plan using the columns as strata and sampling one observation at random from each of the ten columns, and a row-sampling plan using the rows as strata and sampling one observation at random from each row. Box plots of the 30 sample means for each design are shown in Display 4.5. Compare the performances of the three sampling plans.

Cement plant near San Antonio, Texas.

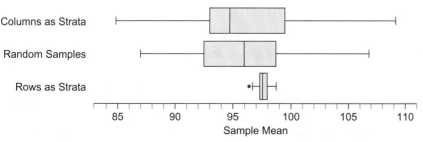

Display 4.5 Plots of 30 sample means from each of three sampling plans.

Solution
All three plans produce fairly symmetric distributions of sample means. All three centers are relatively close to the true population mean, 97.7. The difference is in the variability of the estimates. If you stratify by row (streets), the sample mean is likely to be much closer to the population mean of 97.7 than if you used one of the other sampling plans. Stratifying by row guarantees that you will get values in your sample that are spread over the entire range of PSI measurements. If you stratify by column, you might get all large or all small PSI readings in your sample. In choosing strata, the strata should differ from each other as much as possible but should have as little variation as possible within each stratum. A poor choice of strata (like choosing the columns of the grid as strata) is worse than no stratification at all!

Make strata means as different as possible.

From the preceding example, it certainly looks as though the stratification pays off in producing estimates of the mean with smaller variation. This will be true generally if the stratum means are quite different. The guiding principle is to choose strata that have very different means, whenever that is possible.

Allocate sample sizes to the strata in proportion to the number of units in the strata.

One good way to choose the relative sample sizes in stratified random sampling is to make them proportional to the stratum sizes (the number of units in the stratum). Thus, if all the strata are of the same size, the samples should all be of the same size. If a population is known to have 65% women and 35% men, then a sample of 100 people stratified on gender should contain 65 women and 35 men. If the sample sizes are proportional to the stratum sizes, then the overall sample mean (the mean calculated from the samples of all the strata mixed together) will be a good estimate of the population mean. Proportional allocation is particularly effective in reducing the variation in the sample means if the stratum standard deviations are about equal. (If the stratum standard deviations differ greatly, a more effective allocation of samples to strata can be found.)

DISCUSSION
Stratified Random Samples

D5. An administrator wants to estimate the average amount of time students spend, on a daily basis, traveling to campus from wherever they live. The plan is to stratify the students according to class year and then take a simple random sample from each class year. What is potentially good and what is potentially bad about this plan?

D6. Your assignment is to estimate the proportion of households in a small city that have at least one HDTV. Discuss how you would set up a stratified random sampling plan to accomplish the task. What are the advantages of using stratification here?

Other Methods of Sampling

Simple random samples and stratified random samples offer many advantages, but you often run into a major practical problem in attempting to carry them out. Sampling individual units from a population one at a time is often too costly, too time-consuming, or simply not possible. Several commonly used sampling designs get around this problem by forming larger sampling units out of groups of population units. (Think of sampling households in a neighborhood rather than individual residents.) If the groups are chosen wisely, these designs are often nearly as good as simple or stratified random sampling.

Cluster Samples

To see how well 4th graders in your state do on an arithmetic test, you might take a simple random sample of children enrolled in the 4th grade and give each child a standardized test. In theory, this is a reasonable plan, but it is not very practical. For one thing, how would you go about making a complete list of all the 4th graders in the state? Even if you had a list, it is likely to change every day. For another, imagine the work required to track down each child in your sample and get him or her to take the test. Instead of taking an SRS of 4th graders, it would be more realistic to take an SRS of elementary schools and then give the test to all the 4th graders in those schools. Getting a list of all the elementary schools in your state is a lot easier than getting a list of all the individuals enrolled in the 4th grade. Moreover, once you've chosen your sample of elementary schools, it's a relatively easy organizational problem to give the test to all 4th graders in those schools. This is an example of cluster sampling, with each elementary school being a cluster of 4th-grade students. A **cluster sample** is an SRS of nonoverlapping clusters of units.

> It is often more efficient to sample clusters of units.

Steps in Choosing a Cluster Sample

1. Create a list of all the clusters in your population.
2. Take a simple random sample of clusters.
3. Obtain data on each individual in each cluster in your SRS.

The situation of the 4th graders illustrated the two main reasons for using cluster samples: You need only a list of clusters rather than a list of individuals, and for some studies it is much more efficient to gather data on individuals grouped by clusters than on all individuals one at a time.

Two-Stage Cluster Samples

The National Assessment of Educational Progress (NAEP) uses a variation on cluster sampling. For example, to get a sample of 4th graders in Illinois, NAEP takes a (stratified) random sample of elementary schools. Then, typically 30 4th-grade students are selected randomly from each school to take the mathematics test. [Source: *nces.ed.gov*.] This is an example of **two-stage sampling** because it involves two randomizations.

> Two-stage sampling involves sampling clusters and then sampling from within clusters.

Steps in Choosing a Two-Stage Cluster Sample

1. Create a list of all the clusters in your population, and then take a simple random sample of clusters.
2. Create a list off all the individuals in each cluster already selected, and then take an SRS from each cluster.

Stratification versus Cluster Sapling

Two-stage cluster samples are useful when it is much easier to list clusters than individuals but still reasonably easy and sufficient to sample individuals once the clusters are chosen. Two-stage cluster sampling might sound like stratified random sampling, but the two are different.

- In stratified random sampling, you want to choose strata so that the units within each stratum don't vary much from each other. Then you sample from every stratum.

- In two-stage cluster sampling, you want to choose clusters so that the variation within each cluster reflects the variation in the population, if possible. Then you sample from within only some of your clusters.

Systematic samples can begin with "counting off."

Systematic Samples with Random Start

To get a quick sample of the students in your class, you might use the common system for choosing teams by counting off. Count off by, say, eights, and then select a digit between 1 and 8 at random. Every student who calls out that digit is in the sample. This is an example of a **systematic sample with random start**.

> ### Steps in Choosing a Systematic Sample with Random Start
>
> 1. By a method such as counting off, divide your population into groups of the size you want for your sample.
> 2. Use a chance method to choose one of the groups for your sample.

Suppose your population units come already lined up, such as people waiting in line to enter a theater, or are named in a list, such as a telephone directory, and you want to select a sample of a certain size. The steps in choosing a systematic sample with random start are equivalent to choosing a random starting point between units 1 and k and then taking every kth unit thereafter. The value of k is determined by the sample size. For example, suppose you want a systematic sample of 20 units selected from a list of 1000 units. Then k is 50, because selecting a random start between 1 and 50 and then taking every 50th unit thereafter will result in a systematic sample of 20 units.

When the units in the population are well mixed before counting off, this method will produce a sample much like an SRS. Systematic samples, like cluster samples, are often easier to take than simple random samples of the same size.

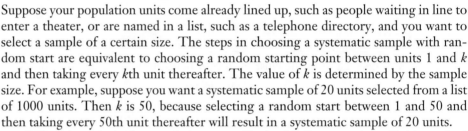

Ducks in a row.

DISCUSSION
Other Methods of Sampling

D7. Why isn't taking a systematic sample with random start equivalent to taking a simple random sample?

D8. The *Los Angeles Times* commissioned an analysis of St. John's wort, an over-the-counter herbal medicine, to determine whether the potency of the pills matched the potency claimed on the bottle. The sampling procedure was described like this: "For the analysis, 10 pills were sampled from each of three containers of one lot of each product." [Source: *Los Angeles Times*, August 31, 1998, p. A10.]

    a. What type of sampling was this?

    b. Can you suggest an improvement in the sampling procedure?

D9. A newspaper reports that about 60% of the cars in your community come from manufacturers based outside the United States. How would you design a sampling plan for gathering the data to substantiate or refute this claim?

D10. Both cluster sampling and stratified random sampling involve viewing the population as a collection of subgroups. Explain the difference between these two types of sampling.

D11. Take a systematic sample, with random start, of five students from your class.

Summary 4.1: Random Sampling: Playing It Safe by Taking Chances

The *population* is the set of units you want to know about. The *sample* is the set of units you choose to examine. A *census* is an examination of all units in the entire population.

    The main reason for using a chance method to choose the individuals for your sample is that randomization protects against sample selection bias, which will be discussed in the next section. And, as you will see, random selection of the sampling units will make inference about the population possible.

There are several ways to take a sample using random selection. The following are the most common methods:

- *Simple random sample.* Number the individuals and use random numbers to select those to be included in the sample.

- *Stratified random sample.* First divide your population into groups of similar units, called strata, and then take a simple random sample from each group.

- *Cluster sample.* First select clusters at random, and then use all the individuals in the clusters as your sample.

- *Two-stage cluster sample.* First select clusters at random, and then select individuals at random from each cluster.

- *Systematic sample with random start.* Count off your population by a number *k* determined by the size of your sample, and select one of the counting numbers between 1 and *k* at random. The units with that number will be in your sample.

Display 4.6 shows schematically how five common sampling designs might look for sampling the blobs in the rectangle.

Simple Random Sampling

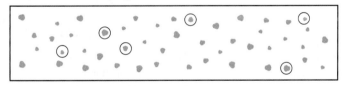

Stratified Random Sampling

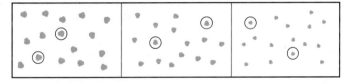

Cluster Sampling

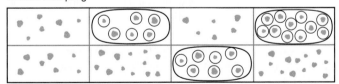

Two-Stage Cluster Sampling

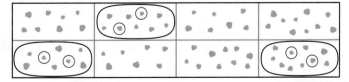

Systematic Sampling with Random Start

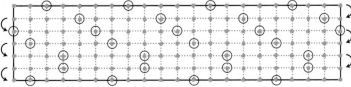

Display 4.6 Various methods of sampling.

Practice

Population and Sample

P1. You want to estimate the average number of TV sets per household in your community.

   a. What is the population? What are the units?

   b. Explain the advantages of sampling over conducting a census.

Simple Random Samples

P2. You are interested in knowing the average number of hours worked by students who live in dorms at a large university. The university has 20 dorms, each of which holds 100 students. You take a sample of students by first taking a random sample of six dorms and then taking a random sample of 10 students from each selected dorm. Does this method give you a simple random sample? Explain why the method does or does not fit the definition of a simple random sample.

P3. Decide whether these sampling methods produce a simple random sample of students from a class of 30 students. Explain why the method does or does not fit the definition of a simple random sample.

   a. Select the first six students on the class roll sheet.

   b. Randomly choose a letter of the alphabet and select for the sample those students whose last name begins with that letter. If no last name begins with that letter, randomly choose another letter of the alphabet.

   c. If the classroom has six rows of chairs with five seats in each row, choose a row at random and select all students in that row.

   d. If the class consists of 15 males and 15 females, assign the males the numbers from 1 to 15 and the females the numbers from 16 to 30. Then use a random digit table to select six numbers from 1 to 30. Select the students assigned those numbers for your sample.

   e. If the class consists of 15 males and 15 females, assign the males the numbers from 1 to 15 and the females the numbers from 16 to 30. Then use a random digit table to select three numbers from 1 to 15 and three numbers from 16 to 30. Select the students assigned those numbers for your sample.

Stratified Random Samples

P4. A psychology student wants to estimate the proportion of undergraduate students on your campus who awaken often from nightmares. Which one of these is *not* a valid design using stratification? Explain why your choice doesn't fit the definition of a stratified random sample.

   A. Two simple random samples are taken: one of the male students and the other of the female students.

   B. Four simple random samples are taken: one in each of the freshman, sophomore, junior, and senior classes.

   C. Three simple random samples are taken: one from students taking a science class, one from students taking a math class, and one from students taking neither of these subjects.

   D. Two random samples are taken: one of students who are married and the other of students who are not married.

P5. Receding gums are most common in people over age 40, but they can occur in younger adults who have periodontal (gum) disease. There does not appear to be a great difference between the rates of males and females with receding gums. You have been hired to estimate the percentage of all adults in your city with receding gums and were given a budget sufficient for a random sample of 1200 adults.

   a. Suppose that, in your city, 43% of adults see their dentist twice a year, 32% once a year, and 25% less frequently than once a year. If you stratify on number of visits to a dentist, how many adults from each stratum should you include in your sample?

   b. In addition to the number of visits to a dentist, which of these variables would be the best to consider stratifying on? Explain your answer.

     • gender (male/female)

     • age (40 and younger/over 40)

P6. Suppose your population is 65% women and 35% men. In a stratified random sample of 100 women and 100 men, you find that 84 women pump their own gas and 69 men pump their own gas. What is the best estimate of the proportion of the entire population who pump their own gas?

Other Methods of Sampling

P7. Suppose 200 people are waiting in line for tickets to a rock concert. You are working for the school newspaper and want to interview a sample of the people in line. Describe how to select a systematic sample, with random start, of

   a. 5% of the people in line

   b. 20% of the people in line

P8. The American Statistical Association directory lists its roughly 17,000 members in alphabetical order. You want a sample of about 1000 members. Describe how to use the alphabetical listing to take a systematic sample with random start.

P9. A direct mail company wants to get a sample of households to test a new advertising campaign. This company advertises to all 45,000 ZIP codes in the U.S., each of which is subdivided into carrier routes.

   a. Describe how to use ZIP codes to take a cluster sample.

   b. Describe how to take a two-stage cluster sample.

P10. *More readability.* Read D4 on page 175, which asked about taking an SRS to estimate the proportion of capital letters in this book.

a. Instead of an SRS, consider taking a cluster sample. What would you use for your clusters? What are the advantages of the cluster sample? Are there any disadvantages?

b. Suppose you want to estimate the average number of capital letters per printed line, so now your sampling

unit is a line. Describe how to take a two-stage cluster sample from this textbook.

c. Now suppose your sampling unit is an individual printed character. Describe how to take a *three*-stage cluster sample of characters from this book.

Exercises

E1. A wholesale food distributor has hired you to conduct a sample survey to estimate the satisfaction level of the businesses he serves, which are mainly small grocery stores and restaurants. A current list of businesses served by the distributor is available for the selection of sampling units.

a. If the distributor wants good information from both the small grocery store owners and the restaurant owners, what kind of sampling plan will you use?

b. If the distributor wants information from the customers who frequent the grocery stores and restaurants he serves, how would you design the sampling plan?

E2. *Cookies.* Which brand of chocolate chip cookies gives you the most chips per cookie? For the purpose of this question, take as your population all the chocolate chip cookies now in the nearest supermarket. Each cookie is a unit in this population.

a. Explain why it would be hard to take an SRS.

b. Describe how to take a cluster sample of chocolate chip cookies.

c. Describe how you would take a two-stage cluster sample. What circumstances would make the two-stage cluster sample better than the cluster sample?

E3. *Haircut prices.* You want to take a sample of students on your campus in order to estimate the average amount students spent on their last haircut. Which sampling method do you think would work best: a simple random sample; a stratified random sample with two strata, males and females; or a stratified random sample with class levels as strata? Give your reasoning.

E4. The *Oxford Dictionary of Quotations*, 3rd edition, has about 600 pages of quotations. Describe how you would take a systematic sample of 30 pages to use for estimating the number of typographical errors per page.

E5. An early use of sampling methods was in crop forecasting, especially in India, where an accurate forecast of the jute yield in the 1930s made some of the techniques (and their inventors) famous. Your job is to estimate the total corn yield, right before harvest, for a county with five farms each with about 200 acres planted in corn. How would you do the sampling?

E6. You are called upon to advise a local movie theater on designing a sampling plan for a survey of patrons on their attitudes about recent movies. About 64% of the patrons are adults, 30% are teens, and 6% are children. The theater has the time and money to interview about 50 patrons. What design would you use?

E7. Quality-control plans in industry often involve sampling items for inspection. A manufacturer of electronic relays (switches) for the TV industry wants to set up a quality-control sampling plan for these relays as they come off its production line. What sampling plan would you suggest if there is only one production line? If there are five production lines?

E8. Suppose that a sample of 25 adults from the U.S. population contains only women. Two explanations are possible: (1) The sampling procedure wasn't random, or (2) a nonrepresentative sample occurred just by chance. In the absence of additional information, which explanation would you be more inclined to believe?

4.2 ▶ Why Take Samples, and How Not To

As you have seen, samples are useful in gathering information about a population without measuring every unit in the population of interest. For example, Gallup periodically polls about 1000 adults in the United States to estimate the approval rating of the president. Similarly, quality assurance engineers in a manufacturing plant maintain the quality of items coming off the production line by periodically selecting a limited number of items (a sample) to be checked carefully for quality. The United States Census Bureau uses sampling to update the census between decennial years, and the Bureau of Labor Statistics uses sampling to estimate the unemployment rate, among other indicators of the health of the economy.

Cost in money or time is a primary reason to use samples. Imagine: It's Sunday night at 8:00, and Nielsen Media Research is gathering data about what proportion of TV sets are tuned to a particular program and what kinds of people are watching that program. To find out how many TV sets are tuned to the program, electronic meters

Sampling saves money and time.

have been hooked up to televisions in a sample of households. To find out who is watching TV, a sample of people are filling out diaries. Why doesn't Nielsen Media Research include everyone in the United States in these surveys? To hook up a meter to every TV set would cost more than anyone would be willing to pay for the information. Also, to try to get a diary from every TV viewer about what they were watching at 8:00 p.m. on Sunday would take so much time that the information would no longer be very useful. So for two reasons—money and time—Nielsen ratings are based on samples.

> Sampling can be destructive.

Sampling lets a cook know how the soup will taste without eating it all just to make sure. A light bulb manufacturer can't test the life of every bulb produced, or there would be none left to sell. Whenever testing destroys the things you test, your only choice is to work with a sample.

> Sampling can provide more information than can a cursory study of all items in the population.

If time and money are limited—and they always are—there's a trade-off between the number of people in your sample and the amount and quality of information you can expect to get from each person. Using a sample allows you to spend more time and money gathering high-quality information from each individual. This often produces greater accuracy in the results than you could get from a quick, but error-prone, study of every individual.

The case for sampling may sound strong and indeed it is, but carrying out a well-designed sampling plan is more difficult than you might imagine at first glance. One of the big issues is bias, which is the subject of the rest of this section.

Bias: A Potential Problem with Survey Data

Samples offer many advantages, but some samples are more trustworthy than others. In this section, you will learn about two ways to get untrustworthy results:

- bias in the way you select your sample
- bias in the way you get a response from the units in your sample

In everyday language, we say an opinion is biased if it unreasonably favors one point of view over others. A biased opinion is not balanced, not objective. In statistics, bias has a similar meaning in that a biased sampling method is unbalanced.

> ### Biased Methods
>
> A sampling method is **biased** if it produces samples such that the estimate from the sample is larger or smaller, on average, than the population parameter being estimated.

> Sampling bias lies in the method, not in the observed sample.

In other words, a sampling method is biased if its results tend to tilt toward one side of truth. There's an important distinction here between the sample itself and the method used for choosing the sample—bias refers to the method of sampling. Estimating the proportion of students on campus who think college athletics are overrated by sampling a crowd leaving a basketball game is a biased method. On average, such samples will underestimate the proportion of students who think college athletics are overrated. However, the estimate obtained by a single lucky sample could be right on track.

Example 4.3	**Binge Drinking**

The Behavioral Risk Factor Surveillance System (BRFSS) of the U.S. Centers for Disease Control (CDC) collects information on health risk behaviors, and many other health-related issues, for all 50 states, the District of Columbia, and U.S. territories by conducting large monthly surveys. Display 4.7 shows recent data on binge drinking as categorized by education level of the person responding. The numbers are the percentages for that column. Binge drinking is defined as drinking with the intention of becoming intoxicated.

	Grade 12 or GED (high school graduate)	College 1 Year to 3 Years (some college or technical)	College 4 Years or more (college graduate)	Combined Education Categories
0 times	67.7	72.3	79.4	72.8
1 time	10.7	10.5	8.9	10.3
2–4 times	13.3	10.5	7.9	10.6
5 or more times	8.3	6.7	3.8	6.3
Total	100.0	100.0	100.0	100.0

Display 4.7 Binge drinking frequency in past month. [Source: CDC, BRFSS.]

a. Describe any patterns you see in the data.

b. Discuss the possible effect of bias when questioning people on a sensitive question like binge drinking.

Solution

a. The percentages for the categories representing one or more episodes within the past month are highest for the lowest-level education group (high school graduate) and decrease as the education level increases. This may not imply that the highest *numbers* of binge drinkers are in the lower education categories because the numbers of people in the various education categories vary substantially.

b. Many people may be reluctant to respond, or respond correctly, to a question about a potentially sensitive question like one on alcohol consumption. Thus, there is considerable potential that those selected for the sample do not respond (nonresponse) or give incorrect responses. Some may like to brag about their drinking while others may want to play down that aspect of their behavior. Further, the tendency for nonresponse and incorrect response may vary depending on the person's educational group.

DISCUSSION
Bias

D12. *Bias in election polling.* In the historic presidential race of 2008 between Barack Obama and John McCain, more preference polls were conducted than ever before in attempts to find trends in voting patterns that would shed light on predicting the winner. Nearly all of these surveys were conducted by telephone, and the traditional method of telephone polling is to use landlines. But many people now rely mainly on their cell phones, and that fact raises interesting questions about the effect of cell phones on election polling. To provide data on the issue, the Pew Research Center for the People & the Press conducted preelection polls with both cell phone and landline samples between the primaries and the election. Some of their data are summarized in Display 4.8.

a. Discuss the differences in the results of the voting preferences for Obama and McCain in the landline survey versus the cell phone survey, as well as possible reasons for the differences.

b. Discuss the differences in party affiliation between the landline and cell phone surveys for the "under age 30" group and possible reasons for these differences.

	Landline Sample (%)	Cell Phone Sample (%)
Obama	45	55
McCain	45	36
Other	10	9
Sample size	1960	176

Registered voters, September 9–14, 2008.

	Landline Sample (%)	Cell Phone Sample (%)
Democratic	54	62
Republican	36	28
Other	10	10
Sample size	390	242

Registered voters under age 30, August and September, 2008.

Display 4.8 Phone surveys and the presidential vote. [Source: http://pewresearch.org/pubs/964/.]

c. Discuss the potential for bias to affect a presidential preference poll if cell phones are not adequately represented in the sample.

d. So, why not just add cell phone numbers to the sample? Two main reasons are that cell phone calls are more expensive (costing the respondent by the minute) and the federal Telephone Consumer Protection Act (TCPA) bans unsolicited calls to a cell phone using automated dialing devices. Discuss how these issues might relate to bias in the results of telephone surveys in general.

Sample Selection Bias

Sample selection bias, or **sampling bias,** is present in a sampling method if samples tend to result in estimates of population parameters that systematically are too high or too low. Various forms of this selection bias can undermine the usefulness of samples and surveys.

Suppose a biologist studying water quality selects lakes for a sample by dropping grains of rice on a map of a state and then selects the lakes that have rice on them. This method of sampling is **size biased** because a randomly dropped grain of rice is more likely to land on a large lake than on a small one. Measuring the length of telephone calls by selecting those in progress at a specific time is also a size-biased method of sampling because long calls are more likely to be interrupted than short ones.

> Size bias is one kind of sample selection bias.

When a television or radio program asks people to call in and take sides on some issue, those who care about the issue will be overrepresented, and those who don't care as much might not be represented at all. Such **volunteer sampling** often leads to **voluntary response bias** and is a second type of sample selection bias.

> Voluntary response bias is another kind of sampling bias.

Here's a simple sampling method: Take whatever's handy. For example, what percentage of the students in your graduating class plan to go to work immediately after graduation? Rather than find a random sample of your graduating class, it would be a lot quicker to ask your friends and use them as your sample—quicker and more convenient, but almost surely biased because your friends are likely to have somewhat similar plans. A **convenience sample** is one in which the units chosen are easy to include. The likelihood of bias makes convenience samples about as worthless as voluntary response samples.

> Convenience sampling is almost always a biased sampling method.

Because voluntary response sampling and convenience sampling tend to be biased methods, you might be inclined to rely on the judgment of an expert to choose a sample that he or she considers representative. As you read in Section 4.1, such samples, not surprisingly, are called **judgment samples.** Unfortunately, though, even experts might overlook important features of a population. In addition, trying to balance several features at once can be almost impossibly complicated. In the early days of election polling, local "experts" were hired to sample voters in their locale by filling certain quotas (so many men, so many women, so may voters over the age of 40, so many employed workers, and so on). The poll takers used their own judgments as to whom they selected for the poll. It took a very close election (the 1948 presidential election, in which polls were wrong in their prediction) for the polling organizations to realize that quota sampling based on judgment was a biased method.

> Judgment sampling, even when done by experts, usually is a biased sampling method.

An unbiased sampling method requires that all units in the population have a known chance of being chosen, so you must prepare a list of population units, called a **sampling frame** or, more simply, a **frame,** from which you select the sample. If you think about enough real examples, you'll come to see that making this list is not something you can take for granted. For the Westvaco employees in Chapter 1 or for the 50 U.S. states, creating the list is not hard, but other populations can pose problems. How would you list all the people using the Internet worldwide or all the ants in Central Park or all the potato chips produced in the United States over a year? For all practical purposes, you can't. There will often be a difference between the population—the set of units you want to know about—and the sampling frame—the list of units you use to create your sample.

> The quality of the sample depends on starting with a good sampling frame.

A sample might represent the units in the *frame* quite well, but how well your sample represents the *population* depends on how well you've chosen your frame. Quite often, a convenient frame fails to cover the population of interest (using a telephone directory to sample residents of a neighborhood, for example). If you start from a bad frame, even the best sampling methods can't save you: bad frame, bad sample.

DISCUSSION
Sample Selection Bias

D13. Identify the type of sampling method used in each of these surveys. Would you expect the estimate of the parameter to be too high or too low?

    a. You use your statistics class to estimate the percentage of students on your campus who study at least 2 hours a night.

    b. You send a survey to all people who have graduated from your college in the past 10 years. You use the mean annual income of those who reply to estimate the mean annual income of all graduates of your school in the past 10 years.

    c. A study was designed to estimate how long people live after being diagnosed with dementia. The researchers took a random sample of the people with dementia who were alive on a given day. The date the person had been diagnosed was recorded, and after the person died the date of death was recorded.

D14. You want to know the percentage of voters who favor state funding for bilingual education. Your population of interest is the set of people likely to vote in the next election. You use as your frame the phone book listing of residential telephone numbers. How well do you think the frame represents the population? Are there important groups of individuals who belong to the population but not to the frame? To the frame but not to the population? If you think bias is likely, identify what kind of bias and how it might arise.

Response Bias

> Bias doesn't always come from the sampling method.

In all the examples so far, bias has come from the method of taking the sample. Unfortunately, bias from other sources can contaminate data even from well-chosen sampling units.

Perhaps the worst case of faulty data is no data at all. It isn't uncommon for 40% of the people contacted to refuse to answer a survey. These people might be different from those who agree to participate. An example of this **nonresponse bias** came from a controversial study that found that left-handers died, on average, about 9 years earlier than right-handers. The investigators sent questionnaires to the families of everyone listed on the death certificates in two counties near Los Angeles asking about the handedness of the person who had died. One critic noted that only half the questionnaires were returned. Did that change the results? Perhaps. [Source: "Left-Handers Die Younger, Study Finds," *Los Angeles Times*, April 4, 1991.]

> Nonresponse bias can occur when people do not respond to surveys.

Statisticians use a variety of methods to minimize nonresponse, ranging from making the questionnaire short and user-friendly to giving people some reward for participating. If the response rate is still unacceptably low, they aggressively follow up on a sample of the nonrespondents to see if they appear to be different from the respondents in a way that might bias the results of the survey.

Nonresponse bias, like bias that comes from the sampling method, arises from who replies. **Questionnaire bias** arises from how you ask the questions. The opinions people give can depend on the tone of voice of the interviewer, the appearance of the interviewer, the order in which the questions are asked, and many other factors. But the most important source of questionnaire bias is the wording of the questions. This is so important that those who report the results of surveys should always provide the exact wording of the questions.

> Questionnaire bias comes from how questions are posed and asked.

For example, *Reader's Digest* commissioned a poll to determine how the wording of questions affected people's opinions. The same 1031 people were asked to respond to these two statements:

1. I would be disappointed if Congress cut its funding for public television.
2. Cuts in funding for public television are justified as part of an overall effort to reduce federal spending.

Agreeing with the first statement is pretty much the same as disagreeing with the second. However, 54% agreed with the first statement, 40% disagreed, and 6% didn't know, while 52% agreed with the second statement, 37% disagreed, and 10% didn't know. [Source: Fred Barnes, "Can You Trust Those Polls?" *Reader's Digest* (July 1995), pp. 49–54.]

"One final question: Do you now own or have you ever owned a fur coat?"

Another problem that polls and surveys have is trying to ensure that people tell the truth. Often, the people being interviewed want to be agreeable and tend to respond in the way they think the interviewer wants them to respond. Newspaper columnist Dave Barry reported that he was called by Arbitron, an organization that compiles television ratings. Dave reports:

So I figured the least I could do, for television, was be an Arbitron household. This involves two major responsibilities:

1. Keeping track of what you watch on TV.
2. Lying about it.

At least that's what I did. I imagine most people do. Because let's face it: Just because you watch a certain show on television doesn't mean you want to *admit* it. [Source: Dave Barry, *Dave Barry Talks Back*, copyright ©1991 by Dave Barry. Used by permission of Crown Publishers, a division of Random House, Inc.]

Bias from **incorrect responses** might be the result of intentional lying, but it is more likely to come from inaccurate measuring devices, including inaccurate memories of people being interviewed in self-reported data. Patients in medical studies are prone to overstate how well they have followed the physician's orders, just as many people are prone to understate the amount of time they actually spend watching TV. Measuring the heights of students with a meter stick that has one end worn off leads to a measurement bias, as does weighing people on a bathroom scale that is adjusted to read on the light side.

> Bias can come from incorrect responses.

DISCUSSION
Response Bias

D15. Like Dave Barry, people generally want to appear knowledgeable and agreeable, and they want to present a favorable face to the world. How might that affect the results of a survey conducted by a school on the satisfaction of its graduates with their education?

D16. Another part of the *Reader's Digest* poll, described on page 188, asked Americans if they agree with the statement that it is not the government's job to financially support television programming. The poll also asked them if they'd be disappointed if Congress cut its funding for public television. Which question do you think brought out more support for public television?

D17. How does nonresponse bias differ from voluntary response bias?

Summary 4.2: Why Take Samples, and How Not To

Important reasons for using a sample in many situations, rather than taking a census of the entire population, are as follows:

- Testing sometimes destroys the items.
- Sampling can save money.
- Sampling can save time.
- Sampling can make it possible to collect more or better information on each unit.

A sampling method is *biased* if it results in estimates that, on average, are too low or too high. This can happen if the method of taking the sample or the method of getting a response is flawed. Here are some sources of bias from the method of taking the sample:

- using a method that gives larger units a bigger chance of being in the sample (size bias)
- letting people volunteer to be in the sample
- using a sample just because it is convenient
- selecting the sampling units based on "expert" judgment
- constructing an inadequate sampling frame

Here are some types of bias derived from the method of getting the response from the sample:

- nonresponse bias
- questionnaire bias
- incorrect response or measurement bias

Practice

Bias

P11. Four people practicing shooting a bow and arrow made these patterns on their targets.

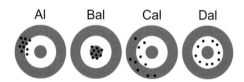

Al   Bal   Cal   Dal

a. Which person had shots that were biased and had low variability?
b. Which person had shots that were biased and had high variability?
c. Which person had shots that were unbiased and had low variability?
d. Which person had shots that were unbiased and had high variability?
e. Do you think it would be easiest to help Al, Cal, or Dal improve?

Sample Selection Bias

P12. What type of sample selection bias would result from each of these sampling methods?

a. A county official wants to estimate the average size of farms in a county in Iowa. He repeatedly selects a latitude and longitude in the county at random and places the farms at those coordinates in his sample. If something other than a farm is at the coordinates, he generates another set of coordinates.

b. In a study about whether valedictorians "succeed big in life," a professor "traveled across Illinois, attending high school graduations and selecting 81 students to participate. . . . He picked students from the most diverse communities possible, from little rural schools to rich suburban schools near Chicago to city schools." [Source: Michael Ryan, "Do Valedictorians Succeed Big in Life?" *Parade*, May 17, 1998, pp. 14–15.]

c. To estimate the percentage of students who passed the most recent AP Statistics Exam, a teacher on an Internet discussion list for teachers of AP Statistics asks teachers on the list to report to him how many of their students took the test and how many passed.

d. To estimate the average length of the pieces of string in a bag, a student reaches in, mixes up the strings, selects one, mixes them up again, selects another, and so on.

e. In 1984, Ann Landers conducted a poll on the marital happiness of women by asking women to write to her.

P13. Suppose the Museum of Fine Arts in Boston wants to estimate what proportion of people who come to Boston from out of town planned their trip to Boston mainly to visit the museum. The sample will consist of all out-of-town visitors to the museum on several randomly selected days. On buying a ticket to the museum, people will be asked whether they came from out of town and, if so, what the main reason for their trip was. Do you expect the museum's estimate to be too high, too low, or just about right? Why? What kind of sampling method is this?

Response Bias

P14. Consider this pair of questions related to gun control:

I. Should people who want to buy guns have to pass a background check to make sure they have not been convicted of a violent crime?

II. Should the government interfere with an individual's constitutional right to buy a gun for self-defense?

Which question is more likely to show a higher percentage of people who favor some control on gun ownership?

P15. In one study, educators were asked to rank Princeton's undergraduate business program. Every educator rated it in the top ten in the country. Princeton does not have an undergraduate business program. What kind of bias is shown in this case? [Source: Anne Roark, "Guidebooks to Colleges Get A's, F's," *Los Angeles Times*, November 21, 1982, Part I, p. 25.]

Exercises

E9. Suppose you want to estimate the percentage of U.S. households that have children under the age of 5 living at home. Each weekday from 9 a.m. to 5 p.m., your poll takers call households in your sample. Every time they reach a person in one of the homes, they ask, "Do you have children under the age of 5 living in your household?" Eventually you give up on the households that cannot be reached.

a. Will your estimate of the percentage of U.S. households with children under the age of 5 probably be too low, too high, or about right?

b. How does this example help explain why poll takers are likely to call at dinnertime?

c. Is this a case of sampling bias or of response bias?

E10. A wholesale food distributor has commissioned a sample survey to estimate the satisfaction level of his customers, who are the owners of small restaurants. The sampling firm takes a random sample from the current list of customers, develops a satisfaction questionnaire, and sends field workers out to interview the owners of the sampled restaurants. The field workers realize that the time does not allow them to interview all the owners selected for the sample, so if the owner is busy when the field worker arrives, the worker moves on to the next business.

a. What is the population? The sample?

b. What kind of bias does this survey have?

c. What kind of restaurants would you expect to be underrepresented in the sample?

E11. *Alcohol consumption.* Attempting to get accurate information on issues like alcohol consumption can be a trying task, as there is great potential for nonresponse to such questions in mailed or telephone surveys. One study that attempted to quantify the bias in such a study compared the responses to a mailed questionnaire (primary respondents) to the responses from some of the nonrespondents to the initial mailed questionnaire (secondary respondents), the latter being obtained by personal interviews. The percentages of responses in each consumption category are given in Display 4.9.

	Primary Respondents ($n = 133$)	Secondary Respondents ($n = 80$)
Frequent excessive drinker	3.8	6.3
Occasional excessive drinker	18.8	16.3
Moderate drinker	50.4	25.0
Abstainer	27.1	52.5

(Alcohol Consumption — row labels)

Display 4.9 Distribution of alcohol consumption of primary respondents and secondary respondents (%). [Source: *Alcohol and Alcoholism*, Vol. 37, 3 (2002), pp. 256–260.]

a. Describe the main ways in which nonresponse appears to have influenced the estimates from the initial mailed questionnaire.

b. The research paper concluded that "There is strong evidence for overrepresentation of nonresponse among abstainers, but weak evidence among frequent excessive drinkers." Do you agree? Do you find this surprising?

E12. *Student engagement.* A doctoral student at a college participating in the National Survey of Student Engagement (NSSE) studied the nonresponse bias of first-year undergraduate students who were in the sample selected to participate in the NSSE. Of the 197 students in the selected sample, only 103 responded (R). Of the 94 nonresponders, the researcher studied a sample of 25 (N) by intensive follow-up methods including personal interviews. For each activity, the involvement of each student was scored on a four-point scale of 1 to 4, with 4 indicating the highest level of engagement. A summary of the results for many of the activities is provided in Display 4.10.

a. In comparing the mean responses for these activities, describe the general pattern of behavior for the R group relative to the N group.

Survey Items	Group	Mean	SD
Active and collaborative learning			
Worked with classmates outside	R	2.11	0.791
of class	N	2.44	0.768
Tutored or taught other students	R	1.54	0.764
	N	1.92	0.909
Student-faculty interaction			
Received prompt feedback from	R	2.51	0.684
faculty on academics	N	3.04	0.790
Work on a research project with	R	1.93	*
a faculty member outisde class	N	3.04	0.978
Enriching educational experience			
Practicum, internship, co-op	R	3.02	0.727
experience, etc.	N	1.76	0.831
Community service or volunteer	R	3.28	0.986
work	N	1.88	1.013
Foreign language coursework	R	2.90	0.928
	N	1.88	0.900
Study abroad	R	1.89	0.803
	N	2.92	0.812
Culminating senior experience	R	1.90	0.934
	N	2.75	0.897

Display 4.10 Summary scores of engagement activities: Respondents (R) and Nonrespondents (N) [Source: *http://nsse.iub.edu*.]

b. How did the nonresponse bias the results? Why might this be important in a national study of student engagement?

E13. *HIV risk.* The Behavioral Risk Factor Surveillance System (BRFSS) of the U.S. Centers for Disease Control collects information on health risk behaviors, and many other health-related issues, for all 50 states, the District of Columbia, and U.S. territories by conducting large monthly surveys. Display 4.11 shows recent data on HIV risk as categorized by age level of the person responding. Risk is gauged mainly by a subject's response to questions about sexual behavior and drug use, along with other health conditions.

		Ages 18 to 24	Ages 25 to 44	Ages 45 to 64
Not at risk	Sample size	6,783	54,081	34,225
	Column %	39.1	51.9	26.3
At risk	Sample size	10,543	50,087	96,136
	Column %	60.9	48.1	73.7
Column total	Sample size	17,326	104,168	130,361
	Column %	100	100	100

Display 4.11 HIV risk percentages. [Source: CDC, BRFSS.]

a. Interpret the number 39.1 in the column for ages 18 to 24.

b. Describe the nature of the association between age group and HIV risk, as portrayed in this table. What are possible reasons for this pattern?

c. Why is there is great potential for bias in the survey procedure?

E14. *Obesity risk.* As stated in E13, the Behavioral Risk Factor Surveillance System (BRFSS) of the U.S. Centers for Disease Control collects information on health risk behaviors, and many other health-related issues, for all 50 states, the District of Columbia, and U.S. territories by conducting large monthly surveys. Display 4.12 shows recent data on obesity risk as categorized by age level of the person responding. The characterization of obesity is primarily in terms of body mass index (BMI), which is calculated according to the following formula:

$$BMI = \frac{weight}{height^2}$$

with weight measured in kilograms and height in meters. The BRFSS surveys depend on self-reported weight and height data.

	Male			
	Ages 18 to 24	Ages 25 to 44	Ages 45 to 64	Ages 65 or Older
Not at risk	53.1	29.2	23.5	32.1
At risk	46.9	70.8	76.5	67.9
Total	100.0	100.0	100.0	100.0

	Female			
	Ages 18 to 24	Ages 25 to 44	Ages 45 to 64	Ages 65 or Older
Not at risk	61.7	48.7	38.9	43.4
At risk	38.3	51.3	61.1	56.6
Total	100.0	100.0	100.0	100.0

Display 4.12 Percentage of adults at risk for obesity. [Source: CDC, BRFSS.]

a. Interpret the numbers 46.9 and 38.3 in the columns for ages 18 to 24.

b. Describe the nature of the association between age group and obesity risk, as portrayed in this table. How does the pattern for males differ from the pattern for females? What are possible reasons for these patterns?

c. Is there potential for bias in the survey procedure? Do you expect the BMIs that form the basic data for these percentages to be too low, about right, or too high? How will this affect the percentage of people in each age group who are at risk of obesity?

E15. Suppose you want to estimate the average response to the question "Do you like math?" on a scale from 1 ("Not at all") to 7 ("Definitely") for all students on your campus. You use your statistics class as a sample. What kind of sample is this? What sort of bias, if any, would be likely? Be as specific as you can. In particular, explain whether you expect the sample average to be higher or lower than the population average.

E16. At a meeting of local Republicans, the organizers want to estimate how well their party's candidate will do in their district in the next race for Congress. They use the people present at their meeting as their sample of voters. What kind of sample is this? Do you expect the estimate to be too favorable, about right, or too pessimistic?

E17. You want to estimate the average number of states that people living in the United States have visited. If you asked only people at least 40 years old, would you expect the estimate to be too high or too low? What if you take your sample only from those living in Rhode Island?

E18. For a study on smoking habits, you want to estimate the proportion of adult males in the United States who are nonsmokers, who are cigarette smokers, and who are pipe or cigar smokers. Tell why it makes more sense to use a sample than to try to survey every individual. What types of bias might show up when you attempt to collect this information?

E19. *Music video games.* Are video games "instrumental" in encouraging students to play real musical instruments? An informal poll by Fender (a guitar manufacturer) and a music education organization called Little Kids Rock contacted 812 music instructors across the nation to ask them, among other questions, if they felt that Rock Band™ and Guitar Hero® video games had increased their guitar enrollments. Among the 517 respondents, 67% indicated that they felt that their guitar program enrollments had increased, 31% indicated that their guitar enrollments had decreased, and 2% indicated these enrollments had stayed the same. [Source: *http://multiplayerblog.mtv.com/2009/ 01/15/fender-survey-shows-music-games-drive-students-to- play-real-instruments/*.]

  a. What kind of sample is this? (There may be more than one possibility.)

  b. Do you trust the results of the survey? Why or why not?

  c. What percentage of the entire population of music educators in the country do you think would say that music video games had a positive effect on their guitar enrollments: more than 67%, less than 67%, or just about 67%? Why?

E20. Are people willing to change their driving habits in the face of higher gasoline prices? At a time of steeply rising gas prices, a Consumers Union poll of Internet users who chose to participate showed these results. [Source: *www.consumersunion.org*.]

As a result of volatile gasoline prices in recent months, I have

Made no changes	42%
Driven less	53%
Carpooled	2%
Relied more on mass transportation	3%

Total Votes: 139

  a. What kind of sample is this?

  b. Do you trust the results of this survey? Why or why not?

  c. What percentage of U.S. drivers do you think drive less as a result of higher gasoline prices: more than

53%, less than 53%, or just about 53%? Why do you think that?

E21. To estimate the average number of children per family in the United States, you use your statistics class as a convenience sample. You ask each student in the sample how many children are in his or her family. Do you expect the sample average to be higher or lower than the population average? Explain why.

E22. Suppose you wish to estimate the average size of English classes on your campus. Compare the merits of these two sampling methods.

  I. You get a list of all students enrolled in English classes, take a random sample of those students, and find out how many students are enrolled in each sampled student's English class.

  II. You get a list of all English classes, take a random sample of those classes, and find out how many students are enrolled in each sampled class.

E23. A Gallup/CNN/*USA Today* poll asked this question:

> As you may know, the Bosnian Serbs rejected the United Nations peace plan and Serbian forces are continuing to attack Muslim towns. Some people are suggesting that the United States conduct air strikes against Serbian military forces, while others say we should not get militarily involved. Do you favor or oppose U.S. air strikes?

On the same day, the ABC News poll worded its question this way:

> Specifically, would you support or oppose the United States, along with its allies in Europe, carrying out air strikes against Bosnian Serb artillery positions and supply lines?

The ABC poll got a larger response favoring air strikes. Explain why that might be the case.

E24. Using the gun control example in P14 on page 191 as a guide, write two versions of a question about a controversial issue, with one version designed to get a higher percentage in favor and the other designed to get a higher percentage opposed. Test your questions on a sample of 20 people.

E25. Display 1.1 (page 3) lists data on the 50 employees of the engineering department at Westvaco. How would you take a sample of 20 employees that is representative of all 50 with regard to all the features listed: pay type, age, seniority, and whether laid off?

E26. Bring to class an example of a survey from the media. Identify the population and the sample, and discuss why a sample was used rather than a census. Do you see any possible bias in the survey method?

4.3 ▶ Experiments and Inference About Cause

The previous two sections showed you how to collect data that will allow you to generalize or infer from the sample you see to a larger but perhaps unseen population. A second kind of inference, maybe even more important than the first, takes you from a pattern you observe to a conclusion about what *caused* the pattern. Does regular exercise cause your heart rate to go down? Does storing coffee beans in the freezer keep them fresher longer than storing them at room temperature? Does passing a course in calculus make a student more likely to be accepted to medical school? Does smoking cigarettes cause lung cancer?

Cause and Effect

Children who drink more milk have bigger feet than children who drink less milk. There are three possible explanations for this association:

I. Drinking more milk causes children's feet to be bigger.

II. Having bigger feet causes children to drink more milk.

III. A lurking variable is responsible for both.

In this case, it is explanation III. The **lurking variable**—one that lies in the background and may or may not be apparent at the outset but, once identified, could explain the pattern between the variables—is the child's overall size. Bigger children have bigger feet, and they drink more milk because they eat and drink more of everything than do smaller children.

But suppose you think that explanation I is the reason. How can you prove it? Can you take a bunch of children, give them milk, and then sit and wait to see if their feet grow? That won't prove anything, because children's feet will grow whether they drink milk or not.

Can you take a bunch of children, randomly divide them into a group that will drink milk and a group that won't drink milk, and then sit and wait to see if the milk-drinking group grows bigger feet? Yes! *Such an experiment is just about the only way to establish cause and effect.*

Suppose you decide to investigate the coffee bean question of the opening paragraph. A more scientific way to pose the question is, "Does storing coffee beans in the freezer *cause* them to stay fresh longer than storing them at room temperature?" You now have two conditions, called **treatments**, to compare, namely "store in freezer" and "store at room temperature." You want to know how the treatments affect the **response** of interest, a measure of freshness in this case, on the **experimental units**, the bag of coffee you choose to use in the study. (When experimental units are people, they are called **subjects**.) Not all of the coffee beans are alike; so the fair way to divide them is by randomly assigning the freezer treatment to some and the room temperature treatment to the remainder. This gives some assurance that the more bitter beans in the bag do not all end up with the same treatment. If the beans stored in the freezer do end up being noticeably fresher for a longer period of time, you can say it was caused by the freezing because other possible causes should have been balanced across the two treatment groups by the randomization.

> Treat the two groups of beans exactly alike except for temperature.

A Real Experiment: Kelly's Hamsters

Here's an example of a real experiment, planned and carried out by Kelly Acampora as part of her senior honors project in biology at Mount Holyoke College. What happens when an animal gets ready to hibernate? This question is too general to answer with a single experiment, but if you know a little biology, you can narrow the question enough for it to serve as the focus of an experiment. Kelly relied on three previously known facts:

1. Golden hamsters hibernate.

2. Hamsters rely on the amount of daylight to trigger hibernation.

3. An animal's capacity to transmit nerve impulses depends in part on an enzyme called Na^+K^+ ATP-ase.

Here are the components of Kelly's experimental design:

Kelly's question: If you reduce the amount of light a hamster gets from 16 hours to 8 hours per day, what happens to the concentration of Na^+K^+ ATP-ase in its brain?

Subjects: Kelly's subjects were eight golden hamsters.

Treatments: There were two treatments: being raised in long (16-hour) days or short (8-hour) days.

Random assignment of treatments: To make her study a true experiment, Kelly randomly selected four hamsters to be raised in 16 hours of daylight, and the other four hamsters were raised in 8 hours of daylight.

Replication: Each treatment was given to four hamsters.

Response variable: Because Kelly was interested in whether a difference in the amount of light causes a difference in the enzyme concentration, she chose the enzyme concentration for each hamster as her response variable.

Results: The resulting measurements of enzyme concentrations (in milligrams per 100 milliliters) for the eight hamsters were

Short Days	12.500	11.625	18.275	13.225
Long Days	6.625	10.375	9.900	8.800

You can imagine Kelly defending her design:

Kelly: I claim that the observed difference in enzyme concentrations between the two groups of hamsters is due to the difference in the number of hours of daylight.

Skeptic: Wait a minute. As you can see, the concentration varies from one hamster to another. Some just naturally have higher concentrations. If you happened to assign all the high-enzyme hamsters to the group that got short days, you'd get results like the ones you got.

Kelly: I agree, and I was concerned about that possibility. In fact, that's precisely why I assigned day lengths to hamsters by using random numbers. The random assignment makes it extremely unlikely that all the high-enzyme hamsters got assigned to the same group. If you have the time, I can show you how to compute the probability.

Skeptic: (*Hastily*) That's okay for now. I'll take your word for it. But maybe you can catch me in Chapter 5.

The characteristics of the plan Kelly used are so important that statisticians try to reserve the word *experiment* for studies like hers that answer a question by comparing the results of treatments assigned to subjects at random.

DISCUSSION
Cause and Effect

D18. Plot Kelly's results in a dot plot, using different symbols for hamsters raised in short days and those raised in long days. In Chapter 11, you'll see a formal method for analyzing data like these to decide whether the observed difference between the long- and short-day hamsters is too big to attribute to chance. This method will be similar to the sort of logic introduced in the *Martin v. Westvaco* case in Chapter 1. For now, just offer your best judgment: Do you think the evidence supports a conclusion that the number of daylight hours causes a difference in enzyme concentration?

D19. Kelly has shown that hamsters raised in less daylight have higher enzyme concentrations than hamsters raised with more daylight. In order for Kelly to show that less daylight *causes* an increase in the enzyme concentration, she must convince us that there is no other explanation. Has she done that?

Confounding in Observational Studies

For sample surveys, selecting the sample at random protects against bias, which otherwise can easily mislead you into jumping to false conclusions. For comparative studies, a threat called *confounding* also can lead to false conclusions. Randomizing, when it is possible, protects against this threat in much the same way it protects against bias in sampling.

In everyday language, *confounded* means "mixed up, confused, at a dead end," and the meaning in statistics is similar.

> **Confounding of Two Variables with the Response**
>
> Two possible influences on an observed response are confounded if they are mixed together in a way that makes it impossible to separate their effects on the response.

Confounding makes it impossible to say which of two explanatory variables causes the response.

Studies that claim to show that review courses increase SAT scores often ignore the important concept of confounding. In one study, students at a large high school were offered an SAT preparation course, and SAT scores of students who completed the course were higher than scores of students who chose not to take the course. The positive effect of the review course was confounded with the fact that the course was taken only by volunteers, who would tend to be more motivated to do well on the SAT. Consequently, you can't tell if the higher scores of those who took the course were due to the course itself or to the higher motivation of the volunteers. As you read the next example, ask yourself which influences are being mixed together. Where's the confounding?

The thymus, a gland in your neck, behaves in a peculiar way. Unlike other organs of the body, it doesn't get larger as you grow—it actually gets smaller. Ignorance of this fact led early-20th-century surgeons to adopt a worthless and dangerous surgical procedure. Imagine yourself in their situation. You know that many infants are dying of what seem to be respiratory obstructions, and in your search for a treatment you begin to do autopsies on infants who die with respiratory symptoms. You've done many autopsies in the past on adults who died of various causes, so you decide to rely on those autopsy results for comparison. What stands out most when you autopsy the infants is that they all have thymus glands that look too big in comparison to their body size. Aha! That must be it! The respiratory problems are caused by an enlarged thymus. It became quite common in the early 1900s for surgeons to treat respiratory problems in children by removing the thymus. In particular, in 1912, Dr. Charles Mayo (one of the two brothers for whom the Mayo Clinic is named) published an article recommending removal of the thymus. He made this recommendation even though a third of the children who were operated on died. Looking back at the study, it's easy to spot the confounded variables.

> This thymus surgery example offers a cautionary tale about confounding.

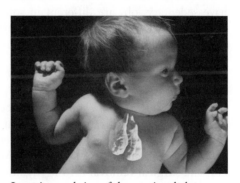

Location and size of thymus in a baby.

		Age	
		Child	Adult
Thymus size	Large	Problems	No evidence
	Small	No evidence	No problems

The doctors couldn't know whether children with a large thymus tend to have more respiratory problems, because they have no evidence about children with a smaller thymus. Age and size of thymus were confounded.

The thymus study is an example of an *observational study*, not an experiment.

Observational Studies Lack Randomization

In an **observational study,** the treatments are not assigned to the units by the experimenter—the conditions of interest are already built into the units being studied.

> Conditions aren't randomly assigned in an observational study.

Observational studies are less desirable than experiments, but often they are the only game in town. Suppose you want to study accident rates at a dangerous intersection to see if the rate on rainy days is higher than the rate on dry days. It turns out that most of the rainy days you have available for study are weekends and almost all the dry days are during the week. You cannot assign rain to a weekday! The best you can do is design your observational study to cover a long enough period so that you are likely to have enough weekdays and weekends in both groups. Otherwise, you have a serious confounding problem.

Don't make the mistake of thinking that observational studies are mere haphazard collections and observations of existing data. Good observational studies, like good experiments, can be designed to help answer specific questions. After researchers take into account every alternative explanation that they can think of, interesting associations can be observed. For example, epidemiologists studying the causes of cancer noticed that lack of exercise seemed to be associated with getting cancer. The best test of a cause-and-effect relationship between amount of exercise and getting cancer would be an experiment in which thousands of people would be assigned randomly to exercise or to not exercise. Researchers would have liked to conduct this experiment, but the practical difficulties were prohibitive and the ethics of such an experiment are questionable. To examine this issue further, one research study selected a group of patients with cancer and then matched them according to as many background variables as they could reasonably measure—except how sedentary their jobs had been—with patients who did not have cancer. Those with cancer turned out to have more sedentary jobs. But cause and effect can not be established based on such an observational study alone. The researchers are worried that a confounding variable is present—that people who have sedentary jobs are different in some other way from those who don't. As one researcher said, "The problem is the things we're not smart enough to know about, the things we haven't even thought of." [Source: Gina Kolata, "But Will It Stop Cancer?" *New York Times*, November 1, 2005.]

For drawing conclusions about cause and effect, a good randomized experiment is nearly always better than a good observational study.

These days, any new medical treatment must prove its value in a **clinical trial**—a randomized comparative experiment. Patients who agree to be part of the study know that a chance method will be used to decide whether they get the standard treatment or the experimental treatment. If Dr. Mayo had used a randomized experiment to evaluate surgical removal for the thymus, he would have seen that the treatment was not effective and many lives might have been spared. However, at the time, randomized experiments were not often used in the medical profession.

> A randomized experiment comparing medical treatments is called a clinical trial.

DISCUSSION
Confounding in Observational Studies

D20. Suppose the surgeons had examined infants without respiratory problems and found that their thymus generally was small.

   a. Have they now proved that a large thymus *causes* respiratory problems in children? If so, why? If not, what is another possible explanation?

   b. Design an experiment to determine whether removal of the thymus helps children with respiratory problems.

D21. What variables might be confounded with amount of exercise in the observational studies on the association of lack of exercise with cancer?

D22. What is the main difference between an experiment and an observational study?

Factors and Levels

Suppose Kelly decides to add two different diets to her hamster experiment—call them light and heavy. Now her experiment has two **factors,** diet and length of day, as shown in the table. Each factor has two **levels:** long and short for length of day, and heavy and light for diet. Choosing one level for length of day and one level for diet gives four possible treatments.

		Factor 1 Type of Diet	
		Light	**Heavy**
Factor 2	**Short**	Light-Short	Heavy-Short
Length of Day	**Long**	Light-Long	Heavy-Long

The Effect of Educational Television

Example 4.4

An educational research organization designed an experiment to test the effects of an educational TV show and the show's website on the learning of mathematical reasoning among upper elementary school students. The website had two types of activities, basic and advanced. The available students were randomly assigned to watch the show or not. Among these same students, some were randomly assigned to skip the website, while others were randomly assigned to use the website. Among the latter, some were randomly assigned to do only the basic activities and the rest to do both the basic and advanced activities. The response measurements were student scores on a variety of mathematical reasoning items. Describe the factors and levels. How many treatments were there? Make a diagram showing the treatments.

Solution
The two factors are TV show and web site. The TV watching is at two levels, none or all. The web site usage was at three levels: none, basic only, or basic and advanced. Thus, there were six treatments, to which students were randomly assigned.

		Website Usage		
		Don't Use	**Use Basic Activities Only**	**Use Basic and Advanced Activities**
TV Show	**Don't Watch**			
	Watch			

In an observational study, nothing is "treated," but the factor and level terminology still works fine. For example, in the thymus study one factor is the size of the thymus (large or small) and a second factor is age (child or adult).

The Importance of Randomizing

Confounding is the main threat to making reliable inferences about cause. You can think of confounding as the two groups you want to compare differing in some important way other than just on the conditions you want to compare. How can you guard against confounding? The best solution, whenever possible, is to randomize: Use a chance device to decide which people, animals, or objects to assign to each set of conditions you want to compare. *If you don't randomize, it's risky to generalize.*

Wait a minute! Are statisticians asking you to believe that if, say, you want to determine if smoking causes cancer, you should assign people to smoke or not on the basis of a coin flip? Of course not. But it *is* true that if you can't use a chance device to assign the conditions you want to compare, for example, smoker and nonsmoker, it becomes very difficult to draw sound conclusions about cause and effect.

Randomizing can protect against confounding.

Often, you can assign treatments at random to available subjects. For example, to evaluate a special course to prepare students for the SAT, you can start with a group of students who want to take the course. Randomly assign half to take the course and the other half to a group that will not get any special preparation. After both groups of students have taken the SAT, compare their scores. This time, if the students who took the course score 40 points better on average, you can be more confident that the difference is due to the course rather than to some confounding trait of the students willing to take a review course.

When you randomly assign treatments to the subjects available, you can infer cause and effect only for the group of subjects you have. Often, that doesn't seem good enough. For example, you would like to conclude that the special course causes an increase in SAT scores for the population of all students who would like a preparation course. In the best of all possible worlds, you would be able to select the subjects at random from that population and then assign the treatments randomly to the random sample. Then you could make the conclusion you want. But in practice such an experiment is impossible unless the population is small and accessible and cooperative. Instead, you must use the group of students that you can round up locally and who are willing to participate in your experiment. You can conclude only that the special course increases SAT scores for this particular group.

Why Randomization Makes Inference Possible

- If you assign treatments to units at random, then there are only two possible causes of any difference in the responses to the treatments: chance or the treatments.

- If the probability is small that chance alone will give you such a difference in the responses, then you can infer that the cause of the difference was the treatment.

A Control or Comparison Group Is Vital

An article on homeopathic medicine in *Time* ["Is Homeopathy Good Medicine?" September 25, 1995, pp. 47–48] began with an anecdote about a woman who had been having pain in her abdomen. She was told to take calcium carbonate, and after two weeks her pain had disappeared. It was reported that her family "now turns first" to homeopathic medicine. This kind of personal evidence—"It worked for me"—is called *anecdotal evidence*.

© Scott Adams/Dist. by United Feature Syndicate, Inc.

Perhaps no one has put anecdotal evidence in its proper place more eloquently than Professor Randy Pausch, who, while dying of cancer, answered a question about various proposed remedies:

I've received 10,000 e-mails—many of them telling me about different remedies. But my first filter is, "Has it been through any kind of clinical study?" **The plural of anecdote is not data,** so if you know three people that did some alternative cure, that's positive, but it's not the same thing as real, clinically proved data. [Source: Randy Pausch, Carnegie-Mellon University, *Time*, April 21, 2008, p. 4.]

Anecdotal evidence can be useful in deciding what treatments might be helpful and so should be tested in an experiment. However, anecdotal evidence cannot *prove*, for example, that calcium carbonate causes abdominal pain to disappear. Why not? After all, the pain did go away.

> Anecdotal evidence isn't proof.

The problem is that pain often tends to go away anyway, especially when a person thinks he or she is receiving good care. When people believe they are receiving special treatment, they tend to do better. In medicine, this is called the **placebo effect.**

A **placebo** is a fake treatment, something that looks like a treatment to the patient but actually contains no medicine. Carefully conducted studies show that a large percentage of people who get placebos, but don't know it, report that their symptoms have improved. The percentage depends on the patient's problem but typically is over 30%. For example, when people are told they are being treated for their pain, even if they are receiving a placebo, changes in their brain cause natural painkilling endorphins to be released. Consequently, if people given a treatment get better, it might be because of the treatment, because time has passed, or simply because they are being treated by someone or something they trust.

How can scientists determine if a new medication is effective or whether the improvement is due either to the placebo effect or to the fact that many problems get better over time? They use a group of subjects who provide a standard for comparison. The group used for comparison usually is called a **control group** if the subjects receive a placebo or a **comparison group** if the subjects receive a standard treatment.

> A control group or a comparison group provides the basis of comparison.

Patients in the **treatment group** are given the drug to be evaluated. The patients given a placebo get a nontreatment carefully designed to look as much like the actual treatment as possible. *The control and treatment groups should be handled exactly alike except for the treatment itself.* If a new treatment is to be compared with a standard treatment, the comparison group receives the standard treatment rather than a placebo.

In order for the control or comparison group and the treatment group to be treated exactly alike, both the subjects and their doctors should be "blind." That is, the patients should not know which treatment they are receiving, and the doctors who evaluate how much the patients' symptoms are relieved should be blind as to what the patients received. If only the patients don't know, the experiment is said to be **blind.** If both the patients and the doctors don't know, the experiment is said to be **double-blind.**

Typical Medical Study Designs

Comparing a treatment and a placebo. To test the stimulus effect of caffeine, 30 male college students were randomly assigned to one of three groups of ten students each. Each group was given one of three doses of caffeine (0, 100, and 200 mg), and 2 hours later the students were given a finger-tapping exercise. The measured response was the number of taps per minute. The 0-mg dose can be considered a placebo. Ideally, the three groups would have been given pills that look exactly alike so that they could not tell which treatment they were receiving. This design allows comparisons to be made among two active treatments and a placebo to sort out the effect of caffeine. You will analyze the results of this experiment in Chapter 13. [Source: Draper and Smith, *Applied Regression Analysis* (John Wiley & Sons, 1981).]

Comparing a new treatment and a standard treatment. In an early study on AIDS-related complex (ARC), the drug zidovudine (commonly known as AZT) combined with acyclovir (ACV) was compared to AZT alone (a standard treatment). A total of 134 patients with ARC agreed to participate in the study, which was designed and run as a double-blind, randomized clinical trial. Each patient was randomly assigned either AZT by itself or a combination of AZT and ACV. One of the outcome measures was how many of the ARC patients developed AIDS during the year of the study. A placebo was not used here—such a "treatment" would be unethical because of the health profession's obligation to administer a treatment thought to be effective. You will analyze the outcome of this experiment in Chapter 9. [Source: David A. Cooper et al., "The Efficacy and Safety of Zidovudine Alone or as Cotherapy with Acyclovir for the Treatment of Patients with AIDS and AIDS-Related Complex: A Double Blind, Randomized Trial," *AIDS*, Vol. 7 (1993), pp. 197–207.]

Comparing a new treatment, a standard treatment, and a placebo. Another experiment compared an undesirable side effect of some antihistamines—drowsiness—for two treatments (meclastine and promethazine) and a placebo. At the time of the study, meclastine was a new drug and promethazine was a standard drug known to cause drowsiness. Each of nine subjects was given each of the three treatments on different days, in random order. The subjects were blind as to which treatment they were receiving. The outcome measure was the *flicker frequency* of patients (the number of flicks of the eyelids per minute). Low flicker frequency is related to drowsiness, because the eyes are staying shut too long. Note that each subject received all treatments. The only randomization was in the order in which he or she received them. You will analyze the results of this experiment in Chapter 9. [Source: D. J. Hand et al., *A Handbook of Small Data Sets* (London: Chapman and Hall, 1994), p. 8.]

To summarize, a good experiment must have both a random assignment of treatments to units and a control or comparison group that is compared to the group getting the treatment of interest. Such an experiment is called a **randomized comparative experiment.**

DISCUSSION
A Control or Comparison Group Is Vital

D23. Why is blinding or double-blinding important in an experiment?

D24. A report about a new study to test the effectiveness of the herb St. John's wort to treat depression says, "The subjects will receive St. John's wort, an antidepressant drug, or a placebo for at least eight weeks and as long as six months." Describe how you would design this study to compare the effects of the three treatments. [Source: *Los Angeles Times*, August 31, 1998, p. A10.]

D25. How would you design an experiment to verify that a placebo effect exists for people who think they are being treated for pain?

Experimental Units and Replication

In educational research, children appear in a group—their classroom—so in order to compare two methods of teaching reading, say, a researcher must assign entire classrooms randomly to the two methods. In such cases, the classroom, not an individual student, is the experimental unit.

Suppose a researcher had six classrooms, each with 25 students, available for her study and assigned three classrooms to each method. The researcher might like to say that each method was replicated on 75 students because the reading ability of 75 students was measured, but, alas, she can claim only that each method was replicated on three classrooms. The students within each classroom were selected as a group and treated in a group setting and so do not contribute independent responses.

Treatments Are Randomly Assigned to the Units

Experimental units are the people, animals, families, classrooms, and so on to which treatments are randomly assigned.

Replication is the random assignment of the same treatment to different units.

Performing many replications in an experiment is the equivalent of having a large sample size in a survey—*the more units you have, the more faith you have in your conclusions.* (Be careful, however, when someone uses the word *replication.* Sometimes it is used to mean that your entire experiment was repeated by someone else who came to the same conclusion you did.)

This box summarizes what you have learned about the characteristics of a well-designed experiment.

Characteristics of a Well-Designed Experiment

- *Compare.* A treatment group is compared to a control group, or two or more treatment groups are compared to each other.
- *Randomize.* Treatments are randomly assigned to the available experimental units.
- *Replicate.* Each treatment is randomized to enough experimental units to provide adequate assessment of how much the responses from the same treatment vary.

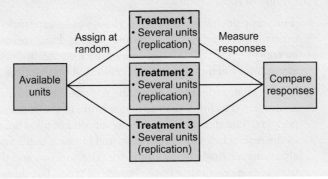

The next example describes an experiment with a design that is typical of experiments done in industrial settings.

Popping Corn

Example 4.5

One Saturday afternoon, you decide to compare the number of kernels left unpopped by a generic and a brand-name popcorn and by your hot air popper and your oil popper. You have the time to pop 20 batches of popcorn before hitting the gym.

a. What are the factors and levels? Describe the treatments.

b. How will you design this study? What are the experimental units? How would randomization be a part of it? Name a variable that might be confounded with your treatments if you didn't randomize.

c. Have you designed a sample survey, an experiment, or an observational study?

Solution

a. The factors are the type of popcorn and the type of popper. There are two levels for type of popcorn: generic and brand name. There are two levels for type of popper: hot air and oil. The four treatments are the four combinations of the two types of popcorn with the two types of popper. The treatments are represented by the four cells in the table.

	Type of Popper	
	Hot Air	Oil
Type of Popcorn — Generic		
Brand Name		

b. You will have time to pop five batches for each treatment. Because you can't select bags of popcorn at random (all the bags in your grocery store probably were produced at the same time), you will have to assume—without much justification—that the bags of popcorn are pretty much all alike within each type. You are using the poppers that you have. They aren't selected at random either.

What you can randomize is the order that you assign to the 20 batches to be popped. You have 20 time periods available to pop batches of popcorn. Your experimental unit is one of these time periods. Place 20 slips of paper in a hat, five with hot air/generic written on them, five with hot air/brand name, five with oil/generic, and five with oil/brand name. Draw them out one at a time, popping the next batch according to the treatment. This type of randomization minimizes confounding the treatment with variables such as how hot the popper gets or the changing humidity in your kitchen.

c. The real world is messy and many designed studies cannot be categorized perfectly as a survey, an observational sturdy, or an experiment. This is one of those studies.

- The study described in part b is closest to an experiment. The experimental units are the 20 time periods. Four treatments were to be compared. Each unit had one of the four treatments assigned to it at random. Each treatment was replicated on five time periods.

  The random assignment over time reduces confounding with variables related to time, such as temperature of the room and of the poppers. The statistics alone, however, do not allow generalization of the results beyond the popcorn and poppers used. Such generalizations are made by assuming that the popcorn and the poppers used are representative of the brands or types, but these assumptions cannot be confirmed by the data generated in the experiment. That is why real-world experiments generally are repeated under varying conditions.

- If you didn't randomize the four treatments to the time period in your answer to part b, there would be no randomization at all and you would have an observational study.

- If you had been able to use random samples of kernels from each of the two different brands of popcorn, the design would be closer to a sample survey than to an experiment. You would be surveying the two populations of popcorn kernels to determine the percentage that would remain unpopped in each of your two poppers.

Giving a name to the type of study isn't important. What is important is that you randomize what you can. In the next section, you will learn more about how to deal with variables that cannot be randomly assigned to the experimental units.

DISCUSSION
Experimental Units and Replication

D26. In the teaching-methods study (page 202), why does it seem reasonable that the experimental unit should be the classroom? Provide an example where the experimental unit might be the school.

D27. To study whether women released from the hospital 1 day after childbirth have more problems than women released 2 days after childbirth, many hospitals in a large city were recruited to participate in an experiment. Each participating hospital was randomly assigned to release all women giving routine births either after 1 day or after 2 days of hospital care. An assessment of problems encountered after the women returned home was then done on a random sample of women released from each hospital. What experimental unit should be used as the basis of this study?

D28. Why do you have more faith in your conclusions if your experiment has many replications?

Summary 4.3: Experiments and Inference About Cause

The goal of an experiment is to compare the responses to two or more treatments. The key elements in any experiment are

- randomization of treatments to units
- replication of each treatment on a sufficient number of units
- a control or comparison group

Randomly assigning treatments to units allows you to make cause-and-effect statements and protects against confounding. If you can't randomize the assignment of treatments to units, you have an observational study, and confounding will remain a threat. Confounding makes it impossible to determine whether the treatment or something else caused the difference in the responses between the groups.

The amount of information you get from an experiment depends on the number of replications. Recognizing experimental units is crucial.

So that you have a basis for comparison, experiments require either a control group receiving no treatment or a comparison group receiving a second treatment.

Practice

Cause and Effect

P16. Research has shown a weak association between living near a major power line and the incidence of leukemia in children. Such a study might measure the incidence of leukemia in children who live near a major power line and compare it to the incidence in children who don't live near a major power line. Typically, the children in the areas near major power lines are matched by characteristics such as age, sex, and family income to children in the areas not near major power lines.

  a. Identify the subjects, treatments, and response variable in such a study.

  b. Is this type of study a true experiment? Explain why or why not.

  c. According to a newspaper article, "While there is a clear association between high-voltage power lines and childhood leukemia, there is no evidence that the power lines actually cause leukemia." [Source: "Power lines tie to cancer unknown," *Paris (Texas) News*, July 30, 2006.] Why might the newspaper come to this conclusion?

P17. Solar thermal systems use heat generated by concentrating and absorbing the sun's energy to drive a generator and produce electrical power. A manufacturer of solar power generators is interested in comparing the loss of energy associated with the reflectivity of the glass on the solar panels. (A measure of energy loss is available.) Three types of glass are to be studied. The company has 12 test sites in the Mojave Desert, in the southwestern United States. Each site has one generator. The sites might differ slightly in the amount of sunlight they receive.

  a. How would you design an experiment to compare the energy losses associated with reflectivity?

  b. What plays the role of "subjects" in your experiment in part a? What are the treatments?

  c. How would you change your design if each site had six generators?

P18. Each pair of variables here is strongly associated. Do you think that I causes II, II causes I, or is a lurking variable responsible for both?

  a. I. Wearing a hearing aid or not
    II. Dying within the next 10 years or not

b. I. The amount of milk a person drinks

II. The strength of his or her bones

c. I. The amount of money a person earns

II. The number of years of schooling

Confounding in Observational Studies

P19. Show the confounding in the SAT study described on page 197 by drawing and labeling a table like the one in the thymus example also on page 203.

P20. Review Example 4.4 on page 199. Suppose the study had been designed by classifying a very large group of upper-elementary students according to which of the six treatments they fell closest to, based on their behavior over the past year. The mathematical reasoning ability of each student was then measured. Give an example of a variable that could be confounded with the treatment in which the student was classified.

Factors and Levels

P21. Does the type of lighting or music in a dentist's waiting room have any effect on the anxiety level of the patient? An experiment to study this question could have nine treatments, represented by this table.

		Type of Music		
		Pop	Classical	Jazz
Brightness of Room	Low			
	Medium			
	High			

a. What are the factors? What are the levels?

b. Describe a possible response variable.

P22–23. Decades ago, when there was less agreement than there is now about the bad effects of smoking on health, a large study compared death rates of three groups of men—nonsmokers, cigarette smokers, and pipe or cigar smokers. The results are show in Display 4.13.

	Deaths (per 1000 men per year)
Nonsmokers	20.2
Cigarette smokers	20.5
Pipe or cigar smokers	35.5

Display 4.13 Death rates of males with various smoking habits. [Source: Paul R. Rosenbaum, *Observational Studies* (New York: Springer-Verlag, 1995).]

P22. The numbers seem to say that smoking a pipe or cigars almost doubles the death rate, from about 20 to 35 per 1000, but that smoking cigarettes is pretty safe.

a. Do you believe that is true? If not, can you suggest a possible explanation for the pattern?

b. Is this study an observational study or an experiment?

c. What is the factor? What are the levels? What is the response variable?

P23. The investigators also recorded the ages of the men in the study, so it was possible to compare the average ages of the three groups.

	Average Age (yr)
Nonsmokers	54.9
Cigarette smokers	50.5
Pipe or cigar smokers	65.9

Display 4.14 Average age of males with various smoking habits.

Does the information in Display 4.14 help you account for the pattern of death rates? If so, tell how. If not, tell why not. Now what are the factors in this study?

P24. *Driving age.* You want to know whether raising the minimum age for getting a driver's license will save lives. You compare the highway death rates for the 50 U.S. states, grouped according to the legal driving age.

a. Is this study observational or experimental?

b. What is the factor? What are the levels? What is the response variable?

c. Think of a variable that might be confounded with legal driving age.

The Importance of Randomizing

P25. Neonatal mortality (death during the first seven days of life) is one of the huge health problems in developing countries. A study in Namibia focused on low birth weight as a possible explanatory variable. Each baby who died (called a *case*) was matched with five *controls*. These controls were the babies who were born immediately before (2 babies) and after (3 babies) the baby who died, but who survived. Each baby then was classified according to low birth weight (LBW) or not. Display 4.15 shows the table for the 281 babies for whom complete information was available.

		Neonatal Deaths (cases)	Surviving Babies (controls)	Total
Low birth weight?	Yes	28	29	57
	No	16	208	224
	Total	44	237	281

Display 4.15 Data for the Namibia neonatal mortality study. [Source: Gotty Muharukua, et al. "Factors Contributing to High Incidence of Neonatal Mortality in Windhoek, Namibia," *Health Systems Research Newsletter/Journal*, World Health Organization (September 1998), pp. 27–29, *www.afronets.org/files/news.22.pdf*.]

a. What is the response variable in this study? What is the explanatory variable?

b. Is this a survey, an observational study, or an experiment? Explain your answer.

c. Compare the percentage of cases who had low birth weight with the percentage of controls who had low birth weight. What appears to be the conclusion?

d. Because the babies could not be randomly assigned to low birth weight or not, many variables could be confounded with birth weight. Name at least three such variables.

e. One possible confounding variable is whether the baby was born prematurely. The two tables in Display 4.16 split the data from the main table into one for premature births and one for full-term births. For each table, compare the percentage of cases who had low birth weight with the percentage of controls who had low birth weight. Explain why premature birth status is a confounding variable.

Premature Births		Neonatal Deaths	Controls	Total
Low birth weight?	Yes	27	17	44
	No	1	5	6
	Total	28	22	50

Full-Term Births		Neonatal Deaths	Controls	Total
Low birth weight?	Yes	1	12	13
	No	15	203	218
	Total	16	215	231

Display 4:16 The data for the Namibia neonatal mortality study split into two tables based on whether the baby was born prematurely.

A Control or Comparison Group is Vital

P26. In Dr. Mayo's thymus studies, described on page 197, what did he use as a control group? Could the placebo effect have been a factor in any successful surgeries he had? Was it possible for the study to be blind? Was it possible for the doctors who evaluated how well the patients were doing after surgery to be blind?

P27. In a study to see if people have a "magnetic sense"—the ability to use Earth's magnetism to tell direction—students were blindfolded and driven around in a van over winding roads. Then they were asked to point in the direction of home. They were able to do so better than could be explained by chance alone. [Source: "Tests Point Away from 6th Sense," *Los Angeles Times*, July 19, 1982.]

a. Are you satisfied with the design of this study? Are there any other possible explanations for the results other than a magnetic sense?

To improve the design, the experimenter placed magnets on the back of some students' heads, which were supposed to confuse their magnetic sense, and nonmagnetic metal bars on the back of other students' heads.

b. What is the control group in the new design?

c. One very important thing wasn't mentioned in the description of the design of this experiment. What is that?

d. Are you satisfied with the new design?

Later it was determined that the magnetic bars tended to stick to the metal wall of the van.

e. Was this a blind experiment? Was this a double-blind experiment?

Experimental Units and Replication

P28. You want to compare two different textbooks for a statistics course. Ten classes with a total of 150 students will take part in the study. To judge the effectiveness of the books, you plan to use the students' scores on a common final exam. The classes are randomly divided into two groups of five classes each. The first group of five classes, with 80 students total, uses the first of the two books. The remaining five classes, with 70 students total, use the other book. Identify the treatments and the experimental units in this study, and then identify the number of units.

P29. You have a summer job working in a greenhouse. The manager says that she has discovered a wonderful new product that will help carnations produce larger blooms, and you decide to design an experiment to check it out. What are your experimental units? How will you use randomization and replication? Do you need a control or comparison group?

P30. On another lazy Saturday afternoon, you decide to compare the strength of two brands of paper towels, Brand A and Brand B, each under the conditions wet and dry. You have one roll of each brand on hand and have time to test 20 towels from each roll before hitting the library. One by one, you stretch each towel in an embroidery hoop, either leaving the towel dry or wetting it with a tablespoon of water, and then place the hoop over a bowl. You place a penny in the center of the hoop, wait 3 seconds, and place another penny on top, continuing until the towel breaks. You record the number of pennies on the paper towel before the towel breaks.

a. What are the factors and levels? What is the response variable?

b. How should you use randomization in this study? What are the experimental units?

c. Is your study most like a survey, an observational study, or an experiment?

Exercises

E27. A psychologist wants to compare children from 1st, 3rd, and 5th grades to determine the relationship between grade level and how quickly a child can solve word puzzles. Two schools have agreed to participate in the study. Would this be an observational study or an experiment? Explain.

E28. An engineer wants to compare traffic flow at four busy intersections in a city. She chooses the times to collect data so that they cover morning, afternoon, and evening hours on both weekdays and weekends. Is this an experiment or an observational study?

E29. *Buttercups.* Some buttercups grow in bright, sunny fields; others grow in woods, where it is both darker and damper. A plant ecologist wanted to know whether buttercups in sunny locations have adapted in a way that makes them less successful in the shade than in the sun. For his study, he dug up ten plants in sunny locations. Five plants were chosen at random from these ten plants and then were replanted in the sun; the remaining five plants were replanted in the shade. At the end of the growing season, he compared the sizes of the plants.

    a. Does this study meet the three characteristics (randomization, replication, control or comparison group) of a true experiment?

    b. Why did the plant ecologist bother to dig up the plants that were just going to be replanted in the sunny location? Use the word *confounded* in your answer.

E30. A metallurgist is studying the properties of copper disks produced by sintering (heating powder until it becomes a solid mass). He has two types of powder available and wants to consider three sintering temperatures. One oven is available for the sintering, and many disks can be made in each run. However, each run can involve only one type of powder.

    a. What are the factors and levels? How many treatments are there?

    b. How should the metallurgist incorporate randomization into the design of his study?

    c. How could you make this study blind?

E31. The college health service at a small residential college wants to see whether putting antibacterial soap in the dormitory bathrooms will reduce the number of visits to the infirmary. In all, 1800 students from 20 dormitories participate. Half the dormitories, chosen at random, are supplied with the special soap; the remaining ones are supplied with regular soap. At the end of one semester, the two groups of students are compared based on the average number of visits to the infirmary per person per semester.

    a. What are the units?

    b. How many units are there?

    c. Is this an observational study or an experiment?

E32. If you've studied chemistry or biology, you may know that different kinds of sugar have different molecular structures. Simple sugars, like glucose and fructose, have six carbon atoms per molecule; sucrose, a complex sugar, has twice as many atoms. A biologist wants to know whether complex sugars can sustain life longer than simple sugars. She prepares eight petri dishes, each containing ten potato leafhoppers. Two dishes are assigned to a control group (no food) and two each to a diet of glucose, fructose, and sucrose. The response variable is the time it takes for half the leafhoppers in a dish to die.

    a. What are the units?

    b. How many units are there?

    c. Is this an observational study or an experiment?

E33. In developing countries, diarrhea is a serious cause of death among babies. A study on the possible deterrents to this health problem identified bottle-feeding versus breastfeeding, exposure to one of two different education programs on healthy babies, and long stay versus short stay in the hospital after birth as conditions to be studied.

    a. Identify the factors and levels that should be used to define the treatments of this study. How many treatments are there?

    b. Describe how researchers might have randomized treatments to the subjects.

    c. Low birth weight is another possible contributing condition to the diarrhea problem. Can this be treated as a factor in designing the experiment? Can it be randomized?

E34. In the 1990s scientists became concerned about the worldwide declines in the population sizes for amphibians. Because many amphibians lay their eggs in shallow water fully exposed to sunlight, a professor at Oregon State University thought that one of the possible causes was the increasing amount of ultraviolet (UV) light to which the eggs were exposed. He and his students collected thousands of eggs of amphibians and put them into enclosures. Each enclosure had a plastic cover that blocked UV rays, a plastic cover that did not block UV rays, or no plastic cover. There were four enclosures of each type, which were randomly placed in shallow water. [Source: A.R. Blaustein, "Amphibians in a Bad Light," *Natural History*, Vol. 103 (1994), pp. 32–37, *http://oregonstate.edu/~blaustea/pdfs/NatHist.pdf.*]

Frog spawn.

a. Describe a response variable that could have been used.

b. Describe the factors and levels for this experiment. How many treatments are there?

c. What are the experimental units?

d. Why was it necessary to have both enclosures with no plastic cover and enclosures with a plastic cover that did not block UV rays?

E35. *Climate and health*. Does living in a colder climate make you healthier? To study the effects of climate on health, imagine comparing the death rates in two states with very different climates, such as Florida and Alaska. (The death rate for a given year tells how many people out of every 100,000 died during that year.) In 2006, Florida's death rate was nearly twice as high as Alaska's. Why is death rate used as a response variable rather than number of deaths? Why might even death rate not be a good choice for a response variable? (Think about the kinds of people who move to the two states.) Is this an experiment or an observational study? [Source: National Center for Health Statistics, *www.cdc.gov*.]

E36. You are to direct a study to compare the graduation rates of the 35 colleges and universities in your state.

a. How will you define "graduation rate"? Why is it important to look at rates rather than the number of graduates?

b. What issues will determine whether you do a census, a sample survey, an observational study, or an experiment?

c. Will you be able to determine the cause of low graduation rates from the data collected in your study?

E37. Suppose you have 50 subjects and want to assign 25 subjects to Treatment A and 25 to Treatment B. You flip a coin. If it is heads, the first subject goes into Treatment A; if it is tails, the first subject goes into Treatment B. You continue flipping the coin and assigning subjects until you have 25 subjects in one of the treatment groups. Then the rest of the subjects go into the other treatment group. Does this method randomly divide the 50 subjects into the two treatment groups? Explain.

E38. Suppose you have recruited 50 subjects and want to assign 25 subjects to Treatment A and 25 to Treatment B. For the first subject to arrive on the day of the study, you flip a coin. If it is heads, the first subject goes into Treatment A; if it is tails, the first subject goes into Treatment B. The second subject to arrive then goes into the other treatment group. For the third subject you flip the coin and again make the random assignment. The fourth subject goes into the opposite treatment from that of the third, and so on. (You flip the coin for the odd-numbered subjects only.) Does this method randomly divide the 50 subjects into the two treatment groups? What good property does this method have?

4.4 ▶ Designing Experiments to Reduce Variability

In the previous section, you learned the basics of a good experiment: Randomly assign one of two or more treatments to each experimental unit, and then handle them as alike as possible except for the treatment itself. Within this framework, however, you have further choices about experimental design. If you can protect yourself against confounding by randomizing, designing a good experiment becomes mainly a matter of managing variability. In the first part of this section, you will learn about two types of variability in experiments: one type that you want because it reveals differences between treatments, and another type that you don't want because it obscures differences between treatments. The rest of the section will show you how a good experimental design can reduce the "bad" kind of variability.

Difference Between Treatments Versus Variability Within Treatments

The results for Kelly's experiment with hamsters raised in short days and hamsters raised in long days (page 195) are given again in Display 4.17. There is a difference in the response *between* the two treatments. The average enzyme concentration for long days is 8.925, and the average concentration for short days is 13.906, a difference of 4.981. There also is variability *within* each treatment—like all living things, hamsters will vary, even when treated exactly alike. In fact, for short days, the difference between the largest and smallest enzyme concentrations (18.275 − 11.625, or 6.65) is even larger than the difference between the two treatment means. Still, because the concentration for each of the short days is larger than the concentration for each of the long days, you probably believe the treatment made a difference.

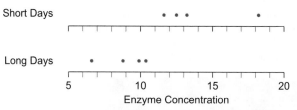

Display 4.17 Dot plots of enzyme concentration in Kelly's hamsters (mg/100ml).

Suppose Kelly had gotten results with more spread in the values but the same means. As Display 4.18 shows, with the values more spread out, it's no longer obvious that the treatment matters.

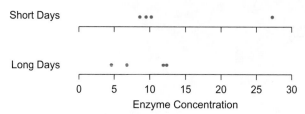

Display 4.18 Altered hamster data: same means, more spread.

To conclude that the treatments make a difference, the difference between the treatment means has to be large enough to overshadow the variation within each treatment.

A good experimental plan can reduce within-treatment variability and will allow you to measure the size of the variability that remains both between and within treatments.

A Design for Every Purpose

The Completely Randomized Design

> In a completely randomized design, treatments are randomly assigned to units without restriction.

In a **completely randomized design (CRD)**, the treatments are assigned to subjects at random, with the only restriction being that the design should be balanced by having each treatment assigned to about the same number of subjects. This can be accomplished, for example, by placing the names of the subjects in a hat and randomly drawing half of them to be assigned Treatment 1, the other half receiving Treatment 2. Kelly Acampora used a completely randomized design for her hamster experiment.

Steps in Creating a Completely Randomized Design (CRD)

1. Number the available experimental units from 1 to n.
2. If you have three treatments, for example, use a random digit table or your calculator to pick $n/3$ integers at random from 1 to n, discarding any repetitions. The units with those numbers will be given the first treatment. Again pick $n/3$ integers, discarding any repetitions. The units with those numbers will be given the second treatment. The remaining units will get the third treatment.

Sit or Stand: Completely Randomized Design   Twenty-two students conducted an experiment to decide if there is a difference in heart rate measured under two treatments:

Treatment 1: standing, with eyes open

Treatment 2: sitting relaxed, with eyes closed

Standing Pulse Rate	Sitting Pulse Rate
60	68
62	68
64	60
88	68
86	60
64	62
72	80
72	58
70	60
48	68
92	74

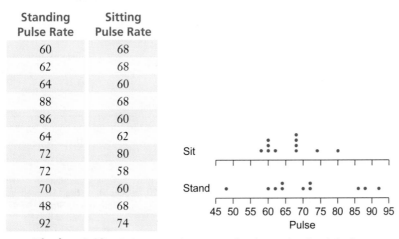

Display 4.19   Pulse rates for a completely randomized design.

The students were randomly assigned to the treatments, so that 11 students were in each treatment group. Display 4.19 shows the results for this completely randomized design. The parallel dot plots show that the pulses for the "stand" group might be a little higher on average, but the large within-treatment variability obscures a clear view of any real differences between the treatment means.

The Randomized Paired Comparison Design   Initially frustrated in their attempt to decide if there is a difference in heart rate when standing and sitting, the students next tried a new design. Each student took his or her pulse rate while sitting. They then paired up, with the two highest sitting pulse rates in the same pair, the next two highest sitting pulse rates in the next pair, and so on. Each pair flipped a coin to decide who would sit and who would stand while their pulse rates were taken again. The two pulse rates within each pair could then be compared. This is an example of a **randomized paired comparison** (or **matched pairs**) design. The pairs of students are called *blocks*.

> Blocks are groups of *similar* units.

Steps in Creating a Randomized Paired Comparison (Matched Pairs) Design

1. Sort your available experimental units into pairs of *similar* units, called *blocks*. The two units in each pair should be enough alike that you expect them to have a similar response to any treatment.

In a randomized paired comparison design, pairs of similar units are randomly assigned different treatments.

2. Randomly decide which unit in each pair is assigned which treatment. For example, you could flip a coin, with heads meaning the first unit gets the first treatment and tails meaning the first unit gets the second treatment. The other unit in the pair gets the other treatment.

Sit or Stand: Matched Pairs Design   Display 4.20 gives the results of the students' matched pairs experiment, where students were paired based on similar sitting pulse rates. Then the treatment of sitting or standing was randomly assigned to the students within each pair.

The upper plot in Display 4.20 compares the pulse rates for each treatment, ignoring the fact that the data were paired on subjects with similar resting pulse rates. As in Display 4.19, this plot shows little difference between the treatments. Because the data are paired, the difference between the standing and sitting pulse rates, as shown in the lower plot, is a reasonable measure of the different effect of the treatments. Over half of these differences are greater than zero, and the mean of the differences is well above zero.

Pair	Standing Pulse Rate	Sitting Pulse Rate	Difference: Stand − Sit
1	64	42	22
2	66	52	14
3	58	64	−6
4	60	58	2
5	80	82	−2
6	80	68	12
7	66	62	4
8	62	64	−2
9	54	64	−10
10	66	60	6
11	64	50	14

Display 4.20   Difference in pulse rates for a matched pairs design. In the upper plot, the pulse rates are shown separately for each treatment. The lower plot shows the difference of the pulse rates within each pair, *stand – sit*.

In a **repeated measures design**, the same subject gets several treatments, in random order.

Sit or Stand: Repeated Measures Design   Finally, the students conducted the sitting/standing experiment using a third design. This time, each student flipped a coin to determine whether he or she would sit or stand first. Then each student took his or her pulse rate twice, either sitting or standing first depending on the result of the coin flip. This *repeated measures design* is also a type of randomized paired comparison design.

Display 4.21 gives the results of the students repeated measures experiment, where both sitting and standing were assigned, but in random order, to each student.

As in the matched pairs design, the plot on the left in Display 4.21 compares the pulse rates for both treatments, ignoring the blocking on subjects. As before, no real difference between the treatments can be detected because there is so much variability among the subjects. While the individual pulse rates jump around a good bit, the differences between the measurements within a pair are less variable, as shown in the plot on the right. Blocking on subjects appears to be highly effective here, as the differences are almost all positive with less variability than in the matched pairs design.

Student	Standing Pulse Rate	Sitting Pulse Rate	Difference: *stand − sit*	Student	Standing Pulse Rate	Sitting Pulse Rate	Difference: *stand − sit*
1	62	58	4	12	88	88	0
2	70	60	10	13	62	40	22
3	60	56	4	14	84	78	6
4	68	62	6	15	66	62	4
5	78	68	10	16	62	58	4
6	56	56	0	17	42	42	0
7	72	66	6	18	64	70	−6
8	68	66	2	19	66	64	2
9	64	58	6	20	56	52	4
10	72	62	10	21	88	84	4
11	58	54	4	22	78	74	4

Display 4.21 Difference in pulse rates for a repeated measures design. On the left, the pulse rates are shown separately for each treatment. The plot on the right shows the difference in pulse rates for each student, *stand − sit*.

DISCUSSION
Sit or Stand

D29. Do the randomized paired comparison designs or the completely randomized design provide stronger evidence that the pulse rate tends to be higher when standing than when sitting?

D30. Which of the two randomized paired comparison designs provides stronger evidence that the pulse rate tends to be higher when standing than when sitting?

STATISTICS IN ACTION 4.3 ▶ Sit or Stand

Now that you have seen some of the general ideas behind designing sound experiments, it is time to make them more explicit by working through a real experimental situation. Consider an experiment to decide if there is a detectable difference in heart rate measured under two treatments:

- Treatment 1: standing, with eyes open
- Treatment 2: sitting relaxed, with eyes closed

You will take your pulse several times during this activity, as directed here.

a. Have your instructor time you for 30 seconds. When your instructor says, "Go," start counting beats until your instructor says, "Stop."

b. Double your count to get your heart rate in beats per minute.

(*continued*)

Part A: Completely Randomized Design

In Part A, you and your classmates will be assigned randomly to one treatment or the other.

1. *Random assignment.* Your instructor will pass around a box with slips of paper in it. Half say "stand" and half say "sit." When the box comes to you, mix up the slips and draw one. Depending on the instruction you get, either stand with your eyes open, or sit with your eyes closed and relax.

2. *Measurement.* Take your pulse.

3. *Record data.* Record your heart rate and that of the other students in your treatment group.

4. *Summaries.* Display the data using side-by-side stemplots for the two treatments. Then compute the mean and standard deviation for each group. Do you think the treatments affect the response differently?

Part B: Randomized Paired Comparison Design (Matched Pairs)

In Part B, you and your classmates will first be sorted into pairs based on an initial measurement. Then, within each pair, one person will be randomly chosen to stand and the other will sit.

1. *Initial measurement.* Take your pulse sitting with your eyes open.

2. *Forming matched pairs.* Line up in order, from fastest heart rate to slowest, and pair off, with the two fastest in a pair, the next two fastest in a pair, and so on.

3. *Random assignment within pairs.* Either you or your partner should prepare two slips of paper, one that says "sit," and one that says "stand." One of you should then mix the two slips and let the other person choose one. Thus, within each pair, one of you randomly ends up sitting and the other ends up standing.

4. *Measurement.* Take your pulse.

5. *Record data.* Calculate the difference *standing* minus *sitting* for each pair of students.

6. *Summaries.* Display the set of differences in a stemplot. Then compute the mean and standard deviation of the differences. What should the mean be if the treatments do not affect the response differently? Do you think the observed difference is real or simply due to variation among individuals?

Part C: Randomized Paired Comparison Design (Repeated Measures)

This time each person is his or her own matched pair. Each of you will take your pulse under both treatments, and standing and sitting. You'll flip a coin to decide the order.

1. *Random assignment.* Flip a coin. If it lands heads, you will sit first and then stand. If it lands tails, you will stand first.

2. *First measurement.* Take your pulse in the position chosen by your coin flip.

3. *Second measurement.* Take your pulse in the other position.

4. *Record data.* Record your heart rates in a table with the heart rates of other students. Calculate the difference *standing* minus *sitting* for each student.

5. *Summaries.* Display the set of differences in a stemplot. Then compute the mean and standard deviation of the differences. Do you think the treatments affect the response differently?

Which of the three designs do you think is best for studying the effect of position on heart rate? Explain what makes your choice better than the other two designs.

The Randomized Block Design

A randomized paired comparison design is a special type of **randomized block design.** In the matched pairs design, each pair of subjects matched on resting pulse rate was a block; in the repeated measures design, each subject in the study was a block. In a generalized block design, more than two experimental units may be in each block.

Steps in Creating a Randomized Block Design

1. Sort your available experimental units into groups (blocks) of *similar* units. The units in each block should be enough alike that you expect them to have a similar response to any one treatment. This is called **blocking**.

2. Randomly assign one treatment to each unit in the first block. (It's usually best if the same number of units is assigned to each treatment.) Then go to the second block and randomly assign a treatment to each unit in this block. Repeat for each block.

> In a randomized block design, treatments are randomly assigned within blocks of similar units.

Is Obesity Related to Stress?

Example 4.6

In a recent laboratory study, some mice were subjected to stress and then fed either a standard diet or a high-fat diet. To make proper comparisons, some non-stressed mice were fed the same diets. There were, then, four treatments: no stress with standard diet, no stress with high-fat diet, stress with standard diet, and stress with high-fat diet. The response was a measure of obesity for the mice. Now, not all laboratory mice are the same, but healthy mice from the same litter are the same age and tend to be similar on other genetically related variables. How could you use blocking to set up this experiment?

Solution
A good design for this experiment would be to identify a number of litters that contain at least four mice each, select four mice from each litter, and randomly assign one of the four treatments to each of the four mice within a litter. The number of litters (blocks) needed depends on the variability within the blocks and the size of the difference in obesity that the scientists want to detect; the short, practical answer is to take as many litters as are available and practically convenient.

The use of the term "block" in statistics comes from the fact that many of the concepts of experimental design (as well as other areas of statistics) originated in agriculture, the area in which R. A. Fisher, one of the great geniuses of modern statistics, happened to work. In comparing four varieties of wheat, for example, to see which produces the highest yield, you would want to eliminate as much variability due to soil conditions, water, sun, and so on, as possible. To do this, you may have to take a small plot of land (a block, if you will) that is fairly uniform in all of these background variables and divide it into four subplots so that each variety could be planted within the block, in a randomly assigned subplot. Since these blocks tend to be small, you would need to use a number of them to get good data on the yields of the wheat.

The effectiveness of blocking depends on how similar the units are in each block and how different the blocks are from each other. Here "similar units" are units that would tend to give similar values for the response if they were assigned the same treatments. *The more similar the units within a block, the more effective blocking will be.*

> Similar units in a block make for effective blocking.

DISCUSSION
Randomized Designs

D31. What is the main difference between a completely randomized design and a randomized block design?

D32. Why is blocking sometimes a desirable feature of a design? Give an example in which you might want to block and an example in which you might not want to block.

Summary 4.4: Designing Experiments to Reduce Variability

Random assignment of treatments to experimental units—the fundamental principle of good design—can be accomplished through two basic plans:

- For a completely randomized design, randomly assign a treatment to each experimental unit while keeping the number of units given each treatment as equal as possible.

- For a randomized block design, place similar experimental units into groups, called blocks, and randomly assign the treatments to the units within each block.

The reason for the randomized block design is that, in any experiment, responses vary not only because of different treatments (between-treatment variation) but also because different subjects respond differently to the same treatment (within-treatment variation). In a well-designed experiment, the experimenter should minimize the amount of variability within the treatment groups because it can obscure the variability resulting from differences between treatments. By taking the blocks into account in analyzing the data, the effect of the within-treatment variability can be minimized.

You have learned two special cases of block design:

- A randomized paired comparison (matched pairs) design involves randomly assigning two treatments within pairs of similar units, such as to twins or to left and right feet.

- A randomized paired comparison (repeated measures) design involves the assignment of all treatments, in random order, to each unit so that comparisons can be made on the same units. An example is a study in which each patient in a clinic is assigned each of three treatments for asthma, in random order.

Practice

Difference Between Treatments Versus Variability Within Treatments

P31. Review the antibacterial soap experiment in E31.

  a. List at least two sources of within-treatment variability.

  b. Is the point of randomization to reduce the within-treatment variability or to equalize it between the treatment groups?

A Design for Every Purpose

P32. To test a new drug for asthma, both the new drug and the standard treatment will be administered in random order, to each subject in the study.

  a. What kind of design is this?

  b. An observant statistician cries, "No, no! Use two similar subjects in each pair, randomized to each treatment." What kind of design is this?

  c. Why did the statistician made this suggestion?

P33. Suppose you want to compare two different methods of teaching young adults to ice skate. The response is how many feet the person can skate without falling after the instruction is complete. Would it be better to block on sex (male/female), age $(18 - 23/24 - 29)$ or whether the person can roller skate (yes/no)? Explain.

P34. For each experiment, describe the within-treatment variability that might obscure any difference between treatments. Then describe an experimental design that includes blocking, and define a response variable.

  a. To determine whether studying with the radio on helps or hurts the ability to memorize, there will be two treatments: listening to the radio and not listening to the radio. The subjects available are all the students in your statistics class.

  b. To determine whether adding MSG to soup makes customers eat more soup, a large restaurant will assign two treatments: adding MSG to the soup and not adding MSG. The subjects available are all customers during one evening.

P35. In the experiment to compare drowsiness caused by two antihistamines (Section 4.3, page 202), a new drug, meclastine, was compared to a standard drug, promethazine, and to a placebo. Each subject was given each treatment. Such a design can be used only because the effect of the antihistamine wears off in a relatively short period of time, allowing more than one treatment to be applied to each subject in the study. The response was the average number of eyelid flicks per minute, because low flicker frequency is related to drowsiness. Results are given in Display 4.22.

a. Do the parallel dot plots show much evidence that one drug is performing better than the other? Is it easy to answer the question from these plots? Explain.

b. Display 4.23 shows the scatterplot for the paired data on Treatments A and C. The line on the scatterplot is the line $y = x$. Does it now look as if Treatment A is better than Treatment C? Is it easier to answer this question from the scatterplot that shows the paired data?

Patient	Treatment A (medastine)	Treatment B (placebo)	Treatment C (promethazine)
1	31.25	33.12	31.25
2	26.63	26.00	25.87
3	24.87	26.13	23.75
4	28.75	29.63	29.87
5	28.63	28.37	24.50
6	30.63	31.25	29.37
7	24.00	25.50	23.87
8	30.12	28.50	27.87
9	25.13	27.00	24.63

Treatment A

Treatment B

Treatment C

23 24 25 26 27 28 29 30 31 32 33 34
Eyelid Flicks per Minute

Display 4.22 Average number of eyelid flicks per minute.

c. How else might you compare the responses for Treatments A and C?

d. What is the design of this study? Explain why this design is better than a completely randomized design.

P36. *Bears in Space.* Congratulations! You have just been appointed director of research for Confectionery

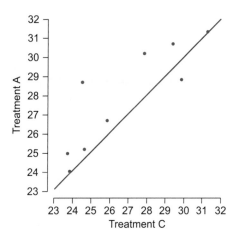

Display 4.23 Average number of eyelid flicks per minute for subjects given both Treatments A and C.

A gummy bear being launched.

Ballistics, Inc., a company that specializes in launching gummy bears from a ramp using a launcher made of tongue depressors and rubber bands.

The CEO has asked you to study a variety of factors that are thought to affect launch distance. Your first assignment is to study the effect of the color of the bears.

Display 4.24 shows actual data for bears launched by members of a statistics class at Mount Holyoke College. Each launch team did ten launches, the first set of five launches using red bears and the second set of five launches using green bears.

	Launch	Team 1	Team 2	Team 3	Team 4	Team 5	Team 6	Median	Mean	SD
	1	15	44	18	13	10	125	16.5	37.5	44.6
	2	24	40	2	39	35	147	37.0	47.8	50.6
Red Bears	3	48	51	10	16	24	35	29.5	30.7	16.8
	4	25	100	41	41	22	81	41.0	51.7	31.7
	5	19	37	88	41	65	125	53.0	62.5	38.9
	6	17	52	46	31	45	27	38.0	36.3	13.4
	7	19	102	72	41	70	187	71.0	81.8	58.9
Green Bears	8	21	86	53	55	21	84	54.0	53.3	28.6
	9	31	89	38	14	33	105	35.5	51.7	36.4
	10	74	120	92	33	37	174	83.0	88.3	53.4
Median		22.5	69.0	43.5	36.0	34.0	115.0			
Mean		29.3	72.1	46.0	32.4	36.2	109.0			
SD		18.4	30.5	31.1	14.0	19.2	53.6			

Display 4.24 Sample launch distance data, in inches, for red and green gummy bears.

a. Which color bear tended to go farther?

b. Your CEO at Confectionery Ballistics, Inc., wants to reward you for discovering the secret of longer launches. Explain why his enthusiasm is premature.

c. Your CEO is adamant: Color is the key to better launches. In vain you argue that color and launch order are confounded. Finally, your CEO issues an executive order: "Prove it. Show me data." Determined to meet the challenge, you remember the basic rule of statistics: *Plot your data first.* Discuss what the plots in Display 4.25 show.

d. How would you change the design of the experiment to eliminate confounding?

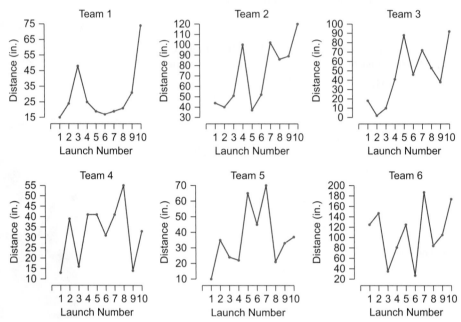

Display 4.25 Plots of launch distances for red and green gummy bears.

Exercises

E39. The SAT discussion from Section 4.3 ended with a revised plan that was truly experimental (see page 200). To which of the two types does that design belong, completely randomized or randomized block? Summarize the design by giving the experimental units, factors, levels, response variable, and blocks (if any).

E40. School academic programs often are evaluated on what are called "gain scores," the gain in academic achievement (as shown by standardized test scores) over the course of an academic year. Such scores often are reported at the school level, not the student level. That is, each school will receive only the mean score of all its children in a given grade. Suppose ten schools have agreed to participate in a study to evaluate the effect of calculator use in teaching mathematics (extensive use versus limited use).

a. In the first design, five participating schools are randomly assigned one of the two treatments and the other five schools are assigned the other treatment. The gain scores in mathematics for the two treatment groups are to be compared at the end of the year. Is this a completely randomized or a randomized block design? Identify the treatments, experimental units, and blocks (if any).

b. In the second design, the participating schools first are paired according to last year's gain scores in mathematics. Within each pair of schools, one school is randomly assigned to extensive use of calculators and the other to limited use. The gain scores in mathematics are to be compared at the end of the year. Is this a completely randomized design or a randomized block design? If the latter, what type of randomized block design is it? Identify the treatments, experimental units, and blocks (if any).

E41. PKU (phenylketonuria) is a disease in which the shortage of an enzyme prevents the body from producing enough dopamine, a substance that helps transmit nerve impulses. To some extent, you can relieve the symptoms by eating a restricted diet, one low in the amino acid phenylalanine. A study was designed to measure the effect of such a diet on the levels of dopamine in the body. The subjects were ten PKU patients. Each patient was measured twice, once after a week on the low phenylalanine diet and again after a week on a regular diet. Identify blocks (if any), treatments, and experimental units, and describe the design used in this study.

E42. In P17, you considered this problem: Solar thermal systems use heat generated by concentrating and absorbing the sun's energy to drive a generator and produce electric power. A manufacturer of solar power generators is interested in comparing the loss of energy associated with the reflectivity of the glass on the solar panels. (A measure of energy loss is available.) Three types of glass are to be studied. The company has 12 test sites in the Mojave Desert, in the southwestern United States. Each site has one generator. The sites might differ slightly in terms of the amount of sunlight they receive.

Now the manufacturer wants to expand the test sites to include generators that are available in the mountains of Tennessee, the desert of northern Africa, and central Europe. Each of these locations has 12 test sites, with one generator at each site. Further, there are two designs of panels in which to place the glass. Describe how you would design this study, identifying treatments, experimental units, and blocks (if any).

Solar panels in the desert.

E43. *Finger tapping.* Many people rely on the caffeine in coffee to get them going in the morning or to keep them going at night. But wouldn't you rather have chocolate? Chocolate contains theobromine, an alkaloid quite similar to caffeine both in its structure and in its effects on humans. In 1944, C. C. Scott and K. K. Chen reported the results of a study designed to compare caffeine, theobromine, and a placebo. Their design used four subjects and randomly assigned the three treatments to each subject, one treatment on each of three different days. Subjects were trained to tap their fingers in a way that allowed the rate to be measured—presumably, this training eliminated any practice effect. The response was the rate of tapping 2 hours after taking a capsule containing either one of the drugs or the placebo. Are treatments assigned to the experimental units completely at random or using a block design? Identify the response, experimental units, treatments, and blocks (if any). [Source: C. C. Scott and K. K. Chen, "Comparison of the Action of 1-ethyl Theobromine and Caffeine in Animals and Man," *Journal of Pharmacologic Experimental Therapy*, Vol. 82 (1944), pp. 89–97.]

E44. *Walking babies.* The goal of this experiment was to compare four 7-week training programs for infants to see whether special exercises could speed up the process of learning to walk. One of the four training programs was of particular interest: It involved a daily 12-minute set of special walking and placing exercises. The second program (the exercise control group) involved daily exercise for 12 minutes but without the special exercises. The third and fourth programs involved no regular exercise (parents were given no instructions about exercise) but differed in their follow-up: Infants in the third program were checked every week, whereas those in the fourth program

were checked only at the end of the study. Twenty-three 1-week-old babies took part, and each was randomly assigned to one of the groups. The response was the time, in months, when the baby first walked without help.

Tell how the treatments are assigned to the experimental units—completely at random or using a block design. Identify the response, experimental units, treatments, and blocks (if any).

E45. *Sawdust for lunch?* Twenty-four tobacco hornworms served as subjects for this experiment, which was designed to see how worms raised on low-quality food would compare with worms raised on a normal diet. The two dozen worms were randomly divided into two groups, and the lucky half was raised on regular worm food. The unlucky half got a mixture of 20% regular food and 80% cellulose. Cellulose has no more food value for a hornworm than it has for you—neither you nor a hornworm can digest the stuff. The experimenter kept track of how much each hornworm ate and computed a response value based on the total amount eaten in relation to body weight.

Tell how treatments were assigned to experimental units in this experiment—completely at random or using a block design. Identify the response, experimental units, treatments, and blocks (if any).

Tobacco hornworm.

E46. Arthritis is painful, and those who suffer from this disease often take pain relief and anti-inflammatory medication for long periods of time. One of the side effects of such medications is that they often cause stomach damage such as lesions and ulcers. The goal of research, then, is to find an anti-inflammatory drug that causes minimal stomach damage. An experiment was designed to test two treatments for arthritis pain (old and new) against a placebo by measuring their effect on lesions in the stomachs of laboratory rats. Rats with similar stomach conditions at the start of the experimental period were randomly assigned to one of the three treatments. Total length of stomach lesions (in millimeters) was measured in each rat after a 2-week treatment period.

a. Was the assignment of treatments in this experiment completely at random or in blocks? Does this seem like the best way to make an assignment in this case?

b. What are the experimental units? What is the response measurement?

c. Is the use of a placebo essential here?

▶ Chapter Summary

This introductory chapter on designs for statistical studies has focused on two basic types of chance-based designs: sample surveys used to estimate population characteristics and experiments used to compare treatments and establish cause. Sampling methods, studied in the first part of this chapter, use chance to choose the individuals to be studied. Typically, you choose individuals in order to ask them questions, as in a Gallup poll. Thus, samples and surveys often go together. Experiments, introduced in the second part of this chapter, are comparative studies that use chance to assign the treatments you want to compare. An experiment should have three characteristics: random assignment of treatments to units, two or more treatments to compare, and replication of each treatment on at least two subjects.

The purposes of sampling and experimental design are quite different. Sample surveys are used to estimate the parameters of fixed, well-defined populations. Experiments are used to establish cause and effect by comparing treatments.

Display 4.26 summarizes the differences between a survey and an experiment.

	What You Examine	Ultimate Goal	Role of Randomization	How You Control Variation	Threats to Inference
Sample Survey	Population	Describe some characteristic of the population	Take a random sample from the population	Stratify	Sampling bias, response bias
Experiment	Treatments	See whether different treatments cause different results	Assign treatments at random to available units	Block	Confounding

Display 4.26   Differences between a sample survey and an experiment.

Randomization is the fundamental principle in both types of study because it allows the use of statistical inference (developed in later chapters) to generalize results. In addition, random selection of the sample reduces bias in surveys, and random assignment of treatments to subjects reduces confounding in experiments.

A careful observational study can *suggest* associations between variables, but the possibility of a lurking variable being responsible for the association means causal inferences cannot be made.

Display 4.27 shows how sample surveys, experiments, and observational studies fit together. Any statistical study has at its core the selection of units on which measurements will be made and the possible assignment of conditions (or treatments) to those units. The sample surveys discussed in this chapter fall into the upper-right box, the experiments into the lower-left box, and the observational studies into the lower-right box. Studies of the type described in the upper-left box, which use both random sampling and random assignment of treatments, have not been discussed in this chapter.

		Assignment of Conditions or Treatments to Units		Inference to Population Appropriate?
		At Random	**Not at Random**	
Selection of Units	At Random	Experiment with broad scope of inference: A random sample of units is selected from a population; treatments are randomly assigned to the units.	Sample survey: A random sample of units is selected from a population; the conditions are already built into the units.	Inference to the population can be drawn.
	Not at Random	Experiment with narrow scope of inference: A group of available units is used; treatments are randomly assigned to the units.	Observational study: A group of available units is used; the conditions are already built into the units.	Inferences are limited to only the units included in the study.
Causal Inference Appropriate		Casual inferences can be drawn.	Associations may be observed but no causal inferences drawn.	

Display 4.27   Inference from designed statistical studies. *[Adapted from F. L. Ramsey and D. W. Schafer, The Statistical Sleuth, 2nd ed., (Duxbury), p. 9, and the notes of Linda Young, University of Florida, 2002.]*

Review Exercises

E47. For each situation, tell whether it is better to take a sample or a census, and give reasons for your answer.

Characteristic of interest	Population of Interest
a. Average life of a battery	Alkaline AAA batteries
b. Average age	Current U.S. senators
c. Average price per gallon	Purchases of regular-octane gasoline sold at U.S. stations next week

E48. You want to estimate the percentage of people with heart disease in your area who also smoke cigarettes. The people in your area who have heart disease make up your population. You take as your frame all records of patients hospitalized in area hospitals within the last 5 years with a diagnosis of heart disease. How well do you think this frame represents the population? If you think bias is likely, identify what kind of bias it would be and explain how it might arise.

E49. Consider using your statistics class as a convenience sample in each of these situations. For each, tell whether you think the sample will be reasonably representative and, if not, in what way(s) you expect your class to differ from the given population of interest.

Characteristic of interest	Population of Interest
a. Percentage who can curl their tongues	U.S., age 12 or older
b. Average age	U.S., all adults
c. Average blood pressure	U.S., students your age
d. Percentage who prefer science to English	U.S., students your age
e. Average blood pressure	U.S., all adults

E50. The U.S. Bureau of Labor Statistics collects data on occupational variables from a nationwide sample of households. This paragraph is from its news bulletin "Union Members in 1996" [January 31, 1997, p. 3]: "The data also are subject to nonsampling error. For example, information on job-related characteristics of the worker, such as industry, occupation, union membership, and earnings, are sometimes reported by a household member other than the worker. Consequently, such data may reflect reporting error by the respondent. Moreover, in some cases, respondents might erroneously report take-home pay rather than gross earnings, or may round up or down from actual earnings."

 a. Would reporting take-home pay rather than gross earnings increase or decrease the estimate of the earnings of various types of workers?

 b. What other sources of nonsampling error might this study contain?

E51. A friend who wants to be in movies is interested in how much actors earn and has decided to gather data using a simple random sample. The *World Almanac and Book of Facts* has a list of actors that your friend plans to use as a sampling frame. Would you advise against using that list as a frame? What sort of bias do you expect?

E52. What is the average number of representatives per state in the U.S. House of Representatives? If you really want to know, you should use a census rather than a sample, but because you like statistics so much you've decided to use a random sample. Tell which of these two sampling methods is biased and describe the bias. Will estimates tend to be too high or too low?

Method I. Start with a list of all the current members of the House. Take a simple random sample of 80 members, and for each representative chosen, record the number of representatives from that person's state. Then take the average.

Method II. Start with a list of all 50 states. Take a simple random sample of five states, and, for each state chosen, record the number of representatives from that state. Then take the average.

E53. For an article on used cars, a writer wants to estimate the cost of repairs for a certain 2008 model. He plans to take a random sample of owners who bought that model from a compilation of lists supplied by all U.S. dealers of that make of car. He tells his research assistant to send letters to all the people in the sample, asking them to report their total repair bills for last year. His research assistant tells him, "You might want to think again about your sampling frame. I'm afraid your plan will miss an important group of owners." What group or groups?

E54. If you look up Shakespeare in just about any book of quotations, you'll find that the listing goes on for several pages. If you look up less famous writers, you find much shorter listings. What's the average number of quotations per author in, say, the *Oxford Dictionary of Quotations*, 3rd edition? Suppose you plan to base your estimate on a random sample taken by this method: You take a simple random sample of pages from the book. For each page chosen, you find the first quotation in the upper-left corner of that page and record the number of quotations by that author. Then you find the average of these values for all the pages in your sample. What's wrong with this sampling method?

E55. You've just been hired as a research assistant with the state of Maine. For your first assignment, you're asked to get a representative sample of the fish in Moosehead Lake, the state's largest and deepest lake. Your supervisor, who knows no statistics and can't tell a minnow from a muskellunge, tells you to drag a net with 1-in. mesh (hole size) behind a motorboat, up and down the length of the lake. "No," you tell him. "That method is biased!" What is the bias? What kinds of fish will tend to be overrepresented? Underrepresented? (You *don't* have to know the difference between a minnow and a muskie to answer this. You *do* have to know a little statistics.)

E56. You are asked to provide a forestry researcher with a random sample of the trees on a 1-acre lot. Your supervisor gives you a map of the lot showing the location of each tree and tells you to choose points at random on the map and, for each point, to take the tree closest to the point for your sample. "Sorry, sir," you tell him. "That sampling method was once in common use, but then someone discovered it was biased. It would be better to number all the trees in the map and use random numbers to take an SRS." Draw a small map, with about ten trees, that you could use to convince your supervisor that his method is biased. Put in several younger trees, which grow close together, and two or three older trees, whose large leaf canopies discourage other trees from growing nearby. Then explain which trees are more likely to be chosen by the method of random points and why.

E57. You're working as an assistant to a psychologist. For subjects in an experiment on learning, she needs a representative sample of 20 adult residents of New York City. She tells you to run an ad in the *New York Times*, asking for volunteers, and to randomly choose 20 from the list of volunteers. What's the bias?

E58. Give an example in which nonresponse bias is likely to distort the results of a survey.

E59. Tell how to carry out a four-stage sample of voters in the United States, with voters grouped by precinct, precincts grouped by congressional district, and congressional districts grouped by state.

E60. Tell how to carry out a multistage sample for estimating the average number of characters per line for the books on a set of shelves.

E61. A researcher designs a study to determine whether young adults who follow a Mediterranean diet (lots of fruits, vegetables, grains, and olive oil; little red meat) end up having fewer heart attacks in middle age than those who follow an average American diet. She randomly selects 1000 young adults in Greece and 1000 young adults in the United States. These 2000 young adults are categorized according to whether they follow a Mediterranean diet or a typical American diet. Thirty years later, she finds that a lower percentage of those who followed the Mediterranean diet have had a heart attack.

a. For the purpose of determining whether a Mediterranean diet results in fewer heart attacks than a typical American diet, is this an observational study or an experiment? Explain.

b. Explain why the researcher cannot conclude that following a Mediterranean diet results in a lower chance of having a heart attack than following an average American diet.

c. What factors might be confounded with diet?

E62. A mother does not keep cola in her house because she is convinced that cola makes her 7-year-old daughter hyperactive. Nevertheless, the daughter gets a cola whenever she has fast food or goes to a birthday party, and the mother has noticed that she then acts hyperactive.

a. Has the mother done an experiment or an observational study?

b. Name two factors that might be confounded with cola, and explain how they are confounded with cola.

c. Design a study that would allow the mother to determine whether giving her daughter cola makes her hyperactive. Are there any potential problems with such a study?

E63. For each science project described,

a. classify the project as a survey, an observational study, or an experiment

b. if a survey, name the explanatory variables; if an observational study or experiment, name the factors and levels

c. name the response variable

d. if a survey, give the sample size; if an observational study, give the number of observed units; if an experiment, give the number of experimental units

i. A student wanted to study the effect of social interaction on weight gain in mice. A large number of baby mice from different mothers were weaned, weighed, and then randomly divided into four groups. One group lived with their mother and siblings, one group lived with their mother but not their siblings, one group lived only with their siblings, and each mouse in the last group lived in isolation. All groups were otherwise treated alike and were given as much food as they wanted. After 90 days, the mice were weighed again.

ii. A student picked four different sites on an isolated hillside at random. At each site, he measured off a 10- × 10-foot square. At each site, a sample of soil was taken and the amounts of ten different nutrients were measured. The student counted the number of species of plants at each site, hoping to be able to predict the number of species from the amounts of the nutrients.

iii. A college professor helped his daughter with a 2nd-grade science project titled "Does Fruit Float?" They tested 15 different kinds of fruit and classified each as "floater" or "nonfloater." For the display, they took a photograph of a watermelon floating in a swimming pool and a grape sitting at the bottom. The daughter wrote her own explanation of what determines whether a particular type of fruit floats.

E64. You want to test whether your company's new shampoo protects against dandruff better than the best-selling brand. Unfortunately, the best-selling brand has a very distinctive color, which you are unable to duplicate in your new shampoo. Design a randomized paired comparison experiment to test your shampoo. Discuss any confounding you expect to encounter.

E65. Investigators found seven pairs of identical twins in which one twin lived in the city and the other in the country, and both were willing to participate in a study to determine how quickly their lungs cleared after they inhaled a spray containing radioactive Teflon particles. Display 4.28 gives the percentage of radioactivity remaining 1 hour after inhaling the spray.

a. What is the factor? What are its levels? Identify a block.

b. Is this an observational study or an experiment? Explain.

c. Is there more variation in response within the urban twins, within the rural twins, or in the differences? Do you believe the study demonstrates that environment makes a difference in this case?

d. Why were twins used? What kind of variation is reduced by using identical twins?

	Environment		
Twin Pair	Rural	Urban	Difference
1	10.1	28.1	−18.0
2	51.8	36.2	15.6
3	33.5	40.7	−7.2
4	32.8	38.8	−6
5	69.0	71.0	−2
6	38.8	47.0	−8.2
7	54.6	57.0	−2.4

Display 4.28 Percentage of radioactivity remaining after 1 hour. [Source: Per Camner and Klas Phillipson, "Urban Factor and Tracheobronchial Clearance," *Archives of Environmental Health*, Vol. 27 (1973), p. 82. Reprinted in Richard J. Larson and Morris L. Marx, *An Introduction to Mathematical Statistics and Its Applications*, 2nd ed. (Englewood Cliffs, NJ: Prentice Hall, 1986).]

E66. Design a taste comparison test. Suppose you want to rate three brands of chocolate chip cookies on a scale from 1 (terrible) to 10 (outstanding). Tell how to run this test as a randomized block design.

E67. *Exercise bikes.* You work for a gym and have been asked to design an experiment to decide which of two types of exercise bikes will get the most use. Your exercise bike room has space for eight bikes, which you will place as in Display 4.29. People enter the room through the door at the top of the diagram. A counter on each bike records the number of hours it has been used.

Design an experiment to compare the number of hours the two different types of exercise bike are used that takes into account the fact that some bikes are in locations that make them more likely to be used than others when the gym isn't full.

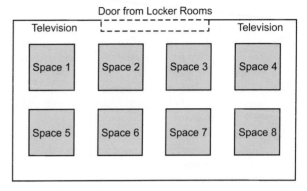

Display 4.29 Diagram of spaces for the eight exercise bikes.

E68. Read one of these articles from *Statistics: A Guide to the Unknown* (Brooks/Cole, 2006), and write a paragraph explaining how randomization and control or comparison groups were used in the study reported in the article.

a. Jennifer Hill, "Evaluating School Choice Programs"

b. William F. Eddy and Margaret L. Smykla, "The Last Frontier: Understanding the Human Mind"

c. William Kahn and Leonard Roseman, "Advertising as an Engineering Science"

Concept Test

C1. Researchers want to estimate the mean number of children per family, for all families that have at least one child enrolled in one of ten similarly sized county high schools. Which sampling plan is biased?

(A) Randomly select 100 families from those families in the county that have at least one child in high school. Compute the mean number of children per family.

(B) Randomly select one high school, and compute the mean number of children in each family with a child or children in that school.

(C) Randomly select 20 students from each of the ten high schools, and ask each one how many children are in his or her family. Compute the mean number of children per family.

(D) From a list of the families in all ten high schools, choose a random starting point and then select every tenth family. Compute the mean number of children per family.

(E) None of the above is a biased plan.

C2. A radio program asked listeners to call in and vote on whether the notorious band, The Rolling Parameters, should perform at the Statistics Day celebration. Of the 956 listeners who responded, 701 answered that The Rolling Parameters should perform. Which type of sampling does this example use?

(A) stratified random   **(B)** cluster
(C) systematic   **(D)** quota
(E) voluntary response

C3. To select students to explain homework problems, a professor has students count off by 5's. She then randomly selects an integer from 1 through 5. Every student who counted off that integer is asked to explain a problem. Which type of sampling plan is this?

(A) convenience   **(B)** systematic
(C) simple random   **(D)** stratified random
(E) cluster

C 4. To conduct a survey to estimate the mean number of minutes adults spend exercising, researchers stratify by age before randomly selecting their sample. Which of the following is not a good reason for choosing this plan?

(A) Without stratification by age, age will be confounded with the number of minutes reported.

(B) Researchers will be sure of getting adults of all ages in the sample.

(C) Researchers will be able to estimate the mean number of minutes for adults in various age groups.

(D) Adults of different ages may exercise different amounts on average, so stratification will give a more precise estimate of the mean number of minutes spent exercising.

(E) All of these are good reasons.

C5. A movie studio runs an experiment in order to decide which of two previews to use for its advertising campaign for an upcoming movie. One preview features the movie's romantic scenes and is expected to appeal more to women. The other preview features the movie's action scenes and is expected to appeal more to men. Sixteen subjects take part in this experiment, eight women and eight men. After viewing one of the previews, each person will rate how much he or she wants to see the movie. Which of the following best describes how blocking should be used in this experiment?

(A) Use blocking, with the men in one block and the women in the other.

(B) Use blocking, with half the men and half the women in each block.

(C) Do not block, because the preview that is chosen will have to be shown to audiences consisting of both men and women.

(D) Do not block, because the response will be confounded with gender.

(E) Do not block, because the number of subjects is too small.

C6. A recent study tried to determine whether brushing or combing hair results in healthier-looking hair. Forty male volunteers were randomly divided into two groups. One group only brushed their hair and the other group only combed their hair. Other than that, the volunteers followed their usual hair care procedures. After two months, an evaluator who did not know the treatments the volunteers used, scored each head of hair by how healthy it looked. There was almost no difference in the scores of the two treatment groups. Which statement best summarizes this study?

(A) If a male wants healthy looking hair, it probably doesn't matter whether he brushes or combs it.

(B) You can't tell whether brushing or combing is better, because the treatments are likely to be confounded with variables such as which kind of shampoo a male uses.

(C) You can't come to a conclusion, because the study wasn't double-blind.

(D) You can't come to a conclusion, because only volunteers were included.

(E) The sample size is too small for any conclusion to be drawn.

C7. Which of the following is not a necessary component of a well-designed experiment?

(A) There must be a control group that receives a placebo.

(B) Treatments are randomly assigned to experimental units.

(C) The response variable is the same for all treatment groups.

D There are a sufficient number of units in each treatment group.

E All units are handled as alike as possible, except for the treatment.

C8. In a clinical trial, a new drug and a placebo are administered in random order to each subject, with six weeks between the two treatments. Which best describes this design?

A completely randomized with blocking

B completely randomized with no blocking

C randomized paired comparison (matched pairs)

D randomized paired comparison (repeated measures)

E two-stage randomized

Investigations

C9. *Needle threading.* With your eye firmly fixed on winning a Nobel prize, you decide to make the definitive study of the effect of background color (white, black, green, or red) on the speed of threading a needle with white thread. Design three experiments—one that uses no blocks, one that creates blocks by grouping subjects, and one that creates blocks by reusing subjects. Tell which of the three plans you consider most suitable, and why.

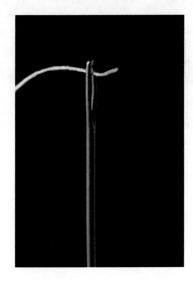

C10. The Census Bureau has considered using sampling to adjust for undercount in the U.S. decennial census. Here is a simplified version of the plan: The Census Bureau collects the information mailed in by the residents of a region. Some residents, however, did not receive forms or did not return them for some reason; these are the uncounted persons. The bureau now selects a sample of blocks (neighborhoods) from the region and sends field workers to find all residents in the sampled blocks. The residents the field workers found are matched to the census data, and the number of residents uncounted in the original census is noted. The census count for those blocks is then adjusted according to the proportion uncounted. (For example, if one-tenth of the residents were uncounted, the original census figures are adjusted upward by 11%.) In addition, the same adjustment factor is used for neighboring regions that have characteristics similar to the region sampled. Comment on the strengths and weaknesses of this method.

Chapter 5
Probability Models

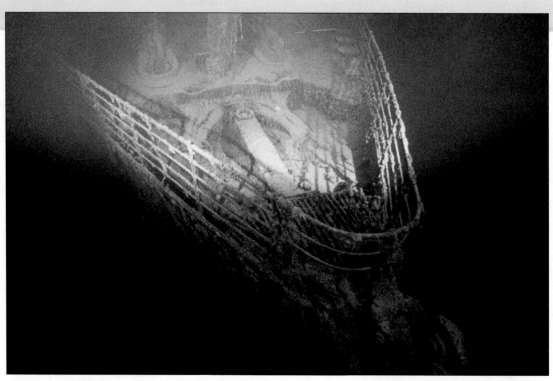

Were some types of people on the *Titanic* more likely to go down with the ship than others? Basic concepts of probability can help you decide.

The idea of probability is not new to you—you have been using it for most of your life to describe your chance of winning a game, or your chance of getting accepted to the college of your choice, or perhaps even your chance of getting an A in this course.

In the study of *Martin v. Westvaco* in Chapter 1, a question was this: If you select three workers at random for layoff from ten workers ages 25, 33, 35, 38, 48, 55, 55, 55, 56, and 64, what is the probability that the mean of the three ages will be 58 or more? This probability was estimated by repeating the process of randomly selecting three ages and computing the mean age for each selection. Because under this random process all possible sets of three ages are equally likely, the probability was calculated by dividing the number of times the mean age was 58 or more by the total number of repetitions in the simulation:

$$\text{probability} \approx \frac{\text{number of repetitions with mean age 58 or more}}{\text{total number of repetitions}}$$

The model—selecting three workers at random—lets you predict how things are supposed to behave. The real data from the court case—the ages of those who actually got laid off—tells you how things actually did behave. The key question, then and for the rest of this book, is this: Are the data consistent with the model? Or should you scrap the model and look for some other way to account for the data? As in Chapter 1, the answer will use the language of probability.

Before going any further into statistics, you need to learn more about buildings and using probability models. Chapter 1 made use of simulation to get an approximate distribution of all possible average ages when three workers are picked at random for layoff. Often, however, you can use the rules of probability to construct exact distributions. That's what you'll learn to do in this chapter and the next.

In this chapter, you will learn to

▷ list all possible outcomes of a chance process in a systematic way

▷ use the Addition Rule to compute the probability that event *A* or event *B* (or both) occurs

▷ use the Multiplication Rule to compute the probability that event *A* and event *B* both occur

▷ compute conditional probabilities, the probability that event *B* occurs given that event *A* occurs

▷ distinguish between mutually exclusive (disjoint) and independent events

5.1 ▶ Models of Random Behavior

Food companies routinely conduct taste tests to decide if a new product tastes at least as good as the competing products. For example, Starbucks announced its newly developed instant coffee and was hitting the airwaves with advertising that emphasized how much it tastes like brewed coffee. "Starbucks has found a way to offer a truly great cup of coffee that you can prepare by just adding water," stated one of the advertisements. The CEO of Starbucks admitted that Starbucks customers have actually been blind taste-testing the product for months as it was slipped into locations in place of regular brew, stating that nobody ever stopped and said, "Hey, something's wrong with this coffee!" [Source: *www.walletpop.com/blog/2009/02/17/the-starbucks-via-taste-test-is-is-really-better-than-your-mo/.*]

Even before the fuss over Starbucks instant coffee, *Consumer Reports* reported that trained taste testers compared coffee from four major retailers and decided that McDonald's beat Starbucks, Dunkin' Donuts, and Burger King. [Source: *www.consumerreports.org/cro/food/beverages/coffee-tea/coffee-taste-test-3-07/overview/0307_coffee_ov_1.htm.*]

Perhaps you should take a slow sip of coffee and contemplate all of this, because *Science Magazine* reports that "taste tests reveal that [people] sometimes can't distinguish gourmet from cheap imitation. So in those cases, we know that people are just fooling themselves." [Source: *www.sciencemag.org/cgi/content/full/323/5917/1006b.*]

How can probability help researchers make rational decisions in these situations? Let's begin with a simple question. Suppose *n* people are each given samples of a gourmet and ordinary coffee and are asked to choose the one that is the gourmet coffee. If the tasters are, in fact, "all fooling themselves," what is the probability that they all make the correct choice?

Why are we concerned with this question? Compare the following two situations: A research firm conducts a taste test with one randomly selected coffee drinker, who correctly identifies the gourmet brand. Another research firm conducts a taste test with 100 randomly selected coffee drinkers, and they all correctly identify the gourmet brand. Which study provides the best evidence that people are able to identify gourmet coffee?

It is the taste test with 100 coffee drinkers and the reasoning goes like this. Even if the single taster can't tell the difference, there is a fifty-fifty chance that he or she chooses correctly just by chance. So the first research firm is unlikely to convince a skeptic that people can choose gourmet coffee. On the other hand, it is extremely unlikely that each of 100 coffee drinkers correctly chooses the gourmet coffee if each person is merely guessing. If the study was well designed, the most hard-core skeptic would be forced to believe that people can pick the gourmet coffee. Even if only 90 of the 100 coffee drinkers correctly choose the gourmet coffee, the skeptic would have to believe that people aren't just guessing. It is just too unlikely to get as many as 90 out of 100 correct if people are merely guessing.

The calculations that will help you decide whether evidence is convincing depend on several basic facts about probability, beginning with those described next.

Event *not A* can be denoted as $\bar{A}$, A^c, or $\sim A$, so $P(\bar{A}) = 1 - P(A)$.

Fundamental Facts About Probability

- An event is a set of possible outcomes from a random situation.

- Probability is a number between 0 and 1 (or between 0% and 100%) that tells how likely it is for an event to happen. At one extreme, events that can't happen have probability 0. At the other extreme, events that are certain to happen have probability 1.

- If the probability that event *A* happens is denoted $P(A)$, then the probability that event *A* doesn't happen is $P(not\ A) = 1 - P(A)$. The event *not A* is called the *complement* of event *A*.

- If you have a list of all possible outcomes and all outcomes are equally likely, then the probability of a specific outcome is

$$\frac{1}{the\ number\ of\ equally\ likely\ outcomes}$$

and the probability of an event is

$$\frac{the\ number\ of\ outcomes\ in\ the\ event}{the\ number\ of\ equally\ likely\ outcomes}$$

In formulating a model for a taste test, the place to begin is with an individual taster. Suppose that the taster cannot identify the gourmet brand, but is merely guessing. Then the taster has two equally likely choices and the probability of choosing the gourmet coffee (call this event G) has probability 0.5. Thus, the probability of a taster guessing correctly can be written as $P(G) = 0.5$. The probability of choosing the ordinary coffee (call this event O) is, then, $P(O) = 0.5$ because there are only two choices.

What if there are two tasters? Assuming that neither can identify the gourmet brand, what is the probability that both guess correctly? Serious academic discussion of problems like this began over 300 years ago. It took an amazingly long time, nearly 150 years, for mathematicians to agree on a systematic, mathematically rigorous approach to probability theory.

The probability of G when coffee tasters can't tell the difference is equivalent to the probability of flipping a coin and observing a head. In a 1750s text, a famous French mathematician, Jean d'Alembert (1717–1783), wrote that the probability of getting two heads in two flips of a coin was 1/3 because the three equally likely outcomes were:

- The first flip is tails.
- The first flip is heads, the second is tails.
- The first flip is heads, the second is heads.

Shortly after the text was published, a Swiss mathematician, Louis Necker (1730–1804), objected to this reasoning. Although agreeing that d'Alembert had a complete list of outcomes, Necker stated that they could not be equally likely because it seemed that the first (the first flip is tails) must be twice as likely as either of the other two.

Necker proposed this list of outcomes:

- The first flip is tails; the second is tails.
- The first flip is tails; the second is heads.
- The first flip is heads; the second is tails.
- The first flip is heads; the second is heads.

Display 5.1 shows the values that d'Alembert's and Necker's models assign to the probability of getting 0, 1, and 2 heads in the flip of two coins. After the following discussion of where probabilities come from, you will see how to determine which model is correct.

Number of Heads	d'Alembert's Probability	Necker's Probability
0	1/3	1/4
1	1/3	1/2
2	1/3	1/4

Display 5.1 Probability models for two coin flips.

DISCUSSION
Probability Models

D1. Why does Necker's model result in the probabilities as shown in Display 5.1?

D2. What is the sum of the probabilities in each of the models in Display 5.1? Why must this be so?

D3. Who do you think is right, d'Alembert or Necker? Give your reason(s). How could your class decide which of them is right?

Where Do Probabilities Come From?

To determine whether d'Alembert or Necker made the correct choice of probabilities it is necessary to think carefully about how probabilities can be assigned to random events. Basically, probabilities come from three sources:

- *Observed data* (long-run relative frequencies). For example, observation of thousands of births has shown that about 51% of newborns are boys. You can use these data to say that the probability of the next newborn being a boy is about 0.51.

- *Symmetry* (equally likely outcomes). If you flip a coin and catch it in the air, nothing about the physics of coin flipping suggests that one side is more likely than the other to land facing up. Based on symmetry, it is reasonable to think that heads and tails are equally likely, so the probability of heads is 0.5. Symmetry also is the reasoning employed in saying that $P(G) = 0.5$ if a taste tester is guessing.

- *Subjective estimates.* What's the probability that you'll get an A in this statistics class? That's a reasonable, everyday kind of question. The use of probability is meaningful in this situation, but you can't gather data or list equally likely outcomes. However, you *can* make a subjective judgment about the probability of such events.

Both d'Alembert and Necker appealed to symmetry in trying to justify their models. Observed data can be used to decide the argument between the two different uses of symmetry. A computer-generated simulation of 3000 flips of two coins resulted in the relative frequencies given in Display 5.2. Comparing the relative frequencies with the probabilities in Display 5.1, it looks like Necker is the winner!

Number of Heads	Frequency	Relative Frequency
0	782	0.26
1	1493	0.50
2	725	0.24
Total	3000	1.00

Display 5.2 Results of 3000 simulated coin flips.

Sample Spaces

Both d'Alembert and Necker used the same principle: Start by making a list of possible outcomes. Over the years, mathematicians realized that such a list of possible outcomes, called a sample space, must satisfy specific requirements.

> A **sample space** for a chance process is a complete list of disjoint outcomes. All of the outcomes in a sample space must have a total probability equal to 1.

It is helpful if the outcomes are equally likely, but this isn't always possible.

Complete means that every possible outcome is on the list. **Disjoint** means that two different outcomes on the list can't occur on the same opportunity. Sometimes the term **mutually exclusive** is used instead of *disjoint*. This book will alternate between the two terms so you can get used to both of them. When possible, a sample space should list equally likely outcomes because that simplifies the calculation of probabilities.

Deciding whether two outcomes are disjoint sounds easy enough, but be careful. You have to think about what an outcome means in your situation. If your outcome is the result of a single coin flip, your sample space is heads (*H*) and tails (*T*). These two outcomes are mutually exclusive because you can't get both heads and tails on a single flip. But suppose you are thinking about what happens when you flip a coin three times. Now your sample space includes outcomes like *HHT* and *TTT*. Even though you get tails on the third flip in both *HHT* and *TTT*, these are disjoint outcomes because the random process in question is a series of three flips. Your sample space consists of triples of flips and these aren't the same triple.

D'Alembert's list has outcomes that are complete and disjoint, but they aren't equally likely because *the first flip is tails* has higher probability than the other two listed outcomes when flipping two coins. He has a legitimate sample space. He has simply assigned the wrong probabilities.

Getting back to the taste testers, Display 5.3 lists a sample space for two tasters attempting to select the gourmet coffee (*G*) over the ordinary coffee (*O*). These four outcomes will be equally likely, by symmetry, if both tasters are merely guessing between the two choices.

Taste Tester A	Taste Tester B
G	*G*
G	*O*
O	*G*
O	*O*

Display 5.3   Sample space for two taste testers.

If the tasters are just guessing, there is a 1/4 chance that they both will guess correctly. Because this probability is so high, a skeptic wouldn't be convinced that the tasters can distinguish the gourmet coffee.

Breaking the Tie for Town Council

Example 5.1

When the new town of Edgewood, New Mexico, held its first election, three candidates tied for a seat on the town council. Such a result is not as unusual as you might think. There are many small towns in the United States that hold many elections in which few people vote. The New Mexico state constitution requires tie votes to be broken by a "game of chance." Suppose that three candidates—A, B, and C—for chair of the town council have received equal numbers of votes, so the rules call for putting the three names in a hat and randomly drawing a name to be town council chair and then drawing a second name to be the vice-chair. List the possible outcomes in the sample space for this random process. Are the outcomes equally likely? [Source: *www.mountainviewtelegraph.com/156592mtnview03-11-04.htm.*]

Solution
The outcomes must take the order of the selection into account, and they can be listed as on Display 5.4. Under random selection of the names from the hat, these outcomes are equally likely by symmetry.

First Name Selected (chair)	Second Name Selected (vice-chair)
A	B
A	C
B	A
B	C
C	A
C	B

Display 5.4   Sample space for selecting city council chair and vice-chair.

DISCUSSION
Sample Spaces

D4. Suppose that a taste test has three tasters, each choosing between a gourmet brand of coffee and ordinary coffee. List the possible outcomes in a sample space for this study. Under what conditions are these outcomes equally likely? Suppose that all three correctly select the gourmet coffee. Are three tasters enough to convince a skeptic that people can distinguish between gourmet and ordinary coffee?

D5. Suppose the tasters in D4 are not all of equal ability and Taster A really can identify the gourmet brand most of the time. Are the outcomes listed in the sample space now equally likely? If not, which ones will have the higher probability?

STATISTICS IN ACTION 5.1 ▶ Spinning Pennies

This activity will help you see some of the difficulties involved in coming up with a realistic probability model.

What You'll Need: one penny per student

In flipping a penny, heads and tails have the same probability. Is the same probability model true for spinning a penny?

1. Use one finger to hold your penny on edge on a flat surface, with Lincoln's head right side up, facing you. Flick the penny with the index finger of your other hand as you let go so that it spins around many times on its edge. When it falls over, record whether it lands heads up or tails up.

2. Repeat, and combine results with the rest of your class until you have a total of at least 500 spins.

3. Are the data consistent with a model in which heads and tails are equally likely outcomes, or do you think the model can safely be rejected?

4. Suppose you spin a penny three times and record whether it lands heads up or tails up.
   a. How many possible outcomes are there?
   b. Are these outcomes equally likely? If not, which is most likely? Least likely?

The Law of Large Numbers

The observed data in Display 5.2 on page 230 convince most people that Necker had a better model than d'Alembert for coin flipping because the relative frequencies from the 3000 simulated coin flips agree with the probabilities in Necker's model but do not agree with the probabilities from d'Alembert's model. People trust that they can estimate probabilities by observing a large number of trials is in accordance with the mathematical principle known as the Law of Large Numbers, which was established in the 1600s by Swiss mathematician Jacob Bernoulli (1654–1705).

The **Law of Large Numbers** says that in random sampling, the larger the sample, the closer the proportion of successes in the sample tends to be to the proportion of successes in the population. In other words, the difference between a sample proportion and the population proportion must get smaller (except in rare instances) as the sample size gets larger. You can see a demonstration of this law in Display 5.5. Think of spinning a coin in such a way that it comes up heads with probability 0.4. The graph in Display 5.5 shows a total of 150 spins, with the proportion of heads accumulated after each spin plotted against the number of the spin. Notice that the coin came up heads for the first few spins, so the proportion of heads starts out at 1. This proportion quickly decreases, however, as the number of spins increases, and it ends up at about 0.38 after 150 spins.

> When sampling is random, the Law of Large Numbers guarantees that the sample proportion converges to the true probability as the sample size increases.

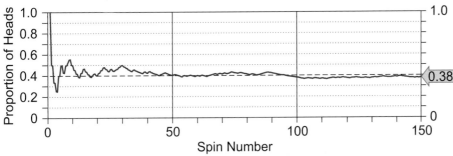

Display 5.5   Proportion of heads for the given number of spins of a coin, with *P(heads)* = 0.4.

Most people intuitively understand the Law of Large Numbers. If they want to estimate a proportion, they know it is better to take a larger sample than a smaller one. After seeing the results from 3000 flips of two coins, most observers immediately reject the d'Alembert model that there are three equally likely outcomes. If there had been only 10 pairs of coin flips, they could not be so sure that his model was wrong.

DISCUSSION
The Law of Large Numbers

D6. An opinion pollster says, "All I need to do to ensure the accuracy of the results of my polls is to make sure I have a large sample."

A casino operator says, "All I need to do to ensure that the house will win most of the time is to keep a large number of people flocking into my casino."

A manufacturer says, "All I need to do to keep my proportion of defective light bulbs low is to manufacture a lot of light bulbs."

Comment on the correctness of each of these statements. In particular, do the people speaking understand the Law of Large Numbers?

D7. Flip a coin 20 times, keeping track of the cumulative number of heads and the cumulative proportion of heads after each flip. Plot the cumulative *number* of heads versus the flip number on one graph. Plot the cumulative *proportion* of heads versus the flip number on another graph. On each graph, connect the points. Repeat this process five times so that you have five sets of connected points on each plot.

Comment on the differences on the patterns formed by the five lines on each graph.

The Fundamental Principle of Counting

As the number of taste testers gets larger, it becomes increasingly difficult to keep track of all the outcomes that should be listed in a sample space. Thus, having a schematic way of listing outcomes and some rule for counting how many there should be are quite helpful. A tree diagram like the one in Display 5.6 is useful both in listing the specific outcomes and in counting how many there are.

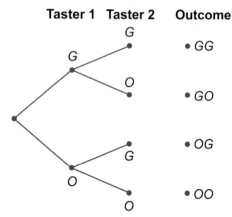

Display 5.6   A tree diagram for all possible outcomes with $n = 2$ tasters.

There are two main branches on the tree, one for each of the two ways the first person can answer. Each main branch has two secondary branches, one for each way the second person can answer after the first person has answered. In all, there are 2×2, or 4, possible outcomes. For $n = 3$, you would add two branches to the end of each of the four branches in Display 5.6. This would give a total of $2 \times 2 \times 2$, or 8, possible outcomes. This idea is called the **Fundamental Principle of Counting**.

Fundamental Principle of Counting

For a two-stage process with n_1 possible outcomes for stage 1 and n_2 possible outcomes for stage 2, the number of possible outcomes for the two stages taken together is $n_1 n_2$.

More generally, if there are k stages, with n_i possible outcomes for stage i, then the number of possible outcomes for all k stages taken together is $n_1 n_2 n_3 \cdots n_k$.

When a process has only two stages, it is often more convenient to list the possible outcomes on a two-way array, as illustrated in Example 5.2.

Example 5.2	Rolling Two Dice

Use the Fundamental Principle of Counting to find the number of possible outcomes when you roll two fair dice. Then make a two-way array that shows all of the possible outcomes. What is the probability that you get doubles (both dice show the same number)?

Solution

Because there are six faces on each die and the dice are fair, the Fundamental Principle of Counting gives 6 × 6, or 36, equally likely outcomes. These outcomes are shown in the array in Display 5.7. Six of these outcomes are doubles, so the probability of rolling doubles is 6/36.

		Second Roll				
	1	**2**	**3**	**4**	**5**	**6**
1	1, 1	1, 2	1, 3	1, 4	1, 5	1, 6
2	2, 1	2, 2	2, 3	2, 4	2, 5	2, 6
3	3, 1	3, 2	3, 3	3, 4	3, 5	3, 6
4	4, 1	4, 2	4, 3	4, 4	4, 5	4, 6
5	5, 1	5, 2	5, 3	5, 4	5, 5	5, 6
6	6, 1	6, 2	6, 3	6, 4	6, 5	6, 6

First Roll labels the rows.

Display 5.7 The 36 possible outcomes when rolling two dice.

DISCUSSION
The Fundamental Principle of Counting

D8. Suppose you flip a fair coin seven times.
  a. How many possible outcomes are there?
  b. What is the probability that you will get seven heads?
  c. What is the probability that you will get heads six times and tails once?

D9. Suppose five taste testers are comparing three brands of coffee (Starbucks, McDonald's, and Dunkin' Donuts) and are asked to choose the one they prefer or respond that they cannot tell any difference. How many possible outcomes are on the list for this taste test? Is it reasonable to assume that these outcomes would be equally likely?

Summary 5.1: Models of Random Behavior

A probability model is a sample space together with an assignment of probabilities. The sample space is a complete list of disjoint outcomes for which these properties hold:

- Each outcome is assigned a probability between 0 and 1.
- The sum of all the probabilities is 1.

Often you can rely on symmetry to recognize that outcomes are equally likely. If they are, then you can compute probabilities simply by counting outcomes: The probability of an event is the number of outcomes that make up the event divided by the total number of possible outcomes. In statistics, the main practical applications of equally likely outcomes are in the study of random samples and randomized experiments. As you saw in Chapter 4, in a survey, all possible simple random samples are equally likely. Similarly, in a completely randomized experiment, all possible assignments of treatments to units are equally likely.

The only way to decide whether a probability model is a reasonable fit to a real situation is to compare probabilities derived from the model with probabilities estimated from observed data.

The Fundamental Principle of Counting says that if you have a process consisting of k stages with n_i outcomes for stage i, the number of outcomes for all k stages taken together is $n_1 n_2 n_3 \cdots n_k$.

Practice

Where Do Probabilities Come From?

P1. Display 5.8 gives the actual low temperature (to the nearest 5°F) in Oklahoma City on 27 days when the National Weather Service forecast was for a low temperature of 30°F.

Actual Low Temperature (°F)	Frequency
20	2
25	8
30	13
35	3
40	1

Display 5.8   Temperatures in Oklahoma City on days with a forecasted low temperature of 30°F. [Source: Harold E. Brooks et al., "Verification of Public Weather Forecasts Available via the Media," *Bulletin of the American Meteorological Society*, Vol. 78 (October 1997), pp. 2167–2177.]

a. Suppose the forecast for tomorrow is for a low temperature of 30°F. What is your estimate of the probability that the low temperature really will be approximately 30°F? That it really will be approximately 25°F?

b. Does the temperature tend to be warmer or colder than predicted?

Sample Spaces

P2. Suppose you flip a coin five times and note whether it is heads or tails.

a. List all possible outcomes.

b. Are these outcomes equally likely?

c. If you are counting the number of heads, what is the probability that you get 0 heads? 1 head? 2 heads? 3 heads? 4 heads? 5 heads?

d. What is the probability that you get at most 4 heads?

P3. Suppose you flip a coin and then roll a die. If you get heads and a 3, then your outcome is H3.

a. List a sample space.

b. Are all outcomes in your sample space equally likely?

c. What is the probability that you get heads and a 3?

P4. You randomly choose two workers to be laid off from a group of workers ages 28, 35, 39, 47, and 55.

a. List a sample space.

b. Are all outcomes in your sample space equally likely?

c. What is the probability that the two youngest people are the ones laid off?

d. What is the probability that the mean age of those laid off is 40 or more?

P5. Suppose that you select two students at random without replacement from Ami, Barbara, Carl, and Darilyn.

a. List a sample space.

b. What is the probability that Ami is among those selected?

P6. Suppose you pick four students at random from your college and check whether they are left-handed or right-handed.

a. Can you list a sample space?

b. Can you determine the probability that all four students are right-handed?

The Law of Large Numbers

P7. The results of 50 spins of a penny are plotted in Display 5.9. The horizontal axis gives the number of the spin, and the vertical axis gives the *cumulative* proportion of spins so far that were heads.

a. Was the first spin heads or tails? The second? The third? The fourth? The 50th?

b. Use these data to estimate the probability that this penny will land heads up when it is spun.

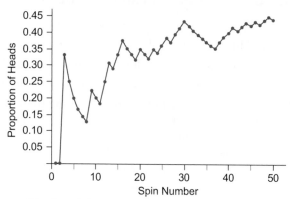

Display 5.9   Results of 50 spins of a penny.

The Fundamental Principle of Counting

P8. You are evaluating a new "improved" version of your company's strawberry ice cream. You will ask each person in a panel of consumers to rate it as good, okay, or poor on flavor and as acceptable or unacceptable on price.

a. On a tree diagram, show all possible outcomes for one person.

b. How many possible outcomes are there?

c. Are all the outcomes equally likely?

P9. A study of different treatments for anxiety randomly assigns each newly diagnosed subject to one of three levels of exercise (mild, moderate, strenuous) and to one of seven different medications.

a. How many different treatments are there?

b. What is the probability the next subject diagnosed with anxiety will be assigned to moderate exercise and the third medication?

c. Illustrate your answer in part a with a two-way array.

d. Illustrate your answer in part a with a tree diagram.

Exercises

E1. Suppose *Consumer Reports* convenes a taste test panel of four coffee drinkers to see if they prefer the taste of Starbucks coffee over that of McDonald's coffee.

   a. List all possible outcomes for the four choices, using *S* if the person chooses Starbucks and *M* if the person chooses McDonald's.

   b. Assume that none of the four can taste any difference between the coffees, and so their choice is equivalent to a random selection of one of the brands. What is the probability that all four people choose Starbucks?

   c. Now suppose that *Consumer Reports* convenes another panel of coffee drinkers. This time, you don't know whether any of the four can tell the difference or not. All four choose Starbucks. Is this evidence convincing enough to enable *Consumer Reports* to claim that Starbucks is preferred over McDonald's by this panel? Explain.

E2. About 85% of those injured snowboarding are male. [Source: National Ski Areas Association, *Facts About Skiing/Snowboarding Safety*, updated October 1, 2008, *www.nsaa.org/nsaa/press/facts-ski-snbd-safety.asp*.] Suppose you take a random sample of five injured snowboarders for a study of how their injuries occurred and note whether the person is male or female.

   a. List all possible outcomes.

   b. Are these outcomes equally likely?

   c. If not, which of the outcomes is most likely? Which is least likely?

E3. Refer to the sample space for rolling two dice shown in Display 5.7 on page 235. Determine each of these probabilities.

   a. not getting doubles

   b. getting a sum of 5

   c. getting a sum of 7 or 11

   d. a 5 occurring on the first die

   e. getting at least one 5

   f. a 5 occurring on both dice

   g. getting 5 as the absolute value of the difference of the two numbers

E4. A tetrahedral die has four sides, with the numbers 1, 2, 3, and 4. (The die on the left has landed as 2 and the one on the right as 3.)

   a. How many equally likely outcomes are there when two fair tetrahedral dice are rolled?

   b. Using a two-way array, show all possible equally likely outcomes.

   c. What is the probability of getting doubles?

   d. What is the probability that the sum is 2?

   e. What is the probability that the larger number is a 2? (If you roll doubles, the number is both the smaller number and the larger number.)

E5. Suppose you roll a tetrahedral die twice (see E4). Determine whether each proposed sample space is disjoint and complete. If the sample space is disjoint and complete, assign probabilities to the given outcomes. If it is not, explain why not.

   a. {no 4 on the two rolls, one 4, two 4's}

   b. {the first roll is a 4, the second roll is a 4, neither roll is a 4}

   c. {the first 4 comes on the first roll, the first 4 comes on the second roll}

   d. {the sum of the two rolls is less than 2, the sum is more than 2}

   e. {the first roll is a 4, neither roll is a 4}

E6. Jean d'Alembert was coauthor (with another Frenchman, Denis Diderot, 1713–1784) of a 35-volume *Encyclopédie*. In it, he wrote that the probability of getting heads at least once in two flips of a fair coin is 2/3. This time, he said that these three outcomes were equally likely:

   • heads on the first flip

   • heads on the second flip

   • heads on neither flip

   a. Is this list of outcomes complete?

   b. Are the outcomes disjoint? Explain.

   c. Are the three outcomes equally likely? Explain.

   d. Is d'Alembert correct about the probability of getting heads at least once?

E7. The results of 50 rolls of a pair of dice are plotted in Display 5.10. The horizontal axis gives the number of the roll, and the vertical axis gives the *cumulative* proportion of rolls so far that were doubles.

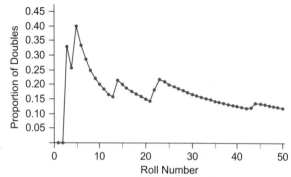

Display 5.10 Cumulative proportion of doubles for 50 rolls of a pair of dice.

a. Was the first roll a double? The second? The third? The fourth?

b. How many rolls were doubles?

c. Use this sample to estimate the probability of rolling doubles with a pair of dice.

d. Why do the lengths of the line segments tend to get shorter as the number of rolls increases?

E8. Jack and Jill decide to investigate the probability that a penny they found will land heads up when spun.

a. Jack spins the penny 500 times and gets 227 heads. What is his estimate of the probability that the penny will land heads up when spun?

b. Jill spins the penny 50 times. Would you expect Jack or Jill to have an estimate closer to the true probability of the penny landing heads up? Why?

c. Jack and Jill now flip the penny 10,000 times. Their results are recorded in Display 5.11. Fill in the missing percentages. Do their results illustrate the Law of Large Numbers? Explain.

d. Does the number of heads get closer to or farther away from half the number of flips?

Number of Flips	Number of Heads	Percentage of Heads
10	6	60
100	56	56
1,000	524	—?—
10,000	5,140	—?—

Display 5.11 Percentage of heads for various numbers of coin flips.

E9. Miguel is designing an experiment to test which combination of colors makes the type on computer screens easiest to read. His two factors are the color of the background (blue, green, yellow, white, or beige) and the color of the type (brown, black, navy, or gray).

a. How many different possible treatments are there?

b. Show the possible treatments on a tree diagram.

c. Suppose Miguel adds a third factor, brightness, to his experiment. He will have two levels of brightness

on the screen, high and low. How many possible distinct treatments are there?

d. Make a tree diagram showing this new situation.

E10. A company that produces clothing for travelers plans to promote a washable, nonwrinkle travel wardrobe. It will recommend that travelers buy six shirts (blue, green, red, yellow, and two white—one long-sleeved, one short-sleeved) and four coordinating pairs of pants (brown, black, blue, and gray).

a. The company wants to advertise the number of different outfits that can be worn. Use the Fundamental Principle of Counting to find this number.

b. Show all possible outfits in a two-way array.

c. If a traveler selects a shirt and a pair of pants at random, determine the probability that he or she will wear

    i. a white shirt

    ii. the gray pants

    iii. a white shirt and the gray pants

    iv. a white shirt or the gray pants

E11. In a Culver City, California, election, the two candidates for a seat on the Board of Education tied with 1141 votes each. To avoid a long and costly runoff election, the city code called for a chance process to select the winner. The city officials made up two bags. One candidate was given a bag that contained eight red marbles and one white marble. The other candidate was given a bag that contained eight blue marbles and one white marble. Simultaneously, the candidates drew a marble. If neither marble was white, the candidates continued to draw marbles simultaneously, one marble at a time. The first candidate to draw a white marble was the winner. The winner drew a white marble on the fourth round. [Source: *Los Angeles Times*, December 2, 2003, pp. B1, B7.]

a. Is this a fair process?

b. How could this process once again result in a tie?

c. What's the chance of a tie on the first round?

d. How could you change the process to reduce the probability of a tie?

E12. You have 27 songs on your MP3 player and randomly select one to play. It's your favorite. When it finishes, you again randomly select a song to play, and it's the same one!

a. What is the chance of this happening?

b. What is the chance that both random selections will be the same song?

E13. *Equally likely outcomes?*

a. Use the Fundamental Principle of Counting to find the number of outcomes for the situation of flipping a fair coin six times. Can you find the probability that you will get heads all six times?

b. Use the Fundamental Principle of Counting to find the number of outcomes for the situation of rolling a fair die six times. Can you find the probability that you will get a 3 all six times?

c. Use the Fundamental Principle of Counting to find the number of outcomes for the situation of picking 1200 U.S. residents at random and asking if they go to school. Can you find the probability that all 1200 will say yes?

E14. How many three-digit numbers can you make from the digits 1, 2, and 7? You can use the same digit more than once. If you choose the digits at random, what is the probability that the number is less than 250?

5.2 ▶ The Addition Rule and Disjoint Events

In this section you will learn to analyze situations like the following: Ann and Bill are planning their engagement party. The room will hold 200 people; so they agree that Ann will invite 100 of her nearest and dearest friends and Bill will invite 100 of his nearest and dearest friends. Everyone invited shows up to the party, but there were only 140 guests there! Ann and Bill are mystified. How could this happen?

The Addition Rule for Disjoint (Mutually Exclusive) Events

You can't roll two dice and get doubles and a sum of 3. You can't draw a card and get an ace and a jack. You can't select a student at random and get a junior and a senior. Two events that can't happen on the same opportunity are called **disjoint** or **mutually exclusive**.

> Two events are disjoint if they can't happen on the same opportunity.

A Property of Disjoint (Mutually Exclusive) Events

If event A and event B are disjoint, then

$$P(A \text{ and } B) = 0$$

> $P(A \text{ and } B)$ may be written $P(A \cap B)$ or $P(AB)$.

If two events are disjoint, then you can add their probabilities to find $P(A \text{ or } B)$. This rule seems natural and you used it, for example, in E3 on page 237 to compute that if you roll a pair of dice, the probability of getting a sum of 7 or a sum of 11 is $\frac{6}{36} + \frac{2}{36} = \frac{8}{36}$.

The Addition Rule for Disjoint (Mutually Exclusive) Events

If event A and event B are disjoint, then

$$P(A \text{ or } B) = P(A) + P(B)$$

> $P(A \text{ or } B)$ is sometimes written $P(A \cup B)$, read "A union B."

The Addition Rule with Dice

Example 5.3

Use the Addition Rule for Disjoint Events to find the probability that if you roll two dice, you get a sum of 7 or a sum of 8.

Solution
Let A be the event *sum of 7*. Let B be the event *sum of 8*. On the same roll, these two events are mutually exclusive, so

$$P(A \text{ and } B) = 0.$$

Then $P(A \text{ or } B) = P(A) + P(B)$. Using the sample space of equally likely outcomes

$$P(sum\ of\ 7\ or\ sum\ of\ 8) = P(sum\ of\ 7) + P(sum\ of\ 8)$$
$$= \frac{6}{36} + \frac{5}{36}$$
$$= \frac{11}{36}$$

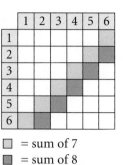

□ = sum of 7
■ = sum of 8

| Example 5.4 | **The Addition Rule with Coffee** |

Use the Addition Rule for Disjoint Events to compute the probability that exactly one person out of two correctly chooses the gourmet coffee when each person is guessing between gourmet coffee (G) and ordinary coffee (O).

Solution
The sample space of equally likely outcomes was: GG, GO, OG, OO. Thus, $P(exactly\ one\ G\ among\ two\ tasters)$

$$= P[(1st\ chooses\ G\ and\ 2nd\ chooses\ O) \text{ or } (1st\ chooses\ O\ and\ 2nd\ chooses\ G)]$$
$$= P(GO\ or\ OG)$$
$$= \frac{1}{4} + \frac{1}{4} = \frac{1}{2}$$

You can add the probabilities because the two outcomes GO and OG are disjoint.

The Addition Rule for Disjoint Events can be generalized. For example, if each pair of events A, B, and C is disjoint, then

$$P(A \text{ or } B \text{ or } C) = P(A) + P(B) + P(C)$$

DISCUSSION
The Addition Rule for Disjoint (Mutually Exclusive) Events

D10. Statistical data often are presented in tables like the ones below. For each table, decide whether the categories are disjoint. How do you know?

a. Display 5.12 gives the status of all noninstitutionalized adults in the United States who were employed or seeking employment. For each person, only the principal source of employment was counted.

Noninstitutional Population	Number of People (in millions)
Employees in nonagricultural industries	144
Employees in agricultural and related industries	2
Unemployed but seeking employment	7
Not in the labor force	79
Total	232

Display 5.12 The U.S. labor force as of January 2008. Institutional includes military, colleges and universities, medical facilities, prisons, and so on. [Source: *Statistical Abstract of the United States, 2009,* Table 581.]

b. Display 5.13 lists some common leisure activities and the percentage of the 221 million U.S. adults who participated in the activity at least once in the prior 12 months.

Activity	Percentage of U.S. Adults Who Engaged in Activity at Least Once in the Prior 12 Months
Dining out	49
Reading books	39
Computer games	20
Going to the beach	24

Display 5.13 Percentage of U.S. adults who engage in various leisure activities. [Source: *Statistical Abstract of the United States*, 2009, Table 1200.]

D11. Suppose you select a person at random from your campus. Are these pairs of events mutually exclusive?

a. has ridden a roller coaster; has ridden a Ferris wheel

b. owns a classical music CD; owns a jazz CD

c. is a senior; is a junior

d. has brown hair; has brown eyes

e. is left-handed; is right-handed

f. has shoulder-length hair; is male

D12. Suppose there is a 20% chance of getting a mosquito bite each summer evening that you go outside. Can you use the Addition Rule for Disjoint Events to compute the probability that you will get bitten if you go outside on three summer evenings? If you go outside on six summer evenings?

The Addition Rule

One of the most common words in the English language—*or*—is often misunderstood. This is because *or* can have two different meanings. For example, if a menu says that soup or salad comes with your entrée, you might not be positive whether you must pick only one or whether it would be acceptable to ask for both. In statistics, as in mathematics, the meaning of *or* allows *both* as a possibility: You can have a salad, the soup, or both. The next example shows how to compute $P(A \text{ or } B)$ when there is overlap in the two events.

> In mathematics, "A or B" includes the possibility that both A and B happen.

Weight and Gender

Example 5.5

The percentages of U.S. adults in various weight categories, as defined and estimated by the Behavioral Risk Factor Surveillance System of the Centers for Disease Control and Prevention, are shown in Display 5.14. Suppose that you pick a U.S. adult at random.

a. What is the probability that the person is overweight or obese?

b. What is the probability that the person is overweight or male?

Category		Male	Female	Total
Weight	Neither Overweight nor Obese (*BMI* < 25)	15.4	23.3	38.7
	Overweight (25 ≤ *BMI* < 30)	21.9	14.9	36.8
	Obese (*BMI* ≥ 30)	12.3	12.2	24.5
	Total	49.6	50.4	100

Display 5.14 Percentages by weight and gender category. *BMI* is body mass index. [Source: Centers for Disease Control and Prevention; BRFSS.]

Solution

a. Because being overweight and being obese are disjoint categories, the probability that a randomly selected adult will be overweight or obese is the probability the person is overweight plus the probability the person is obese: 0.368 + 0.245 = 0.613.

b. If you add the proportion, 0.496, of people who are male to the proportion, 0.368, of people who are overweight you will include the proportion, 0.219, of people who are overweight males twice. The correct computation is

$$P(overweight \text{ or } male) = P(overweight) + P(male) - P(overweight \text{ and } male)$$
$$= 0.368 + 0.496 - 0.219$$
$$= 0.645$$

You can verify this computation by adding the four cells that include anyone who is male or overweight: 0.154 + 0.219 + 0.123 + 0.149 = 0.645.

The computation in part b of Example 5.5 is an example of the Addition Rule.

> **Addition Rule**
>
> For any two events *A* and *B*,
>
> $$P(A \text{ or } B) = P(A) + P(B) - P(A \text{ and } B)$$

Using set notation, $P(A \cup B) = P(A) + P(B) - P(A \cap B)$, where $A \cap B$ is read "*A* intersect *B*."

Example 5.6

Doubles or a Sum of 8

Use the Addition Rule to find the probability that if you roll two dice, you get doubles or a sum of 8.

Solution

Let *A* be the event *getting doubles*, and let *B* be the event *getting a sum of 8*. These two events are not mutually exclusive, because they have a common outcome (4, 4), which occurs with probability 1/36. From the Addition Rule,

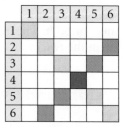

$$P(A \text{ or } B) = P(A) + P(B) - P(A \text{ and } B)$$
$$P(doubles \text{ or } sum \text{ of } 8)$$
$$= P(doubles) + P(sum \text{ of } 8) - P(doubles \text{ and } sum \text{ of } 8)$$
$$= \frac{6}{36} + \frac{5}{36} - \frac{1}{36} = \frac{10}{36}$$

□ = doubles
■ = sum of 8

Computing *P(A* and *B)*

Example 5.7

You will be assigned to do a case study of the insurance history of one small business in your community. In your community, 80% of the small businesses carry liability insurance (*L*) or worker's compensation insurance (*W*). Forty percent carry liability insurance and 50% carry worker's compensation insurance. If your small business is to be chosen at random from those in your community, find the probability that the selected business carries both types of insurance.

Solution

Events *L* and *W* aren't disjoint, so the general form of the Addition Rule applies:

$$P(L \text{ or } W) = P(L) + P(W) - P(L \text{ and } W)$$
$$0.80 = 0.40 + 0.50 - P(L \text{ and } W)$$

Solving, $P(L \text{ and } W) = 0.10$. The probability that the small business carries both types of insurance is 0.10.

DISCUSSION
The Addition Rule

D13. The diagrams in Display 5.15 are called Venn diagrams. They can be used to illustrate the two forms of the Addition Rule.

I.
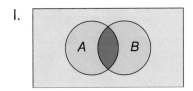
II.
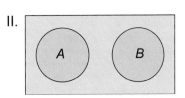

Display 5.15   Venn diagrams illustrating the two forms of the Addition Rule.

a. Which diagram illustrates mutually exclusive events?

b. Use the Venn diagrams to resolve Ann and Bill's situation in the first paragraph of this section, on page 239. Which is the diagram they thought applied? Which actually did? How many friends fall into each section of the diagram?

c. A Roper poll found that 46% of the Americans polled said yes to this question: "Should state governments give legal recognition to marriages between couples of the same sex?" In the same poll, 46% said yes to this question: "Should the federal government give legal recognition to marriages between couples of the same sex?" Draw a Venn diagram that you think illustrates this situation. [Source: AP-National Constitution Center Poll conducted by GfK Roper Public Affairs & Media. September 3–8, 2009. *n* = 1001 adults nationwide.]

D14. In a situation where *A* and *B* are mutually exclusive, what happens if you use the general form of the Addition Rule?

$$P(A \text{ or } B) = P(A) + P(B) - P(A \text{ and } B)$$

Summary 5.2: The Addition Rule and Disjoint Events

In statistics, $P(A$ or $B)$ indicates the probability that event A occurs but B doesn't, event B occurs but A doesn't, or both occur in the same random situation. The word *or* always allows for the possibility that both events occur. In this section, you learned the Addition Rule for any events A and B:

$$P(A \text{ or } B) = P(A) + P(B) - P(A \text{ and } B)$$

Two events are disjoint (mutually exclusive) if they cannot occur on the same opportunity. In the case of disjoint events, $P(A$ and $B) = 0$, so the Addition Rule simplifies to

$$P(A \text{ or } B) = P(A) + P(B)$$

Two-way tables often are helpful in clarifying the structure of a probability problem and in answering questions about the joint behavior of two events..

Practice

The Addition Rule for Disjoint (Mutually Exclusive) Events

P10. Display 5.16 shows the U.S. resident population categorized by age and sex. You are working for a polling organization that is about to select a random sample of U.S. residents.

a. What is the probability the first person selected will be age 65 or older?

b. Which of these pairs of events are disjoint?

   i. getting a female; getting a person age 85 or older

   ii. getting a male age 17 and under; getting a male age 85 and older

   iii. getting a person age 85 or older; getting a person age 65 or older

Age (yr)	Males (millions)	Females (millions)	Total
17 and Under	37.7	36.0	73.7
18 to 29	25.5	24.2	49.7
30 to 64	68.2	69.8	138.0
65 to 84	14.1	18.2	32.3
85 and older	1.5	3.4	4.9
Total	147.0	151.6	298.6

Display 5.16 Resident population of the United States by age and sex (excluding the population living in institutions, college dormitories, and other group quarters). [Source: *American Community Survey*, U.S. Census Bureau, 2007.]

P11. Display 5.17 shows the U.S. college population categorized by age and sex.

a. Suppose that one college student is selected at random. What is the probability the student is age 24 or younger?

b. Suppose that one male college student is selected at random. What is the probability the student is age 24 or younger?

c. Suppose that one female college student is selected at random. What is the probability the student is age 24 or younger?

Age (yr)	Male Enrollment (thousands)	Female Enrollment (thousands)	Total
19 and under	1782	2176	3958
20 to 24	3171	3670	6841
25 to 29	1033	1278	2311
30 and older	1519	2602	4121
Total	7505	9726	17,231

Display 5.17 College population of the U.S. by age and sex. [Source: U.S. Census Bureau, *Statistical Abstract of the United States, 2009*, Table 274.]

P12. If you roll two dice, are these pairs of events mutually exclusive? Explain.

a. doubles; sum is 8

b. doubles; sum is odd

c. a 3 on one die; sum is 10

d. a 3 on one die; doubles

P13. A researcher will select a student at random from a college population where 33% of the students are freshmen, 27% are sophomores, 25% are juniors, and 15% are seniors.

a. Is it appropriate to use the Addition Rule for Disjoint Events to find the probability that the student will be a junior or a senior? Why or why not?

b. Find the probability that the student will be a freshman or a sophomore.

P14. A tetrahedral die has the numbers 1, 2, 3, and 4 on its faces. Suppose you roll a pair of tetrahedral dice.

a. Make an array of all 16 possible outcomes (or use the one you made in E4 on page 237).

b. Use the Addition Rule for Disjoint Events to find the probability that you get a sum of 6 or a sum of 7.

c. Use the Addition Rule for Disjoint Events to find the probability that you get doubles or a sum of 7.

d. Why can't you use the Addition Rule for Disjoint Events to find the probability that you get doubles or a sum of 6?

The Addition Rule

P15. Display 5.18 gives information about all reportable crashes on state-maintained roads in North Carolina in a recent year.

	No Alcohol Involved	Alcohol Involved	Total
Not Speed Related	137,881	6,492	144,373
Speed Related	82,650	4,436	87,086
Total	220,531	10,928	231,459

Display 5.18 Information on reportable crashes in North Carolina. [Source: North Carolina Crash Data Query, *www.hsrc.unc.edu/crash/.*]

a. Are the events *crash involved alcohol* and *crash was speed related* mutually exclusive? How can you tell?

b. Use the Addition Rule to compute the probability that a randomly selected crash involved alcohol or was speed related.

P16. Use the Addition Rule to compute the probability that if you roll two six-sided dice,

a. you get doubles or a sum of 4

b. you get doubles or a sum of 7

c. you get a 5 on the first die or you get a 5 on the second die

P17. Use the Addition Rule to compute the probability that if you flip two fair coins, you get heads on the first coin or you get heads on the second coin.

P18. Use the Addition Rule to find the probability that if you roll a pair of dice, you do not get doubles or you get a sum of 8.

Exercises

E15. Suppose you flip a fair coin five times. Are these pairs of events disjoint?

a. You get five heads; you get four heads and one tail.

b. The first flip is heads; the second flip is heads.

c. You get five heads; the second flip is tails.

d. You get three heads and two tails; the second flip is tails.

e. The first four flips are heads; the first three flips are heads.

f. Heads first occurs on the third flip; the fourth flip is heads.

g. Heads first occurs on the third flip; three of the flips are heads.

E16. Suppose you select two people at random from a nearby health clinic. Are these pairs of events disjoint?

a. One person has health insurance; the other person has health insurance.

b. Both people have health insurance; only one person has health insurance.

c. One person is over age 65; the other person is under age 19.

d. One person is over age 95; the other person is the first person's mother.

E17. Of the 30 million people in the United States who fish, 1.4 million fish in the Great Lakes, 25 million fish in other freshwater, and 7.7 million fish in saltwater. [Source: *Statistical Abstract of the United States, 2009,* Table 1216.]

a. In categorizing people who fish, are these three categories disjoint? How can you tell?

b. Suppose you randomly select a person from among those who fish. Can you find the probability that the person fishes in saltwater?

c. Suppose you randomly select a person from among those who fish. Can you find the probability that the person fishes in freshwater?

d. The number of people who fish in freshwater is 25.4 million. How many people fish in both saltwater and freshwater?

E18. Display 5.19 categorizes the child support received by custodial parents with children under age 21 in the United States.

Child Support Status by Custodial Parents, 2005	Number (1000)
With child support agreement or award	7802
Supposed to receive payments in year shown	6809
Actually received payments in year shown	5259
Received full amount	3192
Received partial payments	2067
Did not receive payments in year shown	1550
Child support not awarded	5803

Display 5.19 Custodial parents and court-ordered child support. [Source: U.S. Census Bureau, *Statistical Abstract of the United States, 2009,* Table 549.]

a. Revise the table so that the categories are disjoint. Note that one category wasn't included: Custodial parents with a child support agreement or award who were not supposed to receive payments in 2005 (but maybe get payments in some other year).

b. If one of the custodial parents is selected at random, what is the probability that he or she was supposed to receive payments and received the full amount or received a partial amount?

E19. Display 5.20 describes students in a large class.

a. Fill in the missing cells and marginal totals.

b. What is the probability that a student randomly selected from this class doesn't receive financial aid or doesn't own a laptop?

		Owns a Laptop?		
		Yes	No	Total
Receives	Yes	—?—	46	72
Financial Aid?	No	112	—?—	—?—
	Total	—?—	—?—	216

Display 5.20 Table of laptop ownership and financial aid receipt.

E20. Display 5.21 classifies crashes in a recent year in Virginia. Some cells are missing because they weren't given directly by the source.

	Driver Violated a Traffic Law	Driver Didn't Violate a Traffic Law	Total
Fatality	521	—?—	837
No Fatality	—?—	—?—	—?—
Total	145,288	—?—	153,907

Display 5.21 Characteristics of crashes in Virginia. [Source: Virginia Department of Motor Vehicles, *www.dmv.state.va.us*.]

a. Fill in the missing cells and marginal totals.

b. What proportion of crashes involved a fatality and a traffic law violation?

c. What proportion of crashes involved a fatality or a traffic law violation?

d. What proportion of fatal crashes involved a driver who violated a traffic law? Is this more or less than the proportion of nonfatal crashes that involved a driver who violated a traffic law?

E21. Suppose that 80% of a population of laboratory rats have genetic mutation A, 70% have genetic mutation B, and 55% have both mutations. What percentage of the rats have mutation A or mutation B?

a. Construct a table to help you answer this question.

b. Use the Addition Rule to answer this question.

E22. After diagnosis with a life-threatening illness, 12% of patients were abandoned by their spouses. Of all the patients, 53% were women and 54% were either abandoned or a woman. Use this information to complete a two-way table (*men/women* versus *stayed together/abandoned*). Were men patients or women patients more likely to be abandoned? [Source: Michael J. Glantz et al., "Gender Disparity in the Rate of Partner Abandonment in Patients with Serious Medical Illness," *Cancer*, American Cancer Society, published online 7/30/2009.]

E23. In the United States, about 7,243,000 people ages 16 to 24 have completed high school and aren't enrolled in college. There are 3,766,000 high school dropouts in this same age group. Of the high school graduates, 71% are employed, 11% are unemployed, and 18% are not in the labor force. Of the high school dropouts, 53% are employed, 14% are unemployed, and 33% are not in the labor force. If you randomly select a person age 16 to 24 who isn't enrolled in college or high school, what is the probability that he or she is employed? Make a two-way table to help you answer this question. [Source: *Statistical Abstract of the United States, 2006*, Table 261.]

E24. Polls of registered voters often report the percentage of Democrats and the percentage of Republicans who approve of the job the president is doing. Suppose that in a poll of 1500 randomly selected voters 860 are Democrats and 640 are Republicans. Overall, 937 approve of the job the president is doing, and 449 of these are Republicans. Assuming the people in this poll are representative, what percentage of registered voters are Republicans or approve of the job the president is doing? First answer this question by using the Addition Rule. Then make a two-way table showing the situation.

E25. Jill computes the probability that she gets heads at least once in two flips of a fair coin:

P(at least one heads in two flips)

$$= P(heads\ on\ first\ flip\ or\ heads\ on\ second)$$

$$= P(heads\ on\ first) + P(heads\ on\ second)$$

$$= \frac{1}{2} + \frac{1}{2} = 1$$

She defends her use of the Addition Rule because getting heads on the first flip and heads on the second flip are mutually exclusive—both can't happen at the same time. What would you say to her?

E26. Rose computes the probability that she gets heads exactly once in two flips of a fair coin:

P(exactly one heads in two flips)

$$= P((tails\ on\ first\ flip\ and\ heads\ on\ second)$$
$$or\ (heads\ on\ first\ flip\ and\ tails\ on\ second))$$

$$= P((tails\ on\ first\ flip\ and\ heads\ on\ second)$$
$$+ P(heads\ on\ first\ flip\ and\ tails\ on\ second))$$

$$= \frac{1}{4} + \frac{1}{4} = \frac{1}{2}$$

She defends her use of the Addition Rule because *HT* and *TH* are mutually exclusive. What would you say to her?

E27. When is this statement true?

$$P(A\ or\ B\ or\ C) = P(A) + P(B) + P(C)$$

E28. Suppose events *A*, *B*, and *C* are three events where $P(A\ and\ B) \neq 0$, $P(A\ and\ C) \neq 0$, $P(B\ and\ C) \neq 0$, and $P(A\ and\ B\ and\ C) \neq 0$.

a. Draw a Venn diagram to illustrate this situation.

b. Use the Venn diagram from part a to help you write a rule for computing $P(A\ or\ B\ or\ C)$.

5.3 ▶ Conditional Probability and the Multiplication Rule

The *Titanic* sank in 1912 without enough lifeboats for the passengers and crew. Almost 1500 people died, most of them men. Was that because a man was less likely than a woman to survive, or did more men die simply because men outnumbered women by more than 3 to 1 on the *Titanic*? You might turn to an old campfire song to decide what might have happened. One line is about loading the lifeboats: "And the captain shouted, 'Women and children first.'" Statisticians aren't opposed to songs. But statisticians also know that stories and theories, even stories set to music, are no substitute for data. Do the data in Display 5.22 support the lyrics of the song?

Titanic survivors nearing rescue ship.

		Gender		
		Male	Female	Total
Survived?	Yes	367	344	711
	No	1364	126	1490
	Total	1731	470	2201

Display 5.22 *Titanic* survival data. [Source: *Journal of Statistics Education*, Vol. 3, 3 (1995).]

Although the numbers alone can't tell you who got to go first on the lifeboats, they do show that $\frac{344}{470}$, or about 73%, of the females survived, while only $\frac{367}{1731}$, or about 21%, of the males survived. Thus, the data are fully consistent with the hypothesis that the song explains what happened.

Overall, $\frac{711}{2201}$, or approximately 32.3%, of the people survived, but the survival rate for females was much higher and that for males much lower. The chance of surviving depended on the *condition* of whether the person was male or female. This commonsense notion that probability can change if you are given additional information is called **conditional probability.**

Conditional Probability from the Sample Space

Often you can calculate conditional probabilities directly from the sample space.

The *Titanic* and Conditional Probability

Example 5.8

Suppose you pick a person at random from the list of people aboard the *Titanic*. Let S be the event that this person survived, and let F be the event that the person was female. Find $P(S|F)$, read "the probability of S given that F is known to have happened."

> The symbol for the conditional probability that A happens given that B happens is $P(A|B)$.

Solution
The conditional probability $P(S|F)$ is the probability that the person survived *given the condition* that the person was female. To find $P(S|F)$, restrict your sample space to only the 470 females (the outcomes for whom the condition is true). Then compute the probability of survival as the number of favorable outcomes—344, the women surviving—divided by the total number of outcomes—470—in the restricted sample space.

$$P(S|F) = P(survived|female) = \frac{344}{470} \approx 0.732$$

Overall, only 32.3% of the people survived, but 73.2% of the females survived.

| **Example 5.9** | **Sampling Without Replacement** |

When you sample without replacement from a small population, the probabilities for the second selection depend on what happens on the first selection. For example, imagine a population of four recent graduates (that is, $N = 4$), two from the west (W) and two from the east (E). Suppose you randomly choose one person from the population and he or she is from the west. What is the probability that the second person selected is also from the west? In symbols, find

$$P(W \text{ selected 2nd} | W \text{ selected 1st})$$

Solution

If you start with a population $\{W, W, E, E\}$ and the first selection is W, that leaves the restricted sample space $\{W, E, E\}$ from which to choose on the second selection. The conditional probability that the second selection is from the west is

$$P(W \text{ selected 2nd} | W \text{ selected 1st}) = \frac{1}{3}$$

DISCUSSION
Conditional Probability from the Sample Space

D15. When you compare sampling with and without replacement, how does the size of the population affect the comparison? Conditional probability lets you answer the question quantitatively. Imagine two populations of students, one large ($N = 100$) and one small ($N = 4$), with half the students in each population male (M). Select random samples of size $n = 2$ from each population.

  a. First, consider the small population. Find $P(2nd \text{ is } M | 1st \text{ is } M)$, assuming you sample without replacement. Then find the probability again, this time assuming you sample with replacement.

  b. Do the same for the larger population.

  c. Based on these calculations, how would you describe the effect of population size on the difference between the two sampling methods?

The Multiplication Rule for *P(A and B)*

You will now investigate a general rule for finding $P(A \text{ and } B)$ for any two events A and B. Suppose, for example, you choose one person at random from the passenger list of the *Titanic*. What is the probability that the person was female (F) and survived (S)? You already know how to answer this question directly by using the data in Display 5.22. Out of 2201 passengers, 344 were women and survived, so

$$P(F \text{ and } S) = \frac{344}{2201}$$

You can also find the answer using conditional probability. It sometimes helps to show the order of events in a tree diagram, such as the one in Display 5.23 on the next page.

You can interpret the top branch of the tree diagram, for example, as

$$P(F) \cdot P(S|F) = \frac{470}{2201} \cdot \frac{344}{470} = \frac{344}{2201} = P(F \text{ and } S)$$

The tree diagram leads to the following rule for finding $P(A \text{ and } B)$.

The Multiplication Rule

The probability that event A and event B both happen is given by

$$P(A \text{ and } B) = P(A) \cdot P(B|A)$$

Alternatively,

$$P(A \text{ and } B) = P(B) \cdot P(A|B)$$

$P(A \text{ and } B)$ is sometimes written as $P(A \cap B)$ or $P(AB)$.

Using set notation,
$P(A \cap B) = P(A) \cdot P(B|A)$ or
$P(A \cap B) = P(B) \cdot P(A|B)$.

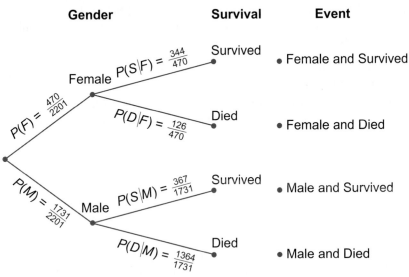

Display 5.23   Tree diagram for the *Titanic* data.

Display 5.24 shows the Multiplication Rule on the branches of a tree diagram.

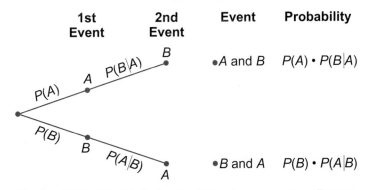

Display 5.24   The Multiplication Rule, shown on a tree diagram.

Females on the *Titanic*

Example 5.10

Use the Multiplication Rule to compute the percentage of passengers on the *Titanic* who were females and survived.

Solution
Twenty-one percent of all *Titanic* passengers were female, and 73% of female passengers survived. So 73% of 21%, or 15.33%, of all passengers were female and survived.

$$P(\text{female and survived}) = P(\text{female}) \cdot P(\text{survived}|\text{female}) = (0.21)(0.73)$$

| Example 5.11 | Coincidences |

The following article describes what appears to be an amazing coincidence.

Man, Wife Beat Odds in Moose-Hunt Draw
The Washington State Game Department conducted a public drawing last week in Olympia for three moose hunting permits. There were 2,898 application cards in the wire mesh barrel. It was cranked around several times before the first name was drawn: Judy Schneider of East Wenatchee, WA. The barrel was spun again. Second name drawn: Bill Schneider, Judy's husband. Result: groans of disbelief from hopeful moose hunters in the auditorium. [Source: *Los Angeles Times*, July 18, 1978.]

a. What is the probability that Judy's name would be the first drawn and her husband's name would be second?

b. What assumptions are you making, when you answer this question?

c. Comment on why coincidences like this seem to happen so often.

Solution

a. $P(\text{Judy first and Bill second}) = P(\text{Judy first}) \cdot P(\text{Bill second}|\text{Judy first})$
$$= \tfrac{1}{2898} \cdot \tfrac{1}{2897} \approx 0.000000119$$

b. The assumptions are that Bill and Judy had only one card each in the barrel and that the cards were well mixed.

c. The chance that Judy's name and then her husband's name will be called is about one chance in 10 million. However, there is also the chance that Bill's name will be called first and then Judy's. This doubles their chances of being the first two names drawn to 0.000000238—still only about one chance in 5 million. However, suppose all the names in the barrel were those of couples. Then the probability that the first two names drawn will be those of a couple is $\frac{1}{2897}$, or about 0.000345, because the first name can be anyone and that person's partner is one of the 2897 left for the second draw. This probability is still small, but now it is 3 chances in 10000. Finally, thousands of lotteries take place in the United States every year, so it is virtually certain that coincidences like this will happen occasionally and be reported in the newspaper.

As time goes on, more coincidences have had an opportunity to appear. For example, since 1978 when the New York State Lotto game began, three different players have won a jackpot twice. [Source: *www.nylottery.org*.] As a result, the public has tended to become less astonished by coincidences. And rightly so: Statisticians have computed that, for example, the probability is more than 50% that, within the next year, a person somewhere in the United States will win the lottery twice. [Source: Gina Kolata, "1-in-a-Trillion Coincidence, You Say? Not Really, Experts Find," *New York Times*, February 27, 1990.]

DISCUSSION
The Multiplication Rule for *P(A and B)*

D16. Use the tree diagram for the *Titanic* data in Display 5.23 on page 249 to find *P(male and survived)* and *P(male and died)*.

D17. Use the *Titanic* data in Display 5.22 on page 247 to show that

$$P(M \text{ and } S) = P(S) \cdot P(M|S) = P(M) \cdot P(S|M)$$

D18. Use the frequencies in the table below to show that

$$P(A \text{ and } B) = P(A) \cdot P(B|A) = P(B) \cdot P(A|B)$$

is true in general.

		Event A Present?	
		Yes	No
Event B Present?	Yes	c	d
	No	e	f

The Definition of Conditional Probability

It's time to define conditional probability formally.

> ## The Definition of Conditional Probability
>
> For any two events A and B such that $P(B) > 0$,
>
> $$P(A|B) = \frac{P(A \text{ and } B)}{P(B)}$$

Using set notation,
$$P(A|B) = \frac{P(A \cap B)}{P(B)}$$

The Multiplication Rule is a consequence of the formal definition of conditional probability. Simply solve the equation in the definition for $P(A \text{ and } B)$.

Conditional Probability **Example 5.12**

Suppose you roll two dice. Use the definition of conditional probability to find the probability that you get a sum of 8 given that you rolled doubles.

Solution
Thinking in terms of the restricted sample space, you already know that the answer is 1/6 because there are six equally likely ways to roll doubles and only one of these is a sum of 8. Using the definition of conditional probability, the computation looks like this.

$$P(\text{sum } 8 | \text{doubles}) = \frac{P(\text{sum } 8 \text{ and } \text{doubles})}{P(\text{doubles})} = \frac{\frac{1}{36}}{\frac{6}{36}} = \frac{1}{6}$$

Conditional Probability and Medical Tests

In medicine, *screening tests* give a quick indication of whether a person is likely to have a particular disease. (For example, the ELISA test is a screening test for HIV, and a chest X-ray is a screening test for lung cancer.) Because screening tests are intended to be relatively quick and noninvasive, they often are not as accurate as other tests that take longer or are more invasive. (For example, a biopsy is more accurate than a chest X-ray if you want to know whether a lung tumor is cancerous.)

A two-way table like the one in Display 5.25 is often used to show the four possible outcomes of a screening test.

		Test Result		
		Positive	Negative	Total
Disease	Present	a	b	$a + b$
	Absent	c	d	$c + d$
	Total	$a + c$	$b + d$	$a + b + c + d$

Display 5.25 Possible results of a screening test.

The effectiveness of screening tests is judged using conditional probabilities. These four terms are commonly used. In each case, higher values are better.

positive predictive value (PPV) $= P(\textit{disease present} | \textit{test positive}) = \dfrac{a}{a + c}$

negative predictive value (NPV) $= P(\textit{no disease present} | \textit{test negative}) = \dfrac{d}{b + d}$

sensitivity $= P(\textit{test positive} | \textit{disease present}) = \dfrac{a}{a + b}$

specificity $= P(\textit{test negative} | \textit{no disease}) = \dfrac{d}{c + d}$

Example 5.13	A Relatively Rare Disease

Relatively rare diseases have tables similar to Display 5.26. This test has 50 **false positives;** that is, 50 people who don't have the disease tested positive. It has only 1 **false negative**—a person with the disease who tested negative. Find the sensitivity, specificity, PPV, and NPV for this test.

		Test Result		
		Positive	Negative	Total
Disease	Present	9	1	10
	Absent	50	9,940	9,990
	Total	59	9,941	10,000

Display 5.26 Hypothetical results of a screening test for a rare disease.

Solution
In some ways, this is pretty good test. It finds 9 out of the 10 people who have the disease, for a sensitivity of 0.9. It correctly categorizes 9940 out of the 9990 people who don't have the disease, for a specificity of 0.99. The NPV is 9940 out of 9941, or 0.9999. However, notice that only 9 out of 59, or 15%, of the people who test positive for the disease actually have it! The PPV is quite low because there are so many false positives.

The previous example shows what can happen when the population being screened is mostly disease free, even with a test of high specificity. Most of the people who test positive do not, in fact, have the disease. On the other hand, if the population being screened has a high incidence of the disease, then there tend to be many false negatives and the negative predictive value tends to be low.

Because the positive predictive value and the negative predictive value depend on the population as well as the screening test, statisticians prefer to judge a test based on the other pair of conditional probabilities: sensitivity and specificity.

Conditional Probability and Statistical Inference

To calculate a probability, you must work from a model. For example, suppose you want to calculate the probability of observing an even number of dots on a roll of a die. Your model is that the die is fair. Using this model, you know that

$$P(\text{even number}|\text{fair die}) = \frac{3}{6}$$

Now suppose you have a die that you suspect isn't fair. If you roll it and get an even number, you cannot calculate $P(\text{fair die}|\text{even number})$. So how can you discredit a model?

If you roll the die 20 times and you get an even number every time, you can compute (by the method of the next section)

$$P(\text{even number on all 20 rolls}|\text{fair die}) = \left(\frac{3}{6}\right)^{20} \approx 0.000001$$

This outcome is so unlikely that you can feel justified in abandoning the model that the die is fair. But you still didn't—and can't—compute

$$P(\text{fair die}|\text{even number on all 20 rolls})$$

Next, let's see how these ideas apply to the age discrimination case of Chapter 1, where three of ten hourly workers were laid off. Their average age was 58. The dialogue that follows is invented and did not actually occur in the *Westvaco* case, but it is based on real conversations one of your authors had on several occasions with a number of different lawyers as they grappled with conditional probabilities.

Statistician: Suppose you draw three workers at random from the set of ten hourly workers. This establishes random sampling as the model for the study.

Lawyer: Okay.

Statistician: It turns out that there are 120 possible samples of size 3, and only 6 of them give an average age of 58 or more.

Lawyer: So the probability is 6/120, or 0.05.

Statistician: Right.

Lawyer: There's only a 5% chance that the company didn't discriminate and a 95% chance that it did.

Statistician: No, that's not true.

Lawyer: But you said . . .

Statistician: I said that *if* the age-neutral model of random draws is correct, then there's only a 5% chance of getting an average age of 58 or more.

Lawyer: So the chance that the company is guilty must be 95%.

Statistician: Slow down. If you start by assuming the model is true, you can compute the chances of various results. But you're trying to start from the results and compute the chance that the model is right or wrong. You can't do that.

Here is the analysis: The lawyer is having trouble with conditional probabilities. The statistician has computed

$$P(average\ age \geq 58 | random\ draws) = 0.05$$

The lawyer wants to know

$$P(no\ discrimination | average\ age = 58)$$

The statistician has computed $P(data | model)$. The lawyer wants to know $P(model | data)$. Finding the probability that there was no discrimination given that the average age was 58 is not a problem that statistics can solve. A model is needed in order to compute a probability.

DISCUSSION
Conditional Probability and Statistical Inference

D20. There's a parallel between statistical testing and medical testing. Write expressions for conditional probabilities—in terms of finding a company guilty or not guilty of discrimination—that correspond to a PPV and an NPV. (You might think of the true medical status as comparable to the true state of affairs within the company.)

Summary 5.3: Conditional Probability and the Multiplication Rule

In this section, you saw these important uses of conditional probability in statistics:

- to compare sampling with and without replacement
- to study the effectiveness of medical screening tests
- to describe probabilities of the sort used in statistical tests of hypotheses

You have learned the definition of conditional probability and the Multiplication Rule and how to use these two key concepts in solving practical problems. When you are solving conditional probability problems, by using the Multiplication Rule or otherwise, tree diagrams and two-way tables help you organize the data and see the structure of the problem.

- The definition of conditional probability is that for any two events A and B, where $P(B) > 0$, the probability of A given the condition B is

$$P(A|B) = \frac{P(A\ and\ B)}{P(B)}$$

- The Multiplication Rule is

$$P(A\ and\ B) = P(A) \cdot P(B|A)\ or\ P(B) \cdot P(A|B)$$

Practice

Conditional Probability from the Sample Space

P19. For the *Titanic* data in Display 5.22 on page 247, let S be the event a person survived and F be the event a person was female. Find and interpret these probabilities.
- **a.** $P(F)$
- **b.** $P(F|S)$
- **c.** $P(not\ F)$
- **d.** $P(not\ F|S)$
- **e.** $P(S|not\ F)$

P20. Display 5.27 gives the hourly workers in the United States, classified by race and by whether they were paid at or below minimum wage or above minimum wage. You select an hourly worker at random.
- **a.** Find $P(worker\ is\ paid\ at\ or\ below\ minimum\ wage)$.
- **b.** Find $P(worker\ is\ paid\ at\ or\ below\ minimum\ wage | worker\ is\ white)$.
- **c.** What does a comparison of the two probabilities in parts a and b tell you?
- **d.** Find $P(worker\ is\ black)$.

		Number Paid at or Below Minimum Wage (thousands of people)	Number Paid Above Minimum Wage (thousands of people)	Total
Race	White	1,420	59,641	61,061
	Black	205	9,760	9,965
	Asian	50	2,680	2,730
	Total	1,675	72,081	73,756

Display 5.27   Hourly wages of workers (in thousands), by race. [Source: U.S. Bureau of Labor Statistics, *Employment and Earnings,* January 2007; U.S. Census Bureau, *Statistical Abstract of the United States, 2009*, Table 631.]

e. Find *P(worker is black | worker is paid at or below minimum wage)*.

f. What does a comparison of the two probabilities in parts d and e tell you?

P21. Suppose Jack draws marbles at random, without replacement, from a bag containing three red and two blue marbles. Find these conditional probabilities.

a. *P(2nd draw is red | 1st draw is red)*

b. *P(2nd draw is red | 1st draw is blue)*

c. *P(3rd draw is blue | 1st draw is red and 2nd draw is blue)*

d. *P(3rd draw is red | 1st draw is red and 2nd draw is red)*

P22. Suppose Jill draws a card from a standard 52-card deck. Find the probability that

a. it is a club, given that it is black

b. it is a jack, given that it is a heart

c. it is a heart, given that it is a jack

The Multiplication Rule for *P(A and B)*

P23. Look again at the *Titanic* data in Display 5.22 on page 247.

a. Make a tree diagram to illustrate this situation, this time branching first on whether the person survived.

b. Write these probabilities as unreduced fractions.

   i. *P(survived)*

   ii. *P(female | survived)*

   iii. *P(survived and female)*

c. Now write a formula that tells how the three probabilities in part b are related. Compare it to the computation on page 249.

d. Write two formulas involving conditional probability to compute *P(male and survived)*.

P24. Use the Multiplication Rule to find the probability that if you draw two cards from a deck without replacing the first before drawing the second, both cards will be hearts. What is the probability if you replace the first card before drawing the second?

P25. Suppose you take a random sample of size *n* = 2, without replacement, from the population {*W, W, M, M*}. Find these probabilities: *P(W chosen 1st)* and *P(W chosen 2nd | W chosen 1st)*. Now find *P(W chosen 1st and W chosen 2nd)*.

P26. Use the Multiplication Rule to find the probability of getting a sum of 8 and doubles when you roll two dice.

The Definition of Conditional Probability

P27. Suppose you roll two dice. Use the definition of conditional probability to find *P(doubles | sum 8)*. Compare this probability with *P(sum 8 | doubles)*.

P28. Suppose you know that, in a class of 30 students, 10 students have blue eyes and 20 have brown eyes. Twenty-four of the students are right-handed, and six are left-handed. Of the left-handers, two have blue eyes. Make and fill in a table showing this situation. Then use the definition of conditional probability to find the probability that a student randomly selected from this class is right-handed, given that the student has brown eyes.

P29. In 1999, Elizabeth Dole was a candidate to become the first woman president in U.S. history, and many observers assumed that she would show particular strength among women voters. According to a Gallup poll, "She did slightly better among Republican women than among Republican men, but this strength was not nearly enough to enable her to challenge Bush. In the October poll, Dole received the vote of 16% of Republican women, compared to 7% of Republican men." Given the additional information that the Republican Party is about 60% male, find the probability that a Republican randomly selected from the October survey would have voted for Dole. [Source: *www.gallup.com*.]

Conditional Probability and Medical Tests

P30. A laboratory screening test for the detection of a certain disease gives a positive result 6% of the time for people who do not have the disease. The test gives a negative result 0.5% of the time for people who do have the disease. Large-scale studies have shown that the disease occurs in about 3% of the population.

a. Fill in this two-way table, showing the results expected for every 100,000 people.

		Test Result		
		Positive	Negative	Total
Disease	Yes	—?—	—?—	—?—
	No	—?—	—?—	—?—
	Total	—?—	—?—	100,000

b. What is the probability that a person selected at random tests positive for this disease?

c. What is the probability that a person selected at random who tests positive for the disease does not have the disease?

P31. When a pregnant woman has been injured by blunt trauma to the abdomen, her physician must check for the presence of pelvic free fluid. Will a fast ultrasound test serve as a good screening test or must the physician use a more accurate ("gold standard") test that is slower to perform and potentially has serious side effects? The data in Display 5.28 is from a study of 328 such women and gives the results of the ultrasound tests, which were checked by one of the "gold standard" tests. [Source: E. L. Ormsby, J. Geng, J. P. McGahan, and J. R. Richards, "Pelvic Free Fluid: Clinical Importance for Reproductive Age Women with Blunt Abdominal Trauma," *Ultrasound in Obstetrics & Gynecology*, September 2005, pp. 271–278.]

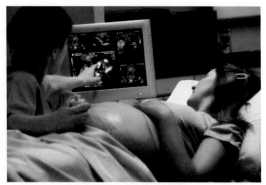

Pregnant woman receiving an ultrasound.

		Result of Ultrasound		
		Positive for Fluid	Negative for Fluid	Total
Fluid Actually	Present	14	9	23
	Absent	15	290	305
	Total	29	299	328

Display 5.28 Results of ultrasound for 328 women.

a. Use the data to estimate the four conditional probabilities defined for screening tests on page 252.

b. In your opinion, is the test a good one? Why or why not?

Conditional Probability and Statistical Inference

P32. Rosa has two pennies. One is an ordinary fair penny, but the other came from a magic shop and has two heads. Rosa chooses one of these coins, but she does not choose it randomly. She also doesn't tell you how she chose it, so you don't know the probabilities *P(coin is two-headed)* or *P(coin is fair)*. She flips the coin once. For each probability, give a numerical value if it is possible to find one. If it is not possible to compute a probability, explain why.

a. *P(coin lands heads | coin is fair)*

b. *P(coin lands heads | coin is two-headed)*

c. *P(coin is fair | coin lands heads)*

d. *P(coin is two-headed | coin lands heads)*

e. *P(coin is fair | coin lands tails)*

f. *P(coin is two-headed | coin lands tails)*

Exercises

E29. Display 5.29 (at the bottom of this page) gives a breakdown of the U.S. population age 16 and older by age and whether the person volunteers time to charitable activities. You are working for a polling organization that is about to select a random sample of U.S. residents age 16 and older. Find these probabilities for the first person selected in the random sample.

a. *P(person is a volunteer)*

b. *P(person is a volunteer | person is age 16 to 24)*

c. *P(person is age 16 to 24 | person is a volunteer)*

d. *P(person is age 16 to 34 | person is not a volunteer)*

e. Which age group has the highest percentage of volunteers? The lowest?

E30. Joseph Lister (1827–1912), British surgeon at the Glasgow Royal Infirmary, was one of the first to believe in Louis Pasteur's germ theory of infection. He experimented with using carbolic acid to disinfect operating rooms during amputations. When carbolic acid was used, 34 of 40 patients lived. When carbolic acid was

not used, 19 of 35 patients lived. Suppose a patient is selected at random. Make a two-way table and find

a. *P(patient died | carbolic acid used)*

b. *P(carbolic acid used | patient died)*

c. *P(carbolic acid used and patient died)*

d. *P(carbolic acid used or patient died)*

E31. A student collects information about the residents of a large nursing home, including age and whether the person receives some sort of medical assistance beyond Medicare. The table gives the proportion of residents in each category.

a. If possible, fill in the rest of this table.

		Receives Additional Medical Assistance?		
		Yes	No	Total
Age	Under 90	0.44	—?—	—?—
	90 or older	—?—	—?—	0.08
	Total	—?—	0.48	1.00

b. If a resident of this nursing home is chosen at random, what is the probability that the resident is 90 or older, given that the resident receives additional medical assistance?

	Age						
	16 to 24	25 to 34	35 to 44	45 to 54	55 to 64	65 and Over	Total
Volunteer	7,798	9,019	12,902	13,136	9,316	8,667	60,838
Not a Volunteer	29,692	30,888	29,399	30,505	23,486	27,748	171,718
Total	37,490	39,907	42,301	43,641	32,802	36,415	232,556

Display 5.29 Persons (in thousands) who performed unpaid volunteer activities in the last year. [Source: U.S. Census Bureau, *Statistical Abstract of the United States, 2009*, Table 566.]

c. What is the probability that a randomly selected resident doesn't receive additional medical assistance and is 90 or older?

E32. Suppose the only information you have about the *Titanic* disaster appears in this two-way table. Is it possible to fill in the rest of the table? Is it possible to find the probability that a randomly selected person aboard the *Titanic* was female and survived? Either calculate the numbers, or explain why it is not possible to do so without additional information and tell what information you would need.

		Gender		
		Male	Female	Total
Survived?	Yes	—?—	—?—	711
	No	—?—	—?—	1490
	Total	1731	470	2201

E33. As you have seen, a useful way to work with the Multiplication Rule is to record the probabilities along the branches of a tree diagram. To practice this, suppose you draw two marbles at random, without replacement, from a bucket containing three red (*R*) and two blue (*B*) marbles. Compute the probabilities in parts a–f and then use these probabilities to fill in and label the branches in a copy of the tree diagram in Display 5.30.

a. P(*R on 1st draw*)

b. P(*B on 1st draw*)

c. P(*R on 2nd draw | R on 1st draw*)

d. P(*B on 2nd draw | R on 1st draw*)

e. P(*R on 2nd draw | B on 1st draw*)

f. P(*B on 2nd draw | B on 1st draw*)

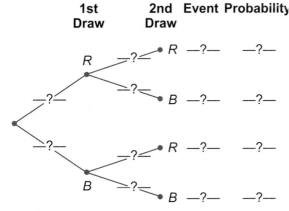

Display 5.30 Tree diagram for drawing marbles.

E34. Your sock drawer hasn't been organized in a while. It contains six identical brown socks, three identical white socks, and five identical black socks. You draw two socks at random (without replacement).

a. Make a tree diagram similar to that in E33 to show the possible results of your draws.

b. Find P(*first sock is white*).

c. Find P(*second sock is white | first sock is white*).

d. Find P(*second sock is white | first sock is brown*).

e. What is the probability that you get two socks that match?

E35. Suppose you draw two cards from a 52-card deck.

a. What is the probability that both cards are aces if you replace the first card before drawing the second?

b. What is the probability that both cards are aces if you don't replace the first card before drawing the second?

c. What is the probability that the first card is an ace and the second is a king if you replace the first card before drawing the second?

d. What is the probability that the first card is an ace and the second is a king if you don't replace the first card before drawing the second?

e. What is the probability that both cards are the same suit if you replace the first card before drawing the second?

f. What is the probability that both cards are the same suit if you don't replace the first card before drawing the second?

E36. A Harris poll found that 36% of all adults said they would be interested in going to Mars. Of those who were interested in going, 62% were under age 25. Write a question for which the solution would be to compute (0.36)(0.62).

E37. Suppose that if rain is predicted, there is a 60% chance that it will actually rain. If rain is not predicted, there is a 20% chance that it will rain. Rain is predicted on 10% of days. On what percentage of days does it rain?

E38. As of July 1 of a recent season, the Los Angeles Dodgers had won 53% of their games. Eighteen percent of their games had been played against left-handed starting pitchers. The Dodgers won 36% of the games played against left-handed starting pitchers. What percentage of their games against right-handed starting pitchers did they win? [Source: *Los Angeles Times*, July 1, 2000, p. D8.]

E39. A laboratory technician is being tested on her ability to detect contaminated blood samples. Among 100 samples given to her, 20 are contaminated, each with about the same degree of contamination. Suppose the technician makes the correct decision 90% of the time.

a. Make a two-way table showing what you would expect to happen.

b. Compute and interpret her PPV.

c. Compute and interpret her NPV.

d. How would these rates change if she were given 100 samples with 50 contaminated?

E40. A screening test for bureaucratic pomposity disorder (BPD) has reasonably good sensitivity and specificity, say 90% for both. (This means that, on average, nine out of every ten people who have BPD test positive and nine out of every ten who don't have BPD test negative.) Consider testing two different populations.

a. Officiousville has a population of 1000. Half the people in Officiousville have BPD. Suppose everyone in Officiousville gets tested. Fill out a table like Display 5.26 on page 252. Then use it to compute the PPV and NPV for this test.

b. Mellowville also has a population of 1000, but only 10 people in the whole town have BPD. Fill out a table to show how the test would perform if you used it to screen Mellowville. Then compute the PPV and NPV for this test.

E41. A new test for HIV, the virus that causes AIDS, uses a mouth swab instead of a blood test. The president of the company that produces this test was quoted as saying that overall the test was more than 99% accurate. The president claimed that, in the last year, the company received reports of only 107 false positives out of 28,436 tests. [Source: *New York Times*, December 10, 2005.]

   a. Can you determine the positive predictive value (PPV) for this test? Explain.

   b. The article gave no information about false negatives. Why do you think no information was given?

   c. How many false negatives would there have to be in order for the test to produce correct results in 99% of the 28,436 tests?

E42. Make a table and do a computation to illustrate that if the population being screened has a high percentage of people who have a certain disease, the NPV will tend to be low.

E43. Return to the situation in P32 in which Rosa has two pennies. One is an ordinary fair penny, but the other came from a magic shop and has two heads. Rosa

chooses one of these coins, this time at random. She flips the coin once. For each probability, give a numerical value if it is possible to find one. If it is not possible, explain why not.

   a. *P(coin lands heads | coin is fair)*

   b. *P(coin lands heads | coin is two-headed)*

   c. *P(coin is fair | coin lands heads)*

   d. *P(coin is two-headed | coin lands heads)*

   e. *P(coin is fair | coin lands tails)*

   f. *P(coin is two-headed | coin lands tails)*

E44. One of these two statements is true, and one is false. Which is which? Make up an example using hypothetical data to illustrate your decision.

$$P(A) = P(A|B) + P(A|not\ B)$$
$$P(A) = P(A\ and\ B) + P(A\ and\ not\ B)$$

5.4 ▶ Independent Events

Because of the health problems associated with obesity, researchers are interested in identifying people who are at risk for becoming obese. Known risk factors include lack of physical activity, poor diet, and age. Is gender one of the risk factors?

The table in Display 5.31 is based on the table in Display 5.14 on page 242. Two weight categories have been combined so that there are now just the two categories of interest, obese and not obese.

		Male (%)	Female (%)	Row Total (%)
Weight Category	Not Obese	38	38	76
	Obese (*BMI* ≥ 30)	12	12	24
	Column Total	50	50	100

Display 5.31 Weight categories by gender, rounded to nearest percentage. [Source: Centers for Disease Control; BFSSC.]

Now, you can see that it makes no difference whether you are talking about males or females—the percentage who are obese is the same. In fact, this is even clearer if you consider the conditional probabilities associated with selecting a person at random from the population of adults:

$$P(obese\,|\,male) = \frac{P(obese\ and\ male)}{P(male)} = \frac{12}{50} = 0.24$$

Since the male and female columns are the same, it is also the case that

$$P(obese\,|\,female) = \frac{P(obese\ and\ female)}{P(female)} = \frac{12}{50} = 0.24$$

The two columns of conditional probabilities, as shown in Display 5.32, are identical and are identical to the overall relative frequency of obese adults shown in the row totals of Display 5.31. Thus, the conditional probability of obesity among males is the same as the unconditional probability of obesity in general. In probability terminology, the event *obese* is said to be **independent** of the event *male*.

		Male	Female
Weight Category	Not Obese	0.76	0.76
	Obese (*BMI* ≥ 30)	0.24	0.24
	Column Total	1.00	1.00

Display 5.32 Weight categories by gender; probabilities by column.

Display 5.32 shows that gender is not a risk factor—males and females are equally likely to become obese. Perhaps educational level is associated with obesity. Display 5.33 shows the conditional percentages of obese and not obese U.S. adults across two categories of education.

		College or Technical School Graduate (%)	Not a College or Technical School Graduate (%)
Weight Category	Not Obese	81	73
	Obese (*BMI* ≥ 30)	19	27
	Column Total	100	100

Display 5.33 Weight categories by education; percentages by column.
[Source: Centers for Disease Control; BFSSC.]

The 19% in the second column is the percentage of obese adults among college or technical school graduates. If a person is chosen at random from the adult population

$$P(obese \mid college \ or \ technical \ school \ graduate) = 0.19$$

Similarly

$$P(obese \mid not \ a \ college \ or \ technical \ school \ graduate) = 0.27$$

Both of these conditional probabilities differ from the unconditional probability of randomly selecting an obese person, which is 0.24. Here, the event *obese* is **dependent** on the education level.

In practical terms, independence is quite important in decision making. If, for example, a government agency is deciding on a campaign to reduce obesity, it need not be concerned about gender differences, as the rate of obesity appears to be independent of gender, but it does need to give more consideration to noncollege graduates than to college graduates.

More formally, events A and B are independent if the probability of event A doesn't depend on whether event B happens. That is, knowing that B happens doesn't change the probability that A happens.

Definition of Independent Events

Suppose $P(A) > 0$ and $P(B) > 0$. Then events A and B are independent events if and only if $P(A|B) = P(A)$ or, equivalently, $P(B|A) = P(B)$.

In E61, you will show that $P(A|B) = P(A)$ if and only if $P(B|A) = P(B)$. In E62, you will show that if events A and B are independent, then so are events A and *not B* (event *not B* is the complement of event B, consisting of all possible outcomes that are not in B).

Example 5.14	**Choosing Gourmet Coffee and Vice-Chair**

Displays 5.3 and 5.4 from page 231 are reprinted here. They show sample spaces of equally likely outcomes, the former for two taste testers attempting to select a gourmet coffee (G) at random and the second for random selections of two candidates from three (A, B, and C) as town council chair and vice-chair.

a. In the taste tester scenario, is *Tester B selects G* independent of *Tester A selects G*?

Taste Tester A	Taste Tester B
G	G
G	O
O	G
O	O

Sample space for two taste testers.

b. In the town council scenario, is *B selected vice-chair* independent of *A selected chair*?

First Name Selected (chair)	Second Name Selected (vice-chair)
A	B
A	C
B	A
B	C
C	A
C	B

Sample space for selecting city council chair and vice-chair.

Solution

a. From the sample space, $P(\textit{Tester B selects G}) = 1/2$. Also,

$P(\textit{Tester B selects G} \mid \textit{Tester A selects G})$

$$= \frac{P(\textit{Tester B selects G and Tester A selects G})}{P(\textit{Tester A selects G})} = \frac{\frac{1}{4}}{\frac{2}{4}} = \frac{1}{2}$$

Because the probabilities are equal, these two events are independent.

b. From the sample space, $P(\textit{B selected vice-chair}) = 2/6 = 1/3$. But

$P(\textit{B selected vice-chair} \mid \textit{A selected chair})$

$$= \frac{P(\textit{B selected vice-chair and A selected chair})}{P(\textit{A selected chair})} = \frac{\frac{1}{6}}{\frac{2}{6}} = \frac{1}{2}$$

Because the probabilities aren't equal, these two events are dependent.

DISCUSSION
Independent Events

D21. Refer to Example 5.14.

a. Choosing different events from those used in the example, show that an event for Tester A is independent of an event for Tester B.

b. Choosing different events from those used in the example, show that an event for choice of chair is not independent of an event for choice of vice-chair.

D22. Suppose you choose a student at random from your college. In each case, does knowing that event A happened increase the probability of event B, decrease the probability of event B, or leave the probability of event B unchanged?

a. A: The student is a football player; B: The student weighs less than 120 lb.

b. A: The student has long fingernails; B: The student is female.

c. A: The student is a freshman; B: The student is male.

d. A: The student is a freshman; B: The student is a senior.

Multiplication Rule for Independent Events

In Section 5.3, you learned how to use the Multiplication Rule, $P(A \text{ and } B) = P(A) \cdot P(B|A)$. If A and B are independent, then $P(B|A) = P(B)$, so the Multiplication Rule reduces to a simple product.

> **Multiplication Rule for Independent Events**
>
> Two events A and B where $P(A) > 0$ and $P(B) > 0$ are independent if and only if
>
> $$P(A \text{ and } B) = P(A) \cdot P(B)$$
>
> More generally, events $A_1, A_2, \ldots, A_n$ are independent if and only if
>
> $$P(A_1 \text{ and } A_2 \text{ and } \ldots \text{ and } A_n) = P(A_1) \cdot P(A_2) \cdot \cdots \cdot P(A_n)$$

You can use this rule to decide whether two events are independent.

If $P(A \text{ and } B) = P(A) \cdot P(B)$, then A and B are independent.

Four Flips   **Example 5.15**

If you flip a fair coin four times, what is the probability that all four flips result in heads?

Solution
The outcomes of the first flips don't change the probability on the remaining flips, so the flips are independent. For a sequence of four flips of a fair coin,

$P(\text{heads on 1st flip} \text{ and } \text{heads on 2nd} \text{ and } \text{heads on 3rd} \text{ and } \text{heads on 4th})$

$= P(\text{heads on 1st}) \cdot P(\text{heads on 2nd}) \cdot P(\text{heads on 3rd}) \cdot P(\text{heads on 4th})$

$= \left(\frac{1}{2}\right)\left(\frac{1}{2}\right)\left(\frac{1}{2}\right)\left(\frac{1}{2}\right) = \frac{1}{16}$

Example 5.16	**Multiplication Rule and Health Insurance**

According to *Health Highlights*, published by the U.S. Department of Health and Human Services, about 30% of young American adults ages 18 to 24 don't have health insurance. [Source: *www.healthfinder.gov*, May 2007.] What is the chance that if you choose two American adults from this age group at random, the first has health insurance and the second doesn't?

Solution

Because the sample is selected from the large number of young adults in the United States, you can model the outcomes of the two trials as independent events. Then

$$P(\text{1st has insurance and 2nd doesn't have insurance}) = (0.7)(0.3) = 0.21$$

You can conveniently list the probabilities of all four possible outcomes for two young adults in a "multiplication" table, as in Display 5.34.

		Second Young Adult		
		No Insurance	Insurance	Total
First Young Adult	No Insurance	$(0.3)(0.3) = 0.09$	$(0.3)(0.7) = 0.21$	0.30
	Insurance	$(0.7)(0.3) = 0.21$	$(0.7)(0.7) = 0.49$	0.70
	Total	0.30	0.70	1.00

Display 5.34   Two-way table for the health insurance example.

Example 5.17	**Computing the Probability of "At Least One"**

About 30% of young American adults ages 19 to 29 don't have health insurance. Suppose you take a random sample of ten American adults in this age group. What is the probability that at least one of them doesn't have health insurance?

Solution

This question is asking for the probability that the first young adult doesn't have health insurance or the second young adult doesn't have health insurance or ... or the tenth young adult doesn't have health insurance. You cannot use the Addition Rule for Mutually Exclusive Events and compute $0.3 + 0.3 + \cdots + 0.3$, because the events aren't mutually exclusive (and note that you'll get a probability larger than 1). However, you can easily compute the complement, the probability that all the young adults have health insurance:

$P(\text{at least one doesn't have insurance})$

$\quad = 1 - P(\text{all have insurance})$

$\quad = 1 - P(\text{1st has insurance and 2nd has insurance} \ldots \text{and 10th has insurance})$

$\quad = 1 - P(\text{1st has insurance}) \cdot P(\text{2nd has insurance}) \cdot \cdots \cdot P(\text{10th has insurance})$

$\quad = 1 - (0.7)(0.7) \cdot \cdots \cdot (0.7)$

$\quad = 1 - (0.7)^{10}$

$\quad \approx 1 - 0.028$

$\quad = 0.972$

You can use the Multiplication Rule because the very large number of young adults in the United States means that the probability of getting someone with insurance does not change (by any noticeable amount) depending on who has already been chosen.

In practice, computing $P(A \text{ and } B)$ for independent events is straightforward: Use the definition, which tells you to multiply. What's not always straightforward in practice is recognizing whether independence is a good model for a real situation. The basic idea is to ask whether one event has a bearing on the likelihood of another. This is something you try to decide by thinking about the events in question. As always, models are simplifications of the real thing, and you'll often use independence as a simplifying assumption to enable you to calculate probabilities, at least approximately.

A Matter of Death and Life

If some of the examples of probability in this chapter seem remote to your own life, consider the story of Sally Clark of England. Mrs. Clark's 11-month-old son died unexpectedly in his sleep. The death was attributed to sudden infant death syndrome (SIDS), and doctors, friends, and family consoled the Clark family for their terrible loss. About 1 year later the horrible nightmare happened again—a second son died in his crib at age 8 months. This time, no sympathy came from outside the family. Rather, Sally Clark was accused of murdering both children.

In the trial that followed, Mrs. Clark was convicted and sentenced to life in prison. After serving 3 years, her conviction was overturned. Why was she imprisoned and why was she released? Perhaps the most serious part of the evidence against her was based on a probability argument given by a pediatrician. Records of crib deaths in England indicated that the chance of a randomly selected baby dying from this cause would be about $\frac{1}{1303}$. For an affluent, nonsmoking family like the Clarks with a mother over the age of 26, this chance decreases to about $\frac{1}{8500}$. The pediatrician simply squared the latter number and testified that the chances of two crib deaths in the same family were about 1 in 73 million—a very small chance.

Many people with knowledge of crib deaths and of mathematics, including the Royal Statistical Society, came to the defense of Mrs. Clark. It was pointed out that crib deaths in the same family cannot be regarded as independent (so the squaring of the probability was not justified). Further data analysis suggested that the chances of a second crib death in the same family was closer to $\frac{1}{100}$, giving an estimated probability of two such deaths in the same family of $\left(\frac{1}{1303} \cdot \frac{1}{100}\right)$, or about 1 in 130,300. Of the approximately 650,000 babies born each year in England, about 220,000 are second births. So the expected number of two-death families is between one and two a year, which is in line with what death records show. Once these arguments became clear to the courts, along with further medical evidence, Mrs. Clark was released from prison.

Sally Clark was a young mother whose life was destroyed in large measure by faulty use of probability, both in its calculation and its interpretation. The story has an even more tragic ending; Sally Clark died 4 years after being released from prison. Her family says that she never recovered from the "appalling miscarriage of justice."

[Sources: *http://plus.maths.org/issue21/features/clark/, www.guardian.co.uk/, http://news.bbc.co.uk.*]

DISCUSSION
Multiplication Rule

D23. Are these situations possible? Explain your answers.

    a. Events A and B are disjoint and independent.

    b. Events A and B are not disjoint and are independent.

    c. Events A and B are disjoint and dependent.

    d. Events A and B are not disjoint and are dependent.

D24. Suppose that you randomly and independently select three houses from those sold recently in the Chicago area. Find $P(all\ three\ are\ below\ the\ median\ price)$. Find $P(at\ least\ one\ is\ below\ the\ median\ price)$.

(Continued)

D25. Refer to the Sally Clark story in *A Matter of Death and Life*.

a. Show how the figure 1 in 73 million was calculated. Was the assumption behind this calculation met? Explain why or why not.

b. Show how the expected number of two-death families a year was calculated.

c. Explain why even if the probability calculation of 1 in 73 million were correct, this does not establish the probable guilt of Mrs. Clark.

Independence with Real Data

You have seen that data often serve as basis for establishing a probability model or for checking whether an assumed model is reasonable. In using data to check for independence, however, you have to be careful. In the obesity by gender data in Display 5.31 on page 258 the percentage of obese males and the percentage of obese females are both about 12%, but not exactly if you include more decimal places. The independence can, however, be used as a convenient and nearly correct model, whereas independence would be far from a correct model for obesity as related to education. It is almost impossible to find perfectly independent events in practice, but using independence as a model is often quite satisfactory.

Example 5.18	**Independence and Baseball**

By July 1, the Los Angeles Dodgers had won 41 games and lost 37 games. The breakdown by whether the game was played during the day or at night is shown in Display 5.35. If one of these games is chosen at random, are the events *win* and *day game* independent?

		Won the Game?		
		Yes	No	Total
Time of Game	Day	11	10	21
	Night	30	27	57
	Total	41	37	78

Display 5.35 Dodger record of wins and losses by time of game.
[Source: *Los Angeles Times*, July 1, 2000, p. D8.]

Solution
First, check the probabilities:

$$P(win) = \frac{41}{78} = 0.526$$

$$P(win \,|\, day\ game) = \frac{11}{21} = 0.524$$

Because these two probabilities aren't exactly equal, you must conclude that the events *win* and *day game* are not independent. Yet, given that the Dodgers played 21 day games, the percentage of day games won couldn't be any closer to 0.526.

So, while the strict *mathematical* definition of independence isn't met, the data are consistent with how the numbers of wins and losses might turn out if, in fact, whether the game is a day game or a night game doesn't affect the probability that the Dodgers will win. In Chapter 12, you will learn a *statistical* test of independence that takes the variability in random situations into account.

STATISTICS IN ACTION 5.2 ▶ **Independence with Real Data**

This Statistics in Action further illustrates the difficulty of establishing independence with real data.

What You'll Need: one penny per student

1. Collect and study data on eyedness and handedness.

   a. Determine whether you are right-eyed or left-eyed: Hold your hands together in front of you at arm's length. Make a space between your hands that you can see through. Through the space, look at an object at least 15 ft away. Now close your right eye. Can you still see the object? If so, you are left-eyed. Now close your left eye. Can you still see the object? If so, you are right-eyed.

   b. Would you expect being right-handed and right-eyed to be independent?

   c. Complete a two-way table for members of your class, showing the frequencies of eyedness and handedness.

   d. For a randomly selected student in your class, are being right-handed and right-eyed independent?

2. Collect and study data on the results of two coin flips.

   a. With other members of your class, flip a coin twice until you have 100 pairs of flips. Place your results in a two-way table with one way showing the result of the first flip and the second way showing the result of the second flip.

   b. Would you expect the results of the first flip and the second flip to be independent?

   c. From your data for two coin flips, do the results of the first flip and the second flip appear to be independent?

3. Did you get the results you expected in steps 1 and 2? Explain.

Summary 5.4: Independent Events

The word *independent* is used in several related ways in statistics. First, two events are independent if knowledge that one event occurs does not affect the probability that the other event occurs. Second, in sampling units from a population with replacement, the outcome of the second selection is independent of what happened on the first selection (and the same is true for any other pair of selections). Sampling without replacement leads to dependent outcomes on successive selections. (However, if the population is large compared to the sample, you can compute probabilities as if the selections were independent without much error.) In both situations, the same definition of independence applies: Two events A and B are independent if and only if $P(A) = P(A|B)$—the probability of A does not change with knowledge that B has happened. If $P(A) = P(A|B)$, then it is also true that $P(B) = P(B|A)$.

You learned in this section that the Multiplication Rule simplifies to

$$P(A \text{ and } B) = P(A) \cdot P(B)$$

if and only if A and B are independent events.

Practice

Independent Events

P33. Suppose you select one person at random from the *Titanic* passengers and crew in Display 5.22 on page 247. Use the definition of independent events to determine whether the events *didn't survive* and *male* are independent. Are any two events in this table independent?

P34. Suppose you draw a card at random from a standard deck. Use the definition of independent events to determine which pairs of events are independent.

   a. getting a heart; getting a jack

   b. getting a heart; getting a red card

   c. getting a 7; getting a heart

Multiplication Rule for Independent Events

P35. About 42% of people have type O blood. Suppose you select two people at random and check whether they have type O blood.

   a. Make a table like Display 5.34 on page 262 to show all possible results.

   b. What is the probability that exactly one of the two people has type O blood?

   c. Make a tree diagram that illustrates the situation.

P36. Suppose you select ten people at random. Using the information in P35, find the probability that

   a. at least one of them has type O blood

   b. at least one of them doesn't have type O blood

P37. After taking college placement tests, freshmen sometimes are required to repeat high school work. Such work is called "remediation" and does not count toward a college degree. About 11% of college freshmen have to take a remedial course in reading. Suppose you select two freshmen at random and check to see if they have to take remedial reading. [Source: C. Adelman, *Principal Indicators of Student Academic Histories in Postsecondary Education*, 1972–2000 (2004), Tables 7.1 and 7.2, preview.ed.gov. Data from U.S. Department of Education, NCES, National Education Longitudinal Study of 1988 (NELS:88/2000), "Fourth Follow-up, 2000."]

   a. Find the probability that both freshmen have to take remedial reading.

   b. Show how to use a table like Display 5.34 on page 262 to find the probability that exactly one freshman needs to take remedial reading.

   c. Show how to use a tree diagram to answer the question in part b.

Exercises

E45. Use the definition of independent events to determine which of these pairs of events are independent when you roll two dice.

   a. rolling doubles; rolling a sum of 8

   b. rolling a sum of 8; getting a 2 on the first die rolled

   c. rolling a sum of 7; getting a 1 on the first die rolled

   d. rolling doubles; rolling a sum of 7

   e. rolling a 1 on the first die; rolling a 1 on the second die

E46. Which of these pairs of events, *A* and *B*, do you expect to be independent? Give a reason for your answer.

   a. For a test for tuberculosis antibodies:

     A: The test is positive.

     B: The person has a relative with tuberculosis.

   b. For a test for tuberculosis antibodies:

     A: The test is positive.

     B: The last digit of the person's Social Security number is 3.

   c. For a randomly chosen state in the United States:

     A: The state lies east of the Mississippi River.

     B: The state's highest elevation is more than 8000 ft.

E47. Display 5.36 gives the handedness and eyedness of a group of 100 people. Suppose you select a person from this group at random.

   a. Find each probability.

     i. *P(left-handed)*

     ii. *P(left-eyed)*

	Right-Eyed	Left-Eyed	Total
Right-Handed	57	31	88
Left-Handed	6	6	12
Total	63	37	100

Display 5.36 Eyedness and handedness of 100 people.

     iii. *P(left-eyed | left-handed)*

     iv. *P(left-handed | left-eyed)*

   b. Are being left-handed and being left-eyed independent events? Explain.

   c. Are being left-handed and being left-eyed mutually exclusive events? Explain.

E48. Display 5.37 gives the decisions on all applications to two of the largest graduate programs at the University of California, Berkeley, by gender of the applicant. Suppose you pick an applicant at random.

	Admit	Deny	Total
Man	650	592	1242
Woman	220	263	483
Total	870	855	1725

Display 5.37 Application decisions for the two largest graduate programs at the University of California, Berkeley, by gender. [Source: David Freedman, Robert Pisani, and Roger Purves, *Statistics*, 3rd ed. (Norton, 1997); data from Graduate Division, University of California, Berkeley.]

a. Find each probability.

   i. $P(admitted)$

   ii. $P(admitted \mid woman)$

   iii. $P(admitted \mid man)$

b. Are being admitted and being a woman independent events?

c. Are being admitted and being a woman mutually exclusive events?

d. Is there evidence of possible discrimination against women applicants? Explain.

Display 5.38 gives the admissions data categorized by which program the applicant applied to.

Program A		Admit	Deny	Total
	Man	512	313	825
	Woman	89	19	108
	Total	601	332	933

Program B		Admit	Deny	Total
	Man	138	279	417
	Woman	131	244	375
	Total	269	523	792

Display 5.38 Application decisions for two of the largest graduate programs at the University of California, Berkeley, by program and gender.

e. Does Program A show evidence of possible discrimination against women applicants? Explain.

f. Does Program B show evidence of possible discrimination against women applicants? Explain.

g. These data illustrate *Simpson's paradox*. What is the paradox?

E49. When a baby is expected, there are two possible outcomes: a boy and a girl. However, they aren't equally likely. About 51% of all babies born are boys.

a. List the four possible outcomes for a family that has two babies (no twins).

b. Which of the outcomes is most likely? Which is least likely?

c. What is the probability that both children will be boys in a two-child family? What assumption are you making?

d. Sometimes people say that "girls run in our family" or "boys run in our family." What do they mean? If they are right, how would this affect your answers in parts a–c?

E50. A Harris poll estimated that 25% of U.S. residents believe in astrology. [Source: *www.sciencedaily.com*.] Your newspaper, which publishes a daily horoscope, will be conducting interviews with readers to see if it should continue this practice. To pilot-test the questionnaire for the interviews, you begin by getting a sample of four readers and ask each whether he or she believes in astrology or not.

a. List the possible outcomes.

b. Assuming that 25% of your readers believe in astrology, which of the outcomes is most likely? Which is least likely?

c. Compute the probability that none of the four believes in astrology. How must you select the people in order to make the computation legitimate?

E51. Forty-two percent of people in a town have type O blood, and 5% are Rh-positive. Assume blood type and Rh type are independent.

a. Find the probability that a randomly selected person has type O blood and is Rh-positive.

b. Find the probability that a randomly selected person has type O blood or is Rh-positive.

c. Make a table that summarizes this situation.

E52. Schizophrenia affects 1% of the U.S. population and tends to first appear between the ages of 18 and 25. Today, schizophrenia accounts for the fifth highest number of years of work lost due to disability among Americans ages 15–44. Roughly 30% of people with this illness are currently employed. Suppose you select a person from the U.S. population at random. State whether each question can be answered using the information given. If the question can be answered, answer it. [Source: National Institutes of Health, *www.nih.gov*.]

a. What is the probability that the person is schizophrenic?

b. What is the probability that the person is unemployed?

c. What is the probability that the person is unemployed and schizophrenic?

d. What is the probability that the person is unemployed or schizophrenic?

e. What is the probability that the person is unemployed given that he or she is schizophrenic?

f. What is the probability that the person is schizophrenic given that he or she is unemployed?

E53. A construction company wants to test two types of cement, H and J, for use on highways. It has two different highways available for the test, each with two identical sections. The company makes up two batches of each type of cement, with one batch being enough for one section of highway. It then randomly selects two of the resulting four batches to go to the first highway.

a. What is the probability that the first highway gets two batches of type J?

b. How would you suggest that the company decide which batches go to which highway?

E54. A committee of three students is to be formed from six juniors and five seniors. If three names are drawn at random, what is the probability that the committee will consist of all juniors or all seniors?

E55. As stated in P37, about 11% of college freshmen have to take a remedial course in reading. Suppose you take a random sample of 12 college freshmen.

a. What is the probability that all 12 will have to take a remedial course in reading?

b. What is the probability that at least one will have to take a remedial course in reading?

E56. As stated in E52, about 1% of the U.S. population is schizophrenic. Suppose you take a random sample of 50 U.S. residents.

a. What is the probability that none will be schizophrenic?

b. What is the probability that at least one will be schizophrenic?

E57. In the "Ask Marilyn" column of *Parade* magazine, this question appeared:

Suppose a person was having two surgeries performed at the same time. If the chances of success for surgery A are 85% and the chances of success for surgery B are 90%, what are the chances that both would fail?

Write an answer to this question for Marilyn. [Source: November 27, 1994. © 1994 Marilyn vos Savant. Initially published in *Parade* magazine. All rights reserved.]

E58. Joaquin computes the probability that if two dice are rolled, then exactly one die will show a 4 as

$$P(4 \text{ on one die and } not \text{ 4 on the other}) = \left(\frac{1}{6}\right)\left(\frac{5}{6}\right)$$
$$= \frac{5}{36}$$

Joaquin is wrong. What did he forget?

E59. A batter hits successfully 23% of the time against right-handed pitchers but only 15% of the time against left-handed pitchers. Suppose the batter faces right-handed pitchers in 80% of his at bats and left-handed pitchers in 20% of his at bats. What percentage of the time does he hit successfully?

E60. In *Dave Barry Talks Back*, Dave relates that he and his wife bought an Oriental rug "in a failed attempt to become tasteful." Before going out to dinner, they admonished their dogs Earnest and Zippy to stay away from the rug.

"NO!!" we told them approximately 75 times while looking very stern and pointing at the rug. This proven training technique caused them to slink around the way dogs do when they feel tremendously guilty but have no idea why. Satisfied, we went out to dinner. I later figured out, using an electronic calculator, that this rug covers approximately 2 percent of the total square footage of our house, which means that if you (not you personally) were to have a random diarrhea attack in our home, the odds are approximately 49 to 1 against your having it on our Oriental rug. The odds against your having four random attacks on this rug are more than five million to one. So we had to conclude that it was done on purpose.

Do you agree with Dave's arithmetic? Do you agree with his logic? [Source: Excerpt from "Just Say No to Rugs," from DAVE BARRY TALKS BACK by Dave Barry, copyright © 1991 by Dave Barry. Used by permission of Crown Publishers, a division of Random House, Inc.]

E61. Two events *A* and *B* are independent if any of these statements holds true:

$$P(A) = P(A|B)$$
$$P(B) = P(B|A)$$
$$P(A \text{ and } B) = P(A) \cdot P(B)$$

Use the definition of conditional probability to show that these three statements are equivalent as long as $P(A)$ and $P(B)$ don't equal 0. That is, show that if any one is true, so are the other two.

E62. Show that if events *A* and *B* are independent, then events *A* and *not B* are also independent.

E63. $P(A) = P(A|B) \cdot P(B) + P(A|\overline{B}) \cdot P(\overline{B})$

sometimes is called the law of total probability.

a. Prove the law of total probability.

b. Use this law to find *P*(*Dodgers win*) from the information in Display 5.35 on page 264.

c. Prove that

$$P(B|A) = \frac{P(A|B) \cdot P(B)}{P(A|B) \cdot P(B) + P(A|\overline{B}) \cdot P(\overline{B})}$$

This result is called Bayes' Theorem.

d. Use Bayes' Theorem to find *P*(*Dodgers win*|*day game*).

E64. Two screening tests are used on patients chosen from a certain population. Test I, the less expensive of the two, is used about 60% of the time and produces false positives in about 10% of cases. Test II produces false positives in only about 5% of cases.

a. Fill in this table of percentages.

		Test I	Test II	Total %
Results	False Positives	—?—	—?—	—?—
	Other Results	—?—	—?—	—?—
	Total %	60	40	100

b. A false positive is known to have occurred in a patient tested by one of these two tests. Find the approximate probability that Test I was used, first by direct observation from the table and then by use of Bayes' Theorem in E63, part c.

Chapter Summary

Probability is the study of random behavior. The probabilities used in statistical investigations often come from a model based on equally likely outcomes. In other cases, they come from a model based on observed data. While you do not know for sure what the next outcome will be, a model based on many observations gives you a pretty good idea of the possible outcomes and their probabilities.

Two important concepts were introduced in this chapter: disjoint (mutually exclusive) events and independent events.

- Two events are disjoint if they can't happen on the same opportunity. If A and B are disjoint events, then $P(A \text{ and } B)$ is 0.

- Two events are independent if the occurrence of one doesn't change the probability that the other will happen. That is, events A and B are independent if and only if $P(A) = P(A|B)$.

You should know how to use each of these rules for computing probabilities:

- The probability that event A doesn't happen is

$$P(\text{not } A) = 1 - P(A)$$

This rule often is used to find the probability of at least one success:

$$P(\text{at least one success}) = 1 - P(\text{no successes})$$

- The Addition Rule gives the probability that event A or event B or both occur:

$$P(A \text{ or } B) = P(A) + P(B) - P(A \text{ and } B)$$

If the events A and B are disjoint, this rule simplifies to

$$P(A \text{ or } B) = P(A) + P(B)$$

- The Multiplication Rule gives the probability that events A and B both occur:

$$P(A \text{ and } B) = P(A) \cdot P(B|A) \text{ or } P(B) \cdot P(A|B)$$

If A and B are independent events, this rule simplifies to

$$P(A \text{ and } B) = P(A) \cdot P(B)$$

Review Exercises

E65. Suppose you roll a fair four-sided (tetrahedral) die and a fair six-sided die.

   a. How many equally likely outcomes are there?

   b. Show all the outcomes in a two-way array or in a tree diagram.

   c. What is the probability of getting doubles?

   d. What is the probability of getting a sum of 3?

   e. Are the events *getting doubles* and *getting a sum of 4* disjoint? Are they independent?

   f. Are the events *getting a 2 on the tetrahedral die* and *getting a 5 on the six-sided die* disjoint? Are they independent?

E66. Backgammon is one of the world's oldest games. Players move counters around the board in a race to get "home" first. The number of spaces moved is determined by a roll of two dice. If your counter is able to land on your opponent's counter, your opponent has to move his or her counter to the "bar" where it is trapped until your opponent can free it. For example, suppose your opponent's counter is five spaces ahead of yours. You can "hit" that counter by rolling a sum of 5 with both dice or by getting a 5 on either die.

   a. Use the sample space for rolling a pair of dice to find the probability of being able to hit your opponent's counter on your next roll if his or her counter is five spaces ahead of yours.

   b. Can you use the Addition Rule for Disjoint Events to compute the probability that you roll a sum of 5 with both dice or get a 5 on either die? Either do the computation or explain why you can't.

E67. Jorge has an alarm clock on his MP3 player. When it's time for him to wake up, the MP3 player randomly selects one song to play from a particular album. There are nine songs on the album and Jorge's favorite song is the third one. What is the probability that Jorge will hear his favorite song at least once in the next week?

E68. This problem appeared in the "Ask Marilyn" column of *Parade* magazine and stirred up a lot of controversy. Marilyn devoted at least four columns to it.

A woman and a man (unrelated) each have two children. At least one of the woman's children is a boy, and the man's older child is a boy. Do the chances that the woman has two boys equal the chances that the man has two boys? [Source: July 27, 1997. © 1997 Marilyn vos Savant. Initially published in *Parade* magazine. All rights reserved.]

E69. The American Society of Microbiology periodically estimates the percentage of people who wash their hands after using a public restroom. (Yes, these scientists spy on people in public restrooms, but it's for a good cause.) [Source: *www.harrisinteractive.com.*] The Society reports that about 75% of men and 90% of women wash their hands after using a public restroom. Suppose you observe one man and one woman at random. Compute each of these probabilities.

   a. P(*both the man and the woman wash their hands*)
   b. P(*neither washes his or her hands*)
   c. P(*at least one washes his or her hands*)
   d. P(*man washes his hands* | *woman washes her hands*)

E70. As you saw in E50 on page 267, about 25% of U.S. residents claim to believe in astrology. Suppose you select two U.S. residents at random. Compute each of these probabilities.

   a. P(*both believe in astrology*)
   b. P(*neither believes in astrology*)
   c. P(*at least one believes in astrology*)
   d. P(*the second person believes in astrology* | *the first person believes in astrology*)

E71. Among recent graduates of mathematics departments, half intend to teach high school. A random sample of size 2 is to be selected from the population of recent graduates.

   a. If mathematics departments had only four recent graduates total, what is the chance that the sample will consist of two graduates who intend to teach high school?
   b. If mathematics departments had 10 million recent graduates, what is the chance that the sample will consist of two graduates who intend to teach high school?
   c. Are the selections technically independent in part a? Are they technically independent in part b? In which part can you assume independence anyway? Why?

E72. You work for a home mortgage company that offers jumbo (large) mortgages and adjustable-rate mortgages (ARMs).

   a. Among jumbo mortgages, 80% have never had a late payment and 45% have borrowers with excellent credit ratings. But 15% of jumbo mortgages have had a late payment and don't have borrowers with excellent credit. What percentage of jumbo mortgages fall into the category of never having had a late payment or of having a borrower with an excellent credit rating?
   b. Thirty percent of ARMs have never had a late payment even though they don't have a borrower with an excellent credit rating. Forty-five percent have a borrower with an excellent credit rating but have had a late payment. Only 20% of ARMs have had a late pay-

ment and don't have a borrower with excellent credit. What percentage of ARMs have never had a late payment or the borrower has an excellent credit rating?

E73. A web site at Central Michigan University collects data about college students. Students were asked their political party preference and which candidate they voted for in the last presidential election. Display 5.39 shows the results for the first 42 students who were 22 years old or older and had either Democratic or Republican Party preference. Suppose you pick one of these 42 students at random.

Party Preference	Voted for Democrat	Voted for Republican	Did Not Vote	Total
Democratic	19	1	6	26
Republican	1	12	3	16
Total	20	13	9	42

Display 5.39 Results of voting and political party preference survey. [Source: Central Michigan University, *stat.cst.cmich.edu.*]

   a. Find each of these probabilities.
      i. P(*Republican*)
      ii. P(*voted Republican*)
      iii. P(*voted Republican and Republican*)
      iv. P(*voted Republican or Republican*)
      v. P(*Republican* | *voted Republican*)
      vi. P(*voted Republican* | *Republican*)
   b. Are *being Republican* and *voting Republican* independent events?
   c. Are *being Republican* and *voting Democratic* mutually exclusive events?

E74. Display 5.40 gives probabilities for a randomly selected adult in the United States.

		Age		
		18 to 25	25 or Older	Total
Has a Cell Phone?	Yes	p	q	$p + q$
	No	r	s	$r + s$
	Total	$p + r$	$q + s$	1

Display 5.40 Cell phone ownership, by age.

Write an expression for the probability that a randomly selected adult

   a. has a cell phone
   b. is 25 or older, given that the person has a cell phone
   c. has a cell phone, given that the person is age 18 to 25
   d. is 25 or older and has a cell phone
   e. is 25 or older or has a cell phone

E75. Consider a situation with these three probabilities:

$$P(A) = \frac{1}{4} \quad P(B|A) = \frac{1}{2} \quad P(B|not\ A) = \frac{1}{4}$$

   a. Fill in a two-way table that has A and *not A* for one way and B and *not B* for the other way.
   b. Are events A and B independent?

E76. Suppose that you conduct a taste test and find that 60 out of 100 randomly selected coffee drinkers correctly choose the gourmet coffee rather than ordinary coffee. You realize that you should be concerned that perhaps everyone was simply guessing and that 60 of them made the correct selection simply by chance. So you flip a fair coin 100 times and count the number of heads. You do this over and over again. Out of thousands of repetitions, you find that you get 60 or more heads about 2.5% of the time. Write a conclusion for your taste test.

E77. Many people who have had a stroke get depressed. These patients tend to have poorer outcomes than stroke patients who aren't depressed. A standard, but lengthy, test for depression was given to 79 stroke patients and 43 were found to be clinically depressed. Researchers wondered if a single question, "Do you often feel sad or depressed?" would be just as good as the longer test. Of the 43 patients found to be clinically depressed, 37 answered "yes," and 6 answered "no." Of the 36 patients classified as not clinically depressed, 8 answered "yes" and 28 answered "no." [Source: C. Watkins et al., "Accuracy of a Single Question in Screening for Depression in a Cohort of Patients After Stroke: Comparative Study," *BMJ*, Vol. 323 (November 17, 2001), p. 1159.] Does the single question appear to be a good diagnostic test for clinical depression in stroke patients? First make a table that summarizes this situation. Then use the ideas of sensitivity, specificity, positive predictive value, and negative predictive value in your answer.

E78. A telephone area code is a three-digit number that begins with a digit from 2 through 9. The second digit can be any number from 1 through 8 or 0. The third number must be different from the second. How many area codes are possible, according to these rules?

E79. A few years ago, the U.S. Post Office added the "zip plus 4" to the original five-digit zip codes. For example, the zip code for the national headquarters of the American Red Cross is 20006-5310.

For this problem, assume that a zip code can be any five-digit number except 00000 and that the four-digit "plus 4" can be any number except 0000.

   a. Assume that the U.S. population is roughly 300 million. If all possible five-digit zip codes are used, what is the average number of people per zip code?

   b. If all possible "zip plus 4" codes are used, what is the average number of people per "zip plus 4"?

E80. RNA is a single strand of genetic code that works with DNA to tell living organisms how to build amino acids. RNA is made up of codons. Codons in turn are made up of the four nucleotides: adenine (A), guanine (G), cytosine (C), and uracil (U). Each codon in RNA consists of a triplet of three nucleotides, such as AAG or UCA, where order matters and the same nucleotide can be repeated two or even three times. How many different codons are there? (There are only 20 different amino acids, so there is quite a bit of redundancy.)

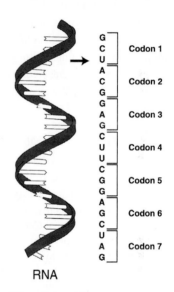

RNA

Ribonucleic acid

Concept Test

C1. A student argues that extraterrestrials will either abduct her statistics instructor before tomorrow's test or they will not, and so there's a 1 out of 2 chance for each of these two events. Which of the following best explains why this reasoning is incorrect?

Ⓐ The two events are not independent.
Ⓑ The two events are mutually exclusive.
Ⓒ The two events are not equally probable.
Ⓓ The two events are complements.
Ⓔ There are more than two events that need to be considered.

C2. Suppose you roll two dice. Which of the following are independent events?

Ⓐ getting a sum of 8; getting doubles
Ⓑ getting a sum of 3; getting doubles
Ⓒ getting a sum of 2; getting doubles
Ⓓ getting a 1 on the first die; getting a sum of 5
Ⓔ getting a 1 on the first die; getting doubles

C3. Suppose you roll two dice. Which of the following are mutually exclusive (disjoint) events?

Ⓐ getting a sum of 8; getting doubles
Ⓑ getting a sum of 3; getting doubles
Ⓒ getting a sum of 2; getting doubles
Ⓓ getting a 1 on the first die; getting a sum of 5
Ⓔ getting a 1 on the first die; getting doubles

C4. A catastrophic accident is one that involves severe skull or spinal damage. The National Center for Catastrophic Sports Injury Research reports that over the last 21 years, there have been 101 catastrophic accidents among female high school and college athletes. Fifty-five of these resulted from cheerleading. [Source: www.unc.edu.] Suppose you want to study catastrophic accidents in more detail, and you take a random sample, without replacement, of 5 of these 101 accidents. What is the probability that none of the 5 accidents resulted from cheerleading?

Ⓐ $\left(\frac{46}{101}\right)^5$

Ⓑ $1 - \left(\frac{55}{101}\right)^5$

Ⓒ $\left(\frac{46}{101}\right)\left(\frac{45}{100}\right)\left(\frac{44}{99}\right)\left(\frac{43}{98}\right)\left(\frac{42}{97}\right)$

Ⓓ $\left(\frac{46}{101}\right)\left(\frac{45}{101}\right)\left(\frac{44}{101}\right)\left(\frac{43}{101}\right)\left(\frac{42}{101}\right)$

Ⓔ None of these is correct.

C5. In a statistics classroom, 50% of the students are female and 30% of the students got an A on the most recent test. What is the probability that a student picked at random from this classroom is a female who got an A on the most recent test?

Ⓐ 0.15　Ⓑ 0.20　Ⓒ 0.40　Ⓓ 0.65
Ⓔ cannot be determined from the information given

C6. In the fictional country of Valorim, 4% of adults are smokers who get lung cancer, 8% are nonsmokers who get lung cancer, 22% are smokers who do not get lung cancer, and the remaining 66% are nonsmokers who do not get lung cancer. What is the probability that a randomly selected adult smoker gets lung cancer?

Ⓐ $\frac{4}{100}$　Ⓑ $\frac{4}{30}$　Ⓒ $\frac{4}{26}$　Ⓓ $\frac{3}{25}$　Ⓔ $\frac{4}{12}$

C7. Seventy percent of Valorites can answer the previous question correctly. Suppose you select three Valorites at random. What is the probability that at least one of the three can answer correctly?

Ⓐ 0.100　Ⓑ 0.343　Ⓒ 0.900　Ⓓ 0.973　Ⓔ 2.100

C8. A new space probe, the Mars Crashlander, has a main antenna and a backup antenna. The probe will arrive at Mars and crash into the surface. The main antenna has a 50% chance of surviving the crash. The backup has a 90% chance of surviving if the main antenna survives, and a 20% chance of surviving if the main antenna fails. What is the probability that the Crashlander will have at least one working antenna after the crash?

Ⓐ 0.10　Ⓑ 0.275　Ⓒ 0.55　Ⓓ 0.6　Ⓔ 0.7

Investigative Tasks

C9. "So they stuck them down below, where they'd be the first to go." This line from the song about the *Titanic* refers to the third-class passengers. The first two-way table in Display 5.41 gives the number of surviving *Titanic* passengers in various classes of travel. The second and third tables break down survival by class and by gender. Use the ideas about conditional probability and independence that you have learned to analyze these tables.

		First	Second	Third	Total
		Class			
Survived?	Yes	203	118	178	499
	No	122	167	528	817
	Total	325	285	706	1316

Females:

		First	Second	Third	Total
		Class			
Survived?	Yes	141	93	90	324
	No	4	13	106	123
	Total	145	106	196	447

Males:

		First	Second	Third	Total
		Class			
Survived?	Yes	62	25	88	175
	No	118	154	422	694
	Total	180	179	510	869

Display 5.41　*Titanic* survival data, by class. [Source: *Journal of Statistics Education*, Vol. 3, 3 (1995).]

272

Chapter 6
Probability Distributions

Forty percent of people have type A blood. A blood bank is in dire need of two donors with type A blood. If it tests ten donors, will it be almost certain of finding at least two donors with type A blood? A probability distribution describes the chances of the possible outcomes.

In Chapter 4 you learned about sampling, and in Chapter 5 you learned about probability. What is the connection between these two topics? Sampling reasons from a sample to a population, and probability reasons from a population to a sample:

- Sampling enables you to estimate an unknown characteristic of your population by taking a sample from that population. A random sample of households from your community helps you estimate the percentage of all households in your community that own a motor vehicle.

- Probability enables you to calculate the chance of getting a specified outcome in a sample taken from a known population. If you know that 91% of households in your community own a motor vehicle, then you can calculate the probability that the three households in a random sample all own a motor vehicle.

One of the most important questions statistics tries to answer is "What are the characteristics of this population?" A census, which would answer this question, is usually impossible or impractical. Fortunately, early-20th-century statisticians had this extraordinary insight: Use probability to learn what samples from populations with known distributions tend to look like. Then take a sample from the unknown population and compare the results to those from known populations. The populations for which your sample appears to fit right in are the plausible homes for your sample. This chapter begins to formalize this profound idea, which will be developed more fully in the remaining chapters of the book.

In this chapter, you will learn

▷ to construct probability distributions

▷ the concept and uses of expected value

▷ how to compute and interpret the standard deviation of a probability distribution

▷ the rules for combining means and variances

▷ to recognize and apply the binomial distribution

6.1 ▶ Probability Distributions and Expected Value

A **probability distribution** describes the possible numerical outcomes of a chance process and allows you to find the probability of any set of possible outcomes. Sometimes a probability distribution is defined by a table, like the one in Display 5.1 on page 229. Sometimes a probability distribution is defined by a graph, like the distribution of male heights in the relative frequency histogram distribution in Display 2.38 on page 43. As you will learn in Section 6.3, sometimes a probability distribution can be defined by a formula. This chapter is about *discrete* probability distributions, those with a finite or a countable number of outcomes. You will return to the story of the *continuous* normal distribution in Chapter 7.

Constructing Probability Distributions

The first probability distributions in this section arise from the familiar situation of rolling two dice.

| **Probability Distributions from Rolling Two Dice** | **Example 6.1** |

In Chapter 5, you constructed an array like Display 6.1, which lists all possible outcomes when two dice are rolled. From this array, you can construct different probability distributions, depending on the numerical summary that interests you.

		Second Die					
		1	**2**	**3**	**4**	**5**	**6**
First Die	**1**	1, 1	1, 2	1, 3	1, 4	1, 5	1, 6
	2	2, 1	2, 2	2, 3	2, 4	2, 5	2, 6
	3	3, 1	3, 2	3, 3	3, 4	3, 5	3, 6
	4	4, 1	4, 2	4, 3	4, 4	4, 5	4, 6
	5	5, 1	5, 2	5, 3	5, 4	5, 5	5, 6
	6	6, 1	6, 2	6, 3	6, 4	6, 5	6, 6

Display 6.1   The 36 outcomes when rolling two dice.

For example, if you are playing Monopoly, the number of spaces you can move depends on the sum of the two dice. If you are playing backgammon, you might be more interested in the larger of the two numbers. Construct each of these probability distributions. Use the appropriate distribution to find the probability that, when you roll two dice, the sum is 3. Then find the probability that the larger number is 3.

Solution

Display 6.2 shows these two probability distributions, one of the sum of two dice and one of the larger number on the two dice. (In the case of doubles, the larger number and the smaller number are the same.)

The sum of the probabilities is 1 in both tables. This should always be the case, subject to round-off error.

Sum of Two Dice, x	Probability, P		Larger Number, x	Probability, p
2	1/36		1	1/36
3	2/36		2	3/36
4	3/36		3	5/36
5	4/36		4	7/36
6	5/36		5	9/36
7	6/36		6	11/36
8	5/36		**Total**	**1**
9	4/36			
10	3/36			
11	2/36			
12	1/36			
Total	**1**			

Display 6.2   Probability distributions for the roll of two dice: the sum of the two numbers and the larger number.

When you roll two dice, the probability that the sum of the numbers is 3 is 2/36. The probability that the larger number is 3 is 5/36.

Capital *X* represents the random variable itself; a lowercase *x* represents a particular value that the random variable takes on.

The language of probability distributions reflects the chance involved in random outcomes. The variable of interest, X, is called a **random variable** because its numerical values vary from trial to trial. For example, if the random variable X represents the sum of the roll of two dice, then $P(X = 7) = 1/6$, because the probability of rolling a sum of 7 is 6/36, or 1/6.

Example 6.2 | Sampling Lung Cancer Patients

Display 6.3 shows data on the proportion of lung cancer cases caused by smoking. Suppose two lung cancer patients are randomly selected from the large population of people with that disease. Construct the probability distribution of X, the number of patients with lung cancer caused by smoking.

Lung Cancer Cases	Proportion
Smoking responsible	0.87
Smoking not responsible	0.13

Display 6.3 Lung cancer cases caused by smoking.
[Source: American Lung Association, 2009.]

Solution

There are four possible outcomes for the two patients. With "yes" representing "caused by smoking" and "no" representing "not caused by smoking," the possibilities are

no for 1st patient and *no for 2nd patient*

no for 1st patient and *yes for 2nd patient*

yes for 1st patient and *no for 2nd patient*

yes for 1st patient and *yes for 2nd patient*

The fact that the patients are selected at random from a large population implies that the outcomes for the two selections may be considered independent. Using the Multiplication Rule for Independent Events of Chapter 5,

$P(\text{no for 1st patient and no for 2nd patient}) = P(\text{no for 1st}) \cdot P(\text{no for 2nd})$
$= (0.13)(0.13) = 0.0169$

$P(\text{no for 1st patient and yes for 2nd patient}) = (0.13)(0.87) = 0.1131$

$P(\text{yes for 1st patient and no for 2nd patient}) = (0.87)(0.13) = 0.1131$

$P(\text{yes for 1st patient and yes for 2nd patient}) = (0.87)(0.87) = 0.7569$

The first of these outcomes results in $X = 0$, the second and third each result in $X = 1$, and the fourth results in $X = 2$. Because the second and third outcomes are disjoint, their probabilities can be added to get $P(X = 1)$. The probability distribution of X is given by the table in Display 6.4.

Number Caused by Smoking, *x*	Probability, *p*
0	0.0169
1	0.1131 + 0.1131 = 0.2262
2	0.7569

Display 6.4 Probability distribution for the number of patients whose lung cancer was caused by smoking in a random sample of two patients.

| Enough Parking Spaces? | Example 6.3 |

You are working for a contractor who is building 500 new single-family houses and wants to know how many parking spaces will be needed per house. Your job is to estimate the number of vehicles per household. Fortunately, before starting the work of taking a survey, you find a distribution of the number of vehicles per household (Display 6.5) from the U.S. Census Bureau. Assuming that these national results are typical of your community, your work is done! You can report to your boss, for example, that you expect that about 0.381 + 0.201, or 0.582 (more than half) of the households will have two vehicles or more.

Vehicles per Household, x	Proportion of Households, p
0	0.087
1	0.331
2	0.381
3	0.201

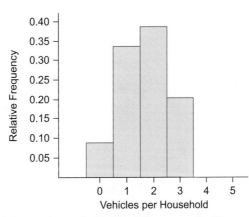

Display 6.5 The number of motor vehicles per household. The "3" represents "3 or more." (The proportion of households with more than three vehicles is very small.) The graph has bars with a width of 1 even though getting a household with, say, 1.54 vehicles is impossible. Because the width is 1, the proportion of households in each category is represented both by the height of the bar and by the area of the bar (width of 1 times the height). Identifying the probability with the area of the bar can be helpful when estimating probabilities. [Source: U.S. Census Bureau, American Community Survey, 2007. Released 2008.]

Your boss is pleased that you now have so much extra time and gives you a new project. Your construction company is also building 500 duplexes (a building with two housing units), and your boss wants a table of probabilities for the total number of vehicles that will be parked by the two households occupying a duplex.

Solution
After some thought, you realize that you can compute the probability that two randomly selected households have no vehicles using the methods of Chapter 5:

P(no vehicles in a duplex) = *P(no vehicles in the 1st household* and *no vehicles in the 2nd household)*

= *P(no vehicles in the 1st household)* · *P(no vehicles in the 2nd household)*

= 0.087 · 0.087

≈ 0.008

This calculation depends on the assumptions that households living in duplexes mirror households in general with respect to the number of vehicles per household and that the number of vehicles in one household is independent of the number of vehicles in the neighboring household. How good the approximation is depends on how well these assumptions hold up. This is typical of all modeling problems—the accuracy of the result depends on the appropriateness of the model.

Computing the probability of exactly one vehicle per duplex is just a little more complicated:

P(one vehicle in a duplex) = P(one vehicle in the 1st household and no vehicles in the 2nd household or no vehicles in the 1st household and one vehicle in the 2nd household)

$\qquad$ = P(one vehicle in the 1st household) · P(no vehicles in the 2nd household) + P(no vehicles in the 1st household) · P(one vehicle in the 2nd household)

$\qquad$ = 0.331 · 0.087 + 0.087 · 0.331

$\qquad$ ≈ 0.058

Continuing in this way, you construct the probability distribution in Display 6.6.

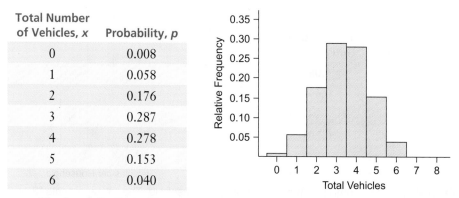

Total Number of Vehicles, x	Probability, p
0	0.008
1	0.058
2	0.176
3	0.287
4	0.278
5	0.153
6	0.040

Display 6.6 Probability distribution of total vehicles in two households.

You can now report to your boss, for example, that there is a 0.087 chance that a single-family household will have no vehicles but only a 0.008 chance that the occupants of a duplex will have no vehicles.

Example 6.4	**Sampling Hockey Teams**

Display 6.7 gives the capacity of the home arenas of the five teams in the Atlantic Division of the National Hockey League. Suppose you pick two teams at random to play a pair of exhibition games, with one game played in each home arena. Construct the probability distribution of X, the total number of people who could attend the two games. Then find the probability that the total possible attendance could be at least 36,000.

Team	Arena Seating (thousands)
New Jersey Devils	19
New York Islanders	16
New York Rangers	18
Philadelphia Flyers	18
Pittsburgh Penguins	17

Display 6.7 Hockey team home arena capacities.

Solution
There are ten ways to pick two teams from the five teams: $_5C_2 = \binom{5}{2} = 10$. These ten pairs of teams, with the total possible attendance, are shown in Display 6.8.

$$\binom{n}{k} = \frac{n!}{k!(n-k)!}$$

Teams	Maximum Attendance (thousands)
Devils/Islanders	$19 + 16 = 35$
Devils/Rangers	$19 + 18 = 37$
Devils/Flyers	$19 + 18 = 37$
Devils/Penguins	$19 + 17 = 36$
Islanders/Rangers	$16 + 18 = 34$
Islanders/Flyers	$16 + 18 = 34$
Islanders/Penguins	$16 + 17 = 33$
Rangers/Flyers	$18 + 18 = 36$
Rangers/Penguins	$18 + 17 = 35$
Flyers/Penguins	$18 + 17 = 35$

Display 6.8   Total maximum attendance for pairs of exhibition games.

Using the fact that the random choice makes the outcomes equally likely, the probability distribution of the total possible attendance for two games is given in Display 6.9.

Total Possible Attendance (thousands), x	Probability, p
33	0.1
34	0.2
35	0.3
36	0.2
37	0.2

Display 6.9   Probability distribution of total possible attendance at two games.

The probability that the possible total attendance is at least 36,000 is $0.2 + 0.2$, or 0.4.

DISCUSSION
Constructing Probability Distributions

D1. Compare the shape, center, and spread of the distribution in Display 6.5 with that in Display 6.6.

D2. Your boss finds it odd that there is a fairly good chance (0.201) that a single household will need to park three (or more) vehicles while there is very little chance (0.040) that the two households in a duplex will need to park six (or more) vehicles. Explain to your boss why this is reasonable.

D3. Verify that the probability given in Display 6.6 is correct for a total of two vehicles in a duplex.

Expected Value

When you are making decisions, it is often helpful to know the mean value that you can expect from a probabilistic situation. For example, in some board games where the number of spaces you move is determined by the roll of two dice, it is helpful to know that the mean of the probability distribution of the sum of the two dice is 7. So the number of spaces you will be able to move on one roll, on average, is 7. Further, to make a complete trip around a board of 36 spaces, you would expect to require 36/7, or 5½, rolls. The mean of a probability distribution has a special name, although continuing to call it the "mean" is not wrong.

> You can say either *mean* or *expected value* or *expected number*.

The mean of a probability distribution for the random variable X is called its **expected value** and is denoted by μ_X or $E(X)$.

Formula for the Mean of a Probability Distribution

The expected value (mean) μ_X of a discrete probability distribution may be found using this formula:

$$E(X) = \mu_X = \sum x \cdot p$$

Here, p is the probability that the random variable X takes on the specific value x.

To find the expected value of a probability distribution given in a table, you can use a method similar to that on page 59 in Chapter 2 for the mean of a distribution given in a frequency table.

Example 6.5 | **Expected Number of Vehicles per Household**

Your boss has looked at the table in Display 6.10 and would like to know the average number of vehicles that would need parking spaces for single-family houses.

Vehicles per Household, x	Proportion of Households, p
0	0.087
1	0.331
2	0.381
3	0.201

Display 6.10 The number of motor vehicles per household. The "3" represents "3 or more." (The proportion of households with more than three vehicles is very small.) [Source: U.S. Census Bureau, American Community Survey, 2007. Released 2008.]

Solution

Assuming that your community is like the entire United States, you can find the expected number of vehicles per household using the formula in the preceding box. You expect 8.7% of households to have no vehicles, 33.1% to have one vehicle, and so on. Thus, the expected number of vehicles per household is

$$\mu_X = 0(0.087) + 1(0.331) + 2(0.381) + 3(0.201) = 1.696$$

You can report to your boss that the average number of vehicles per household is 1.696. Notice that the mean, 1.696, is the balance point of the graph in Display 6.5.

Adding a third column to the table, as in Display 6.11, is an easy way to organize your work when computing an expected value.

Vehicles per Household, x	Proportion of Households, p	Contribution to the Mean, x · p
0	0.087	0
1	0.331	0.331
2	0.381	0.762
3	0.201	0.603
	Sum	1.696

Display 6.11 Computing expected value from a probability distribution table.

In many real-life situations, you can determine the best course of action by considering expected value.

Burglary Insurance

Example 6.6

A woman comes to your insurance agency and asks you to insure her household so that if it is burglarized, you would pay her $5000. How much you will charge her per year depends on how likely it is that her household will be burglarized. Insurance companies and the mathematicians they employ, called actuaries, keep careful records of various crimes and disasters so that they can know the probabilities that these will occur. Your company doesn't have an actuary, and all the information you can find is the nationwide rate of 26.9 burglaries per 1000 households. What should you charge for the insurance in order to expect just to break even?

Solution
Display 6.12 gives the two possible outcomes, their payouts, and their probabilities.

Outcome	Payout, x	Probability, p	x · p
No burglary	0	0.9731	0
Burglary	5000	0.0269	134.50

Display 6.12 Table of payouts and their probabilities. [Source: U.S. Bureau of Justice Statistics, *Criminal Victimization, 2007, www.ojp.usdoj.gov.*]

The expected payout per policy is

$$E(X) = \mu_X = 0(0.9731) + 5000(0.0269) = 134.50$$

You expect to break even if you charge $134.50 for the insurance. But as a good businessperson, you will actually charge more to cover your costs of doing business and to earn a profit.

Example 6.7	**The Wisconsin Lottery**

The Wisconsin lottery had a scratch-off game called "Big Cat Cash." It cost $1 to play. Wisconsin gave the probabilities of winning various amounts that are listed in Display 6.13. Do the probabilities sum to 1? If not, add to the table so that the sum of the probabilities is 1. Find and interpret the expected value when playing one game.

Winnings, x	Probability, p
$1	1/10
$2	1/14
$3	1/24
$18	1/200
$50	1/389
$150	1/20,000
$900	1/120,000

Display 6.13   Probabilities for Wisconsin scratch-off game.
[Source: *www.wilottery.com.*]

Solution

The probabilities don't sum to 1—it's not even close. That's because the most likely outcome is winning nothing. So add another row with $0 for "Winnings" and 0.7793 for "Probability." Using the expected-value formula, you can verify that the expected value for this probability distribution is 0.6014. This means that if you spend $1 to play this game, you "expect" to get 60.14¢ back in winnings. Of course, you can't get this amount on any one play, but over the long run that would be the average return. Another way to understand this is to imagine playing the game 1000 times. You expect to get back $601.40, but you will have spent $1000.

The expected value may not be of much importance to an individual player (unless he or she is going to play many times), but it is of great importance to the State of Wisconsin, which can expect to pay out only $601.40 for every $1000 it takes in.

> Don't routinely round expected values to the nearest whole number. The expected value is an average, so it doesn't have to be one of the possible outcomes.

The solution to the Wisconsin lottery example used the following rule, which seems natural because it is similar to the fact that $\bar{x} = \frac{\Sigma x}{n}$ or $\Sigma x = n \cdot \bar{x}$. For example, if, on average, you spend 37 minutes commuting each day you come to campus, you expect to spend a total of $5 \cdot 37$, or 185, minutes commuting on five randomly selected days you come to campus. Note that this estimate wouldn't be very good if, instead of five randomly selected days, you picked only Mondays when traffic is heaviest.

Mean of a Sum of *n* Values of a Random Variable

Suppose that you have *n* values of a random variable that has mean, μ_x. Then the expected value of their sum is $n\mu_x$.

| **Ten Wisconsin Lottery Tickets** | **Example 6.8** |

A single ticket in the Wisconsin lottery game has an expected value of $0.6014. Suppose a person is given ten tickets. What is the expected value of the person's total winnings?

Solution

The expected winnings are $n\mu_X = 10(\$0.6014) = \$6.014$.

DISCUSSION
Expected Value

D4. In Chapter 2, you used the following formula to compute the mean, $\bar{x}$, of the values in a frequency table, where x is the value, f its frequency, and $n = \Sigma f$:

$$\bar{x} = \frac{\Sigma x \cdot f}{n}$$

a. Use this formula to compute the mean of the values in this frequency table.

Value, x	Frequency, f
5	12
6	23
8	15

b. Suppose the data had been given in a relative frequency table like this one, which shows the proportion of times each value occurs. Fill in the rest of the second column.

Value, x	Proportion, f/n
5	0.24
6	?
8	?

c. Show that you can find the mean, $\bar{x}$, using the formula

$$\Sigma\left(x \cdot \frac{f}{n}\right)$$

d. How does the formula in part c relate to the formula for the expected value, μ_X?

D5. This sentence appeared in the British humor magazine *Punch*: "The figure of 2.2 children per adult female was felt to be in some respects absurd, and a Royal Commission suggested that the middle classes be paid money to increase the average to a rounder and more convenient number." [Source: M. J. Moroney, *Facts from Figures* (Baltimore, MD: Penguin Books, 1951).] What lesson does this quotation teach about expected values?

D6. Describe the shapes of the distributions of the two random variables in Display 6.2 on page 275. Find their expected values, using each in a sentence.

Summary 6.1: Probability Distributions and Expected Value

A probability distribution describes the possible numerical outcomes, x, of a chance process and allows you to find the probability of any set of possible outcomes. A probability distribution table for a discrete random variable lists each of these outcomes in one column and its associated probability in the other column. You can compute the expected value (mean) of a discrete probability distribution using the formula

$$E(X) = \mu_X = \sum x \cdot p$$

Finding the expected value has many real-world applications. For example, you can figure out expected losses from gambling or the break-even price for insurance.

Practice

Constructing Probability Distributions

P1. Suppose you roll two dice. Construct the probability distribution of the smaller of the two numbers. (In the case of doubles, such as rolling two 3's, the larger number and the smaller number both are 3.)

P2. The best estimate of the percentage of all email that is spam (electronic junk mail) is about 70%. Suppose you randomly select two email messages. Construct the probability distribution of the random variable, X, defined as the number of messages in your sample that are spam.

P3. Five equally qualified students (Betina, Charlotte, Max, Alisa, and Shaun) have applied to serve as student representatives on the campus housing committee. Alisa and Shaun live in the dorms and the other students live off campus. You will randomly select three of the five applicants to serve on the committee.

  a. List all possible random samples of size 3 from this group of five students.

  b. Each of your samples from part a is equally likely to occur. Construct the probability distribution of the random variable X, defined as the number of student committee members who live in the dorms.

  c. What is the probability that the committee has at least one student who lives in the dorms?

Expected Value

P4. Refer to your work in P2. What is the expected number of messages that would be spam?

P5. The distribution in Display 6.14 gives the number of children per family in the United States.

Number of Children	Proportion of Families
0	0.524
1	0.201
2	0.179
3	0.070
4 or more	0.026

Display 6.14 The number of children per family. [Source: *Statistical Abstract of the United States, 2004–2005, www.census.gov/prod/www/statistical-abstract-04.html.*]

  a. Compute the mean number of children per family. Count "4 or more" as 4.

  b. Write a summary sentence using the mean.

  c. What is the expected number of children in 10 randomly selected families?

P6. Construct the probability distribution table for the situation of flipping a coin five times and counting the number of heads. Compute the expected value of the distribution, and construct a graph of the probability distribution.

P7. Suppose you roll two tetrahedral dice, each with faces numbered 1, 2, 3, and 4. (The die on the left has landed as 2 and the one on the right as 3.)

  a. Make a probability distribution table for the sum of the numbers on the two dice. What is the probability that the sum is 3?

  b. What is the expected value of the probability distribution in part a?

P8. For a raffle, 500 tickets will be sold.

    a. What is the expected value of a ticket if the only prize is worth $600?

    b. What is the expected value of a ticket if there is one prize worth $1000 and two prizes worth $400 each?

P9. The burglary rate in your town is 23.2 burglaries per 1000 households. You want to take out insurance that would pay you $5000 if your home were burglarized. What price would a company charge for such insurance in order for it to expect to break even?

Exercises

E1. Beethoven composed five piano concertos, which are all on your iPod. The length of time each concerto takes to play is given in the following table. You use the shuffle feature of your iPod to select two different concertos at random.

Concerto	Number of Minutes
1	38
2	30
3	35
4	33
5	40

Ludwig van Beethoven, 1770–1827.

    a. Construct the probability distribution of the total number of minutes the two randomly selected concertos will take to play.

    b. You have 74 minutes before you have to leave for class. What is the probability that you have finished listening to both concertos before you have to leave?

E2. Suppose a quality-control inspection plan calls for randomly sampling two computer monitors from each group of six that comes off the assembly line and inspecting them for defects. On a particularly bad Monday morning, the first six monitors coming off the line include three monitors with defects.

    a. List all possible samples of size 2 that could be selected from the six monitors. Then make a probability distribution table for X, the number of sampled monitors with defects.

    b. If either of the two sampled monitors shows any defects, the company's policy calls for shutting down the line for an investigation of the cause of the defects. What is the probability that the line will be shut down after the inspection of two randomly selected monitors from the first six that come off the line this Monday morning?

E3. According to a recent government report, 16% of occupants of pickup trucks don't wear seatbelts. Suppose you randomly sample three occupants of pickup trucks. [Source: *Traffic Safety Facts: Seat Belt Use in 2009—Overall Results, www-nrd.nhtsa.dot.gov.*]

    a. Construct the probability distribution of the random variable, X, defined as the number of people in your sample who don't wear seatbelts.

    b. Would you tend to doubt the 16% estimate if none of the three were wearing seatbelts?

E4. The report cited in E3 says that 84% of drivers now use seatbelts regularly. Suppose a police officer at a road check randomly stops three vehicles to check for seatbelt use.

    a. Construct the probability distribution of X, the number of drivers using seatbelts.

    b. Would you tend to doubt the 84% estimate if none of the drivers was wearing a seatbelt?

E5. Investigators wanted to determine whether getting a speeding ticket deters speeding. They found that 11% of people in Maryland who received a speeding ticket got at least one more speeding ticket within a year. [Source: Saranath Lawpoolsri, Jingyi Li, and Elisa R. Braver, "Do Speeding Tickets Reduce the Likelihood of Receiving Subsequent Speeding Tickets? A Longitudinal Study of Speeding Violators in Maryland," *Traffic Injury Prevention*, Vol. 8, 1 (March 2007), pp. 26–34.]

    a. Suppose you randomly sample three Maryland with a speeding ticket. Construct the probability distribution of the random variable, X, defined as the number of sampled people who get at least one more speeding ticket within the next year.

    b. Does the information given here show that getting a speeding ticket does not deter speeding? If so, explain. If not, what additional information would you need to know?

E6. As stated in Example 6.2 on page 276, 87% of lung cancer cases are caused by smoking. Suppose three lung cancer patients are randomly selected from the large population of people with that disease.

    a. Construct the probability distribution of X, the number of patients with cancer caused by smoking.

    b. Would you tend to doubt the 87% estimate if none of the three could have had their lung cancer caused by smoking?

E7. Display 6.15 gives the percentage of U.S. households with various numbers of color televisions. Suppose a duplex is occupied by two randomly selected households and you are interested in the random variable X, the total number of color TV sets in the duplex.

Number of Color TVs	Percentage of Households
0	1.2
1	27.4
2	35.9
3	21.8
4	9.5
5	4.2

Display 6.15 Color television sets per household. [Source: U.S. Census Bureau, *Statistical Abstract of the United States, 2006*, Table 963.]

a. Use the rules of probability from Chapter 5 and the information in Display 6.15 to construct the distribution of X.

b. Find the expected value of X and use it in a sentence.

c. Compute the expected value of the number of color televisions in a single randomly selected household. Compare this expected value to the one you computed in part b.

E8. Refer to Display 6.6 on page 278.

a. Verify that the probabilities are correct for duplexes with 3, 4, 5, and 6 vehicles.

b. Find the expected value of this distribution.

c. Compare the expected value from part b to the one for Display 6.5 on page 277. What do you notice?

E9. Refer to Display 6.5 on page 277. If your boss provides two parking spaces per single-family house, how many vehicles do you estimate will have to be parked on the street in the new neighborhood of 500 single-family houses?

E10. Refer to Displays 6.5 and 6.6 on pages 277 and 278. Your boss suggests providing two parking spaces for each of the 500 single-family houses and, to keep things fair, four spaces for each of the 500 duplexes. If this plan is followed, which neighborhood do you expect will have a larger proportion of vehicles parked on the street?

E11. Refer to Example 6.6 on page 281.

a. If you charge a householder $134.50 per year for insurance, what is the largest profit you could earn on that one policy for the year? What is the largest possible loss for the year?

b. If you want to earn an expected yearly profit of $5000 per 1000 customers, how much should you charge per customer?

c. Factors besides the location of the household affect the probability of a burglary. What other factors might an insurance company take into account?

E12. The passenger vehicle with the highest theft loss is the Cadillac Escalade EXT, with 20.2 claims for theft per 1000 insured vehicles per year and an average payment of $14,939 per claim. How much would you charge an owner of a Cadillac Escalade EXT for theft insurance per year if you simply wanted to expect to break even? [Source: Highway Loss Data Institute news release, October 19, 2004, *www.hwysafety.org*.]

E13. For each million tickets sold, the original New York Lottery awarded one $50,000 prize, nine $5,000 prizes, ninety $500 prizes, and nine hundred $50 prizes.

a. What was the expected value of a ticket?

b. The tickets sold for 50¢ each. How much could the state of New York expect to earn for every million tickets sold?

c. What percentage of the income from the lottery was returned in prizes?

E14. A scratch-off card at Burger King gave the information contained in Display 6.16. All cards were potential winners. In order to actually win the prize, the player had to scratch off a winning path without making a false step. The probability of doing this was given as 1/5 for the food item and 1/10 for any nonfood item. So, for example, to win the prize worth $1 million, you must first get the card for $1 million and then scratch off the correct path without making a mistake.

Value of Prize ($)		Number of Game Cards
	1,000,000	1
	500,000	2
(house)	200,000	1
	100,000	10
(car)	16,619	75
	10,000	50
(cruise)	2,900	150
	1,000	200
	100	1,000
	10	10,000
(food item)	0.73	196,000,000

Display 6.16 Values of prizes, with frequencies.

a. What is the total value of all the potential prizes?

b. What total amount did Burger King expect to pay out?

c. What is the expected value of a card?

E15. You are trying to decide whether to buy a $250 service contract for your new computer. If you buy the service contract, you get an unlimited number of free repairs. If you don't buy the service contract, each repair will cost $150. You investigate and find the information in Display 6.17 about the number of repairs and their probabilities.

Number of Repairs	Probability
0	0.40
1	0.30
2	0.15
3	0.10
4	0.05

Display 6.17 Probability distribution of number of repairs.

a. Find the expected cost of each option.

b. What would be an advantage of buying the service contract? Of not buying it?

E16. Display 6.18 gives information about four commonly used cellular phone service plans.

Service	Monthly Charge	Monthly Minutes	Each Additional Minute
AT&T	$49.99	600	$0.45
Verizon	$49.00	500	$0.45
Voicestream	$39.99	500	$0.25
Sprint PCS	$49.99	500	$0.30

Display 6.18 Cell phone plans. [Source: New York City Department of Consumer Affairs, "The Buyers' Guide to Better Cellular Service," April 2002; *www.nyc.gov*.]

a. If you talk for 500 minutes or less each month, which plan will be the cheapest?

b. Some months you have a lot to talk about, and you estimate that the number of minutes you talk each month follows the probability distribution in Display 6.19. Rank the service plans according to how much you would expect them to cost per month.

Number of Minutes per Month	Probability
500 or less	0.4
550	0.3
600	0.2
700	0.1

Display 6.19 Probability distribution of number of minutes using cell phone.

6.2 ▶ Rules for Means and Variances of Probability Distributions

In the previous section, you learned how to compute and interpret the expected value (mean) of a probability distribution. As you have learned, a good summary of a distribution also includes a measure of spread, typically the standard deviation. Often the variance is used instead because some of the rules you will see in this section have a simpler form with the variance.

Variance of a Probability Distribution

As did the formula for the expected value, the formula for the variance of a probability distribution is similar to that for a frequency distribution of data.

> The variance of a distribution is the square of its standard deviation.

Formula for the Variance of a Probability Distribution

The variance of a probability distribution listed in a table is given by

$$Var(X) = \sigma_X^2 = \sum (x - \mu_X)^2 \cdot p$$

Here, p is the probability that the random variable X takes on the specific value x and μ_X is the expected value. To get the standard deviation, take the square root of the variance.

> The variance is the expected value of the squared deviations from the mean.

Standard Deviation of the Distribution of Number of Vehicles per Household

Example 6.9

In Example 6.5, Expected Number of Vehicles per Household on page 280, you found that you can report to your boss that the expected number of vehicles per household is 1.696. However, you realize that you should give your boss an estimate of how much households are likely to differ from this average. Thus you decide to calculate the standard deviation of the number of vehicles per household. The distribution reappears in Display 6.20 as a probability distribution.

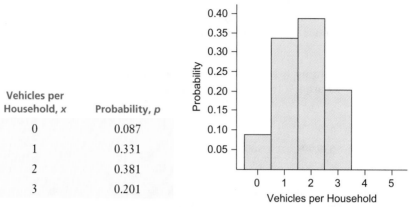

Vehicles per Household, x	Probability, p
0	0.087
1	0.331
2	0.381
3	0.201

Display 6.20 The number of motor vehicles per household.

Solution
The variance of the probability distribution is given by

$$\sigma_X^2 = (0 - 1.696)^2\,(0.087) + (1 - 1.696)^2\,(0.331) + (2 - 1.696)^2\,(0.381) \\ + (3 - 1.696)^2\,(0.201)$$
$$\approx 0.7876$$

The standard deviation is $\sigma_X = \sqrt{0.7876} \approx 0.8875$. You tell your boss that the neighborhood of single-family houses can expect to have 1.696 motor vehicles per house (on average), but quite often a house will have about 0.9 vehicles more or less than this mean.

Example 6.10 — Standard Deviation for the Wisconsin Lottery

The table of payoffs for the Wisconsin Lottery example in Display 6.13 is repeated in Display 6.21. The expected value of the distribution turned out to be $0.6014. Compute and interpret the standard deviation. Why is the standard deviation so large?

Winnings, x	Probability, p
$0	0.7793
$1	1/10
$2	1/14
$3	1/24
$18	1/200
$50	1/389
$150	1/20,000
$900	1/120,000

Display 6.21 Probabilities for Wisconsin scratch-off lottery game.
[Source: *www.wilottery.com*.]

Solution
The variance is

$$\sigma_X^2 = \sum (x - \mu_X)^2 \cdot p$$
$$= (1 - 0.6014)^2 \cdot \frac{1}{10} + (2 - 0.6014)^2 \cdot \frac{1}{14} + (3 - 0.6014)^2 \cdot \frac{1}{24}$$

$$+ (18 - 0.6014)^2 \cdot \frac{1}{200} + (50 - 0.6014)^2 \cdot \frac{1}{389} + (150 - 0.6014)^2 \cdot \frac{1}{20,000}$$

$$+ (900 - 0.6014)^2 \cdot \frac{1}{120,000} + (0 - 0.6014)^2(0.7793)$$

$$\approx 16.32039$$

Taking the square root, the standard deviation, σ_X, is approximately \$4.040.

The expected amount won per ticket is \$0.6014 with a standard deviation of \$4.040. The standard deviation is so large because of the small chance of winning a relatively large amount. The fact that the standard deviation is much larger than the mean also indicates that the distribution of winnings must have a long tail toward the larger values.

DISCUSSION
Variance of a Probability Distribution

D7. *Excessive compensatory damages.* The media occasionally report a case of a runaway jury that awards an excessive amount of money to a person who suffered little harm. To gain information about how widespread this actually is, researchers surveyed all 389 Texas district court judges. The judges were asked this question: "In what percentage of cases tried before you as presiding judge during the past 48 months, in which the jury awarded compensatory damages, do you believe that the jury's verdict on compensatory damages was disproportionately high given the evidence presented during the trial?" The results, from the 235 judges who responded to this question, are presented in Display 6.22 as a probability distribution. For example, if you pick one of the 235 judges at random, the probability is 0.834 that he or she saw no such cases in the past 48 months. Make a reasonable estimate of the center of each interval and then compute and interpret the expected value and standard deviation of this distribution.

Percentage of Cases	Probability
0	0.834
1–25	0.149
26–50	0.009
51–75	0.004
76–100	0.004
Total	1.00

Display 6.22 Percentage of jury cases where excessive compensatory damages were awarded by judge. [Source: Larry Lyon Bradley et al., "Straight from the Horse's Mouth: Judicial Observations of Jury Behavior and the Need for Tort Reform," *Baylor Law Review*, Vol. 59, 2 (Spring 2007), pp. 419–434.]

Rules for Means and Variances

In Section 2.3, you learned that if you recenter a distribution—that is, add the same number c to all the values—it doesn't change the shape or standard deviation but adds c to the mean. If you rescale a distribution—that is, multiply all the values by the same nonzero number d—it doesn't change the basic shape but multiplies the mean by d and the standard deviation by $|d|$ where $|d|$ indicates the absolute value of d. These rules for distributions of data also apply to probability distributions.

> Adding a constant recenters, or shifts, a distribution, while multiplying by a constant stretches, or shrinks it.

> **Linear Transformation Rule: The Effect of a Linear Transformation of X on μ_X and σ_X**
>
> Suppose you have a probability distribution for random variable X, with mean μ_X and standard deviation σ_X. If you transform each value by multiplying it by d and then adding c, where c and d are constants, then the mean and the standard deviation of the transformed values are given by
>
> $$\mu_{c+dX} = c + d\mu_X$$
> $$\sigma_{c+dX} = |d|\sigma_X$$

Example 6.11

Tripling the Wisconsin Lottery

Suppose that for a special promotion, the prizes are tripled for the Wisconsin scratch-off game "Big Cat Cash" described in Display 6.13 on page 282.

a. What are the expected winnings for a person who plays one game? What is the standard deviation?

b. Taking into account that the cost of a lottery ticket is $1, what is the expected profit and the standard deviation of that profit for a person who plays one game during the special promotion?

Solution

a. The most efficient method of doing this problem is to transform the expected value, 0.6014, and standard deviation, 4.040 (from Example 6.10), for the original game using the linear transformation rule. Here, $c = 0$ and $d = 3$, so the expected value of the tripled prize is 3(0.6014), or $1.804, with a standard deviation of 3(4.040), or $12.12.

A second method of doing this problem is to rewrite the table in Display 6.22 using the winnings during the special promotion, as shown in Display 6.23.

Original Winnings, x	Winnings in Special Promotion, $3x$	Probability, p
$0	$0	0.7793
$1	$3	1/10
$2	$6	1/14
$3	$9	1/24
$18	$54	1/200
$50	$150	1/389
$150	$450	1/20,000
$900	$2700	1/120,000

Display 6.23 Winnings during the special promotion.

Then use the formulas for the expected value and variance from pages 280 and 287 as before:

$$E(X) = 0(0.7793) + 3(1/10) + 6(1/14) + 9(1/24) + 54(1/200) + 150(1/389) + 450(1/20,000) + 2700(1/120,000)$$
$$\approx 1.804$$

$$Var(X) = (0 - 1.804)^2(0.7793) + (3 - 1.804)^2(1/10) + (6 - 1.804)^2(1/14)$$
$$+ (9 - 1.804)^2(1/24) + (54 - 1.804)^2(1/200)$$
$$+ (150 - 1.804)^2(1/389) + (450 - 1.804)^2(1/20,000)$$
$$+ (2700 - 1.804)^2(1/120,000)$$
$$\approx 146.89$$

The expected winnings are \$1.804, with a standard deviation of $\sqrt{146.89}$, or \$12.12. This standard deviation is so large because the distribution is skewed, with small probabilities of winning the larger prizes.

b. The expected profit will be \$1.804 − \$1, or \$0.804. This would be a good game to play because it has a positive expectation, even when taking into account the cost of a ticket. (So don't expect any state ever to offer such a game!)

The standard deviation of the profit will not change because subtracting \$1 from each of the winnings in Display 6.23 does not change the spread of the distribution.

Suppose that, instead of tripling the winnings, the special promotion gives out three tickets for the price of one. What are the mean and standard deviation of the winnings now? You saw in the previous section that the mean may be computed using the rule on page 282. There is a similar rule for computing the standard deviation. Both rules appear in the following box.

Mean and Variance of a Sum of *n* Independent Values of a Random Variable

Use this rule when you are adding random variables from the *same* distribution.

Suppose that you have n independent values of a random variable that has mean, μ_X, and variance, σ_X^2. Then the expected value of their sum is $n \cdot \mu_X$. The variance of their sum is $n \cdot \sigma_X^2$, and the standard deviation is $\sqrt{n} \cdot \sigma_X$.

Getting Three Lottery Tickets

Example 6.12

A single ticket in the Wisconsin lottery game has an expected value of \$0.6014 with standard deviation \$4.040. Suppose you are given three different tickets. What are your expected total winnings and the standard deviation of your total winnings?

Solution
Your expected winnings are $n \cdot \mu_X = 3(\$0.6014) = \$1.804$. The standard deviation is $\sqrt{n} \cdot \sigma_X = \sqrt{3} \cdot 4.040 \approx 6.997$.

The formula for the mean works without independence, but the formula for the standard deviation is guaranteed to be correct only when the values of the tickets are independent. Independent means, for example, that the probability that the third ticket is a \$150 winner remains 1/20,000, no matter what the value of the first two tickets were. That is approximately the case here, if there were a great number of lottery tickets sold.

Compare the previous two examples. In the first, there was *one* randomly selected ticket and its value was tripled. In the second, there were *three* randomly selected tickets and their values were added. The expected values are the same, $1.804, but there is more variability when the winnings from a single ticket are tripled.

The previous rule is a special case of the next rule. You may use the previous rule when the random selections are made from the *same* distribution. You must use the following rule when the random selections are made from *different* distributions. (You also may use it when the two distributions are the same.)

Addition and Subtraction Rules for Random Variables

Use this rule when you are combining random variables from *different* distributions.

If X and Y are random variables, then

$$\mu_{X+Y} = \mu_X + \mu_Y$$
$$\mu_{X-Y} = \mu_X - \mu_Y$$

If X and Y are independent, then

$$\sigma^2_{X+Y} = \sigma^2_X + \sigma^2_Y$$
$$\sigma^2_{X-Y} = \sigma^2_X + \sigma^2_Y$$

The Addition Rule generalizes in the obvious way when there are more than two random variables.

Example 6.13   Two Different Lotteries

Suppose that you buy tickets in two different lotteries. The first has an expected value of $0.50 with a standard deviation of $0.25. The second has an expected value of $0.75 with a standard deviation of $0.32. Not counting the cost of a ticket, what are your expected total winnings, and what is the standard deviation of this total?

Solution
Your expected winnings are

$$\mu_{X+Y} = \mu_X + \mu_Y$$
$$= 0.50 + 0.75$$
$$= 1.25$$

Because the outcomes of the two lotteries are independent, you can use the Addition Rule for Random Variables to compute the variance of the total.

$$\sigma^2_{X+Y} = \sigma^2_X + \sigma^2_Y$$

$$= (0.25)^2 + (0.32)^2$$

$$= 0.1649$$

The standard deviation is then $\sigma_{X+Y} = \sqrt{0.1649} \approx 0.41$.

The expected value of your winnings is $1.25, but you shouldn't be surprised to get $0.41 more or less than this.

Most of the rules in this section may seem fairly natural, with one exception. The most surprising rule says that to get the variance of the *difference* of two independently selected random variables, you *add* the individual variances. Why add and not subtract? The following example will help you see the reason.

The Sum and Difference of Two Dice

Example 6.14

If you roll a single die, you get 1, 2, 3, 4, 5, or 6, with equal probability. It's straight-forward to compute that the mean roll is $\mu = 3.5$, with a variance of $\sigma^2 = 2.917$. The probability distribution for the *sum* of two dice is shown on the left in Display 6.24. It is triangular, with mean $\mu_1 + \mu_2 = \mu + \mu = 3.5 + 3.5 = 7$ and variance $\sigma_1^2 + \sigma_2^2 = \sigma^2 + \sigma^2 = 2.917 + 2.917 = 5.834$.

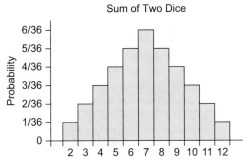
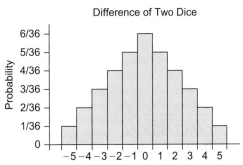

Display 6.24 Probability distribution for the sum (on the left) and the difference (on the right) of the roll of two dice.

a. Construct a probability distribution for the *difference* when a die is rolled twice (first roll minus second roll). Then compare the distribution of the difference to that of the sum.

b. Compute the mean μ_X and the variance σ_X^2 of the distribution of the difference using the formulas

$$\mu_X = \sum x \cdot p \quad \text{and} \quad \sigma_X^2 = \sum (x - \mu_X)^2 \cdot p$$

c. Using the fact that the mean and variance of the distribution of the roll of one die are 3.5 and 2.917, verify that the mean of the probability distribution of the difference is equal to $\mu_1 - \mu_2$, and the variance is $\sigma^2 + \sigma^2$, or $2\sigma^2$, where μ and σ are the mean and standard deviation of the distribution of the outcome of a single roll of a die.

Solution

a. The probability distribution for the difference (first roll minus second roll) is:

Difference	Probability
−5	1/36
−4	2/36
−3	3/36
−2	4/36
−1	5/36
0	6/36
1	5/36
2	4/36
3	3/36
4	2/36
5	1/36

This distribution is shown on the right in Display 6.24.

The shapes and spreads of the two distributions are identical. (This won't always be the case with a sum and difference of two independent random variables.) Only the means are different: 7 for the sum and 0 for the difference.

b. Using the formulas and the probability distribution table in part a, the mean is 0, and the variance is approximately 5.834.

c. The mean of the difference is equal to the difference of the mean of the roll of the first die and the mean of the roll of the second die:

$$\mu - \mu = 3.5 - 3.5 = 0$$

The variance of the difference is indeed the sum of the variances of the roll of each die:

$$\sigma^2 + \sigma^2 = 2.917 + 2.917 \approx 5.834$$

This is the same variance as for the sum of two dice.

DISCUSSION
Rules for Means and Variances of Probability Distributions

D8. If you expect to work 10 hours next week with a standard deviation of 2 hours and you expect to study 15 hours next week with a standard deviation of 3 hours, what is the total number of hours you expect to spend working or studying? Use the rule for adding variances to estimate the standard deviation of the total number of hours spent studying or working. Is it reasonable to use the rule in this case? Explain.

D9. Provide an intuitive argument as to why the variability in a sum should be larger than the variability in either random variable making up the sum. Do the same for the variability in a difference.

Summary 6.2: Rules for Means and Variances of Probability Distributions

You can compute the expected value (mean) and variance of a probability distribution with values denoted by x and corresponding probability by p using the formulas

$$E(X) = \mu_X = \sum x \cdot p$$
$$Var(X) = \sigma_X^2 = \sum (x - \mu_X)^2 \cdot p$$

If you take a single value, x, at random from a probability distribution with mean, μ_X and standard deviation σ_X, and multiply it by d and then add c, then the expected value of this new random variable is $\mu_{c+dX} = c + d\mu_X$, with standard deviation $\sigma_{c+dX} = |d| \, \mu_X$.

If you take n values randomly from a probability distribution with mean μ_X and standard deviation σ_X and add them, then the expected value of this sum is $n \cdot \mu_X$. If the values are taken independently, the standard deviation is $\sqrt{n} \cdot \sigma_X$.

Now suppose that you have two random variables, X and Y. That is, you take one value at random from distribution X and one from distribution Y. Then the expected values of the sum and difference are given by

$$\mu_{X+Y} = \mu_X + \mu_Y$$
$$\mu_{X-Y} = \mu_X - \mu_Y$$

If X and Y are independent, then both the variance of the sum and the variance of the difference are given by

$$\sigma_{X+Y}^2 = \sigma_{X-Y}^2 = \sigma_X^2 + \sigma_Y^2$$

Practice

Variance of a Probability Distribution

P10. A large class was assigned a difficult homework problem. Display 6.25 shows the scores the students received and the proportion of students who received each score. Suppose you select a student at random. Compute and interpret the expected score and the standard deviation of the scores.

Score	Proportion of Students
0	0.45
1	0.09
2	0.00
3	0.04
4	0.15
5	0.27

Display 6.25 Distribution of homework scores.

P11. Display 6.26 gives the ages of students enrolled at the University of Texas, San Antonio. Suppose you select a student at random. Make a reasonable estimate of the center of each age group, and compute and interpret the expected age and standard deviation of the ages.

Age	Percentage
17–22	44.2
23–29	31.4
30 +	24.4

Display 6.26 Ages of students. [Source: University of Texas, *www.utsa.edu.*]

P12. The payout information for the Texas Lottery game called "Lucky Dice" is given in Display 6.27. There are approximately 8,054,375 tickets in Lucky Dice.

a. Verify the following statement from the information page about the lottery: The overall chances of winning any prize in Lucky Dice are 1 in 4.76 including break-even prizes. (The game costs $2 to play, so winning $2 is a break-even prize.)

Amount	Number Printed
$20,000	10
$2,000	31
$200	6,498
$50	41,490
$20	48,255
$10	112,930
$5	96,600
$4	740,879
$2	644,597

Display 6.27 The Texas lottery game, Lucky Dice. [Source: *www.txlottery.org.*]

b. Add a row to the table for the event of winning nothing. Taking into account the cost of a ticket, compute the expected value of a single ticket, and the standard deviation. Interpret these values.

P13. A neighborhood hardware store rents a steam cleaner for 3 hours at a fee of $30.00 or for a day (up to 8 hours) at a fee of $50.00. Sometimes the cleaner is rented out for two 3-hour periods in the same day. By studying rental records, the manager of the store comes up with approximate probabilities for renting the cleaner on a typical Saturday, as shown in Display 6.28.

Rental Period	Probability
One 3-hour rental	0.4
One full-day rental	0.3
Two 3-hour rentals	0.1
Not rented	0.2

Display 6.28 Steam cleaner rental probabilities.

a. Find the expected rental income for the steam cleaner on a typical Saturday.

b. Find the standard deviation of the rental income on a typical Saturday. Given that the standard deviation is supposed to measure a typical deviation from the mean, does this value seem reasonable?

Rules for Means and Variances of Probability Distributions

P14. The distribution in Display 6.29 gives the probability that a randomly selected family in the United States will have the given number of children.

a. Compute the expected value and standard deviation of this population. Count "4 or more" as 4. (You may have already computed the expected value in P5 on page 284.)

b. Suppose you take a random sample of 30 households. Compute and interpret the expected number of children in the 30 households and the standard deviation of the number of children.

Number of Children	Probability
0	0.524
1	0.201
2	0.179
3	0.070
4 or more	0.026

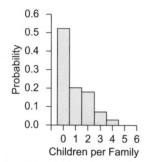

Display 6.29   The number of children per family.
[Source: *Statistical Abstract of the United States, 2004–2005.*]

P15. The amount you save varies from week to week, but averages $35 with a standard deviation of $15.

   a. What is the expected amount you would save in a year (52 weeks)? What is the standard deviation of your yearly savings?

   b. If your grandmother adds $10 a week to your savings, what is the expected amount that you would save in a year? What is the standard deviation?

P16. Refer to P7, where you constructed the probability distribution for the sum of two tetrahedral dice.

   a. Compute the mean μ_X and the variance σ_X^2 of the distribution of the sum of two tetrahedral dice using your distribution from P7 and the formulas

$$\mu_X = \sum x \cdot p \quad \text{and} \quad \sigma_X^2 = \sum (x - \mu_X)^2 \cdot p$$

   (You may have the mean from P7.)

   b. Find the mean and the variance of the roll of a single tetrahedral die (where the numbers 1, 2, 3, and 4 each occur with probability $\frac{1}{4}$).

   c. Verify that the mean of the probability distribution of the sum from part a is equal to the sum of the mean of the first roll and the mean of the second roll.

   d. Verify that the variance of the probability distribution of the sum from part a is equal to the sum of the variance of the first roll and the variance of the second roll.

P17. Refer to Displays 6.5 and 6.6 on pages 277 and 278.

   a. Find the expected value and standard deviation of the total number of vehicles per duplex directly from Display 6.6.

   b. Verify that the mean of the probability distribution of the total number of vehicles in a duplex is equal to $\mu + \mu$, and the variance is $\sigma^2 + \sigma^2$, or $2\sigma^2$, where μ and σ are the mean and standard deviation of the distribution of the number of vehicles in a single household.

Exercises

E17. Your construction company plans to build apartment buildings, each with one common parking lot. Your boss would like an estimate of how many spaces would be needed for a building with 200 households.

   a. Using the mean number of vehicles per household, 1.696, from Display 6.11 on page 281, find the expected number of vehicles for an apartment building with 200 households.

   b. You realize that a randomly selected group of 200 households probably won't have exactly the expected number of vehicles. How far off is your estimate likely to be?

   c. What assumptions are you making in your answers to parts a and b? Do you think they are reasonable for apartment dwellers?

E18. Display 6.30 gives the probability that a randomly selected family in the United States will have the given number of color televisions.

   a. Compute and interpret the expected value and standard deviation of this distribution. (You may have already computed the expected value in E7 on pages 285–286.)

   b. Suppose you take a random sample of 30 households. Compute the expected number of color

televisions in the 30 households and the standard deviation of the number of color televisions.

Number of Color TVs	Probability
0	0.012
1	0.274
2	0.359
3	0.218
4	0.095
5	0.042

Display 6.30   Color television sets per household.
[Source: *Statistical Abstract of the United States, 2006*, Table 963.]

E19. The amount of tips you get each week in your job as a server varies widely, according to the probability distribution in Display 6.31.

   a. What are your expected total weekly tips? What is the standard deviation?

   b. Now suppose you get a base salary of $60 a week, in addition to your tips. What are your total expected weekly earnings? What is the standard deviation?

Amount of Tips ($)	Probability
200	0.1
300	0.3
400	0.4
500	0.2

Display 6.31 Probability distribution of amount of tips.

c. Further, suppose that each week you share 20% of your tips with other restaurant employees. What are your expected total weekly earnings now? What is the standard deviation?

E20. Working as part-time salesperson in the large-screen television department of an electronics store, you get a commission of $25 for each large-screen television you sell. The probability distribution of the number of televisions you sell in a week is given in Display 6.32.

Number Sold	Probability
0	0.10
5	0.40
10	0.25
15	0.15
20	0.10

Display 6.32 Probability distribution of number of televisions sold.

a. What is your expected total weekly commission? What is the standard deviation?

b. Now, suppose you get a base salary of $150 a week, in addition to your commission. What are your total expected weekly earnings? What is the standard deviation?

c. Each week, your employer withholds 30% of your total earnings for taxes. What are your expected total weekly take-home earnings? What is the standard deviation?

E21. Many mechanical and electronic systems are built to include backups to major components (much like the spare tire in a car). If the main component fails, the backup kicks in. Often, however, the backup component is for emergency use only and is not built to the same specifications as the main component. Suppose a main pump in a city water system works without failure for 1 month with probability 0.1, 2 months with probability 0.3, and 3 months with probability 0.6. The pump's backup works without failure for 1 month with probability 0.2 and for 2 months with probability 0.8. The main pump is used alone until it fails; then the backup pump kicks in and is used alone until it fails. You can assume that the main pump and its backup operate independently of each other.

a. Find the probability distribution of the total working time of the main pump followed by its backup.

b. Find the expected value and standard deviation of the total working time of the main pump and its backup.

Pumping station in Orleans Parish, Louisiana.

c. Show that the expected total time in part b is the sum of the expected working times of the main pump and of its backup.

E22. You pay $1 to play Game A, which generates a payoff of $0, $1, $2, or $3 with respective probabilities 0.4, 0.3, 0.2, and 0.1. You also pay $2 to play Game B, which generates a payoff of $0, $2, or $4 with respective probabilities 0.7, 0.2, and 0.1. The games are operated independently of each other.

a. Construct the probability distribution of your total gain from playing the two games. (Your gain is your winnings minus the cost of playing the game.)

b. Find the expected value and standard deviation of your total gain.

c. Show that your expected total gain in part b is the sum of your expected gains from each of the two games.

E23. A regular die is rolled, and then a tetrahedral die is rolled. The difference is calculated.

a. Construct the probability distribution of the difference of the number on the regular die and the number on the tetrahedral die.

b. Verify that the mean of the probability distribution is $\mu_1 - \mu_2$ and the variance is $\sigma_1^2 + \sigma_2^2$, where μ_1 and σ_1^2 are, respectively, the mean and variance of the distribution of the roll of a single six-sided die and μ_2 and σ_2^2 are, respectively, the mean and variance of the distribution of the roll of a single four-sided die.

E24. To give a small child practice in subtracting, you generate problems of the form *larger number − smaller number*. To get the larger number, you have the child roll a die. To get the smaller number, you use 1 less than the larger number. For example, if the child rolls a 3, you give him the problem 3 − 2.

a. Construct the probability distribution for the larger number and compute its mean, μ_{larger}, and variance, σ_{larger}^2.

b. Construct the probability distribution for the smaller number and compute its mean, $\mu_{smaller}$, and variance, $\sigma_{smaller}^2$.

c. Construct the probability distribution for the difference, *larger number − smaller number*. Compute its mean and variance using the formulas

$$\mu_X = \sum x \cdot p \quad \text{and} \quad \sigma_X^2 = \sum (x - \mu_X)^2 \cdot p$$

d. Is the mean of the probability distribution for the difference equal to $\mu_{larger} - \mu_{smaller}$?

e. Is the variance of the probability distribution for the difference equal to $\sigma^2_{larger} + \sigma^2_{smaller}$. Why should this be the case?

E25. Refer to Exercise E19, part c.

a. What are your expected total yearly earnings (for 52 weeks' work)? What is the standard deviation?

b. Would it be unusual for you to earn more than $20,000 in a year? Explain.

E26. Working as part-time salesperson in the large-screen television department of an electronics store, you get paid $150 per week, plus a commission of $25 for each large-screen television you sell. For each customer you talk to, the probability is 0.1 that you will be able to sell him or her a large-screen television.

a. Make a probability distribution table for the amount of commission you receive for talking to one customer.

b. What are your expected total weekly earnings if you talk to only one customer? What is the standard deviation?

c. What are your expected total weekly earnings if you talk to 100 customers? What is the standard deviation?

d. Would it be unusual for you to earn less than $300 if you talk to 100 customers? Explain.

6.3 ▶ The Binomial Distribution

Some types of probability distributions occur so frequently in practice that it is important to know their names and characteristics. Among these is the binomial distribution, which closely (but not perfectly) reflects many real-world situations.

Binomial Probabilities

Many random variables seen in practice amount to counting the number of successes in n independent trials. Examples include:

- counting the number of doubles in four rolls of a pair of dice (on each roll either you get doubles or you don't)

- counting the number of children who have been diagnosed as autistic in a random sample of 1000 children (each child either has been diagnosed as autistic or not)

- counting the number of defective items in a sample of 20 items (either an item is defective or it isn't)

These situations are called binomial because each trial has two possible outcomes.

In Chapter 5, you studied a binomial situation when you wanted to know the answer to a question like this: "If you ask four people which is the gourmet coffee and none of them can tell gourmet coffee from ordinary coffee, what is the probability that two of the four will guess correctly?" The following example reviews how to list all possible outcomes in order to find this probability and introduces a formula for counting how many outcomes there are.

Example 6.15	Equally Likely Outcomes: Guessing Which Is the Gourmet Coffee

Each of four people will be presented with gourmet coffee and ordinary coffee and asked to identify the gourmet coffee. None of the four can tell the difference and are merely guessing.

a. List all 16 possible outcomes. Use the list to find the probability that exactly two people correctly identify the gourmet coffee.

b. Show that the number of outcomes in which exactly two people out of four correctly identify the gourmet coffee may be found using the formula

$_nC_k = \binom{n}{k} = \dfrac{n!}{k!(n-k)!}$, where n is the number of tasters and k is the number of tasters who guess correctly.

c. Use the result from part (b) and the Multiplication Rule for Independent Events to find the probability that exactly two people correctly identify the gourmet coffee.

Solution

a. The 16 possible outcomes are organized in the table in Display 6.33.

Number Who Correctly Identify Gourmet Coffee	Outcomes	Count of Outcomes
0	OOOO	1
1	OOOG OOGO OGOO GOOO	4
2	OOGG OGOG GOOG OGGO GOGO GGOO	6
3	OGGG GOGG GGOG GGGO	4
4	GGGG	1

> The counts are the fifth row of Pascal's triangle.
>
> ```
>         1
>       1   1
>     1   2   1
>   1   3   3   1
> 1   4   6   4   1
> ```
> Pascal's triangle

Display 6.33 All possible outcomes when four people are asked which of two cups of coffee is the gourmet coffee. G means choosing correctly and O means choosing incorrectly.

Each of the 16 outcomes is equally likely if none of the four people can tell the difference and are merely guessing. Six outcomes represent two of the four people guessing correctly. Thus, the probability of getting exactly two people who select the gourmet coffee is 6/16.

b. Use the formula $\binom{n}{k} = \dfrac{n!}{k!(n-k)!}$, with $n = 4$ and $k = 2$.

$$\binom{4}{2} = \frac{4!}{2!(4-2)!} = \frac{4!}{2! \cdot 2!} = \frac{24}{2 \cdot 2} = 6$$

Note that if you wish to verify the other counts in the table, you will need to use the facts that $0! = 1$ and $1! = 1$. For example, for $k = 0$

$$\binom{4}{0} = \frac{4!}{0!(4-0)!} = \frac{4!}{1 \cdot 4!} = 1$$

For $k = 3$

$$\binom{4}{3} = \frac{4!}{3!(4-3)!} = \frac{4!}{6 \cdot 1} = \frac{24}{6} = 4$$

> Reminder:
> $n! = n \cdot (n-1) \cdot (n-2) \cdots 3 \cdot 2 \cdot 1$.

> The symbol $\binom{n}{k}$ is read "n choose k."

c. Using the Multiplication Rule for Independent Events, the probability of getting any one particular outcome where two of the four people guess correctly, say, GOOG, is $\frac{1}{2} \cdot \frac{1}{2} \cdot \frac{1}{2} \cdot \frac{1}{2}$, or $\frac{1}{16}$. From part b, there are six outcomes where two people guess correctly and these are mutually exclusive. So the probability of getting two people who correctly select the gourmet coffee is

$$6 \cdot \frac{1}{16}, \text{ or } \frac{6}{16}.$$

| Example 6.16 | **When Outcomes Aren't Equally Likely: College Graduates** |

The proportion of adults age 25 and older in the United States with at least a bachelor's degree is 0.29. [Source: U.S. Census Bureau, *Statistical Abstract of the United States, 2009*, Table 221.] Suppose you pick seven adults at random. What is the probability that exactly three will have a bachelor's degree or higher?

Solution

In this situation, the outcomes aren't equally likely. For example, the outcome

grad   grad   grad   grad   grad   grad   grad

is far less likely than the outcome

not   not   not   not   not   not   not

While you can't find the probability using the method in part (a) of the previous example, there is nothing about the method in parts (b) and (c) that requires equally likely outcomes.

For example, one outcome with exactly three college graduates is

grad   not   grad   grad   not   not   not

By the Multiplication Rule, the probability of this particular outcome is

$$(0.29)(0.71)(0.29)(0.29)(0.71)(0.71)(0.71) = (0.29)^3(0.71)^4$$

Another outcome with exactly three college graduates is

not   grad   grad   not   not   not   grad

The probability of this particular outcome is

$$(0.71)(0.29)(0.29)(0.71)(0.71)(0.71)(0.29) = (0.29)^3(0.71)^4$$

which is exactly the same. That's because the computation of the probability of each outcome with exactly three college graduates has the same factors but in a different order.

The number of different outcomes with exactly three college graduates is

$$\binom{7}{3} = \frac{7!}{3!(7-3)!} = 35$$

Thus, the probability of getting exactly three college graduates is

$$35(0.29)^3(0.71)^4 \approx 0.2169$$

A summary of the method developed in the two previous examples is given next.

| **The Binomial Probability Distribution** |

Suppose you have a series of trials that satisfy these conditions:

B: They are binomial—each trial must have one of two different outcomes, one we will call a "success" and the other a "failure."

I: Each trial is independent of the others.

N: There is a fixed number, n, of trials.

S: The probability, p, of a success is the same on each trial.

Then the distribution of the random variable, X, that counts the number of successes is called a **binomial distribution**. Further, the probability that you get exactly $X = k$ successes is

$$P(X = k) = \binom{n}{k} p^k (1 - p)^{n-k}$$

Here

$$\binom{n}{k} = \frac{n!}{k!(n - k)!} \quad \text{and} \quad n! = n(n - 1)(n - 2) \cdots 3 \cdot 2 \cdot 1$$

> Notation: p is the probability of a success on any one trial, n is the number of trials, and k is the number of successes.

> Reminder:
> $\binom{n}{n} = 1$
> $\binom{n}{0} = 1$

Counting Bachelor's Degrees

Example 6.17

You pick seven adults (age 25 and older) from your city at random to participate in a focus group about proposed changes in library service. As you saw in Example 6.16, 29% of adults in the United States have at least a bachelor's degree and that is also true in your community. Construct the probability distribution table and graph for the number in your focus group with a bachelor's degree. What is the probability that five or more have a bachelor's degree?

Solution

The distribution in the table and graph in Display 6.34 was constructed using the formula in the preceding box with $n = 7$, $p = 0.29$, and $k = 0$ through 7.

Number of College Graduates, k	Computation	Probability, $P(X = k)$
0	$\binom{7}{0} (0.29)^0 (1 - 0.29)^7$	0.091
1	$\binom{7}{1} (0.29)^1 (1 - 0.29)^6$	0.260
2	$\binom{7}{2} (0.29)^2 (1 - 0.29)^5$	0.319
3	$\binom{7}{3} (0.29)^3 (1 - 0.29)^4$	0.217
4	$\binom{7}{4} (0.29)^4 (1 - 0.29)^3$	0.089
5	$\binom{7}{5} (0.29)^5 (1 - 0.29)^2$	0.022
6	$\binom{7}{6} (0.29)^6 (1 - 0.29)^1$	0.003
7	$\binom{7}{7} (0.29)^7 (1 - 0.29)^0$	0.000

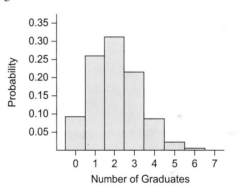

Display 6.34 Probability distribution of number of college graduates in a sample of size 7.

The graph is skewed to the right, which is typical when $p < 0.5$. It is more likely to get smaller numbers of college graduates than larger.

The probability that five or more adults in the focus group have a bachelor's degree is $0.022 + 0.003 + 0.000 = 0.025$.

You might have noticed that the trials in the previous example aren't really independent. The first adult selected has probability 0.29 of being a college graduate. If that person is a college graduate, the probability that the next person selected is a college graduate is a bit less. However, the change in probability is so small that you can safely ignore it. For example, if there are 150,000 adults age 25 and older in your city, then 43,500 of them would be college graduates. The probability that the first adult selected at random is a college graduate is 43,500/150,000, or 0.29. If that person turns out to be a college graduate, the probability that the second adult selected is a college graduate is 43,499/149,999, or 0.2899953, which is very close to 0.29.

You can treat your random sample as a binomial situation as long as the sample size, n, is small compared to the population size, N. A simple guideline that works well in practice is that n should be less than 10% of the size of the population, or $n < 0.10N$.

> It is generally safe to regard trials as independent if the condition $n < 0.10N$ is satisfied.

DISCUSSION
Binomial Probabilities

D10. Make a probability distribution table for the number of people who *aren't* college graduates in a random sample of seven adults if 29% of adults are college graduates. Make a graph of this distribution and compare with the one in Display 6.34.

D11. Show why the population must be "large" in order to use the binomial probability formula by pretending that there are 12 adults in your family, of whom 4 are chocoholics. Compute the exact probability that in a random sample of 7 adults, none will be chocoholics. Compare your results to those you get using the binomial formula.

STATISTICS IN ACTION 6.1 ▶ Can People Identify the Tap Water?

In this activity, you'll conduct an investigation to determine whether people can distinguish between tap water and bottled water. This will give you practice in statistical decision making as you decide how many subjects will have to correctly select the tap water before you are convinced that people can tell the difference.

What You'll Need: small paper or plastic cups, a container of tap water, a container of bottled water, 20 volunteer subjects

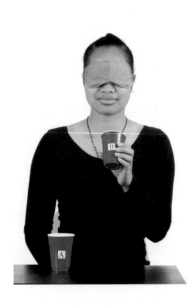

1. Design a study to see whether people can identify which of two cups of water contains the tap water. Refer to the principles of good experimental design in Chapter 4. How will you use randomization in your design?

2. Suppose none of your 20 subjects actually can tell the difference. Their choice is equivalent to selecting one of the cups at random. Construct a probability distribution of the number who correctly identify the tap water.

3. If only ten people correctly select the tap water, you have no reason to conclude that people actually can tell the difference. Why?

4. Decide how many people will have to make the correct selection before you are reasonably convinced that people can tell the difference.

5. Conduct your investigation and write a conclusion.

Shape, Center, and Spread of a Binomial Distribution

The fact that the binomial distribution is represented by such a simple and elegant formula might lead you to hope that there is a simple formula for the expected value and standard deviation as well. That is indeed the case. For the coffee-tasting experiment, the distribution of X, the number of the four tasters who correctly guessed the gourmet coffee, is shown in Display 6.35.

The expected value for this distribution is

$$E(X) = 0(1/16) + 1(4/16) + 2(6/16) + 3(4/16) + 4(1/16) = 2$$

Notice that $E(X)$ can be written as $4 \cdot 1/2$, which is np. In other words, you would expect about half of the four to guess correctly given two equally likely choices.

k	$P(X = k)$
0	1/16
1	4/16
2	6/16
3	4/16
4	1/16

Display 6.35   Distribution of number of correct guesses.

Is this just coincidence, or is it a general rule? Let's try one more example to see if we have any hope of the latter. The discussion of college graduates in Example 6.17 on page 301 stated that about 29% of U.S. adults have at least a bachelor's degree. If seven adults are randomly selected to participate in a focus group, the distribution of X, the number with at least a bachelor's degree, was shown in Display 6.34 on page 301. The expected value of this distribution is

$$
\begin{aligned}
E(X) &= 0(0.091) + 1(0.260) + 2(0.319) + 3(0.217) + 4(0.089) + 5(0.022) \\
&\quad + 6(0.003) + 7(0) \\
&= 2.03 \\
&= 7(0.29) \\
&= np
\end{aligned}
$$

A simple formula does seem to be a possibility, and that indeed is the case. As a bonus, there also is a formula for the standard deviation that turns out to be just about as simple as the one for the expected value.

The Characteristics of a Binomial Distribution

If a random variable X has a binomial distribution with n trials and probability of success p, the mean (expected value) and standard deviation for the distribution are

$$E(X) = \mu_X = np \quad \text{and} \quad \sigma_X = \sqrt{np(1 - p)}$$

The shape of the distribution becomes more normal as n increases.

These formulas hold for all sample sizes n.

Multiracial Americans

Example 6.18

Whether a person should be allowed to designate himself or herself as multiracial has been controversial throughout U.S. history, with positions shifting from one view to the other among various groups. For the first time, the 2000 U.S. census allowed people to "Mark one or more races to indicate what this person considers himself/herself to be." Overall, 2.4% of the population chose two or more races, designating themselves as multiracial. Gallup and other polling organizations randomly sample about 1200 U.S. residents for political polls. In a random sample of 1200 U.S. residents who participated in the 2000 census, what is the expected number who designated themselves as multiracial? What is the standard deviation of the number in the sample who designated themselves as multiracial?

Solution

The expected number is $E(X) = 1200(0.024)$, or about 28.8 who designated themselves as multiracial, with a standard deviation of

$$\sigma_X = \sqrt{np(1-p)} = \sqrt{1200 \cdot 0.024(1 - 0.024)} \approx 5.30.$$

Plots of binomial distributions for various values of n and p are shown in Display 6.36. Examine how the shape, center, and spread of these binomial distributions change as p increases and as n increases.

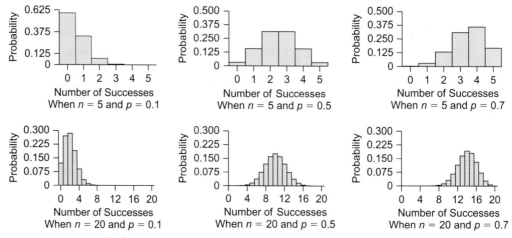

Display 6.36 Binomial distributions for different values of n and p.

DISCUSSION
Shape, Center, and Spread of a Binomial Distribution

D12. Refer to Display 6.36. How do the shape, center, and spread of a binomial distribution change as p increases for fixed values of n? How do they change as n increases for fixed values of p?

D13. Ten MP3 players of a discontinued model are selling for $100 each, with a double-your-money-back guarantee if they fail in the first month of use. Suppose the probability of such failure is 0.08. What is the expected net gain for the retailer after all ten MP3 players are sold? Ignore the original cost of the MP3 players to the retailer.

D14. For a random sample of three adults, the expected number of college graduates is $(3)(0.29)$, or 0.87, less than one whole person. Provide a meaningful interpretation of this number.

Summary 6.3: The Binomial Distribution

The random variable X is said to have a binomial distribution if X represents the number of successes in n independent trials, where the probability of a success is p on each trial. For example, X might represent the number of successes in a random sample of size n from a large population, with probability of success p on each selection.

In Section 6.1, to find the shape, expected value, and standard deviation of a random variable, typically you first had to construct the entire distribution, which was a lot of work. For the binomial distribution, and some of the other most commonly used random variables, mathematical formulas will give you the distribution directly, as well as its expected value and standard deviation.

The formulas for a binomial distribution are relatively simple.

- The probability of getting exactly $X = k$ successes is given by the formula

$$P(X = k) = \binom{n}{k} p^k (1 - p)^{n-k}$$

- The expected value, or mean, is $E(X) = \mu_X = np$.
- The standard deviation is $\sigma_X = \sqrt{np(1 - p)}$.

Practice

Binomial Probabilities

P18. Suppose you flip a coin eight times.

    a. What is the probability that you'll get exactly three heads?

    b. What is the probability you get exactly 25% heads?

    c. What is the probability you get at least seven heads?

P19. You roll a pair of dice five times. Find the probability of each outcome.

    a. You get doubles exactly once.

    b. You get exactly three sums of 7.

    c. You get at least one sum of 7.

    d. You get at most one sum of 7.

P20. To test whether elephants recognize themselves in a mirror, a visible mark was placed on the forehead of Happy, a 34-year-old female Asian elephant at the Bronx Zoo. After observing the mark in the mirror, Happy touched the side of her forehead that had the visible mark with her trunk 12 times and each time came directly in contact with the mark or close to it. She never touched the other side of her forehead, which had a sham (nonvisible) mark. [Source: Joshua M. Plotnik, Frans B. M. de Waal, and Diana Reiss, "Self-Recognition in an Asian Elephant," *Proceedings of the National Academy of Sciences*, Vol. 103, 45 (November 7, 2006), pp. 17053–17057.]

    a. If Happy touches one side of her forehead at random, what is the probability she touches the side with the mark?

    b. If Happy touches one side of her forehead at random on 12 different occasions, what is the probability that she touches the side with the mark each time?

    c. What is a reasonable conclusion?

P21. On the first page of this chapter, you read that 40% of people have type A blood and that a blood bank is in dire need of two donors with type A blood. The blood bank will test 10 random donors and count the number with type A blood.

    a. What is the probability that the blood bank gets only 0 or 1 donor with type A blood?

    b. If the blood bank tests 10 random donors, will it be almost certain of finding at least two with type A blood?

Shape, Center, and Spread of a Binomial Distribution

P22. About 11% of people ages 18–24 are high school dropouts, persons who have not completed high school and are not enrolled. Suppose you pick five people at random from this age group and count the number of dropouts. [Source: *Statistical Abstract of the United States, 2009*, Table 262.]

    a. Make a probability distribution table for this situation.

    b. Make a graph of the distribution and describe its shape.

    c. What is the expected number of dropouts? What is the standard deviation?

    d. If your random sample contained five dropouts, would you have reason to doubt the government estimate that 11% of people this age are high school dropouts? Explain.

P23. According to a recent government report, 84% of drivers now use seatbelts regularly. Suppose a police officer at a road check randomly stops four cars to check for seatbelt usage. Construct the probability distribution of X, the number of drivers using seatbelts. [Source: National Highway Traffic Safety Administration, *Traffic Safety Facts*, September 2009.]

    a. Make a probability distribution of the number of drivers who wear seatbelts. Then make a graph of the distribution and describe its shape.

    b. Compute and interpret the mean of the distribution.

    c. Compute and interpret the standard deviation of the distribution.

    d. If none of the four drivers were wearing a seatbelt, would you have cause to doubt the government estimate of 84%?

Exercises

E27. Suppose you ask six people who can't tell the difference between gourmet coffee and ordinary coffee to identify which is the gourmet coffee.

   a. Make a probability distribution of the number who correctly identify the gourmet coffee. Then make a graph of the distribution and describe its shape.

   b. Compute and interpret the mean of the distribution.

   c. Compute and interpret the standard deviation of the distribution.

   d. Now suppose you ask six randomly selected people who claim they can tell the difference between gourmet coffee and ordinary coffee to identify which is the gourmet coffee and they all get it right. Do you have cause to believe that people who think they can tell the difference really can?

E28. According to a recent United States Census Bureau report, 13.2% of the population live below the poverty level. You plan to randomly sample 25 Americans. [Source: U.S. Census Bureau, 2008 American Community Survey, issued September 9, 2009, *www.census.gov*.]

   a. What is the probability that your sample will include at least two people with incomes below the poverty level?

   b. What are the expected value and standard deviation of the number of people in your sample with incomes below the poverty level?

   c. You want to see if the percentage of 13.2% holds for the region in which you live. You take a random sample of 25 people from your region. How many people in your sample would have to live below the poverty level before you have reason to believe that the percentage in your region is higher than 13.2%?

E29. According to the U.S. Census Bureau, about 15% of residents have no health insurance. You are to randomly sample 20 residents for a survey on health insurance coverage. [Source: U.S. Census Bureau, 2008 American Community Survey, issued September 9, 2009, *www.census.gov*.]

   a. What is the probability that your sample will include at least three people who do not have health insurance?

   b. Compute and interpret the expected value and standard deviation of the number of people in your sample without health insurance.

   c. If you plan to interview the residents in your sample without health insurance coverage and each interview will take 45 minutes, how many hours do you expect to spend interviewing? What is the standard deviation of the time spent interviewing?

   d. If your sample contains five or more people who do not have health insurance, do you have reason to doubt the estimate of 15% given by the Census Bureau?

E30. The median annual household income for U.S. households is about $50,000. [Source: U.S. Census Bureau, 2008 American Community Survey, issued September 9, 2009, *www.census.gov*.]

   a. What is the probability that a randomly selected household has a yearly income less than $50,000?

   b. Among five randomly selected U.S. households, find the probability that four or more have incomes less than $50,000 per year.

   c. Consider a random sample of 16 U.S. households.

   i. What is the expected number of households with annual incomes of less than $50,000 per year?

   ii. What is the standard deviation of the number of households with annual incomes of less than $50,000 per year?

   iii. What is the probability of getting at least 10 out of the 16 households with annual incomes of less than $50,000 per year?

   d. In a sample of 16 U.S. households, suppose none had annual incomes of less than $50,000 per year. What might you suspect about this sample?

E31. You neglected to study for a true-false quiz on the government of Botswana, so you will have to guess all the answers. You need to have at least 60% of the answers correct to pass. Would you rather have a 5-question quiz or a 20-question quiz? Explain your reasoning, referring to Display 6.36.

E32. You want to survey randomly selected students on your campus to estimate the proportion who voted in the last election for student body president. You think the proportion is around 50%. Use Display 6.36 to explain why it is better to have a sample of size 20 than a sample of size 5.

E33. You buy 15 lottery tickets for $1 each. With each ticket, you have a 0.06 chance of winning $10. Taking into account the cost of the tickets,

   a. what is the probability that you will gain $10 or more?

   b. what is your expected gain (or loss) on this purchase?

   c. what is the standard deviation of your gain?

E34. An oil exploration firm is to drill ten wells, each in a different location. Each well has a probability of 0.1 of producing oil. It will cost the firm $60,000 to drill each well. A successful well will bring in oil worth $1 million. Taking into account the cost of drilling,

   a. what is the firm's expected gain from the ten wells?

   b. what is the standard deviation of the firm's gain for the ten wells?

   c. what is the probability that the firm will lose money on the ten wells?

   d. what is the probability that the firm will gain $1.5 million or more from the ten wells?

E35. A home alarm system has one detector for each of the n zones of the house. Suppose the probability is 0.7 that the detector sounds an alarm when an intruder passes through its zone and that this probability is the same for each detector. The alarms operate independently. An intruder enters the house and passes through all the zones.

a. What is the probability that at least one alarm sounds if $n = 3$?

b. What is the probability that at least one alarm sounds if $n = 6$? Is the probability from part a doubled?

c. Estimate how large must n be so that the probability of at least one alarm sounding is about 0.99.

E36. Complex electronic systems for which failure could be catastrophic (such as airplane electronic systems) are built with a number of parallel backup components that automatically take over in case of failure. Suppose that at a certain key juncture there are n electronic switches (the main one and $n - 1$ backups) that operate independently of one another under standard conditions. The probability that any one switch works properly is 0.92.

a. What is the probability that at least one switch works properly if $n = 2$?

b. What is the probability that at least one switch works properly if $n = 3$?

c. An engineer wants the probability that at least one switch works properly to be $1 - 10^{-5}$, or 0.99999. Estimate the number of backup switches that she should place in the system.

E37. A potential buyer will sample DVDs from a large lot of new DVDs. If she finds at least one defective DVD, she'll reject the entire lot. In each case, find the sample size n for which the probability of detecting at least one defective DVD is 0.50.

a. 10% of the DVDs are defective.

b. 4% of the DVDs are defective.

E38. Scientists often study the conditions of populations of fish (and other animals) by tagging some and then recapturing a few tagged ones, along with some untagged ones, at a later date. Suppose a random sample of fish from a large population are tagged and then a random sample of n fish is to be selected from that population. Find n such that the probability of recapturing at least one tagged fish is 0.80 if it is estimated that

a. 5% of the population was tagged.

b. 2% of the population was tagged.

E39. The table shows a binomial distribution with $n = 1$ trial.

Number of Successes	Probability
0	$1 - p$
1	p
Total	1

a. Find the mean and standard deviation of this distribution.

b. Use the appropriate rules from Section 6.2 to show that the mean and standard deviation for the probability distribution of the number of successes in n trials are $E(X) = \mu_X = np$ and $\sigma_X = \sqrt{np(1 - p)}$.

E40. The sum of the probabilities in a binomial distribution must equal 1. To show that this is true, first let $q = 1 - p$ so that $p + q = 1$. The table shows that for a binomial distribution with $n = 1$ trial, the sum of the probabilities is 1.

Number of Successes	Probability
0	q
1	p
Total	1

a. Make a probability distribution table for the binomial situation when $n = 2$ and the probability of a success is p. Show that the sum of the probabilities is $p^2 + 2pq + q^2$. Factor this sum to show that it is equal to 1.

b. Make a probability distribution table for the binomial situation when $n = 3$ and the probability of a success is p. Find the sum of the probabilities. Factor this sum to show that it is equal to 1.

Montana Fish and Game Department tagging fish in the Yellowstone River.

Chapter Summary ◂

In this chapter, you have learned that random variables are variables with a probability attached to each possible numerical outcome. Probability distributions, like data distributions, are characterized by their shape, center, and spread. You can use formulas for the mean and standard deviation that are similar to those for relative frequency distributions of data.

The binomial distribution is a model for the situation in which you count the number of successes in a random sample of size n from a large population of successes and failures. A typical question is, "Suppose that 53% of voters approve of the job the president is doing. If you select 20 voters at random, what is the probability of getting exactly 6 who approve of the job the president is doing?"

Beginning in Chapter 8, you will learn to turn the reasoning around. A typical question will be: "Suppose that 53% of the voters in a sample of size 1000 approve of the job the president is doing. What percentages are plausible as the percentage of *all* voters who would approve?"

Review Exercises

E41. Suppose you roll a dodecahedral (12-sided) die twice. Your summary statistic will be the sum of the two rolls.

   a. What is the probability that the sum is 3 or less?

   b. Compute the mean and standard deviation of the distribution of outcomes of a single roll of one die.

   c. Compute the mean and standard deviation of the probability distribution of the sum of two rolls.

E42. Suppose you and your partner each roll two dice. Each of you computes the average of your own two rolls. The you find the difference between your average and your partner's average. Describe the probability distribution of these differences.

E43. Two different pumping systems on levees have pumps numbered 1, 2, 3, and 4, configured as in Display 6.37. In System I, water will flow from A to B only if *both* pumps are working. In System II, water will flow from A to B if *either* pump is working. Assume that each pump has this lifelength distribution: 1 month with probability 0.1, 2 months with probability 0.3, and 3 months with probability 0.6. "Lifelength" refers to the length of time the pump will work continuously without repair. Assume that the pumps operate independently of each other. (You are interested only in whether water will flow, not the amount of water flowing.)

   a. Find the distribution of the length of time water will flow for System I.

   b. Find the distribution of the length of time water will flow for System II.

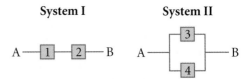

Display 6.37 Two pumping systems.

   c. Find the expected time that water will flow for each system and compare. Which system would you recommend?

E44. Suppose an insurance company wants to compute the premium to charge customers in order to break even, considering only payouts and premiums collected. Display 6.38 gives the chances that an adult (age 18 or older) living in the United States will become a victim of the given events.

Event	Rate per 1000 Adults per Year
Injury associated with stairs or steps	3.81
Injury associated with a bicycle	1.63
Injury associated with jewelry	0.28
Struck by lightning	0.0025

Display 6.38 Rates of accidents or death per 1000 adults per year. [Source: National Weather Service, 2009, *www.lightningsafety.noaa.gov/medical.htm; Statistical Abstract of the United States, 2009*, Table 193, Injuries Associated With Selected Consumer Products: 2006, *www.census.gov*.]

   a. If there are 300,000,000 people in the United States, about how many suffer each kind of accident each year?

   b. What should a randomly selected U.S. adult be charged yearly for an insurance policy that pays him or her $100,000 if he or she has an injury associated with stairs or steps?

   c. What should a randomly selected U.S. adult be charged yearly for a policy that pays him or her $10,000 if he or she is struck by lightning?

   d. The insurance company determines that people are willing to pay $50 a year for a policy that pays them a set amount if they have an injury associated with jewelry. What should this set amount be?

E45. A blood bank knows that only about 10% of its regular donors have type B blood. The technician is able to check ten donations today. What is the chance that at least one will be type B? What assumptions are you making?

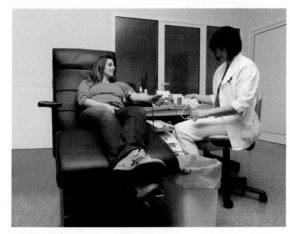

Donating blood.

E46. "One in ten high school graduates in the state of Florida sends an application to the University of Florida," says the director of admissions there. If Florida has approximately 120,000 high school graduates next year, what are the mean and standard deviation of the number of applications the university can expect to receive from Florida high school students? Do you see any possible weaknesses in using the binomial model here?

E47. In the United States, 33% of women and 26% of men aged 25 to 29 have attained a bachelor's degree or higher. [Source: *Educational Attainment in the United States: 2007, www.census.gov/Press-Release/www/releases/archives/education/011196.html.*] You take a random sample of 20 women and 20 men from your community.

　　a. What is the probability that more than half the women in the sample have attained a bachelor's degree? What assumptions are you making?

　　b. What is the probability that more than half the men in the sample have attained a bachelor's degree? What assumptions are you making?

　　c. If instead of 20 women and 20 men, you take one sample of 40 people at random, what is the approximate probability that more than half of those sampled will have attained a bachelor's degree? What assumptions are you making?

E48. About 15% of Americans have no health insurance. A polling organization randomly samples 500 Americans to ask questions about their health. [Source: U.S. Census Bureau, 2008 American Community Survey, issued September 9, 2009, *www.census.gov.*] If it costs $40 to interview a respondent with health insurance and $20 to interview a respondent without health insurance, what is the expected cost of conducting the 500 interviews?

E49. Airlines routinely overbook popular flights because they know that not all ticket holders show up. If more passengers show up than there are seats available, an airline offers passengers $100 and a seat on the next

flight. A particular 120-seat commuter flight has a 10% no-show rate.

　　a. If the airline sells 130 tickets, make a probability distribution table for the number of passengers who show up.

　　b. On average, how much will the airline pay out per flight if it sells 130 tickets?

E50. This question demonstrates one reason why the mean and variance are considered so important. Suppose you select one book at random to read from List A and one from list B. The numbers given are the pages in each book.

List A	List B
200	100
225	250
600	700

　　a. Find the total number of pages in each of the nine different possible pairs of books. Compute the standard deviation, σ, and variance, σ^2, of these nine totals.

　　b. Compute the variance of the number of pages in the three books in List A. Compute the variance of the number of pages in the three books in List B. What relationship do you see between these two variances and the variance in part a?

　　c. Does the same relationship as in part b hold for the three standard deviations?

　　d. Can you add the mean of the numbers in List A to the mean of the numbers in List B to get the mean of the nine totals?

　　e. Can you add the median of the numbers in List A to the median of the numbers in List B to get the median of the nine totals?

E51. A few years ago, Taco Bell had a scratch-off card promotion that contained four separate games. The player could play only one of them and could choose which one. The probability of winning Game A was 1/2, and the prize was a drink worth $0.55. The probability of winning Game B was 1/4, and the prize was a food item worth $0.69. The probability of winning Game C was 1/8, and the food prize was worth $1.44. The probability of winning Game D was 1/16, and the food prize was worth $1.99. Assuming you like all the prizes, which is the best game to play? This question is open to different interpretations, but one reasonable way to decide is to calculate your expected winnings on each game.

　　You can organize the information for each game in a table. For example, for Game A, the table looks like this:

Value of Prize	Probability
$0.55	0.5
$0.00	0.5

　　a. Compute the expected winnings from Games A, B, C, and D. On which game are your expected winnings greatest?

　　b. For what reasons would people choose to play the other games, even if they knew the expected values were lower?

E52. Suppose you earn $12 an hour for tutoring but spend $8 an hour for dance lessons. You save the difference between what you earn and the cost of your lessons. The number of hours you spend on each activity in a week varies independently according to the probability distributions below. Find your expected weekly savings and the standard deviation of your weekly savings.

Hours of Tutoring, y	Probability, p
1	0.3
2	0.3
3	0.2
4	0.2

Hours of Dance Lessons, x	Probability, p
0	0.4
1	0.3
2	0.3

E53. Suppose you roll a balanced die seven times. What is the probability that you will get an even number exactly two times? More than half the time?

E54. Suppose you select five numbers at random from 10 through 99, with repeats allowed. Find the probability of each outcome.

a. Exactly three of the numbers are even.

b. Exactly one of the numbers has digits that sum to a number greater than or equal to 9.

c. At least one of the numbers has digits that sum to a number greater than or equal to 9.

C1. The table gives the percentage of women who ultimately have a given number of children. For example, 19% of women ultimately have 3 children. What is the probability that two randomly selected women will have a total of exactly 2 children?

Number of children	0	1	2	3	4	5 or more
Percentage of women	18%	17%	35%	19%	7%	4%

Ⓐ 0.0289 **Ⓑ** 0.1549 **Ⓒ** 0.17 **Ⓓ** 0.34 **Ⓔ** 0.70

C2. For every 100 people who enter a summer fun run, 10 will be selected at random to get a $5 gift certificate. What is the standard deviation of the amount won for a person who enters the fun run?

Ⓐ $0.50 **Ⓑ** $1.50 **Ⓒ** $2.25 **Ⓓ** $2.50 **Ⓔ** $4.50

C3. Ray and June are the same age now. Ray's life expectancy is 78 years with a standard deviation of 4 years. June's life expectancy is 86 years with a standard deviation of 3 years. June expects to live 8 years longer than Ray, but with what standard deviation?

Ⓐ 1 year
Ⓑ 5 years
Ⓒ 7 years
Ⓓ 8 years
Ⓔ 25 years

C4. Every morning, Russell rolls a fair die. If the result is a 1 or 2, he doesn't go jogging that day, and if the result is a 3, 4, 5, or 6, he goes jogging. What is the probability that he goes jogging on exactly two of the next five days?

Ⓐ 0.111
Ⓑ 0.400
Ⓒ 0.165
Ⓓ 0.329
Ⓔ 0.671

C5. Dawna walks 10 miles starting at 1:00 p.m. every day. On average she finishes at 3:50 p.m., with a standard deviation of 10 minutes. Jeanne leaves 30 minutes later, and every day she runs 10 miles exactly twice as fast as Dawna walks on that day. What is the mean time Jeanne will finish and the standard deviation of the number of minutes it takes Jeanne to finish?

Ⓐ mean 2:25 p.m., standard deviation 5 minutes
Ⓑ mean 2:25 p.m., standard deviation 20 minutes
Ⓒ mean 2:25 p.m., standard deviation 35 minutes
Ⓓ mean 2:55 p.m., standard deviation 5 minutes
Ⓔ mean 2:55 p.m., standard deviation 35 minutes

C6. According to the World Bank, in the year 2000 approximately 35% of all the people in India spent less than $1 per day. Suppose that you repeatedly take random samples of 100 Indians and record the number who spent less than

$1 per day. What are the approximate mean and standard deviation of this distribution?

Ⓐ mean 0.35, standard deviation 0.2275
Ⓑ mean 35, standard deviation 0.2275
Ⓒ mean 35, standard deviation 0.4770
Ⓓ mean 35, standard deviation 4.7697
Ⓔ mean 35, standard deviation 22.75

Investigations

C7. Random sampling from a large lot of a manufactured product yields a number of defectives, X, with an approximate binomial distribution, with p being the true proportion of defectives in the lot. A sampling plan consists of specifying the number, n, of items to be sampled and an acceptance number, a. After n items are inspected, the lot is accepted if $X \le a$ and is rejected if $X > a$.

   a. For $n = 5$ and $a = 0$, calculate the probability of accepting the lot for values of p equal to 0, 0.1, 0.3, 0.5, and 1.

   b. Graph the probability of lot acceptance as a function of p for this plan. This is the *operating characteristic curve for the sampling plan.*

Now, a quality-control engineer is considering two different lot acceptance sampling plans:

$$(n = 5, a = 1) \text{ and } (n = 25, a = 5).$$

   c. Construct operating characteristic curves for both plans.

   d. If you were a seller producing lots with proportions of defectives between 0% and 10%, which plan would you prefer?

   e. If you were a buyer wishing high protection against accepting lots with proportions of defectives over 30%, which plan would you prefer?

C8. Nationwide, 57% of undergraduate students are women. [Source: *The Condition of Education, 2009*, National Center for Education Statistics.] Suppose you start selecting undergraduates at random until you get a man.

   a. What is the probability that the first undergraduate is a man?

   b. What is the probability that the first man is the second undergraduate you select?

   c. What is the probability that the first man is the third undergraduate you select?

   d. Write a formula that gives the probability that the first man is the nth undergraduate you select.

   e. Use your formula to construct a probability distribution for $n = 1, 2, 3, 4, 5, 6, 7,$ and 8. Graph this distribution, which is an example of a *geometric distribution*, and describe its shape.

Chapter 7
Sampling Distributions

What would happen if you could take random samples over and over again from your population? Sampling distributions show how much your results might vary from sample to sample, as when you use a random sample of only 40 people to estimate what percentage of the world's population has a cell phone.

You have studied methods that are good for exploration and description, but for inference—going beyond the data in hand to conclusions about the population or the process that created the data—you need to collect the data by using a random sample (a survey) or by randomly assigning treatments to subjects (an experiment). The promise of Chapter 4 was that if you used those methods to produce a data set, you then could use your data to draw sound conclusions. Randomized data production not only protects against bias and confounding but also makes it possible to imagine repeating the data production process so that you can estimate how the summary statistic you compute from the data would vary from sample to sample. To oversimplify, but only a little, a sampling distribution is what you get by repeating the process of producing the data and computing the summary statistic.

In this chapter, you will learn

▶ to generate sampling distributions of common summary statistics such as the sample mean and the sample proportion

▶ to describe the shape, center, and spread of the sampling distributions of common summary statistics without actually generating them

▶ to use the sampling distribution to determine which results are reasonably likely and which would be considered rare

7.1 ▶ Generating Sampling Distributions

Sometimes it is possible to imagine repeating the process of data collection. A **sampling distribution** is the distribution of the summary statistics you get from taking repeated random samples. For example, Chapter 1 contained a case study of possible age discrimination in layoffs at the Westvaco Corporation. In one round of the downsizing, three workers were chosen from ten workers to be laid off. The ages of all ten workers were

A sampling distribution answers the question, "How does my summary statistic behave when I repeat the process many times?"

$$25 \quad 33 \quad 35 \quad 38 \quad 48 \quad 55 \quad 55 \quad 55 \quad 56 \quad 64$$

The ages of the three workers actually laid off were 55, 55, and 64. To decide whether there is any evidence of possible age discrimination, the actual layoffs must be compared to what might happen from an age-blind process of selecting people for lay off.

There is no process more age-blind than selecting three workers at random. The distribution in Display 7.1 shows the results of selecting three different workers at random, computing their mean age (the summary statistic), and repeating the process until there are hundreds of sample means.

The mean age of the three workers actually laid off is $\frac{55 + 55 + 64}{3}$, or 58. As you can see from Display 7.1 (on page 314), it's rather hard to get an average age that large just by chance. While this doesn't prove there was age discrimination, it does mean that Westvaco had some explaining to do.

The Westvaco analysis illustrates the four steps in using simulation to generate an approximate sampling distribution of the mean age of three workers:

1. Take a random sample of a fixed size n from a population.
2. Compute a summary statistic.
3. Repeat steps 1 and 2 many times.
4. Display the distribution of the summary statistics.

These steps describe how to use simulation to make a *simulated sampling distribution*. Alternatively, you could construct an exact sampling distribution by listing all possible samples of 3 workers who could have been laid off from among the 10 workers. There would be

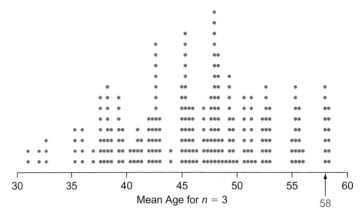

Display 7.1 A simulated sampling distribution of the mean age from random samples of three people who could have been laid off at Westvaco.

$$\binom{n}{k} = \frac{n!}{k!(n-k)!}$$

$\binom{10}{3} = {}_{10}C_3 = 120$ samples in your list. Then, for each of the 120 samples, compute the mean age. The distribution of the 120 sample means is called the *sampling distribution*.

Using a Sampling Distribution

In the Westvaco analysis, a sampling distribution was used to compare what actually did happen in the layoffs to what should have happened if the workers had been selected at random. A sampling distribution also is useful when you want to estimate the mean of a population by using the mean of a random sample taken from that population, as illustrated in Example 7.1.

Example 7.1	**Estimating the Mean NBA Salary**

The salaries of players in the National Basketball Association are public information. The distribution of these salaries, shown in Display 7.2, is highly skewed with mean $4.6 million and standard deviation $4.7 million.

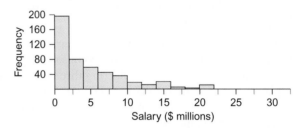

Display 7.2 Salaries of players in the NBA, 2008–2009.

Now suppose that the salaries were not made public. An agent representing an average player intends to estimate the mean salary by getting a random sample of ten players and finding their salaries. The agent knows that the population of salaries must be highly skewed, and so is concerned about whether the sample mean will be a good estimate of the population mean. Should he be concerned?

Solution
Because you have the population of all salaries, you can determine whether the agent should be concerned. Take random samples of size 10, compute the sample mean for each sample, and display the resulting values, as in Display 7.3. This simulated sampling distribution shows the sample means from 200 random samples, each of size 10.

The distribution of sample means is almost mound-shaped. The sample means are centered at about the population mean of $4.6 million and have a standard deviation of about $1.5 million. From what you know about the normal distribution, this means that the agent can be assured that he has a 95% chance that the sample mean from a random sample of size 10 will be within about $3 million of the population mean.

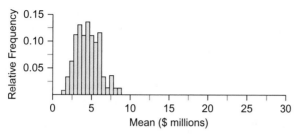

Display 7.3 Simulated sampling distribution of the means of 200 samples of size 10 taken from NBA player salaries.

Example 7.1 shows how sampling distributions are useful in practice. The agent doesn't really have the entire population—that's why he needs to take a sample to get an estimate of the population mean. But because he has some idea about the shape, center, and spread of the sampling distribution of sample means, he can make some estimates about how close his sample mean is likely to be to the population mean.

There are now two quite different standard deviations in the picture—the standard deviation of the population, $4.7 million, and the smaller standard deviation of the sampling distribution, $1.5 million. To tell them apart, these two types of standard deviation have different names. The standard deviation of the population, σ, is called, naturally enough, the *population standard deviation*. The standard deviation of the sampling distribution is commonly called the **standard error,** or *SE* for short.

> Two standard deviations: standard deviation of the population and standard error of the sampling distribution.

Computing the Population Standard Deviation

When computing the population standard deviation, σ, divide by the size of the population, N, rather than $N - 1$:

$$\sigma = \sqrt{\frac{\sum(x - \mu)^2}{N}}$$

Here, μ is the mean of the N values of x in the population.

Values that lie in the middle 95% of a sampling distribution are called **reasonably likely events**. For example, from Display 7.3, getting a mean salary of $3 million is a reasonably likely event. Values that lie in the lower 2.5% or in the upper 2.5% of a sampling distribution are called **rare events**. For example, getting a mean salary of $8 million would be a rare event. When the distribution is normal, rare events are those values that lie more than 1.96 standard deviations from the mean.

> Two events: reasonably likely and rare.

Notice that if the agent selects one salary at random, he wouldn't be at all surprised to get one as high as $8 million. You can see from Display 7.2 that about 20% of the salaries are $8 million or more. However, if the agent selects 10 salaries at random, he should be very surprised to get a mean as high as $8 million. From Display 7.3, a mean that high happened only a couple of times out of 200 samples.

DISCUSSION
Using a Sampling Distribution

D1. As you saw in Example 7.1, the agent has a 95% chance that his sample mean will be within about $3 million of the population mean. Suppose $3 million either way isn't precise enough for him. What should he do?

D2. Compare Displays 7.2 and 7.3.

    a. Why is the largest value in Display 7.2 larger than the largest value in Display 7.3? Why is the smallest value smaller?

    b. Would you expect the means of the distributions in Display 7.2 and 7.3 to be equal? Why or why not?

    c. Would you expect the two standard deviations to be equal? Why or why not?

    d. How does the shape of the sampling distribution of sample means for samples of size 10 compare to the shape of the distribution of the population of salaries?

D3. In the sampling distribution of sample means in Display 7.3, approximately what values of the sample mean for samples of size 10 would be reasonably likely? Which would be rare events?

Exact Sampling Distributions

When the population size is very small, you can construct sampling distributions exactly by listing all possible samples.

Example 7.2	Utah's National Parks

Simulation isn't necessary here.

Utah has five national parks. Your company has been hired to search each park in an effort to locate all natural bridges and arches. You decide to select two parks at random and begin work on them. Use Display 7.4 to construct the sampling distribution for the total number of square miles in the two parks on which you begin work. You think that you can search 600 square miles in a year. What is the probability that you don't finish the first two parks in your first year?

National Park	Area (sq mi)
Arches (A)	119
Bryce Canyon (B)	56
Canyonlands (C)	527
Capitol Reef (R)	378
Zion (Z)	229

Display 7.4   Sizes of Utah's national parks.　　Delicate Arch, Arches National Park.

Solution

There are $\binom{5}{2} = {}_5C_2$, or 10, equally likely ways to select two national parks, as shown in Display 7.5. The dot plot shows the sampling distribution of the sum. The probability that the two parks have more than 600 square miles, and so you don't finish the first two parks this year, is 4/10.

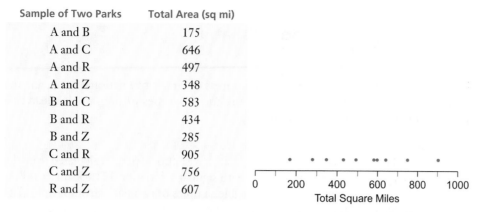

Sample of Two Parks	Total Area (sq mi)
A and B	175
A and C	646
A and R	497
A and Z	348
B and C	583
B and R	434
B and Z	285
C and R	905
C and Z	756
R and Z	607

Display 7.5 Total number of square miles in two of Utah's national parks.

Properties of Point Estimators

Suppose you want to know the mean of the NBA salaries plotted in Display 7.2. Because of time or money constraints, you cannot find them all. You know what to do: Take a random sample from the population of salaries; compute the sample mean salary, $\bar{x}$; use $\bar{x}$ as your estimator of the population mean salary, μ. This process seems so natural that it is easy to overlook just how powerful it is. Two very fortunate things are happening in the background:

- Although an individual sample mean, $\bar{x}$, usually is smaller or larger than the population mean, μ, in the long run, over many samples, the average value of $\bar{x}$ turns out to be exactly equal to μ.

- The value of $\bar{x}$ tends to be pretty close to the value of μ, and, as you will see in the next section, the larger the sample size, the closer to μ it tends to be.

Even though in practice you can't repeat the sampling process, it is nice to know whether the estimator you are using generally works out well. For example, from Display 7.3, you can see that while a single estimate $\bar{x}$ from a sample of size 10 might be too small or too large, the estimates are centered around the population mean. Display 7.3 also gives you an idea of how close your estimate is likely to be to the population mean if you sample only ten salaries.

When you use a summary statistic from a sample to estimate a population parameter, such as using the sample mean as an estimate of the population mean, the summary statistic is called a **point estimator**. Two desirable properties of point estimators are given next.

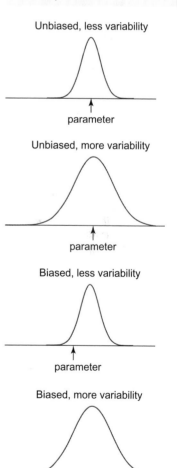

μ is the population mean. $\bar{x}$ is the sample mean.

Unbiased, less variability

↑ parameter

Unbiased, more variability

↑ parameter

Biased, less variability

↑ parameter

Biased, more variability

↑ parameter

Desirable Properties of Point Estimators

When you use a summary statistic from a random sample to estimate a parameter of a population, you would like the summary statistic to have two properties:

- The summary statistic should be **unbiased**. That is, the mean of the sampling distribution is equal to the value you would get if you computed the summary statistic using the entire population. More formally, an estimator is unbiased if its expected value equals the parameter being estimated.

- The summary statistic should have as little variability as possible (be more *precise* than other estimates) and should have a standard error that decreases as the sample size increases.

| Example 7.3 | The Sample Mean as an Estimator of the Population Mean |

Use the data in Display 7.4, which gives the areas of Utah's five national parks, and the samples of size 2 in Display 7.5 to illustrate the properties of the sample mean as a point estimator of the population mean.

Solution
The mean of the population of the areas of Utah's five national parks, calculated from the areas in Display 7.4 is 261.8 square miles, with standard deviation 171.85 square miles.

Display 7.6 shows each of the 10 possible samples of two parks from Display 7.5. It also gives the total area of the two parks and the mean area of the two parks.

Sample of Two Parks	Total Area (sq mi)	Mean Area (sq mi)
A and B	175	87.5
A and C	646	323.0
A and R	497	248.5
A and Z	348	174.0
B and C	583	291.5
B and R	434	217.0
B and Z	285	142.5
C and R	905	452.5
C and Z	756	378.0
R and Z	607	303.5
	Mean	261.8

Display 7.6 Total area and mean area for all possible samples of two Utah national parks.

The mean of the ten sample means that make up this sampling distribution is 261.8 square miles, exactly the same as the mean of the original five areas. This illustrates that, in random sampling, the sample mean is an unbiased estimator of the population mean.

The standard error of the sampling distribution of means is 105.23, much smaller than the standard deviation of the five population values, 171.85. This illustrates that the standard error of the sampling distribution of the sample mean is smaller with a sample size of 2 than with a sample size of 1 (the population). The two dot plots in Display 7.7 have the same mean, but the second has a smaller spread.

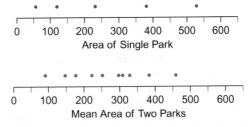

Display 7.7 Dot plots of the areas of the five national parks of Utah and of the mean area of all ten samples of two national parks.

Not all estimators are unbiased or have a relatively small standard error, as Example 7.4 illustrates.

The Sample Maximum as an Estimator of the Population Maximum

Example 7.4

Your observatory is given the names of 31 galaxies. Your job is to estimate the distance to Earth from the farthest of the 31 galaxies. You have the resources to measure the distance of only 10 galaxies, which you will select at random. Unbeknownst to you, the distances, in megaparsecs (Mpc), are

0.008	0.76	0.81	7.2	7.5	9.7	10.6	11.2	11.4	11.6
11.7	13.2	15.0	15.0	15.3	15.5	15.7	16.1	16.1	16.8
16.8	20.9	22.9	22.9	24.1	25.9	26.2	29.2	31.6	58.7	93

The maximum distance, 93 Mpc, is the population parameter you hope to estimate.

Because you want to estimate the maximum distance from Earth, you decide to use the maximum distance in the sample as your summary statistic. For example, if your sample of size 10 is

NASA image of the M81 spiral galaxy.

0.81	7.5	11.6	11.7	13.2	16.1	16.8	24.1	26.2	58.7

you will estimate the population maximum to be 58.7 Mpc. This is smaller than the maximum in the population of 93 Mpc, so you have underestimated.

Display 7.8 shows a simulated sampling distribution of the maximum for samples of size 10. What tends to happen when you use the maximum of the sample as an estimator of the maximum of this population?

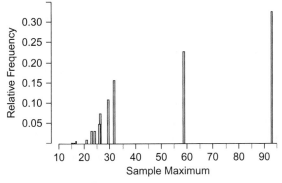

Display 7.8 Simulated sampling distribution of the maximum for samples of size 10 taken from the population of 31 galaxies.

Solution

The population maximum is 93 Mpc, and you have a probability of about 0.32 of getting that as your estimate. In the rest of the samples, the estimates all would be too small.

The mean of the sampling distribution is only 56 Mpc. If you could repeat your process of estimating the population maximum by using the maximum in the sample, on average your estimate would be too small. In other words, the sample maximum is a biased estimator of the population maximum. That's not too surprising, because the maximum of a sample can never be larger than the maximum of its population.

The sample maximum can never be too big!

Further, there is a great deal of variability in the sample maximum. Your estimate of the population maximum might be as high as 93 Mpc (the population maximum), but about half the time your estimate will be less than 32 Mpc, way too small. Using the sample maximum as an estimator of the population maximum doesn't work out very well.

STATISTICS IN ACTION 7.1 ▶ The Return of the Random Rectangles

What You'll Need: a copy of Display 4.1, a method of producing random digits

In this activity, you will construct simulated sampling distributions of various summary statistics using the population of rectangles in Display 4.1 on page 173. The distribution in Display 7.9 shows the areas of all 100 rectangles in this population, which have a mean of 7.42 square units and a standard deviation of 5.20 square units.

1. Generate five distinct random numbers between 00 and 99.

2. Find the rectangles in Display 4.1 that correspond to your random numbers. (The rectangle numbered 100 can be called 00.) This is your random sample of five rectangles.

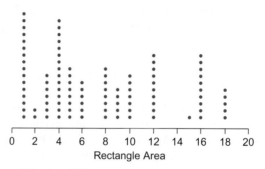

Display 7.9 Areas of all 100 rectangles.

3. Determine the areas of the rectangles in your random sample and find the sample median. (Save these five areas for step 7.)

4. Combine your sample median with those of other students in the class, and repeat steps 1 through 3 until you have 200 sample medians. Construct a plot showing the distribution of the 200 sample medians.

5. Describe the shape, mean, and standard error of this distribution of sample medians. How do these compare to the shape, mean, and standard deviation of the population of areas above?

6. Repeat steps 4 and 5, but this time your summary statistic will be the maximum area of the five rectangles in each sample. For example, if the areas in your sample are 1, 8, 4, 8, and 3, the maximum area is 8. (You can use the same samples as before.)

7. Compare the population in the dot plot above to your simulated sampling distributions.

   a. What is the median of the population of 100 rectangle areas? Is the sample median a good estimator of the median of the areas of all 100 rectangles?

   b. What is the maximum of the population of 100 rectangle areas? Is the sample maximum a good estimator of the largest area of the 100 rectangles?

DISCUSSION
Properties of Point Estimators

D4. How are the concepts of a biased estimator and of bias in sampling similar? How are they different?

Summary 7.1: Generating Sampling Distributions

A sampling distribution answers the question, "How would my summary statistic behave if I could repeat the process of collecting data using random samples?"

In subsequent chapters, you will compare the summary statistic you observed in a random sample to the sampling distribution for that statistic from a known population. If the observed summary statistic doesn't fit comfortably with those in the sampling distribution, you will conclude that your sample belongs to a different population.

You can generate a simulated sampling distribution of any sample statistic by following these steps:

1. Take a random sample of size n from a population.
2. Compute a summary statistic.
3. Repeat steps 1 and 2 many times.
4. Display the distribution of the summary statistics.

You can sometimes create an exact sampling distribution by listing all possible samples. In many more situations, you will be able to describe the shape, mean, and standard error of the sampling distribution without simulation and without listing samples. You will learn how to do this for sample means and sample proportions in Sections 7.2 and 7.3.

Here are several key facts about sampling distributions:

- A simulated, or approximate, sampling distribution is the distribution of the sample statistic for a large number of repeated random samples.
- Sampling distributions, like probability distributions and data distributions, are best described by shape, center, and spread.
- Sampling distributions don't necessarily have the same shape as the population from which the random samples were taken.
- If the mean of the sampling distribution is equal to the population parameter being estimated, then the summary statistic you are using is an unbiased estimator of that parameter.
- The standard deviation of a sampling distribution is called the "standard error."
- For normal distributions, reasonably likely outcomes are those that fall within approximately two standard errors of the mean.

Practice

Using a Sampling Distribution

P1. The following table gives the number of moons for each planet in our solar system. [Source: NASA, *solarsystem.nasa.gov*.] As one of the summer interns working for NASA, you are given three different planets, selected at random, and are to write a description of each of the planets' moons for NASA's web site.

Planet	Number of Moons
Mercury	0
Venus	0
Earth	1
Mars	2
Jupiter	63
Saturn	56
Uranus	27
Neptune	13

a. What is the smallest number of moons that you might have to describe? The largest?

b. Display 7.10 (on the next page) shows the sampling distribution for the total number of moons you might have to describe.

There are $\binom{8}{3} = {}_8C_3 = 56$ different samples of three moons. Locate the point on the sampling distribution that represents the sample of Saturn, Uranus, and Neptune.

c. What is the probability that you will have to describe three moons or fewer?

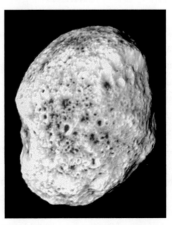

Hyperion, a moon of Saturn.

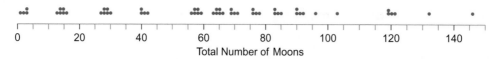

Display 7.10 Sampling distribution of the total number of moons for three randomly selected planets.

P2. Display 7.11 shows the distribution of exam scores for 192 students in an introductory statistics course at the University of Florida. From this population, simulated sampling distributions were generated for the minimum, maximum, and median for samples of size 10. Match each histogram in Display 7.12 to its summary statistic.

  I. the minimum

  II. the maximum

  III. the median

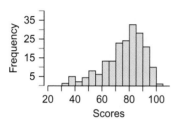

Display 7.11 A distribution of exam scores.

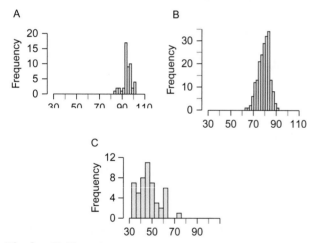

Display 7.12 Histograms of different summary statistics from random samples of size 10 of statistics exam scores.

Exact Sampling Distributions

P3. Every year, *Forbes* magazine releases a list of the top-earning dead celebrities. In 2008, the top six and their yearly earnings were

  1. Elvis Presley, $52 million
  2. Charles M. Schulz, $33 million
  3. Heath Ledger, $20 million
  4. Albert Einstein, $18 million
  5. Aaron Spelling, $15 million
  6. Dr. Seuss (Theodor Geisel), $12 million

[Source: *www.forbes.com.*]

Your talent agency gets an opportunity to represent two of these dead celebrities, to be selected at random. You will be paid 10% of their earnings.

  a. What is the most you could be paid? The least?

  b. Construct the sampling distribution of your total possible earnings.

  c. What is the probability that you will be paid $7 million or more?

Properties of Point Estimators

P4. A quality control plan adopted by the painting department of an automobile repair facility calls for randomly sampling two cars out of every six painted and inspecting them thoroughly for paint defects. The mean number of defects per sampled car is plotted on the plant's quality control chart as a long-run check on quality. For simplicity, suppose the next six cars painted have 0, 1, 2, 3, 4, and 5 paint defects, respectively. These six values form the population.

  a. Find the mean and standard deviation of the six values.

  b. List the 15 samples of size 2 that can be selected from this population of six cars, and calculate the mean number of defects in each of the samples.

  c. Verify that the sample mean is an unbiased estimator of the population mean. (In other words, verify that the mean of the sampling distribution of means is equal to the mean of the population.)

  d. Verify that the standard error of the sampling distribution of means is smaller than the standard deviation of the population.

P5. Refer to Example 7.2 on page 316. Your company's contract has changed. You now are to estimate the range (largest minus smallest) of the areas of Utah's national parks. Your procedure will be to select three of the five parks at random and use the range of their three areas as the estimate.

  a. What is the value of the population parameter that you are estimating?

  b. Refer to Display 7.5 and complete a copy of the table in Display 7.13.

  c. Construct a plot of the exact sampling distribution of the sample range. What is the mean of this sampling distribution?

  d. Is the sample range an unbiased estimator of the population range? Explain.

  e. Is there much variability in the estimates?

Sample of Three Parks	Range of the Areas (sq mi)
A, B, C	527 − 56 = 471
A, B, R	378 − 56 = 322
⋮	⋮

Display 7.13 Range of areas of three randomly chosen national parks in Utah.

P6. Three very small populations are given, each with a mean of 30.

  I. 10 50

  II. 10 20 30 40 50

  III. 20 30 40

a. Match each population to the sampling distribution of the sample mean (Display 7.14) for a sample of size 2 (taken with replacement).

b. What is the mean of each sampling distribution?

c. Which has the smallest standard error?

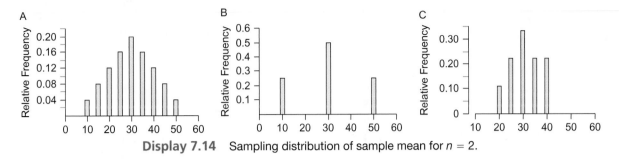

Display 7.14 Sampling distribution of sample mean for n = 2.

Exercises

E1. In Chapter 4, you read about Kelly Acampora's hamster experiment. The four hamsters raised in short days (with long nights) had enzyme levels

> 12.500 11.625 18.275 13.225

The four hamsters raised in long days (with short nights) had enzyme levels

> 6.625 10.375 9.900 8.800

You'll use a sampling distribution to decide whether this is persuasive evidence that hamsters raised in short days have higher enzyme levels than hamsters raised in long days.

a. Your summary statistic will be the mean enzyme level of hamsters raised in short days, $\bar{x}$. What is the value of the summary statistic $\bar{x}$ for Kelly's experiment?

b. Suppose the length of a day makes no difference in enzyme levels; that is, suppose the eight enzyme levels would have been the same no matter which treatment the hamster received. If Kelly had randomly assigned the hamsters to the long and short days as follows, what would the value of the summary statistic $\bar{x}$ have been?

| Short days | 6.625 | 10.375 | 13.225 | 8.800 |
| Long days | 12.500 | 11.625 | 18.275 | 9.900 |

c. Display 7.15 shows the sampling distribution of $\bar{x}$ for all 70 possible random assignments of the eight hamsters to the two treatments. What is the probability of getting a value of $\bar{x}$ as large as that from the actual experiment?

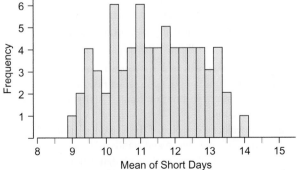

Display 7.15 Sampling distribution of the mean for short days for all 70 possible random assignments of treatments to hamsters.

d. Is Kelly's sample statistic of $\bar{x}$ a reasonably likely outcome if the length of a day makes no difference in the enzyme level? What can you conclude?

E2. The histogram in Display 7.16 shows the distribution of exam scores for 192 students in an introductory statistics course at the University of Florida.

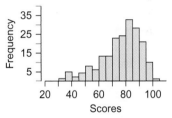

Display 7.16 A distribution of exam scores.

a. Match each histogram in Display 7.17 to its description.

   I. a simulated sampling distribution of the mean of the scores of 100 random samples of 4 students

   II. a simulated sampling distribution of the mean of the scores of 100 random samples of 30 students

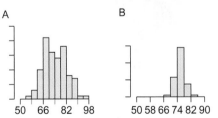

Display 7.17 Histograms for simulated sampling distributions of exam scores.

b. The second-hour class of 30 students, whose exam scores are part of the data set, had a class average of 86 on this exam. The instructor says that this class is just a random assortment of typical students taking this course. Do you agree?

E3. The areas of all 50 U.S. states, in millions of acres, are given in Display 7.18.

a. Describe the shape of the distribution.

b. Select a random sample of five areas. What is the mean of your five areas?

State	Area (millions of acres)	State	Area (millions of acres)
AL	33.1	MT	94.1
AK	366.0	NC	33.7
AR	34.0	ND	45.2
AZ	73.0	NE	49.5
CA	101.6	NH	5.9
CO	66.6	NJ	5.0
CT	3.2	NM	77.8
DE	1.3	NV	70.8
FL	37.5	NY	31.4
GA	37.7	OH	26.4
HI	4.1	OK	44.8
IA	36.0	OR	62.1
ID	53.5	PA	29.0
IL	36.1	RI	0.8
IN	23.1	SC	19.9
KS	52.6	SD	49.4
KY	25.9	TN	27.0
LA	30.6	TX	170.8
MA	5.3	UT	54.3
MD	6.7	VA	26.1
ME	21.3	VT	6.1
MI	37.4	WA	43.6
MN	54.0	WI	35.9
MO	44.6	WV	15.5
MS	30.5	WY	62.6

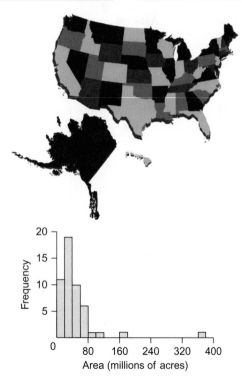

Display 7.18 Areas of the 50 U.S. states in millions of acres (not including inland bodies of water). [Source: *World Almanac and Book of Facts 2001*, p. 621.]

c. The plot in Display 7.19 shows the mean area for 25 random samples, each of size 5, just like the sample you generated in part b. Compare this simulated sampling distribution to the population.

d. From what particular sample could the largest value in the plot in Display 7.19 have come?

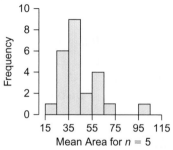

Display 7.19 A simulated sampling distribution of the sample mean of five state areas.

E4. The mean and the median are only two of many possible measures of center. Another is the midrange, defined as the midpoint between the minimum and the maximum in a data set. The rectangles in Display 4.1 on page 173 have areas ranging from 1 to 18, so the population midrange is $\frac{1+18}{2}$, or 9.5. The histogram in Display 7.20 shows a simulated sampling distribution of the midrange based on 1000 random samples of size 5 from the population of 100 rectangles. The mean of this distribution is 8.05, and the estimated standard error is 2.17.

a. Take a random sample of five rectangles; compute the midrange of their areas.

b. Describe the sampling distribution of the midrange. How does it compare to the sampling distribution of the sample mean in the lower plot in Display 4.2 on page 174?

c. What advantages or disadvantages do you see in using the midrange as a measure of center for the rectangle area population?

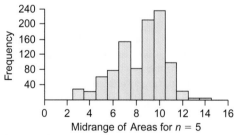

Display 7.20 A sampling distribution of the sample midrange ($n = 5$).

E5. The five tennis balls in a can have diameters 62, 63, 64, 64, and 65 mm. Suppose you select two of the tennis balls at random, without replacing the first before selecting the second.

a. Construct a dot plot of the five population values and find their mean and standard deviation.

b. List all possible sets of size 2 that can be chosen from the five balls. There are $\binom{5}{2} = {_5}C_2$, or 10, possible sets of two balls.

c. List all 10 possible sample means, and construct a dot plot of the sampling distribution of the sample mean. Compute the mean and standard error of this distribution. Compare these to the mean and standard deviation of the population.

d. Construct a dot plot of the sampling distribution of the maximum diameter for samples of size 2.

e. Construct a dot plot of the sampling distribution of the range (*maximum* minus *minimum*) for samples of size 2.

E6. As part of a statistics project at Iowa State University, a student tested how well a bike with treaded tires stopped on concrete. In six trials, these lengths of skid marks (in centimeters) were produced: 365, 374, 376, 391, 401, and 402. Suppose instead that the student had done only three trials, which could have been any three selected from this population of six, each with equal probability of being selected. [Source: Stephen B. Vardeman, *Statistics for Engineering Problem Solving* (Boston: Prindle, Weber & Schmidt, 1994), p. 349.]

a. Construct a dot plot of the six population values and find its mean.

b. List all possible sets of three skid lengths the student could get. There are $\binom{6}{3} = {}_6C_3$, or 20, possible sets.

c. Construct a dot plot of the sampling distribution of the sample mean. Find its mean. Is the mean equal to the mean of the population?

d. Construct a dot plot of the sampling distribution of the sample median. Find its median. Is it equal to the median of the population?

e. Construct a dot plot of the sampling distribution of the sample minimum. Describe the shape of this distribution. Is the sample minimum a good estimator of the population minimum of all six lengths? Explain.

E7. Refer to your exact sampling distributions in E5. For each of the following summary statistics, explain why the summary statistic is or is not an unbiased estimator of the population parameter. If it is a biased estimator, does it tend to be too big or too small, on average?

a. sample mean (see part c)

b. sample maximum (see part d)

c. sample range (see part e)

E8. Refer to your exact sampling distributions in E6. For each summary statistic here, explain why the summary statistic is or is not an unbiased estimator of the population parameter. If it is a biased estimator, does it tend to be too big or too small, on average?

a. sample mean (see part c)

b. sample median (see part d)

c. sample minimum (see part e)

E9. Joel has heard that the sample range is a biased estimator of the population range. He tests this using the population of numbers {0, 1, 2, 3, . . . , 98, 99, 100} and sample sizes of 8, taken with replacement.

a. What is the value of the population parameter Joel is trying to estimate?

b. The distribution of sample ranges from Joel's tests is given in Display 7.21. Use this distribution to explain why the sample range is a biased estimator of the population range.

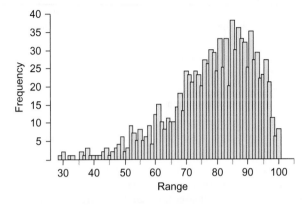

Variable	Mean	Median	SD	Min	Max	Q1	Q3
Range	78.725	81	13.52	29	100	71	89

Display 7.21 Simulated sampling distribution of the range for samples of size 8.

c. Joel has a new idea. Instead of using the sample range to estimate the population range, he will double the sample interquartile range (*IQR*) and use that as an estimate of the population range. To test his estimator, Joel takes 10,000 samples of size 8, computes the *IQR* of each one, and doubles it. His results are shown in Display 7.22. Does Joel's estimator of the range appear to be biased or unbiased? Do you have any other concerns about it?

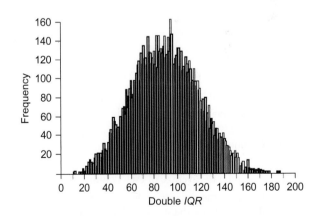

Variable	Mean	Median	SD	Min	Max	Q1	Q3
Double IQR	89.82	89	28.00	11	187	70	109

Display 7.22 Simulated sampling distribution of 2 · *IQR* for samples of size 8.

E10. An inspector is called to see how much damage has been caused to a warehouse full of frozen fish by a recent power failure. The fish are stored in cartons containing 24 one-pound packages. There are hundreds of cartons in the warehouse.

The inspector decides he needs a sample of 48 one-pound packages of fish in order to assess the damage. He selects two cartons at random and treats this as a sample of 48 one-pound packages. The inspector doesn't realize it, but if one package of fish in a carton spoils, very rapidly the whole carton spoils. So it is safe to assume that each carton in the warehouse is either completely spoiled or not spoiled at all.

a. If the inspector is using pounds of spoiled fish in the sample as a statistic, what will the sampling distribution of the statistic look like? Describe this distribution as completely as you can.

b. Did the inspector choose a good sampling plan? If not, explain how he could have chosen a better one.

For E11–E12: In Chapter 2, you might have wondered why the denominator of the sample standard deviation is $n - 1$ rather than n. Now that you know about sampling distributions, you can use them to answer this question in true statistical fashion. If a sample standard deviation is to be a good estimator of the population standard deviation, σ, its sampling distribution should be centered at σ, or at least near that value. In E11 and E12, you will see whether that turns out to be the case.

E11. Suppose you have a population that consists of only the three numbers 2, 4, and 6. You take a sample of size 2, replacing the first number before you select the second. Use parts a–e to complete the table in Display 7.23.

a. In the first column, list all nine possible samples. (Consider 2, 4 as different from 4, 2; that is, order matters.)

b. Compute the mean of each sample and write it in the second column.

Sample	Mean	Variance, Dividing by $n = 2$	Variance, Dividing by $n - 1 = 1$
2, 2	2	0	0
2, 4	3	1	2
⋮	⋮	⋮	⋮
Average	—?—	—?—	—?—

Display 7.23 Variance calculations.

c. Compute the variance of each sample, dividing by $n = 2$, and enter it in the third column. (You should be able to do this in your head.)

d. Compute the variance of each sample, dividing by $n - 1 = 1$, and enter it in the fourth column.

e. Compute the average of each column.

f. Compute the variance of the population {2, 4, 6} using the formula for the standard deviation on page 315. Compare it to your answers in part e.

g. Do you get an unbiased estimate of the variance of the population if you divide by n or by $n - 1$? Explain how you know.

h. If you divide by n rather than by $n - 1$, does the variance tend to be too large or too small, on average?

E12. The rectangle areas of Display 4.1 on page 173 have a population mean area of 7.42, and a population standard deviation, σ, of 5.20.

a. Take a random sample of five rectangles. Compute the standard deviations of the areas of these rectangles, dividing by $n - 1$, or 4. Then compute the standard deviation again, but divide by n, or 5. Which standard deviation is larger?

b. The first histogram in Display 7.24 shows the sample standard deviations from 1000 random samples of five rectangles. The mean of these standard deviations is 5.02. If you divide by 5 rather than by 4 (that is, by n instead of $n - 1$), the results produce the distribution shown in the second histogram. The mean of this distribution is 4.49. Explain why dividing by $n - 1$ gives a better estimate of the population standard deviation.

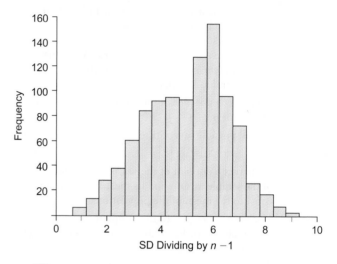

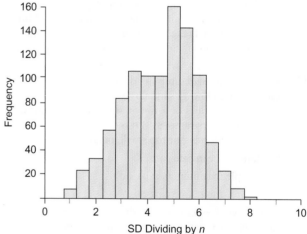

Display 7.24 Simulated sampling distributions of sample standard deviation ($n = 5$).

E13. As you saw in Example 7.4 on page 319, if you use the sample maximum as an estimate of the population maximum, it tends to be too small. Here is one possible rule for adjusting it:

$$(population\ maximum) \approx \frac{n + 1}{n}\ (sample\ maximum)$$

As always, n is the sample size. This rule works best when the population is uniform.

a. Suppose you take a random sample of four numbers from the uniform distribution on $[0, N]$, where N is the population maximum. On average, what size gap would you expect between two adjacent numbers in your sample? The dot plot in Display 7.25 might help. It shows a sample of size 4 taken from a uniform distribution with minimum 0 and maximum N.

b. Where would you expect the maximum of a sample of size 4 to be?

c. If you know the sample maximum of a sample of size 4 taken from a uniform distribution on $[0, N]$, what would you expect N to be?

d. Test how well this estimator works using the sample of size 10 from Example 7.4 on page 319: 0.81 7.5 11.6 11.7 13.2 16.1 16.8 24.1 26.2 58.7.

Display 7.25 Four numbers randomly chosen from a uniform distribution with minimum 1.

E14. With symmetric or nearly symmetric distributions, the variation in the sampling distribution of the sample mean is smaller than the variation in the sampling distribution of the sample median. Suppose the distribution for your population is highly skewed (for example, the areas of the U.S. states given in E3).

a. Might the sampling distribution of the sample median have less variation than the sampling distribution of the sample mean? Explain your reasoning.

b. Might the shape of the sampling distribution of the sample median be more normal than that of the sample mean?

c. The plots in Display 7.26 show simulated sampling distributions for random samples of size 10 taken from the 50 areas of the states. Does this simulation help confirm your answers to parts a and b? Do they prove your answers?

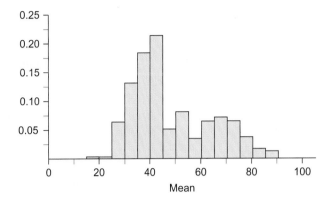

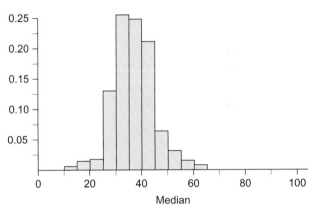

Display 7.26 Simulated sampling distributions for sample mean and median of the areas of the 50 states.

7.2 ► Sampling Distribution of the Sample Mean

You have seen that the sample mean (average) is a very important statistic. In fact, you knew that the mean is important before you ever took a statistics course because you see averages used all around you—grade point average, average income, average age, batting average, and on and on. In this section, you will learn to predict the mean, standard deviation, and shape of the sampling distribution of the sample mean. Because its properties are so well understood, the sample mean is used almost exclusively as the measure of center in statistical inference.

STATISTICS IN ACTION 7.2 ► Cents and Center

In this activity, you will watch the normal distribution emerge from a very non-normal population.

What You'll Need: 25 pennies collected from recent day-to-day change, 1 nickel, 1 dime, 1 quarter

(Continued)

1. If you were to construct a histogram of the ages of all the pennies from all the students in your class, what do you think the shape of the distribution would look like?

2. Find the age of each of your pennies by subtracting the date on the penny from the current year. Construct a histogram of the ages of all the pennies in the class.

3. Estimate the mean and standard deviation of the distribution. Then confirm these estimates by actual computation.

4. Take a random sample of size 5 from the ages of your class's pennies, and compute the mean age of your sample.

5. If you were to construct a histogram of all the mean ages computed by the students in your class in step 4, do you think the mean of the values in this histogram would be larger than, smaller than, or the same as the mean for the population of the ages of all the pennies? Regardless of your choice, try to make an argument to support each choice. Estimate what the standard deviation of the distribution of the mean ages will be.

6. Construct the histogram of the mean ages, using nickels, and determine the mean and standard deviation. Which of the three choices in step 5 appears to be correct?

7. Repeat steps 4 through 6 for samples of size 10, making the plot using dimes, and size 25, using quarters.

8. Look at the four histograms that your class has constructed. What can you say about the shape of the histogram as *n* increases? About the center? About the spread?

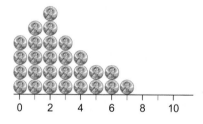

Shape, Center, and Spread of the Sampling Distribution of $\bar{x}$

> For a finite population, the sampling distribution of the sample mean is the set of means from all possible samples of a specific size.

Some of the distributions of data that you have studied so far have had a roughly normal shape, but many others were not at all normal. Part of what makes the normal distribution important is its tendency to emerge when you create sampling distributions.

Display 7.27 shows the distribution of the number of children per family in the United States. This isn't a sampling distribution. Rather, it's the population distribution you will use to build a sampling distribution. If you count all families with four or more children as having four children, this highly skewed population has a mean of about 0.9 and a standard deviation of about 1.1.

You work for the county superintendent of schools, who wants to estimate the total number of children per family in your county. The number of families is known from the number of housing units, so if you can get a good estimate of the mean number of children in the families, you can multiply the mean by the number of families to get an estimate of the total number of children.

Your plan is to take a random sample of families and compute the mean number of children per family in your sample. You wonder what sample size will be necessary to get a good estimate of the population mean and are considering samples of size 1, 4,

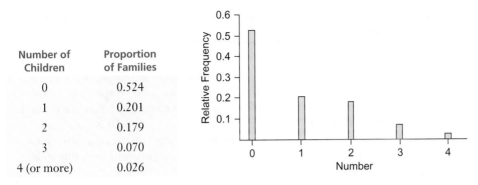

Number of Children	Proportion of Families
0	0.524
1	0.201
2	0.179
3	0.070
4 (or more)	0.026

Display 7.27 The number of children per family in the United States. [Source: U.S. Census Bureau, *Statistical Abstract of the United States, www.census. gov.*]

10, 20, and 40. You don't know it, but the distribution of the number of children per family in your county is the same as nationally, as given in Display 7.27. Display 7.28 gives the five sampling distributions of the sample mean for your sample sizes of 1, 4, 10, 20, and 40. Note that the exact sampling distribution of the mean for samples of size 1 is identical to the population distribution.

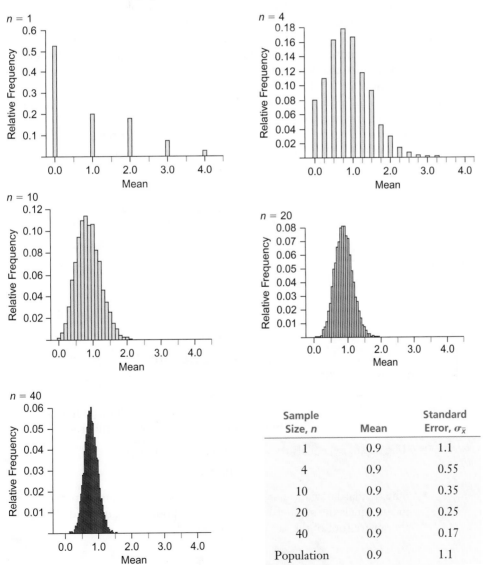

Sample Size, n	Mean	Standard Error, $\sigma_{\bar{x}}$
1	0.9	1.1
4	0.9	0.55
10	0.9	0.35
20	0.9	0.25
40	0.9	0.17
Population	0.9	1.1

Display 7.28 Sampling distributions of the sample mean and summary table for samples of size 1, 4, 10, 20, and 40.

The plots of these sampling distributions and the summary table in Display 7.28 reveal some interesting patterns:

- The population started out with only five values and a large skew toward the larger numbers. The sampling distribution for samples of size 1 is identical to the distribution of the population. The sampling distribution for samples of size 4 is still somewhat skewed but has more possible values. The sampling distribution for samples of size 10 is even less skewed. The sampling distributions for samples of size 20 and 40 look nearly normal. (This progression toward normality will turn out to be very important.)

- The means of all five sampling distributions are equal, about 0.9. This is the mean of the population.

- The standard errors of the five sampling distributions decrease from 1.1 for samples of size 1 (which is the same as the population standard deviation) to 0.55 for samples of size 4, to 0.35 for samples of size 10, and so on. Just as you would predict, the sample mean tends to be closer to the population mean with larger samples than it is with smaller samples.

There is a 95% chance that $\bar{x}$ is within 1.96 *SE* of μ.

Looking at the summary table, you can see that if you take a random sample of 20 families, you have a 95% chance that your estimate of the mean number of families will be a reasonably likely event and so be within 1.96(0.25), or 0.49, of the population mean of 0.9. If that's not close enough and you take a random sample of 40 families, you have a 95% chance that your estimate will be within 1.96(0.17), or about 0.33, of the population mean of 0.9. For either sample size, the distribution is approximately normal so you can compute probabilities using z-scores and the standard normal distribution in Table A on page 760.

The patterns in Display 7.28 suggest a set of rules for describing the behavior of sampling distributions of the sample mean. These rules appear in the box following this table of commonly used symbols.

	Common Symbols		
	Population Parameter	Sample Statistic	Sampling Distribution
Mean	μ	$\bar{x}$	$\mu_{\bar{x}}$
Standard Deviation	σ	s	$\sigma_{\bar{x}}$ or SE
Size	N	n	

Properties of the Sampling Distribution of the Sample Mean, $\bar{x}$

Center.

Shape.

If a random sample of size n is selected from a population with mean μ and standard deviation σ, then the sampling distribution of $\bar{x}$ has these properties.

- The mean, $\mu_{\bar{x}}$, equals the mean of the population, μ:

$$\mu_{\bar{x}} = \mu$$

- The standard deviation, $\sigma_{\bar{x}}$, sometimes called the standard error of the mean, equals the standard deviation of the population, σ, divided by the square root of the sample size n:

$$\sigma_{\bar{x}} = \frac{\sigma}{\sqrt{n}}$$

Center ... Spread ... Shape.

- The shape will be approximately normal if the population is approximately normal. For other populations, the sampling distribution becomes more normal as *n* increases. (This property is called the **Central Limit Theorem**.)

All three properties are of great importance in statistics, and all three depend on *random* samples. The first property, $\mu_{\bar{x}} = \mu$, says that the means of random samples are centered at the population mean. This might seem obvious, and it is obvious for symmetric populations. But as you saw in the NBA salary study of Example 7.1, it's also true for skewed populations.

The second property, $\sigma_{\bar{x}} = \sigma/\sqrt{n}$, is the main reason for using the standard deviation to measure spread: You can find the standard error of the sample mean *without simulation* if you know the standard deviation of the population. This result validates our intuitive feeling that large samples are better: The larger the sample, the closer its mean tends to be to the population mean.

The third property, the Central Limit Theorem, deals with the shape of the sampling distribution and will indeed be central to your work in the rest of this course. It helps you decide which outcomes are reasonably likely and which are not.

DISCUSSION
Shape, Center, and Spread of the Sampling Distribution of $\bar{x}$

D5. Why is it the case that the sampling distribution of the mean for samples of size 1 is identical to the population distribution?

D6. The scatterplot in Display 7.29 shows the standard error, *SE*, plotted against the sample size, *n*, for the table in Display 7.28 on the page 329. This plot illustrates that while the standard error decreases as the sample size increases, it decreases more and more slowly. Use the pattern in the plot to estimate the standard error when the sample size is 30. What is the equation of the curve that goes through these points?

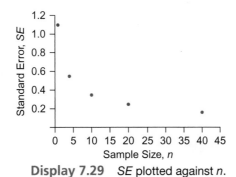

Display 7.29 *SE* plotted against *n*.

D7. Justify the comment "Large samples are better, because the sample mean tends to be closer to the population mean."

Finding Probabilities Involving Sample Means

As you will see in the next example, you can solve problems involving the sample mean by using the three properties of the sampling distribution in combination with what you learned about normal distributions in Chapter 2.

Example 7.5	**Average Number of Children**

What is the probability that a random sample of 20 families in the United States will have an average of 1.5 children or fewer?

Solution

By looking at the sampling distribution in Display 7.28 for $n = 20$, you can see that the sampling distribution of the sample mean is approximately normal. The mean of the population is about 0.9, and the standard deviation of the population is 1.1. So, for the sampling distribution, we have

$$\mu_{\bar{x}} = \mu \approx 0.9$$

$$\sigma_{\bar{x}} = \frac{\sigma}{\sqrt{n}} = \frac{1.1}{\sqrt{20}} \approx 0.25$$

Display 7.30 shows a sketch of this situation.

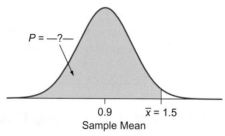

Display 7.30 Sampling distribution of $\bar{x}$ when $\mu_{\bar{x}} = 0.9$, $\sigma_{\bar{x}} = 0.25$, and $n = 20$. The shaded area shows the probability that the sample mean is less than 1.5.

The z-score for the value 1.5 is

$$z = \frac{\bar{x} - mean}{standard\ deviation} = \frac{\bar{x} - \mu_{\bar{x}}}{\sigma_{\bar{x}}} = \frac{\bar{x} - \mu}{\sigma/\sqrt{n}} \approx \frac{1.5 - 0.9}{0.25} \approx 2.4$$

You can use Table A or technology to find that the area under a standard normal curve to the left of 1.5, 2.4 is about 0.9918, which is the probability that the sample mean will fall below 1.5. In a random sample of 20 families, it is almost certain that the average number of children per family will be less than 1.5.

Sometimes situations are stated in terms of the total number in the sample rather than the average number.

Example 7.6	**Total Number of Children**

What is the probability that there are 30 or fewer children in a random sample of 20 families in the United States?

Solution

> When you are given the sample total rather than $\bar{x}$.

First, find the equivalent average number of children, $\bar{x}$, by dividing the total number of children, 30, by the sample size, 20.

$$\bar{x} = \frac{30}{20} = 1.5$$

Now you can use the same formulas and procedure as in the previous example.

Reasonably Likely Averages

Example 7.7

You take a random sample of 20 families. How close is the sample mean likely to be to the population mean of 0.9?

Solution

From Example 7.5 on the previous page, the distribution of the mean number of children is approximately normal with mean 0.9 and standard error 0.25. Reasonably likely outcomes are those within 1.96 standard errors of the mean. This interval is

$$0.9 \pm 1.96(0.25) = 0.9 \pm 0.49$$

There is a 95% chance that $\bar{x}$ will be within 0.49 of the population mean of 0.9.

Corn Yields

Example 7.8

Corn has become an important commodity in the modern world because of its central role in both food and fuel. Estimating the yield of corn is, then, a very important undertaking. A standard way of producing such estimates is to begin with randomly selected plots of size $\frac{1}{1000}$ of an acre, which translates into about 18 feet of one row of corn. Within each plot, the number of kernels of corn is approximated by a fairly accurate technique honed over years of practice. For typical farms in Indiana, the yields per plot have a mean of about 15,000 kernels with a standard deviation of about 2,000 kernels.

a. Suppose twenty-five $\frac{1}{1000}$-acre plots are to be randomly chosen on a typical farm in Indiana. What is the approximate probability that the mean number of kernels per plot will exceed 16,000 kernels?

b. What are the reasonably likely outcomes for the mean yield of the 25 sample plots?

Solution

a. Assuming the distribution of the number of kernels per plot is not highly skewed, the sample size of 25 should be large enough to use the normal distribution as an approximate sampling distribution for the mean counts. The z-score for a sample mean of 16,000 is

$$z = \frac{\bar{x} - \mu_x}{\sigma/\sqrt{n}} = \frac{16,000 - 15,000}{2000/\sqrt{25}} = 2.5$$

The probability of a sample mean exceeding 16,000 kernels is 0.0062.

b. The reasonably likely outcomes for the sample mean of the 25 plots is

$$15,000 \pm 1.96 \cdot \frac{2000}{\sqrt{25}} = 15,000 \pm 784$$

or about 14,216 to 15,784 kernels per plot.

DISCUSSION
Finding Probabilities Involving Sample Means

D8. What is the probability that a random sample of nine U.S. families will have an average of 1.5 children or fewer? Can this problem be done like the example for 20 families? Why or why not?

Using the Properties of the Sampling Distribution of the Mean

At this stage, you must have a few questions about how to use the properties in the box on page 330.

1. *When can I use the property that the mean of the sampling distribution of the mean is equal to the mean of the population, $\mu_{\bar{x}} = \mu$?* Good news. In random sampling, it is always true that $\mu = \mu_{\bar{x}}$. The shape of the population doesn't matter, nor does how large the sample is, how large the population is, or whether you sample with replacement or without replacement.

2. *When can I use the property that the standard error of the sampling distribution of the mean, $\sigma_{\bar{x}}$, is equal to $\sigma/\sqrt{n}$?* Good news here too, mostly. You can use this formula with a population of any shape and with any sample size as long as you randomly sample with replacement or you randomly sample without replacement and the sample size is less than 10% of the population size. In practice, random samples are generally taken without replacement, but the populations are usually large compared to the sample size. So, almost always, you can use this formula. However, in the rare instance when you are taking a random sample without replacement from a small population, you should use the formula given in E29.

3. *To compute probabilities, I find a z-score and then use Table A to get the probability. This doesn't work unless the sampling distribution of the mean is approximately normal. When can I assume that the sampling distribution is approximately normal?* Here's where the trouble starts. As you saw in Display 7.28, for example, when the population is skewed, the sampling distribution also is skewed for small sample sizes. With a large enough sample, the sampling distribution is approximately normal. The problem is that there is no hard-and-fast rule to follow in knowing what sample size is "large enough." The following box gives you some guidelines.

When Can the Sampling Distribution of the Mean Be Considered Approximately Normal?

- If the population is approximately normally distributed, you can assume that the sampling distribution of $\bar{x}$ is approximately normal too, no matter what the sample size.

- If the sample size is 40 or more, it's pretty safe to assume that the sampling distribution of $\bar{x}$ is approximately normal.

- Using the normal approximation may be reasonably accurate with smaller sample sizes if the population isn't too badly skewed.

For example, if the population looks like the one in Display 7.28, you can see that the sampling distribution of $\bar{x}$ is approximately normal for samples of size 40 or larger, and a normal approximation wouldn't be too bad for samples of size 20 or larger.

4. *Isn't the size of the population really important? Surely a random sample of 200 households would provide more information about a city of size 20,000 households than about a city with 2 million households.* Not really. As long as the sample was randomly selected and as long as the population is much larger than the sample, it doesn't matter how large the population is. Perhaps an extreme example will help. Suppose you want to know the percentage of all Skittles that are yellow. You start sampling Skittles. After checking 100,000 pieces, you find that about 20% are yellow. With such a large random sample, you should have a lot of faith that this estimate is close to the real population percentage. You shouldn't have less faith in this estimate if you know that 100 million Skittles have been produced than if you know 10 million pieces have been produced.

DISCUSSION
Using the Properties of the Sampling Distribution of the Mean

D9. If you select 100 households at random, would the standard error of the sampling distribution of the mean number of children be larger if the population size, N, is 1000 or if it is 10,000?

D10. Consider the sampling distribution of the mean of a random sample of size n taken from a population of size N with mean μ and standard deviation σ.

   a. For a fixed N, how does the mean of the sampling distribution change as n increases?

   b. For a fixed N, how does the standard error change as n increases?

   c. If $N = n$, what are the mean and standard error of the sampling distribution?

Summary 7.2: Sampling Distribution of the Sample Mean

For a finite population, the sampling distribution of the sample mean is the set of means from all possible samples of a specific size.

Because averages are so important, it seems fortunate that the mean and standard error of the sampling distribution of $\bar{x}$ have such simple formulas. It seems doubly fortunate that, for large sample sizes, the shape is approximately normal. This allows you to use z-scores and the standard normal distribution to determine whether a given sample mean is reasonably likely or a rare event. But fortune has nothing to do with it! It's precisely because these formulas are so simple that the mean and standard deviation are so important.

Here is a summary of the characteristics of the sampling distribution of $\bar{x}$. They apply only if the samples are selected at random.

- The mean of the sampling distribution equals the mean of the population, or $\mu_{\bar{x}} = \mu$, and this fact does not depend on the shape of the population or on the sample size.

- The standard error of the sampling distribution equals the population standard deviation divided by the square root of the sample size, or $\sigma_{\bar{x}} = \sigma/\sqrt{n}$, and this fact does not depend on the shape of the population or on the sample size as long as you sample with replacement or your sample size is no more than 10% of the size of the population.

- If the population is normally distributed, the sampling distribution is normally distributed for all sample sizes n. If the population is not normally distributed, then as the sample size increases, the sampling distribution will become more normal. This is called the Central Limit Theorem.

- If the population is at least ten times the size of the sample, whether you sample with or without replacement is of little consequence.

- The population size does not have much effect on the analysis unless the sample size is greater than 10% of the population size.

What rule can you use to decide whether the shape of your sampling distribution is approximately normal? Unfortunately, there isn't one. The required size of *n* depends on how close the population itself is to normal and how accurate you want your approximations to be. But generally you are safe using the normal distribution as an approximation to the sampling distribution of the sample mean as long as the sample size is 40 or more.

Practice

Shape, Center, and Spread of the Sampling Distribution of x̄

P7. The distribution of the number of motor vehicles per household in the United States (the population) and sampling distributions of the mean for samples of size 4 and 10 are shown in Display 7.31.

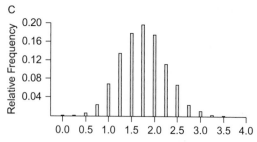

Display 7.31   Three distributions.

a. Which distribution is which? Make a rough estimate of the mean and standard deviation of each distribution.

b. The mean of the population is about 1.7. Theoretically, what is the mean of the sampling distribution for samples of size 4? Of size 10? Are these computed means consistent with your estimates of the means in the histograms?

c. The standard deviation of the population is about 1. Theoretically, what is the *SE* of the sampling distribution for samples of size 4? Of size 10? Are these

computed *SEs* consistent with your estimates of the *SEs* in the histograms?

d. Compare the shapes of the three distributions. Are the shapes consistent with the Central Limit Theorem?

P8. From 1910 through 1919, the single-season batting averages of individual Major League Baseball players had a distribution that was approximately normal, with mean .266 and standard deviation .037. Suppose you construct the sampling distribution of the mean batting average for random samples of 15 players. What are the shape, mean, and standard error of this sampling distribution? [Source: Stephen Jay Gould, *The Spread of Excellence from Plato to Darwin* (New York: Harmony Books, 1996).]

Babe Ruth (1895–1948).

Finding Probabilities Involving Sample Means

P9. The process for manufacturing a ball bearing results in weights that have an approximately normal distribution with mean 0.15 g and standard deviation 0.003 g.

a. If you select one ball bearing at random, what is the probability that it weighs less than 0.148 g?

b. If you select four ball bearings at random, what is the probability that their mean weight is less than 0.148 g?

c. If you select ten ball bearings at random, what is the probability that their mean weight is less than 0.148 g?

P10. The distribution of the number of children for families in the United States has mean 0.9 and standard deviation 1.1. Suppose a television network selects a random sample of 1000 families in the United States for a survey on TV viewing habits.

a. Describe the sampling distribution of the possible values of the average number of children per family.

b. What average numbers of children are reasonably likely in the sample?

c. What is the probability that the average number of children per family in the sample will be 0.8 or less?

d. What is the probability that the average number of children per family in the sample will be between 0.8 and 1.0?

P11. The distribution of the number of motor vehicles per household in the United States is roughly mound-shaped, with mean 1.7 and standard deviation 1.0.

a. If you pick 15 households at random, what is the probability that they have at least 30 motor vehicles among them?

b. If you pick 20 households at random, what is the probability that they have between 25 and 30 motor vehicles among them?

Using the Properties of the Sampling Distribution of the Mean

P12. Refer to the population of NBA salaries in Display 7.2 on page 314, which has mean 4.6 and standard deviation 4.7 (both in millions of dollars). You want to estimate the mean salary by taking a sample of size 5.

a. Compute the standard error of the sampling distribution of the mean for samples of size 5, $\sigma_{\bar{x}} = \sigma/\sqrt{n}$. Then use a z-score and the table of the normal distribution to estimate the probability that you will get a mean that is less than 1.5.

b. Display 7.32 shows a simulated sampling distribution of the mean for samples of size 5. Use it to estimate the probability you will get a mean that is less than 1.5.

c. Is your estimate in part b consistent with your computation from part a? Why or why not?

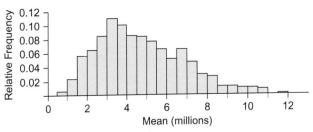

Display 7.32 Simulated sampling distribution of the mean NBA salary for samples of size 5.

P13. A typical political opinion poll in the United States questions about 1500 people. How would the analysis change if this were a random sample of 1500 people from Arizona rather than from the entire United States?

Exercises

E15. A college entrance examination has scores that are approximately normally distributed with mean 500 and standard deviation 100.

a. If you select one score at random, what is the probability that it is 510 or greater?

b. If you select four scores at random, what is the probability that their average is 510 or greater?

c. If you select 25 scores at random, what is the probability that their average is 510 or greater?

d. If you select 25 scores at random, what is the probability that their average is between 490 and 510?

E16. Last January 1, Jenny thought about buying individual stocks. Over the next year, the mean of the percentage changes in individual stock prices is 6.5% and the standard deviation of these percentage changes is 12.8%. The distribution of price changes is approximately normal.

a. If Jenny had picked one stock at random, what is the probability that it would have gone down in price?

b. If Jenny had picked four stocks at random, what is the probability that their mean percentage changes would be negative?

c. If Jenny had picked eight stocks at random, what is the probability that their mean percentage change would be between 8% and 10%?

d. If Jenny had picked eight stocks at random, what mean percentage changes in price are reasonably likely?

E17. An emergency room in Oxford, England, sees a mean of 67.4 children ages 7–15 per summer weekend, with

standard deviation 10.4. The numbers of children are approximately normally distributed. [Source: Stephen Gwilym et al., "Harry Potter Casts a Spell on Accident Prone Children," *British Medical Journal*, Vol. 331 (December 2005), pp. 1505–1506.]

a. How many children in all would you expect to be seen in the emergency room over two randomly selected summer weekends?

b. Find the probability that the emergency room will see a mean of 36.5 (total of 73) or fewer children over two randomly selected summer weekends.

c. Over the two summer weekends when Harry Potter books were released, a mean of 36.5 (total of 73) children were seen in the emergency room. What can you conclude?

E18. Illinois has nine riverboat casinos. In 2007, they admitted a total of 16,525,437 people. The mean casino win (customer loss) per admission was $120.02. Assume that the standard deviation is about $100. [Source: Center for Gaming Research, University of Nevada, Las Vegas, *gaming.unlv.edu/abstract/il_annual.html*.]

a. What were the total winnings on gambling for all nine riverboat casinos that year? What was the average win on gambling per casino?

b. If you randomly select 100 people who were admitted, what is the probability that the mean win per person for the casino was over $125? What is the probability that the total winnings for the casinos from these 100 people was over $12,500?

c. If you randomly select 100 people who were admitted, what is the probability that the mean win per person for the casino would be less than $0 (i.e., the casino lost money)?

E19. Consider the skewed population shown in Display 7.33. Display 7.34 shows simulated sampling distributions of the sample mean for samples of size 2, 4, and 25.

a. Match the theoretical summary information with the correct simulated sampling distribution (A, B, or C) and the correct sample size (2, 4, or 25).

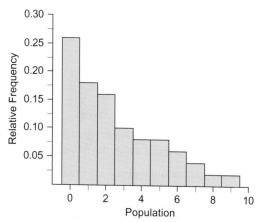

Display 7.33   A skewed population.

I.   mean 2.50: standard error 0.48
II.   mean 2.50: standard error 1.20
III.   mean 2.50: standard error 1.70

b. Does the rule for computing the standard error of the mean from the standard deviation of the population appear to hold in all three situations?

c. Does the sampling distribution appear to be approximately normal in all cases? If not, explain how the given shape came about.

d. For which sample sizes would it be reasonable to use the rule stating that 95% of all sample means lie within approximately two standard errors of the population mean?

E20. Consider the M-shaped population in Display 7.35. Display 7.36 shows simulated sampling distributions of the sample mean for samples of size 2, 4, and 25.

a. Match the theoretical summary information with the correct simulated sampling distribution (A, B, or C) and the correct sample size (2, 4, or 25).

I.   mean 4.50: standard error 1.75
II.   mean 4.50: standard error 0.70
III.   mean 4.50: standard error 2.47

b. Does the rule for computing the standard error of the mean from the standard deviation of the population appear to hold in all three situations?

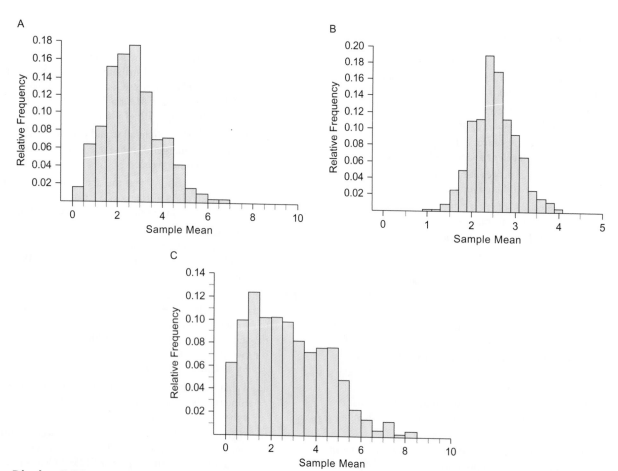

Display 7.34   Three sampling distributions of the sample mean (for $n = 2$, 4, and 25, not necessarily in that order).

c. Does the sampling distribution appear to be approximately normal in all cases? If not, explain how the given shape came about.

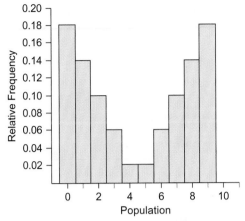

Display 7.35   An M-shaped population.

d. For which sample sizes would it be reasonable to use the rule stating that 95% of all sample means lie within approximately two standard errors of the population mean?

E21. Suppose police records in a small city show that the number of automobile accidents per day for 1045 days has the frequency distribution shown in Display 7.37. The relative frequencies, in order, are 0.36, 0.37, 0.17, 0.09, and 0.01.

a. Would it be a rare event to see two accidents on a randomly selected day in this city? The two histograms in Display 7.38 show simulated sampling distributions of the mean number of accidents per day for random samples of 4 days and 8 days. Each distribution has 200 sample means.

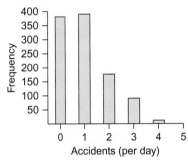

Display 7.37   Frequency distribution of the number of automobile accidents per day.

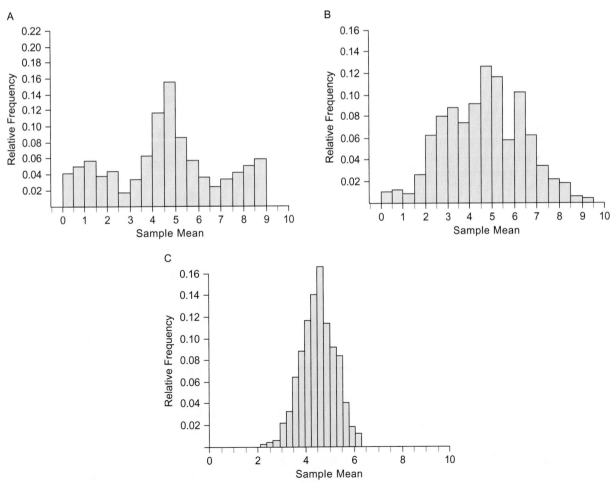

Display 7.36   Three sampling distributions (for $n = 2$, 4, and 25, not necessarily in that order).

A

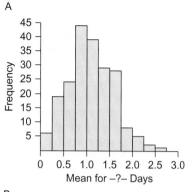

B

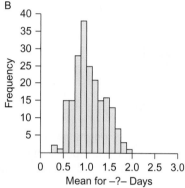

Display 7.38 Simulated sampling distributions (for *n* = 4 and *n* = 8, not necessarily in that order).

b. Which histogram is for samples of 4 days? Which is for samples of 8 days?

c. Would it be more unusual to see an average of 1.75 accidents per day for a random sample of 4 days or a random sample of 8 days?

d. Suppose the city has seven accidents over the next 8 days. Would this be a rare event? What assumptions must you make in order to answer this question? Do they seem reasonable for this situation?

E22. The distribution of grades on the very first AP Statistics Exam are given in Display 7.39.

Grade	Percentage Receiving Grade
5	13.5
4	21.3
3	24.6
2	18.7
1	21.9

Display 7.39 Distribution of exam grades.

a. Construct a plot of the distribution of the population of scores.

b. The plots in Display 7.40 show simulated sampling distributions of the mean for random samples of sizes 1, 5, and 25. Match each histogram to its sample size.

c. A teacher reports that her class averaged 3.6 on this exam. Do you think she has a class of 5 students or a class of 25 students? Explain your reasoning.

A

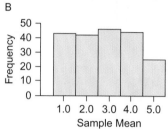

B

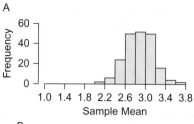

C

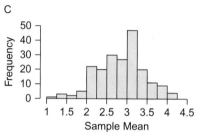

Display 7.40 Simulated sampling distributions (for *n* = 1, 5, and 25, not necessarily in that order).

E23. A college entrance examination has scores that are approximately normally distributed with mean 500 and standard deviation 100.

a. If you pick 40 scores at random, what is the probability that their mean will be within 10 points of the population mean of 500?

b. How large a random sample of scores would you need in order to be 95% sure that the sample mean would be within 10 points of the population mean of 500?

E24. Refer to E16.

a. If Jenny picked ten stocks at random, what is the probability that her mean percentage change in price is more than 7%?

b. What is the smallest number of stocks Jenny should have picked at random so that the probability is 95% or more that her mean percentage change in price would have been more than 5%?

E25. Refer to Display 7.27 on page 329. Suppose a television network selects a random sample of 1000 families in the United States for a survey on TV viewing habits.

a. Is it reasonably likely that a sample of 1000 households will produce at least 1000 children? Explain your reasoning.

b. Suppose the network changes to a random sample of 1200 households. Does this dramatically improve the chances of seeing at least 1000 children in the sampled households? Explain.

E26. Suppose the network in E25 repeatedly uses one of these four sample sizes and computes the mean number of children per family. What interval should contain the middle 95% of the sample means? That is, for each sample size, determine what sample means are reasonably likely.

a. 25

b. 100

c. 1000

d. 4000 (This is the approximate sample size for the Nielsen television ratings.)

E27. The advertising agency for a new family-oriented theme park has run a promotion that promises 100 local people free admission for a year. In preparation for a public drawing, the address of each household in the community is placed in a bin. Winning addresses will be drawn, and each member of the household at those addresses will get free admission for a year. The director of the advertising agency suddenly realizes a flaw in his scheme: He doesn't know how many addresses to draw in order to end up with 100 people. (The number of people in each household won't be known until the winners are notified.) His budget allows for 100 free admissions and he doesn't want to go over this, as it will cut into his profit. He finds from census data that the distribution of the number of people per household in the community is mound-shaped with a mean of 4.3 and a standard deviation of 1.4.

a. What is the smallest number of addresses that should be drawn in order to have a 98% chance that there will be at least 100 people total in the households? (You can use your calculator to estimate the solution to the resulting equation to the nearest whole number of names.)

b. The cost of one free admission for a year is $250. If the director draws the number of addresses you recommended in part a, how much extra will he be expected to have to pay because of his poor planning?

E28. Display 7.41 gives the distribution of the number of households in the United States that own various numbers of color televisions. Suppose a government agency wants to check 1000 color televisions to assess whether televisions in homes have a tendency to overheat. How large a random sample of households should the agency take to have a 95% chance of getting 1000 color televisions or more among the households in its sample?

Number of Color Televisions	Percentage of Households
0	1.2
1	27.4
2	35.9
3	21.8
4	9.5
5	4.2

Display 7.41 Number of color televisions per household. [Source: *Statistical Abstract of the United States, 2006*, Table 963.]

E29. If you have a small population and you sample *without replacement*, the formula for the standard error of the sampling distribution of the mean must be adjusted:

$$SE = \sigma_{\bar{x}} = \frac{\sigma}{\sqrt{n}} \cdot \sqrt{\frac{N - n}{N - 1}}$$

a. Compute the value of the adjustment factor, $\sqrt{\frac{N-n}{N-1}}$, when the population size, N, is 100 and the sample size, n, is

i. 10

ii. 50

iii. 100

b. Describe the how the adjustment factor affects the standard error. Why is this sensible?

E30. You are allowed to drop your two lowest grades on the five tests in your statistics class. There will be no makeup tests. If you show up for all five tests, your scores will be 68, 75, 82, 90, and 95. This population has a mean μ of 82 and a standard deviation σ of 9.78. You pick two test days at random to miss class. You will get a 0 on these two tests, and these two 0s will be the grades you drop.

a. Construct the exact sampling distribution of your mean test score for the remaining three tests. Sample without replacement, so that you have $n = 10$ sample means. Compute the mean and standard error of the population of 10 values in the sampling distribution. Verify that the mean of the sampling distribution is equal to the mean of the population, 82.

b. Use each of the following "shortcut" formulas to compute $\sigma_{\bar{x}}$. The first is the usual one, which may be used when the sample size is small compared to the size of the population. The second, from E29, should be used when sampling without replacement and the sample size is large compared to the size of the population:

$$\sigma_{\bar{x}} = \frac{\sigma}{\sqrt{n}} \quad \text{and} \quad \sigma_{\bar{x}} = \frac{\sigma}{\sqrt{n}} \cdot \sqrt{\frac{N - n}{N - 1}}$$

c. Which formula in part b gives the same standard error as in part a?

Use this information for E31–E34: Section 6.2 gives the rules for finding the mean and variance when you add or subtract two random variables (see the box on page 292). The same rules apply when working with sampling distributions.

E31. In 2009, SAT critical reading scores had a mean of 501 and a standard deviation of 112. The SAT math scores had a mean of 515 and a standard deviation of 116. Each distribution was approximately normal. Suppose one SAT critical reading score is selected at random and one SAT math score is independently selected at random and the scores are added. [Source: *2009 College Bound Seniors: Total Group Profile Report, www.collegeboard.com.*]

a. Find the mean and standard error of the sampling distribution of this sum.

b. Find the probability that the sum of the scores is less than 800.

c. What total SAT scores are reasonably likely?

d. What is the probability that the math score is at least 100 points higher than the critical reading score?

e. Now suppose instead that one student is selected at random and his or her SAT critical reading score and math score are added. If you can, describe the mean and standard error of the sampling distribution of this sum.

E32. The heights of males in a population are approximately normally distributed with mean 69.3 inches and standard deviation 2.92. The heights of females in the same population are approximately normally distributed with mean 64.1 inches and standard deviation 2.75.

a. Suppose one male from this age group is selected at random and one female is independently selected at random and their heights added. Find the mean and standard error of the sampling distribution of this sum.

b. Find the probability that the sum of the heights is less than 125 inches.

c. What total heights are reasonably likely?

d. What is the probability that the male is at least 2 inches taller than the female?

e. Suppose that one male is selected at random and his height is added to that of his closest female relative. If you can, describe the mean and standard error of the sampling distribution of this sum.

E33. You saw in Chapter 6 that, when n values are taken at random and independently from the same population with mean μ and standard deviation σ^2, the distribution

of their sum has mean $\mu + \mu + \cdots + \mu$, or $n\mu$, and variance $\sigma^2 + \sigma^2 + \cdots + \sigma^2$, or $n\sigma^2$. These rules also apply to sampling distributions.

a. If you roll a die three times, what is the mean of sampling distribution of the sum? The variance?

b. If you roll a die seven times, what is the mean of the sampling distribution of the sum? The variance?

c. In 2009, SAT critical reading scores had a mean 501 and standard deviation 112. If 20 scores are selected at random, what are the mean and variance of their sum? What is the probability that the sum is less than 10,000?

E34. F. Paca weighed rams in Lima, Peru, and found that the distribution was approximately normal with mean 75.4 kg and standard deviation 7.38 kg. [Source: F. Paca, New Mexico State University, 1977.]

a. Suppose two rams are selected at random. Describe the sampling distribution of the sum of their weights.

b. What is the probability that the sum of their weights is less than 145 kg?

c. What is the probability that the sum of the weights of ten randomly selected rams is more than 750 kg? (Use the rules given in E33.)

d. What is the probability that the first ram selected is more than 5 kg heavier than the second ram selected?

7.3 ▶ Sampling Distribution of the Sample Proportion

A "success" is whatever characteristic you are counting.

The goal of many surveys, polls, and observational studies is to estimate the proportion of a population with a certain characteristic (that are a "success"): the proportion of adult Americans who approve of the job the president is doing, the proportion of U.S. households in which there is spousal abuse, the proportion of college students who have difficulty sleeping, and on and on. These proportions are estimated using the proportion of "successes" in a sample. But that estimate depends on which sample you get. Knowing how much the sample proportion varies from sample to sample is the key to understanding how close the estimate from a sample is likely to be to the population proportion. That is what you will study in this section. The material here will seem very familiar because the properties of sample proportions parallel those of sample means.

Shape, Center, and Spread for Sample Proportions

The following question is typical of those you will learn how to answer in this section.

The United Nations reports that the number of cell phone subscriptions worldwide is around 4.1 billion. This means that 60% of all people in the world have a cell phone. If you take a random sample of 40 people, what is the probability that 75% or more of the people in the sample have a cell phone? [Source: *www.guardian.co.uk/technology/2009/mar/03/mobile-phones1; www.Itu.int/newsroom/press_releases/2009/07.html.*]

Some notation will make your work more efficient. Suppose you count the number of successes in a random sample of size n taken from a population in which the true proportion of successes is given by p. The symbol for the sample proportion is $\hat{p}$ (read "p-hat"). That is,

$$\hat{p} = \frac{number\ of\ successes}{sample\ size}$$

For example, suppose your sample of 40 people contains 26 who have a cell phone. Then $n = 40$, $p = 0.60$, and $\hat{p} = \frac{26}{40} = 0.65$.

Display 7.42 shows the sampling distributions of $\hat{p}$ for samples of size 10, 20, and 40 drawn from a population with 60% successes. For example, look at the distribution for $n = 10$. You can picture this being constructed by drawing a random sample of size 10 from the people of the world and counting the number who have a cell phone. If the number who have a cell phone, is, say, 4, then $\hat{p} = 0.4$, which belongs in the bar above 0.4. Repeating this over and over and adding new values of $\hat{p}$ would produce the distribution shown. (Another way to construct these distributions is to use the formula for the binomial distribution on page 298 in Section 6.3, and that's how these were constructed.)

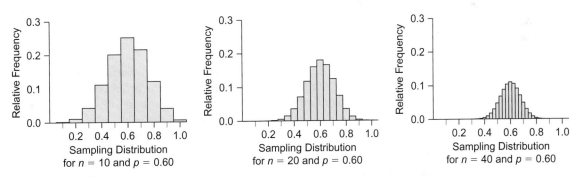

Display 7.42   Exact sampling distributions of $\hat{p}$ for samples of size 10, 20, and 40 when $p = 0.60$.

As Display 7.42 illustrates, the sampling distribution for the proportion of successes has the same characteristics as does the sample mean:

- As the sample size increases from 10 to 20 to 40 people, the shape becomes more approximately normal.
- The mean does not change depending on the sample size. No matter how large the sample size, the mean of the sampling distribution is equal to 0.60.
- The spread decreases as the sample size increases.

These characteristics hold because you can think of the sample proportion as a mean if you replace all of the successes in the population with a 1 and all of the failures with a 0. Then

$$\hat{p} = \frac{number\ of\ \text{``successes''}}{sample\ size} = \frac{\Sigma(0s\ and\ 1s\ from\ sample)}{n}$$

The latter is identical to the formula for the mean of the values (0s and 1s) in the sample.

The properties of the sampling distribution of the sample proportion are summarized in this box, followed by an example showing how to use them.

Properties of the Sampling Distribution of the Sample Proportion, $\hat{p}$

If a random sample of size n is selected from a population with proportion of successes p, then the sampling distribution of $\hat{p}$ has these properties:

- The mean, $\mu_{\hat{p}}$, is equal to the proportion, p, of successes in the population, or $\mu_{\hat{p}} = p$.

- The standard error is equal to the standard deviation of the population divided by the square root of the sample size:

$$\sigma_{\hat{p}} = \frac{\sigma}{\sqrt{n}} = \frac{\sqrt{p(1-p)}}{\sqrt{n}} = \sqrt{\frac{p(1-p)}{n}}$$

- As the sample size gets larger, the shape becomes more normal and will be approximately normal if n is large enough.

As a guideline, if both np and $n(1-p)$ are at least 10, then using the normal distribution as an approximation of the shape of the sampling distribution will give reasonably accurate results.

Example 7.9 | Describing the Sampling Distribution of a Sample Proportion

Drivers in the Northeast and Mid-Atlantic states had the highest failure rate, 20%, on the GMAC Insurance National Driver's Test. (They also were the drivers most likely to speed.) [Source: *Insurance Journal, www.insurancejournal.com.*] Describe the shape, center, and spread of the sampling distribution of the proportion of drivers who would fail the test in a random sample of 60 drivers from these states. What are the reasonably likely proportions of drivers who would fail the test?

Solution
Both $np = 60(0.2) = 12$ and $n(1-p) = 60(0.8) = 48$ are at least 10, so the sampling distribution of $\hat{p}$ is approximately normal. The histogram in Display 7.43 shows the exact distribution and normal approximation.

The mean of the sampling distribution of $\hat{p}$ is $\mu_{\hat{p}} = p = 0.2$ and the *SE* is

$$\sigma_{\hat{p}} = \sqrt{\frac{p(1-p)}{n}} = \sqrt{\frac{0.2(1-0.2)}{60}} \approx 0.05$$

Thus, the reasonably likely proportions are $0.2 \pm 1.96(0.05)$, or between about 0.1 and 0.3. This means that with a sample of only 60 people, you have a 95% chance of getting an estimate from your sample that is within 0.10 of the true population proportion of 0.20.

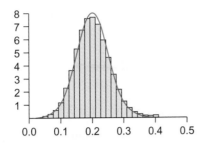

Display 7.43 Exact sampling distribution and normal approximation for $p = 0.2$ and $n = 60$.

STATISTICS IN ACTION 7.3 ▶ Sixty Percent of the World Use Cell Phones

Sixty percent of the people of the world have a cell phone. In this activity, you will simulate taking random samples of people and discovering how the proportion who have cell phones varies from sample to sample.

What You'll Need: a table of random digits

1. Describe how to use a table of random digits to simulate selecting a random sample of ten people from the world's population and asking each person whether he or she has a cell phone.

2. Use your method to select a random sample of size 10. Count the number who have a cell phone and compute the proportion who have a cell phone.

3. Repeat step 2 until your class has 100 samples of size 10.

4. Plot your 100 sample proportions on a dot plot.

5. Calculate the mean and standard deviation of your simulated sampling distribution of the sample proportion.

6. Repeat steps 2 through 5 for samples of size 20 and size 40.

7. How do the shape, mean, and standard deviation differ for the simulated sampling distributions of sizes 10, 20, and 40?

DISCUSSION
Shape, Center, and Spread for Sample Proportions

D11. Use Display 7.42 on page 343 to answer these questions.

 a. Which distribution is most normal in shape? Least normal?

 b. About 60% of people have a cell phone. On average, what proportion of people would you expect to have a cell phone in a sample of size 10? Size 20? Size 40?

 c. Which distribution in Display 7.42 has the largest spread? The smallest?

 d. Would you be most likely to find that 75% or more of the people in the sample have a cell phone if the sample size is 10, 20, or 40?

D12. Look again at Display 7.42.

 a. Compute the mean and standard error for each distribution ($n = 10, 20$, and 40) using the formulas from this section.

(Continued)

b. Estimate the means and standard errors simply from looking at the histograms. Do the computed means and standard errors in part a match your estimates?

c. From Chapter 6, the expected value (mean) μ_x and variance σ_x^2 of a discrete probability distribution are

$$E(X) = \mu_x = \sum x_i p_i \quad \text{and} \quad Var(X) = \sigma_x^2 = \sum (x_i - \mu_x)^2 p_i$$

where p_i is the probability that the random variable takes on the specific value x_i. The standard deviation is the square root of the variance.

The table in Display 7.44 gives the exact sampling distribution for samples of size 10, with $p = 0.6$. Compute the mean and *SE* using the formulas from Chapter 6. Do these formulas give the same mean and *SE* as the formulas you used in part a?

$\hat{p}$	Probability
0.0	0.000105
0.1	0.001573
0.2	0.010617
0.3	0.042467
0.4	0.111477
0.5	0.200658
0.6	0.250823
0.7	0.214991
0.8	0.120932
0.9	0.040311
1.0	0.006047

Display 7.44 Exact sampling distribution for samples of size 10 with $p = 0.6$.

Finding Probabilities Involving Proportions

You can now find answers to specific questions by using your knowledge of the properties of the sampling distribution of a proportion.

Example 7.10

Using the Properties of a Sampling Distribution of a Proportion to Find Probabilities

About 60% of people have a cell phone. Suppose your class conducts a survey of 40 randomly selected people.

    a. What is the chance that 75% or more of those selected have a cell phone?

    b. Would it be quite unusual to find that fewer than 10 of the people selected have a cell phone?

Solution

From the developments in this section, you know that $\hat{p}$ has a sampling distribution with mean and standard error

$$\mu_{\hat{p}} = p = 0.6$$

$$\sigma_{\hat{p}} = \sqrt{\frac{p(1-p)}{n}} = \sqrt{\frac{0.6(1-0.6)}{40}} = 0.0775$$

Furthermore, the sampling distribution of $\hat{p}$, shown in Display 7.45, is approximately normal because np and $n(1-p)$ are both at least 10:

$$np = 40(0.6) = 24 \quad \text{and} \quad n(1-p) = 40(1-0.6) = 16$$

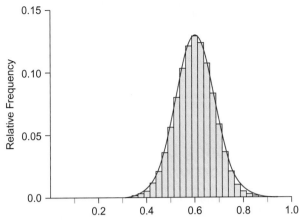

Display 7.45   Exact sampling distribution of $\hat{p}$ and normal approximation for $p = 0.6$ and $n = 40$.

a.   The z-score for the value $\hat{p} = 0.75$ in this distribution is

> Using the normal distribution as an approximation

$$z = \frac{\hat{p} - \mu_{\hat{p}}}{\sigma_{\hat{p}}} = \frac{\hat{p} - p}{\sqrt{\dfrac{p(1-p)}{n}}} = \frac{0.75 - 0.6}{0.0775} \approx 1.935$$

The proportion of the standard normal distribution that lies below this value of z is 0.9735. It follows that the probability of getting 75% or more people who have a cell phone in a sample of size 40 is only $1 - 0.9735$, or 0.0265.

b.   You are given the number of successes, 10. To use the formulas, you first must convert to the proportion of successes, $\hat{p} = 10/40 = 0.25$. Next, compute the z-score for $\hat{p} = 0.25$.

$$z = \frac{\hat{p} - p}{\sqrt{\dfrac{p(1-p)}{n}}} = \frac{0.25 - 0.6}{\sqrt{\dfrac{0.6(1-0.6)}{40}}} \approx -4.52$$

The sample proportion of 0.25 is about 4.5 standard deviations below the mean of the sampling distribution. So such a result would be unusual indeed!

Coffee Taste Test

Example 7.11

In a taste test of preference among Starbucks, McDonald's and Dunkin' Donuts coffee, 45 randomly chosen coffee drinkers were given samples of each (in random order) and asked to state a preference. If each of the 45 taste testers had no preference and was just choosing a "favorite" at random, what is the probability that 20 or more would choose Starbucks?

Solution
Under random guessing, the chance of selecting Starbucks is $p = \frac{1}{3}$ for any one taster. The value of the sample proportion of interest is $\hat{p} = \frac{20}{45}$. Conditions are met for the

normal distribution as an approximation of the shape of the sampling distribution of $\hat{p}$ to give reasonably accurate results. The z-score for this sample proportion is

$$z = \frac{\hat{p} - p}{\sqrt{\dfrac{p(1-p)}{n}}} = \frac{\left(\frac{20}{45}\right) - \left(\frac{1}{3}\right)}{\sqrt{\dfrac{\left(\frac{1}{3}\right)\left(\frac{2}{3}\right)}{45}}} \approx 1.58$$

The probability of exceeding this value is 0.0571. Because this is a fairly small probability, the random choice hypothesis might come into question if this result actually occurred in practice.

Summary 7.3: Sampling Distribution of the Sample Proportion

Sample proportions, like sample means, are among the most commonly used summary statistics and knowledge of their behavior is fundamental to the study of statistics.

The formulas in this section about the proportion of successes are derived from the fact that sample proportions behave like sample means. You compute a z-score and use the normal distribution to approximate the probability of getting specified results from a random sample. To do this, you use these facts:

- The samples must be selected randomly.
- The mean, $\mu_{\hat{p}}$, of the sampling distribution of $\hat{p}$ is equal to the proportion p of successes in the population, or $\mu_{\hat{p}} = p$.
- The standard deviation $\sigma_{\hat{p}}$ of the sampling distribution of $\hat{p}$, also called the standard error of a sample proportion, is given by the formula

$$\sigma_{\hat{p}} = \sqrt{\frac{p(1-p)}{n}}$$

- The shape of the sampling distribution becomes more normal as n increases. You can use a normal approximation for computing probabilities as long as np and $n(1 - p)$ are both at least 10.

Practice

Shape, Center, and Spread for Sample Proportions

P14. In the 2000 U.S. Census, 53% of the population over age 30 were women. [Source: U.S. Census Bureau, *www.census.gov*.]

a. Describe the shape, mean, and standard error of the sampling distribution of $\hat{p}$, the proportion of women in the sample, for random samples of size 100 taken from this population. Make an accurate sketch, with a scale on the horizontal axis of this distribution.

b. To be a member of the U.S. Senate, you must be at least 30 years old. In 2000, 9 of the 100 members of the U.S. Senate were women. Is this a reasonably likely event if gender plays no role in whether a person becomes a U.S. Senator?

Members of the United States Congress.

P15. According to the U.S. Census Bureau, 15% of the U.S. population age 5 and older are disabled.

    a. What proportions of people who are disabled would be reasonably likely to occur in a random sample of 400 people age 5 and older? How close is your estimate from the sample reasonably likely to be to the true proportion of 0.15?

    b. Suppose you take a random sample of 4000 people age 5 and older from the U.S. population to estimate the proportion who are disabled. What proportions who are disabled would be reasonably likely to occur in your sample? How close is your estimate from the sample likely to be to the true proportion of 0.15?

Finding Probabilities Involving Proportions

P16. In 2008, about 38% of the citizens of the United States considered themselves Democrats. [Source: Harris Poll, *www.pollingreport.com*.]

    a. If 435 citizens are selected at random, what is the probability of getting 257 (59%) or more who are Democrats?

    b. In the 2008 elections, 435 members were elected to the U.S. House of Representatives and 257 of these were Democrats. Is this percentage higher than could reasonably occur by chance?

    c. If so, what are some possible reasons for this unusually large percentage?

P17. A survey of hundreds of thousands of college freshmen found that 61% are attending their first-choice college. [Source: Higher Education Research Institute, UCLA, *The American Freshman, National Norms for Fall 2008*.] Suppose you take a random sample of 100 of the freshmen surveyed for additional questioning. You would like to have at least 50 students in your sample who are attending their first-choice college. Can you be fairly certain this will happen?

P18. About 23% of the people of the world use the Internet. [Source: *www.itu.int/newsroom/press_releases/2009/07.html*.]

    a. If you take a random sample of 1000 people from the world, what is the probability that fewer than 18% of the people in your sample use the Internet?

    b. If you take a random sample of 1000 people from the world, what is the probability that more than 24% of the people in your sample use the Internet?

    c. If you take a random sample of 1000 people from the billions in the world to estimate the proportion who use the Internet, how close is your estimate from the sample reasonably likely to be to the true proportion of 0.23?

Exercises

E35. The ethnicity of about 92% of the population of China is Han Chinese. Suppose you take a random sample of 1000 Chinese. [Source: *CIA World Factbook*.]

    a. Make an accurate sketch, with a scale on the horizontal axis, of the sampling distribution of the proportion of Han Chinese in your sample.

    b. What is the probability of getting 90% or fewer Han Chinese in your sample?

    c. What is the probability of getting 925 or more Han Chinese?

    d. What proportions of Han Chinese would be rare events (not reasonably likely)?

E36. According to the U.S. Census Bureau, 22.3 percent of the Spanish-surnamed population in the United States have one of these surnames: Garcia, Martinez, Rodriguez, Lopez, Hernandez, Gonzalez, Perez, Sanchez, Rivera, Ramirez, Torres, Gonzales. Suppose you take a random sample of 500 Spanish-surnamed people in the United States. [Source: David L. Word and R. Colby Perkins Jr., *Building a Spanish Surname List for the 1990's—A New Approach to an Old Problem*, Technical Working Paper no. 13, March 1996.]

    a. Make an accurate sketch of the sampling distribution of the proportion of people in your sample who have one of these surnames.

    b. What is the probability of getting 20% or fewer with one of these surnames in your sample?

    c. What is the probability of getting 105 or more people with one of these surnames?

    d. What proportions of people with one of these surnames would be rare events (not reasonably likely)?

E37. Refer to the situation in E35. This time, suppose you take a random sample of 100 Chinese rather than 1000.

    a. Describe how the shape, center, and spread of the sampling distribution of the proportion of Han Chinese in your sample will be different from your sketch in E35, part a.

    b. With a sample of size 100, will the probability of getting 90% or fewer Han Chinese in your sample be larger or smaller than the probability you computed in E35, part b? Explain.

E38. Refer to the situation in E36. This time, suppose you take a random sample of 100 Spanish-surnamed people rather than 500.

  a. Describe how the shape, center, and spread of the sampling distribution of the proportion of people in your sample with one of the given surnames will be different from your sketch in E36, part a.

  b. With a sample of size 100, will the probability of getting 20% or fewer with one of the given surnames be larger or smaller than the probability you computed in E36, part b? Explain.

E39. In 1991, the median age of residents of the United States was 33.1 years. [Source: U.S. Census Bureau, *www.census.gov*.]

  a. What is the probability that one person, selected at random that year, would be under the median age?

  b. In a random sample of 50 people that year, what is the probability of getting 10 or fewer under the median age?

  c. Of the 50 Westvaco employees listed in Display 1.1 on page 3, 10 were under the median age. Is this about what you would expect from a random sample of 50 residents of the United States, or should you conclude that this group is special in some way? If you think the group is special, what is special about it?

E40. In fall 2008, 37% of the 50,187 first-year students attending the California State University system needed remedial work in mathematics. [Source: California State University, *www.asd.calstate.edu/performance/proficiency.shtml*.]

  a. Suppose you select 4478 students at random. Make an accurate sketch, with a scale on the horizontal axis, of the sampling distribution of the proportion who need remedial work.

  b. What is the probability of getting 58% or more who need remedial work in a random sample of 4478 students?

  c. Of the 4478 students entering California State University, Northridge, 58% needed remedial work. Is this result about what you would expect from a random sample, or should you conclude that this group is special in some way?

E41. About 60% of married women are employed. If you select 75 married women, what is the probability that between 30 and 40 of them are employed? What assumptions underlie your computation? [Source: *Statistical Abstract of the United States, 2006*.]

E42. Suppose 80% of a certain brand of computer disk contain no bad sectors. If 100 such disks are inspected, what is the approximate chance that 15 or fewer contain bad sectors? What assumptions underlie this approximation?

E43. The GMAC Insurance National Driver's Test found that about 10% of drivers ages 16 to 65 would fail a written driver's test if they had to take one today. [Source: *Insurance Journal*, 2005, *www.insurancejournal.com*.] The sampling distributions in Display 7.46 give the proportion of drivers who would fail for samples of sizes 10, 20, 40, and 100 taken from a population where $p = 0.10$.

  a. Which distribution is most normal in shape? Least normal? Does the guideline that both np and

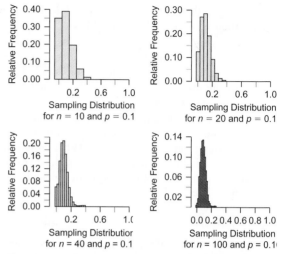

Display 7.46 Exact sampling distributions of $\hat{p}$ for samples of size 10, 20, 40, and 100 when $\hat{p} = 0.10$.

$n(1 - p)$ should be at least 10 work well in predicting approximate normality?

  b. On average, what proportion of drivers would you expect to fail a driving test in a sample of size 10? Size 20? Size 40? Size 100?

  c. Which distribution has the largest spread? The smallest?

  d. Would you be most likely to find that 20% or more of a sample of drivers fail a driving test if the sample size is 10, 20, 40, or 100?

E44. Refer to the sampling distributions in E43.

  a. Compute the mean and standard error for each distribution ($n = 10, 20, 40,$ and 100) using the formulas in this section.

  b. Estimate the means and standard errors simply from looking at the histograms. Do the computed means and standard errors in part a match your estimates?

  c. The table in Display 7.47 gives the exact sampling distribution for samples of size 10, with $p = 0.1$. Compute the mean and SE using the formulas given in D12 on page 345. Do these formulas give the same mean and SE as the formulas you used in part a?

$\hat{p}$	Probability
0.0	0.348678
0.1	0.387420
0.2	0.193710
0.3	0.057396
0.4	0.011160
0.5	0.001488
0.6	0.000138
0.7	0.000009
0.8	0.000000
0.9	0.000000
1.0	0.000000

Display 7.47 Exact sampling distribution for $n = 10$ and $p = 0.1$.

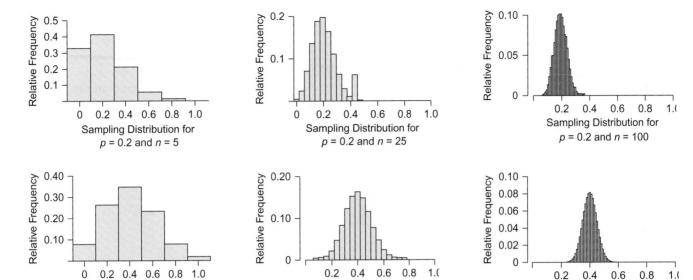

Display 7.48 Sampling distributions of $\hat{p}$ for $p = 0.2$ and $p = 0.4$ for samples of sizes $n = 5, 25,$ and 100.

E45. The histograms in Display 7.48 are sampling distributions of $\hat{p}$ for samples of sizes 5, 25, and 100, first for a population with $p = 0.2$ and then for a population with $p = 0.4$.

   a. How do the means of the sampling distributions change as p increases from 0.2 to 0.4? As n increases?

   b. How do the standard errors of the sampling distributions change as p increases from 0.2 to 0.4? As n increases?

   c. How do the shapes of the sampling distributions change as p increases from 0.2 to 0.4? As n increases?

   d. For which combinations(s) of p and n would you be willing to use the rule that roughly 95% of the values lie within two standard errors of the mean?

E46. The guideline states that it is appropriate to use the normal distribution as an approximation for the sampling distribution of a sample proportion if both np and $n(1 - p)$ are greater than or equal to 10. To check this out, generate simulated sampling distributions for values of p equal to 0.90 and 0.98. Use sample sizes of 50, 100, and 500 with each value of p. Does the guideline appear to be reasonable?

E47. In this exercise, you will learn why the introduction to this section said that the properties of sample proportions parallel the properties of sample means in the previous section. Recall that 60% of people have a cell phone. Imagine the population of the world as consisting of a barrel containing one piece of paper per person. Those people who have a cell phone are represented by a piece of paper with the number 1 on it. Those people who don't have a cell phone are represented by a piece of paper with the number 0 on it. Display 7.49 shows this population.

   a. Instead of taking a random sample of 40 actual people, you take a random sample of 40 slips of paper out of this barrel. Suppose you get 26 slips of paper with a 1 on them and 14 slips of paper with 0 on them. Compute the sample mean using the formula

$\bar{x} = \frac{\Sigma x}{n}$. How does this mean compare to the proportion of successes, $\hat{p}$ in the sample? Why is that the case?

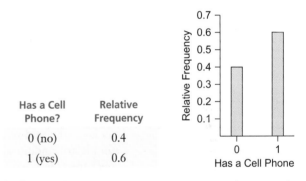

Has a Cell Phone?	Relative Frequency
0 (no)	0.4
1 (yes)	0.6

Display 7.49 Relative frequency histogram for a population with $p = 0.6$.

   b. Find the mean of this population (Display 7.49) using the formula $\mu_x = \Sigma x_i p_i$. Compute the mean using the formula in the box on page 344. How do the means compare?

   c. Describe what you have learned in this exercise about the similarities of a sample proportion and a sample mean.

E48. Refer to E47, where a population with 60% successes is thought of as a population with 60% 1s and 40% 0s.

   a. Compute the standard error of the population in Display 7.49 using the formula

$$\sigma_x = \sqrt{\Sigma(x_i - \mu_x)^2 p_i}$$

   b. Compute the standard error of the population in Display 7.49 using the formula in the box on page 344. How does this compare with the standard error you computed in part a?

Chapter Summary

In this chapter, you have learned to create a sampling distribution of a summary statistic. To create a simulated sampling distribution, you first define a process for taking a random sample from the given population. You then take this random sample and compute the summary statistic you are interested in. Finally, you generate a distribution of values of the summary statistic by repeating the process. For some simple situations, you were able to list all possible samples and get the exact distribution of the summary statistic.

Some summary statistics have easily predictable sampling distributions.

- If a random sample of size n is taken from a population with mean μ and variance σ^2, the sampling distribution of the *sample mean*, $\bar{x}$, has mean and standard error

$$\mu_{\bar{x}} = \mu \quad \text{and} \quad \sigma_{\bar{x}} = \frac{\sigma}{\sqrt{n}}$$

- If a random sample of size n is taken from a population with percentage of successes p, the sampling distribution of the *sample proportion*, $\hat{p}$, has mean and standard error

$$\mu_{\hat{p}} = p \quad \text{and} \quad \sigma_{\hat{p}} = \sqrt{\frac{p(1 - p)}{n}}$$

These formulas are summarized in Display 7.50

In the case of means, the sampling distribution becomes more and more normal in shape as the sample size increases. This is called the Central Limit Theorem. If the population is normally distributed to begin with, the sampling distributions of means, sums, and differences are normal for all sample sizes.

For proportions, you can use the normal distribution as an approximation to the sampling distribution of $\hat{p}$ as long as np and $n(1 - p)$ are both at least 10.

It might still be unclear to you why sampling distributions are so important. If so, it might help you to review the discussion of the *Westvaco* case at the beginning of Section 7.1. That sampling distribution helped you decide whether the ages of the workers laid off could reasonably be attributed to the variation that occurs from random sample to random sample, or whether you should investigate to see if some other mechanism is at work.

There's another reason why sampling distributions are so important. Suppose a pollster takes a random sample of 1500 voters to find out what percentage, p, of voters approve of some issue. The proportion $\hat{p}$ in the sample who approve almost certainly won't be exactly equal to the proportion p in the population who approve. The difference is called sampling error. How large is the sampling error likely to be? That's what sampling distributions can tell us. That might seem a bit backward. In this chapter, you always knew p and then found the reasonably likely values of $\hat{p}$. But the pollster has $\hat{p}$ and wants to find plausible values of p. This isn't hard to do now that you know what values of $\hat{p}$ tend to come from which values of p, but the details will have to wait until Chapter 8.

	Mean	SE
All populations	$\mu_{\bar{x}} = \mu$	$\sigma_{\bar{x}} = \frac{\sigma}{\sqrt{n}}$
Special case of binomial population	$\mu_{\hat{p}} = p$	$\sigma_{\hat{p}} = \sqrt{\frac{p(1 - p)}{n}}$

Display 7.50 Formulas for the mean and standard error of sampling distributions.

Review Exercises

E49. Twenty of the 30 most polluted cities in the world are in China. [Source: *news.bbc.co.uk/2/hi/business/7972125.stm.*] In the past, *China Daily*, an English-language newspaper published in the People's Republic of China, reported the Air Pollution Index (API) for 32 major Chinese cities (Display 7.51).

City	API
Beijing	123
Changchun	74
Changsha	77
Chengdu	89
Chongquing	77
Fuzhou	86
Guangzhou	98
Guiyang	85
Haikou	41
Hangzhou	62
Harbin	138
Hefei	42
Hohhot	164
Jinan	128
Kunming	119
Lanzhou	500
Nanchang	56
Najing	61
Nanning	75
Shanghai	75
Shengyang	114
Shenzhen	42
Shijiazhuang	218
Taiyuan	346
Tianjin	156
Urumqi	94
Wuhan	83
Xi'an	224
Xining	500
Yantai	76
Yinchuan	328
Zhengzhou	157

Display 7.51 API for Chinese cities. [Source: *China Daily*, March 13, 1999.]

Shanghai, China.

The mean of this distribution, shown in the dot plot in Display 7.52, is 140.9, with standard deviation 119.4.

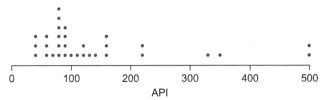

Display 7.52 Dot plot of the Air Pollution Index (API) for 32 major Chinese cities.

Suppose China had the resources to monitor air quality in only three of these cities.

a. Find the mean and standard error of the sampling distribution of the sample mean $\bar{x}$ for randomly selected samples of three cities.

b. Can the sampling distribution be considered approximately normal?

E50. Refer to the situation in E49 of selecting three Chinese cities at random from the list of 32 cities. Air quality is considered good if the API is less than 100.

a. Find the mean and standard error of the sampling distribution of the proportion of cities with good air quality for samples of size 3.

b. Can the sampling distribution be considered approximately normal?

E51. About 68% of the people in China live in rural areas. Suppose a random sample of 200 Chinese people is taken.

a. Describe the shape, mean, and standard error of the sampling distribution of $\hat{p}$, the proportion of people in the sample who live in rural areas.

b. What is the probability that 75% or more in the sample live in rural areas?

c. What values of $\hat{p}$ would be rare events?

d. What is the probability that 130 or more in the sample live in rural areas?

E52. The spine widths of the books in a university library have a distribution that is skewed slightly to the right, with mean 4.7 cm and standard deviation 2.1 cm.

a. If you select 50 books at random from this library, what is the probability that they will fit on a shelf that is 240 cm in length?

b. Does your answer to part a imply that the 50 books in the philosophy section probably will fit on one 240-cm shelf?

E53. *Forbes* magazine found that the mean age of the chief executive officers (CEOs) of America's 500 largest companies is 56 years, with standard deviation 6.3 years. The ages are roughly mound-shaped, as shown in Display 7.53.

a. What do you find interesting about this histogram?

b. If you select ten of these CEOs at random, what is the probability that their mean age is between 55 and 60?

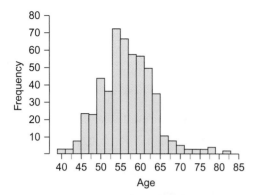

Display 7.53 Histogram of ages of 500 CEOs. [Source: *www.forbes.com, 2009.*]

E54. W. J. Youden weighed many new pennies and found that the distribution is approximately normal with mean 3.11 g and standard deviation 0.43 g. What are the reasonably likely weights of a roll of 50 randomly selected new pennies? [Source: W. J. Youden, *Experimentation and Measurement* (National Bureau of Standards, 1984), pp. 107–109.]

E55. Young men in a population have an average height of about 68 in. with standard deviation 2.7 in. Young women have an average height of about 64 in. with standard deviation 2.5 in. Both distributions are approximately normal. A young man and a young woman are selected at random.

   a. Describe the sampling distribution of the difference of their heights.

   b. Find the probability that he is 2 in. or more taller than she is.

   c. What is the probability that she is taller than he is?

E56. Two buses always arrive at a bus stop between 11:13 and 11:23 each day. Over a sample of 5 days, the average time between the times the two buses arrived was 7 minutes. Are the buses arriving at random in the 10-minute interval? Follow these steps to arrive at an answer you think is reasonable.

   a. The population of interest is the set of possible times between the two arrivals (interarrival times). Describe a way to generate five of these interarrival times, assuming the two buses arrive at random and independent times during the 10-minute interval. Then take such a sample.

   b. What was the mean interarrival time for your sample of size 5?

   c. The dot plot in Display 7.54 shows the results of 100 repetitions of the process in parts a and b. What is the largest mean interarrival time recorded? Give an example of five pairs of arrival times that would give that mean.

   d. Based on the sampling distribution, do you think it's reasonable to conclude that the buses in question were arriving at random?

E57. You select two digits at random from 0 through 9 (with replacement) and take their average. Your opponent also selects two digits at random and takes their average. You win an amount equal to the difference between your average and your opponent's average. (If the difference

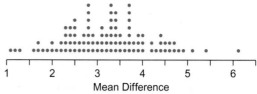

Display 7.54 Dot plot of the mean difference between arrival times for two buses for samples of 5 days.

is negative, you lose that amount.) Describe the distribution of all possible outcomes.

E58. Random samples are taken from the population of random digits 0 through 9, with replacement.

   a. Each histogram in Display 7.55 is a simulated sampling distribution of the sample mean. Match each sampling distribution to the sample size used: 1, 2, 20, or 50.

   b. Compare the means. How does the mean of the sampling distribution depend on the sample size?

   c. Compare the spreads. How does the spread of the sampling distribution depend on the sample size?

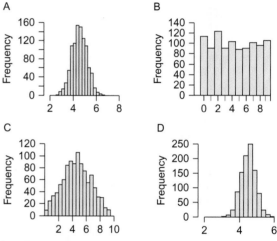

Display 7.55 Histograms of sample means of random digits.

E59. Suppose you take a sample of size *n* with replacement from a binomial population. As *n* increases, describe what happens to the shape, mean, and standard error of the sampling distribution of the sample proportion.

E60. As the sample size *n* increases, describe what happens to the mean and standard error of the sampling distribution of the sample mean.

E61. The histogram in Display 7.56 shows the number of airline passengers (in millions) departing from or arriving at the 30 largest world airports during a recent year.

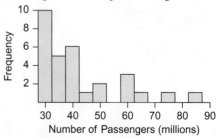

Display 7.56 Number of passengers at the 30 largest world airports. [Source: Airports Council International, *www.airports.org.*]

Sample Size	Mean (millions)	Estimated Standard Error (millions)
5	42.296	6.085
10	42.441	3.790
20	42.526	1.909

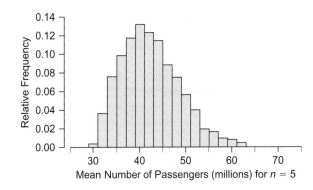

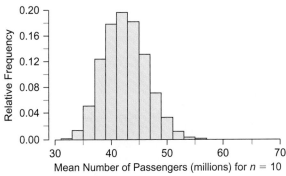

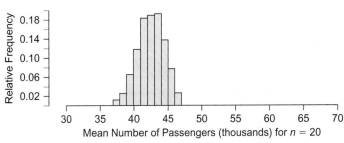

Display 7.57 Summary statistics and histograms of the distribution of sample means (in millions) for samples of sizes 5, 10, and 20.

The only outlier is Atlanta, with 83,606,583 passengers. The mean of this population is 42,520,000, and the standard deviation is 14,758,000.

The histograms and summary statistics in Display 7.57 show simulated distributions of 5000 sample means for samples of sizes 5, 10, and 20, selected without replacement from the numbers of passengers.

Check how well the shapes, means, and estimated standard errors of the simulated sampling distributions agree with what the theory says they should be. Do you see any reason why the theory you have learned should not work in any of these cases?

E62. In this problem, you will review sampling with and without replacement and the rules of probability from Chapter 5. Suppose you take a random sample of two students from the population of 100 students in Display 7.58.

Number of Term Papers	Number of Students
0	53
1	20
2	17
3	7
4 (or more)	3
Total	100

Display 7.58 Population of students and the number of term papers assigned this semester.

a. If you sample with replacement, what is the probability that both students have exactly one term paper to write?

b. If you sample without replacement, what is the probability that both students have exactly one term paper to write?

c. If you sample with replacement, what is the probability that neither student has any term papers to write?

d. If you sample without replacement, what is the probability that neither student has any term papers to write?

E63. Bottle caps are manufactured so that their inside diameters have a distribution that is approximately normal with mean 36 mm and standard deviation 1 mm. The distribution of the outside diameters of the bottles is approximately normal, with mean 35 mm and standard deviation 1.2 mm. If a bottle cap and a bottle are selected at random (and independently), what is the probability that the cap is not too small to fit on the bottle?

E64. Refer to E63. Suppose a cap is too loose if it is at least 1.1 mm larger than the bottle. If a cap and a bottle are selected at random, what is the probability that the cap is too loose?

E65. A student drives to campus in the morning and drives home in the afternoon. She finds that her commute time depends on the day of the week and whether it is morning or afternoon. Her data are shown in Display 7.59.

Day	Morning Commute Time (minutes)	Afternoon Commute Time (minutes)
Monday	18	9
Tuesday	14	8
Wednesday	13	5
Thursday	11	7
Friday	10	11

Display 7.59   Commute times.

a. What are the mean μ and variance σ^2 of the morning commute times? Of the afternoon commute times?

b. If the student selects a day at random and finds the total commute time for that day, what are the mean and variance of the sampling distribution of this total commute time?

c. Are your answers in part b equal to the sum of those in part a? Explain why they should or shouldn't be equal.

E66. Refer to Example 7.1 on page 314. Now suppose that the basketball agent is representing a very good player and wants to estimate the 75th percentile (third quartile) of NBA salaries from a random sample of 10 players. In fact, about 25% of NBA players earn more than $6.3 million and 75% earn less. The simulated sampling distribution in Display 7.60 shows the third quartile of the salaries in 200 random samples, each of size 10. Discuss whether you think the agent can get a good estimate of the 75th percentile from a sample of size 10.

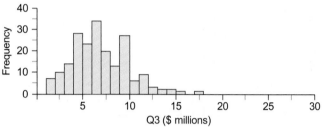

Display 7.60   Simulated sampling distribution of the third quartile of 200 samples of size 10 taken from NBA player salaries.

C1. Five English majors were asked how many novels they read last month, and the numbers were 1, 1, 1, 2, 2. Random samples of size two are taken from this population (without replacement). What is the median of the sampling distribution of the median?

Ⓐ 1

Ⓑ 1.4

Ⓒ 1.5

Ⓓ 2

Ⓔ none of the above

C2. With random sampling, which of the following is the best reason *not* to use the sample maximum as an estimator for the population maximum?

Ⓐ The sample maximum has too much variability.

Ⓑ The sample maximum is a biased estimator.

Ⓒ The sample maximum is difficult to compute.

Ⓓ The sample range is an unbiased estimator.

Ⓔ The sample maximum does not have a normally distributed sampling distribution.

C3. In computing the sample standard deviation, the formula calls for a division by $n - 1$. Which of the following is the best reason for dividing by $n - 1$ instead of n?

Ⓐ For averages, always divide by n, and for standard deviations, always divide by $n - 1$.

Ⓑ You use only $n - 1$ data values when computing standard deviations.

Ⓒ Dividing by $n - 1$ gives less variation in the sampling distribution of the population standard deviation.

Ⓓ Dividing by $n - 1$ makes the sample variance an unbiased estimator of the population variance in random sampling.

Ⓔ The sampling distribution of the sample standard deviation is closer to normal when dividing by $n - 1$.

C4. The distribution of the population of the millions of household incomes in New York is skewed to the right. Which of the following best describes what happens to the sampling distribution of the sample mean when the size of a random sample taken from this population increases from 10 to 100?

Ⓐ Its mean gets closer to the population mean, its standard deviation gets closer to the population standard deviation, and its shape gets closer to the population's shape.

Ⓑ Its mean gets closer to the population mean, its standard deviation gets smaller, and its shape gets closer to normal.

Ⓒ Its mean stays constant, its standard deviation gets closer to the population standard deviation, and its shape gets closer to the population's shape.

Ⓓ Its mean stays constant, its standard deviation gets smaller, and its shape gets closer to normal.

Ⓔ None of the above.

C5. The scores on a standardized test are normally distributed with mean 500 and standard deviation 110. In a randomly selected group of 100 test takers, what is the probability that the mean test score is less than 510?

Ⓐ less than 0.0001

Ⓑ 0.1817

Ⓒ 0.4638

Ⓓ 0.5362

Ⓔ 0.8183

C6. A statistics professor claims to be able to guess, with better than 25% accuracy, which of four symbols (circle, wavy lines, square, or star) is printed on a card. To test this claim, you ask her to guess the symbol on 40 cards. If you use the normal approximation to the binomial to compute the probability that she would get 15 (37.5%) or more correct just by random guessing, which of the following is closest to your z-score?

Ⓐ 0.183

Ⓑ 0.38

Ⓒ 1.25

Ⓓ 1.83

Ⓔ 1067

C7. In a large city, 20% of adults favor a new, stricter recycling program. Which of the following is closest to the probability that more than 24% of 1200 adults randomly selected for a survey will be in favor of the new recycling program?

Ⓐ less than 0.0001

Ⓑ 0.0003

Ⓒ 0.0029

Ⓓ 0.38

Ⓔ 0.9997

C8. A study found that infants get a mean of 290 minutes of sleep between midnight and 6 a.m. with a standard deviation of 30 minutes. A parent wants to know the probability that two randomly and independently selected infants will sleep a total of more than 600 minutes between midnight and 6 a.m., but a statistics student tells the parent that there's not enough information here to perform that calculation. Which of the following is the best reason for the student's conclusion?

Ⓐ The mean of the total cannot be determined.

Ⓑ Statistical techniques can be used for the mean but not for the total.

Ⓒ The standard deviation of the total cannot be determined.

Ⓓ The shape of the sampling distribution of the total is unknown.

Ⓔ The value of z cannot be determined.

C9. On an average day, about 246,000 vehicles travel east on the Santa Monica Freeway in Los Angeles to the interchange with the San Diego Freeway. Assume that a randomly selected vehicle is equally likely to go straight through the interchange, go south on the San Diego Freeway, or go north on the San Diego Freeway. [Source: Caltrans, *www.dot.ca.gov.*]

a. What is the best estimate of the number of vehicles that will go straight through the interchange on an average day?

b. What numbers of vehicles are reasonably likely to go straight through?

c. On an average day, Caltrans found that 138,300 vehicles went straight through the interchange, 46,800 went north on the San Diego Freeway, and 60,900 went south. What can you conclude?

C10. You estimate that the people using an elevator in an office building have an average weight of 150 lb with a standard deviation of 20 lb. The elevator is designed for a 2000-lb weight maximum. This maximum can be exceeded on occasion but should not be exceeded on a regular basis. Your job is to post a sign in the elevator stating the maximum number of people to ensure safe use. Keep in mind that it is inefficient to make this number too small but dangerous to make it too large.

a. What number would you use for maximum occupancy? Explain your reasoning and assumptions.

b. Otis Elevator Company sells the Holed Hydraulic Elevator, which lists a capacity of 2000 lb, or 13 people in the United States and 12 people in Canada. Are these numbers close to the number you decided on in part a? Explain why you would or would not expect that to be the case. [Source: *www.otis.com*, 2009.]

Chapter 8
Inference for a Proportion

A survey found that 55% of young singles aren't in a committed relationship and aren't actively looking for a romantic partner. How accurate are such percentages likely to be?

In this chapter, you will learn the two basic techniques of statistical inference: confidence intervals and significance testing.

If you are familiar with polls, you have seen the idea of a confidence interval. Suppose that a polling organization reports that 54% of Americans approve of the job the president is doing, with a margin of error of ± 3%. This statement is equivalent to a confidence interval of 51% to 57%, an interval of plausible values for the true percentage of all Americans who approve of the job the president is doing.

If you read reports of medical studies, you have seen the idea of significance testing. When a report says that the difference in the success rates of two treatments is "statistically significant," it means that the difference cannot reasonably be attributed just to chance.

The ideas of confidence intervals and significance testing developed in this chapter are fundamental to work in statistical inference. They will form the basis of the inferential procedures in the chapters to follow.

In this chapter, you will learn

▶ to construct and interpret a confidence interval for the proportion of successes in a population

▶ to conduct a test of significance for a proportion and interpret the *P*-value of the test

▶ to use a test of significance to make a decision

▶ to understand how you can plan your study to increase the chance that the decision is the correct one

8.1 ▶ A Confidence Interval for a Proportion

Open the daily newspaper, listen to a news broadcast, or read the news on the Internet, and you are likely to see the results of a public opinion poll. For example, a national survey found that 55% of singles ages 18–29 say that they aren't in a committed relationship and are not actively looking for a romantic partner. (We will call these young singles "unattached.") The survey didn't ask all young singles in the United States, only 1068 of them, and so reported a margin of error of 3%. That means that unless there were some special difficulties in getting a random sample or with the wording of the question, the researchers are 95% confident that the error in the percentage is less than 3% either way. In other words, they are 95% confident that if they were to ask all young singles in the United States, the percentage who would report that they are unattached would fall somewhere in the interval 55% ± 3%, or between 52% and 58%. How can they be so sure about this from so small a sample? What do they, in fact, mean by the phrase "95% confident"? The answer involves the concept of sampling distributions. [Source: Pew Research Center, *Not Looking for Love: The State of Romance in America*, February 13, 2006, *www.pewresearch.org*.]

> You can quantify how confident you are in an estimate from a sample.

Plausible Values for the Population Proportion, *p*

> The population proportion *p* is the unknown parameter.

One way to interpret the interval of 52% to 58% is to say that these percentages are *plausible* values of the proportion, *p*, of all young singles who are unattached. For example, 56% is a plausible value for *p* because if it were true that $p = 0.56$, it wouldn't be unusual to get $\hat{p} = 0.55$ in a sample of size 1068. On the other hand, 20% isn't a plausible value for p because if it were true that $p = 0.20$, it would be quite unusual to get $\hat{p} = 0.55$ in a sample of size 1068.

Plausible Values for *p*

You have a random sample from a population with an unknown proportion of successes, p. If the sample proportion, $\hat{p}$, falls in the middle of the sampling distribution of $\hat{p}$ for the population with proportion of successes p_0, then p_0 is a plausible value for the proportion of successes in the population. If $\hat{p}$ falls in the tails, then p_0 is not plausible as the proportion of successes in the population. "Middle" typically refers to the middle 95%, but may be 90%, 99%, or other values depending on the situation.

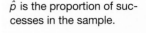
$\hat{p}$ is the proportion of successes in the sample.

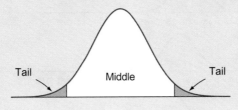

The next example will show you how to determine whether a proportion is a plausible value for p. It will rely on the facts that you learned in Chapter 7 about the sampling distribution of $\hat{p}$:

- If both np and $n(1 - p)$ are at least 10, the sampling distribution is approximately normal in shape.

- If the sampling distribution is approximately normal, then, for example, 95% of all values of $\hat{p}$ lie in the interval $p \pm 1.96 \sqrt{\frac{p(1-p)}{n}}$.

Plausible Values for the Proportion of All Young Singles Who Are Unattached

Example 8.1

Of the 1068 young singles surveyed, 55% are unattached. If you could ask all young singles in the United States if they are unattached, then you would know the true value of the population parameter, p (and it's likely that p won't be exactly 55%). For parts a and b, determine whether $\hat{p} = 0.55$ lies in the middle 95% of the sampling distribution for the proposed value of p.

    a. Is it plausible that p would be 0.57?

    b. Is it plausible that p would be 0.51?

    c. Refer to Display 8.1. Which proportions are plausible values for p?

Solution

    a. If p were 0.57, then the middle 95% of the sampling distribution of $\hat{p}$ would be in the interval

$$p \pm 1.96 \sqrt{\frac{p(1-p)}{n}} = 0.57 \pm 1.96 \sqrt{\frac{0.57(1 - 0.57)}{1068}} \approx 0.57 \pm 0.02969$$

or between about 0.54 and 0.60. Because $\hat{p} = 0.55$ lies in this interval, 57% is a plausible value for p. Display 8.1 shows that $\hat{p} = 0.55$ lies in the middle of the sampling distribution of $\hat{p}$ for $p_0 = 0.57$.

b. If p were 0.51, then the middle 95% of the sampling distribution of $\hat{p}$ would be in the interval

$$p \pm 1.96 \sqrt{\frac{p(1-p)}{n}} = 0.51 \pm 1.96 \sqrt{\frac{0.51(1-0.51)}{1068}} \approx 0.51 \pm 0.02998$$

or between 0.48 0.54. Because $\hat{p} = 0.55$ from the sample lies in the tail of the distribution, not in the middle, 51% is not a plausible value for p. Display 8.1 shows that $\hat{p} = 0.55$ lies in the tail of the sampling distribution of $\hat{p}$ for $p_0 = 0.51$.

c. From Display 8.1, $\hat{p} = 0.55$ lies in the middle of the sampling distribution for all proportions from 52% to 58%. These are the plausible values for p.

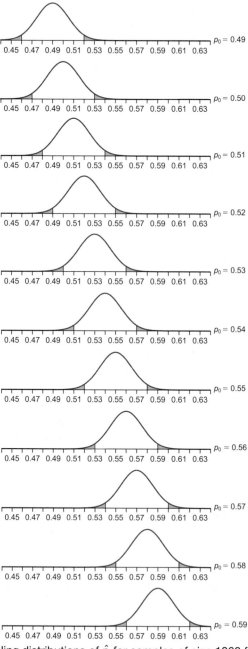

Display 8.1   Sampling distributions of $\hat{p}$ for samples of size 1068 for $p_0 = 0.49$ to $p_0 = 0.59$.

In Example 8.1, you may have noticed that the two values of

$$1.96 \sqrt{\frac{p(1-p)}{n}}$$

are almost the same for $p = 0.57$ and $p = 0.51$. As you will see next, this turns out to be the key to getting a formula that gives an interval that contains all plausible values for p.

STATISTICS IN ACTION 8.1 ▶ The Vulcan Salute

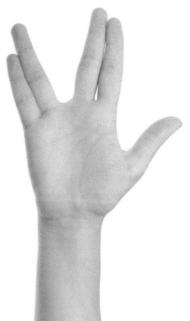

In this activity, you will collect data about the proportion of students who can make the Vulcan salute. If your sample of students can be considered random, you will then be able to make an inference about the proportion of *all* students who can make the Vulcan salute.

The Vulcan salute from *Star Trek*, which means "live long and prosper," originated as part of a priestly blessing in a Jewish ceremony. Your goal is to estimate what proportion of students can make this salute, clearly and easily, with both hands.

1. Without giving anyone the chance to practice, ask at least 40 students to make the Vulcan salute with both hands at once. Count the number who are successful.

2. Suppose you can reasonably consider your sample a random sample of all students on your campus. From the data you gathered, is it plausible that 10% of all students could make this salute? Is it plausible that the true percentage is 60%? What percentages are plausible?

A Formula for a Confidence Interval for a Proportion

The polling organization that conducted the study of young singles was 95% confident that if they were to ask all young singles in the United States, the percentage who would report that they are unattached would fall somewhere in the interval 55% ± 3%, or between 52% and 58%. This *95% confidence interval* contains the plausible values for p, the proportion of all young singles who would report that they are unattached.

> Think of a confidence interval as a list of all of the plausible homes for your orphan sample.

Concept of a Confidence Interval

A **confidence interval** for the proportion of successes in a population contains all of the values for p that are plausible, given that the proportion of successes in the sample was $\hat{p}$.

Read the follow dialog to see how confidence intervals may be computed without checking all values of p_0 individually.

Statistician: Suppose that you take a random sample of size n from a population where the proportion p of successes is unknown. The proportion of successes in your sample turns out to be $\hat{p}$. You don't know what p is, but you know that in 95% of all random samples, the distance between $\hat{p}$ and p is less than $1.96\sqrt{\frac{p(1-p)}{n}}$. That means that in 95% of random samples, p will be between

$$\hat{p} - 1.96\sqrt{\frac{p(1-p)}{n}} \quad \text{and} \quad \hat{p} + 1.96\sqrt{\frac{p(1-p)}{n}}$$

$\sqrt{\frac{p(1-p)}{n}}$ is the *SE*.

Student: That gives me an interval, but how can I use that formula when I don't know what p is?

Statistician: Although $\hat{p}$ may not be all that close to p, the products $p(1-p)$ and $\hat{p}(1-\hat{p})$ are likely to be very close to each other anyway.

Student: I saw that idea in Example 8.1. But how does that help?

Statistician: You can approximate p using $\hat{p}$, which gives this formula for a 95% confidence interval:

$\sqrt{\frac{\hat{p}(1-\hat{p})}{n}}$ is an estimate of the *SE*.

$$\hat{p} \pm 1.96\sqrt{\frac{p(1-p)}{n}} \approx \hat{p} \pm 1.96\sqrt{\frac{\hat{p}(1-\hat{p})}{n}}$$

Statistician: Do you see how to change the formula if you want, say, 99% confidence rather than 95% confidence?

Student: Sure, instead of 1.96, I would use 2.576, because in 99% of random samples, p and $\hat{p}$ are no more than 2.576 standard errors apart. I'm ready for a summary box.

A Confidence Interval for a Population Proportion

A confidence interval for the proportion of successes p in the population is given by the formula

z^* is called the critical value.

$$\hat{p} \pm z^*\sqrt{\frac{\hat{p}(1-\hat{p})}{n}}$$

larger z^* = greater confidence

Here n is the sample size and $\hat{p}$ is the proportion of successes in the sample. The value of z^* depends on how confident you want to be that p will be in the confidence interval. If you want 95% confidence, use $z^* = 1.96$; for 90% confidence, use $z^* = 1.645$; for 99% confidence, use $z^* = 2.576$. Table A or technology will give you the value of z^* for other confidence levels.

You have the given level of confidence that p is in your interval when three conditions are met:

• The sample is a random sample from a binomial population.
• Both $n\hat{p}$ and $n(1-\hat{p})$ are at least 10.
• The size of the population is at least 10 times the size of the sample.

The first two conditions listed in the box are necessary for you to be able to use the normal distribution as an approximation to the binomial distribution. If the third condition isn't met, your confidence interval will be wider than it needs to be. See E25 on page 379 for the formula you should use in this case.

Sleep Patterns of College Students

Example 8.2

Those interested in the health of the American public have become alarmed because of recent studies that show that most Americans, including children, do not get as much sleep as they need. Here is part of the authors' abstract of such a study, about the sleep patterns of college students:

> The authors' purpose in this study was to determine the sleep patterns of college students to identify problem areas and potential solutions. Participants: A total of 313 students returned completed surveys. Methods: A sleep survey was emailed to a random sample of students at a North Central university. . . . 43% woke more than once nightly. [Source: LeAnne M. Forquer et al., "Sleep Patterns of College Students at a Public University," *Journal of American College Health*, Vol. 56, 5 (March–April 2008), pp. 563–565.]

a. What is the population in this situation? Describe the parameter being estimated.

b. Are conditions met for constructing a confidence interval?

c. Construct a 95% confidence interval for the proportion of all college students at this university who woke up more than once nightly.

d. Interpret this interval.

Solution

a. The population consists of all of the students at this university. The parameter is the proportion of all students at this university who would report that they woke more than once nightly.

b. All three conditions are met for making an inference from the sample to the population:

- You are told you have a random sample (assuming most students returned the survey).
- Both $n\hat{p} = 313(0.43) = 134.59$ and $n(1 - \hat{p}) = 313(1 - 0.43) = 178.41$ are at least 10.
- The number of students at a large university certainly will be at least $10(313) = 3130$.

c. A 95% confidence interval for the proportion of all college students at this university who woke up more than once nightly is

$$\hat{p} \pm z^* \sqrt{\frac{\hat{p}(1 - \hat{p})}{n}} = 0.43 \pm 1.96 \sqrt{\frac{0.43(1 - 0.43)}{313}} \approx 0.43 \pm 0.055$$

d. You are 95% confident that, if all of the students at this university had been asked, between 37.5% and 48.5% of them would have reported that they woke up more than once nightly.

Statistical software will compute confidence intervals for you. Display 8.2 shows Minitab output for the confidence interval of Example 8.2. The 95% confidence interval in the top output uses the same method as in Example 8.2. The interval is slightly different because Minitab requires that you enter the number of successes in the sample, not $\hat{p}$ (which was rounded in Example 8.2). The second interval, called an "exact" interval, does not use the normal distribution as an approximation of the sampling distributions of $\hat{p}$. It uses the formula on page 301 in Section 6.3. The second interval,

Your calculator also may want the number of successes, *x*, not $\hat{p}$, as on the TI-84 screens on the next page.

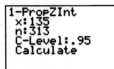

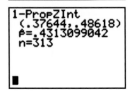

which is not centered at $\hat{p} = 0.431310$, is the better one to use if you have access to statistical software.

Test and CI for One Proportion

Sample	X	N	Sample p	95% CI
1	135	313	0.431310	(0.376443, 0.486177)

Using the normal approximation.

Test and CI for One Proportion

Sample	X	N	Sample p	95% CI
1	135	313	0.431310	(0.375726, 0.488206)

Display 8.2 Minitab output for a confidence interval for $\hat{p} = 0.43$ and $n = 313$, as in Example 8.2.

Polls in the media that track, say, the president's approval rating almost always report the proportion of successes in the sample (for example, 54%) along with a margin of error (±3%) rather than giving the confidence interval (51% to 57%).

Margin of Error

When estimating a population proportion from the proportion $\hat{p}$ of sucesses in a random sample of size n, the quantity

$$E = z^* \sqrt{\frac{\hat{p}(1 - \hat{p})}{n}}$$

is called the **margin of error.** It is equal to half the width of the confidence interval.

Example 8.3

A 95% Confidence Interval for the Unattached Singles

As you read at the beginning of this section, a national survey found that 55% of young singles say that they are unattached. The sample size was 1068, and the margin of error reported was 3%. Because this survey was done by a reliable polling organization, you may assume that the sample was randomly selected. Use the formula for a 95% confidence interval for p to verify this margin of error. Then describe specifically what p stands for in this situation.

Solution

You can use the formula for the margin of error because the sample is random and both $n\hat{p} = 1068(0.55) = 587.4$ and $n(1 - \hat{p}) = 1068(1 - 0.55) = 480.6$ are at least 10. With $n = 1068$ and $\hat{p} = 0.55$, the margin of error for 95% confidence is

$$E = 1.96\sqrt{\frac{\hat{p}(1 - \hat{p})}{n}} = 1.96\sqrt{\frac{0.55(1 - 0.55)}{1068}} = 0.0298$$

The margin of error of 0.0298 rounds to 3% and so was reported correctly.

In this situation, the population proportion, p, is the proportion of all singles ages 18–29 in the United States who would say they are unattached. You are 95% confident that p lies in the interval (0.52, 0.58).

| **College Student Engagement** | **Example 8.4** |

The National Survey of Student Engagement helps universities and colleges assess whether their students believe the campus offers academic challenge, active and collaborative learning, student-faculty interaction, enriching educational experiences, and a supportive environment. The results are based on responses from 184,457 first-year students who were randomly sampled from 722 four-year colleges and universities in the United States. One of the results was that 17% of first-year students report spending more than 20 hours per week preparing for class (studying, reading, writing, doing homework or lab work, analyzing data, rehearsing, and other academic activities). [Source: National Survey of Student Engagement, *Promoting Engagement for All Students: The Imperative to Look Within, 2008 Results.*]

   a. What is the population in this situation?

   b. Are conditions met for making an inference to the population from the sample?

   c. Construct and interpret a 95% confidence interval for the proportion of first-year students who report spending 20 or more hours per week preparing for class.

   d. The National Survey of Student Engagement also surveyed 194,858 seniors who were randomly sampled from the 722 four-year colleges and universities. Twenty percent of the seniors report spending 20 or more hours per week preparing for class. Construct a 95% confidence interval for the proportion of seniors who report spending more than 20 hours per week preparing for class.

   e. Compare the two margins of error and the two confidence intervals from parts c and d. What can you conclude?

Solution

   a. The population consists of all first-year students in these 722 four-year colleges and universities in the United States.

   b. Conditions are met for making an inference to the population from the sample:

- You are told that the students were selected at random.

- Both $n\hat{p} = 184{,}457(0.17) = 31{,}357.69$ and $n(1 - \hat{p}) = 184{,}457 (1 - 0.17) = 153{,}099.31$ are at least 10.

- The number of first-year students at these 722 colleges and universities should be at least $10(184{,}457) = 1{,}844{,}570$, which would be an average of 2555 first-year students each. So this condition may or may not have been met. But if not, the confidence interval will be slightly wider than it needs to be.

   c. The 95% confidence interval for the proportion of first-year students who report spending more than 20 hours per week preparing for class is

$$\hat{p} \pm z^* \sqrt{\frac{\hat{p}(1 - \hat{p})}{n}} = 0.17 \pm 1.96 \sqrt{\frac{0.17(1 - 0.17)}{184{,}457}} \approx 0.17 \pm 0.0017$$

You are 95% confident that, if you could ask all first-year students at these 722 four-year colleges and universities, the proportion who would report spending more than 20 hours per week preparing for class would be between 0.1683 and 0.1717.

   d. The 95% confidence interval for the proportion of seniors who would report spending more than 20 hours per week preparing for class is

$$\hat{p} \pm z^* \sqrt{\frac{\hat{p}(1 - \hat{p})}{n}} = 0.20 \pm 1.96 \sqrt{\frac{0.20(1 - 0.20)}{194{,}858}} \approx 0.20 \pm 0.0018$$

or 0.1982 to 0.2018.

smaller n = wider interval

e. The sample sizes are very large, so the margins of error for first-year students and seniors are only 0.0017 and 0.0018, respectively.

While the proportions from the two samples are quite close, 0.17 and 0.20, the two confidence intervals of 0.1683 to 0.1717 for first-year students and 0.1982 to 0.2018 for seniors don't overlap. Thus, there is no population proportion that is plausible for both groups. You can conclude that the proportion of all seniors at these two colleges who report spending more than 20 hours per week preparing for class is larger than the proportion of all first-year students who would report this.

DISCUSSION
A Formula for a Confidence Interval for a Proportion

D1. Why do we say we want a confidence interval for p rather than saying we want a confidence interval for $\hat{p}$?

D2. What, in words, does $n\hat{p}$ represent in Example 8.4 on page 367? What, in words, does $n(1 - \hat{p})$ represent in this example? What do these two quantities represent in general?

D3. In the survey described in Example 8.4, the average response rate for colleges that administered the survey on the Web was 39% and for colleges that administered the survey on paper was only 32%. Nevertheless, the final sample sizes were very large, 184,457 first-year students and 194,858 seniors. Do you think the low response rates had any affect on the results? If so, in what way?

D4. With the same sample size and confidence level, do confidence intervals get wider and narrower as the sample proportion, $\hat{p}$, moves away from 0.5?

D5. Sometimes polls report the "error attributable to sampling" instead of the "margin of error." Explain what this means.

D6. Most legitimate polling organizations publish details on how their polls are conducted. Many include a statement similar to this: "In addition to sampling error, the practical difficulties of conducting a survey of public opinion may introduce other sources of error." List at least three other possible sources of error, and explain why they are not included in the sampling error measured by statistical formulas.

The Meaning of "95% Confident"

Suppose you pick a method of constructing confidence intervals and a level of confidence (typically 95%) and use this method repeatedly to estimate a population parameter. Sometimes your confidence interval will capture the true population proportion, p, and sometimes it won't. In the long run, however, 95% of your 95% confidence intervals should capture the population parameter.

Confidence Level (Capture Rate)

The **confidence level** (or **capture rate**) of a method of constructing confidence intervals is the proportion of confidence intervals that contain the population parameter in repeated use of the method.

| The Capture Rate | Example 8.5 |

Each of 200 students in a large section of introductory statistics at a very large university is assigned to get a random sample of students on campus, ask them if they have borrowed money to attend college, and construct a 95% confidence interval for the proportion of all students on campus who have borrowed money to attend college. The statistics students don't know it, but the true proportion of all students on campus who have borrowed money is 53%. How many of the confidence intervals do you expect to contain the true proportion of 53%?

Solution
You expect that 95% of the 200 confidence intervals, 190 of them, will capture the true proportion of 53%. What about the other 10 confidence intervals? The true proportion of 53% will not be in these confidence intervals because the sample proportion, $\hat{p}$, by chance, is more than 1.96 standard deviations from 53%.

While the statistics students in Example 8.5 know that 190 of the confidence intervals are expected to capture the true proportion of students who have borrowed money to attend college, each student has no way of knowing whether he or she has one of these confidence intervals.

The meaning, then, of "95% confident" is that you expect the true value of p to be in 95 out of every 100 of the 95% confidence intervals you construct. However, it is not correct to say, after you have found a confidence interval, that there is a 95% *probability* that p is in that confidence interval. Your confidence is in your method, not in any particular interval, which may or may not contain p.

You are 95% confident in your method.

| A Survey of Young Singles | Example 8.6 |

The Pew Research Center survey reported that 55% of young singles are unattached. The size of the random sample was 1068. This survey had a margin of error of 3%, so the 95% confidence interval was 52% to 58%.

   a. What is the interpretation of the confidence interval of 52% to 58%?
   b. What is the meaning of "95% confidence"?
   c. For 99% confidence, would you need a wider or narrower interval?

Solution
   a. You are 95% confident that if you could ask all singles ages 18–29 if they are unattached, between 52% and 58% would say yes.
   b. If you were to take 100 random samples of young singles and compute the 95% confidence interval from each sample, then you can expect 95 of the confidence intervals to contain the proportion of all young singles ages 18–29 who are unattached.
   c. To have more confidence that p is captured by the confidence interval, the interval must be wider. For 99% confidence, you would use $z^* = 2.576$; so the interval would be

$$\hat{p} \pm z^* \sqrt{\frac{\hat{p}(1-\hat{p})}{n}} = 0.55 \pm 2.576\sqrt{\frac{0.55(1-0.55)}{1068}} = 0.55 \pm 0.0392$$

The margin of error, 0.0392, is larger than the margin of error, 0.0298, for 95% confidence.

STATISTICS IN ACTION 8.2 ▶ The Capture Rate

In this activity, you will test whether it is indeed the case that 95% of all confidence intervals, constructed from random samples, contain the population proportion p.

What You'll Need: computer or one table of random digits per student, a copy of Display 8.3

1. Your instructor will assign you a group of 40 random digits, or you will generate 40 random digits using technology.
2. Count the number of even digits in your sample of 40 random digits.
3. Compute a 95% confidence interval for the proportion of random digits that are even.
4. Each member of your class should draw his or her confidence interval as a vertical line segment on a copy of Display 8.3.
5. What is the true proportion of all random digits that are even?
6. What percentage of the confidence intervals captured the true proportion? Is this what you expected? Explain.
7. Was every student's confidence interval the same? Is this how it should be? Explain.

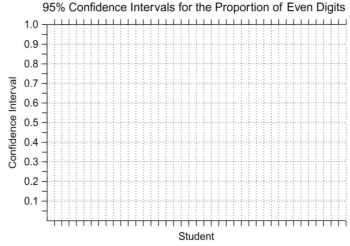

95% Confidence Intervals for the Proportion of Even Digits

Display 8.3 Chart for recording a sample of 95% confidence intervals.

DISCUSSION
The Meaning of "95% Confident"

D7. What is the relationship between the width of the confidence interval and the capture rate if the sample size, n, and the proportion of successes, $\hat{p}$, in the sample are held constant?

Margin of Error and Sample Size

The 95% confidence intervals for large sample sizes are narrower than those for small sample sizes. This makes sense—the larger the sample size, the closer $\hat{p}$ should be to p. Let's see how this works in a specific example.

The Effect of Sample Size on the Margin of Error	Example 8.7

Suppose you take a random sample of first-year college students nationwide and find that 70% of the students in your sample would like to spend a semester abroad.

a. If your sample size is 100, find the 95% confidence interval for the proportion of all first-year students who would like to spend a semester abroad, and state the margin of error.

b. Describe the difference in the confidence interval and margin of error if your sample size had been 400 instead of 100.

Solution

a. The 95% confidence interval for a sample size of 100 is

$$\hat{p} \pm 1.96\sqrt{\frac{\hat{p}(1-\hat{p})}{n}} = 0.7 \pm 1.96\sqrt{\frac{0.7(1-0.7)}{100}}$$
$$\approx 0.7 \pm 1.96(0.0458)$$
$$\approx 0.7 \pm 0.0898$$

The 95% confidence interval is (0.6102, 0.7898), and the margin of error is approximately 0.0898.

b. If you quadruple your sample size to 400, your 95% confidence interval would be

$$\hat{p} \pm 1.96\sqrt{\frac{\hat{p}(1-\hat{p})}{n}} = 0.7 \pm 1.96\sqrt{\frac{0.7(1-0.7)}{400}}$$
$$\approx 0.7 \pm 1.96(0.0229)$$
$$\approx 0.7 \pm 0.0449$$

The 95% confidence interval is (0.6551, 0.7449), half as wide as with a sample size of 100. The margin of error is approximately 0.0449, or half what it was before.

> Quadrupling the sample size cuts the margin of error in half.

DISCUSSION
Margin of Error and Sample Size

D8. With the same confidence level and proportion of successes, $\hat{p}$, in the sample, should the margin of error for a sample of size 80 be larger or smaller than the margin of error for a sample of size 40? Examine the formula for a confidence interval, and explain why it is true that with the same sample proportion and the same confidence level, the margin of error is smaller with a larger sample size than with a smaller one.

What Sample Size Do You Need?

People who conduct surveys often ask statisticians, "What sample size should I use?" The simple answer is that the larger the sample size, n, the more precise the results will be. With the same sample proportion and the same confidence level, the margin of error is smaller with a larger sample size than with a smaller one. However, researchers always have limited time and money; so for practical reasons they have to limit the

> How large the sample should be depends on how small a margin of error you want.

size of their samples. Usually the researcher has an idea of the largest margin of error that would be acceptable. If so, the formula for the margin of error, E, can be solved for the needed sample size, n:

$$E = z^* \sqrt{\frac{p(1-p)}{n}}$$

$$E^2 = (z^*)^2 \left(\frac{p(1-p)}{n} \right)$$

$$n = (z^*)^2 \left(\frac{p(1-p)}{E^2} \right)$$

To use this formula, you have to know the margin of error, E, that is acceptable. You have to decide on a level of confidence so that you know what value of z^* to use, and you have to have an estimate of p. If you have a good estimate of p from previous experience, use it in this formula. If you do not have a good estimate of p, use $p = 0.5$. Using 0.5 as the estimate for p might give you a sample size that is a bit too large, but it will never give one that is too small. (This is because the largest possible value of $p(1-p)$ occurs when $p = 0.5$. See E23.)

> ## Computing the Needed Sample Size for a Given Margin of Error
>
> If you want to estimate the proportion, p, of "successes" in a binomial population to within a margin of error of $\pm E$ (where E is given as a proportion), then the sample size, n, that you should use is
>
> $$n = (z^*)^2 \frac{p(1-p)}{E^2}$$
>
> In this formula, the value of z^* depends on the confidence level you would like. You don't know the value of p, of course, but substitute the closest estimate you can get for it into this formula. If you have no idea what p might be, use $p = 0.5$.

| Example 8.8 | **Estimating the Needed Sample Size** |

Two years ago, the city council passed a law requiring that low-flow showerheads be installed in all city housing units within 12 months. The consumption of water has not gone down. The council suspects that households have not complied with the law. You have been asked to estimate, to within 3%, the proportion of housing units in your city that have complied and have low-flow showerheads. You have no idea what that proportion might be. What sample size should you use for a survey of housing units if you want the margin of error to be at most 3% with 95% confidence?

To get a margin of error of 3% with 95% confidence is one reason why national polling organizations use sample sizes of 1000 to 1500.

Solution

Because you do not have an estimate of p, use $p = 0.5$. Then

$$n = (z^*)^2 \left(\frac{p(1-p)}{E^2} \right) = 1.96^2 \left(\frac{0.5(1-0.5)}{0.03^2} \right) \approx 1067.111$$

Because you must have a sample size of at least 1067.111, round up to 1068. You must survey 1068 randomly selected households.

DISCUSSION
What Sample Size Do You Need?

D9. Suppose it costs $5 to survey each person in your sample. You estimate that p is about 0.5. What will your survey cost if you want a 95% confidence interval with a margin of error of about 10%? About 1%? About 0.1%?

Summary 8.1: A Confidence Interval for a Proportion

When it is impossible or impractical to examine the entire population, you can use a random sample from the population to estimate the value of p, the proportion of successes in the population. You can estimate how close to p your sample proportion $\hat{p}$ is likely to be because, with a sufficiently large sample size, for example, 95% of all sample proportions fall within 1.96 standard deviations of the population proportion.

The formula for a confidence interval has the form

statistic $\pm$ (*critical value*) · (*estimated standard error*)

In the case of estimating the proportion of successes in a binomial population by taking a random sample taken from that population, the confidence interval is given by

$$\hat{p} \pm z^* \sqrt{\frac{\hat{p}(1 - \hat{p})}{n}}$$

Here, $\hat{p}$ is the observed sample proportion and n is the sample size. This interval contains all proportions that are plausible as the population proportion p, given that the proportion of successes in the sample was $\hat{p}$.

With $z^* = 1.645$, you expect that 90% of the confidence intervals constructed using this method capture the population proportion, p; with $z^* = 1.96$, you expect that 95% will do so; and with $z^* = 2.576$, you expect that 99% will do so.

This confidence interval has about the advertised capture rate when three conditions are met:

- The sample was a random sample from a binomial population.
- Both $n\hat{p}$ and $n(1 - \hat{p})$ are at least 10.
- The size of the population is at least 10 times the size of the sample.

There are two parts to giving an interpretation of a confidence interval: describing what is in the confidence interval and describing what is meant by "confidence." For a 95% confidence interval, for example, this interpretation would include:

- You are 95% confident that, if you could examine each unit in the population, the proportion of successes, p, in this population would fall in the confidence interval.
- If you were able to repeat this process 100 times and construct the 100 resulting confidence intervals, you'd expect 95 of them to contain the population proportion of successes, p. In other words, 95% of the time the process results in a confidence interval that captures the true value of p.

Remember that there can be sources of error other than sampling error. For example, if the samples aren't randomly selected or if there is a problem such as a poorly worded questionnaire, the capture rates don't apply.

To estimate the sample size, *n*, needed for a given margin of error *E*, use the formula

$$n = (z^*)^2 \frac{p(1-p)}{E^2}$$

In this formula, use a rough estimate for *p* if you have it; if not, use *p* = 0.5.

Practice

Plausible Values for the Population Proportion, *p*

P1. A study of 40 kleptomaniacs (mentally ill people who steal compulsively) found that 14 had onset of the behavior before age 11. Assume that these patients can be considered a random sample of all kleptomaniacs. [Source: Elias Aboujaoude, MD, Nona Gamel, MSW, and Lorrin M. Koran, MD, "Overview of Kleptomania and Phenomenological Description of 40 Patients," *Journal of Clinical Psychiatry*, Vol. 6, 6 (2004), pp. 244–247, *www.pubmedcentral.nih.gov/articlerender.fcgi?artid=535651.*]

   a. Is it plausible that 25% of all kleptomaniacs had onset before age 11?

   b. Is it plausible that 30% of all kleptomaniacs had onset before age 11?

   c. Is it plausible that 55% of all kleptomaniacs had onset before age 11?

P2. In a poll conducted in October by Roper, 8% of a random sample of 1166 pet owners planned to dress up a pet for Halloween. [Source: *www.ap-gfkpoll.com/pdf/AP-GfK_Petside_Poll_10.14.09.pdf.*]

   a. Is it plausible that 10% of all pet owners planned to dress up a pet for Halloween?

   b. Is it plausible that 5% of all pet owners planned to dress up a pet for Halloween?

   c. The sample size of 1166 is close to that used to construct the distributions in Display 8.1. Are any of the proportions 0.49 to 0.59 plausible values for the proportion of all pet owners who planned to dress up a pet for Halloween?

A Formula for a Confidence Interval for a Proportion

P3. In a famous experiment in psychology that would be considered unethical today, 40 volunteer male subjects were recruited for what they thought was an experiment on learning. Assume that these male volunteers can be considered a random sample of all male volunteers. They were told to give electric shocks to a likeable 47-year-old man whenever he made a mistake learning pairs of words. Twenty-five of the 40 subjects continued to give shocks up to a level that was marked "Danger: Severe Shock." In fact, the "learner" was acting and received no shocks at all. [Source: *www.experimentresources.com/stanley-milgram-experiment.html.*].

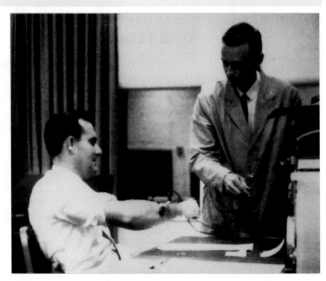

Milgram obedience experiment.

   a. Find a 95% confidence interval for the population proportion *p*.

   b. You are 95% confident that *p* is in your confidence interval. Define specifically what *p* stands for in this situation.

P4. Although the media, college officials, and many parents tend to obsess about the ranking of top colleges, students seem to pay relatively little attention when selecting a college. In a recent survey, only 20% of the randomly sampled 500 college-bound students reported reading any articles or reports that ranked colleges. [Source: *www.artsci.com/StudentPOLL/v5n1/question1.htm.*]

   a. What is the population in this situation? Describe the parameter being estimated. Is 20% the value of *p* or of *p̂*?

   b. Are conditions met for computing a confidence interval?

   c. Compute and interpret a 90% confidence interval for the proportion of college-bound students nationwide who read college rankings as part of their selection process.

The Meaning of "95% Confident"

P5. Every day Gallup conducts a poll to track the proportion of Americans who approve of the job the president is doing. It reports the results of each of these daily *tracking polls* using a 95% confidence interval. Over the course of 365 days, 365 different random samples are taken and used to construct 365 different 95% confidence intervals. How many of these confidence intervals do you expect to contain the proportion of all Americans who approved of the job the president is doing on the day the poll was taken?

P6. Suppose that each student in a large lecture section of 350 students is assigned to flip a penny a large number of times. Each student then is to construct a 95% confidence interval to estimate the probability that their penny lands heads. How many of the resulting 350 confidence intervals do you expect to capture the true proportion of 0.5?

P7. Some medical procedures, such as abortion or risky elective cosmetic surgery, are perfectly legal to perform, but some doctors may have religious or moral objections to them. Whether such doctors may refuse to provide legal medical treatments based on their objections is highly controversial. To study this issue, researchers sent a questionnaire by mail to a stratified random sample of 1820 U.S. physicians. One question was this:

> If a patient requests a legal medical procedure, but the patient's physician objects to the procedure for religious or moral reasons, does the physician have an obligation to refer the patient to someone who does not object to the requested procedure?

A total of 1144 of 1820 physicians (63%) responded to the survey. Seventy-one percent said that the physician had an obligation to refer the patient to another clinician who does not object to the requested procedure. [Source: Farr A. Curlin et al., "Religion, Conscience, and Controversial Clinical Practices," *New England Journal of Medicine*, Vol. 356 (2007), pp. 593–600.]

a. Check that the three conditions are met for computing a confidence interval for the population proportion, discussing any reservations you have. (For this purpose, you may consider the stratified random sample as equivalent to a random sample.)

b. Regardless of your answer to part a, compute the 95% confidence interval.

c. Interpret this confidence interval.

d. Explain the meaning of "95% confident."

Margin of Error and Sample Size

P8. The last NBC/*Wall Street Journal* poll of likely voters before the 2008 presidential election predicted that Barack Obama would win with 51% of the vote. The poll reported a margin of error of ± 3.1%. [Source: *www.msnbc.msn.com/id/27488250/*.]

a. Barack Obama actually got 53% of the vote. Was this true percentage within the margin of error of the predicted percentage?

b. The sample size for this poll was 1011. Verify that the margin of error was reported correctly.

P9. The last Fox News poll of likely voters before the 2004 presidential election predicted that George Bush would get 46% of the vote. The poll reported a margin of error of ± 3%. [Source: *www.foxnews.com/projects/pdf/110104_poll.pdf*.]

a. George Bush actually got 50% of the vote. Was this true percentage within the margin of error of the predicted percentage?

b. The sample size for this poll was 1400. Verify that the margin of error was reported correctly.

P10. You will take a random sample of 200 apples from a large shipment from abroad and will construct an 80% confidence interval to estimate the proportion of apples with the residue of any pesticide that has been banned by the U.S. Food and Drug Administration.

a. What value of z^* should you use for an 80% confidence interval?

b. Will the margin of error for 80% confidence be larger or smaller than for 95% confidence? Explain your answer.

P11. If all other values remain constant, will the width of the confidence interval increase or decrease with the following change?

a. The sample size is doubled from 400 to 800.

b. The confidence level is decreased from 95% to 90%.

c. The critical value, z^*, is changed from 1.96 to 1.645.

What Sample Size Do You Need?

P12. A Gallup poll of a random sample of 1011 American adults found that 36% reported that they do not drink alcohol. Gallup asks this question every year. What sample size should Gallup use next year for a 95% confidence interval with margin of error of 3% and survey as few people as possible? [Source: The Gallup Poll, June 29, 2009, *www.gallup.com/*.]

P13. The *USA Today*/CNN/Gallup poll generally reports a 3% margin of error with 95% confidence. Approximately what sample size does it use?

P14. The accuracy of opinion polls is examined closely in presidential election years. Suppose your large polling organization is planning a final poll before the presidential election.

a. What sample size should you use to have a margin of error of 2% with 95% confidence?

b. What sample size should you use to have a margin of error of 1% with 99% confidence?

c. What sample size should you use if you want a margin of error of 0.5% with 90% confidence?

Exercises

E1. The major polling organizations sponsored by news media periodically take surveys to determine how well Americans think the president is performing. The lowest approval rating ever given to a president since the polls began in the 1930s was 19%. This rating was for George W. Bush, in an American Research Group poll taken in September 2008, during a financial crisis and wars in Iraq and Afghanistan. The sample size was 1100 randomly selected American adults. [Source: *www.ropercenter.uconn.edu/ data_access/data/presidential_approval.html*.]

   a. Find a 95% confidence interval for the proportion of all American adults who approved of the job President Bush was doing.

   b. Locate your confidence interval on the following Minitab output. Compute the margin of error directly from the output.

Test and CI for One Proportion

Sample	X	N	Sample p	95% CI
1	209	1100	0.190000	(0.166817, 0.213183)

Using the normal approximation.

E2. Refer to E1. President George W. Bush also received the highest approval rating ever given to a president since the presidential approval polls began in the 1930s. Right after the September 11, 2001 terrorist attacks on New York City and Washington, he received an approval rating of 92% in the ABC news poll of 1009 randomly selected American adults. [Source: *www.ropercenter.uconn.edu/ data_access/data/presidential_approval.html*.]

   a. Find a 95% confidence interval for the proportion of all American adults who approved of the job President Bush was doing.

   b. Locate your confidence interval on the following Minitab output. Compute the margin of error directly from the output.

Test and CI for One Proportion

Sample	X	N	Sample p	95% CI
1	928	1009	0.919722	(0.902957, 0.936488)

Using the normal approximation.

E3. The National Survey of America's College Students (NSACS) examines the literacy of graduating U.S. college students.

After testing a random sample of about 1000 graduating students from four-year colleges, the NSACS found that 34% tested proficient in quantitative literacy. [Source: *The Literacy of America's College Students, American Institutes for Research*, January 2006. *www.air.org/news/documents/ The%20Literacy%20of%20Americas% 20College%20Students_final%20report.pdf*.]

   a. What is the population in this situation? Describe the parameter being estimated. Is 0.34 the value of p or of $\hat{p}$?

   b. Are conditions met for computing a confidence interval?

   c. Compute and interpret a 95% confidence interval for the percentage of all graduating four-year college students who would score proficient in quantitative literacy.

   d. NSACS also tested a random sample of about 800 graduating students from two-year colleges, and 18% scored proficient in quantitative literacy. Compute and interpret a 95% confidence interval for the percentage of all graduating two-year college students who would score proficient in quantitative literacy.

   e. Are you convinced that the percentages are different for the four-year college students and two-year college students? Explain.

E4. Refer to E3. What is the margin of error for the two-year college students? What is the margin of error for the four-year college students? What two variables account for the difference in widths?

E5. The correct dosage of some medicines is determined by a person's weight. Other medical procedures or advice can depend on a person's height or body mass index (BMI). In such situations, medical personnel often rely on people's own reports of their height and weight. A study was carried out to determine how risky that can be. The researchers randomly selected 10% of the participants in the Oxford cohort of the European Prospective Investigation into Cancer and Nutrition (EPIC–Oxford). They asked people to report their height and weight on a questionnaire, and later a nurse measured them. Based on their self-reported height and weight, 40.9% of the 279 randomly selected obese men (BMI > 30) would have been classified as being no more than overweight (BMI 25–30). [Source: Elizabeth A. Spencer at al., "Validity of Self-Reported Height and Weight in 4808 EPIC–Oxford Participants," *Public Health Nutrition*, Vol. 5, 4 (2002), pp. 561–565.]

   a. What is the population in this situation? What is the parameter being estimated?

   b. Check to see if the three conditions for computing a confidence interval are met.

   c. Compute and interpret a 95% confidence interval for the population proportion.

   d. Compute a 90% confidence interval for the population proportion.

   e. Which confidence interval, 90% or 95%, is wider? Why should that be the case?

E6. In the study described in E5, based on their self-reported height and weight, 27.0% of the 433 randomly selected obese women would have been classified as being overweight at most.

   a. What is the population in this situation? What is the parameter being estimated?

   b. Check to see if the three conditions for computing a confidence interval are met.

   c. Compute and interpret a 95% confidence interval for the population proportion.

d. How does the margin of error here compare to the margin of error for men from part c in E5? What is the reason for this?

e. Compare the confidence intervals for men (part c in E5) and women (part c). Are you confident that the parameters are different?

E7. In a survey consisting of a randomly selected national sample of 600 teens aged 13–17, 65% responded that their school was helping them discover what type of work they would love to do as a career. [Source: Gallup, *Teens: Schools Help Students Find Career Path, poll .gallup.com.*]

a. Check to see if the three conditions for computing a confidence interval are met.

b. Find a 95% confidence interval for the percentage of all teens aged 13–17 in the United States who would respond that their school is helping them discover a career path. What is the margin of error?

c. Find a 90% confidence interval for the percentage of all teens aged 13–17 in the United States who would respond that their school is helping them discover a career path. What is the margin of error?

d. Which confidence interval, 90% or 95%, is wider? Why should that be the case?

E8. In the same survey as in E7, 4% of the 600 teens responding gave their school a D rating (on a scale of A, B, C, D, F).

a. Check to see if the three conditions for computing a confidence interval are met.

b. Find a 95% confidence interval for the percentage of all teens aged 13–17 in the United States who would give his or her school a D rating.

c. How does the width of the confidence interval in part b compare to that of the confidence interval in part b of E7? What is the reason for this?

E9. Researchers wanted to estimate the annual rate of sexual infidelity in married women. They interviewed the equivalent of a random sample of 4884 married women. When the women were asked in a face-to-face interview, 1.08% admitted to sexual infidelity in the last year. When the same women answered an identical question on a computer-assisted self-interview, the rate was 6.13%. [Source: M. A. Whisman and D. K. Snyder, "Sexual Infidelity in a National Survey of American Women: Differences in Prevalence and Correlates as a Function of Method of Assessment," *Journal of Family Psychology*, Vol. 21, 2 (June 2007), pp. 147–154.]

a. Compute and interpret a 95% confidence interval for the percentage of all married women who would admit to sexual infidelity if asked in a face-to-face interview.

b. Compute and interpret a 95% confidence interval for the percentage of all married women who would admit to sexual infidelity if asked in a computer-assisted self-interview.

c. Based on the intervals in parts a and b, are you convinced that the two parameters are different? Explain.

E10. To study the effect of a cash incentive on the response rates to a mail survey, a group of 4675 Medicaid enrollees received $2 in cash along with the survey. Another group the same size received the same survey, but no money. The initial response rates were 47% and 39%, respectively. You may consider each group a random sample of Medicaid enrollees. [Source: Timothy Beebe et al., "The Effect of a Prepaid Monetary Incentive Among Low Income and Minority Populations," paper presented at the annual meeting of the American Association for Public Opinion Research, Pointe Hilton Tapatio Cliffs, Phoenix, AZ, May 11, 2004. downloaded from *www.allacademic.com/meta/p115965_index.html* on February 6, 2009.]

a. Compute and interpret a 95% confidence interval for the percentage of all Medicaid enrollees who would return the survey if they receive $2.

b. Compute and interpret a 95% confidence interval for the percentage of all Medicaid enrollees who would return the survey if they receive no cash incentive.

c. Based on the intervals in parts a and b, are you convinced that the two parameters are different? Explain.

E11. The Black Youth Project collected data on random samples of youth between the ages of 15 and 25, sampling 634 black youths, 567 white youths, and 314 Hispanic youths. For a question on whether or not they believe they have the skills to participate in politics, 74% of blacks, 70% of whites, and 66% of Hispanics responded in the affirmative. [Source: *http://blackyouthproject. uchicago.edu/writings /research_summary.pdf.*]

a. For each group, construct a 95% confidence interval for the proportion of all youths in that group who believe they have the skills to participate in politics.

b. Do any of the three groups appear to differ significantly from the others?

c. Explain the meaning of "95% confidence" with respect to the confidence interval for the black youths.

d. The sampled participants were selected by random digit dialing, and the methodology for the survey called for paying eligible respondents $20 for completing the basic telephone interview and an additional $50 for participating in a face-to-face interview. How might these practices have biased the results?

E12. The following article describes differences between teens and adults on what technology might bring in the next decade.

Study: Teens Optimistic About Innovation

Brian Bergstein, Associated Press

BOSTON—Teenagers have some seemingly high expectations about what technology might bring over the next decade, according to a new Massachusetts Institute of Technology study.

For example, 33 percent of teens predicted that gasoline-powered cars will go the way of the horse and buggy by 2015. Just 16 percent of adults agreed. Meanwhile, 22 percent of teenagers predicted desktop computers will become obsolete a decade from now, while only 10 percent of adults agreed. Adults, on the other hand, were far more certain about the demise of the landline telephone by 2015 (45 percent made that prediction) than teenagers (17 percent). . . .

But he [the director of the survey] also wonders whether enough of today's teens are in a position to invent such solutions, noting that engineering was teens' third-most attractive career choice, picked by 14 percent as the field that most interested them—and just 4 percent of girls. Only 9 percent of all teens said they were leaning toward science. The top two career choices: Arts and medicine, each picked by 17 percent of the kids surveyed.

The Lemelson–MIT program, which focuses on encouraging young people to pursue innovation, commissioned its "invention index" in November, interviewing 500 teens and 1,030 adults nationwide. The margin of sampling error was plus or minus 4 percentage points for teens and 3 for adults. [Source: *www.siliconvalley.com*, June 11, 2006.]

a. What would you like to know about the survey design that is not described here?

b. Are the conditions for constructing confidence intervals met for teens' responses? For adults' responses?

c. Do the computations of the margins of error for both teens and adults, using the formula in this section, agree with that given in the press release?

d. Assuming the conditions have been met to construct a 95% confidence interval for the proportion of teenagers who think engineering is an attractive career choice, interpret this confidence interval.

e. Explain the meaning of "95% confidence" with respect to the interval in part d.

E13. A *U.S. News & World Report* survey of a random sample of 1000 U.S. adults reported that 81% thought TV contributes to a decline in family values.

a. Compute and interpret a 95% confidence interval for the population proportion. What is the margin of error for this survey?

b. Another part of the *U.S. News & World Report* survey went to Hollywood leaders because the magazine also wanted to see what Hollywood leaders thought. Of 6059 mailed surveys, only 570 were returned. Among the returned surveys, 46% thought TV contributed to a decline in family values. The magazine does not report a margin of error for this part of the survey. Should a margin of error be reported? Explain.

E14. A nationwide survey of 19,441 teens about their attitudes and behaviors toward epilepsy found that only 51% knew that epilepsy is not contagious. The executive summary from the Epilepsy Foundation for this survey gave this technical information:

The two-page survey was distributed to teens nationally by 20 affiliates of the Epilepsy Foundation from March 2001 through July 2001 in schools selected by each affiliate. A total of 19,441 valid surveys were collected. Mathew Greenwald & Associates, Inc., edited the surveys and performed the data entry. Greenwald & Associates was also responsible for the tabulation, analysis, and reporting of the data. The data were weighted by age and region to reflect national percentages.

The margin of error for this study (at the 95% confidence level) is plus or minus approximately 1%.

The survey was funded by the Centers for Disease Control and Prevention. [Source: *www.efa.org*, April 16, 2002.]

a. Do you have any reservations about the design of this study?

b. Are the conditions for constructing a confidence interval met?

c. Does a computation of the margin of error using the formula in this section agree with that given in the press release?

d. Assuming the conditions to construct a confidence interval have been met, what is it that you are 95% sure is in the confidence interval?

E15. You want to determine the percentage of seniors who drive to campus. You take a random sample of 125 seniors and ask them if they drive to campus. Your 95% confidence interval turns out to be from 0.69 to 0.85. Select *each* correct interpretation of this situation. There might be no, one, or more than one correct statement.

A. 77% of the seniors in your sample drive to campus.

B. 95% of all seniors drive to campus from 69% to 85% of the time, and the rest drive more frequently or less frequently.

C. If the sampling were repeated many times, you would expect 95% of the resulting samples to have a sample proportion that falls in the interval from 0.69 to 0.85.

D. If the sampling were repeated many times, you would expect 95% of the resulting confidence intervals to contain the proportion of all seniors who drive to campus.

E. You are 95% confident that the proportion of seniors in the sample who drive to campus is between 0.69 and 0.85.

F. You are 95% confident that the proportion of all seniors who drive to campus is in the interval from 0.69 to 0.85.

G. All seniors drive to campus an average of 77% of the time.

H. A 90% confidence interval would be narrower than the interval given.

E16. To prepare for a nationwide advertising campaign, a survey of a random sample of U.S. adults was conducted to determine what cell phone services adults prefer. The results of the survey showed that 73% of the adults wanted email service, with a margin of error of plus or minus 4%. Which of these sentences explains most accurately what is meant by "plus or minus 4%"?

A. They estimate that 4% of the population surveyed might change their minds between the time the poll is conducted and the time the survey is published.

B. There is a 4% chance that the true percentage of adults who want email service is not in the confidence interval from 69% to 77%.

C. Only 4% of the population was surveyed.

D. To get the observed sample proportion of 73% would be unlikely unless the actual percentage of all adults who want email service is between 69% and 77%.

E. The probability that the sample proportion is in the confidence interval is 0.04.

E17. You take a random sample of 200 apples from a shipment from abroad and construct a 95% confidence interval to estimate the proportion that have the residue of any pesticide that has been banned by the FDA. You get a margin of error that is three times larger than you would like. What sample size should you have used to obtain the desired margin of error? Justify your answer.

E18. If you want a margin of error that is one-quarter of what you estimate it will be with your current sample size, by what factor should you increase the sample size?

E19. Suppose you know that a population proportion, p, is 0.60. Now suppose 80 different researchers select independent random samples of size 40 from this population. Each researcher constructs a 90% confidence interval based on the value of $\hat{p}$ from his or her sample. How many of the resulting confidence intervals would you expect to include the population proportion, p, of 0.60?

E20. A quality-control plan in a plant that manufactures computer mice calls for taking 20 different random samples of mice during a week and estimating the proportion of defective mice in a 95% confidence interval for each sample. If the percentage of mice that are defective is maintained at 4% throughout the week, how many of the confidence intervals would you expect *not* to capture this value?

E21. Explain the difference between p and $\hat{p}$. Which is the parameter? Which is always in the confidence interval?

E22. If all other values remain constant, what happens to the width of a confidence interval

a. as the sample size, n, increases?

b. as the level of confidence increases?

E23. If you don't have an estimate for p when computing the needed sample size, you use $p = 0.5$. This is the safest value to use because $p = 0.5$ requires the largest sample size, all other things being equal. Answer parts a–d to prove that the largest possible value of $p(1 - p)$ is achieved when $p = 0.5$.

a. Sketch the graph of $y = p(1 - p)$, with p on the horizontal axis. What is the name of this type of function?

b. In the context of this situation, what does p represent? What does y represent? What should be the domain of p?

c. What is the largest possible value of y? At what value of p does this occur?

d. State what you have just shown.

E24. Explain why it is not appropriate to use the method of constructing a confidence interval for a population proportion described in this section when the condition that both $n\hat{p}$ and $n(1 - \hat{p})$ are at least 10 is violated. You can use examples or a simulation to support your arguments.

E25. When the sample size is more than 10% of the population size, the formula for a confidence interval should be adjusted with a correction factor:

$$\hat{p} \pm z^* \sqrt{\frac{\hat{p}(1 - \hat{p})}{n}} \cdot \sqrt{\frac{N - n}{N - 1}}$$

Here, N is the size of the population and n is the sample size.

In a study of how well doctors explain the benefits and risks of screening procedures to patients, a 60-year-old physician went undercover as a patient to 20 urologists and asked for advice on PSA screening. (The PSA is a test that can indicate the presence of prostate cancer.) The 20 urologists were drawn randomly from 135 urologists in Berlin. Suppose that 10 of the urologists could not answer most of the "patient's" questions correctly. (It actually was 14!) [Source: "Urologen im Test: Welchen Nutzen hat der PSA-Test?" ["Testing Urologists: What Are the Benefits of a PSA Test?"], *Stiftung Warentest* (the German equivalent of *Consumer Reports*), February 2004, pp. 86–89. As reported in Gerd Gigerenzer et al., "Helping Doctors and Patients Make Sense of Health Statistics," *Psychological Science in the Public Interest*, Vol. 8, 2 (2008), p. 68.]

a. Is the condition met that the sample size is at most 10% of the population size?

b. Compute a 95% confidence interval for the proportion of all urologists in Berlin who could not answer most of the questions correctly using the formula

$$\hat{p} \pm z^* \sqrt{\frac{\hat{p}(1 - \hat{p})}{n}}$$

c. Now compute a 95% confidence interval for the proportion of all urologists in Berlin who could not answer most of the questions correctly using the formula with the correction factor

$$\hat{p} \pm z^* \sqrt{\frac{\hat{p}(1 - \hat{p})}{n}} \sqrt{\frac{N - n}{N - 1}}$$

d. Compare the two intervals. Which is shorter? Why does this make sense in this situation?

e. When N is very big compared to n, what, approximately, is the value of the correction factor?

E26. Refer to the formula in E25 for constructing confidence intervals when the sample size, n, is more than 10% of the population size, N.

Complaints about the problems with jury trials frequently appear in the media. Generally, the complaints are of the form, "The system is rife with frivolous lawsuits, unethical behavior by plaintiffs', attorneys, and runaway juries." There seems to be a feeling that juries, influenced by sharp attorneys, are prone to reach ridiculous verdicts, like the famous $2.7 million judgment against McDonald's for burns from hot coffee (later reduced to $480,000).

To check these claims, an investigator asked all 389 district court judges in Texas (known as a prodefendant state) about the nature of judgments they had seen juries render over a 4-year period. One question asked of the judges was, "In what percentage of cases tried before you as presiding judge during the past 48 months, in which the jury awarded compensatory damages, do you believe that the jury's verdict on compensatory damages was disproportionately high given the evidence presented during the trial?" Of 235 judges who replied to this question, 196 said that this had not happened in a single case. While this may not be so, you may assume for this question that the 235 judges are a random sample from the 389 judges. [Source: "Straight from the Horse's Mouth: Judicial Observations of Jury Behavior and the Need for Tort Reform," *Baylor Law Review*, Vol. 59, 2 (Spring 2007), pp. 419–434.]

a. Is the condition met that the sample size is at most 10% of the population size?

b. Use the formula

$$\hat{p} \pm z^* \sqrt{\frac{\hat{p}(1 - \hat{p})}{n}}$$

to compute a 95% confidence interval for the proportion of all district court judges in Texas who would say that not a single jury awarded disproportionately high compensatory damages in the last 48 months.

c. Now use the formula with the correction factor

$$\hat{p} \pm z^* \sqrt{\frac{\hat{p}(1 - \hat{p})}{n}} \cdot \sqrt{\frac{N - n}{N - 1}}$$

to compute a 95% confidence interval for the proportion of all district court judges in Texas who would say that not a single jury awarded disproportionately high compensatory damages in the last 48 months.

d. Compare the two intervals. Which is shorter? Why does this make sense in this situation?

8.2 ▶ A Significance Test for a Proportion: Interpreting a *P*-Value

Investigators often collect data for the purpose of comparing the results from the sample to some predetermined standard. In such tests of significance the goal is to test the significance of the difference between the results in the sample and the results predicted by the standard. If the difference is small, the data are consistent with the standard. When the difference is so large that it can't reasonably be attributed to chance, the investigator has strong evidence that the standard does not hold. In other words, the difference between the sample data and the standard is **statistically significant.**

The following two examples illustrate the reasoning of a significance test. The first describes a survey where the results are statistically significant. The second describes an experimental study where they are not.

Tests of significance look for departures from a standard.

| Example 8.9 | **Test of Radiation and Mutation** |

Solution

In 1986, the nuclear reactor at Chernobyl, in Ukraine, leaked radioactivity, causing concern that the radiation might cause mutations in the genes of humans and animals that would be passed on to offspring. About 2% of barn swallows have white feathers in places where the plumage is normally blue or red. The white feathers are caused by genetic mutations. In a sample of 266 barn swallows captured around Chernobyl between 1991 and 2000, about 16% had white feathers in places where the plumage is normally blue or red. What is the standard against which the results from the sample should be compared? Does the survey provide convincing evidence that there was a higher rate of these mutations in the barn swallows around Chernobyl than among barn swallows in general?

Solution

Researchers compared the proportion of mutations from the sample, $\hat{p} = 0.16$, with the standard proportion, which is 0.02. Display 8.4 shows the sampling distribution of

$\hat{p}$ for $n = 266$ and $p = 0.02$. Assuming that you can consider the captured barn swallows a random sample of barn swallows from the Chernobyl area, the sampling distribution shows that getting a sample proportion as high as 0.16 is extremely unlikely when $n = 266$ and $p = 0.02$.

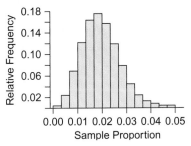

Display 8.4   Sampling distribution of $\hat{p}$ for $n = 266$ and $p = 0.02$.

Chernobyl barn swallow with white feathers indicating genetic mutation (above) and normal barn swallow (below). (Photograph by T. A. Mousseau. Published with permission.)

   In other words, if the overall percentage of barn swallows with this mutation in the Chernobyl area was still only 2%, it is not at all likely that 16% of the barn swallows in a sample of size 266 would have these white feathers. So the researchers came to the conclusion that there was an increased proportion of barn swallows with genetic mutations in the Chernobyl area. [Source: A. P. Moller and T. A. Mousseau, "Albinism and Phenotype of Barn Swallows (*Hirundo rustica*) from Chernobyl," *Evolution*, Vol. 55 (2001), pp. 2097–2104.]

Test of the Validity of Natal Charts

Example 8.10

Scientists and others are deeply concerned about the fact that, in the 21st century, there still are people who believe in paranormal phenomena such as psychic readers and astrology columns. In an attempt to debunk astrology, a well-designed study tested the astrological proposition that

> the position of the "planets" (all planets, the Sun and Moon, plus other objects defined by astrologers) at the moment of birth can be used to determine the subject's general personality traits and tendencies in temperament and behavior, and to indicate the major issues which the subject is likely to encounter.

   This study is important partly because it had the full cooperation of the National Council of Geocosmic Research, an organization of astrologers, which approved the procedure and recommended the astrologers involved.

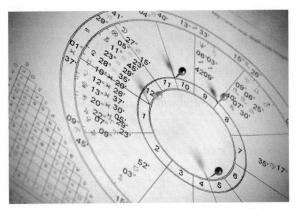

Natal chart shows positions of the sun, moon, and planets at the time of birth.

The astrologers prepared natal charts (horoscopes based on a person's exact time and place of birth) for 83 volunteer subjects. Each subject then was given three natal charts (their own and two randomly chosen charts that were made for other people). They were asked to pick out the one that most correctly described them. Twenty-eight out of 83 selected their own chart. What is the standard against which the results should be compared? Does the experiment provide convincing evidence that a person's own natal chart describes them better (in their own opinion) than randomly selected ones? [Source: Shawn Carlson, "A Double-Blind Test of Astrology," *Nature*, Vol. 318 (December 5, 1985), pp. 419–425.]

Solution

If their own natal chart doesn't describe a person any better than the randomly selected ones, the probability that a person will select their own chart is $p = 1/3$. This is the standard against which the results from the experiment should be compared.

The sampling distribution in Display 8.5 shows the proportion of times each possible value of $\hat{p}$ occurs when $n = 83$ and $p = \frac{1}{3}$.

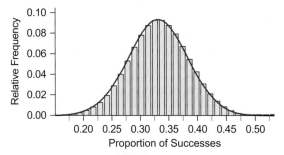

Display 8.5 Sampling distribution of $\hat{p}$ for $n = 83$ and $p = \frac{1}{3}$.

The sample proportion from this experiment is $\hat{p} = \frac{28}{83} \approx 0.337$, which lies near the middle of this distribution. This means that their own natal charts don't describe these people any better than randomly selected charts. So if the probability a person selects their own chart is $\frac{1}{3}$, getting 28 correct selections out of 83 tries is reasonably likely. In fact, it's just about exactly what you would expect. This certainly isn't convincing evidence that their own chart describes these people any better than randomly selected charts.

If you are like most people, this logic above seems natural: If someone wants to make a case for astrology, she has to show you that astrology does a better job of prediction than anybody can do just by guessing. The burden of proof is on the "geocosmic researchers" (astrologers). Similarly, if someone wants to make the case that barn swallows around Chernobyl have a higher rate of mutation than barn swallows elsewhere, he must show not just that the proportion of mutations in his sample is higher than 2%, but that it is so much higher than 2% that it is unlikely to have occurred just by chance. Keep that logic in mind while learning the more formal terminology in the rest of the section.

The two studies you have just read use the same logic, but the conclusions are different. The Chernobyl study was a sample survey. The barn swallows were, we are assuming, a random sample from a larger population. Thus, the results can be generalized to the larger population of all barn swallows around Chernobyl. The natal charts study involves randomization only in that the charts should have been presented to the subjects in random order. The subjects were not selected at random, but were volunteers. Thus, the results cannot be generalized to any larger population. Astrology failed its test, but only with these subjects.

Hypotheses

In a formal research study, the standard to which the result from the sample is compared is stated in the *null hypothesis*. The null hypothesis typically states that nothing has changed, that one population is identical to another in some aspect, that the process is fair, or that all choices are equally likely to occur. The null hypothesis is paired with an *alternative* (or *research*) *hypothesis* that states the nature of the change the researcher is trying to establish in the study.

> "Null" means amounting to nothing.

The Null Hypothesis and the Alternative Hypothesis

If you are doing a test of significance for a proportion and wish to compare the results from a sample to a standard value, p_0, begin by writing the **null hypothesis**.

The null hypotheses for a sample survey may be worded using either form below:

H_0: The proportion, p, of successes in the population from which the sample was taken is equal to the hypothesized, or standard, value, p_0.

H_0: $p = p_0$, where p is the proportion of successes in the population from which the sample was taken.

The null hypothesis for a study involving the probability of a success would be worded this way:

H_0: The probability, p, of a success on any one trial is equal to the hypothesized, or standard, value, p_0.

The **alternative hypothesis** has three forms, depending on what you want to establish. The following are the three forms for sample surveys.

H_a: $p \neq p_0$ The proportion of successes, p, in the population is not equal to the hypothesized value p_0.

H_a: $p > p_0$ The proportion of successes, p, in the population is greater than the hypothesized value p_0.

H_a: $p < p_0$ The proportion of successes, p, in the population is less than the hypothesized value p_0.

The first form defines a **two-sided test**. The second and third define **one-sided tests**.

Hypotheses for Test of Radiation and Mutation

Example 8.11

State the null hypothesis and alternative hypothesis for the test of whether barn swallows were more likely to have the white feather mutation in the Chernobyl area than worldwide. Do the results of the test provide evidence against the null hypothesis?

Solution

The standard was that, in the Chernobyl area, the proportion of barn swallows with the white feather mutation was 0.02, the same as worldwide. Thus, the null hypothesis is

H_0: $p = 0.02$, where p stands for the proportion of barn swallows in the Chernobyl area with the white feather mutation.

In this situation, only a sample proportion larger than 2% would provide evidence of an association between mutation and radiation, so this is a one-sided test. The alternative hypothesis is

H_a: $p > 0.02$, where p stands for the proportion of barn swallows in the Chernobyl area with the white feather mutation.

The results provide strong evidence against the null hypothesis. From Display 8.4, you can see that it would be very unlikely to get 16% of the birds in the sample having mutations if it is the case that $p = 0.02$.

| Example 8.12 | **Hypotheses for Test of the Validity of Natal Charts** |

State the null hypothesis and alternative hypothesis for the test of whether a person's natal chart describes him or her better (in that person's own opinion) than two randomly selected charts. Does the result that 28 out of 83 subjects selected their own chart provide evidence against the null hypothesis?

Solution
The standard was that, given three choices, the probability that a person selects his or her own natal chart is $\frac{1}{3}$. Thus, the null hypothesis is

H_0: $p = \frac{1}{3}$, where p stands for the probability that a person selects his or her own natal chart.

In this situation, only a sample proportion larger than $\frac{1}{3}$ would provide evidence of the validity of natal charts, so this is a one-sided test. The alternative hypothesis is:

H_a: $p > \frac{1}{3}$, where p stands for the probability that a person selects his or her own natal chart.

The results do not provide evidence against the null hypothesis. From Display 8.5, you can see that 28 out of 83 subjects correctly choosing their own chart , or $\hat{p} \approx 0.337$, is a reasonably likely outcome under the null hypothesis that $p = \frac{1}{3}$.

In the previous example, there was no evidence against the null hypothesis. This doesn't mean, however, that you can conclude that the null hypothesis is correct that $p = \frac{1}{3}$. There are other values of p that also are plausible for a sample proportion of $\hat{p} = 28/83 \approx 0.3373$. One of those, certainly, is $p = 0.3373$. R. A. Fisher, who developed the idea of tests of significance put it this way: "... the null hypothesis is never proved or established, but is possibly disproved, in the course of experimentation. Every experiment may be said to exist only in order to give the facts a chance of disproving the null hypothesis." [Source: R. A. Fisher, *The Design of Experiments* (Edinburgh, UK: Oliver and Boyd, 1935).]

[Source: S. Adams, January 13, 1998. © 1998 United Features Syndicate, Inc.]

In the test of the barn swallow mutations, there are three different proportions to keep straight:

p: the population proportion

- p stands for the proportion of *all* barn swallows in the Chernobyl area with mutations. No one knows this value for sure because not all barn swallows have been examined.

- $\hat{p}$ stands for the proportion of barn swallows *in the sample* with mutations. For this sample, $\hat{p} = 0.16$.

- p_0 stands for the *hypothesized* proportion of barn swallows in the Chernobyl area with mutations. In the example, the value of p_0 is 0.02.

$\hat{p}$: the sample proportion

p_0: the hypothesized value of the population proportion

DISCUSSION
Hypotheses

D10. For each of these situations, give the value of the standard p_0, say whether the situation calls for a one-sided or two-sided test, write the null and alternative hypotheses, and give the value of the sample proportion, $\hat{p}$.

a. You want to see if people can identify the gourmet coffee from three cups of coffee containing the gourmet coffee, ordinary coffee, and instant coffee. You give 100 randomly selected people a taste of each (in random order) and 52 people correctly choose the gourmet coffee.

b. You suspect gender discrimination in hiring in your local police department. Forty percent of the applicants are women. A random sample of employees finds that only 15% are women.

STATISTICS IN ACTION 8.3 ► Spinning and Flipping Pennies

People tend to believe that pennies are balanced. They generally have no qualms about flipping a penny to make a fair decision. Is it really the case that penny flipping is fair? What about spinning pennies?

In this activity, you will record the results of 400 flips of a penny and 400 spins of a penny. You will use the data to test whether you have evidence that spinning and flipping pennies aren't fair.

What You'll Need: pennies (at least one per student)

1. Begin spinning your penny. To spin, hold the penny upright on a table or the floor with the forefinger of one hand and flick the side edge with a finger of the other hand. The penny should spin freely, without bumping into anything before it falls. Spin pennies until your class has a total of 400 spins. Count the number of heads and compute $\hat{p}$.

2. Begin flipping your penny. Let it rotate many times in the air, catch it in the air while it is still rotating, and then open your hand. Record whether it is heads or tails. Flip pennies until your class has a total of 400 flips. Count the number of heads and compute $\hat{p}$.

3. Perform a test of the significance of the results of your penny spins by answering the following questions:

   a. When investigating whether spinning a penny is fair, what is the standard?

   b. What is your sample proportion, $\hat{p}$, of heads from step 1?

   c. Locate your sample proportion, $\hat{p}$, on the sampling distribution in Display 8.6.

(Continued)

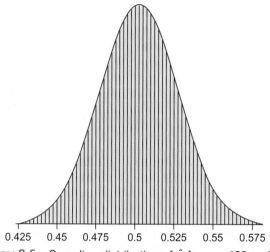

Display 8.6   Sampling distribution of $\hat{p}$ for $n = 400$ and $p = 0.5$.

    d. What is your conclusion? Do your data provide evidence against the standard?

4. Now use your penny-flipping results from step 2 to answer the questions in step 3.

A Test Statistic

For both the natal chart and barn swallow examples, it was easy to see where the sample proportion $\hat{p}$ fell in the sampling distribution by looking at Display 8.4 or 8.5. You could then decide whether you had strong evidence against the null hypothesis. To determine this without constructing the entire sampling distribution, for some binomial situations (when $np_0 \geq 10$ and $n(1 - p_0) \geq 10$) you will be able to compute a z-score and use the fact that the sampling distribution for $\hat{p}$ is approximately normal.

| A *z*-score! |

A Test Statistic for Testing a Proportion

| The z-score is an example of a test statistic. |

To determine where $\hat{p}$ lies in the sampling distribution for the given standard p_0, compute the value of the **test statistic**

$$z = \frac{statistic - parameter}{standard\ error} = \frac{\hat{p} - p_0}{\sqrt{\dfrac{p_0(1 - p_0)}{n}}}$$

The value of z tells you how many standard errors the sample proportion $\hat{p}$ lies from the standard, p_0.

Example 8.13 | A Test Statistic for the Natal Charts

Compute the test statistic for Example 8.10 on page 381, where 28 out of 83 subjects selected their own natal chart when given three choices.

Solution

The sampling distribution of $\hat{p}$ for $p = \frac{1}{3}$ and $n = 83$ in Display 8.5 is approximately normal because $np_0 = 83\left(\frac{1}{3}\right) = 27.7 \geq 10$ and $n(1 - p_0) = 83(1 - \frac{1}{3}) = 55.3 \geq 10$. This means you can locate $\hat{p}$ by computing a z-score and using facts you know about the normal distribution. For the sample proportion, $\hat{p}$, of $\frac{28}{83}$:

$$z = \frac{\hat{p} - p_0}{\sqrt{\frac{p_0(1 - p_0)}{n}}}$$

$$= \frac{\frac{28}{83} - \frac{1}{3}}{\sqrt{\frac{\frac{1}{3}\left(1 - \frac{1}{3}\right)}{83}}}$$

$$\approx 0.08$$

The test statistic of $z = 0.08$ means that the sample proportion is a mere 0.08 standard errors above the hypothesized mean of $\frac{1}{3}$. Thus, the result from the sample is just about exactly what you would expect to happen if people cannot pick their own natal chart with probability any greater than $\frac{1}{3}$.

DISCUSSION
A Test Statistic

D11. What does it mean if the test statistic is equal to 0?

D12. If everything else remains constant, what will happen to the test statistic z when

    a. the sample size n increases?

    b. $\hat{p}$ gets farther from p_0?

D13. Could the researchers have come to the wrong conclusion when they concluded that people are not able to select their own natal chart at a rate any better than pure chance? Explain.

P-Values and Statistical Significance

Good statistical practice requires reporting a probability along with (or instead of) the test statistic. This probability, called a *P-value*, corresponds to the area under the sampling distribution for $\hat{p}$ (constructed assuming that the null hypothesis is true) that represents possible outcomes that are more extreme than the value of $\hat{p}$ from the sample.

> ### The *P*-Value for a Test of Significance
>
> The **P-value** for a test is the probability of seeing a result that is as extreme as or more extreme than the result you got from your sample *if the null hypothesis is true*.
> The *P*-value measures the strength of the evidence against the null hypothesis. The closer the *P*-value is to 0, the stronger the evidence against the null hypothesis (and in favor of the alternative hypothesis). The closer the *P*-value is to 1, the weaker the evidence against the null hypothesis.

Here is how to find the *P*-value under the assumption that the null hypothesis is correct that $p = p_0$.

- Be sure you have a random sample, and check that the sampling distribution of $\hat{p}$ is approximately normal: $np_0 \geq 10$ and $n(1 - p_0) \geq 10$.
- Compute the test statistic (*z*-score) for your sample.

- Use the table of the normal distribution to find the area that falls outside of that z score. Whether you include the area in only one or in both tails depends on the form of the alternative hypothesis:

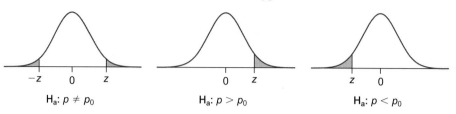

$H_a: p \neq p_0$ $H_a: p > p_0$ $H_a: p < p_0$

Example 8.14 | Computing *P*-Values

It has been estimated that 60% of university students are anxious when they have to deal with mathematics. [Source: University of Granada, "Six Out of 10 University Students Have Math Anxiety, Spanish Study Finds," *Science Daily*, April 21, 2009, *www.sciencedaily.com/releases/2009/04/090401103123.htm*.] You want to test whether this percentage is different on your campus. Your null hypothesis is $p = 0.6$, where p is the proportion of all students on your campus who are anxious when they have to deal with mathematics. You take a random sample of 100 students from your campus. For each of the following situations, write the appropriate null hypothesis, draw a sketch of each situation, find the *P*-value, and interpret it.

    a. You believe that the proportion is higher on your campus. Your test statistic turns out to be $z = 1.84$.

    b. You believe that the proportion is lower on your campus. Your test statistic turns out to be $z = -1.84$.

    c. You don't have a hypothesis about whether the proportion is higher or lower on your campus. Your test statistic turns out to be $z = 1.84$.

    d. You don't have a hypothesis about whether the proportion is higher or lower on your campus. Your test statistic turns out to be $z = -1.84$.

Solution

First, the sampling distribution of $\hat{p}$ is approximately normal because both $np_0 = 100(0.6) = 60$ and $n(1 - p_0) = 100(1 - 0.6) = 40$ are at least 10. You can use the table of the normal distribution to find probabilities.

    a. For this one-sided test, evidence against the null hypothesis occurs only when $\hat{p}$ is greater than 0.6. The alternative hypothesis is $p > 0.6$. From Table A, the area above $z = 1.84$ is $1 - 0.9671$, or 0.0329. If you take a random sample from a population where $p = 0.6$, then the probability of getting a value of z equal to 1.84 or even larger is 0.0329.

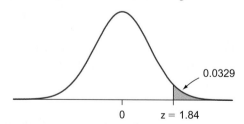

0.0329

0 $z = 1.84$

    b. For this one-sided test, evidence against the null hypothesis only occurs when $\hat{p}$ is less than 0.6. The alternative hypothesis is $p < 0.6$. From Table A, the area below $z = -1.84$ is 0.0329. If you take a random sample from a population where $p = 0.6$, then the probability of getting a value of z less than or equal to -1.84 is 0.0329.

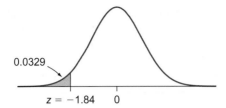

c. For this two-sided test, evidence against the null hypothesis occurs when $\hat{p}$ is less than or greater than 0.6. The alternative hypothesis is $p \neq 0.6$. From Table A, the area above $z = 1.84$ is $1 - 0.9671$, or 0.0329, and the area below $-z = -1.84$ is 0.0329. The total area is 2(0.0329), or 0.0658. If you take a random sample from a population where $p = 0.6$, then the probability of getting a value of z at least as extreme as 1.84 or -1.84 is 0.0658.

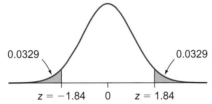

d. The alternative hypothesis, computations, *P*-value, and interpretation are the same as in part *c*.

Interpreting a *P*-Value

Example 8.15

Chimpanzees were given the choice of pulling one of two heavy rakes toward them. Pulling one of the rakes would bring food with it. Pulling the other rake would not bring in food. Display 8.7 shows a typical setup for this study. The round object represents the food. The researchers found that the correct rake was chosen significantly more often than would be expected by chance (*P*-value = 0.01). Interpret this *P*-value in the context of the situation. [Source: Victoria Horner and Andrew Whiten, "Causal Knowledge and Imitation/Emulation Switching in Chimpanzees (*Pan troglodytes*) and Children (*Homo sapiens*)," *Animal Cognition*, Vol. 8 (2005), pp. 164–181.]

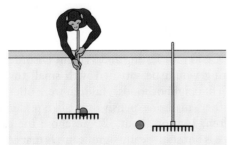

Display 8.7   A chimpanzee about to rake in some food. If the chimpanzee had pulled in the other rake, it wouldn't have gotten the food.

Solution

The *P*-value of 0.01 means that *if the chimps were selecting a rake at random*, which gives them a 0.5 chance of selecting the one that rakes in the food, the probability is only 0.01 that the chimps would rake in the food as often as or more often than they did. The researchers therefore can conclude that the chimps were able to deliberately choose the rake that got them the food.

"Statistically significant" means the value of $\hat{p}$ is unlikely to occur by chance under the null hypothesis.

The word "significantly" in the previous example has a technical meaning. "Significant" or "statistically significant" means that a difference as large as that between $\hat{p}$ and $p_0 = 0.5$ would be unlikely to occur just by chance if the probability a chimpanzee selects the right rake is 0.5.

The following box gives a rule for deciding when a result from a sample is statistically significant.

Statistical Significance

A sample proportion is said to be statistically significant if the *P*-value places it in an outer tail of the sampling distribution constructed under the assumption that the null hypothesis is true. In many statistical studies, the term "statistically significant" is used only when the *P*-value is less than 0.05. Sometimes the *level of significance* will be 0.01 or 0.001, or another value.

Example 8.16 | Statistical Significance for the Test of Natal Charts

Is the result from the test of natal charts statistically significant at the 0.05 level of significance?

Solution

No, getting 28 volunteers out of 83 who correctly select their own natal chart isn't statistically significant. From Example 8.13, the test statistic, z, is 0.08. For a one-sided test, the *P*-value is 0.4681, which is greater than 0.05. (Using a calculator without rounding intermediate steps, the *P*-value will be 0.4691.) The result from the sample is consistent with the null hypothesis that the probability a person selects his or her own natal chart is $\frac{1}{3}$.

Example 8.17 | Halloween Treats

Researchers at the Yale Center for Eating and Weight Disorders wanted to see if children out trick-or-treating would be satisfied with small toys instead of candy. In households in Connecticut neighborhoods, children were offered two bowls: one contained candy and the other small, inexpensive toys like plastic bugs that glow in the dark. Of the 283 children, 148, or about 52.3%, chose the candy. The researchers report that the difference is not statistically significant. Is that correct? You may assume that the children are a random sample from all children trick-or-treating in Connecticut neighborhoods that year. [Source: "Trends: Halloween, for Skinnier Skeletons," *New York Times*, October 21, 2003, report of a study published in *The Journal of Nutrition Education and Behavior*, (July–August 2003).]

Solution

Hypotheses

The researchers wanted to determine whether it makes any difference to children whether they get candy or small toys, rather than preferring one over the other, so this will be a two-sided test of significance.

H_0: $p = 0.5$, where p is the proportion of *all* children out trick-or-treating in Connecticut neighborhoods who would have picked the candy. (You could just as well test that the proportion of children who would have picked the toy is 0.5.)

H_a: $p \neq 0.5$

Computations

The test statistic is

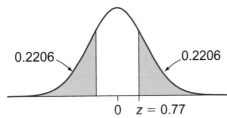

$$z = \frac{\hat{p} - p_0}{\sqrt{\frac{p_0(1 - p_0)}{n}}}$$

$$= \frac{\frac{148}{283} - 0.5}{\sqrt{\frac{0.5(1 - 0.5)}{283}}}$$

$$\approx 0.77$$

0.2206   **0.2206**

0   z = 0.77

Display 8.8   Standard normal distribution.

You can use a table of the normal distribution to find the *P*-value illustrated in Display 8.8 because $np_0 = 283(0.5) \geq 10$ and $n(1 - p_0) = 283(1 - 0.5) \geq 10$. From Table A, a *z*-score of 0.77 has an area of 0.2206 above it. Because this is a two-sided test, the *P*-value is 2(0.2206), or 0.4412. (Using a calculator, as in the margin, the *P*-value will be 0.4397.)

The following Minitab output gives the same value of *z* and *P*-value as the calculator, along with other information.

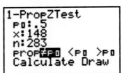

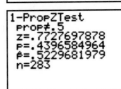

Test and CI for One Proportion

Test of p = 0.5 vs. p not = 0.5

Sample	X	N	Sample p	95% CI	z-value	P-value
1	148	283	0.522968	(0.464776, 0.581161)	0.77	0.440

Using the normal approximation.

Interpretation of *P*-value

The *P*-value is more than 0.05, so this is not a statistically significant result. If children are equally as likely to pick the candy as the toy, getting 148 out of 283 randomly selected children who pick the candy is reasonably likely. There is no statistically significant evidence that, overall, children out trick-or-treating have a preference one way or the other. (However, as the researchers point out, the children may have selected the toy because toys were a novelty at Halloween. If every household offered toys, they might not be so satisfied with them.)

Statistical significance doesn't mean the same thing as "practical" significance. If the sample size is large enough, a result can be statistically significant even though the difference between $\hat{p}$ and p is too small to make any difference in the real-life situation. For example, a study of a huge number of births in the United States found that 51.2% were boys and 48.8% were girls. This is a statistically significant difference, meaning

> Just because a result is statistically significant doesn't mean the difference is of any practical significance.

that the difference between the proportions of boys and girls can't reasonably be attributed to chance. However, it would be of no practical significance to a pair of prospective parents who were hoping for a boy or for a girl. They would not rush out and buy blue booties because, for them, the probability that the baby will be a boy is no different than the probability that it will be a girl. [Source: Centers for Disease Control, 2005, *www.cdc.gov*; *Trend Analysis of the Sex Ratio Birth in the United States*, *National Vital Statistics Reports*, Vol. 53, 20 (June 14, 2005), p. 1.]

DISCUSSION
P-Values and Statistical Significance

D14. Compute the *P*-value for each situation in D10. Which results are statistically significant at the 0.05 level of significance?

D15. Why is the phrase "if the null hypothesis is true" necessary in the interpretation of a *P*-value?

D16. Does the *P*-value give the probability that the null hypothesis is true?

D17. What does it mean if the *P*-value is 1?

D18. Refer to the preceding discussion about the fact that 51.2% of all babies born in the United States are boys and 48.8% are girls.

    a. Give an example of a situation or an industry to which this difference does have some practical significance.

    b. Suppose that parents had four children. Is there much difference in the probability that they would all be boys and the probability that they would all be girls?

Summary 8.2: A Significance Test for a Proportion— Interpreting a *P*-Value

Suppose you want to decide whether it is reasonable to assume that your random sample comes from a population where the proportion of successes is p_0. A test of significance for a proportion tells you whether the results from your sample are so different from what you would expect that you have statistically significant evidence that p_0 is not a plausible value for the proportion of successes in the population. If the sample proportion, $\hat{p}$, is relatively far from p_0, you cannot reasonably attribute the difference to chance variation and should conclude that p_0 is not among the plausible values for the proportion of successes in the population. If the sample proportion, $\hat{p}$, is relatively close to p_0, you can reasonably attribute the difference to chance variation and must consider p_0 as being among the plausible values for the proportion of successes in the population.

You measure how close the sample proportion $\hat{p}$ is to p_0 by computing the test statistic,

$$z = \frac{\hat{p} - p_0}{\sqrt{\dfrac{p_0(1 - p_0)}{n}}}$$

The *P*-value for a test is the probability of seeing a value of the test statistic, z, that is as extreme as or more extreme than the test statistic computed from your random sample *if the null hypothesis is true*.

The larger in absolute value the test statistic, $|z|$, the smaller the *P*-value will be. The closer the *P*-value is to 0, the more evidence you have against the null hypothesis. If the *P*-value is large, typically defined as larger than 0.05, you don't have statistically significant evidence against the null hypothesis.

When you have a study where the subjects aren't a random sample from a larger population, but you want to test whether the difference between the results and the standard can reasonably be attributed to chance alone, you can use the same reasoning as above. You will be testing the null hypothesis that $p = p_0$, where p stands for the probability of a "success" for the subjects involved in your study.

Practice

Hypotheses

P15. This year, half of the graduating seniors wanted extra tickets for their graduation ceremony. To anticipate whether there might be a change in that percentage next year, the junior class took a random sample of 40 juniors and found that 16 will want extra tickets.

a. What is the standard, p_0?

b. Should this be a one-sided or two-sided test? Write the null and alternative hypotheses.

c. What is the sample proportion, $\hat{p}$?

d. Use Display 8.9 to determine whether there is strong evidence of a change.

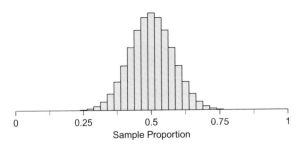

Display 8.9 Sampling distribution of $\hat{p}$ for $n = 40$ and $p = 0.5$.

P16. A student took a 40-question true-false test and got 30 answers correct. The student says, "That proves I was not guessing at the answers."

a. What is the standard, p_0?

b. Should this be a one-sided or two-sided test? Write the null and alternative hypotheses.

c. What is the sample proportion, $\hat{p}$?

d. Use Display 8.9 to assess whether there is strong evidence that the student wasn't guessing.

e. Does the answer to part d *prove* that the student was not guessing?

The Test Statistic

P17. Find the values of the test statistic for the situations in P15 and Pl6.

P18. Forty-five dogs and their owners were photographed separately. A judge was shown a picture of each owner and pictures of two dogs and asked to pick the dog that went with the owner. The judge was right 23 times. [Source: Based on a study published in *Psychological Science*, 2004.]

a. What is the standard, p_0, in a test that the judge did better than could reasonably be expected if he was just guessing?

b. Should this be a one-sided or two-sided test? Write the null and alternative hypotheses.

c. What is the sample proportion, $\hat{p}$?

d. What is the value of the test statistic?

P-Values and Statistical Significance

P19. Suppose you recreate the chimpanzee experiment in Example 8.15 on page 389 with an apparatus that tests whether 2-year-olds can select the rake that results in the food. You test 50 two-year-olds and find that 28 select the rake that pulls in the food. Find the *P*-value for this test and interpret it.

P20. A 2008 National Center for Education Statistics report found that two-thirds of undergraduates get some form of financial aid. You wonder if this proportion is different today. You take a random sample of 400 undergraduates from the United States and find that 302 receive some form of financial aid. [Source: *http://nces.ed.gov/pubs2009/2009166.pdf.*]

a. What is the standard, p_0?

b. Should this be a one-sided or two-sided test? Write the null and alternative hypotheses.

c. What is the sample proportion, $\hat{p}$?

d. What is the value of the test statistic? What is the *P*-value?

e. Assess the strength of the evidence against the null hypothesis. In particular, is the result statistically significant at the 0.05 level of significance?

P21. Inattentional blindness occurs when people don't see something that is right in front of their eyes. In a study on inattentional blindness in dynamic scenes, researchers asked subjects to watch one of several versions of a video and count the number of passes of a basketball among three players. The subjects needed to concentrate because the counting was complicated by the presence of a second team of three players (dressed in a different color) who also were passing a basketball. About 46 seconds into the video, an unexpected event occurred. Either a tall woman with an open umbrella or a person dressed as a gorilla walked through the scene, staying onscreen for about 5 seconds. The researchers conclude: "Out of all 192 observers across all conditions, 54% noticed the unexpected event and 46% failed to notice the unexpected event, revealing a substantial level of sustained inattentional blindness for a dynamic event . . . ."

Figure provided by Daniel Simons [Source: Daniel J. Simons and Christopher F. Chabris, "Gorillas in Our Midst: Sustained Inattentional Blindness for Dynamic Events," *Perception*, Vol. 28, (1999), pp. 1059–1074. See the videos at viscog.beckman.illinois.edu/flashmovie/15.php.]

a. Suppose your definition of *substantial level* is 50%. Is it plausible that (at least) half of all people would fail to notice the unexpected event? Use the alternative hypothesis that less than 50% would fail to notice the unexpected event.

b. What proportions are plausible?

P22. At the beginning of this section, you read that 2% of barn swallows have white feathers in places where the plumage is normally blue or red. However, about 16% of barn swallows captured around Chernobyl after 1991 had such genetic mutations. The sample captured around Chernobyl was relatively large—266 barn swallows. Display 8.4, reproduced in Display 8.10, shows the sampling distribution of the sample proportion, $\hat{p}$, for $n = 266$ and $p = 0.02$.

a. Is the normal distribution a good approximation to the binomial distribution in Display 8.10?

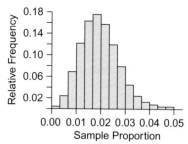

Display 8.10   Sampling distribution of $\hat{p}$ for $n = 266$ and $p = 0.02$.

b. If you were to estimate the P-value for the barn swallows using a z-score and Table A, would that estimate differ much from the true value in this case?

Exercises

E27. A psychologist was struck by the fact that at square tables many pairs of students sat on adjacent sides rather than across from each other and wondered if people prefer to sit that way. The psychologist collected some data, observing 50 pairs of students seated in the student cafeteria of a California university. The tables were square and had one seat available on each of the four sides. The psychologist observed that 35 pairs sat on adjacent sides of the table, while 15 pairs sat across from each other. He wanted to know if the evidence suggested a preference by students for adjacent sides, as opposed to simply random behavior. [Source: Joel E. Cohen, "Turning the Tables," in Frederick Mosteller et al., eds., *Statistics by Example: Exploring Data*, (Reading, MA: Addison-Wesley, 1973), pp. 87–90.]

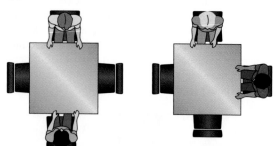

a. Suppose that two students sit down randomly at a square table. Explain why the probability that they sit on adjacent sides is 2/3.

b. What is an appropriate null hypothesis to test whether students' seating pattern is random, as opposed to their having a preference to sit on adjacent sides?

c. Is this a one-sided or two-sided test? What is the appropriate alternative hypothesis?

d. What is the test statistic?

e. Find the P-value for this test.

f. Write a conclusion for the psychologist.

E28. Suppose that the psychologist in E27 mistakenly thought that the probability that two randomly seated students sit on adjacent sides is $\frac{1}{2}$.

a. What is his null hypothesis for testing whether students' seating pattern is random, as opposed to their having a preference to sit on adjacent sides?

b. Is this a one-sided or two-sided test? What is the psychologist's alternative hypothesis?

c. What is his test statistic?

d. Find the P-value for his test.

e. What conclusion would the psychologist come to? Is the conclusion valid?

E29. According to the U.S. Census Bureau, about 69% of houses across the country are occupied by their owners. Your class randomly samples 50 houses in your community and finds that 30 houses are occupied by the owners. Follow these steps to test whether your community differs from the nation as a whole as to the percentage of houses that are owner occupied.

a. State the null hypothesis to be tested. What is the alternative hypothesis?

b. From the following Minitab output, find the value of the test statistic.

c. Find the P-value for this test and explain what it means.

d. Write a conclusion, based on your analysis, in the context of the problem.

Test and CI for One Proportion

Test of p = 0.69 vs p not = 0.69

Sample	X	N	Sample p	95% CI	z-Value	P-value
1	30	50	0.600000	(0.464210, 0.735790)	−1.38	0.169

Using the normal approximation.

E30. You wonder whether a majority of black youths in the country think that black youths receive a poorer education than do white youths. You discover that the Black Youth Project has collected data on a random sample of 634 black youths between the ages of 15 and 25. One of the reported results is that 54% of the black youths interviewed think that black youth receive a poorer education, on average, than do white youth. [Source: *http://black-youthproject.uchicago.edu/writings/research summary.pdf*.]

a. State the null hypothesis to be tested. What is the alternative hypothesis?

b. From the following Minitab output, find the value of the test statistic.

c. Find the *P*-value for this test and explain what it means.

d. Write a conclusion, based on your analysis, in the context of the problem.

Test and CI for One Proportion

Test of $p = 0.5$ vs. $p > 0.5$

Sample	X	N	Sample P	95% Lower Bound	z-Value	P-Value
1	342	634	0.539432	0.506871	1.99	0.024

Using the normal approximation.

E31. In a second test in the astrology experiment described in Example 8.10 on page 381, the astrologers were given a natal chart and asked which of three California Psychological Inventory (CPI) test results belonged to the subject of the chart. The astrologers chose the correct CPI test results for 40 of the 116 charts. They had predicted that they would be able to chose the correct results half of the time. Test the strength of the evidence against the astrologers' claim.

a. What is the standard, p_0?

b. Should this be a one-sided or two-sided test? Write the null and alternative hypotheses.

c. What is the sample proportion, $\hat{p}$?

d. What is the value of the test statistic? What is the *P*-value?

e. Assess the strength of the evidence against the null hypothesis.

E32. In a second test in the astrology experiment in Example 8.10 on page 381, the astrologers were given a natal chart and asked which of three California Psychological Inventory (CPI) test results belonged to the subject of the chart. The astrologers chose the correct CPI test results for 40 of the 116 charts. Test whether there is a statistically significant difference between the actual results and random guessing. [Source: Shawn Carlson, "A Double-Blind Test of Astrology," *Nature*, Vol. 318 (December 5, 1985) pp. 419–425.]

a. What is the standard, p_0?

b. Should this be a one-sided or two-sided test? Write the null and alternative hypotheses.

c. What is the sample proportion, $\hat{p}$?

d. What is the value of the test statistic? What is the *P*-value?

e. Assess the strength of the evidence against the null hypothesis.

E33. For an Associated Press–Ipsos poll, a random sample of 1013 adults answered this question: "In general, do you believe in or do you not believe in ghosts?" Thirty-four percent responded that they believe in ghosts. Test whether it is plausible that, if you could ask all adults in the United States this question, half would say they believe in ghosts. Use the alternative hypothesis that less than half would say this. [Source: *www.polling report.com*, October 16–18, 2009.]

E34. "Americans increasingly favor raising the driving age, a *USA Today*/CNN/Gallup poll has found. Nearly two-thirds—61%—say they think a 16-year-old is too young to have a driver's license. Only 37% of those polled thought it was okay to license 16-year-olds, compared with 50% who thought so in 1995. A slight majority, 53%, thinks teens should be at least 18 to get a license. The poll of 1002 adults was conducted Dec 17-19, 2004." If you asked all adult residents of the United States in 2004 if they thought that a 16-year-old is too young to have a driver's license, is it plausible that only 55% would say yes? You may assume that the sample is equivalent to a random sample. [Source: *USA Today* polls, *Is 16 Too Young to Drive a Car?* March 2, 2005.]

E35. Kevin has heard that if you spin a penny by flicking it with your finger and letting it spin until it falls onto the table, it is less likely to land heads up than tails up. He spins a penny 40 times and gets 10 heads. Is this statistically significant evidence that a spun penny is less likely to land heads up than tails up?

a. What is the standard, p_0?

b. Should this be a one-sided or two-sided test? Write the null and alternative hypotheses.

c. What is the sample proportion, $\hat{p}$?

d. What is the value of the test statistic? What is the *P*-value?

e. Assess the strength of the evidence against the null hypothesis.

E36. For a project in her psychology class, Maya presented 40 subjects with two horoscopes, in random order. (The horoscopes were selected from an out-of-town newspaper that the subjects would not have seen.) Each subject was then asked which horoscope more closely matched the type of day he or she had yesterday. One of the horoscopes actually was yesterday's horoscope for that person's astrological sign and the other horoscope was for a completely different astrological sign. Twenty-three of the subjects chose their own horoscope rather than one for a different sign of the zodiac. Is this statistically significant evidence that horoscopes can predict the type of day a person will have?

a. What is the standard, p_0?

b. Should this be a one-sided or two-sided test? Write the null and alternative hypotheses.

c. What is the sample proportion, $\hat{p}$?

d. What is the value of the test statistic? What is the *P*-value?

e. Assess the strength of the evidence against the null hypothesis.

E37. Suppose 28 students out of a random sample of 50 students carry a backpack around campus. Test the claim that 60% of the students on campus carry backpacks to class.

   a. State the null hypothesis in words and in symbols.

   b. Calculate the value of the test statistic. Calculate the P-value for the test. Use this P-value in a sentence that explains what it represents.

   c. What is your conclusion? Explain it in the context of this problem.

E38. Follow the three steps in E37 for a random sample of 50 students, but select the students from your own campus.

E39. Which of statements A–E is the best explanation of what is meant by the P-value of a test of significance?

   A. Assuming that you had a random sample and the other conditions for a significance test are met, the P-value is the probability that H_0 is true.

   B. Assuming that H_0 is true, the P-value is the probability of observing a value of a test statistic, z, at least

as far out in the tails of the sampling distribution as is the value of z from your sample.

   C. The P-value is the probability that H_0 is false.

   D. Assuming that the sampling distribution is normal, the P-value is the probability that H_0 is true.

   E. Assuming that H_0 is true, the P-value is the probability of observing the same value of z that you got in your sample.

E40. Which of statements A–E is the best explanation of what is meant by statistical significance?

   A. The result from the sample, or one more extreme, is unlikely to have happened by chance.

   B. The result from the sample, or one more extreme, is unlikely to have happened if the null hypothesis is true.

   C. The difference between p_0 and $\hat{p}$ is less than 0.05.

   D. The difference between p_0 and $\hat{p}$ is more than 0.05.

   E. The result from the sample is pretty much what you would expect if the null hypothesis is true.

8.3 ▶ A Significance Test for a Proportion: Making a Decision

In the previous section, you learned how to find and interpret a P-value. But sometimes, researchers must do more than assess the strength of the evidence against the null hypothesis. They must make a decision. Here are three common situations where significance testing helps people decide on a course of action:

Child with Down syndrome.

- *Quality control.* Five percent of the windshields produced by the long-time supplier for a car manufacturer don't meet specifications. The car manufacturer is considering switching to a new supplier and inspects a random sample of 300 windshields produced by this supplier. Three percent don't meet specifications. Based on this evidence, would you recommend a switch?

- *Medicine.* Current tests available for the first trimester of a pregnancy detect Down syndrome in 82% to 87% of the cases. Your company would like to market a new test, but only if it exceeds the 87% detection rate. In a random sample of 200 women whose babies were later verified to have Down syndrome, 182 of the cases were detected by the new test. Should your company market the new test?

- *Political funding.* A labor union has a $2,000,000 fund held in reserve. It plans to spend the money on advertising on behalf of a new bond issue, but only if statewide support for the bond issue falls below 55%. A random sample of 700 likely voters finds current support at 53%. Should the labor union begin its advertising campaign?

Beginning in 1926, Jerzy Neyman and Egon Pearson laid out the foundations of the "hypothesis" testing approach to statistical inference, which allows you to make decisions in situations such as those above. You will study their approach in this section.

The Level of Significance

You saw in the last section that results from a sample often are called statistically significant when the P-value is less than 0.05. The proportion, 0.05 in this case, is called **the level of significance** and is denoted α. Other proportions can be used depending on the level of significance desired.

| **Fast Forward** | **Example 8.18** |

Advertisers are deeply worried that people who fast-forward through a commercial will miss the advertiser's message. Researchers thought that fast-forwarders would respond better to commercials with heavy branding (brand information is on screen for 12 seconds) than to commercials with limited branding (brand information is on screen for 3 seconds).

An experiment was designed where subjects viewed commercials for two unfamiliar chocolate bars: Flake and Aero. For each chocolate bar, two versions of each 30-second commercial were created, one with heavy branding and one with limited branding. Each subject watched a nature show, fast-forwarding through the two commercials, one for Flake and one for Aero. One commercial, for a chocolate bar that was randomly selected, had heavy branding and the other had limited branding. As they were leaving, the subjects were told that they were free to take a chocolate bar from a basket filled with Flake and Aero bars. Eighteen of 27 chose the bar corresponding to the commercial they had seen that had heavy branding. [Source: S. Adam Brasel and James Gips, "Breaking Through Fast-Forwarding: Brand Information and Visual Attention," *Journal of Marketing*, Vol. 72 (November 2008), pp. 31–48.]

a. Should this be a one-sided or two-sided test?

b. What are the hypotheses?

c. What is the *P*-value for this test?

d. Based on this evidence and a level of significance of 0.05, can you reject the null hypothesis? What if the level of significance were 0.01?

e. What decision should advertisers make about the type of commercial they should use for people who fast-forward through commercials?

Solution

a. This will be a one-sided test because the researchers expected the fast-forwarding subjects to respond better to the commercial with the heavy branding.

b. If the type of branding makes no difference, the probability is 0.5 that the subject would pick the bar with heavy branding. Thus, the hypotheses are

H_0: $p = 0.5$, where p stands for the probability that the subjects would pick the bar with heavy branding.

H_a: $p > 0.5$

c. The test statistic is

$$z = \frac{\hat{p} - p_0}{\sqrt{\frac{p_0(1 - p_0)}{n}}}$$

$$= \frac{\frac{18}{27} - 0.5}{\sqrt{\frac{0.5(1 - 0.5)}{27}}}$$

$$\approx 1.73$$

> The test statistic for a one-sided test is the same as for a two-sided test.

0.0418

0   z = 1.73

> The *P*-value for a one-sided test will be half as large as it would be for a two-sided test.

> Reject a null hypothesis that is not likely to have produced the sample proportion.

From Table A, a *z*-score of 1.73 has an area of 0.0418 above it. Because this is a one-sided test, the *P*-value is 0.0418. (From a calculator, the *P*-value is 0.0416.)

d. The *P*-value is less than 0.05, so you reject the null hypothesis. If the type of commercial makes no difference in which chocolate bar fast-forwarders would select, there is only a 0.0418 chance that so many fast-forwarders would pick the chocolate bar corresponding to the commercial with heavy branding. If the level of significance were 0.01, you would not reject the null hypothesis because the *P*-value of 0.0418 is larger than 0.01.

e. Using a 0.05 level of significance, advertisers would decide to use heavy branding for people who fast-forward through commercials. (The experiment also included a group of people who didn't fast-forward through commercials. They picked the heavily branded chocolate bar at about the same rates as the fast-forwarders, so you can expect to see heavy branding in commercials.)

The following box summarizes the use of the level of significance.

The Level of Significance and Decision Making

If the *P*-value computed from your sample is less than the level of significance, α, that you have chosen, you have evidence against the null hypothesis. Reject the null hypothesis and say that the result is statistically significant.

If the *P*-value is greater than α, you do not reject the null hypothesis. The result is not statistically significant.

> $\alpha = 0.05$ is the default.

If a level of significance isn't specified, it is usually safe to assume that $\alpha = 0.05$.

DISCUSSION
The Level of Significance

D19. If everything else remains constant, is it easier to reject the null hypothesis when the level of significance, α, is larger or smaller?

D20. Select the correct words in each sentence.

  a. A larger level of significance, α, makes it harder/easier to reject the null hypothesis.

  b. A smaller level of significance, α, makes it harder/easier to reject the null hypothesis.

  c. A one-sided test (on the correct side) makes it harder/easier to reject the null hypothesis than a two-sided test.

The Components of a Fixed-Level Test of Significance for a Proportion

Formal terminology for a *fixed-level* test of significance for a proportion appears in the following box. The term **fixed-level** means that you fix the level of significance, α, in advance. You will examine each component thoroughly in the rest of this section.

Components of a Fixed-Level Significance Test for a Proportion

1. **State the hypotheses, defining any symbols.** When testing the significance of a proportion for a random sample taken from a larger population, the null hypothesis, H_0, is

   H_0: The proportion of successes, p, in the population from which the sample came is equal to p_0. The alternative hypothesis, H_a, has three possible forms:

   H_a: The proportion of successes, p, in the population from which the sample came is not equal to p_0.

   H_a: The proportion of successes, p, in the population from which the sample came is greater than p_0.

   H_a: The proportion of successes, p, in the population from which, the sample came is less than p_0.

   With a one-sided alternative, such as $p < p_0$, the null hypothesis may be stated as $p \geq p_0$. The null hypothesis for an experimental study involving the probability of a success would be worded as follows, with corresponding alternative hypothesis.

   H_0: The probability p of a success among this group of subjects is p_0.

2. **Check the conditions for using the test.**

   - To make an inference about a larger population, you must have a random sample from that population.

   - To use the normal approximation to the binomial distribution, both np_0 and $n(1 - p_0)$ must be at least 10.

   - The population size must be at least 10 times the sample size.

3. **Compute the test statistic, z, and find the P-value.** Include a sketch that illustrates the situation. The test statistic is

$$z = \frac{\hat{p} - p_0}{\sqrt{\dfrac{p_0(1 - p_0)}{n}}}$$

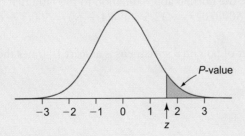

4. **Write a conclusion.** There are two parts to stating a conclusion:

   - Compare the P-value to the predetermined level of significance α. You reject the null hypothesis if the P-value is smaller than α. Otherwise, you say that you do not reject the null hypothesis.

   - Write a sentence giving your conclusion in context.

Example 8.19	Drug Testing in the Workplace

Random tests during the last year on millions of workers in the United States who weren't likely to expect random drug testing found that 5.3% had used illicit drugs within the previous three days. The American Civil Liberties Union (ACLU) and others oppose such random (i.e., without cause) testing as an unconstitutional violation of privacy. Another objection is the relatively high false positive rate for such tests.

One commonly used test has been found to have a false positive rate of 2.5% for cocaine. That means that 2.5% of the samples that test positive for cocaine are not in fact positive. The false positive rate can be determined because, typically, positive readings are retested with a different test with higher specificity.

Suppose that a pharmacologist who works for the company that makes the text has modified the test in an effort to reduce the false positive rate (without increasing the false negative rate). The CEO would like to advertise that the test has been improved, but only if he believes that the false positive rate is now less than 2.5%, with a 5% level of significance. He orders a random sample of 500 specimens that have tested positive for cocaine with the modified test and, checking the results using the test with higher specificity, finds that 9 are false positives.

Based on this evidence, would you advise the CEO to go ahead with the advertising campaign? [Sources: Quest Diagnostics, *www.questdiagnostics.com/employersolutions/dti/2009_05/dti_index.html*; Christy Visher, *A Comparison of Urinalysis Technologies for Drug Testing in Criminal Justice*, U.S. Department of Justice, 1991, p. 15.]

Solution

State your hypotheses first.

Hypotheses:

This should be a one-sided test because the company can advertise that the test has been improved only if the false positive rate is less than before. The hypotheses are:

H_0: The false positive rate among all specimens that test positive remains 2.5%. ($p = 0.025$)

H_a: The false positive rate among all specimens that test positive is now less than 2.5%. ($p < 0.025$)

Always check the conditions or randomness and normality.

Conditions:

- You were told that the sample is a random sample from a larger population.
- You can use the normal approximation to the binomial distribution because both $np_0 = 500(0.025) = 12.5$ and $n(1 - p_0) = 500(1 - 0.025) = 487.5$ are at least 10.
- The number of potential specimens is at least 10 times the sample size of 500.

Compute the test statistic, draw a sketch, and find the P-value.

Computations:

The test statistic is

$$z = \frac{\hat{p} - p_0}{\sqrt{\frac{p_0(1 - p_0)}{n}}}$$

$$= \frac{\frac{9}{500} - 0.025}{\sqrt{\frac{0.025(1 - 0.025)}{500}}}$$

$$\approx -1.00$$

The *P*-value for the one-sided test is 0.1580, as illustrated in Display 8.11.

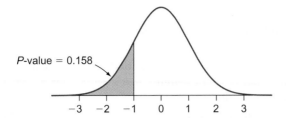

Display 8.11 The *P*-value of 0.1580 is larger than $\alpha = 0.05$.

Conclusion:

Because the *P*-value of 0.1580 is higher than the level of significance of 0.05, you cannot reject the null hypothesis. (You should never say that you "accept" the null hypothesis, because if you constructed a confidence interval for *p*, all the other values in it—not just 0.025—are also plausible values of *p*.) You don't have statistically significant evidence that the test has been improved. You should advise the CEO that he should not go ahead with the advertising campaign.

> Write a conclusion that is linked to the *P*-value and is stated in the context of the situation.

Planning a Test of Significance

You will more easily remember the structure of a test of significance if you keep in mind what you should do *before you look at the data from the sample.*

> What to decide before looking at the data.

- *Decide whether you have a one-sided or two-sided test.* When testing the effectiveness of a new drug the investigator must establish that the new drug has a *better* cure rate than the older treatment (or that there are *fewer* side effects). He or she isn't interested in simply rejecting the null hypothesis that the new drug has the same cure rate as the older treatment but instead needs to know if it is *better*. In such clear-cut cases where the investigator has an indication of which way any change from the standard should go, you would plan a one-sided test of significance. If you were interested in detecting a change from the standard in either direction, you would plan a two-sided test. If you aren't sure which to use, always choose a two-sided test.

> A two-sided test is the default.

- *Write your hypotheses.* The hypotheses should be based on the research question to be investigated, not on the data, so the investigator sets the hypotheses before the data are collected. This means that *the value of $\hat{p}$ should not appear in your hypotheses.*

- *Check the conditions for the test.* Checking the conditions to be met for a significance test for a proportion requires only knowledge of how the sample was collected, the sample size, and the value of the hypothesized standard, p_0.

- *Decide on the level of significance.* The level of significance, α, is set by the investigator before the data are collected. Some fields have set standards for the level of significance, typically $\alpha = 0.05$.

You will use the data from the sample (number of successes and value of $\hat{p}$) only to compute the test statistic and to write your conclusion.

DISCUSSION
The Components of a Test of Significance for a Proportion

D21. What is the difference in the meaning of the symbols p, p_0, and $\hat{p}$? Of the three, which varies depending on the sample you select? Which of the three is unknown?

(Continued)

D22. Suppose the null hypothesis is true and you have a large random sample of size n.

    a. What is the approximate probability that the test statistic will be greater than 1.96? What is the probability that it will either be larger than 1.96 or be less than -1.96?

    b. What is the probability that the test statistic will exceed 1.645? What is the probability that it will either be larger than 1.645 or be less than -1.645?

    c. Why do you need to know that H_0 is true to answer part a and part b?

D23. If everything else remains the same, is it easier to reject a false null hypothesis with a one-sided test or with a two-sided test?

Summary 8.3: A Significance Test for a Proportion— Making a Decision

The four components of a test of significance are listed here, with some suggestions about how to interpret each one.

    1. *Writing hypotheses.* When you write the hypotheses, you are not supposed to have seen the sample yet. Thus, the value from the sample, $\hat{p}$, does not appear in the null hypothesis (nor does it affect the direction of the alternative hypothesis).

        If you use the symbol p in the null hypothesis, always say in words what it represents in the context of the problem.

        Using a one-sided alternative hypothesis makes it easier to reject a false null hypothesis—if you use the correct side. But you can use a one-sided alternative hypothesis only if, before looking at the sample, you have an indication of which direction from the standard the true proportion falls. When in doubt about which type of test to use, use a two-sided test.

    2. *Checking conditions.* In real surveys, it is almost always the case that the sample was not a simple random sample from the population. If the sample comes from a more complicated design that uses randomization, inference is possible. In other cases, convenience samples are used—people are chosen for a sample because they were easy to find. In such cases, confidence intervals and significance tests generally should not be used.

    3. *Doing computations.* The test statistic measures how compatible the result from the sample is with the null hypothesis. Test statistics typically come in this form:

$$\frac{statistic - parameter}{standard\ error}$$

In the case of testing a proportion, this becomes

$$z = \frac{\hat{p} - p_0}{\sqrt{\dfrac{p_0(1 - p_0)}{n}}}$$

    4. *Writing a conclusion.* If z is large and thus the P-value is smaller than some significance level, α, that you have decided on in advance, this is evidence against the null hypothesis and so you should reject it. If the P-value is larger than or equal to the predetermined value α, do not reject the null hypothesis.

        If you reject the null hypothesis, you are saying that the difference between $\hat{p}$ and p_0 is so large that you can't reasonably attribute it to chance variation. If you don't reject the null hypothesis, you are saying that the difference between $\hat{p}$ and p_0 is small enough so that it looks like the kind of variation you would expect from a binomial situation in which the proportion of successes is p_0.

Do not simply write "Reject the null hypothesis" or "Do not reject the null hypothesis." Explain what your conclusion means in terms of the context of the problem.

You should never write "Accept the null hypothesis." The reason is that the sample can tell you only whether the value of p in the null hypothesis is plausible. If you get a very small P-value, then it isn't reasonable to assume that the null hypothesis is true and you should reject it. However, if you get a large P-value, then the null hypothesis is plausible but still might not give the exact value of p. Other values of p will be plausible, too.

Practice

The Level of Significance

P23. *Quality control.* Five percent of the windshields produced by the long-time supplier for a car manufacturer don't meet specifications. The car manufacturer is considering switching to a new supplier and inspects a random sample of 300 windshields produced by this supplier. Three percent don't meet specifications. You will do a one-sided test of significance, with null hypothesis that the defective rate for the new supplier is (at least) 5%. You decide to use a level of significance of 0.01.

    a. What is the alternative hypothesis?

    b. What is the P-value for this test?

    c. Based on this evidence, would you recommend a switch to the new supplier?

P24. *Medicine.* Current tests available for the first trimester of a pregnancy detect Down syndrome in 82% to 87% of the cases. Your company would like to market a new test, but only if it exceeds the 87% detection rate. In a random sample of 200 women whose babies were later verified to have Down syndrome, 182 of the cases were detected by the new test. [Source: The American College of Obstetricians and Gynecologists Education Pamphlet AP165, Screening for Birth Defects, March 2007, *www. acog.org/publications/patient_education/hp165.cfm.*]

    a. Should this be a one-sided or two-sided test?

    b. What are the hypotheses?

    c. What is the P-value for this test?

    d. Based on this evidence and a 5% level of significance, would you recommend marketing the new test?

P25. *Political funding.* A labor union has a $2,000,000 fund held in reserve. It plans to spend the money on advertising on behalf of a new bond issue, but only if statewide support for the bond issue falls below 55%. A random sample of 700 likely voters finds current support at 53%.

    a. Should this be a one-sided or two-sided test?

    b. What are the hypotheses?

    c. What is the P-value for this test?

    d. Based on this evidence and a 1% level of significance, would you recommend that the labor union begin its advertising campaign?

The Components of a Test of Significance for a Proportion

P26. For each situation, state appropriate hypotheses for a test of a proportion.

    a. You wonder if a student is guessing on a multiple-choice test of 60 questions where each question has five possible answers.

    b. You wonder if the percentage of people age 18 and older who are veterans is different in your county from the national percentage of 10.4%.

    c. You wonder if people who wash their cars once a week are most likely to wash them on Saturday.

P27. The United States 2000 Census found that 4% of all households were multigenerational (consisting of three or more generations of parents and their children). You want to test the null hypothesis that this percentage is the same this year as it was in 2000. You write the null hypothesis as $p = 0.04$. Which of these statements best describes what p stands for? [Source: *www. census.gov.*]

    A. the proportion of all households that were multigenerational in 2000

    B. the proportion of all households that are multigenerational this year

    C. the proportion of multigenerational households in the sample in 2000

    D. the proportion of multigenerational households in the sample this year

    E. the proportion of all multigenerational households in both 2000 and this year

P28. American roulette wheels have 38 numbers, of which 18 are red. You are checking a roulette wheel to see if red occurs with the right relative frequency. You observe 500 spins, in which the ball lands on red 194 times. Carry out the four steps in a test of the null hypothesis that the probability of getting red is 18/38. Use $\alpha = 0.10$.

P29. Wildlife scientists noticed that bears seem to prefer to break into minivans. Gathering information from incident reports, they found that over a 2-year period, bears broke into 412 vehicles in Yosemite National Park and that 120 of the vehicles were minivans. Yet only 7% of the vehicles parked overnight at the park were minivans. For your analysis, assume that the 412 vehicles may be considered a random sample of the bear's breaking-and-entering preferences and that 7% is the true proportion of the vehicles parked overnight that are minivans. Carry out the four steps in a test of the null hypothesis that the probability that a bear will select a minivan to break into is 0.07 against the alternative that the probability is greater than 0.07. [Source: Stewart W. Breck, Nathan Lance, and Victoria Seher, "Selective Foraging for Anthropogenic Resources by Black Bears: Minivans in Yosemite National Park," *Journal of Mammalogy*, Vol. 90, 5 (2009), pp.1041–1044.]

P30. You claim that the percentage of teens in your community who know that epilepsy is not contagious is larger than the national percentage, 51%. You take a random sample of 169 teens in your community and find that 55% know that epilepsy is not contagious. Carry out the steps of a test of significance to check your claim. Use $\alpha = 0.05$.

Exercises

E41. A Gallup poll asked a random sample of 1012 adult Americans how much they worry about the pollution of drinking water. Fifty-nine percent chose the reply "a great deal." Following the four steps of a significance test, can you reject the null hypothesis that half of Americans worry a great deal about the pollution of drinking water? Use $\alpha = 0.01$. [Source: Gallup, March 25, 2009, *www.gallup.com.*]

E42. A Harris poll asked, "Do you agree or disagree with this statement? Protecting the environment is so important that requirements and standards cannot be too high, and continuing environmental improvements must be made regardless of cost." Of the random sample of 1217 adults surveyed nationwide, 74% agreed. Following the four steps of a significance test, can you conclude that the true proportion of the nation's adults who agree with the statement is different from two-thirds? Use $\alpha = 0.01$. [Source: The Harris Report, August 2005, *www.pollingreport.com.*]

E43. The following Minitab output is for the situation in E41. How could you come to a decision about whether to reject the null hypothesis based only on the confidence interval? Explain your reasoning.

Test and CI for One Proportion

Test of p = 0.5 vs p not = 0.5

Sample	X	N	Sample p	99% CI	z-Value	P-Value
1	597	1012	0.589921	(0.550096, 0.629746)	5.72	0.000

Using the normal approximation.

E44. The following Minitab output is for the situation in E42. How could you come to a decision about whether to reject the null hypothesis based only on the confidence interval? Explain your reasoning.

Test and CI for One Proportion

Test of p = 0.666667 vs. p not = 0.666667

Sample	X	N	Sample p	99% CI	z-Value	P-Value
1	901	1217	0.740345	(0.707972, 0.772718)	5.45	0.000

Using the normal approximation.

E45. In a poll of 1000 randomly sampled adults in the United States, 49% said they were satisfied with the quality of K–12 education. Can you reject the null hypothesis that (at least) half of adult residents are satisfied with the quality of education? [Source: Based on a Gallup poll.]

   a. What is the null hypothesis? What is the most appropriate alternative hypothesis? Explain your reasoning.

   b. Verify that the conditions for the test are satisfied.

   c. Calculate a test statistic and the corresponding *P*-value.

   d. Using a 0.05 level of significance, write a conclusion in context.

E46. In the same poll of adults as described in E45, 51% of the 1000 respondents said that they were dissatisfied with the quality of K–12 education in schools today. Can you reject the null hypothesis that (at least) half of adult residents are dissatisfied with the quality of education? Work through the same steps as in E45. Then reconcile your conclusion with that in E45.

E47. In 2009, 45% of a random sample of 1004 Americans polled by the *Washington Post*/ABC News poll approved of the way President Barack Obama was handling the situation in Afghanistan. Suppose that you are were writing a headline for an article in the campus newspaper reporting this result. Is it fair to say, "Fewer than half of Americans approve of the president's handling of Afghanistan"? [Source: *www.washingtonpost.com*, polling done October 15–19, 2009.]

E48. The editors of a magazine have noticed that people seem to believe that a successful life depends on having good friends. They would like to have a story about this and use a headline such as "Most Adults Believe Friends Are Important for Success." So they have commissioned a survey to ask a random sample of adults whether a successful life depends on having good friends. In a random sample of 1027 adults, 53% said yes. Should the editors go ahead and use their headline?

E49. From a random sample of 1012 adult Americans, a Gallup poll found that 39% approved of the way Congress was handling its job. The margin of error was given as 3%. Which of these is not a true statement? [Source: Gallup, 2008 *www.gallup.com*.]

   A. You can reject the hypothesis that 31% of all adult Americans approved of the way Congress was handling its job.

   B. You cannot reject the hypothesis that 80% of all Americans would not have said that they approved of the way Congress was handling its job.

   C. The true proportion of all Americans who approved of the way Congress was handling its job on that date must lie between 0.36 and 0.42.

   D. If, in fact, 35% of all Americans approved of the way Congress was handling its job on that date, then the observed sample result is reasonably likely.

   E. If, in fact, 75% of all Americans would not have said that they approved of the way Congress was handling its job on that date, then the observed sample result is not reasonably likely.

E50. A random sample of dogs was checked to see how many wore a collar. The 95% confidence interval for the percentage of all dogs that wear a collar turned out to be 0.82 to 0.96. Which of these is *not* a true statement?

   A. You can reject the hypothesis that 75% of all dogs wear a collar.

   B. You cannot reject the hypothesis that 90% of all dogs wear a collar.

   C. If 90% of all dogs wear a collar, then you are reasonably likely to get a result like the one from this sample.

   D. If 75% of all dogs wear a collar, then you are reasonably likely to get a result like the one from this sample.

   E. The true proportion of dogs that wear a collar may or may not be between 0.82 and 0.96.

E51. In this exercise, data are known for the entire population of interest. A confidence interval makes no sense in this situation. However, the computations of a test of significance can be useful in deciding whether you can conclude that the results can reasonably be explained by chance alone or whether some other explanation is needed.

   Do Rock Stars Die Young? A study of the survival rates of the 196 most famous North American rock and pop stars found that 39 died within 30 years of the beginning of their fame. Based on the stars' age, sex, and ethnicity, the expected number of deaths (compared to matched general populations) was 11.3. [Source: Mark A Bellis et al., "Elvis to Eminem: Quantifying the Price of Fame Through Early Mortality of European and North American Rock and Pop Stars," *Journal of Epidemiology and Community Health*, Vol. 61 (2007), pp. 896–901.]

   a. Should this be a one-sided or two-sided test?

   b. Write appropriate null and alternative hypotheses to test whether the fact that rock-and-roll pop stars died more frequently can reasonably be attributed to chance.

   c. Can the results reasonably be attributed to chance?

E52. Read the first paragraph of the previous exercise.

   Does the Taller Man Win? In an election between two men, does the taller man tend to win more often?

   Display 8.12 gives the results of the U.S. presidential elections from 1900 to 2008

   a. Plot the heights so you can compare the heights of the winners and losers. (You may want to consider that in 2000, the taller candidate received more popular votes, but the shorter candidate won the Electoral College vote and became president.)

   b. Write appropriate null and alternative hypotheses to test whether the fact that the taller man has won more often can reasonably be attributed to chance.

   c. Can the results reasonably be attributed to chance?

	Winning Candidate		Runner-up				Winning Candidate		Runner-up	
Year	Candidate	Height	Candidate	Height		Year	Candidate	Height	Candidate	Height
1900	McKinley	5'7"	Bryan	6'0"		1960	Kennedy	6'0"	Nixon	5'11.5"
1904	T. Roosevelt	5'10"	Parker	6'0"		1964	Johnson	6'3"	Goldwater	6'0"
1908	Taft	6'0"	Bryan	6'0"		1968	Nixon	5'11.5"	Humphrey	5'11"
1912	Wilson	5'11"	T. Roosevelt	5'10"		1972	Nixon	5'11.5"	McGovern	6'1"
1916	Wilson	5'11"	Hughes	5'11"		1976	Carter	5'9.5"	Ford	6'0"
1920	Harding	6'0"	Cox	NA		1980	Reagan	6'1"	Carter	5'9.5"
1924	Coolidge	5'10"	Davis	6'0"		1984	Reagan	6'1"	Mondale	5'10"
1928	Hoover	5'11"	Smith	NA		1988	G. Bush	6'2"	Dukakis	5'8"
1932	F. Roosevelt	6'2"	Hoover	5'11"		1992	Clinton	6'2"	G. Bush	6'2"
1936	F. Roosevelt	6'2"	Landon	5'8"		1996	Clinton	6'2"	Dole	6'2"
1940	F. Roosevelt	6'2"	Willkie	6'1"		2000	G. W. Bush	5'11"	Gore	6'1"
1944	F. Roosevelt	6'2"	Dewey	5'8"		2004	G. W. Bush	5'11"	Kerry	6'4"
1948	Truman	5'9"	Dewey	5'8"		2008	Obama	6'1.5"	McCain	5'9"
1952	Eisenhower	5'10.5"	Stevenson	5'10"						
1956	Eisenhower	5'10.5"	Stevenson	5'10"						

Display 8.12 Heights of presidential candidates. [Sources include: Paul M. Sommers, "Presidential Candidates Who Measure Up," *Chance*, Vol. 9, 3 (1996), pp. 29–31.]

8.4 ▶ Types of Errors and the Power of a Test

Whether you reject the null hypothesis or do not reject the null hypothesis, you might be making a error. This section is about these types of errors and how to minimize the chance of making them.

Types of Errors

The reasoning of significance tests often is compared to that of a jury trial. The possibilities in such a trial are given in the following diagram. In the American system of justice, convicting an innocent person is considered worse than letting a guilty person go.

		Defendant Is Actually	
		Innocent	Guilty
Jury's Decision	Not Guilty	Correct	Error
	Guilty	Worse error	Correct

Similarly, there are two types of errors in significance testing. You could convict an innocent (true) null hypothesis or you could fail to convict a guilty (false) null hypothesis. Like convicting an innocent person, the error of rejecting a true null hypothesis is considered serious and so a null hypothesis isn't rejected unless the evidence against it is convincing beyond a reasonable doubt.

		Null Hypothesis Is Actually	
		True	False
Your Decision	Don't Reject H_0	Correct	Type II error
	Reject H_0	Type I error	Correct

The following box defines the two types of error.

> **Types of Errors**
>
> A **Type I** error occurs when you reject a true null hypothesis.
> A **Type II** error occurs when you don't reject a false null hypothesis.

Example 8.20	An ESP Test and a Type I Error

Several Internet sites allow you to take a test to see if you have extrasensory perception (ESP). Ann took one of these tests. She was presented with the back of one of the five different cards shown in Display 8.13. She guessed which of the five cards it was. In 100 tries with different cards, she guessed right 29 times. The web site said that her result was "significant" and the evidence for ESP was "fair." What is the P-value for this test? Discuss what type of error might have been made.

Display 8.13: Zener cards are used in EPS experiments.

Solution
The null hypothesis is $p = 0.2$, where p is the probability that Ann names a randomly selected card correctly. Because evidence of ESP occurs only when $\hat{p}$ is greater than 0.20, this should be a one-sided test.

Conditions have been met to use a normal approximation to the binomial distribution because Ann was presented with a random sample of cards and both $np_0 = 100(0.2) = 20$ and $n(1 - p_0) = 100(1 - 0.2) = 80$ are at least 10.

The test statistic is

$$z = \frac{\hat{p} - p_0}{\sqrt{\frac{p_0(1 - p_0)}{n}}} = \frac{0.29 - 0.20}{\sqrt{\frac{0.20(1 - 0.20)}{100}}} = 2.25$$

The corresponding P-value is 0.01, as illustrated in Display 8.14 , so reject the null hypothesis.

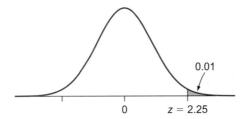

Display.8.14 The P-value for a z-score of 2.25.

Ann knows that she does not have ESP and, in fact, didn't even try to discern what the card was, instead selecting her choices rapidly and at random. In other words, the null hypothesis is true that, in the long run, she will guess the correct card 20% of the time. A Type I error of rejecting a true null hypothesis has been made. Out of every 100 people who take such a test, we expect that only one of them will get 29 or more cards right. Ann was the lucky one. (Yet, amazed at her previously unsuspected skill, Ann tried the ESP test again—and correctly identified only 17 cards, a little bit worse than the 20 she would expect just by chance.)

> A Type I error is rejecting a true H_0.

Drug Testing and a Possible Type II Error

Example 8.21

In Example 8.19 on page 400, 9 out of 500, or 1.8%, of the specimens in the sample were false positive using the modified test. Although this was lower than the 2.5% false positive rate of the original test, the P-value of 0.1580 was higher than the level of significance of 0.05. Thus, the difference wasn't statistically significant and the CEO was not able to reject the null hypothesis that the modified test had a false positive rate equal to that of the original test. What type of error might he be making? What is the possible consequence to the company?

Solution
The CEO may have made the error of not rejecting a null hypothesis that is false. If so, he has made a Type II error. It is possible that the modified test does indeed have a lower rate of false positives than the original test, but the difference was not big enough to be detected by the significance test. The consequence to the company would be that it fails to market a modified test that is an improvement over the original test in that it has a lower false positive rate.

> A Type II error is failing to reject a false H_0.

> A Type I error usually is more serious than a Type II error.

The company in Example 8.21 may have made a Type II error (failing to market a better test), but would consider this less serious than a Type I error (marketing a test that isn't better). If a clinical trial fails to show that a new treatment is better than the old but the company is convinced that it is better, then research will continue and the truth will eventually come out. If, however, the new treatment is not better than the old, and the clinical trial says it is, then millions of dollars will be spent manufacturing and marketing a new product that serves no useful purpose (and may make no profit for the company).

The factors to consider when you make a decision about a null hypothesis are given in the following box. The first part is for a situation like Example 8.18 on page 397, where you reject the null hypothesis. The second part is for a situation like Example 8.19, on page 400 where you don't reject the null hypothesis.

What to Consider When You Reject the Null Hypothesis

If you reject the null hypothesis, then there are several possibilities to consider:

> Making an error doesn't mean you did anything wrong. Blame the sample.

- You are making the correct decision. The null hypothesis isn't true, and that's why the sample proportion, $\hat{p}$, was so far from p_0.
- You are making a Type I error—rejecting H_0 even though H_0 is actually true. It was just bad luck that resulted in $\hat{p}$ being so far from p_0.
- The sampling process or method of getting a response from the sample was biased in some way; so the value of $\hat{p}$ is itself suspicious.

What to Consider When You Don't Reject the Null Hypothesis

If you don't reject the null hypothesis, then there are also several possibilities to consider:

- You are making the correct decision, The null hypothesis is true, and you got just about what you would expect in the sample.
- You are making a Type II error—not rejecting H_0 even though H_0 is false. It was just by chance that $\hat{p}$ turned out to be close to p_0.
- The sampling process or method of getting a response from the sample was biased in some way, so the value of $\hat{p}$ is itself suspicious.

The Probability of Making a Type I Error

Suppose the null hypothesis is true. What is the chance that you will reject it, making a Type I error? The only way you can make a Type I error is for your value of $\hat{p}$ to be a rare event. For example, if you are using a significance level of 0.05 in a two-sided test, you would make a Type I error if you get a P-value less than 0.05. This happens when the null hypothesis is true only 5% of the time no matter what the sample size. Thus, the probability of a Type I error is 0.05. If you used a significance level of 0.01, then the probability of making a Type I error would be only 0.01.

Probability of a Type I Error

If the null hypothesis is true, the probability of a Type I error is equal to the level of significance α. If you are worried about making a Type I error, then your best strategy is to have a low level of significance.

However, if the null hypothesis is false, setting a low level of significance increases the probability of making a Type II error.

> If H_0 is true, the probability of a Type I error is α.

You may have been wondering why 0.05 was set as the standard level of significance. Like many statistical practices, it goes back to R. A. Fisher, who put it this way:

> It is convenient to draw the line at about the level at which we can say: "Either there is something in the treatment, or a coincidence has occurred such as does not occur more than once in twenty trials." . . . If one in twenty does not seem high enough odds, we may, if we prefer it, draw the line at one in fifty (the 2 per cent point), or one in a hundred (the 1 per cent point). Personally, the writer prefers to set a low standard of significance at the 5 per cent point, and ignore entirely all results which fail to reach this level. [Source: R. A. Fisher, "The Arrangement of Field Experiments," *Journal of the Ministry of Agriculture of Great Britain*, Vol. 33 (1926), pp. 503–513.]

There is some empirical evidence that 1 in 20 is the level where people get very uncomfortable with the null hypothesis. If a teacher flips a coin in class, hands suddenly go up all over the room after five heads or tails in a row. [Source: Frederick Mosteller, quoted in Victor Cohn, *News and Numbers* (Iowa State University Press, 1989).]

DISCUSSION
Types of Errors

D24. Suppose you use a level of significance of 0.02 and the null hypothesis is actually true. What is the probability that you will get a sample that results in rejecting the null hypothesis? Explain. What type of error is this?

D25. To avoid Type I errors, why not always use a very small level of significance?

Power of a Test

Sometimes your sample size isn't large enough to enable you to reject a false null hypothesis.

A Two-Headed Silver Dollar   Example 8.22

Unbeknownst to you, your uncle has purchased a two-headed silver dollar from a store that sells magic tricks. He arrives for Sunday dinner and offers to leave it to chance whether you or he does the dishes. He pulls the silver dollar from his pocket and says, "Heads you do the dishes and tails I do the dishes." The coin lands heads and you do the dishes. After the fourth Sunday where you do the dishes, you begin to feel a bit suspicious. Do you have enough evidence to accuse your beloved uncle of cheating? How should you proceed?

Solution
You don't have enough evidence for an accusation. The null hypothesis is that the silver dollar is fair. If your uncle were using a fair silver dollar, then the probability of

getting four heads in a row is $(0.5)^4$, or 0.0625. This is larger than the usual level of significance, 0.05. You cannot reject the null hypothesis that the coin is fair.

What do you do? You have to get a larger sample size. A sample size of 4 is too small for your test to detect a statistically significant difference between the proportion of heads in your sample, four out of four or 1.00, and the value under the null hypothesis that the coin is fair, 0.5. The probability of five heads in a row is only $(0.5)^5$, or 0.03125, so if the silver dollar comes up heads next Sunday, you will have statistically significant evidence against the hypothesis that the silver dollar is fair.

The probability of a Type II error is denoted β.

In Example 8.22, your initial conclusion, after four weeks, resulted in a Type II error. You failed to reject a false null hypothesis. Another way to state this is to say that your test did not have enough *power* to be able to detect that the true proportion of heads, $p = 1.00$, is different from the proportion hypothesized, $p_0 = 0.5$.

Power of a Test

The **power** of a test is the probability of rejecting the null hypothesis. When the null hypothesis is false

$$power = 1 - probability\ of\ a\ Type\ II\ error$$

It follows that if the probability of a Type II error is small, then the power of the test is large.

Bigger n, more power.

If the null hypothesis is false and should be rejected, a larger sample size increases the probability that you will be able to reject it. A larger sample size can give a statistically significant result while a smaller sample size will not because of the difference in the spread of the two sampling distributions. The spread is controlled by the sample size: bigger sample, less spread.

However, if the null hypothesis is true, increasing the sample size has no effect on the probability of making a Type I error. The only way you can decrease the probability of making a Type I error is to make the level of significance, α, smaller. As you'll see in the next example, if the null hypothesis is actually false, this strategy results in a higher probability of a Type II error and lower power!

Example 8.23	**You Lose Power and Make an Error About Your Uncle's Silver Dollar**

Recall that your beloved uncle is cheating you by using a two-headed coin to decide who does the dishes, but you do not know that. (Refer to Example 8.22.) On the fifth Sunday, the coin is heads once again. The difference now is statistically significant at the 0.05 level. But you are worried about making a Type I error—accusing your beloved uncle of cheating when he was not. Consequently, you decide to use a 0.01 level of significance. That is, if it is true that his silver dollar is fair, you want to have only a 0.01 chance of deciding wrongly that it isn't fair. What is your conclusion now?

Solution

Bigger α, less power.

The probability of getting 5 heads in a row with a fair coin is $(0.5)^5$, or 0.03125. At the 0.01 level of significance, you cannot reject the false null hypothesis that your uncle's silver dollar is fair. You will make a Type II error because a 0.01 level of significance gives you less power than a 0.05 level of significance.

Example 8.23 illustrates the difficulty in balancing the risk of making a Type I error and the risk of making a Type II error. Deciding on the correct balance depends on the real-word situation and the consequences of each type of error. As Neyman and Pearson put it:

> These two sources of error can rarely be eliminated completely; in some cases it will be more important to avoid the first, in others, the second. . . . Is it more serious to convict an innocent man or to acquit a guilty? That will depend on the consequences of the error; is the punishment death or fine; what is the danger to the community of released criminals; what are the current ethical views on punishment? From the point of view of mathematical theory all that we can do is to show how the risk of errors may be controlled and minimised. The use of these statistical tools in any given case, in determining just how the balance should be struck, must be left to the investigator. [Source: J. Neyman and E. Pearson, "On the Problem of the Most Efficient Tests of Statistical Hypotheses," *Philosophical Transactions of the Royal Society of London, Ser. A*, 231, 1933, pp. 289–337.]

Here are three important properties of power:

- Power increases as the sample size increases, all else being held constant.
- Power increases as the value of α increases, all else being held constant.
- Power increases when the population proportion, p, is farther from the hypothesized value, p_0. When p is farther from p_0, then $\hat{p}$ tends to be farther from p_0, so the test statistic tends to be larger and the P-value smaller. (For a one-sided test, this assumes that $\hat{p}$ lies in the direction of the alternative hypothesis.)

Only the first of these typically is an acceptable way for you to increase the power of your significance test. The following box summarizes what you should understand about types of error and power.

Types of Error and Power

Type I error. When the null hypothesis is true and you reject it, you have made a Type I error. The probability of making a Type I error is equal to the significance level, α, of the test. To decrease the probability of a Type I error, make α smaller. Changing the sample size has no effect on the probability of a Type I error.

If the null hypothesis is false, you can't make a Type I error.

Type II error. When the null hypothesis is false and you fail to reject it, you have made a Type II error. The probability of making a Type II error can be estimated using statistical software to compute *power curves*, which will not be covered in this text. To decrease the probability of making a Type II error, get a larger sample or make the significance level, α, larger.

If the null hypothesis is true, you can't make a Type II error.

Power. Power is the probability of rejecting the null hypothesis.

When the null hypothesis is false, you want to reject it and therefore you want the power to be large. To increase power, theoretically you could either take a larger sample or make α larger.

In practice, you should use a larger sample size.

DISCUSSION
Power of a Test

D26. Explain why an increase in sample size increases the power of a test, all else remaining unchanged.

D27. What happens to the power of a test as the population proportion, p, moves farther away from the hypothesized value, p_0, all else remaining unchanged?

D28. Can a statistical test of the type discussed in this chapter ever have a power of 1? Can a statistical test of the type discussed in this chapter ever have a power of 0? If so, would either be desirable from a practical point of view?

D29. How many times will you have to do the dishes on Sunday before you have enough power to confront the fairness of your uncle's silver dollar at the 0.01 level of significance?

Summary 8.4: Types of Errors and the Power of a Test

If you reject the null hypothesis in a test of significance, it's always possible that you have made an error. Rejecting a true null hypothesis is called a Type I error. The value you set for the significance level, α, of the test controls the probability that you will make a Type I error. If the null hypothesis is true, the probability of a Type I error is α. So to limit the probability of making a Type I error, have a small value of α.

Perversely, when you reduce α, you then increase the probability of making a Type II error. A Type II error is failing to reject a null hypothesis that is false. The best way to limit the probability of making a Type II error is to increase the power of your test by having as large a sample size as practicable.

Practice

Types of Errors

P31. Hila is rolling a pair of dice to test whether they land doubles 1/6 of the time. She doesn't know it, but the dice are fair. She will use a significance level, α, of 0.05. Hila rolls the dice 100 times and gets doubles 25 times.

    a. What is the P-value for this test? What conclusion should Hila come to?

    b. Did Hila make an error? If so, which type?

P32. In the New York State Lotto, players pick six numbers from 1 through 59. Jack wants to test whether the New York State Lotto is fair with respect to his favorite number, 1. If all numbers are equally likely, which they are, any one of the 59 numbers has a probability of 6/59 ≈ 0.102 of coming up in any one game. In the last 104 New York State Lotto games, the number 1 has come up 18 times. [Source: *www.nylottery.org*.]

    a. What is the P-value for this test? What conclusion should Jack come to?

    b. Did Jack make an error? If so, which type?

P33. Researchers wanted to determine if people could pick out the dog food from five choices consisting of the dog food, duck liver mousse, pork liver pâté, liverwurst, and Spam, all of which had been blended to the same consistency. Of 18 subjects, only three were able to correctly identify the dog food. The researchers concluded the results "did not support the hypothesis that the distribution of guesses was significantly different from random." [Source: John Bohannon, Robin Goldstein, and Alexis Herschkowitsch, "Can People Distinguish Pâté From Dog Food?" American Association of Wine Economists, Working Paper No. 36, April 2009, *www.wine-economics.org*.]

    a. What is the null hypothesis in this situation? Did the researchers reject it or fail to reject it?

    b. What type of error might the researchers be making?

    c. Explain what that type of error means in this context.

P34. You are studying a random sample of children who have lived for at least 5 years within 0.5 mile of a freeway in Los Angeles. You plan to conduct a test of significance with the null hypothesis that the proportion of all such children who have asthma is the same as the (known) proportion of all children in Los Angeles who have asthma. Your research hypothesis is that the proportion is greater.

    a. Describe what a Type I error would be in this situation. Name at least one serious consequence that might result from a Type I error.

    b. Describe what a Type II error would be in this situation. Name at least one serious consequence that might result from a Type II error.

Power of a Test

P35. A statistical test is designed with a significance level of 0.05 and a sample size of 100. A similar test of the same null hypothesis is designed with a significance level of 0.10 and a sample size of 100. If the null hypothesis is false, which test has the greater power? Explain.

P36. A statistical test is designed with a significance level of 0.05 and a sample size of 100. A similar test of the same null hypothesis is designed with a significance level of 0.05 and a sample size of 200. If the null hypothesis is false, which test has the greater power? Explain.

P37. Suppose a medical researcher wants to test the hypothesis that the proportion of patients who develop undesirable side effects from a certain medication is 6%, as reported by the pharmaceutical company. The researcher is pretty sure that the true proportion is between 8% and 10%. She has a random sample of patients who will use the medication and chooses a 5% level of significance. Is the power of the test larger if the true proportion is nearer 8% or nearer 10%? Explain.

Exercises

E53. When flipping a penny, the probability of getting heads is 0.5. Suppose 240 people perform a test of significance with this penny. Each person plans to reject the null hypothesis that $p = 0.5$ if their P-value is less than 0.05.

   a. How many of these people do you expect to make a Type I error?

   b. How many do you expect to make a Type II error?

E54. Fifty laboratories across the country are conducting similar tests on the possible side effects of a certain medication. Each laboratory tests the null hypothesis that no more than 15% of the subjects taking the medication develop undesirable side effects, and each test is done at the 10% significance level. One of these two questions can be answered. Answer that one, and explain why you cannot answer the other.

   a. How many of the laboratories do you expect to reject the null hypothesis if it is true?

   b. How many of the laboratories do you expect to reject the null hypothesis if it is false?

E55. Jeffrey and Taline each want to test whether the proportion of adults in their neighborhood who have graduated from high school is 0.94, as claimed in the newspaper. Jeffrey takes a random sample of 200 adults and uses $\alpha = 0.05$. Taline takes a random sample of 500 adults and uses $\alpha = 0.05$. Suppose the newspaper's percentage is actually right.

   a. Is it possible for Jeffrey or Taline to make a Type I error? If so, who is more likely to do so?

   b. Is it possible for Jeffrey or Taline to make a Type II error? If so, who is more likely to do so?

E56. Jeffrey and Taline each want to test whether the proportion of adults in their state who have graduated from college is 0.6, as claimed in the newspaper. Jeffrey takes a random sample of 200 adults and uses $\alpha = 0.01$. Taline takes a random sample of 200 adults and uses $\alpha = 0.05$. Suppose the newspaper's percentage is actually right.

   a. Is it possible for Jeffrey or Taline to make a Type 1 error? If so, who is more likely to do so?

   b. Is it possible for Jeffrey or Taline to make a Type II error? If so, who is more likely to do so?

E57. Jeffrey and Taline each want to test whether the proportion of adults in their neighborhood who took chemistry in high school is 0.25, as claimed in the newspaper. Jeffrey takes a random sample of 200 adults and uses $\alpha = 0.05$. Taline takes a random sample of 500 adults and uses $\alpha = 0.05$. Suppose the newspaper's percentage is actually wrong.

   a. Is it possible for Jeffrey or Taline to make a Type I error? If so, who is more likely to do so?

   b. Is it possible for Jeffrey or Taline to make a Type II error? If so, who is more likely to do so?

E58. Jeffrey and Taline each want to test whether the proportion of adults in their state who have gone to nursery school is 0.55, as claimed in the newspaper. Each takes a random sample of 200 adults. Jeffrey uses $\alpha = 0.01$. Taline uses $\alpha = 0.05$. Suppose the newspaper's percentage is actually wrong.

   a. Is it possible for Jeffrey or Taline to make a Type I error? If so, who is more likely to do so?

   b. Is it possible for Jeffrey or Taline to make a Type II error? If so, who is more likely to do so?

E59. A psychology professor flashed a pair of head shots before subjects and asked them to identify the face that displayed more competence. The subjects didn't know it, but the head shots were the two opposing candidates in 600 recent House of Representative elections. It turned out that 401 of the candidates who were perceived as more competent by the subjects won the election. The article says that the statistical test used "tests the proportion of correctly predicted races against the chance level of 50%." [Source: A. Todorov, A. N. Mandisodza, A. Goren, and C. C. Hall, "Inferences of Competence from Faces Predict Election Outcomes," *Science*, Vol. 308 (2005), pp. 1623–1626.]

   a. What is $\hat{p}$? What is p_0?

   b. The test statistic given in the article is 8.25. Is this correct?

   c. The P-value reported was "$P < 0.001$." Is this correct for a two-sided test?"

   d. Does the sample size make an important contribution to the power of the test in this situation? Explain.

   e. What type of error might the professor be making? What are the consequences of such an error?

E60. A psychology professor flashed a pair of head shots before subjects and asked them to identify the face that displayed more competence. The subjects didn't know it, but the head shots were the two opposing candidates in the 32 U.S. Senate elections of 2004. It turned out that 22 of the candidates who were perceived as more competent by the subjects won the election. [Source: A. Todorov, A. N. Mandisodza, A. Goren, and C. C. Hall, "Inferences of Competence from Faces Predict Election Outcomes, *Science*, Vol. 308 (2005), pp. 1623–1626.]

a. To determine the statistical significance of this result, should a one-sided or two-sided test be used?

b. What is $\hat{p}$? What is p_0?

c. What is the value of the test statistic?

d. The *P*-value reported was "$P < 0.034$." Is this correct for a two-sided test?

e. Does the sample size make an important contribution to the power of the test in this situation? Explain.

f. What type of error might the professor be making? What are the consequences of such an error?

E61. In P18 on page 393 of Section 8.2, you computed a test statistic for this study: Forty-five dogs and their owners, chosen at random, were photographed separately. A judge was shown a picture of each owner and pictures of two dogs and asked to pick the dog that went with the owner. The judge was right 23 times.

a. If you use a significance level of 0.05, do you reject the null hypothesis that, in the long run, the judge will select the correct dog half the time?

b. What type of error might you be making? What are the consequences of such an error?

c. How would you suggest that the investigator get more power for this test?

E62. Suppose that 42 out of 80 randomly selected students prefer hamburgers to hot dogs and that 38 prefer hot dogs to hamburgers.

a. Test the null hypothesis that the proportion of all students who prefer hamburgers is 0.55.

b. Test the null hypothesis that the proportion of all students who prefer hot dogs is 0.55.

c. How can you reconcile your conclusions in parts a and b?

Now suppose you go out and get a larger random sample of 800 students. Suppose that 420 of these students prefer hamburgers to hot dogs and that 380 prefer hot dogs to hamburgers.

d. Test the null hypothesis that the proportion of all students who prefer hamburgers is 0.55.

e. Test the null hypothesis that the proportion of all students who prefer hot dogs is 0.55.

f. How can you reconcile your conclusions in parts a, b, d, and e?

E63. Read the following excerpt from the abstract of an article that evaluated research reporting experiments concerning surgery of the spine:

A literature search was conducted of MED-LINE, PubMed, and Cochrane databases, using the key words, "spine" and "surgery" between 1967 and 2002. Trials were included if they were of a 2-group randomized controlled trial design, which reported a nonsignificant difference in the primary outcome. . . . A total of 37 studies satisfied the inclusion criteria. The mean type II error (beta error) was 82%. CONCLUSION: The spine surgical literature is plagued with a high potential for type II error. A trial's methodology should be scrutinized to prevent misinterpretation of the results. [Source: C.S. Bailey, C.G. Fisher, and M. F. Dvorak, "Type II Error in the Spine Surgical Literature," *Spine*, Vol. 29, 10 (May 15, 2004), pp. 1146–1199, *www. ncbi.nih.nlm.gov/pubmed/15131445*.]

a. Did each of the 37 studies reject the null hypothesis or fail to reject it?

b. The abstract says that the mean Type II error was 82%. Which of the following is the best explanation of what that must mean?

    A. 82% of the studies made a Type II error.

    B. On average, if the null hypothesis were false, the study had an 82% chance of being able to reject it.

    C. On average, if the null hypothesis were false, the study had an 82% chance of failing to reject it.

E64. Read the following excerpt from the abstract of an article that evaluated research reports concerning cardiovascular experiments. You will learn about *t*-tests in Chapter 10. For now, think of them as similar to the *z*-test.

Frequently in biomedical literature, measurements are considered "not statistically different" if a statistical test fails to achieve a *P* value that is < or = 0.05. This conclusion may be misleading because the size of each group is too small or the variability is large, and a type II error (false negative) is committed. In this study, we examined the probabilities of detecting a real difference (power) and type II errors in unpaired *t*-tests in Volumes 246 and 266 of the *American Journal of Physiology: Heart and Circulatory Physiology*. In addition, we examined all articles for other statistical errors. The median power of the *t*-tests was similar in these volumes (approximately 0.55 and approximately 0.92 to detect a 20% and a 50% change, respectively). In both volumes, approximately 80% of the studies with nonsignificant unpaired *t*-tests contained at least one *t*-test with a type II error probability > 0.30. Our findings suggest that low power and a high incidence of type II errors are common problems in this journal. [Source: J. L. Williams, C. A. Hathaway, K. L. Kloster, and B. H. Layne, "Low Power, Type II Errors, and Other Statistical Problems in Recent Cardiovascular Research," *American Journal of Physiology: Heart and Circulatory Physiology*, Vol. 273 (1997), pp. H487–H493, Copyright 1997 by American Physiological Society. Reproduced with permission of American Physiological Society in the format Textbook via Copyright Clearance Center.]

a. What two reasons does the abstract give for why a Type II error can occur?

b. Did the studies with "nonsignificant" tests reject the null hypothesis or fail to reject the null hypothesis?

c. The abstract says that approximately 80% of the studies with nonsignificant results contained at least one test where the Type II error probability was greater than 0.30. Which of the following is the best explanation of what that must mean?

    A. 30% of the studies made a Type II error.

    B. For these studies, if the null hypothesis were false, the study had at most a 30% chance of being able to reject it.

    C. For these studies, if the null hypothesis were false, the study had at most a 30% chance of failing to reject it.

E65. A friend wants to do a two-sided test of whether spinning pennies is fair. His null hypothesis is that the percentage of heads when a penny is spun is 50%. He plans to do 20 spins and use a 5% level of significance.

    a. Use the binomial probability formula on page 300 to construct the binomial distribution for $n = 20$ and $p = 0.5$.

    b. Use your distribution from part a to find the numbers of heads that would result in rejecting the null hypothesis.

c. You are pretty sure that the true percentage of heads is about 40%. Supposing that is the case, use exact binomial probabilities to compute the probability that your friend will be able to reject the null hypothesis.

d. What should you suggest to your friend?

E66. A quality improvement plan for a business office calls for selecting a sample of ten invoices that came in during a specified week and counting the number that were not paid on time. Initially, management wants to test to see if the proportion of invoices not paid on time exceeds 30%.

    a. Suppose the plan calls for rejecting the null hypothesis if the number of invoices not paid on time is six or greater out of the sampled ten. Use the binomial probability formula on page 300 to construct the binomial distribution for $n = 10$ and $p = 0.3$. Then use this distribution to find the significance level for this test.

    b. Suppose that, in fact, the true rate of invoices not paid on time is 40%. What is the power of the test described in part a to detect this alternative? Use exact binomial probabilities to calculate this power.

Chapter Summary ◀

To use the confidence interval and significance test of this chapter, you need a random sample from a population that is made up of successes and failures. (When no randomness is involved, you proceed with the test only if you state clearly the limitations of what you have done. If you reject the null hypothesis in such a case, all you can conclude is that something happened that can't reasonably be attributed to chance.)

You use a confidence interval if you want to find a range of plausible values for p, the proportion of successes in a population. The confidence interval has the form

$$\text{statistic} \pm (\text{critical value}) \cdot (\text{estimated standard error})$$

You conduct a test of significance when you want to assess the plausibility of a null hypothesis. The significance test includes these steps:

1. State the null hypothesis and the alternative hypothesis.

2. Justify your reasons for choosing this particular test. Discuss whether the conditions are met, and decide whether it is okay to proceed if they are not strictly met.

3. Compute the test statistic and find the P-value. Draw a diagram of the situation.

4. Use the computations to decide whether to reject or not reject the null hypothesis by comparing the P-value to α, the significance level. Then state your conclusion in terms of the context of the situation. (Simply saying "Reject H_0" is not sufficient.) Mention any doubts you have about the validity of your conclusion.

The test statistic, z, is computed using the hypothesized value, p_0, as the mean. So, the P-value weighs the evidence against the null hypothesis found in the data: A small P-value tells you that the sample proportion you observed is quite far away from p_0. Your data aren't behaving in a manner consistent with the null hypothesis. Your results are said to be statistically significant. A large *P*-value tells you that the sample proportion you observed is near p_0; so your result isn't statistically significant.

This table summaries the information in the last paragraph.

P-Value Is	Evidence Against H_0?	Statistically Significant?	Decision	Possible Error
Small	Yes	Yes	Reject H_0.	Type I
Large	No	No	Do not reject H_0.	Type II

Review Exercises

E67. A 6th-grade student, Emily Rosa, performed an experiment to test the validity of therapeutic touch. According to an article written by her RN mother, a statistician, and a physician: "Therapeutic Touch (TT) is a widely used nursing practice rooted in mysticism but alleged to have a scientific basis. Practitioners of TT claim to treat many medical conditions by using their hands to manipulate a 'human energy field' perceptible above the patient's skin." To investigate whether TT practitioners actually can perceive a "human energy field," 21 experienced TT practitioners were tested to see whether they could correctly identify which of their hands was closest to the investigator's hand. They placed their hands through holes in a screen so they could not see the investigator. Placement of the investigator's hand was determined by flipping a coin. Practitioners of TT identified the correct hand in only 123 of 280 trials. [Source: *Journal of the American Medical Association*, Vol. 279 (1998), pp. 1005–1010.]

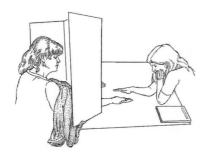

Therapeutic touch.

a. What does "blinded" mean in the context of this experiment? How might it have been done? Why wasn't double-blinding necessary?

b. Should this be a one-sided or a two-sided test? Write appropriate null and alternative hypotheses.

c. Given the results, is it necessary to actually carry out the test?

d. The article says, "The statistical power of this experiment was sufficient to conclude that if TT practitioners could reliably detect a human energy

field, the study would have demonstrated this." What does this sentence mean?

E68. "Most teens are not careful enough about the information they give out about themselves online." In a survey that randomly sampled 971 teenagers who have online access, 78% agreed with this statement. [Source: Pew Internet and American Life Project, 2005, *www.pewinternet.org*.]

a. Compute the 95% confidence interval for the population proportion of teenagers who would agree with the statement.

b. Interpret this confidence interval, making it clear exactly what it is that you are 95% sure is in the confidence interval.

c. Explain the meaning of 95% confidence.

E69. Suppose a null hypothesis is tested at the 0.05 level of significance in 200 different studies in which the null hypothesis is true.

a. How many Type I errors do you expect?

b. Find the probability that no Type I error is made in any of the 200 studies.

E70. Suppose you use the formula for a 90% confidence interval for many different random samples from the same population. What fraction of times should the intervals produced actually capture the population proportion of successes? Explain your reasoning.

E71. What is the best explanation of the use of the term "95% confidence"?

A. We can never be 100% confident in statistics; we can only be 95% confident.

B. The sample proportion will fall in 95% of the confidence intervals we construct.

C. In the long run, the population proportion will fall in 95% of the confidence intervals we construct.

D. We are 95% confident that the confidence interval contains the sample proportion.

E. We are 95% confident that we have made a correct decision when we reject the null hypothesis.

E72. How could this explanation of margin of error be improved?

> Statisticians over the years have developed quite specific ways of measuring the accuracy of samples—so long as the fundamental principle of equal probability of selection is adhered to when the sample is drawn. For example, with a sample size of 1,000 national adults (derived using careful random selection procedures), the results are highly likely to be accurate within a margin of error of plus or minus three percentage points. So, if we find in a given poll that President Clinton's approval rating is 50%, the margin of error indicates that the true rating is very likely to be between 53% and 47%. It is very unlikely to be higher or lower than that. To be more specific, the laws of probability say that if we were to conduct the same survey 100 times, asking people in each survey to rate the job Bill Clinton is doing as president, in 95 out of those 100 polls, we would find his rating to be between 47% and 53%. In only five of those surveys would we expect his rating to be higher or lower than that due to chance error. [Source: Frank Newport, Lydia Saad, and David Moore, "How Polls Are Conducted," in *Where America Stands* (Hoboken, NJ: John Wiley & Sons, Inc., © 1997). Used by permission of John Wiley & Sons, Inc.]

E73. Some people complain that election polls cannot be right, because they personally were not asked how they voted. Write an explanation about how polls can get a good idea of how the entire population will vote by asking a relatively small number of voters.

E74. To perform a significance test for a proportion, you must decide on the hypotheses (null and alternative), the level of significance, and the sample size. Explain the effects of each of these components on the power of the test.

E75. The American Religious Identification Survey is conducted periodically to determine the religious preferences and beliefs of American adults. The survey questioned 54,461 respondents in English or Spanish. You may consider this a random sample of American adults. [Source: *American Religious Identification Survey, Summary Report*, March 2009, *http://b27.cc.trincoll.edu/weblogs/American ReligionSurvey-ARIS/reports/ARIS_Report_2008.pdf.*]

a. Seventy-six percent identified themselves as Christian. Compute a 95% confidence interval for the proportion of all American adults who would identify themselves as Christian.

b. What is the margin of error for the proportion estimated in part a?

c. A subsample of 1000 respondents, which you may also consider a random sample of all American adults, was asked, "Regarding the existence of God, do you think . . . ? Their responses are in the following table. If you are interested in the proportion of Americans who respond "There is definitely a personal God," what is the margin of error for this portion of the poll?

There is no such thing	2.3%
There is no way to know	4.3%
I'm not sure	5.7%
There is a higher power but no personal God	12.1%
There is definitely a personal God	69.5%
Refused to answer	6.1%
Total	100%

E76. Suppose a test rejects this null hypothesis in favor of the alternative:

$$H_0: p = p_0 \qquad \text{versus} \qquad H_a: p > p_0$$

a. Explain why it also would have been reasonable to reject the null hypothesis if it had been of this form: $H_0: p \leq p_0$.

b. Would the null hypothesis necessarily have been rejected if the alternative had been $H_a: p \neq p_0$? Explain.

E77. Determine whether each statement is true or false.

a. Always look at the data before writing the null and alternative hypotheses.

b. All else being equal, using a one-sided test will result in a larger *P*-value than using a two-sided test.

c. The *P*-value is the probability that the null hypothesis is true.

d. A statistically significant result means the *P*-value is small.

C1. Only 33% of students correctly answered a difficult multiple-choice question on an exam given nationwide. Professor Chang gave the same question to her 35 students, hypothesizing that they would do better than students nationwide. Despite the lack of randomization, she performed a one-sided test of the significance of a sample proportion and got a P-value of 0.03. Which is the best interpretation of this P-value?

Ⓐ Only 3% of her students scored better than students nationally.

Ⓑ If the null hypothesis is true that her students do the same as students nationally, there is a 3% chance that her students will do better than students nationally on this question.

Ⓒ Between 30% and 36% of her population of students can be expected to answer the question correctly.

Ⓓ There is a 3% chance that her students are better than students nationally.

Ⓔ There is a 3% chance that a random sample of 35 students nationwide would do as well as or better than her students did.

C2. Researchers constructed a 95% confidence interval for the proportion of people who prefer apples to oranges. They computed a margin of error of ±4%. In checking their work, they discovered that the sample size used in their computation was 1/4 of the actual number of people surveyed. Which is closest to the correct margin of error?

Ⓐ 1%

Ⓑ 2%

Ⓒ 4%

Ⓓ 8%

Ⓔ 16%

C3. A survey of 200 randomly selected students at a large university found that 105 favor a stricter policy for keeping cars off campus. Is this convincing evidence that more than half of all students favor a stricter policy for keeping cars off campus?

Ⓐ Yes, because 105 is more than half of 200.

Ⓑ Yes, by constructing a 95% confidence interval, you can see that it is plausible that more than half of all students favor a stricter policy for keeping cars off campus.

Ⓒ No, by constructing a 95% confidence interval, you can see. that it is plausible that 50% or less of all students favor a stricter policy for keeping cars off campus.

Ⓓ No, because the survey only used a sample of students.

Ⓔ Because students probably have strong opinions that aren't normally distributed, no conclusion can be reached.

C4. In college populations, the annual incidence of infectious mononucleosis has been estimated to be as many as about 50 cases per 1000 students. A university student health service took a survey of students to test whether the rate of mononucleosis on their campus is different from this national rate. With $\alpha = 0.05$, they rejected the null hypothesis. Which is the best interpretation of "$\alpha = 0.05$" in this context? [Source: *www. aafp.org/afp/20041001/1279.html*.]

Ⓐ There's a 5% chance that the rate of mono on this campus is different from the national rate.

Ⓑ There's a 5% chance that the rate of mono on this campus is not different from the national rate.

Ⓒ If the rate of mono is the same on this campus as nationally, there's a 5% chance that the researchers will mistakenly conclude that it is different.

Ⓓ If the rate of mono is different on this campus, there's a 5% chance that the researchers will mistakenly conclude that it is the same.

Ⓔ The study has enough power to detect a difference in the rates of mono of 0.05 or more.

C5. In a pre-election poll, 51% of a random sample of voters plan to vote for the incumbent. A 95% confidence interval was computed for the proportion of all voters who plan to vote for the incumbent. What is the best meaning of "95% confidence"?

Ⓐ If all voters were asked, there is a 95% chance that 51% of them will say they plan to vote for the incumbent.

Ⓑ You are 95% confident that the confidence interval contains 51%.

Ⓒ In 100 similar polls, you expect that 95 of the confidence intervals constructed will contain the percentage of all voters who plan to vote for the incumbent.

Ⓓ In 100 similar polls, you expect that 95 of the confidence intervals constructed will contain 51%, which is the best estimate of the percentage of all voters who plan to vote for the incumbent.

Ⓔ The probability is 0.95 that the confidence interval constructed in this poll contains the percentage of all voters who plan to vote for the incumbent.

C6. Sheldon takes a random sample of 50 U.S. housing units and finds that 30 are owner occupied. Using a significance test for a proportion, he is not able to reject the null hypothesis that exactly half of U.S. housing units are owner occupied. Later, Sheldon learns that the U.S. Census for the same year found that 66.2% of housing units are owner occupied. Select the best description of the type of error in this situation.

Ⓐ No error was made.

Ⓑ A Type I error was made because a false null hypothesis was rejected.

Ⓒ A Type I error was made because a false null hypothesis wasn't rejected.

Ⓓ A Type II error was made because a false null hypothesis was rejected.

Ⓔ A Type II error was made because a false null hypothesis wasn't rejected.

Investigations

C7. Fifty students want to know what percentage of Skittles candies are orange. Each student takes a random sample of 50 candies and constructs a 90% confidence interval using the formula

$$\hat{p} \pm z^* \sqrt{\frac{\hat{p}(1 - \hat{p})}{n}}$$

a. Of all Skittles candies, 20% are orange. How many of their fifty 90% confidence intervals would you expect to capture the population proportion of 0.20?

b. The students' confidence intervals are plotted in Display 8.15 as vertical lines. How many of their confidence intervals capture the population proportion of 0.20? What percentage is this?

c. Are most of the confidence intervals that don't capture the population proportion too close to 0 or too close to 0.5? Do they tend to be shorter or longer than intervals that do capture the population proportion?

d. Unfortunately, what you have discovered here is true in general: If you use the standard formula to construct confidence intervals for a proportion, your actual capture rate will be smaller (and sometimes a lot smaller) than advertised. List several ways in which this formula is based on approximations and so might have an actual capture rate different from that advertised.

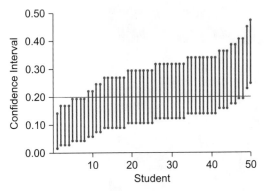

Display 8.15 Fifty 90% confidence intervals for random samples of size 50 with $p = 0.20$ constructed using the standard formula.

C8. *The Plus 4 interval.* In C7, you saw that the standard formula for a confidence interval results in a capture rate that tends to be less than advertised. The capture rate should improve if the more extreme intervals are moved

toward the center and made a little longer, as long as the remaining intervals aren't disturbed too much. There is a method that will do exactly that, called the Plus 4 method: you simply add 2 to the number of observed successes and add 2 to the number of observed failures (which adds 4 to the sample size). That is, if x denotes the number of successes in a sample of n items, the new estimator of the population proportion, p, is

$$\hat{p} = \frac{x + 2}{n + 4}$$

The Plus 4 confidence interval is then

$$\hat{p} \pm z^* \sqrt{\frac{\hat{p}(1 - \hat{p})}{n + 4}}$$

a. Construct 90% confidence intervals for $n = 50$ and $x = 5$ using both the standard formula and the Plus 4 method. What happened to the center of the interval? The length of the interval?

b. Repeat the procedures and questions of part a for $n = 50$ and $x = 20$. Compare your answers to those in part a.

c. The students used the same samples as in C7 to construct Plus 4 confidence intervals. These Plus 4 confidence intervals are given in Display 8.16. How many of the intervals capture the population proportion of 0.20? What percentage is this?

d. Describe how the Plus 4 confidence intervals differ from the standard intervals.

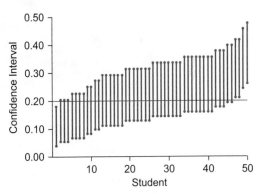

Display 8.16 Fifty 90% confidence intervals for the same random samples as used in Display 8.15 with $n = 50$ with $p = 0.20$, this time constructed using the Plus 4 method.

419

Chapter 9

Comparing Two Populations: Inference for the Difference of Two Proportions

Sixty-three percent of U.S. households answering a recent survey reported that they own a pet, up from 56% in 1994, the year the survey started tracking this information. Are you confident that there has been an increase in the percentage of U.S. households that own a pet? Your answer will depend on how many households were included in the surveys.

In Chapter 8 you learned how to estimate a population proportion and how to test a hypothesis about a proportion. You will now expand that knowledge to cover situations that require the comparison of two population proportions.

Comparing two proportions is probably more common than making an inference about a single proportion and may well be more important. In political polls the difference between the proportions of two groups who favor some issue may be of more concern than the actual proportions. In a medical study comparing an experimental treatment with a standard treatment, the difference in their success rates is the essential factor in the decision as to whether the experimental treatment is approved for general use. In quality improvement programs, the determining factor in management decisions is whether the increase in the proportion of products conforming to the manufacturing standard is worth the cost of implementing a new process.

Thus, when comparing sample two proportions, you will begin by computing their difference. You can then test whether the difference is statistically significant, or you can find a confidence interval estimate for the difference between the population proportions.

In this chapter, you will learn

▶ to construct and interpret a confidence interval for the difference between the proportion of successes in one population and the proportion of successes in another population

▶ to use a test of significance to decide if it is reasonable to conclude that two samples might have been drawn from two populations that have the same proportion of successes

▶ to construct and interpret confidence intervals and tests of significance for experiments and observational studies

9.1 ▶ A Confidence Interval for the Difference of Two Proportions

A recent poll of 29,700 U.S. households found that 63% owned a pet. The percentage in 1994 was 56%. [Source: American Pet Products Manufacturers Association, *www.appma.org.*] The two populations are the households in the United States in 1994 and the households now. The question you will investigate is, "What was the change in the percentage of U.S. households that own a pet?" The obvious answer is that the percentage increased by 7 percentage points, but this is only an estimate because 7% is the difference of two *sample* percentages that probably are not exactly equal to the *population* percentages. In this section, you will learn how to find the margin of error to go with the difference of 7%.

Many decisions are based on comparisons.

A confidence interval for the difference of two proportions, $p_1 - p_2$, where p_1 is the proportion of successes in the first population and p_2 is the proportion of successes in the second population, has this familiar form:

$$(\hat{p}_1 - \hat{p}_2) \pm z^* \cdot \textit{estimated standard error of } (\hat{p}_1 - \hat{p}_2)$$

Here $\hat{p}_1$ and $\hat{p}_2$ are the proportions of successes in the two samples. Substituting $\hat{p}_1 = 0.63$, $\hat{p}_2 = 0.56$, and $z^* = 1.96$ into the formula gives the 95% confidence interval for the difference between the proportion of U.S. households that own pets now and the proportion that owned pets in 1994:

$$0.07 \pm 1.96 \cdot \textit{estimated standard error of } (\hat{p}_1 - \hat{p}_2)$$

The hard part, as always, is estimating the size of the standard error.

The Formula for the Confidence Interval

The statistic of interest is a difference of random variables, and you know from Chapter 6 that the mean of a difference is the difference of the means. If the variables are *independent*, the variance of the difference is the sum of the variances. In symbols,

$$\mu_{\hat{p}_1 - \hat{p}_2} = p_1 - p_2 \quad \text{and} \quad \sigma^2_{\hat{p}_1 - \hat{p}_2} = \sigma^2_{\hat{p}_1} + \sigma^2_{\hat{p}_2}$$

In addition, under the independence assumption, the sampling distribution of the differences will be approximately normal if the sample sizes are suitably large.

Taking the square root of the variance, the standard error is

$$\sigma_{\hat{p}_1 - \hat{p}_2} = \sqrt{\sigma^2_{\hat{p}1} + \sigma^2_{\hat{p}2}}$$

From Section 7.3, the standard error of the distribution of the sample proportion $\hat{p}_1$ is

$$\sigma_{\hat{p}_1} = \sqrt{\frac{p_1(1 - p_1)}{n_1}}$$

which can be estimated by

$$\sqrt{\frac{\hat{p}_1(1 - \hat{p}_1)}{n_1}}$$

where n_1 is the sample size. Similarly, the standard error of the distribution of the sample proportion, $\hat{p}_2$, is

$$\sigma_{\hat{p}_2} = \sqrt{\frac{p_2(1 - p_2)}{n_2}}$$

which can be estimated by

$$\sqrt{\frac{\hat{p}_2(1 - \hat{p}_2)}{n_2}}$$

where n_2 is the sample size.

Substituting these into the formula for the standard error of a difference, an estimate of the standard error of the difference of two proportions is

$$\sigma_{\hat{p}_1 - \hat{p}_2} \approx \sqrt{\frac{\hat{p}_1(1 - \hat{p}_1)}{n_1} + \frac{\hat{p}_2(1 - \hat{p}_2)}{n_2}}$$

Now you can write the complete confidence interval for the difference between two proportions.

Confidence Interval for the Difference of Two Proportions

A confidence interval for the difference, $p_1 - p_2$, of the proportion of successes in one population and the proportion of successes in a second population is

$$(\hat{p}_1 - \hat{p}_2) \pm z^* \sqrt{\frac{\hat{p}_1(1 - \hat{p}_1)}{n_1} + \frac{\hat{p}_2(1 - \hat{p}_2)}{n_2}}$$

where $\hat{p}_1$ is the proportion of successes in a random sample of size n_1 taken from the first population and $\hat{p}_2$ is the proportion of successes in a random sample of size n_2 taken from the second population. (The sample sizes don't have to be equal.)

The conditions that must be met to use this formula are:

- The two samples are taken randomly and independently from two populations.
- Each population is at least 10 times as large as its sample size.
- $n_1\hat{p}_1$, $n_1(1 - \hat{p}_1)$, $n_2\hat{p}_2$, and $n_2(1 - \hat{p}_2)$ are all at least 5.

Here is the interpretation of the interval: If you use $z^* = 1.96$, for example, you are 95% confident that, if you were to examine all units in both populations, the difference between the proportion, p_1, of successes in the first population and the proportion of successes, p_2, in the second population would fall in the confidence interval.

It doesn't matter which sample proportion you call $\hat{p}_1$ and which you call $\hat{p}_2$, as long as you remember which is which. Most students like to assign the larger one to be $\hat{p}_1$ so that the difference, $\hat{p}_1 - \hat{p}_2$, is positive. If there is a good reason to work with negative numbers, do it the other way!

Which sample proportion should be $\hat{p}_1$ and which $\hat{p}_2$?

A Difference in Pet Ownership?

Example 9.1

As described at the beginning of this section, a recent pet ownership survey found that 63% of the 29,700 U.S. households sampled own a pet. A 1994 survey, conducted on behalf of the same organization, found that 56% of the 6786 U.S. households sampled owned a pet. Although the sampling procedure was more complicated, you may consider these two independently selected random samples. Find and interpret a 95% confidence interval for the difference between the proportion of U.S. households that own a pet now and the proportion of U.S. households that owned a pet in 1994.

Solution

The two samples can be considered random samples. They were taken independently from the population of U.S. households in two different years. The number of U.S. households in each year is larger than 10 times 29,700. Finally, you must check that the sample sizes are large enough to ensure that the sampling distribution of $\hat{p}_1 - \hat{p}_2$ is approximately normal. That is the case because each of the following is at least 5.

$$n_1\hat{p}_1 = 29,700(0.63) = 18,711 \qquad n_1(1 - \hat{p}_1) = 29,700(0.37) = 10,989$$
$$n_2\hat{p}_2 = 6,786(0.56) \approx 3,800 \qquad n_2(1 - \hat{p}_2) = 6,786(0.44) \approx 2,986$$

Check conditions.

The 95% confidence interval for the difference of the two population proportions, $\hat{p}_1$ and $\hat{p}_2$, is

Do computations.

$$(\hat{p}_1 - \hat{p}_2) \pm z^* \sqrt{\frac{\hat{p}_1(1 - \hat{p}_1)}{n_1} + \frac{\hat{p}_2(1 - \hat{p}_2)}{n_2}}$$
$$= (0.63 - 0.56) \pm 1.96\sqrt{\frac{(0.63)(1 - 0.63)}{29,700} + \frac{(0.56)(1 - 0.56)}{6,786}}$$
$$= 0.07 \pm 0.013$$

Alternatively, you can write this confidence interval as (0.057, 0.083). You also can calculate this interval using technology by entering each sample size and the number of successes in each sample. (To estimate the number of successes, multiply the sample proportion by the sample size and round to the nearest whole number.)

You are 95% confident that the increase in the rate of pet ownership from 1994 to today is between 0.057 and 0.083. This means that it is plausible that the difference in the percentage of households that own pets now and the percentage in 1994 is 5.7%. It is also plausible that the difference is 8.3%. Note that a difference of 0 does not lie within the confidence interval. This means that if the difference in the proportion of pet owners now and in 1994 actually is 0, getting a difference of 0.07 in the samples is not at all likely. Thus, you are convinced that there was an increase in the percentage of households that own a pet.

The margin of error, 0.013, in the previous example was so small because the sample sizes were very large.

Example 9.2 | Wireless Versus Wired

The *Mobile Difference* is a report aimed at describing how behaviors differ among those embracing wireless connectivity and those who are still happy with traditional desktop computing. Developed by the Pew Internet and American Life Project, the study used random digit dialing to select samples that covered various types of technology users within two main classes, those "motivated by mobility" and those for whom "stationary media will do." Here, four types are studied, two from each main class, with their definitions:

> *Digital Collaborators (DC)*. With the most tech assets, Digital Collaborators use them to work with and share their creations with others. They are enthusiastic... and confident in how to manage digital devices and information.
>
> *Media Movers (MM)*. Media Movers have a wide range of online and mobile habits, and they are bound to find or create an information nugget, such as a digital photo, and pass it on.
>
> *Desktop Veterans (DV)*. This group of older, veteran online users is content to use high-speed connection and a desktop computer to explore the Internet and stay in touch with friends, placing their cell phone and mobile connections in the background.
>
> *Information Encumbered (IE)*. Most people in this group suffer from information overload and think taking time off from the Internet is a good thing.

Within the mobility class, DC covers the high end and MM the midrange users. In the stationary class, DV covers the high end and IE the midrange users. Display 9.1 provides the sample sizes (in parentheses) and observed percentages for a variety of characteristics within each type. With regard to getting health information online, estimate the difference between the population proportions for DC and MM types in a 90% confidence interval. Interpret this interval.

Solution

The samples are selected by random digit dialing, and so there is no problem with the randomness of the initial sample. Those responding to the survey are actually separated

	Motivated by Mobility		Stationary Media Will Do	
	DC (273)	MM (228)	DV (442)	IE (399)
College grad	61	32	41	33
Get health information online	93	90	85	76
Buy a product online	95	81	82	52
Play a video game	48	44	28	14

Display 9.1 Characteristics of wireless and wired users of technology (percent). [Source: *www.pewinternet.org/Reports/2009/5-The-Mobile-Difference-Typology.aspx*.]

into their respective types after the initial sample is completed, but that does not have a serious effect on the independence or the randomness of the samples as long as the sample size is large. Let $\hat{p}_1$ denote the DC sample proportion of 0.93 and $\hat{p}_2$ the MM sample proportion of 0.90. Then the conditions necessary for approximate normality of the sampling distribution of $\hat{p}_1 - \hat{p}_2$ are satisfied because

$$n_1\hat{p}_1 = 273(0.93) = 253.89$$
$$n_1(1 - \hat{p}_1) = 273(1 - 0.93) = 19.11$$
$$n_2\hat{p}_2 = 228(0.90) = 205.2$$
$$n_2(1 - \hat{p}_2) = 228(1 - 0.90) = 22.8$$

are all at least 5.

For a 90% confidence interval, use $z^* = 1.645$:

> Do computations.

$$(\hat{p}_1 - \hat{p}_2) \pm z^* \sqrt{\frac{\hat{p}_1(1 - \hat{p}_1)}{n_1} + \frac{\hat{p}_2(1 - \hat{p}_2)}{n_2}}$$
$$= (0.93 - 0.90) \pm 1.645 \sqrt{\frac{0.93(1 - 0.93)}{273} + \frac{0.90(1 - 0.90)}{228}}$$
$$= 0.03 \pm 0.04$$

The interval is $(-0.01, 0.07)$.

Statistical software will also compute confidence intervals for you. The following Minitab output gives the same confidence interval as the computations above. It also includes the *P*-value for two different significance tests for the difference of two proportions, the topic of Section 9.2.

Test and CI for Two Proportions

Sample	X	N	Sample p
1	254	273	0.930403
2	205	228	0.899123

Difference = p (1) − p (2)
Estimate for difference: 0.0312801
90% CI for difference: (−0.0101690, 0.0727292)
Test for difference = 0 (vs not = 0): Z = 1.26 P-Value = 0.208

Fisher's exact test: P-Value = 0.257

Because this method captures the population difference 90% of the time, you are 90% confident that the difference in the population proportions (for the populations of DC and MM) of those who get health information online is between −0.01 and 0.07. In other words, the population proportion for DC is at most 0.07 higher or at most 0.01 lower than the population proportion for MM, with 90% difference. Any value of the

> Write interpretation in context.

difference in this interval is a plausible value for the actual difference, $p_1 - p_2$. Because the interval includes zero, it is plausible that there is no difference between the proportions in these two populations.

DISCUSSION
The Formula for the Confidence Interval

D1. A survey of 1200 households found that 45% of households own a dog, 34% own a cat, and 20% own both.

    a. What percentage of households in the survey own neither a cat nor a dog?

    b. Are owning a dog and owning a cat independent events?

    c. Should you use the techniques of this section to estimate the difference in the percentage of households that own a dog and the percentage that own a cat? Explain.

D2. In estimating a single proportion, the standard error of the sample proportion, $\sqrt{\frac{p(1-p)}{n}}$, is largest when the population proportion, p, is equal to 0.5. Argue that a quick and easy approximation to the maximum margin of error corresponding to a 95% confidence interval for p is $\frac{1}{\sqrt{n}}$. Develop a similar rule to approximate the maximum margin of error corresponding to a 95% confidence interval for the difference between two population proportions.

STATISTICS IN ACTION 9.1 ▶ Who's Yellow?

In this Statistics in Action, you will explore how the confidence interval for the difference in two proportions behaves from sample to sample when the proportions of successes in two populations are different.

What You'll Need: a sample of 50 Skittles, a sample of 50 Milk Chocolate M&M's (or bags prepared by your instructor)

    1. Count the number of yellow candies in your sample of Skittles and in your sample of M&M's. Compute the two sample proportions.

    2. Find the 95% confidence interval for the difference in the two population proportions: *proportion of Skittles that are yellow − proportion of M&M's that are yellow*. (Every student should subtract the sample proportions in this order, even if the difference is negative.)

    3. Draw your confidence interval on a large copy of Display 9.2. Other members of your class should draw their confidence intervals on the same chart.

    4. Compare the centers of the confidence intervals. Are they close to each other, or do they vary a lot? Why do they vary?

    5. Compare the widths of the confidence intervals. Are the widths fairly equal, or do they vary a lot? Why is there any variability in the width?

    6. How many of the confidence intervals would you expect to capture the true difference in proportions?

    7. Skittles have 20% yellow candies. Milk Chocolate M&M's have 14% yellow candies. How many of the confidence intervals actually captured the true difference in these proportions?

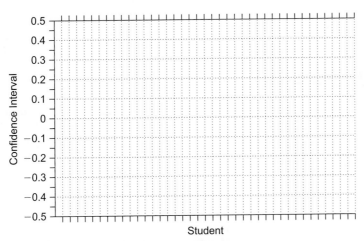

Display 9.2 Chart for recording 95% confidence intervals for the difference in the proportion of yellows in Skittles and M&M's.

Summary 9.1: A Confidence Interval for the Difference of Two Proportions

Let $\hat{p}_1$ be the proportion of successes in a random sample of size n_1 taken from a population with an (unknown) proportion of successes p_1 and let $\hat{p}_2$ be the proportion of successes in an independent random sample of size n_2 taken from a second population with an (unknown) proportion of successes p_2. When the sample sizes, n_1 and n_2, are large enough, the sampling distribution of $\hat{p}_1 - \hat{p}_2$ is approximately normal, is centered at $p_1 - p_2$, and has standard error

$$\sigma_{\hat{p}_1 - \hat{p}_2} = \sqrt{\frac{p_1(1 - p_1)}{n_1} + \frac{p_2(1 - p_2)}{n_2}}$$

In practice, you estimate $\sigma_{\hat{p}_1 - \hat{p}_2}$ using the two sample proportions. This leads to a confidence interval for the difference, $p_1 - p_2$, of the proportion of successes in one population and the proportion of successes in a second population

$$(\hat{p}_1 - \hat{p}_2) \pm z^* \sqrt{\frac{\hat{p}_1(1 - \hat{p}_1)}{n_1} + \frac{\hat{p}_2(1 - \hat{p}_2)}{n_2}}$$

where $\pm z^*$ are the values that enclose an area in a normal distribution equal to the confidence level, which typically is 0.95. Don't forget to check the three conditions on page 429 that must be met to use this confidence interval.

When interpreting, for example, a 95% confidence interval, you should say that you are 95% confident that the difference between the proportion of successes in the first population and the proportion of successes in the second population is in this confidence interval. (However, say this in the context of the situation.) If the confidence interval includes 0, you can't conclude that there is any difference in the proportions in the two populations.

Practice

The Formula for the Confidence Interval

P1. Suppose the surveys on pet ownership described in Example 9.1 on page 429 had used sample sizes of only 100 people. That is, a 1994 survey of 100 randomly selected U.S. households found that 56% owned a pet, and this year a survey of 100 randomly selected U.S. households found that 63% owned a pet.

a. Check the conditions for constructing a confidence interval for the difference of two proportions.

b. Compute and interpret the confidence interval.

c. Is 0 in the confidence interval? What does your answer imply?

P2. A poll of 549 teenagers asked if it was appropriate for parents to install a special device on the car to allow parents to monitor teenagers' driving speeds. Among the 325 13- to 15-year-olds in the sample, 50% responded yes to this question. Among the 224 16- to 17-year-olds, only 28% responded yes. You may consider these independent random samples. [Source: *Teens Slow to Support Speed Monitors*, 2005, *poll.gallup.com.*]

   a. Check the conditions for constructing a confidence interval for the difference of two proportions.

   b. Estimate the difference between the proportion of all 13- to 15-year-olds who would answer yes to this question and the proportion of all 16- to 17-year-olds who would answer yes.

   c. Interpret the resulting interval in the context of the problem.

   d. Is 0 in the confidence interval? What does your answer imply?

P3. The National Survey of America's College Students (NSACS) examines the quantitative literacy of graduating U.S. college students. (See E3 in Chapter 8 on page 382 for further details.) The survey sampled about 1000 students from four-year colleges and about 800 students from two-year colleges. Among the four-year college students, 34% tested proficient in quantitative literacy, while the figure was 18% for the two-year college students. Assuming that the samples were randomly selected, estimate the true difference in proportions of quantitatively literate students for these two populations. Interpret the resulting interval in the context of the problem. [Source: *The Literacy of America's College Students*, American Institutes for Research, January 2006, *www.air.org/news/documents.*]

P4. Refer to Example 9.2. With regard to playing a video game online, estimate the difference between the population proportions for DC and DV types in a 95% confidence interval. Interpret this interval.

P5. Twitter use has been ballooning since late 2008 and now is a common method of communication. The Pew Internet and American Life Project collected a large random sample of users of the World Wide Web and split it into two groups. The first group was a sample of 663 "American adults [who] said they used a service like Twitter or another service that allowed them to share updates about themselves or to see the updates of others." (These will be referred to as "those who Twitter.") The second group was a sample of 1591 Internet users who do not Twitter. (These will be referred to simply as "Internet users.") Among those who Twitter, 35% live in urban areas, while among Internet users only 29% live in urban areas. [Source: *pewresearch.org/pubs/1117/ twitter-tweet-users-demographics.*]

   a. Check the conditions for constructing a confidence interval for the difference of two proportions.

   b. Find a 95% confidence interval for the difference of proportions.

   c. Interpret the resulting interval in the context of the problem.

   d. Is 0 in the confidence interval? Is it fair to say that the percentage who are urban among those who Twitter is greater than among Internet users?

   e. What is meant by the term "95% confident" in this situation?

Exercises

E1. A statistics instructor kept track of students' grades on her final examination and whether they did the exam in pencil or in pen. Of the 75 students who used a pencil, 63 got grades of 70% or higher. Of the 19 students who used a pen, 10 got grades of 70% or higher.

   Assume that these students can be considered random samples from the students who could take this instructor's final examination. Locate her 95% confidence interval for the difference of two proportions in the following output and then decide which of these choices are the best two interpretations of this confidence interval.

Test and CI for Two Proportions

Sample	X	N	Sample p
1	63	75	0.840000
2	10	19	0.526316

Difference = p (1) − p (2)
Estimate for difference: 0.313684
95% CI for difference: (0.0743322, 0.553036)
Test for difference = 0 (vs not = 0): Z = 2.93 P-Value = 0.003

Fisher's exact test: P-Value = 0.011

   A. The instructor must observe more students, because this confidence interval is too wide to be of any possible use.

   B. The instructor has statistically significant evidence that a larger proportion of students who use a pencil score 70% or higher on her final than students who use a pen.

   C. If the instructor could observe all students who could take her final examination, she is 95% confident that the difference in the proportion who use a pencil and get a grade of 70% or higher and the proportion who use a pen and get a grade of 70% or higher is between 0.07 and 0.55.

   D. If the instructor could observe all students who could take her final examination, she is 95% confident that between 7% and 55% of them will get 70% or higher.

E2. Has Dr. Martin Luther King's 1963 "I Have a Dream" speech been fulfilled in the 45 years since its delivery? A January 2009 CNN/Opinion Research Corporation poll of 332 blacks and 798 whites found 69% of blacks said it has been fulfilled. Among whites, only 46% said the dream has been fulfilled. Locate the 95% confidence interval for the difference of two proportions in the following output and then decide which of these choices are the best two interpretations of this confidence interval. [Source: *www.cnn.com/2009/POLITICS/01/19/king.poll.*]

Test and CI for Two Proportions

Sample	X	N	Sample p
1	229	332	0.689759
2	367	798	0.459900

Difference = p (1) − p (2)
Estimate for difference: 0.229859
95% CI for difference: (0.169264, 0.290454)
Test for difference = 0 (vs not = 0): Z = 7.05 P-Value = 0.000

Fisher's exact test: P-Value = 0.000

A. You can be confident that, if you looked at all adults in the country, a larger percentage of blacks than whites would think that the "Dream" has been fulfilled.

B. If all adults were questioned, you can be 95% confident that the true percentage of adults who think the "Dream" has been fulfilled would be between 17% and 29%.

C. This confidence interval would be wider if the sample sizes were smaller, with the same observed percentages.

D. If all adults who did not think the "Dream" had been fulfilled were questioned, you are 95% confident that more of them would be black than white.

E3. Researchers at the University of California, San Diego, found that when given photos of a dog owner and two dogs, judges correctly matched 16 out of 25 purebred dogs with their owners. When the dogs were mutts, only 7 dogs out of 20 were correctly matched with their owners. [Source: *www.newscientist.com.*]

a. Check the conditions for constructing a confidence interval for the difference of two proportions.

b. Find a 95% confidence interval for the difference of proportions.

c. Interpret the resulting interval in the context of the problem.

d. Is 0 in the confidence interval? What does your answer imply?

e. What would you recommend that the researchers do?

f. What is meant by term "95% confident" in this situation?

E4. The USC Annenberg School Center for the Digital Future found that, in 2009, 30.4% of Americans felt uncomfortable with the lack of face-to-face contact when ordering a product online. The corresponding percentage was 24.9% in 2001. Assume the samples were independently and randomly selected and that the sample size was 2000 in both years. [Source: *www.digitalcenter.org.*]

a. Check the conditions for constructing a confidence interval for the difference of two proportions.

b. Find a 99% confidence interval for the difference of proportions.

c. Interpret the resulting interval in the context of problem.

d. Is 0 in the confidence interval? What does your answer imply?

e. What is meant by the term "99% confident" in this situation?

E5. College administrations are continually perplexed about how to deal with hazing issues on their campuses. But how widespread is hazing, and, in fact, what constitutes hazing anyway? One study tried to get an objective handle on this persistent college phenomenon by carrying out a large survey of college and university students on 53 campuses. The study leaders defined hazing as "any activity expected of someone joining or participating in a group that humiliates, degrades, abuses, or endangers them regardless of a person's willingness to participate." Portions of data from this are summarized in the following questions. [Source: *www.hazingstudy.org/publications/hazing_in_view_web.pdf.*]

a. Of the survey responses, 4775 were females associated with campus groups and 2685 were males associated with campus groups. On the overall question of having ever experienced hazing in any form, 61% of the males who participate in campus groups and 52% of the females who participate in campus groups responded positively. Estimate the difference between the proportions of students who have experienced hazing among the males who participate in campus activities as compared to the females who participate in campus activities. Use a 90% confidence interval.

b. This was a Web-based survey with a low response rate of only about 12%. How do you think that might have affected the reported percentages? Even if the reported percentages are biased, might not the difference between the two be a valid comparison? Explain your reasoning.

E6. For the hazing study of E5, the two types of activities receiving the highest rate of participation overall were "participate in a drinking game" and "sing or chant by yourself or with a few selected team members in a public situation that is not related to the event, game, or practice." Display 9.3 shows the sample sizes for those responding and the proportions for each activity across a variety of typical campus groups. In comparing these groups, regard the samples as being independent even though some of the same students may be in more than one group.

Campus Group	n	Drinking Game	Sing or Chant
Varsity athletics	640	0.47	0.27
Social fraternity or sorority	1295	0.53	0.32
Service fraternity or sorority	544	0.26	0.18
Intramural sports	1060	0.28	0.13
Performing arts	818	0.23	0.25
Academic clubs	1061	0.10	0.06

Display 9.3 Proportion of group participating in two types of hazing activity.

a. Estimate the difference, with 95% confidence, between the proportions engaged in drinking games among those in varsity athletics as compared to those in social fraternities or sororities. Do the same for those who have been involved in singing or chanting.

b. Estimate the difference, with 90% confidence, between the proportions engaged in drinking games among those in service fraternities or sororities as compared to those in performing arts. Do the same for those who have been involved in singing or chanting.

E7. A Harris poll asked this question of about 425 men and 425 women: "When you get a sales or customer service phone call from someone you don't know, would you prefer to be addressed by your first name or by your last name, or don't you care one way or the other?" Twenty-three percent of the men and 34% of the women said they would prefer being addressed by their last name. Although this was not quite the case, you can assume that the samples were random and independent. [Source: Harris, November 14, 2001, *www.harrisinteractive.com*.]

a. Check the conditions for constructing a confidence interval for the difference of two proportions.

b. Find a 99% confidence interval for the difference of proportions.

c. Interpret the resulting interval in the context of the problem.

d. Is 0 in the confidence interval? What does your answer imply?

E8. "At what age do you think people should be permitted to have a driver's license?" In 1995, 46% of 1000 randomly sampled adults in the United States responded that 16 was the correct age. In 2004, only 35% of 1000 randomly sampled adults responded this way. [Source: CNN/Gallup poll, December 17–19, 2004.]

a. Check the conditions for constructing a 99% confidence interval for the difference of two proportions.

b. Estimate the true difference between the 1995 and 2004 proportions favoring 16 as the correct age to begin driving.

c. Interpret the resulting interval in the context of the problem.

d. Is 0 in the confidence interval? What does your answer imply?

E9. In a recent national survey, 15,200 high school seniors in 128 schools completed questionnaires about physical activity. Male students were significantly more likely (59%) than female students (48%) to have played on sports teams run by their school during the 12 months preceding the survey. Check the accuracy of the phrase "significantly more likely," assuming that there were equal numbers of male and female students in this survey and that the samples are equivalent to independent simple random samples. [Source: Participation in School Athletics, *www.childtrendsdatabank.org*.]

E10. In the report referenced in E9, the participation of females in school athletics increased from 47% in 1991 to 48% recently. The total sample size (males and females) for 1991 was approximately 15,400. Does this increase represent a significant change in the level of participation for females? Explain your reasoning.

E11. What is the effect of an increase in the sample sizes on the width of a confidence interval for the difference of two proportions? Explain.

E12. Statements I and II are interpretations of a 95% confidence interval for a single proportion. Write similar statements for a 95% confidence interval for a difference of two proportions.

I. A 95% confidence interval consists of those population proportions p for which the proportion from the sample, $\hat{p}$, is reasonably likely to occur.

II. If you construct a hundred 95% confidence intervals, you expect that the population proportion p will be in 95 of them.

E13. The formula for the confidence interval for the difference of two proportions involves z. Why is it okay to use a table of the normal distribution in this case?

E14. What can you conclude if the confidence interval for the differences of two proportions

a. includes 0?

b. does not include 0?

E15. In a recent Associated Press poll of 1008 adult Americans, 75% favored a legal drinking age of 21. A student computed a 95% confidence interval for the difference between the population proportion favoring 21 and the proportion not favoring 21 as follows:

$$(0.75 - 0.25) \pm 1.96 \sqrt{\frac{0.75(1 - 0.75)}{1008} + \frac{0.25(1 - 0.25)}{1008}}$$
$$= 0.50 \pm 0.038$$

Assuming the sample was randomly selected, is the student's method correct? If so, write an interpretation of this interval. If not, do a more appropriate analysis of the data. [Source: *www.icrsurvey.com.*]

E16. In the *USA Today* poll on driving ages described in E8, 35% of 1000 randomly sampled adults chose 16 as the preferred initial driving age while 42% chose 18. Can the difference of the proportion in the population who prefer 16 and the proportion in the population who prefer 18 be estimated by the confidence interval developed in this chapter? Explain why or why not.

9.2 ▶ A Significance Test for the Difference of Two Proportions

In the previous section, you learned how to estimate the size of the difference of two proportions. But sometimes you must decide between two alternatives. For example:

- Are snowboarders or skiers more likely to be seriously injured?
- Are opinions about the general state of the country more positive than 5 years ago?
- Is there a difference between the proportions of pellets (a renewable, clean-burning home heating fuel) that conform to specifications when produced by two different processes?

The size of any difference involved is not the main issue, as it was in Section 9.1. The main concern is whether you have enough evidence to conclude that there is a difference. In this section, you will learn to perform a test of significance so you can decide if the observed difference can reasonably be attributed to chance alone or if the difference is so large that something other than chance variation from sample to sample must be causing it.

> Should the observed difference be attributed to chance alone?

A Sampling Distribution of the Difference

Before you learn how to use a test of significance to answer questions like those above, you will learn about what sizes of differences you should expect in your samples when there is no actual difference in the proportions of successes in the two populations. Approximate sampling distributions (constructed from 5000 pairs of samples) for samples of size 30, 50, and 100 are shown in Display 9.4. In each case, the proportion of successes, p_1, in the first population is 0.2, and the proportion of successes, p_2, in the second population is also 0.2. For each sample size, values of $\hat{p}_1 - \hat{p}_2$ are plotted, where $\hat{p}_1$ is the proportion of successes in a random sample of the given size taken from a population with proportion of successes 0.2 and $\hat{p}_2$ is the proportion of successes in an independent random sample of the given size taken from a second population also with proportion of successes 0.2.

Note three facts about these approximate sampling distributions:

- Each sampling distribution is approximately normal in shape and becomes more so with larger sample sizes.
- The mean of each sampling distribution is $p_1 - p_2 = 0.2 - 0.2 = 0$.
- The formula from Section 9.1 can be used to compute the standard error:

$$\sigma_{\hat{p}_1 - \hat{p}_2} = \sqrt{\frac{p_1(1 - p_1)}{n_1} + \frac{p_2(1 - p_2)}{n_2}} = \sqrt{\frac{0.2(1 - 0.2)}{n_1} + \frac{0.2(1 - 0.2)}{n_2}}$$

Display 9.5 lists the standard errors (SEs) computed from the formula and from the approximate sampling distributions in Display 9.4. They match very closely.

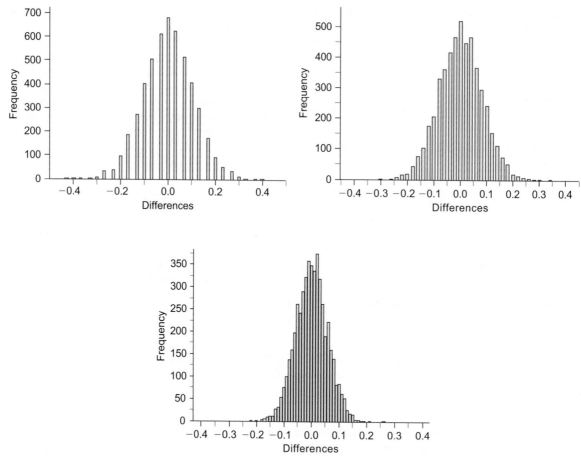

Display 9.4 Simulated sampling distributions of $\hat{p}_1 - \hat{p}_2$ when $p_1 = p_2 = 0.2$ for samples of sizes 30, 50, and 100.

Sample Size for n_1 and n_2	SE When $p_1 = p_2 = 0.2$	SE from Simulation
30	0.103	0.104
50	0.080	0.079
100	0.057	0.058

Display 9.5 Standard errors (SEs) from the formula and from the simulation in Display 9.4.

DISCUSSION
A Sampling Distribution of the Difference

D3. Use technology to take two samples of size 25 from a population with 35% successes. Find the difference, $\hat{p}_1 - \hat{p}_2$. Repeat until you have 200 differences.

a. Construct a plot of the differences of the two sample proportions.

b. From your plot, estimate the mean and standard error of the sampling distribution of $\hat{p}_1 - \hat{p}_2$.

c. Find the theoretical value of the mean and standard error of the sampling distribution of $\hat{p}_1 - \hat{p}_2$. Compare your answers to your estimates in part b.

The Theory of a Significance Test for the Difference of Two Proportions

Let's see how these ideas apply to a real sample survey. Think about the left-handers you know or know about. Queen Victoria was left-handed, and so were Julius Caesar, Leonardo da Vinci, and Benjamin Franklin. Angelina Jolie and Oprah Winfrey are left-handed, and so are Bruce Willis, Keanu Reeves, and Spike Lee. You probably know more left-handed males than females. So you might ask this question: "Is there objective evidence to show that the proportion of males who are left-handed in the country is larger than the proportion of females who are left-handed?" In fact, there are survey data that can help us answer this question.

The government's Health and Nutrition Survey (HANES) of 1976–1980 recorded the handedness of random samples of 1067 males and 1170 females from across the country. The study found that 113 of the males and 92 of the females were left-handed. From these data you can construct a significance test to answer the preceding question. [Source: D. Freedman, R. Pisani, and R. Purves, *Statistics*, 3rd ed. (Norton, 1998), p. 537.]

Using a sampling procedure that is typical of many such surveys, the HANES survey did not randomly sample 1067 males and then randomly sample 1170 females to ask about handedness. It randomly (by a somewhat complex design) selected a large sample of adults from across the country and then sorted out the males and females. The resulting two samples can still be considered to be independent and random, as long as the sample sizes are large.

| **Are Men More Likely to Be Left-Handed?** | **Example 9.3** |

The survey just described found that 113 of 1067 randomly selected males were left-handed while 92 of 1170 randomly selected females were left-handed. Perform a significance test to determine whether males are more likely to be left-handed than females.

Solution

Let p_1 denote the proportion of left-handers among all males in the adult population, and let p_2 denote the proportion among all adult females. The null hypothesis here is

> **Define symbols and state hypotheses.**

H_0: There is no difference between the proportions of males who are left-handed and females who are left-handed. In symbols, $p_1 = p_2$ or $p_1 - p_2 = 0$.

The alternative hypothesis, sometimes called the research hypothesis, is a statement of what the researcher is trying to establish. Here you are looking for evidence that the proportion of left-handers is greater among males, so the alternative hypothesis is

H_a: The proportion of left-handers among male adults is greater than the proportion among female adults. In symbols, $p_1 > p_2$ or $p_1 - p_2 > 0$.

The conditions that must be checked are the same as those for a confidence interval for the difference of two proportions: random samples and normal sampling distribution. You were told that these samples may be considered independent random samples. All these quantities are at least 5:

> **Check conditions of randomness and morality.**

$$n_1\hat{p}_1 = 113 \quad n_1(1 - \hat{p}_1) = 954 \quad n_2\hat{p}_2 = 92 \quad n_2(1 - \hat{p}_2) = 1078$$

Each population size is far larger than 10 times the sample sizes of 1067 and 1170.

The test statistic builds on the best estimates of the population proportions, namely, the sample proportions:

> **Compute the test statistic and *P*-value.**

$$\hat{p}_1 = \frac{113}{1067} \approx 0.106 \quad \text{and} \quad \hat{p}_2 = \frac{92}{1170} \approx 0.079$$

The general form of a test statistic for testing hypotheses is

$$\frac{statistic - parameter}{standard\ error}$$

The difference from the samples (statistic) is $\hat{p}_1 - \hat{p}_2$, or $0.106 - 0.079 = 0.027$. The hypothesized difference (parameter) is 0, the value under the null hypothesis that $p_1 - p_2 = 0$. The standard error is given exactly by

$$\sigma_{\hat{p}_1 - \hat{p}_2} = \sqrt{\frac{p_1(1 - p_1)}{n_1} + \frac{p_2(1 - p_2)}{n_2}}$$

You could do as you did in Section 9.1: Estimate p_1 with $\hat{p}_1$ and estimate p_2 with $\hat{p}_2$. However, here you can do even better. The null hypothesis states that the proportion of males who are left-handed is equal to the proportion of females who are left-handed, that is, $p_1 = p_2$. You can estimate this common value of p_1 and p_2 by combining the data from both samples into a **pooled estimate**, $\hat{p}$. Find this pooled estimate by combining males and females into one group:

$$\hat{p} = \frac{total\ number\ of\ left\text{-}handers}{total\ number\ of\ people} = \frac{113 + 92}{1067 + 1170} = \frac{205}{2237}$$

> $\hat{p}$ is called the pooled estimate of the common proportion of successes.

The standard error of the difference, then, is approximately

$$\sqrt{\frac{p_1(1 - p_1)}{n_1} + \frac{p_2(1 - p_2)}{n_2}} \approx \sqrt{\frac{\hat{p}(1 - \hat{p})}{n_1} + \frac{\hat{p}(1 - \hat{p})}{n_2}}$$

$$= \sqrt{\hat{p}(1 - \hat{p})\left(\frac{1}{n_1} + \frac{1}{n_2}\right)}$$

The test statistic then takes on the value

$$z = \frac{statistic - parameter}{estimated\ standard\ error}$$

$$= \frac{(\hat{p}_1 - \hat{p}_2) - 0}{\sqrt{\hat{p}(1 - \hat{p})\left(\frac{1}{n_1} + \frac{1}{n_2}\right)}}$$

$$= \frac{\left(\frac{113}{1067} - \frac{92}{1170}\right) - 0}{\sqrt{\frac{205}{2237}\left(1 - \frac{205}{2237}\right)\left(\frac{1}{1067} + \frac{1}{1170}\right)}} \approx 2.23$$

> The test statistic is normally distributed.

This test statistic is based on the difference of two approximately normally distributed random variables, $\hat{p}_1$ and $\hat{p}_2$. Such a difference itself has a normal distribution. Thus, you can use Table A on page 759 to find the P-value. The test statistic $z = 2.23$ has a (one-sided) P-value of 0.0129 as shown in Display 9.6.

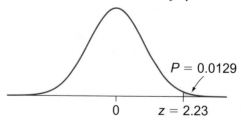

Display 9.6 P-value for a one-sided test with $z = 2.23$.

Your calculator or statistical software may do the computations for you. The following Minitab output gives the same value of z and P-value as the previous calculations. The second significance test, Fisher's exact test, which you may study in Chapter 16, is based on the hypergeometric distribution rather than on a normal approximation. For now, use the "Test for difference."

Test and CI for Two Proportions

Sample	X	N	Sample p
1	113	1067	0.105904
2	92	1170	0.078632

Difference = p(1) − p(2)
Estimate for difference: 0.0272719
95% lower bound for difference: 0.00708203
Test for difference = 0(vs > 0): z = 2.23 P-Value = 0.013
Fisher's exact tests: P-Value = 0.015

This *P*-value is very small, so you have evidence against the null hypothesis. The difference between the rates of left-handedness in these two samples is too large to attribute to chance variation alone. The evidence supports the alternative hypothesis that males have a higher rate of left-handedness.

> State the conclusion in context.

As you have seen, the significance test for the difference of two proportions proceeds along the same lines as the test for a single proportion. The steps in the significance test for the difference of two proportions are given in the following box.

Components of a Significance Test for the Difference of Two Proportions for Surveys

1. *Write a null and an alternative hypothesis.* You can write the null hypothesis in several different ways as long as you define p_1 and p_2:

    H_0: The proportion of successes p_1 in the first population is equal to the proportion of successes p_2 in the second population.

    H_0: $p_1 = p_2$, where p_1 is the proportion of successes in the first population and p_2 is the proportion of successes in the second population.

    H_0: $p_1 - p_2 = 0$, where p_1 is the proportion of successes in the first population and p_2 is the proportion of successes in the second population.

    > Null hypothesis: $p_1 = p_2$.

    The form of the alternative hypothesis depends on whether you have a two-sided or a one-sided test:

    H_a: The proportion of successes p_1 in the first population is not equal in the proportion of successes p_2 in the second population. In symbols, $p_1 \neq p_2$ or $p_1 - p_2 \neq 0$.

    H_a: The proportion of successes p_1 in the first population is greater than the proportion of successes p_2 in the second population. In symbols, $p_1 > p_2$ or $p_1 - p_2 > 0$.

    H_a: The proportion of successes p_1 in the first population is less than the proportion of successes p_2 in the second population. In symbols, $p_1 < p_2$ or $p_1 - p_2 < 0$.

    > Alternative hypothesis: $p_1 \neq p_2$, $p_1 > p_2$, or $p_1 < p_2$.

2. *Check conditions.* A random sample of size n_1 is taken from a large binomial population with proportion of successes p_1. The proportion of successes in this sample is $\hat{p}_1$. A second and independent random sample of size n_2 is taken from a large binomial population with proportion of successes p_2. The proportion of successes in this sample is $\hat{p}_2$. All these quantities must be at least 5:

    > Randomness and normality.

    $$n_1\hat{p}_1 \quad n_1(1 - \hat{p}_1) \quad n_2\hat{p}_2 \quad n_2(1 - \hat{p}_2)$$

    Each population size must be at least 10 times the sample size.

| Computations with a sketch. | **3.** *Compute the test statistic and a P-value.* |

$$z = \frac{(\hat{p}_1 - \hat{p}_2) - 0}{\sqrt{\hat{p}(1-\hat{p})\left(\frac{1}{n_1} + \frac{1}{n_2}\right)}}$$

where

$$\hat{p} = \frac{\text{total number of successes in both samples}}{n_1 + n_2}$$

The *P*-value is the probability of getting a value of z as extreme as or even more extreme than that from your samples if H_0 is true. See Display 9.7.

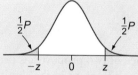 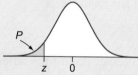 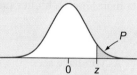

Display 9.7   *P*-values for two-sided and one-sided tests.

4. *Write a conclusion.* State whether you reject or do not reject the null hypothesis. You reject the null hypothesis if the *P*-value is smaller than the level of significance, α. Then write a sentence giving your conclusion in context.

| Example 9.4 | **Has the Percentage of Overweight Adults Changed over Recent Years?** |

Health experts agree that one of the best ways to improve health and extend life among U.S. residents is to reduce the percentage of those overweight. What is not so clear is whether the percentage of overweight adults has changed over recent years. It may seem that more are overweight simply because the issue has garnered much media attention. On the other hand, with the current emphasis on exercise and healthy eating, perhaps the percentage of overweight adults has actually decreased. So the question for investigation is whether the proportion of overweight adults has changed over the past 5 years.

One source of data on this issue is the Harris Poll on lifestyles conducted every year with a random sample of about 1000 adults. In 2004, 30% of adults in the sample were 20% or more overweight, while in 2009 the percentage was 32%. Does this represent a real change in population proportions at the 10% level of significance? [Source: *www. harrisinteractive.com/harris_poll/pubs/Harris_Poll_2009_03_04.pdf.*]

Solution

| Hypotheses | **1.** H_0: The proportion of adults overweight by 20% or more in 2004, p_1, is equal to the proportion of adults overweight by 20% or more in 2009, $p_2 = p_1 = p_2$ or $(p_1 - p_2) = 0$. |

H_a: $p_1 \neq p_2$. This is a two-sided alternative because you are only looking for a change, which could occur in either direction.

| Check conditions of randomness and normality. | **2.** The sampling design used by the Harris Poll is more complex than a simple random sample, but the simple random sample serves as a reasonably good approximation when dealing with proportions. The adults are sampled independently from year to year. Regarding 2004 adults as population |

1 and 2009 adults as population 2, the respective sample proportions are $\hat{p}_1 = 0.30$ and $\hat{p}_2 = 0.32$. Using the sample sizes of approximately 1000 each, all of

$$n_1\hat{p}_1 = 1000(0.30) = 300 \qquad n_1(1 - \hat{p}_1) = 1000(0.70) = 700$$
$$n_2\hat{p}_2 = 1000(0.32) = 320 \qquad n_2(1 - \hat{p}_2) = 1000(0.68) = 680$$

are much greater than 5. The adult population of the country is much greater than 10 times the sample size.

3. The pooled estimate of the common proportion of overweight adults is

Computations with a diagram.

$$\hat{p} = \frac{\textit{total number of successes in both samples}}{n_1 + n_2} = \frac{300 + 320}{1000 + 1000} = 0.31$$

The test statistic is

$$z = \frac{(\hat{p}_1 - \hat{p}_2) - 0}{\sqrt{\hat{p}(1 - \hat{p})\dfrac{1}{n_1} + \dfrac{1}{n_2}}} = \frac{(0.30 - 0.32) - 0}{\sqrt{0.31(1 - 0.31)\left(\dfrac{1}{1000} + \dfrac{1}{1000}\right)}} = -0.97$$

With a test statistic of -0.97, the P-value for a two-sided test is $2(0.1660) = 0.3320$, as seen in Display 9.8.

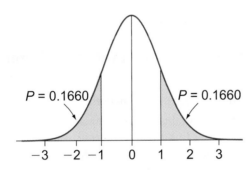

$P = 0.1660 \qquad\qquad P = 0.1660$

Display 9.8 A two-sided test with $z = -0.97$.

4. With a P-value of 0.3320, the difference in the sample proportions is not statistically significant. If the population proportions of overweight adults were equal, a difference in sample proportions as large as larger or than 0.02, in either direction, would occur over 33% of the time in repeated sampling. What you have observed is not an unlikely outcome under the null hypothesis; therefore, the null hypothesis of a zero difference cannot be rejected as a plausible value of the difference in the two population proportions.

Conclusion in context.

Display 9.9 shows computer output for this example.

Display 9.9 Output of the significance test for the overweight example.

Another way to determine that the change in the proportion of adults who are overweight is not statistically significant is to construct a 90% confidence interval for the difference in the proportion of all adults who were overweight in 2009 and the proportion of all adults who were overweight in 2004. That interval is –0.014 to 0.054, as shown below. Because 0 is in this interval, it is plausible that the difference in the proportions is 0. In other words, it is plausible that the proportions of all adults who were overweight in the two years are equal. When the test of significance is two-sided, you can use either method to determine statistical significance.

Attribute (categorical): unassigned

Attribute (categorical or grouping): unassigned

Interval estimate of the difference in proportions
320 out of **1000**, **or 0.32**, of sample of adults in 2009 are overweight.
300 out of **1000**, or **0.3**, of sample of adults in 2004 are overweight.

Confidence level: **90.0** %
```
Estimate:     0.02 +/- 0.0340131
Range:       -0.0140131 to 0.0540131
```

DISCUSSION
The Theory of a Significance Test for the Difference of Two Proportions

D4. In Example 9.3 on page 433, the null hypothesis was rejected.

    a. Which type of error might have been made in the significance test? Is the chance of making this error very large?

    b. How can you check quickly that the condition is met that all of $n_1 \hat{p}_1$, $n_1(1 - \hat{p}_1)$, $n_2 \hat{p}_2$, and $n_2(1 - \hat{p}_2)$ are at least 5?

D5. In Example 9.4, which type of error could have been made? What might be the practical consequences of making this error?

D6. What does it suggest about the two proportions if the significance test for a difference of two proportions

    a. fails to reject the null hypothesis?

    b. rejects the null hypothesis?

D7. Suppose one population has 10% successes and a second population has 20% successes. Generate a simulated sampling distribution of the difference between sample proportions for samples of size 10 from each population. Now repeat the procedure for samples of size 50 from each population. Explain how the two sampling distributions illustrate the need for the condition that $n_1 \hat{p}_1$, $n_1 (1 - \hat{p}_1)$, $n_2 \hat{p}_2$, and $n_2(1 - \hat{p}_2)$ must each be at least 5.

STATISTICS IN ACTION 9.2 ▶ Differences in Proportions of Orange

This Statistic in Action demonstrates chance variation from sample to sample.

What You'll Need: a sample of 50 Skittles, a sample, of 50 Milk Chocolate M&M's (or bags prepared by your instructor)

1. Count the number of orange candies in your sample of Skittles and in your sample of M&M's. Compute the two sample proportions.

2. Compute the difference in the two sample proportions:

   proportion of Skittles that are orange − proportion of M&M's that are orange

   (Every student should subtract in this order, even if the difference is negative.)

3. Each member of your class should mark his or her difference in sample proportions above a number line so that everyone can see the distribution.

4. Repeat with fresh samples until your class has at least 100 differences.

5. What is the shape of the sampling distribution?

6. Where should the sampling distribution be centered? Where is it centered?

7. What, approximately, is the standard error of the sampling distribution?

Summary 9.2: A Significance Test for the Difference of Two Proportions

Suppose that $\hat{p}_1$ is the proportion of successes in a random sample of size n_1 taken from a population with (unknown) proportion of successes p_1 and $\hat{p}_2$ is the proportion of successes in an independent random sample of size n_2 taken from a second population with (unknown) proportion of successes p_2. To get a simulated sampling distribution of the difference, $\hat{p}_1 - \hat{p}_2$, you

- take independent random samples from each population
- compute the proportion of successes in each sample, $\hat{p}_1$ and $\hat{p}_2$
- subtract the proportions of successes in the samples, $\hat{p}_1 - \hat{p}_2$
- repeat this process many times

You learned how to test whether two samples were drawn from populations that have the same proportion of successes. Follow these steps, as with any significance test:

- *Write a null and an alternative hypothesis.* The null hypothesis typically is that the two populations contain the same proportion of successes.

- *Check the conditions of randomness and normality.* You need two sufficiently large random samples selected independently from two different populations.

- *Compute the test statistic and a P-value.* The test statistic is

$$z = \frac{(\hat{p}_1 - \hat{p}_2) - 0}{\sqrt{\hat{p}(1 - \hat{p})\left(\frac{1}{n_1} + \frac{1}{n_2}\right)}}$$

where $\hat{p}$ is computed by combining both samples. The P-value is the probability of getting a value of z as extreme as or even more extreme than that from your samples if H_0 is true.

- *Write a conclusion.* State whether you reject or do not reject the null hypothesis, linking this decision to your P-value. Then restate your conclusion in the context of the situation.

Practice

A Sampling Distribution of the Difference

P6. According to the 2009 *Statistical Abstract of the United States*, 12% of Americans adults visited a zoo in the last year and 12% went to a museum. Suppose you take a random sample of 1000 American adults and compute the proportion, $\hat{p}_1$, who went to the zoo last year. Then you take another random sample of 800 American adults and compute the proportion, $\hat{p}_2$, in that sample who went to a museum last year. You subtract the two proportions to get $\hat{p}_1 - \hat{p}_2$. You repeat this process until you have a distribution of millions of values of $\hat{p}_1 - \hat{p}_2$. (That's why the problem says "suppose.")

   a. What is the expected value of the difference $\hat{p}_1 - \hat{p}_2$?

   b. What is the standard error of the distribution of $\hat{p}_1 - \hat{p}_2$?

   c. Sketch the distribution of $\hat{p}_1 - \hat{p}_2$, with a scale on the horizontal axis.

   d. Compute the probability that the difference will be 0.05 or greater.

P7. Suppose you take two independent samples from two different populations that have the same proportion of successes. You repeat this millions of times and plot the distribution of the differences. Answer true or false to each statement.

   a. It is always the case that $\hat{p}_1 = \hat{p}_2$.

   b. It is true that $p_1 = p_2$.

   c. Sometimes the difference between the two sample proportions can be relatively large.

   d. Theoretically, the plot should be centered exactly at 0.

   e. To have a greater chance of having the difference $\hat{p}_1 - \hat{p}_2$ nearer 0, have larger sample sizes.

The Theory of a Significance Test for the Difference of Two Proportions

P8. A random sample of 600 probable voters was taken at the start of the campaign for mayor, and 321 of the 600 said they favored the challenger over the incumbent. However, it was revealed shortly afterward that the challenger had dozens of outstanding parking tickets. Subsequently, a new random sample of 750 probable voters showed that 382 voters favored the challenger.

   Do these data support the conclusion that there was a decrease in voter support for the challenger after the parking tickets were revealed? Give appropriate statistical evidence to support your answer.

P9. Pellet fuel is a renewable, clean-burning, and cost-efficient home heating fuel produced from wood or biomass. The processes for producing the pellets can, however, result in large proportions of pellets that do not conform to specifications. Two different methods for producing pellets were compared by randomly sampling 100 pellets from each process during a single day's production and comparing the proportions of conforming pellets. With Method A, 38% of the sampled pellets conformed to specifications while with Method B, 29% conformed. Is there statistically significant evidence of a difference between the proportions of pellets that conform to specifications for these two methods? [Source: S. Vardeman, *Statistics for Engineering Problem Solving* (WS Publishing, 1994), p. 335.]

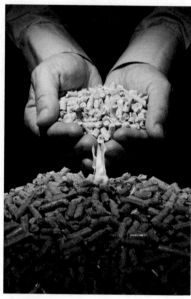

Wood pellet fuel.

   a. State the appropriate null and alternative hypotheses for the problem, defining all symbols used.

   b. Calculate the test statistic.

   c. Report an approximate *P*-value.

   d. Write an appropriate conclusion in the context of the problem, taking care to describe the populations for which the inference is relevant.

P10. Do active Christians behave differently from those with no active religious faith? Responses from independent random samples of 3011 active Christian adults and 1055 adults professing no active religious faith appear in Display 9.10.

	Christian	No Faith
Registered to vote	89%	78%
Active in the community	68%	41%
Suffer from personal addictions	12%	13%
Considered to be overweight	41%	26%
Feel stressed out much of the time	26%	37%

Display 9.10 Poll results for active Christian adults and adults professing no active religious faith. [Source: *www.barna.org/*.]

   a. Which behaviors show a statistically significant difference, at the 5% significance level, between active Christians and those espousing no active religious faith?

b. For which behavior is the evidence of a difference strongest?

c. This study did not control for differences in the demographic characteristics of the two groups. Name a lurking variable that might account for most of these differences.

P11. You work for a life insurance company in Detroit that wants to survey 50 men and 50 women who were married in Wayne County within the last year. (During this year, Michigan did not allow same-sex marriage, and no person was married twice in the last year.) You have copies of all 7211 marriage licenses. For each of the following, explain why the method described does or does not produce independent random samples of 50 men and 50 women. [Source: www.mdch.state.mi.us/pha/osr/Marriage/MarriageTrendNumbers.asp.]

a. Randomly select 50 of the marriage licenses and survey the 50 men and 50 women.

b. Randomly select 50 of the marriage licenses and survey the 50 men listed on them. Mix these licenses back in with the others and randomly select 50 more licenses from the 7211 licenses. Survey the 50 women listed on them.

c. Randomly select 50 of the marriage licenses and survey the 50 men listed on them. Randomly select 50 more licenses from the 7161 remaining licenses. Survey the 50 women listed on them.

Exercises

E17. Milk Chocolate M&M's are 24% blue, and Almond M&M's are 20% blue. Suppose you take a random sample of 100 of each. You compute the proportion that are blue in each sample. You then subtract the proportion, $\hat{p}_2$, of blue in the sample of Almond M&M's from the proportion, $\hat{p}_1$, of blue in the sample of Milk Chocolate M&M's to get $\hat{p}_1 - \hat{p}_2$. You repeat this process until you have a distribution of millions of values of $\hat{p}_1 - \hat{p}_2$.

a. What is the expected value of $\hat{p}_1 - \hat{p}_2$?

b. What is the standard error of the distribution of $\hat{p}_1 - \hat{p}_2$?

c. Sketch the distribution of $\hat{p}_1 - \hat{p}_2$ with a scale on the horizontal axis.

d. Compute the probability that the difference will be 0.05 or greater.

E18. Almond M&M's and Peanut Butter M&M's both are 20% blue. Suppose you take a random sample of 100 of each. You compute the proportion that are blue in each sample. You then subtract the proportion, $\hat{p}_2$, of blue in the sample of Peanut Butter M&M's from the proportion, $\hat{p}_1$, of blue in the sample of Almond M&M's to get $\hat{p}_1 - \hat{p}_2$. You repeat this process until you have a distribution of millions of values of $\hat{p}_1 - \hat{p}_2$.

a. What is the expected value of $\hat{p}_1 - \hat{p}_2$?

b. What is the standard error of the distribution of $\hat{p}_1 - \hat{p}_2$?

c. Sketch the distribution of $\hat{p}_1 - \hat{p}_2$ with a scale on the horizontal axis.

d. Compute the probability that the difference will be 0.05 or greater.

E19. An annual poll of the sleeping habits of Americans found that of a random sample of 177 Americans ages 18–29, 30% slept 8 hours or more on a weekday. Of an independent random sample of 616 people ages 30–49, 24% slept 8 hours or more on a weekday. Is this a statistically significant difference? (When answering this question, include all four steps in your test of significance.) [Source: National Sleep Foundation, 2005, www.sleepfoundation.org.]

E20. "Cyberbullying involves the use of information and communication technologies to support deliberate, repeated, and hostile behavior by an individual or group, that is intended to harm others." The National Crime Prevention Council reports that, in a study of this phenomenon involving 680 men and 698 women, 33% of the men and 36% of the women say they have been victims of cyberbullying. [Source: S. Hinduja and J. W. Patchin, "Cyberbullying: An Exploratory Analysis of Factors Related to Offending and Victimization," Deviant Behavior], 29, 2 (208), pp.129–156.

a. Is this significant evidence that women are cyberbullied more than men? Assume that these are independent random samples.

b. Actually, these were convenience samples rather than random samples. Does the conclusion you came to in part a still have some validity? Explain.

E21. In 1990, the Gallup Poll began to periodically ask adult Americans, "How would you rate the job being done by NASA—the U.S. space agency? Would you say it is doing an excellent, good, only fair, or poor job?" In the first poll, 46% of about 1000 randomly selected adult Americans, gave NASA an excellent or good rating. In 2007, the poll asked 1010 randomly selected adult Americans, and 56% gave NASA a good to excellent rating. Was NASA being looked upon more favorably by the American public in 2007 than in 1999? Would you say that there is strong evidence to support your conclusion? (Always include all four steps in your test of significance.) [Source: Gallup. Americans Continue to Rate NASA Positively, 2007, www.gallup.com/poll/102466/Americans-Continue-Rate-NASA-Positively.aspx.]

E22. The question of lowering the legal drinking age in the United States from 21 to 18 has been greatly debated in recent years, especially because some college presidents have come out in favor of the lower age. In a poll involving a random sample of 1000 adults, of whom 400 were men and 600 were women, 37% of the men and 27% of the women favored lowering the age to 18. You may consider these independent random samples. Among the men, about one-fourth of whom were under the age of 40, 52% of the under-40 group favored the younger drinking age, while only 32% of the 40 and older group favored lowering the age to 18. [Source: *Rasmussen Reports*, August 16, 2008.]

a. In comparing the populations of adult men and women, is there statistically significant evidence that the proportion of men favoring the lower drinking age is greater than that for women? Would you say that the evidence of your conclusion is strong?

b. In comparing the population of men under 40 with the population of men 40 and older, is there statistically significant evidence that the proportion favoring the age of 18 in the younger population is greater than that proportion in the older population? Is the evidence for your conclusion here stronger than the evidence in part a?

c. In the total sample of 1000, 52% favor maintaining the current legal age of 21, while 31% favor lowering the age to 18. Can you test to see if these two sample proportions are significantly different using the methods of this section? Explain why or why not.

E23. The Pew Internet and American Life Project collected information on Twittering by conducting a randomized survey of people who use the World Wide Web, which was then split into two groups, the first being a sample of 663 "American adults [who] said they used a service like Twitter or another service that allowed them to share updates about themselves or to see the updates of others." (These will be referred to as "those who Twitter.") The second group consisted of 1591 Internet users who do not Twitter. (These will be referred to as "Internet users.") [Source: *http://pewresearch.org/pubs/1117/twitter-tweet-users-demographics.*]

a. Among those who Twitter, 35% live in urban areas, while among Internet users only 29% live in urban areas. Is it fair to say that a larger percentage of those who Twitter are urban than those who just use the Internet? Justify your answer with a statistical test.

b. Among those who Twitter, 76% read a newspaper online, while that percentage is only 60% among Internet users. Is it fair to say that those who Twitter read newspapers online at a higher percentage than those who just use the Internet? Justify your answer statistically.

c. In the initial sample of 2254 people, about 19% of those between the ages of 18 and 24 Twitter, while that percentage drops to 10% among those between the ages of 35 and 44. Do you have enough information to conduct a statistical test to compare the proportions of all Web users in these age groups who Twitter? If not, what other information do you need?

E24. How different are liberals and conservatives on matters of religious faith? In studying this issue, the Barna Group sampled 992 adults in the United States who defined themselves as "mostly conservative" and 511 adults who defined themselves as "mostly liberal." Although it actually was more complex, the sampling process may be considered equivalent to one that produces two random and independently selected samples for the basic inference done in this exercise. [Source: *www.barna.org/barna-update/article/13-culture/258-survey-shows-how-liberals-and-conservatives-differ-on-matters-of-faith.*]

a. Of the conservatives, 21% said they were associated with the Catholic Church, while this percentage was 30% for the liberals. In using these sample percentages to test equality of population proportions, which null hypothesis are you testing?

A. The population of conservatives in the United States has the same proportion of Catholics as does the population of liberals in the United States.

B. The population of Catholics in the United States has equal proportions of conservatives and liberals.

b. Carry out the appropriate test for the null hypothesis chosen in part a.

c. The survey found that 12% of the conservatives had participated in at least one short-term mission trip; the percentage was 6% for the liberals. Is there strong evidence that the conservative and liberal populations differ on the rate of participation in mission trips?

d. The conservative and liberal samples came from a larger sample of about 3000 adults, among whom 32% considered themselves to be "mostly conservative" on social and political matters and 17% considered themselves to be "mostly liberal" on such matters. Can you test to see if the difference between these two proportions is statistically significant using the methods of this section? Explain why or why not.

E25. A large survey of college and university students on 53 campuses studied the prevalence of different types of hazing. The study leaders defined hazing as "any activity expected of someone joining or participating in a group that humiliates, degrades, abuses, or endangers them regardless of a person's willingness to participate." [Source: *www.hazingstudy.org/publications/hazing_in_view_web.pdf.*] Of the survey responses, 4775 were females associated with campus groups and 2685 were males associated with campus groups. On the overall question of having ever experienced hazing in any form, 61% of the males who participate in campus groups and 52% of the females who participate in campus groups responded positively.

Use the following printout to decide whether you consider this to be strong evidence that the population of females and the population of males differ on this question.

Test and CI for Two Proportions

Sample	X	N	Sample p
1	1638	2685	0.610056
2	2483	4775	0.520000

Difference = p (1) − p (2)
Estimate for difference: 0.0900559
95% CI for difference: (0.0667932, 0.113319)
Test for difference = 0 (vs not = 0): Z = 7.51 P-Value = 0.000

Fisher's exact test: P-Value = 0.000

E26. In the study described in E25, the two types of activities with the highest rate of participation were "participate in a drinking game" and "sing or chant by self or with others in a public situation that is not related to the event, game, or practice." Display 9.11 shows the sample sizes for those responding and the proportions of students who had participated in each activity across a variety of typical campus groups.

Campus Group	n	Drinking Game	Sing or Chant
Varsity athletics	640	0.47	0.27
Social fraternity or sorority	1295	0.53	0.32
Service fraternity or sorority	544	0.26	0.18
Intramural sports	1060	0.28	0.13
Performing arts	818	0.23	0.25
Academic clubs	1061	0.10	0.06

Display 9.11 Proportions of groups participating in two types of hazing activity.

In comparing these groups, regard the samples as being independent even though some of the same students may be in more than one group.

a. Is there a statistically significant difference between varsity athletics and social fraternities or sororities with regard to the proportions engaged in drinking games? Use the following output to answer the question.

Test and CI for Two Proportions

Sample	X	N	Sample p
1	686	1295	0.529730
2	301	640	0.470313

Difference = p (1) − p (2)
Estimate for difference: 0.0594172
95% CI for difference: (0.0121494, 0.106685)
Test for difference = 0 (vs not = 0): Z = 2.46 P-Value = 0.014

Fisher's exact test: P-Value = 0.016

b. Is there a statistically significant difference between intramural sports and performing arts with regard to the proportions engaged in singing or chanting? Use the following output to answer the question.

Test and CI for Two Proportions

Sample	X	N	Sample p
1	205	818	0.250611
2	138	1060	0.130189

Difference = p (1) − p (2)
Estimate for difference: 0.120423
95% CI for difference: (0.0844734, 0.156372)
Test for difference = 0 (vs not = 0): Z = 6.70 P-Value = 0.000

Fisher's exact test: P-Value = 0.000

E27. The Food and Drug Administration (FDA) of the United States government has the task of monitoring the country's food supply, whether domestically produced or imported. One of the main factors checked on a regular basis is the amount of pesticide residue found on various foods about to be sent to market. Foods that show residue are not necessarily in violation of a standard, as the standards allow small amounts of pesticide residue to enter our food supply. The food samples tested are selected by a rather complicated process involving the experience of experts, but, when doing a test of significance for the difference of two proportions, you may regard the process as producing results similar to independent random samples.

Worker spraying strawberry fields with pesticide.

Display 9.12 shows the results of pesticide residue tests on samples of fruits and vegetables for a recent year.

Pesticide Residue: Domestic Foods	Fruit	Vegetables
Sample size	344	672
% showing no residue	44.2	73.8
% showing residue in violation	0.9	2.4

Pesticide Residue: Imported Foods	Fruit	Vegetables
Sample size	1136	2447
% showing no residue	70.4	60.4
% showing residue in violation	3.6	5.4

Display 9.12 Pesticide residue data for domestic and imported food samples. [Source: Pesticide Program Residue Monitoring 2004–2006, FDA.]

a. Is there a statistically significant difference between the domestic and imported fruits with regard to the proportions showing no residue?

b. Is there a statistically significant difference between the domestic and imported vegetables with regard to the proportions showing no residue?

c. The samples selected for testing are said to be intentionally biased toward the foods that are known from past experience to be potentially high in pesticide residue. Does this fact necessarily negate the previous comparisons made above? Explain your reasoning.

E28. Refer to the data on pesticide testing in E27.

a. Is there a statistically significant difference between the domestic and imported vegetables with regard to the proportions showing residue in violation of a standard? In your conclusion, describe the appropriate populations for which the inference is relevant.

b. Is there a statistically significant difference between the domestic and imported fruits with regard to the proportions showing residue in violation of a standard? In your conclusion, describe the appropriate populations for which the inference is relevant.

c. In part b, one condition for the test is not satisfied. Which condition is it, and do you think that violation might cause your conclusion to be invalidated?

E29. It probably comes as no surprise that Internet usage is increasing, but a Gallup poll sheds some light on how fast this is happening. In a random sample of about 1000 U.S. adults taken in early 2009, 48% reported using the Internet more than 1 hour per day. In a similar and independent sample of 1000 adults taken a year earlier, 43% reported using the Internet more than 1 hour per day. [Source: Gallup Poll, January 2, 2009.]

a. Do these data represent a statistically significant increase in Internet use over the year?

b. Are you justified in concluding that less than a majority of adults used the Internet more than 1 hour per day in January 2009?

E30. In the poll on Internet usage described in E29, 62% of the 18–29 age group reported using the Internet more than 1 hour per day, while 54% of the 30–49 age group reported this level of usage.

a. Assuming there were about 100 people in the 18–29 sample and about 300 people in the 30–49 sample, do these data represent a statistically significant difference in Internet use between these two age groups?

b. In the 18–29 age group, 12% reported using the Internet up to 1 hour daily and 17% reported using it only few times a week. Can you test for a statistically significance difference between these two proportions using the methods of this section? Why or why not?

E31. You feed Diet A to a random sample of cows and Diet B to another random sample. Your null hypothesis is that the two diets produce the same proportion of contented cows. Your two-sided test has a P-value of 0.36. Select the *one* best interpretation of this P-value.

A. If the null hypothesis that there is no difference in the proportion of contented cows produced by the two diets is true, the probability is 0.36 that the proportion of contented cows produced by each diet won't be very different.

B. You can't conclude that the diets make a difference because it's fairly likely to get as big a difference in the proportion of contented cows produced by the two diets as you did even if there is no difference at all in the overall proportion of contented cows that each diet produces.

C. The probability is 0.36 that the two diets differ in the proportion of contented cows that they produce. Because this probability is so high, you can reject the null hypothesis.

D. If the null hypothesis is true, then the probability is 0.36 that the two diets differ in the proportion of contented cows.

E. You can't reject the null hypothesis because if, in fact, the null hypothesis is true, the probability is 0.36 that the difference in the true proportions is the same as in your experiment.

E32. Explain how knowledge of sampling distributions for differences in sample proportions is used in the development of the significance test of the difference of two proportions.

E33. In this section, you have been testing the null hypothesis that $p_1 - p_2$ equals 0. The null hypothesis does not have to state that the difference in the two population proportions is 0. You can test for any difference you want. In the z-statistic for the next two problems, replace the parameter $p_1 - p_2$ with the difference in population proportions you are testing for rather than replacing it with 0. Then, when estimating the SE, $\sqrt{\frac{p_1(1-p_1)}{n_1} + \frac{p_2(1-p_2)}{n_2}}$, you should not use a pooled estimate of a common population proportion. The pooled estimate can be used only under the hypothesis that the two population proportions are equal, which is not what you will be hypothesizing in the following situation.

Return to Example 9.3 on page 433, where a survey found that 113 of 1067 males were left-handed while 92 of 1170 females were left-handed. Perform a test to determine if there is statistically significant evidence that the percentage of males who are left-handed is greater than the percentage of females who are left-handed by at least 2 percentage points.

E34. Read the first paragraph of E33 before answering the following question. You can take either Bus A or Bus B to campus, and you're concerned about being late to class. You check out Bus A on 100 randomly selected mornings and find that you would be late to class five mornings. You check out Bus B on 64 randomly selected mornings and find that you would be late five mornings. Because you live a bit closer to the stop for Bus B, you will use Bus A only if you can conclude that the difference in the proportions of late arrivals is more than 2%. Which bus will you use?

9.3 ► Inference for Experiments and Observational Studies

To use the methods for comparing two proportions discussed thus far—confidence interval and test of significance—the underlying condition was that the data came from random samples selected independently from two different populations. This is the typical situation with sample surveys, such as surveys comparing pet ownership rates for two different years or comparing two processes for making fuel pellets. However, this is not the typical situation in experiments.

Experiments usually require investigators to use those subjects who volunteer or those experimental units available for the study. If there are two treatments, this group of subjects is then split into two groups by random assignment of the treatments to the subjects or units. In such situations, the population is the group of subjects or units being used in the study, and the inferences drawn are directly applicable only to this group. A clinical trial designed to study the effectiveness of a new drug versus a standard drug by randomly assigning the new drug to half of the subjects who volunteered for the study (and the standard drug to the remainder) is a typical randomized experiment.

An experiment uses the available subjects.

If comparisons are to be made on subjects or units without random assignment, such as in comparing cancer rates in a group of smokers with rates in a group of non-smokers, the investigation is called an observational study.

Observational studies don't involve randomization.

Significance Test for a Difference in Proportions from an Experiment

The statistical methods introduced earlier in this chapter for a sample survey (independent random samples taken from two larger populations) can be adapted to experiments, as summarized in the next box. Even though the randomized experiment results in dependent samples from a small population (the available subjects or units), the formula for the standard error is the same as that for sample surveys. This seems like a great piece of luck, but it is actually due to some neat statistical theory.

> ### Sampling Distribution of the Difference in Proportions for an Experiment
>
> The sampling distribution of the differences in proportions, $\hat{p}_1 - \hat{p}_2$, from an experiment in which two treatments are randomly assigned to the available subjects
>
> - is approximately normal when the sample sizes are large enough
> - is centered at 0 when there is no difference in the effect of the two treatments
> - has standard error that can be estimated by
>
> $$\sigma_{\hat{p}_1 - \hat{p}_2} = \sqrt{\frac{\hat{p}_1(1 - \hat{p}_1)}{n_1} + \frac{\hat{p}_2(1 - \hat{p}_2)}{n_2}}$$
>
> Here, $\hat{p}_1$ represents the proportion of successes among the n_1 subjects who received the first treatment, and $\hat{p}_2$ represents the proportion of successes among the n_2 subjects who received the second treatment.

Thus, for a test of the significance of the difference of the two proportions, you can construct a test statistic just as you did in the case of sample survey data. The main difference will be how you state the hypotheses and conclusions. An outline of a significance test for comparing proportions from an experiment is given in the box.

> ## Components of a Significance Test for the Difference of Two Proportions from an Experiment

No larger populations exist.

1. *Write a null and an alternative hypothesis.* The null hypothesis is that if all subjects had received the first treatment, the proportion of successes would be the same as if all subjects had received the second treatment. The alternative hypothesis for a two-sided test is that the two treatments would not have resulted in the same proportion of successes. The alternative hypothesis for a one-sided test is that a specific treatment would have resulted in a higher proportion of successes than the other.

   In symbols,

   H_0: $p_1 = p_2$, where p_1 is the proportion of subjects in this study for which the outcome would have been a success if all subjects had been given the first treatment and p_2 is the proportion of subjects for which the outcome would have been a success if all subjects had been given the second treatment.

   H_a: $p_1 \neq p_2$ for a two-sided test, or $p_1 > p_2$ or $p_1 < p_2$ for a one-sided test.

   You also may state the null hypothesis this way: Which treatment each subject received made no difference in whether the outcome was a success or a failure.

Randomness and normality.

2. *Check conditions.* Two treatments, A and B, are randomly assigned to available experimental units, with n_1 receiving Treatment A and n_2 receiving Treatment B.

   Let $\hat{p}_1$ represent the proportion of successes from Treatment A and $\hat{p}_2$ represent the proportion of successes from Treatment B. Then, all of the following values must be at least 5:

   $$n_1\hat{p}_1 \quad n_1(1 - \hat{p}_1) \quad n_2\hat{p}_2 \quad n_2(1 - \hat{p}_2)$$

3. *Compute the test statistic and a P-value.*

   $$z = \frac{(\hat{p}_1 - \hat{p}_2) - 0}{\sqrt{\hat{p}(1 - \hat{p})\left(\frac{1}{n_1} + \frac{1}{n_2}\right)}}$$

   where

   $$\hat{p} = \frac{total\ number\ of\ success\ in\ both\ groups}{n_1 + n_2}$$

   The *P*-value is the probability of getting a value of z as extreme as or even more extreme than that from your samples if H_0 is true. See Display 9.13.

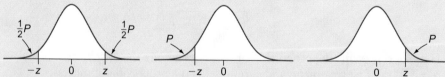

Display 9.13 *P*-values for a two-sided test (left) and one-sided tests (center and right).

4. *Write a conclusion.* State whether you reject or do not reject the null hypothesis. You reject the null hypothesis if the *P*-value is smaller than the level of significance, α. Then write a sentence giving your conclusion in the context of this situation.

So you can see how to apply these ideas to a real experiment, Example 9.5 describes an experiment to determine whether an N95 respirator or a surgical face mask offers nurses better protection against the flu.

Protection from the Flu

Example 9.5

For use during influenza epidemics, two types of respiratory protection are available: surgical masks and N95 respirators. Surgical masks are less expensive than N95 respirators, and in greater supply, but they do not fit tightly and do not protect wearers from smaller particles. An experiment was performed to see which type of mask was more effective. During influenza season, 446 nurses who were caring for patients with respiratory illness including fever were randomly divided into two groups. One group of 225 nurses received surgical masks and 221 received fitted N95 respirators. Display 9.14 gives the number of nurses in each group who got influenza. Is this statistically significant evidence that one type of respiratory protection is better than another?

		Treatment		Total
		Surgical Mask	N95 Respirator	Total
Influenza?	Yes	50	48	98
	No	175	173	348
	Total	225	221	446

Display 9.14 Results from influenza protection experiment. [Source: Mark Loeb et al., "Surgical Mask vs N95 Respirator for Preventing Influenza Among Health Care Workers: A Randomized Trial," *Journal of the American Medical Association*, Vol. 302, 17 (October 2009), *www.jama.ama-assn.org/cgi/content/abstract/2009.1466v1*.]

Solution

The goal was to see which type of protection was better, so this situation calls for a two-sided test.

Write a null and an alternative hypothesis.

H_0: $p_1 = p_2$, where p_1 is the proportion of nurses in this study who would have gotten influenza if all nurses had worn the surgical mask, and p_2 is the proportion who would have gotten influenza if they had all worn the N95 respirator.

H_a: $p_1 \neq p_2$.

The first condition that needs to be met is that the treatments were randomly assigned to the subjects.

Check random assignment and normality.

The nurses weren't randomly selected from the population of all nurses, but were volunteers from eight Ontario hospitals. They should not be regarded as a random sample from any specific population. However, the treatments were randomly assigned to the nurses available. (This type of randomization is typical in medical research, agricultural research, and many other kinds of experiments for which getting a simple random sample from the entire population of interest is impossible.)

You don't need a random sample.

The second condition is that $n_1\hat{p}_1 = 50$, $n_1(1 - \hat{p}_1) = 175$, $n_2\hat{p}_2 = 48$, and $n_2(1 - \hat{p}_2) = 173$ are all 5 or more. Both conditions are met.

The pooled estimate of the proportion of nurses getting influenza if the null hypothesis is true is

Compute the test statistic and P-value.

$$\hat{p} = \frac{total\ number\ getting\ influenza}{n_1 + n_2} = \frac{50 + 48}{225 + 221} = \frac{98}{446} \approx 0.2197$$

The test statistic is

$$z = \frac{(\hat{p}_1 - \hat{p}_2) - 0}{\sqrt{\hat{p}(1 - \hat{p})\left(\frac{1}{n_1} + \frac{1}{n_2}\right)}} = \frac{\left(\frac{50}{225} - \frac{48}{221}\right) - 0}{\sqrt{\frac{98}{446}\left(1 - \frac{98}{446}\right)\left(\frac{1}{225} + \frac{1}{221}\right)}} \approx 0.128$$

This statistic is approximately normally distributed, so the z-score of approximately 0.13 gives a P-value of 2(0.4483) = 0.8966, using Table A of the Appendix, or 0.8980 from technology, as in the following output.

Test and CI for Two Proportions

Sample	X	N	Sample p
1	50	225	0.222222
2	48	221	0.217195

Difference = p (1) − p (2)
Estimate for difference: 0.00502765
95% CI for difference: (−0.0718245, 0.0818798)
Test for difference = 0 (vs not = 0): Z = 0.13 P-Value = 0.898
Fisher's exact test: P-Value = 0.909

Write a conclusion.

Because this P-value is quite large, the data do not provide evidence against the null hypothesis and you don't reject it. There is no evidence that the proportion of the 446 nurses who would have gotten influenza if they all had been given surgical masks would differ from the proportion if they all had been given N95 respirators. The P-value of 0.8980 means that if the difference in the proportions of nurses who would get influenza with the two treatments is actually 0, the probability of getting a difference as large or larger (in absolute value) than the one, $\hat{p}_1 - \hat{p}_2 = \left(\frac{50}{225} - \frac{48}{221}\right) \approx 0.005$, from this randomization is 0.8980. Thus, we aren't convinced that the two types of respiratory protection differ with respect to the proportion of these nurses who would get influenza.

The next example is from an important early clinical trial that compared two treatments for people with HIV and AIDS.

Example 9.6 | Treatment for AIDS

From 1986 to 1988, an important study on the treatment of patients with HIV was run by a collaborative group of researchers from eight European countries and Australia. Part of the study involved 131 patients who had already developed AIDS. They participated a double-blind, randomized clinical trial. Two treatments were to be compared. Each patient was given either zidovudine (commonly known as AZT) by itself or a combination of AZT and acyclovir (ACV).

The group of AIDS patients weren't randomly selected from the population of all AIDS patients but were volunteers who had to give "informed consent." Thus, they might be quite different from the population of all AIDS patients and should not be regarded as a random sample from any specific population. However, the treatments were randomly assigned to the AIDS patients who were available. (This type of randomization is typical in medical research, agricultural research, and many other kinds of experiments for which getting a simple random sample from the entire population of interest is impossible.)

Display 9.15 shows the number of patients who survived and who died during the year-long experiment.

		Treatment		
		AZT	AZT + ACV	Total
Survived?	No	28	13	**41**
	Yes	41	49	**90**
	Total	**69**	**62**	**131**

Display 9.15 Results from AIDS experiment. [Source: David A. Cooper et al., "The Efficacy and Safety of Zidovudine Alone or as Cotherapy with Acyclovir for the Treatment of Patients with AIDS and AIDS-Related Complex: A Double Blind, Randomized Trial," *AIDS*, Vol. 7 (1993), pp. 197–207.]

The sample proportions were $\hat{p}_1 = \frac{41}{69} \approx 0.594$, where $\hat{p}_1$ is the proportion who survived in the group treated with AZT alone, and $\hat{p}_2 = \frac{49}{62} \approx 0.790$, where $\hat{p}_2$ is the proportion who survived in the group treated with AZT plus ACV. For a new therapy (in this case, AZT plus ACV) to become accepted practice among physicians, the research must show that the new therapy is an improvement over the old one. Can that claim be made based on these data?

Solution

Because the researchers are attempting to show that the new therapy is better than the old one, a one-sided significance test is in order. The hypotheses are

> State hypothesis.

H_0: For these subjects, the new therapy is not better than the old, or $p_1 = p_2$, where p_1 is the proportion of patients who would have survived if all 131 patients had been given AZT alone and p_2 is the proportion of patients who would have survived if they all had been given AZT plus ACV.

H_a: The new therapy is better than the old one, or $p_1 < p_2$.

Treatments were randomly assigned to the available subjects, and each of

$$n_1\hat{p}_1 = 41 \quad n_1(1 - \hat{p}_1) = 28 \quad n_2\hat{p}_2 = 49 \quad n_2(1 - \hat{p}_2) = 13$$

is at least 5.

The pooled estimate, $\hat{p}$, of the probability that a patient will survive, under the null hypothesis that the treatment a patient received made no difference in whether he or she survived, is $\frac{90}{131}$, or about 0.687. The test statistic then takes on the value

> Check random assignment and normality.

> Compute the test statistic and the P-value.

$$z = \frac{statistic - parameter}{estimated\ standard\ error}$$

$$= \frac{(\hat{p}_1 - \hat{p}_2) - 0}{\sqrt{\hat{p}(1 - \hat{p})\left(\frac{1}{n_1} + \frac{1}{n_2}\right)}}$$

$$= \frac{(0.594 - 0.790) - 0}{\sqrt{0.687(1 - 0.687)\left(\frac{1}{69} + \frac{1}{62}\right)}} \approx -2.415$$

This test statistic is based on the difference of two approximately normally distributed random variables, $\hat{p}_1$ and $\hat{p}_2$. The test statistic $z = -2.42$ has a (one-sided) P-value of 0.0078, as illustrated in Display 9.16.

Test and CI for Two Proportions

Sample	X	N	Sample p
1	41	69	0.594203
2	49	62	0.790323

Difference = p (1) − p (2)
Estimate for difference: −0.196120
95% upper bound for difference: −0.0669451
Test for difference = 0 (vs < 0): Z = −2.42 P-Value = 0.008

Fisher's exact test: P-Value = 0.012

Display 9.16 *P*-value for a one-sided test with $z = -2.42$.

Conclusion in context.

This *P*-value is very small, so reject the null hypothesis. If all subjects in the experiment had been given the AZT plus ACV treatment, you are confident that there would have been a larger survival rate than if all had received only AZT. The difference between the survival rates for these two treatments is too large to be attributed to chance variation alone for this group of patients.

Confidence intervals aren't usually constructed from data from an experiment.

In this section, you have learned to test whether the difference in the proportions of successes from two treatments is statistically significant. In some situations, it seems that it also would be important to have an estimate of the size of that difference. Yet you may have noticed that this section does not teach you to construct a confidence interval for the difference of two proportions arising from an experiment. That's because the primary purpose of most experiments is to see if there is evidence of differences among the treatments. Standard methods can be used to estimate the size of a difference, but it is important to keep in mind that the results of the testing or estimation procedure do not have statistical generalizations beyond the experimental subjects or units used in the study.

DISCUSSION
Significance Test for a Difference in Proportions from an Experiment

D8. Discuss how the randomization used in a two-population sample survey differs from the randomization used in a two-treatment experiment.

D9. Discuss the differences in stating hypotheses for testing the equality of proportions in a two-population sample survey versus a two-treatment experiment.

D10. Suppose the difference in the proportion of patients who improve with Treatment A and the proportion of patients who improve with Treatment B is 0.05. Does this mean that the majority of patients improve with Treatment A? Does it mean that Treatment B isn't very good?

D11. The data in Example 9.6 on page 448 came from a "double-blind, randomized clinical trial." What is the meaning of each part of this phrase?

D12. In Example 9.5 on page 447, the null hypothesis was not rejected. Which type of error might have been made in the significance test? Discuss the practical ramifications of making this error in this context.

D13. In Example 9.6 on page 448, the null hypothesis was rejected. Which type of error might have been made in the significance test? Discuss the practical ramifications of making this error in this context.

STATISTICS IN ACTION 9.3 ▶ Random Assignment in an Experiment

In this Statistics in Action, you will study the sampling distribution of the difference between the proportions of students who are "cured" by two different treatments.

What You'll Need: a group of students (the number in the group must be divisible by 4; any remaining students can pair up with another student); enough small cards for all students, half of the cards marked "sit" and half marked "stand"; a second deck of small cards for the students, three-fourths of them marked with a "C" and one-fourth marked with a "B."

Suppose that each member of your class is afflicted with boredom.

1. Randomly assign to each student the treatment of *sit* or *stand*. Do this by shuffling the sit/stand cards and giving one to every student. Each student should implement his or her assigned treatment by either sitting or standing.

2. Suppose that neither treatment cures boredom, but three-quarters of you will be cured of your boredom for some other reason (perhaps thinking about Saturday night). To simulate that, shuffle the C/B cards and give one to every student. If you get a C card, you are cured! If you get a B card, you are still bored.

3. For each treatment, find the proportion of students who were cured. Then find the difference in the proportions of students who were cured, $\hat{p}_{sit} - \hat{p}_{stand}$. Here $\hat{p}_{sit}$ represents the proportion of sitting students who happened to be cured, and $\hat{p}_{stand}$ represents the proportion of standing students who happened to be cured. Mark the difference on a dot plot.

4. Because the treatments had no effect on whether you were cured, keep the same C or B card. To see the effect that a different random assignment of the treatments would have had on the difference $\hat{p}_{sit} - \hat{p}_{stand}$, again shuffle the treatment cards and pass them out. Find the difference in the two proportions, $\hat{p}_{sit} - \hat{p}_{stand}$. Mark this second difference on the dot plot. Repeat until you have enough differences to enable you to see the shape, center, and spread of the sampling distribution.

5. Describe the shape, mean, and the standard error of your approximate sampling distribution. Is your distribution centered where you expect it to be?

6. Compare the SE from your approximate sampling distribution to the value you get from the formula for the standard error of the difference in sample proportions.

$$\sigma_{\hat{p}_1 - \hat{p}_2} = \sqrt{\frac{p_1(1-p_1)}{n_1} + \frac{p_2(1-p_2)}{n_2}}$$

For this experiment, n_1 and n_2 are each half the number of students, and p_1 and p_2 are each 0.75, the probability of being cured. What can you conclude?

Inference for an Observational Study

As you saw in Chapter 4, studies in which the conditions of interest are already built into the units being studied are called "observational studies." Sometimes it is impossible, either ethically or practically, to assign treatments to subjects or experimental units. This often is the situation in epidemiological studies of disease, because it is unethical to give a subject a condition that might be harmful, even as part of an experiment.

For example, studies have found an association between cell phone use and auto accidents. Drivers who use cell phones are four times as likely to get into crashes serious enough to injure themselves than those who don't use cell phones while driving. [Source: NHTSA, Insurance Institute for Highway Safety, *www.iihs.org/externaldata/srdata/docs/sr4305.pdf*.] These were observational studies because it would be unethical to randomly assign some drivers to use cell phones and others to refrain from using cell phones. Such studies do not allow you to conclude that cell phone use causes the increase in accidents or to estimate population parameters in a rigorous way.

Even though no randomization is involved, significance testing can still be useful if you are very careful in stating the question and your conclusion. What you are asking is, "Is the difference in the proportion of cell phone users who get into accidents and the proportion of non-cell phone users who get into accidents so small that I can reasonably attribute it to chance alone?" If the answer is yes, then the data don't provide evidence of an association between cell phone use and accidents. If the answer is no, then there is evidence of an association that should be investigated further. (Such an investigation would include analysis of confounding variables, such as the fact that cell phone users are the younger drivers, who are more likely to get into accidents, cell phone use or not. Investigators have categorized drivers by age and sex, and the association is still strong within each group. Also, investigators have gone on to replicate the results in experiments by randomly assigning drivers to use a cell phone or not, and then test their driving in a simulator.)

Automobile simulator. Courtesy of David Strayer, University of Utah Applied Cognition Lab.

| Example 9.7 | **A Plan to Stop Smoking?** |

Do you know someone who is trying to quit smoking? Does the person have a plan to stop? How successful has he or she been? Some have argued that a spur-of-the-moment decision to stop smoking (cold turkey, if you will) is just as successful as the best laid plan to stop smoking. In a study on this question, researchers observed 611 documented attempts to quit smoking. Among the 297 unplanned, spur-of-the-moment attempts, 194 succeeded six months or longer. Among the 314 planned attempts, 133 succeeded six months or longer. [Source: *www.medpagetoday.com*.] Could the difference reasonably be attributed to chance alone?

Solution
The conditions for a significance test for the difference between proportions are not met because there was no randomization in this study. The researchers studied the 611 attempts that were available to them from data collected in a national survey, and the treatments (planned and unplanned attempts to quit smoking) were not randomly assigned. Nevertheless, a test statistic and *P*-value will help answer the question of whether the observed difference in proportions could be chalked up to chance. The test statistic is calculated in the usual way, beginning with the sample proportions and the pooled proportion of successes:

$$\hat{p}_1 = \frac{194}{297} \approx 0.65 \qquad \hat{p}_2 = \frac{133}{314} \approx 0.42$$

$$\hat{p}_1 = \frac{total\ number\ of\ successes\ in\ both\ samples}{n_1 + n_2} = \frac{194 + 133}{297 + 314} = 0.535$$

$$z = \frac{(\hat{p}_1 - \hat{p}_2) - 0}{\sqrt{\hat{p}(1-\hat{p})\left(\frac{1}{n_1} + \frac{1}{n_2}\right)}} = \frac{(0.65 - 0.42) - 0}{\sqrt{0.535(0.465)\left(\frac{1}{297} + \frac{1}{314}\right)}} = 5.70$$

The *P*-value associated with this large *z*-score is very close to zero, so it looks like the observed difference in proportions cannot be attributed to chance alone. Something other than chance must be at work here.

However, this study cannot determine whether the method itself resulted in a higher success rate or some other variable was responsible, such as a difference in the willpower of people who choose the two methods. Nor can this study determine if the result applies to the larger population of smokers who attempt to quit. All that can be concluded is that this difference cannot easily be attributed to chance alone. Thus, this result might suggest to the researchers that they should collect more data on this question, although they probably can never do a randomized study. (They could not randomly assign a person to make an immediate decision to stop smoking.)

The Effectiveness of a Helmet Law

Example 9.8

Does a helmet law have much impact on whether bicycle riders actually wear helmets? Anticipating the passage of a helmet law, the government of North Carolina commissioned a study of helmet usage among bicycle riders at various locations across the state in 1999. This study provided baseline data for studying the effects of the law. In 2002, six months after the law was passed, a similar study was conducted at the same types of locations. Some of the resulting data summaries are shown in Display 9.17.

	% Helmet Use		Sample Size	
Location Type	1999	2002	1999	2002
Local streets	16	19	1116	848
Collector streets	25	30	592	513
Greenways	42	55	404	369
Mountain biking trails	84	90	336	219
Total			2448	1949

Display 9.17   Helmet usage at various locations. [Source: University of North Carolina Highway Safety Research Center, *www.hsrc.unc.edu.*]

Is there sufficient evidence to say that the difference in the percentage using helmets on local streets before the law and after the law is so large that it cannot reasonably be attributed to chance alone?

Solution

Even though some of the sites for observing riders were randomly selected, the riders observed were those who just happened to be there at the time of observation. In addition, there were no randomly assigned treatments and no controls on any of the myriad other variables that could affect helmet use. Nevertheless, a test of significance can shed some light on the possible association between the law and helmet usage. The null hypothesis is that there is no difference in the proportions of helmet users on local streets for 1999 and 2002. The alternative hypothesis of interest is that the 2002 rate of helmet use is higher than the 1999 rate (perhaps because of the law).

Under the null hypothesis of no difference in proportions, the pooled estimate of helmet use on local streets is

$$\hat{p} = \frac{(0.16)(1116) + (0.19)(848)}{1116 + 848} \approx 0.173$$

If you let $\hat{p}_1 = 0.16$ denote the proportion of helmet users observed on local streets in 1999 and $\hat{p}_2 = 0.19$ denote the proportion observed in 2002, the test statistic becomes

$$z = \frac{0.16 - 0.19}{\sqrt{(0.173)(1 - 0.173)\left(\frac{1}{1116} + \frac{1}{848}\right)}} \approx -1.741$$

If this were a formal one-sided test of significance, the *P*-value would be 0.04. This indicates that if there was no change in the percentage of helmet users, there would be very little chance of seeing a difference this large in the percentage wearing helmets in samples of randomly selected bicycle riders. Thus, there is some evidence that there was a higher rate of helmet wearing in 2002, but that could be due to a combination of many factors, one of which might be the law. The law could have coincided with an increased emphasis on safety education, a general trend in the behavior of bicycle riders, the age or experience of the riders, or even a change in the weather. None of these possible factors are measured, controlled, or balanced by randomization.

DISCUSSION
Inference for an Observational Study

D14. Discuss the differences in the types of conclusions that can be reached in a randomized experiment versus those that can be reached in an observational study.

D15. There is a large controversy in educational research about whether randomized experiments can be used to reach conclusions about curriculum materials and methods. For example, suppose a new method for teaching the addition of fractions is to be compared to a standard method that has been in use for years. To be accepted as "better," the new method must be tested against the old in real classroom situations. Should these tests be set up as randomized experiments or as observational studies? What do you suppose is a key argument in favor of randomized experiments? What do you suppose is a key argument against trying to conduct randomized experiments?

Summary 9.3: Inference for Experiments and Observational Studies

Tests of significance proceed for experiments much as they did for sample surveys, but the hypotheses and conclusions must be stated differently. This is because an experiment generally does not result in a comparison based on independent random samples from two populations. Rather, one population of available experimental units is split randomly into two groups by the assignment of treatments, and so the results cannot be generalized beyond the subjects who participated in the experiment. As long as the basic sample size considerations are met, however, the same test statistic used in the analysis of sample surveys applies and provides information on whether the observed statistic could reasonably have been produced by chance.

Tests of significance can be useful when trying to understand observational studies. However, all you can conclude is whether the association can or cannot reasonably be attributed to chance alone.

Practice

Significance Test for a Difference in Proportions from an Experiment

P12. The landmark Physicians' Health Study of the effect of low-dose aspirin on the incidence of heart attacks checked to see if aspirin use was associated with increased risk of ulcers. This was a randomized, double-blind, placebo-controlled clinical trial. Male physician volunteers with no previous important health problems were randomly assigned to an experimental group ($n_1 = 11{,}037$) or a control group ($n_2 = 11{,}034$). Those in the experimental group were asked to take a pill containing 325 mg of aspirin every second day, and those in the control group were asked to take a placebo pill every second day. Of those who took the aspirin, 169 got an ulcer, compared to 138 in the placebo group. Does this provide statistically significant evidence of a difference in the proportions of ulcers associated with the two treatments? [Source: Steering Committee of the Physicians' Health Study Research Group, "The Final Report on the Aspirin Component of the Ongoing Physicians' Health Study," *New England Journal of Medicine*, Vol. 321, 3 (1989), pp. 129–135, *content. nejm.org.*]

P13. The Pique Technique says that people will remember and respond more often to an unusual request than to a standard request because the unusual request will pique their curiosity about why you ask. To test this theory, researchers set up a study in which students divided 144 target adults into two groups of 72 each. Each person in one group was asked for a quarter, and each person in the other group was asked for 17 cents. Assume the division into groups was random. The group from which a quarter was requested had 30.6% success rate while the other group had 43.1% success rate. Do these data support the premise of the Pique Technique? [Source: Fred L. Ramsey and Daniel W. Schafer, *The Statistical Sleuth*, 2nd ed, (Duxbury, 2002), p. 549.]

P14. In 1910, the German Nobel Prize-winning chemist Adolf Windaus first linked high cholesterol and heart disease. In the century since then, further research has led to effective treatments. A randomized clinical trial compared two drugs designed to lower cholesterol (Lipitor and Zocor) with regard to both the cholesterol outcome and the rate of heart attacks over a 5-year period. It is the prevention of heart attacks that is the ultimate goal, and Zocor is much less expensive than Lipitor (because Zocor's patent expired in 2006). High doses of Lipitor won out over low doses of Zocor on lowering cholesterol, but the heart attack issue was much closer. Of 4420 patients on Lipitor, 411 had heart attacks. Of 4452 patients on Zocor, 463 had heart attacks. [Source: *USA Today*, November 16, 2005.]

a. Is this evidence of a statistically significant difference in heart attack rates for the two drugs?

b. Prior to the study, many experts thought that the high doses of Lipitor should be much more effective against heart attacks than the low doses of Zocor. Conduct the appropriate one-sided test to check this claim. What is your conclusion? How do you reconcile it with your conclusion in part a?

P15. In P14, which type of error could have been made in part a? In part b? Which do you think is the more serious error in the practical context of this problem? Explain your reasoning.

P16. To study the effect of reducing television viewing time on children's obesity, researchers conducted a randomized experiment using as subjects 3rd- and 4th-grade students at two schools in the same city. The intervention being tested was to educate children about television watching and to give them incentives to watch less TV. The outcome measures were body mass index and other variables related to obesity.

The randomization, however, was to schools rather than to students, with one school randomly assigned the intervention (for all of its participating students) and the other school being the control (no intervention for any student). This may or may not be a good method of randomization, so the first question to study is, "Do the treatment and control groups look alike before the intervention is applied, or are there important differences that cannot be explained by chance?" Display 9.18 shows a few of the items the researchers measured in that regard.

	Intervention, $n = 92$	Control, $n = 100$
TV in the bedroom of student participant	43.5%	43%
College grads among parents or guardians	45%	21%
Female student participant	45%	48.5%

Display 9.18 Effect of reducing television viewing time on children's obesity. [Source: *http://jama. ama-assn.org/cgi/reprint/282/16/1561.pdf.*]

a. Use statistical inference to check to see if these differences in percentages could be relegated to chance or if other possible causes should be considered.

b. Comment on whether you think the randomization to schools rather than students was a good idea.

Inference for an Observational Study

P17. Do victims of violence exhibit more violence toward others? In a classic study of this question, a researcher found records of 908 people who had been abused as children. Then, matching the demographic characteristics of this group as closely as she could, she found records of 667 individuals who had not been abused as children. Based on further searches of records, she discovered that 103 of the abused children later perpetrated violent crimes, while 53 of the children not abused perpetrated violent crimes. [Source: C. S. Wisdom, "The Cycle of Violence," *Science*, Vol. 244 (1989), p. 160.]

a. What kind of study was this: sample survey, experiment, or observational study?

b. Is there statistically significant evidence that abused children are more likely to perpetrate violent crimes than nonabused children?

c. Write a conclusion in the context of the problem, being careful to consider the type of design that was used. Does this prove that abuse of children leads to violent crime down the road?

P18. Refer to Example 9.8 on page 453. Among the last three locations (collector streets, greenways, and mountain biking trails), which shows the strongest association between helmet use and the law?

Exercises

E35. Nobel laureate Linus Pauling (1901–1994) conducted a famous medical experiment to try to justify his belief that vitamin C prevents colds. His subjects were 279 French skiers who were randomly assigned to receive vitamin C or a placebo. Of the 139 given vitamin C, 17 got a cold. Of the 140 given the placebo, 31 got a cold. [Source: L. Pauling, "The Significance of the Evidence About Ascorbic Acid and the Common Cold," *Proceedings of the National Academy of Sciences*, Vol. 68 (1971), pp. 2678–2681.]

a. Test to see if there is evidence of a statistically significant difference in proportions here, using a 0.05 level of significance.

b. Test to see if there is evidence of a statistically significant difference in proportions here, using a 0.01 level of significance. Does your conclusion change from your interpretation in part a?

Linus Pauling (1901–1994).

E36. A randomized clinical trial on Linus Pauling's claim that vitamin C helps prevent the common cold was carried out in Canada among 818 volunteers, with results reported in 1972. The data showed that 335 of the 411 in the placebo group got colds over the winter in which the study was conducted, while 302 of the 407 in the vitamin C group got colds. [Source: T. W. Anderson, D. B. Reid, and G. H. Beaton, "Vitamin C and the Common Cold," *Canadian Medical Association Journal*, Vol. 107 (1972), pp. 503–508.]

a. Test to see if there is evidence of a statistically significant difference in proportions here, using a 0.05 level of significance.

b. Test to see if there is evidence of a statistically significant difference in proportions here, using a 0.01 level of significance. Does your conclusion change from your interpretation in part a?

E37. In 1954, the largest medical experiment of all time was carried out to test whether the newly developed Salk vaccine was effective in preventing polio. This study incorporated all three characteristics of an experiment: use of a control group of children who received a placebo injection (an injection that felt like a regular immunization but contained only saltwater), random assignment of children to either the placebo injection group or the Salk vaccine injection group, and assignment of each treatment to several hundred thousand children. Of the 200,745 children who received the Salk vaccine, 82 were diagnosed with polio. Of the 201,229 children who received the placebo, 162 were diagnosed with polio. Conduct a test to establish whether the Salk vaccine produced a lower rate of polio than that produced by the placebo. [Source: Paul Meier, "The Biggest Public Health Experiment Ever." in Judith M. Tanur et al., eds., *Statistics: A Guide to the Unknown*, 3rd ed., (Pacific Grove, CA: Brooks/Cole, 1989).]

Jonas Salk (1914–1995).

E38. Two different methods of treating wool with a moth-proofing agent were tested in a laboratory by randomly dividing 40 wool samples into two groups of 20, one group randomly assigned to Method A and the other to Method B. Moth larvae were then attached to each sample of wool and observed for a fixed period of time. For Method A, 8 of the 20 larvae died; for Method B, 12 of the 20 larvae died. Is there evidence of a difference in the effectiveness of these two methods? Answer the question by performing and interpreting a test of significance. [Source: G. E. P. Box, W. G. Hunter, and J. S. Hunter, *Statistics for Experimenters* (Hoboken, NJ: Wiley, 1978), pp. 132–133.]

E39. The Physicians' Health Study was a randomized, double-blind, placebo-controlled clinical trial. (See P12 on page 455.) Male physician volunteers with no previous important health problems were randomly assigned to an experimental group ($n_1 = 11{,}037$) or a control group ($n_2 = 11{,}034$). Those in the experimental group were asked to take a pill containing 325 mg of aspirin every second day, and those in the control group were asked to take a placebo pill every second day. After about 5 years, there were 139 heart attacks in the aspirin group and 239 in the placebo group. Is there statistically significant evidence that aspirin use reduces the rate of heart attacks? Test at the 1% level using the following output.

Test and CI for Two Proportions

Sample	X	N	Sample p
1	239	11034	0.021660
2	139	11037	0.012594

Difference = p (1) − p (2)
Estimate for difference: 0.00906632
99% lower bound for difference: 0.00500537
Test for difference = 0 (vs > 0): Z = 5.19 P-Value = 0.000

Fisher's exact test: P-Value = 0.000

E40. A related part of the landmark study discussed in E39 dealt with the question of whether aspirin reduces the seriousness of a heart attack should one occur. Of the 139 volunteers in the aspirin group who had a heart attack, 10 died. Of the 239 in the placebo group who had a heart attack, 26 died. Is the difference in death rate statistically significant? Test at the 1% level using the following output.

Test and CI for Two Proportions

Sample	X	N	Sample p
1	26	239	0.108787
2	10	139	0.071942

Difference = p (1) − p (2)
Estimate for difference: 0.0368442
99% lower bound for difference: −0.0324011
Test for difference = 0 (vs > 0): Z = 1.18 P-Value = 0.120

Fisher's exact test: P-Value = 0.160

E41. "Exercise Helps Delay Onset of Dementia" read a headline in the *Gainesville Sun* on January 17, 2006. Researchers followed a group of people age 65 and older from 1994 to 2003. Among the 1185 free of dementia at the end of this time, 77% reported exercising three or more times a week. Among the 158 who showed signs of dementia at the end of the period, 67% reported exercising three or more times a week.

a. Is this an experiment, a sample survey, or an observational study?

b. Test to see if the difference in these proportions is statistically significant. Then state your conclusions carefully, in light of your answer to part a. Does this study demonstrate that exercise causes a delay in the onset of dementia?

E42. In one early study (1956) of the relationship between smoking and mortality, Canadian war pensioners were asked about their smoking habits and then followed to see if they were alive 6 years later. Of the 1067 non-smokers, 950 were still alive. Of the 402 pipe smokers, 348 were still alive. [Source: G. W. Snedecor and W. G. Cochran, *Statistical Methods*, 6th ed. (Iowa State Press, 1967), pp. 215–216.]

a. Is this an experiment, a sample survey, or an observational study?

b. Is there statistically significant evidence that pipe smoking increases the rate of mortality? State your conclusions carefully, in light of your answer to part a.

E43. Do you tend to mimic the food preferences of others? In a study designed to help answer this question, college students were told that they were to watch a video of a series of advertisements and then comment on their impressions. The real goal of the study was to see if the students mimicked the snacking pattern of the host on the video. Bowls of goldfish crackers and animal crackers were set in front of both the host and the viewers. In the first setting, the host ate only the goldfish and in that setting, 21 of the 29 viewers selected goldfish as their snack of preference. In the second setting, the host ate only animal crackers, and in that setting, 12 of the 26 viewers selected goldfish as their snack of preference. Assume the 29 students in the first setting and the 26 students in the second setting were randomly assigned to their viewing option. It is important to note that a pretest of snack preferences rated goldfish and animal crackers about equal among these college students. [Source: Robin J. Tanner et al., "Of Chameleons and Consumption: The Impact of Mimicry on Choice and Preferences," *Journal of Consumer Research*, Vol. 34, 6 (April 2008), pp. 754–766.]

a. What would be the null hypothesis that the viewers were not mimicking the eating behavior of the host? What would be the most appropriate alternative hypothesis?

b. Do the results of the experiment provide significant evidence of mimicry?

E44. In a classic 1962 social experiment in Ypsilanti, Michigan, 3- and 4-year-old children were assigned at random to either a treatment group receiving two years of preschool or a control group receiving no preschool. The participants in the study were followed through their teenage years. One variable recorded was the number in each group arrested for a crime by the time they were 19 years old. The data are shown in Display 9.19.

	Arrested	Not Arrested
Preschool	19	42
No preschool	32	30

Display 9.19 Arrests by the age of 19, by preschool attendance. [Source: F. Ramsey and D. Schafer, *The Statistical Sleuth* (Duxbury, 2002), p. 546; *Time*, July 29, 1991.]

a. Are the conditions met for constructing a statistical test on the difference of two proportions?

b. Is there statistically significant evidence that preschool lowers the proportion arrested?

c. Carefully state a conclusion in the context of the problem.

d. This is an experiment, but is a clear cause-and-effect relationship established here? Explain your reasoning.

E45. A study of newly diagnosed lung cancer patients who had enrolled in an Early Lung Cancer Action Program regimen of screening found that 234 of the 258 patients with tumors measuring 15 mm or less had no metastases (spreading to other parts of the body) and that 98 of the 118 patients with tumors measuring 16–25 mm had no metastases. The researchers needed to determine whether the difference in these two proportions is statistically significant. [Source: "Computed Tomographic Screening for Lung Cancer: The Relationship of Disease Stage to Tumor Size," *Archives of Internal Medicine*, Vol. 166 (February 13, 2006), 321–325.]

a. The investigators chose to do a one-sided test of significance. Why does that make sense?

b. What kind of study was this: sample survey, experiment, or observational study?

c. The *P*-value reported in the article was 0.02. Is that correct?

d. Interpret this *P*-value in the context of this situation.

E46. A study at UCLA found that 55% of 100 cosmetic surgery patients take herbal supplements, whereas only 24% of a similar group of 100 people not undergoing surgery take herbal supplements. The researchers needed to determine whether the difference in these two proportions is statistically significant [Source: *Los Angeles Times*, February 20, 2006, p. F4.]

a. Would you have chosen to do a one-sided or a two-sided test of significance? Explain your reasons.

b. What kind of study was this: sample survey, experiment, or observational study?

c. Using the test you chose in part a, find the *P*-value for a significance test of the difference in the two proportions.

d. Interpret the *P*-value you found in part c in the context of this situation.

E47. The great baseball player Reggie Jackson batted .262 during regular season play and .357 during World Series play. He was at bat 9864 times in the regular season and 98 times in the World Series. Can this difference reasonably be attributed to chance, or did Reggie earn his nickname "Mr. October"? (The World Series is played in October.) Use $\alpha = 0.05$. Note that you do not have two

random samples from different populations. You have Reggie's entire record, so all you can decide is whether the difference is about the size you would expect from chance variation or you should look for some other explanation. [Source: Major League Baseball, February 23, 2002, *www.mlb.com.*]

Reggie Jackson.

E48. The Battle of Bunker Hill was one of the first battles of the American Revolution. About 2400 British troops were engaged, of which 1054 were wounded. Of the 1054 wounded, 226 died. Out of 1500 American participants in the battle, losses were estimated at 140 killed and an additional 271 wounded who didn't die of their wounds. (The number of casualties varies somewhat depending on the source.) You have all the necessary information about the two populations. For parts a and b, should the difference in the proportions be attributed to chance variation, or should you look for another explanation? Use $\alpha = 0.05$. [Source: Christopher Ward, *The War of the Revolution* (New York: Macmillan, 1952), p. 96.]

a. the proportion of the American troops who were wounded (including those killed) and the proportion of the British troops who were wounded (including those killed)

b. the proportion of the American troops who were killed and the proportion of the British troops who were killed

Bunker Hill Monument.

Chapter Summary ◀

To use the confidence intervals and significance tests of this chapter for a survey, you need two random samples taken independently from two distinct populations, each made up of successes and failures. This allows you to come to a conclusion about the difference in the proportions of successes in the populations from which the samples were taken. If the study is an experiment, you should have two treatments randomly assigned to the available subjects or units. The conclusion of your test of significance must be limited to the subjects who participated in the experiment. When no randomness is involved in the design, you have an observational study. While computing a test statistic and P-value can be useful in an observational study, if you reject the null hypothesis, all you can conclude is that the association between the condition and the outcome can't reasonably be attributed to chance alone.

The significance tests you studied follow these steps:

1. State the null hypothesis and the alternative hypothesis.

2. Justify your reasons for choosing a particular test. Discuss whether the conditions are met and decide whether it is okay to proceed if they are not strictly met.

3. Compute the test statistic and find the P-value. Draw a diagram of the situation. The test statistic has the usual form:

$$\frac{statistic - parameter}{estimated\ standard\ error}$$

4. Use the computations to decide whether to reject or not reject the null hypothesis by comparing the P-value to α, the significance level. Then state your conclusion in terms of the context of the situation. (Simply saying "Reject H_0" is not sufficient.) Mention any doubts you have about the validity of your conclusion.

If you have a sample survey, you can use a confidence interval if you want to find a range of plausible values for $p_1 - p_2$, the difference between the proportion of successes in one population and the proportion of successes in another population. The confidence interval has the usual form:

$$statistic \pm (critical\ value) \cdot (estimated\ standard\ error)$$

When the situation is two-sided, you can use a confidence interval to determine statistical significance. If 0 is in the confidence interval, the result is not statistically significant.

Review Exercises

E49. Each year Gallup conducts a poll about public education on behalf of Phi Delta Kappa, an association of professional educators. Beginning in 1974, respondents were asked to give a grade to the public schools in their community. That year, 48% of the sample of 1702 adults assigned a grade of A or B. In 2008, the comparable figure was 46%, with a sample size of 1002. Find and interpret a 90% confidence interval for the difference of two proportions. You may consider Gallup's samples as equivalent to random samples of American adults. [Source: *www.pdkintl.org/kappan.*]

E50. A study of all injuries over a 2-year span at the three largest ski areas in Scotland found that of the 531 snowboarders who were injured, 148 had fractures. Of the 952 skiers who were injured, 146 had fractures. (For both groups, most of the other injuries were sprains, lacerations, or bruising.) [Source: *www.ski-injury.com*, March 29, 2002.]

a. Is this a survey, experiment, or observational study?

b. Is this difference statistically significant at the 0.05 level? Write a conclusion, taking into account the type of study design.

c. There are about twice as many skiers as snowboarders. Can you use this fact and the data from this study to determine whether snowboarders are more likely than skiers to be injured?

E51. To study the brain's response to placebos, researchers at UCLA gave 51 patients with depression either an antidepressant or a placebo. The article reports that the "51 subjects then were randomly assigned to receive 8 weeks of double-blind treatment with either placebo or the active medication.... Overall, 52% of the subjects (13 of 25) receiving antidepressant medication responded to treatment, and 38% of those receiving placebo (10 of 26) responded." [Source: A. F. Leuchter et al., "Changes in Brain Function of Depressed Subjects During Treatment with Placebo," *American Journal of Psychiatry*, Vol. 159, 1 (January 2002), pp. 122–129.]

a. Perform a test to determine if the difference in the proportion who responded is statistically significant. Use $\alpha = 0.02$.

b. If neither the patient nor the examining physician knows which treatment the patient is receiving, the experiment is called *double-blind*. See page 207 of Chapter 4. Describe how the experiment could have been made double-blind.

E52. In a random sample of 1000 U.S. adults consisting of 480 men and 520 women, 48% of the men favored legalizing marijuana, whereas only 34% of the women favored legalization. With 90% confidence, estimate the difference between the population proportions of men and women favoring the legalization of marijuana. Do the data support the claim that a higher proportion of men favor legalization? [Source: Rasmussen Reports, February 19, 2009.]

E53. An April 2009 Gallup poll of 1023 adults living in the United States found that 39% favored stricter gun control laws. Gallup first asked this question in 1990, when 78% favored stricter gun control laws. You may consider the samples as equivalent to random samples of American adults. [Source: *www.gallup.com/poll/117361/ Recent-Shootings-Gun-Control-Support-Fading.aspx* and *www.cnn.com/2009/POLITICS/04/08/gun.control.poll.*]

a. The sample size for 1990 is not given in the source. Assuming that it is about the same as in 2009, is there statistically significant evidence of a decline in the proportion favoring stricter gun control laws?

b. Would your conclusion change in part a if the sample size in 1990 had been only 100?

c. Compute confidence intervals for the difference of the two proportions using sample sizes of 1023 and then of 100 for the 1990 sample. How do the lengths compare?

E54. Medical record keeping is an important issue in the discussions of the overall effectiveness and cost of health care in the United States. In response to this interest, the Department of Health Policy and Management of the Harvard School of Public Health conducted a sample survey of hospitals in the country to ask about the availability of electronic health record (EHR) systems.

A summary of their findings is found in Display 9.20 Comprehensive systems are those that cover the entire hospital, while basic systems are those that may be used in just one department of the hospital. The data are reported as percentages along with the standard errors of the percentages (in parentheses). Use the confidence interval on page 428 to answer parts a and b.

Type of Hospital	Comprehensive EHR System (%) (standard error)	Basic EHR System (%) (standard error)
For-profit hospitals	1.0 (0.4)	5.0 (1.1)
Private nonprofit hospitals	1.5 (0.3)	8.0 (0.7)
Public hospitals	2.7 (0.7)	6.9 (1.1)

Display 9.20 Percentages of hospitals having EHR systems. [Source: *http://content.nejm.org/cgi/ reprint/NEJMsa0900592v1.pdf.*]

a. Is there evidence of a difference in the population percentages of hospitals having comprehensive EHR systems between for-profit and public hospitals? How about between private nonprofit and public hospitals?

b. Is there evidence of a difference in the population percentages of hospitals having basic EHR systems between for-profit and public hospitals? How about between private nonprofit and public hospitals?

E55. Blood glucose level is an important measure of health, not only among those suffering from diabetes but also among those suffering from any serious illness. A study of intensive versus conventional control of blood glucose levels in critically ill patients (those in ICU for 3 or more consecutive days) was undertaken in a randomized study of 6104 patients. Of these, 3054 were randomized to the intensive treatment and 3050 to the conventional treatment. The summary response of interest was the number of deaths within 90 days after treatment began. There were 829 deaths among those in the intensive treatment group and 751 deaths among those in the conventional treatment group. Do these data provide evidence of a statistically significant difference in death rates for the two treatments? Would the same conclusion hold if there had been only about 300 patients in each group instead of about 3000, with roughly the same percentages of deaths? [Source: *http://content.nejm.org/cgi/content/short/360/13/1283.*]

E56. To become a lawyer, a person must pass the bar exam for his or her state. Suppose you are searching for possible inequities that result in unequal proportions of males and females passing the bar exam. Each state has its own exam; so you check a random sample of the exam records in each of the 50 states. For each state, you plan to perform a significance test of the difference of the proportion of males and females passing the exam, using

$\alpha = 0.05$. If the difference is statistically significant for any state, you will conclude that there has been some inequity. Do you see anything wrong with this plan?

E57. Election polls typically report the estimated percentages of the vote that the leading candidates will get at a particular point in time. What is of most concern to the candidates, however, is the difference between the percentages for, say, the top two candidates. In a typical polling situation, can the population difference be estimated by the methods of this chapter? Explain why or why not.

E58. Refer to P20 on page 331 in Section 6.3. In that research study, a visible mark was placed on one side of the forehead of Happy, an elephant. The researchers also were interested in whether Happy touched the side of her forehead with her trunk more frequently when at the mirror or recently at the mirror than when she was not at the mirror. Suppose that, while at the mirror, she touched her forehead 12 times and, in an equal span of time while not at the mirror, she touched her forehead 7 times.

a. If being at the mirror makes no difference, what proportion of the head touches would you expect to occur at the mirror?

b. What test could you use to determine if this is a statistically significant difference?

c. Conduct the test you chose.

Concept Test

C1. A Harris poll asked about 1000 U.S. adults to name their favorite sport. Twenty-six percent named professional football in 1998, and 30% named the same sport in 2008. With 90% confidence, estimate the change in the proportion favoring professional football over this decade. Choose the correct statement about the resulting confidence interval.

A The confidence interval includes 0, so it is plausible that there was no real change.

B The confidence interval does not include 0, so there is evidence of a real change.

C The confidence interval does not include 0, so there is no evidence of a real change.

D The samples cannot be independent because they both involve lots of people who like professional football.

E The audience for pro football has increased by 4% over the decade.

C2. In a clinical trial experiment involving two treatments, labeled A and B, for a common knee injury, the response of interest was whether there was full recovery in 1 year. After the data were collected, the difference in proportions was 0.12, with Treatment A having the higher recovery rate. Based on a significance test for the difference between proportions, the null hypothesis of equal proportions of successes for these treatments was not rejected. Later research showed quite clearly that Treatment A was the better treatment for this injury. Select the best description of the error in this earlier experiment.

A No error was made.

B A Type I error was made because a false null hypothesis was rejected.

C A Type I error was made because a false null hypothesis was not rejected.

D A Type II error was made because a false null hypothesis was rejected.

E A Type II error was made because a false null hypothesis was not rejected.

C3. The professor in charge of coordinating the teaching of multiple sections of an introductory statistics course decides to gather data on whether having regular quizzes throughout the term improves the proportion of students receiving a grade of B or better. She knows from past experience that the students are fairly consistent across sections in terms of academic ability, gender, and so on. She chooses two sections taught by an instructor who uses quizzes and two sections taught by a different instructor who does not use quizzes, and she compares the relevant proportions at the end of the term. By doing a test of significance, she concludes that use of quizzes does produce better grades. Choose the best description of this study and its conclusion.

A This is an experiment, so the conclusion is valid.

B This is an experiment, but the conclusion is not valid because cause and effect cannot be determined by such studies.

C This is an observational study, so the professor's conclusion is valid.

D This is an observational study, but the conclusion is not valid because cause and effect cannot be determined by such studies.

E This is a sample survey and the results can be used only to estimate the effect of quizzes for all the sections from which these sections were sampled.

C4. In a study of the effectiveness of two tutors in preparing students for an exam, 50 students were randomly assigned either to Mr. A or to Mr. B. A larger proportion of students tutored by Mr. A passed the exam, resulting in a 95% confidence interval for the difference of two proportions of $(-0.05, 0.45)$. Which is *not* a correct conclusion to draw from this?

A There is statistically significant evidence that Mr. A is the better tutor.

B You cannot reject a null hypothesis of equal tutor effectiveness.

C The difference in the percentage who passed the exam and were tutored by Mr. A and the percentage who passed and were tutored by Mr. B is 20%.

D A Type II error may be made.

E The design of this study didn't have enough power to pick up any but a very large difference in passing rates.

C5. With $\alpha = 0.05$, researchers conducted a test of the difference of two proportions to compare the rate of alcohol use among college students this year and in 1990. The rates for both years are based on large, independent random samples of college students. Which is the best interpretation of "$\alpha = 0.05$" in this context?

A There's a 5% chance that the rate of alcohol use has changed.

B There's a 5% chance that the rate of alcohol use has not changed.

C If the rate of alcohol use has not changed, there's a 5% chance that the researchers will mistakenly conclude that it has.

D If the rate of alcohol use has changed, there's a 5% chance that the researchers will mistakenly believe it hasn't.

E The study has enough power to detect a difference in the rate of alcohol use of 0.05 or more.

C6. A student obtains a random sample of M&M's and a random sample of Skittles. She finds that 7 of the 40 M&M's are yellow and 13 of the 35 Skittles are yellow. Her null hypothesis is that the proportion of yellow candies is equal in both brands. Her alternative hypothesis is that the proportion is higher in Skittles. What is her conclusion, if she uses $\alpha = 0.05$?

Ⓐ Reject the null hypothesis. The difference in the two proportions can reasonably be attributed to chance alone.

Ⓑ Reject the null hypothesis. The difference in the two proportions cannot reasonably be attributed to chance alone.

Ⓒ Do not reject the null hypothesis. The difference in the two proportions can reasonably be attributed to chance alone.

Ⓓ Do not reject the null hypothesis. The difference in the two proportions cannot reasonably be attributed to chance alone.

Ⓔ Accept the null hypothesis because the difference in the two proportions is not statistically significant.

Investigation

C7. A test of significance asks the question, "Is it reasonable to assume that the result I see happened by chance, or should I look for another cause?" In randomized experiments, the answer to this question can be found simply by studying what happens in repeated randomizations, without recourse to standard errors, test statistics, or the normal distribution. The following analysis will show you how that is done.

Thirty-six people with chronic migraines were randomly assigned to receive either acupuncture or a sham acupuncture treatment. After the 3 months of treatment, 12 of the 19 people who received the real acupuncture and 8 of the 17 people who got the sham treatment reported at least a 50% reduction in the frequency of migraines. Call a person who reported at least a 50% reduction in frequency a "success." [Source: J. Alecrim-Andrade et al., "Acupuncture in Migraine Prevention: A Randomized Sham Controlled Study with 6-Months Posttreatment Follow-Up," *Clinical Journal of Pain*, Vol. 24, 2 (2008), pp. 98–105 (Medline: 18209514).]

a. What is the null hypothesis for this study?

b. What is the value of $\hat{p}_1 - \hat{p}_2$, where $\hat{p}_1$ represents the proportion of successes in the acupuncture group and $\hat{p}_2$ represents the proportion of successes in the sham acupuncture group?

c. Now imagine there had been a different random assignment of treatments to these subjects. This is called a *randomization*. If the null hypothesis is true, how many successes would you expect to see in any randomization?

d. Still assuming the null hypothesis is true, here are the results from another possible random assignment of

treatments to these subjects, with successes labeled S and failures labeled F.

Acupuncture	S S S S S S S S S F F F F F F F F F F
Sham	S S S S S S S S S S S F F F F F F

What is the value of $\hat{p}_1 - \hat{p}_2$ for this randomization?

e. The following histogram gives the values of $\hat{p}_1 - \hat{p}_2$ for 5000 randomizations, still assuming that the null hypothesis is true. Locate the value of $\hat{p}_1 - \hat{p}_2$ for the randomization of part d. Locate the value of $\hat{p}_1 - \hat{p}_2$ for the actual experiment.

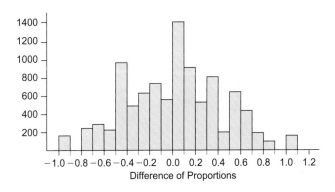

f. Is there any reason to reject the null hypothesis? Explain.

C8. The following steps describe the construction of a randomization test based on the AIDS data in Display 9.15 on page 448. The null hypothesis was that there is no difference in the death rates for the two treatments, AZT and AZT plus ACV.

I. Number 90 chips or slips of paper with a 1 to represent those alive and number 41 chips with a 0 to represent those not alive at the end of the study time.

II. Randomly split the 131 chips into two groups, one with 69 chips (representing those treated with AZT alone) and one with 62 chips (representing those treated with AZT plus ACV).

III. Calculate the difference between the two proportions of 1's (subtracting AZT plus ACV from AZT).

IV. Repeat the process many times and plot the values of the generated differences in proportions. Under the null hypothesis, each of the generated differences has the same chance of occurring.

A plot of 200 runs of this simulation is shown in Display 9.21

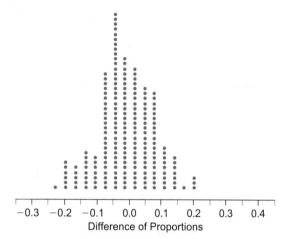

Display 9.21 A plot of 200 runs of a randomization process.

a. The observed difference in the proportions for the data from the real experiment was about −0.20. How many times, out of 200, did the randomization procedure generate a difference this small or smaller? What does this suggest about the chance of seeing a value this small under the null hypothesis?

b. Explain what your conclusion would be regarding the two treatments based solely on the randomization procedure. Does your conclusion agree with the one reached in Example 9.6 on page 448?

Chapter 10
Inference for Means

In 1879, A. A. Michelson made 100 measurements of the velocity of light in air. How can scientists use data like these to estimate the true velocity of light, while accounting for measurement variability? Statistical inference can help.

You now know how to make inferences for proportions based on categorical data. But often the data that arise in everyday contexts are measurement data rather than categorical data. Measurement data, as you know, are often summarized by using the mean. Mean income, mean scores on exams, mean waiting times at checkout lines, and mean housing prices are all commonly used in making decisions that affect your life.

In Chapter 8, you learned how to construct confidence intervals and perform significance tests for a proportion. Although the formulas you'll use for the inferential procedures in this chapter will change a little from those you learned there, the basic concepts remain the same. These concepts are all built around the question, "What are the reasonably likely outcomes from a random sample?" This chapter will follow the same outline as Chapter 8 except that the methods are applied to means rather than proportions.

In this chapter, you will learn

▷ to construct and interpret a confidence interval for estimating an unknown population mean

▷ to compute a P-value for a t-test and interpret it in terms of the strength of the evidence against the null hypothesis

▷ to use a fixed-level significance test (hypothesis test) to decide if it is reasonable to conclude either that your sample might have been drawn from a population with a specified mean or that it was not drawn from such a population

10.1 ▷ A Confidence Interval for a Mean

In this section, you will learn how to construct a confidence interval for a population mean, μ. The structure of the confidence interval will be similar to that in Section 8.1 for a proportion:

$$\hat{p} \pm z^* \sqrt{\frac{\hat{p}(1-\hat{p})}{n}} = statistic \pm critical\ value \cdot estimated\ standard\ error$$

Let's start with an example of a real-life situation in which it is important to locate the mean of a population.

| Example 10.1 | **Average Body Temperature** |

When you are sick and take your temperature, you probably compare your temperature to the so-called normal temperature, 98.6°F. How was that temperature determined? In the 1860s, the German physician Carl Wunderlich measured the temperatures of thousands of healthy people and reported the mean as 37°C (98.6°F). Investigators have questioned his methodology and the quality of his thermometers. To determine an up-to-date average, researchers took the body temperatures of 148 people at several different times during two consecutive days. A portion of these measurements (in °F), for ten randomly selected women, is given here:

97.8   98.0   98.2   98.2   98.2   98.6   98.8   98.8   99.2   99.4

[Source: Allen L. Shoemaker, "What's Normal?—Temperature, Gender, and Heart Rate," *Journal of Statistics Education*, Vol. 4 (1996). Original source: P. A. Mackowiak et al., "A Critical Appraisal of 98.6 Degrees F, the Upper Limit of the Normal Body Temperature, and Other Legacies of Carl Reinhold August Wunderlich," *Journal of the American Medical Association*, Vol. 268 (September 1992), pp. 1578–1580.]

The mean body temperature, $\bar{x}$, for this sample of women is 98.52, and the standard deviation, s, is 0.527. Are these statistics likely to be equal to the mean, μ, and standard deviation, σ, for the population? How can you determine the plausible values of the mean temperature of all women?

Solution

Neither $\bar{x}$ nor s is likely to be exactly equal to the population parameters μ and σ. (This would be true no matter how large the sample.) Plausible values of the mean body temperature of all women, μ, are those values that lie close to $\bar{x} = 98.52$, where "close" is defined in terms of standard error. From Section 7.2, the standard error of the sampling distribution of a sample mean is given by $SE_{\bar{x}} = \sigma/\sqrt{n}$, where σ is the standard deviation of the population and n is the sample size. When the sample size is large enough or the population is normally distributed, in 95% of all samples $\bar{x}$ and μ are no farther part than $1.96 \cdot \frac{\sigma}{\sqrt{n}}$. So plausible values of μ lie in the interval

$$\bar{x} \pm 1.96 \frac{\sigma}{\sqrt{n}} \quad \text{or} \quad 98.52 \pm 1.96 \frac{?}{\sqrt{10}}$$

For the sample of body temperatures, you know $\bar{x}$ and n, but you don't know σ, and you seldom will know in real-life situations. What now? As you might be thinking, the best decision, generally, is to use the sample standard deviation, s, as an estimate of σ.

The Effect of Estimating σ

To compute a confidence interval for the mean, you use the sample standard deviation, s, as an estimate of σ. How will substituting s for σ affect your confidence interval? Some samples give an estimate of s that's too small: $s < \sigma$. In this case, the confidence interval will be too narrow. Others give an estimate that's too large: $s > \sigma$. In this case, the confidence interval will be too wide.

On average, the sampling distribution of s has its center near σ, so, on average, the width of the confidence interval is about right. That's a nice feature, but confidence intervals are judged on their capture rate, not their average width. Does replacing σ with s change your chance of capturing the unknown population mean? If so, how?

The answers to the questions depend on the shape of the sampling distribution of s. The histogram on the left in Display 10.1 shows the values of s for 1000 samples of size 4 taken from a normally distributed population with $\sigma = 107$. For this small sample size, the distribution is quite skewed. The histogram on the right shows the values of s for 1000 samples of size 20. There is little skewness here, so s is smaller than σ about as often as it is larger. Although the average value of s is about equal to σ, s tends to be smaller than σ more often than it is larger, unless the sample size is very large, because the median is smaller than the mean for right-skewed distributions. This causes the confidence intervals to be too narrow more than half the time, so the capture rate will be less than the advertised value of 95%. Fortunately, the sampling distribution of s becomes less skewed and more approximately normal as the sample size increases.

> s is smaller than σ more often than it is larger.

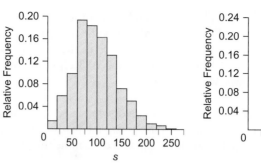

Display 10.1 Approximate sampling distribution of s for samples of size 4 (left) and size 20 (right) for a normally distributed population with $\sigma = 107$.

STATISTICS IN ACTION 10.1 ▶ The Effect of Estimating σ on Measuring Romantic Love

Nerve growth factor (NGF) is a neurotrophin found in blood plasma. Researchers found that the mean NGF concentration of people "truly, deeply and madly in love" for six months or less was significantly higher than that in single people or in people in a long-term relationship. The NGF levels in those who recently fell in love were normally distributed with mean 227 (pg/mL) and standard deviation 107. Assume that these are the population parameters, μ and σ. [Source: Enzo Emanuele et al., "Raised Plasma Nerve Growth Factor Levels Associated with Early-Stage Romantic Love," *Psychoneuroendocrinology*, Vol. 31, 3 (2006), pp. 288–294.]

Suppose you must estimate the population mean from the NGF levels of a random sample of four people who recently fell in love.

1. Use technology to simulate taking a random sample of four NGF levels from the population of those who recently fell in love.

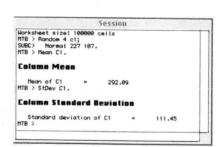

2. Compute the mean, $\bar{x}$, and standard deviation, s, for your sample. Record the value of s.

3. Compute a confidence interval using the formula

    $$\bar{x} \pm 1.96 \frac{s}{\sqrt{n}}$$

    Does your confidence interval capture the population mean, μ, of 227?

4. Repeat steps 1 through 3 until your class has generated 100 samples. You will have 100 values of s and 100 decisions as to whether the confidence interval captured the population mean, μ.

5. Make a dot plot of the 100 values of s and describe its shape, mean, and spread. How close to σ is the mean of the values of s? How many times is s smaller than σ? Larger?

6. What percentage of your confidence intervals capture μ? What can you conclude about the effect of using s to estimate σ?

DISCUSSION
The Effect of Estimating σ

D1. When you use s to estimate σ, the capture rate is too small unless you make further adjustments. If an interval's capture rate is smaller than what you want it to be, do you need to use a wider or a narrower interval to get the capture rate you want?

D2. Consider two extreme situations, $n = 10$ and $n = 1000$. If you substitute s for σ in the formula, $\bar{x} \pm 1.96\, \sigma/\sqrt{n}$, which sample size—10 or 1000—do you expect to give a capture rate closer to 95%? Why?

D3. A firm has given an aptitude test to a large number of applicants for a training program for computer technicians. The scores have a standard deviation of 10 points. Test scores are now available for a random sample of 12 applicants from among the many who applied this month, and the standard deviation of these 12 scores is 8. You are to estimate the mean score for all of this month's applicants using the formula

$$\bar{x} \pm 1.96 \frac{?}{\sqrt{n}}$$

   a. Describe the advantages and disadvantages of using 10 in the place of $?$.

   b. Describe the advantages and disadvantages of using 8 in place of $?$.

How to Adjust for Estimating σ

As you have seen, using s as an estimate of σ in a confidence interval lowers the overall capture rate because s tends to be smaller than σ more often than it is larger. To compensate, you will increase the width of the confidence interval by replacing z^* with a larger value, called t^*:

| Replacing σ with s tends to make the interval too narrow. Compensate by replacing z^* with t^*. |

$$\bar{x} \pm t^* \frac{s}{\sqrt{n}}$$

Let's listen in on a discussion about t^*.

Student: Where does the value of t^* come from?

Statistician: In principle, you could find it using simulation. Set up an approximately normal population, take a random sample, and compute the mean and standard deviation. Do this thousands of times. Then use the results to figure out the value of t^* that gives a 95% capture rate for intervals of the form $\bar{x} \pm t^* \frac{s}{\sqrt{n}}$.

Student: Wouldn't that take a lot of work?

Statistician: Yes, especially if you went about it by hand. Fortunately, this work has already been done, long ago. Statistician W. S. Gosset (1876–1937), who worked for the Guinness Brewery, actually did this back in 1915. Four years later, geneticist and statistician R. A. Fisher (1890–1962) figured out how to find values of t^* using calculus. If the population is nearly normal, it turns out the value of t^* depends on just two things: how many observations you have and the capture rate you want.

Student: So t^* doesn't depend on the unknown mean or unknown standard deviation?

Statistician: No, it doesn't, which is very handy because in practice you don't know these numbers. Suppose, for example, you have a sample of size 5 and you want a 95% confidence interval. Then you can use $t^* = 2.776$ no matter what the values of μ and σ are.

Student: Where did you get that value for t^*?

Statistician: From a *t*-table, although I could have found it from a calculator or statistical software. A brief version of the table is shown in Display 10.2. Table B on page 761 is more complete. For 95% confidence, you go to the column with 95% at the bottom. For the row, you need to know the degrees of freedom, *df* for short.

df	Tail Probability p		
	.05	.025	.02
1	6.314	12.71	15.89
2	2.920	4.303	4.849
3	2.353	3.182	3.482
4	2.132	2.776	2.999
5	2.015	2.571	2.757
	90%	95%	96%
	Confidence level C		

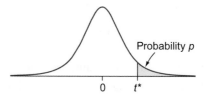

Display 10.2 An abbreviated version of a *t*-table.

Student: Degrees of freedom? What's that?

Statistician: There's a short answer, a longer answer, and a very long answer. The longer answer will come in E44 on page 504. The very long answer is for another course. For the moment, here's the short answer: The number of **degrees of freedom** is one less than the sample size. (It is the number you use for the denominator when you calculate the standard deviation, *s*.) So for these confidence intervals, $df = n - 1$, where *n* is the sample size. When $n = 5$, $df = 4$, and you look in that row of the table. If you look in the row with $df = 4$ and the column with confidence level 95%, you'll find the value 2.776 for t^*.

> **Finding the Value of t^* to Use in Your Confidence Interval**
>
> Turn to Table B on page 761. Find your confidence level in the last row of the table. Find your *df*, $n - 1$, where *n* is the sample size, in the column at the left. Your value of t^* is at the intersection of this row and column. If your *df* falls between two of those listed in the column on the left, use the smaller *df*.

Example 10.2 | **Who Is Hotter?**

Is normal body temperature the same for men and women? As you saw in the previous example, medical researchers interested in this question collected data from a large number of men and women. Random samples from that data are recorded in Display 10.3. (The women's temperatures are the same as those given in the previous example.)

   a. Use a 95% confidence interval to estimate the mean body temperature of men.

   b. Use a 95% confidence interval to estimate the mean body temperature of women.

Body Temperature (°F)	
Male	**Female**
96.9	97.8
97.4	98.0
97.5	98.2
97.8	98.2
97.8	98.2
97.9	98.8
98.0	98.6
98.1	98.8
98.6	99.2
98.8	99.4

Display 10.3   Two random samples of body temperatures. [Source: *Journal of Statistics Education Data Archive*, April 15, 2002, *www. amstat.org.*]

Solution

First, you need to check that these are random samples from relatively large populations and that it's reasonable to assume that the populations are normal. The problem states that these are random samples selected from a group of men and women examined by the researchers; so your final result generalizes only to this population. The researchers reported that the distributions are normal. You should always plot the data, however, and the stemplots in Display 10.4 give you no reason to suspect that the populations are not normally distributed.

```
Male Temperatures          Female Temperatures

  96 |                        97 |
   • | 9                       • | 8
  97 | 4                      98 | 0222
   • | 5889                    • | 688
  98 | 01                     99 | 24
   • | 68
```

97 | 8 represents 97.8°.

Display 10.4   Stemplots of body temperature data.

For a sample of size 10 (9 degrees of freedom) and a confidence level of 95%, Table B on page 761 gives a value of t^* of 2.262.

a. For the males, the sample mean is 97.88, and the sample standard deviation is 0.555. That gives a 95% confidence interval of

$$\bar{x}_m \pm t^* \frac{s_m}{\sqrt{n}} \quad \text{or} \quad 97.88 \pm 2.262 \frac{0.555}{\sqrt{10}}$$

or (97.48, 98.28). Any population of body temperatures with a mean in this interval could have produced a sample mean of 97.88 as a reasonably likely outcome.

b. For the females, the sample mean is 98.52, and the sample standard deviation is 0.527. You are 95% confident that the mean body temperature of women in this population is in the interval

$$\bar{x}_f \pm t^* \cdot \frac{s_f}{\sqrt{n}} \quad \text{or} \quad 98.52 \pm 2.262 \cdot \frac{0.527}{\sqrt{10}}$$

or (98.14, 98.90).

From the sample means it might appear that the average body temperature of females is higher than that of males, but beware! There is some overlap in the confidence intervals, so that conclusion might not be valid. You'll come back to the question of comparing two means in Chapter 11.

DISCUSSION
How to Adjust for Estimating σ

D4. What parameter does the standard deviation, s, of a random sample tend to approach as the sample size gets larger? Look at Table B on page 761. What happens to the value of t^* as the sample size increases? Find the row where df equals ∞. Do those numbers look familiar?

D5. Overall, the researchers found a mean body temperature of about 98.2°F, which is lower than Wunderlich's 98.6°F that has become the standard. Was Wunderlich wrong? Wunderlich worked in the metric system. He rounded his mean body temperature to the nearest degree to get 37°C. What is the lowest his mean could have been in degrees Celsius? The highest? Convert the lowest possible mean, the reported mean (37°C), and the highest possible mean to degrees Fahrenheit. The conversion formula is $F = \frac{9}{5}C + 32$. What can you conclude?

Constructing a Confidence Interval for a Mean

Whenever you construct a confidence interval for a mean, you must check three conditions. Officially, you need:

- a random sample
- an approximately normally distributed population or a sufficiently large sample
- a population that is at least ten times the size of the sample

The fact that you must always have a random sample should be no surprise. A normally distributed population is (theoretically) required because the values of t^* in Table B were computed by assuming that the population is normally distributed. However, you can get away with less than a normally distributed population if the sample size is large enough. If the sample size is more than about 10% of the population size, the estimated standard error of the mean used in this chapter will tend to be larger than it needs to be for the specified confidence level.

That's the last of the new things to learn about constructing a confidence interval for a mean. Here's a summary of what you've learned.

> **A Confidence Interval for a Mean**
>
> 1. *Check conditions.* You must check three conditions:
>   - *Randomness.* In the case of a survey, the sample must have been randomly selected.
>   - *Normality.* The sample must look like it's reasonable to assume that it came from a normally distributed population *or* the sample must be sufficiently large. There is no exact rule for determining whether a normally distributed population is a reasonable assumption or what constitutes a "large enough" sample. Think carefully about the possible shape of the population. If

it is reasonable to assume it to be highly skewed, then a rough guideline is that the sample size should be at least 40.
- *Sample size.* In the case of a sample survey, the population size should be at least ten times as large as the sample size.

2. *Do computations.* A confidence interval for the population mean, μ, is given by

$$\bar{x} \pm t^* \frac{s}{\sqrt{n}}$$

where n is the sample size, $\bar{x}$ is the sample mean, s is the sample standard deviation, and t^* depends on the confidence level desired and the degrees of freedom, $df = n - 1$.

3. *Give interpretation in context.* A good interpretation is of this form: "I am 95% confident that the population mean, μ, is in this confidence interval." Of course, when you interpret a confidence interval, you will do it in context, describing the population you are talking about.

A 95% confidence interval may also be interpreted this way: A 95% confidence interval for a population mean, μ, includes the plausible values of μ. A population with any one of these values for μ could have produced the observed sample mean as a reasonably likely outcome.

The Atomic Weight of Silver

Example 10.3

The National Institute of Standards and Technology (NIST) conducts research that helps U.S. industry measure, test, and standardize products and services. To reduce uncertainty in the atomic weight of silver, a NIST researcher measured the atomic weight of a reference sample of silver. The 24 measurements from one mass spectrometer are given in Display 10.5. Note that all 24 measurements begin with 107.868. Thus, it will easier to work with the coded values, the last decimal places.

Construct and interpret a 95% confidence interval for the mean measurement (using this mass spectrometer) of the atomic weight of silver.

Measurement	Coded Measurement	Measurement	Coded Measurement
107.8681079	1079	107.8681198	1198
107.8681344	1344	107.8681482	1482
107.8681513	1513	107.8681334	1334
107.8681197	1197	107.8681609	1609
107.8681604	1604	107.8681101	1101
107.8681385	1385	107.8681512	1512
107.8681642	1642	107.8681469	1469
107.8681365	1365	107.8681360	1360
107.8681151	1151	107.8681254	1254
107.8681082	1082	107.8681261	1261
107.8681517	1517	107.8681450	1450
107.8681448	1448	107.8681368	1368

Display 10.5 Twenty-four measurements of a sample of silver. [Source: L. J. Powell, T. J. Murphy, and J. W. Gramlich, "The Absolute Isotopic Abundance & Atomic Weight of a Reference Sample of Silver," *NBS Journal of Research*, Vol. 87 (1982), pp. 9–19, *www.itl.nist.gov*.]

Check conditions.

Solution

You can consider these measurements to be a random sample of all possible measurements by this mass spectrometer. Display 10.6 shows that the distribution of measurements from the sample is reasonably symmetric with no outliers, so it is reasonable to assume that they could have come from a normal distribution.

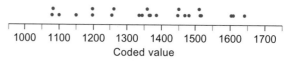

Display 10.6   A dot plot of the sample data.

Do computations.

The mean and standard deviation of the measurements are 1363.54 and 169.017, respectively. For 95% confidence and *df* equal to 24 − 1, or 23, use $t^* = 2.069$. The 95% confidence interval is

$$\bar{x} \pm t^* \frac{s}{\sqrt{n}} = 1363.54 \pm 2.069 \cdot \frac{169.017}{\sqrt{24}}$$

$$\approx 1363.54 \pm 71.38$$

or about (1292, 1435).

Give interpretation in context.

You are 95% confident that the mean measurement of the atomic weight of silver using this mass spectrometer is in the interval from 107.8681292 to 107.8681435. In other words, all values from 107.8681292 to 107.8681435 are plausible values for the mean measurement of the atomic weight of silver.

You also can use your calculator or computer to calculate the confidence interval for a mean. The following Minitab output was computed from the original data. The TI-84 output was computed from the summary statistics.

 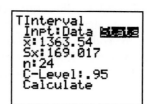 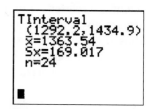

One-Sample T: Coded Atomic Weight

Variable	N	Mean	StDev	SE Mean	95% CI
Coded Atomic Weight	24	1363.5	169.0	34.5	(1292.2, 1434.9)

DISCUSSION
Constructing a Confidence Interval for a Mean

D6. A student said, "Let's be sure and capture μ by producing an interval with a 100% confidence level." Is this possible? Would it be wise to shoot for a 99.5% confidence level?

D7. You have produced a confidence interval for your supervisor, and she says, "This interval is too wide to be of any practical value." What are your options for producing a narrower interval in the next study of this same type? Which option would you choose if the study might have consequences that are vital to the future of the firm?

More on Interpreting a Confidence Interval

As in Chapter 8, the **capture rate** of a confidence interval is the proportion of random samples for which the resulting confidence interval captures the (population) parameter. You can think of it as the chance that the method used to produce a confidence interval actually will work correctly.

The Capture Rate of a Confidence Interval for a Mean

The proportion of intervals of the form $\bar{x} \pm t^* s/\sqrt{n}$ that capture μ is equal to the confidence level; that is, a 95% confidence interval will have a 95% capture rate, a 90% confidence interval will have 90% capture rate, and so on. This will be true provided that

- the sample is random
- the population is normally distributed
- the size of the population is at least 10 times the sample size

The capture rate will be approximately correct for non-normally distributed populations as long as the sample size is large enough.

The capture rate is the same as the confidence level.

Remember that the capture rate is a property of the method, not of the sample. After the sample is selected, either the resulting interval will capture the population mean or it won't. So it is incorrect to say about a specific interval that there is a 95% probability (or 95% chance) that the true mean falls in this interval. All you can say is that you are 95% confident that μ is in your interval, meaning that the method produces a confidence interval that captures μ, the true mean, 95% of the time.

The Meaning of "95% Confident"

Saying you are "95% confident" means that if you could take random samples repeatedly from this population and compute a confidence interval from each sample, in the long run 95% of these different intervals would contain (or capture) μ.

In practice, samples might not be random, distributions might not be normal, and survey questions might be worded so that people answer incorrectly. In such situations, the capture rate may be far below what is advertised.

DISCUSSION
More on Interpreting a Confidence Interval

D8. Explain why it is incorrect to say the probability is 0.95 that the population mean is in the confidence interval that you just constructed.

D9. "The capture rate equals the confidence level." Explain why this statement depends on

    a. the randomness of the sampling

    b. the normality of the population

D10. a. Suppose you take a single random sample from a population and construct a 95% confidence interval and a 99% confidence interval for a mean. Is it possible that the population mean is inside both intervals? Inside the 95% confidence interval but not the 99% confidence interval? Inside the 99% confidence interval but not the 95% confidence interval? Inside neither?

    b. Now, suppose you take two (independent) random samples from your population and construct a 95% confidence interval from the first sample and a 99% confidence interval from the second sample. Is it possible that the population mean is inside both intervals? Inside the 95% confidence interval but not the 99% confidence interval? Inside the 99% confidence interval but not the 95% confidence interval? Inside neither?

Margin of Error

The quantity

$$E = t^* \frac{s}{\sqrt{n}}$$

> The margin of error is half the width of the confidence interval.

is called the **margin of error**. It is half the width of the confidence interval. When your samples are random, larger samples provide more information than do smaller ones. So, in general, as the sample size gets larger, the margin of error gets smaller.

Example 10.4	**Margin of Error for the Mean Atomic Weight of Silver**

Find the margin of error for the mean atomic weight of silver in Example 10.3 on page 473.

Solution
The confidence interval for the mean atomic weight was

$$\bar{x} \pm t^* \frac{s}{\sqrt{n}} = 1363.54 \pm 2.069 \cdot \frac{169.017}{\sqrt{24}}$$

$$\approx 1363.54 \pm 71.38$$

So the margin of error is 71.38.

Summary 10.1: A Confidence Interval for a Mean

In this section, you have learned how to construct a confidence interval for a population mean, μ. A confidence interval for μ is given by

$$\bar{x} \pm t^* \frac{s}{\sqrt{n}}$$

where $\bar{x}$ is the sample mean, s is the sample standard deviation, and n is the sample size. Because you are using s to estimate σ, you must use values of t^* rather than values of z^*. Find the value of t^* in Table B on page 761 or with technology, using $df = n - 1$. For a given confidence level and sample size, the value of t^* is slightly larger than the corresponding value of z^*, but this difference becomes smaller with larger sample sizes. You must check that you have a random sample from a relatively large population (or a random assignment of treatments to units) and that the sample came from a distribution that is approximately normal or, if not, that the sample is large enough.

Here is how to interpret a 95% confidence interval:

- You are 95% confident that the population mean, μ, is in this confidence interval.

- This method of constructing a 95% confidence interval ensures that the chance of getting a value of $\bar{x}$ whose interval captures μ is 95%. This is called the *capture rate* or *confidence level*.

Practice

The Effect of Estimating σ

P1. *Aldrin in the Wolf River.* Aldrin is a highly toxic organic compound that can cause various cancers and birth defects. Ten samples taken from Tennessee's Wolf River downstream from a toxic waste site once used by the pesticide industry gave these aldrin concentrations, in nanograms per liter:

$$5.17 \quad 6.17 \quad 6.26 \quad 4.26 \quad 3.17$$
$$3.76 \quad 4.76 \quad 4.90 \quad 6.57 \quad 5.17$$

Wolf River, Tennessee.

a. Calculate a confidence interval of the form $\bar{x} \pm 1.96 \cdot s/\sqrt{n}$, using s as an estimate of σ.

b. In general, how do you expect this estimate—just substituting s for σ—to affect the center of a confidence interval? Will it tend to make the interval too wide or too narrow? Will it make the capture rate larger than 95% or smaller than 95%?

How to Adjust for Estimating σ

P2. *Using a t-table.* Use Table B on page 761 to find the value of t^* to use for each of these situations.

a. 95% confidence interval based on a sample of size 10

b. 95% confidence interval based on a sample of size 7

c. 99% confidence interval based on a sample of size 12

d. 90% confidence interval with $n = 3$

e. 99% confidence interval with $n = 44$

f. 99% confidence interval with $n = 82$

P3. For each situation, construct a 95% confidence interval for the unknown mean.

a. $n = 4 \quad \bar{x} = 27 \quad s = 12$

b. $n = 9 \quad \bar{x} = 6 \quad s = 3$

c. $n = 16 \quad \bar{x} = 9 \quad s = 48$

P4. Using the data in P1, construct a 95% confidence interval for the mean level of aldrin. Compare the width of this confidence interval to your interval in P1.

Constructing a Confidence Interval for a Mean

P5. An article in the *Journal of the American Medical Association* included the body temperatures of 122 men. (The data in Example 10.2 on page 472 were a random sample taken from this larger group of men.) The summary statistics for the entire sample of men's temperatures were $\bar{x} = 98.1$, $s = 0.73$, and $n = 122$. [Source: P. A. Mackowiak et al., "A Critical Appraisal of 98.6 Degree F, the Upper Limit of the Normal Body Temperature, and Other Legacies of Carl Reinhold August Wunderlich," *Journal of the American Medical Association*, Vol. 268 (September 1992), pp. 1578–1580.]

a. Construct a 95% confidence interval for the mean of the population from which these data were selected. You can assume that the conditions have been met.

b. Write an interpretation of this confidence interval.

c. Is 98.6°F in this interval? What can you conclude?

P6. If you are lost in the woods and do not have clear directional markers, will you tend to walk in a circle? To study this question, some students recruited 30 volunteers to attempt to walk the length of a football field while blindfolded. Each volunteer began at the middle of one goal line and was asked to walk to the opposite goal line, a distance of 100 yards. None of them made it. Display 10.7 shows the number of yards they walked before they crossed a sideline.

a. Are the conditions met for finding a confidence interval estimate of the mean distance walked before crossing a sideline? Explain any reservations you have.

b. Regardless of your answer to part a, construct a 99% confidence interval for the mean distance walked before crossing a sideline.

c. Interpret your interval in part b.

Yards Walked Before Walker Went Out of Bounds	Yards Walked Before Walker Went Out of Bounds
35	59
37	60
37	60
38	61
40	65
42	68
42	70
48	70
49	71
50	73
52	74
52	75
55	80
56	81
56	95

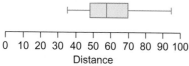

0 10 20 30 40 50 60 70 80 90 100
Distance

Display 10.7 Number of yards walked by blind-folded volunteers before going out of bounds. [Source: Andrea Axtell, "When Direction Vanishes: Walking Straight or in Circles," *STATS*, Vol. 39 (2004), pp. 18–21.]

P7. The data in Display 10.8 are from a survey taken in an introductory statistics class during the first week of the semester. These data are the number of hours of study per week reported by each of the 61 students. Because of the way students register for class, these students can be considered a random sample of all students taking this course.

a. You were told that the sample can be considered a random selection from this large population of students. Do the data look as if they reasonably could have come from a normal distribution?

b. Regardless of your answer to part a, estimate the mean study hours per week for all the students taking this course, with 90% confidence.

c. Explain the meaning of your confidence interval.

	n	MEAN	MEDIAN	STDEV
Study Hours	61	10.26	10	6.22

Stem-and-leaf of Study

N = 61   Leaf Unit = 1.0

```
        0   2222333
        0   444555555
        0   666677777
        0   8888
        1   0000000011
        1   222333
        1   5555555
        1
        1   88
        2   0000
        2   3
        2   5
        2
        2
        3   0
```

1 | 2 = 1.2 h/week

Display 10.8 Number of hours of study per week.

More on Interpreting a Confidence Interval

P8. Refer to the 95% confidence interval you constructed for the aldrin data in P4. Which of statements A–E are correct interpretations of that interval?

A. If you take one more measurement from the Wolf River, you are 95% confident that this measurement will fall in the confidence interval.

B. If you take ten more measurements from the Wolf River, you are 95% confident that the sample mean of the ten measurements will fall in the confidence interval.

C. There is a 95% chance that the mean aldrin level of the Wolf River falls in the confidence interval.

D. You are 95% confident that the mean aldrin level of the Wolf River falls in the confidence interval.

E. You are 95% confident that the sample mean falls in the confidence interval.

P9. A 95% confidence interval for a population mean is calculated for a random sample of weights, and the resulting confidence interval is from 42 to 48 lb. For each statement, indicate whether it is a true or false interpretation of the confidence interval.

a. 95% of the weights in the population are between 42 and 48 lb.

b. 95% of the weights in the sample are between 42 and 48 lb.

c. The probability that the interval includes the population mean, μ, is 95%.

d. The sample mean, $\bar{x}$, might not be in the confidence interval.

e. If 200 confidence intervals were generated using the same process, about ten of these confidence intervals would not include the population mean.

Margin of Error

P10. Refer to the confidence intervals for the mean body temperature of men and women in Example 10.2 on page 472.

a. Find the margins of error for the men and for the women.

b. Which margin of error is larger? Why is it larger?

Exercises

E1. The National Survey of America's College Students (NSACS) examines the literacy of U.S. college students. (See E3 on page 382 in Chapter 8 for more details.)

In a sample of about 1000 students from 4-year colleges, the mean score on quantitative literacy (QL) questions was 330 with a standard deviation of 111. In a sample of about 800 students from two-year colleges, the mean score was 310 with a standard deviation of 79. Answer the following assuming that the samples were randomly and independently selected. [Source: *www.air.org/news/documents/the%20literacy%20of%20Americas%20College%20Students_final%20report.pdf*.]

a. Find a 95% confidence interval for the mean QL score of all four-year college students.

b. Find a 95% confidence interval for the mean QL score of all two-year college students.

c. Compare the confidence intervals in parts a and b. Does it look like there is a difference in the two population means? (You will see a better way to answer this question in Section.11.1.)

E2. Exposure to high levels of solar UV radiation can damage or destroy living cells. Thus, various types of filters (including sunglasses) are designed to reduce the flow of UV radiation into areas of sensitivity. An industrial process involves a machine that measures transmittance of UV radiation through a filter. The operator of the machine suspects that it has a tendency to slide off target. To test this, the amount of radiation was held steady and the measurements made by the machine over 50 equally spaced time periods were recorded (see Display 10.9). One way to analyze these measurements is to compare those from the first 25 time periods (first half) with those from the second 25 time periods (second half).

a. Construct parallel boxplots of the 25 measurements from the first half and from the second half. Does it look like the measures are sliding off target? Does it look like these data are suitable for inference on the means?

b. Assuming that these measurements can be regarded as random samples of all measurements that could have been collected in the first half and second half, construct a confidence interval estimate of the plausible population means for the first half. Do the same for second half.

Time Period	Transmittance	Time Period	Transmittance
1	18	26	14
2	17	27	13
3	18	28	14
4	19	29	15
5	18	30	14
6	17	31	15
7	15	32	16
8	14	33	15
9	15	34	16
10	15	35	19
11	17	36	20
12	18	37	20
13	18	38	21
14	19	39	22
15	19	40	23
16	21	41	24
17	20	42	25
18	16	43	27
19	14	44	26
20	13	45	26
21	13	46	26
22	15	47	27
23	15	48	26
24	16	49	25
25	15	50	24

Display 10.9 Transmittance of UV radiation (rescaled). [Source: *www.itl.nist.gov/div898/handbook/index.htm*.]

c. Do the confidence intervals from part b appear to support the operator's suspicion that the machine is sliding off target? (You will see a better way to answer this question in Chapter 11.)

E3. To avoid any further hill climbing, Jack and Jill have opened a water-bottling factory. The distribution of the number of ounces of water in the bottles is approximately normal. The mean, μ, is supposed to be 16 oz, but the water-filling machine slips away from that amount occasionally and has to be readjusted. Jack and Jill take a random sample of bottles from today's production and weigh the water in each. The weights (in ounces) are

| 15.91 | 16.08 | 16.08 | 15.94 | 16.02 |
| 15.94 | 15.96 | 16.03 | 15.82 | 15.96 |

Should Jack and Jill readjust the machine? Use a statistical argument to support your advice.

E4. *Velocity of light.* In 1879, A. A. Michelson made 100 determinations of the velocity of light in air. These measurements, in kilometers per second, coded by subtracting 299,000 from each, are shown in the following histogram. One way to combine the measurements into an estimate of the "true" velocity of light is to think of them as a random sample from a population that has the true value as its mean. The 100 measurements have a mean of 852.2 and a standard deviation of 78.9. What are the plausible values of the true velocity of light? Estimate them in a 90% confidence interval. [Source: *http://lib.stat.cmu.edu/DASL/Datafiles/Michelson.html.*]

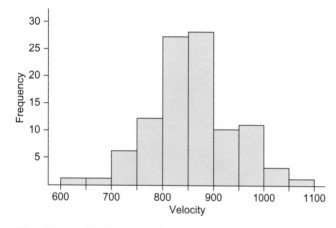

E5. Display 10.10 gives the hot dog prices for three randomly selected National Basketball Association (NBA) teams.

Team	Hot Dog Price (dollars)
Phoenix	3.00
Milwaukee	3.50
Detroit	3.00

Display 10.10 A sample of hot dog prices. [Source: *www.teammarketing.com*, March 2008.]

a. From this very small sample, is there any reason to suspect that the conditions are not met for constructing a confidence interval for the mean hot dog price for all NBA teams?

b. Regardless of your answer in part a, construct a 95% confidence interval.

c. The mean hot dog price for all NBA teams is $3.97. Is that value in your confidence interval? If not, list two reasons why this could have happened. The plots of all 30 prices in Display 10.11 might help you. (The outlier is San Francisco.)

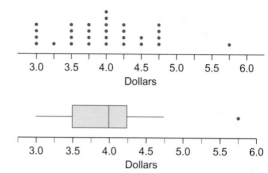

Display 10.11 Dot plot and boxplot of NBA hot dog prices.

d. There are only 30 teams in the NBA. If ten teams had been sampled instead of three, would the inferential methods of this section have worked satisfactorily? If not, where do the potential problems lie?

E6. Refer to E4. When measuring the speed of light, Michelson actually took the measurements in five groups of 20, each successive group coming at a later time. Display 10.12 shows a plot and summary statistics for the first and fourth groups of measurements.

a. With 95% confidence estimate the true velocity of light using each group.

b. It is now known the Michelson's measurements were a little too high due to bias in his measurement process. Does he seem to be improving over time? Explain your reasoning.

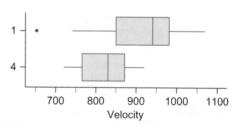

Group	Mean	Standard Deviation
1	909	105
4	820	60

Display 10.12 Velocity of light measurements in two groups.

E7. A statistics class at the University of Wisconsin-Stevens Point decided to estimate the average mass of a small bag of french fries sold at McDonald's. They bought 32 bags during two different time periods on two consecutive days at the same McDonald's and weighed the fries. The data and Minitab output are given in Display 10.13.

a. Are the conditions met for constructing a 95% confidence interval for the mean?

b. Regardless of your answer in part a, give and interpret a 95% confidence interval.

c. The "target value" set by McDonald's for the mass of a small order of fries was 74 g. Is there statistically significant evidence that this McDonald's franchise wasn't meeting that target?

Mass of Small Bag of Fries (g)	Mass of Small Bag of Fries (g)
62.8	92.7
67.2	71
103.6	60.9
71.1	63
64.5	64.7
67.6	83.1
76.5	77.1
84.2	79.6
82.9	77.4
60.2	82.7
60.6	58.7
76.4	72
70.9	76.8
69.2	84.3
62.7	77.6
69.4	87.6

Number of Fries in Small Bag	Number of Fries in Small Bag
30	58
43	53
54	37
70	42
40	39
45	55
46	56
55	44
41	56
40	53
36	36
50	52
52	56
42	66
46	51
43	53

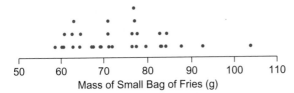

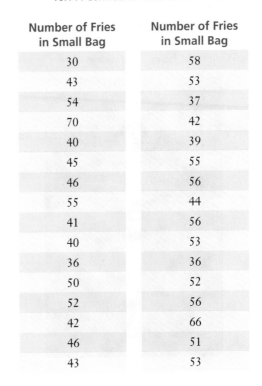

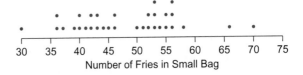

One-Sample T: Mass

Variable	N	Mean	StDev	SE Mean	95% CI
Mass	32	73.72	10.52	1.86	(69.93, 77.51)

Display 10.13 Mass of small bags of McDonald's fries. [Source: Nathan Wetzel, "McDonald's French Fries. Would You Like Small or Large Fries?" *STATS*, Vol. 43 (Spring 2005), pp. 12–14.]

One-Sample T: Number

Variable	N	Mean	StDev	SE Mean	95% CI
Number	32	48.13	8.98	1.59	(44.89, 51.36)

Display 10.14 Number of fries in small bags of McDonald's fries.

E8. The statistics class from E7 decided to estimate the average number of fries in a small bag of McDonald's french fries. They bought 32 bags during two different time periods on two consecutive days at the same McDonald's and counted the fries. The data and Minitab output are given in Display 10.14.

    a. Are the conditions met for constructing a 95% confidence interval for the mean?

    b. Regardless of your answer in part a, give and interpret a 95% confidence interval.

E9. A page was randomly selected from each of 30 brochures for cancer patients published by the American Cancer Society and the National Cancer Institute. The pages were judged for readability, using standard readability tests. The reading grade levels of the 30 pages selected are given in Display 10.15.

    a. Suppose you want to estimate the mean reading level of cancer brochures. Do you have the required random sample?

Reading Grade	Number of Pages
6	3
7	3
8	8
9	4
10	1
11	1
12	4
13	2
14	1
15	2
16	1

Display 10.15 Reading level of pages of cancer brochures. [Source: Thomas H. Short et al., "Readability of Educational Materials for Patients with Cancer," *Journal of Statistics Education*, Vol. 3, 2 (July 1995).]

b. Using the data in Display 10.15, construct a 90% confidence interval.

c. What is the best interpretation of your interval?

E10. Your grade in your literature class isn't all that you had hoped it would be. You took five exams, each consisting of one essay question worth 100 points. Your scores were 52, 63, 72, 41, and 73. Your instructor averages these scores, gets 60.2, and says you have earned a D. You are, however, doing very well in your statistics class and think that what you have learned in this section can convince your literature instructor to give you a C, which is a mean score of between 70 and 79. You remember that your instructor told the class early in the semester that the questions on the exams would be only a sample of the questions that she could ask. Write a paragraph to your instructor giving your argument for a C. You can assume that the instructor knows something about statistics.

E11. Refer to the quantitative literacy data in E1 on page 479.

a. Find the margin of error for the four-year college estimate. Do the same for two-year college estimate.

b. Which margin of error is larger, and why is this so?

E12. Which confidence interval is widest in each of parts a–c? (You don't need to do any computing.)

a. 95% interval with $n = 4$ and $s = 10$;
90% interval with $n = 5$ and $s = 9$;
99% interval with $n = 4$ and $s = 10$

b. 95% interval with $n = 3$ and $s = 10$;
95% interval with $n = 4$ and $s = 10$;
95% interval with $n = 5$ and $s = 10$

c. 90% interval with $n = 10$ and $s = 5$;
95% interval with $n = 10$ and $s = 5$;
95% interval with $n = 10$ and $s = 10$

E13. A large population of measurements has unknown mean and standard deviation. Two different random samples of 50 measurements are taken from the population. A 95% confidence interval for the population mean is constructed for each sample. Which statement would most likely be true of these two confidence means?

A. They would have identical values for the lower and upper limits of the confidence interval.

B. They would have the same margin of error.

C. The confidence intervals would have the same center but different widths.

D. None of the above is true.

E14. What happens to the margin of error if all remains the same except that

a. the standard deviation of the sample is increased?

b. the sample size is increased?

c. the confidence level is increased?

E15. Suppose a large random sample, with $n = 100$, is going to be taken from a population of 6-year-old girls. A 90% confidence interval will be constructed to estimate the population mean height. A smaller random sample, with $n = 50$, will also be taken from the same population of 6-year-old girls, and a 99% confidence interval will be constructed to estimate the population mean height. Which confidence interval has a better chance of capturing the population's mean height? Explain your reasoning.

A. The 90% confidence interval based on a sample of 100 heights has a better chance.

B. The 99% confidence interval based on a sample of 50 heights has a better chance.

C. Both methods have an equal chance.

D. You can't determine which method will have a better chance.

E16. Is the capture rate affected by

a. changes in the sample size?

b. the size of the population standard deviation?

c. the confidence level?

E17. When you do not know the value of σ and have to estimate it with s, the capture rate will be too small if you use z as the multiplier in your confidence interval. Complete this exercise to see part of the reason why.

a. Display 10.16 displays and summarizes 100 sample standard deviations, s, for random samples of size 4 taken from a normal distribution with mean $\mu = 511$ and standard deviation $\sigma = 112$. What is the shape of this distribution? On average, how well does s approximate σ?

b. Which happened more often, a value of s smaller than σ or a value of s larger than σ?

c. How does your answer to part b explain why the capture rate tends to be too small if you use z as the multiplier in your confidence interval?

E18. Suppose you know σ, the standard deviation of the population. (In practice, σ is almost never known.) These steps give a justification for the 95% confidence interval for the population mean.

Step 1: If the population is normally distributed or the sample size is large enough, 95% of all sample means fall within 1.96 standard errors of the population mean.

Step 2: Writing this algebraically, for 95% of all sample means

$$\mu - 1.96\ SE < \bar{x} < \mu + 1.96\ SE$$

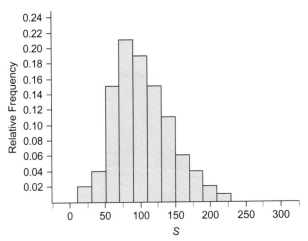

Mean	102.8
Minimum	15.3
First quartile	72.9
Median	98.5
Third quartile	127.3
Maximum	211.3

Display 10.16 Simulated sampling distribution of s for samples of size 4 from a normal distribution with $\sigma = 112$.

(This interval contains the values of $\bar{x}$ that are reasonably likely.)

Step 3: You can write this in another, algebraically equivalent, way. For 95% of all sample means,

$$\bar{x} - 1.96 \, SE < \mu < \bar{x} + 1.96 \, SE$$

(This interval contains all the values of μ that are plausible as the population mean. In other words, each population mean in the interval $\bar{x} \pm 1.96 \cdot SE$ plausibly could produce the observed value of $\bar{x}$.)

Step 4: The inequality in step 3 can be rewritten using the fact that the standard error of the sampling distribution of the sample mean is $\sigma_{\bar{x}}$ or $\sigma/\sqrt{n}$.

a. Show algebraically why the inequality in step 3 is equivalent to that in step 2.

b. Rewrite the inequality based on the information in step 4.

c. Explain why the inequality in part b is equivalent to a 95% confidence interval for the population mean.

10.2 ▶ A Significance Test for a Mean: Interpreting a *P*-Value

Significance tests all rely on the same basic idea: Be suspicious of any model that assigns low probability to what actually happened. In this section you will learn when your data tell you to be suspicious of any claimed value of the population mean.

The basic procedure for testing a hypothesis is always the same: Compute a test statistic from the data and compare it with a known distribution. As in Sections 8.2 and 8.3, you will state the null hypothesis and alternative hypothesis (two-sided or one-sided), check conditions for the test, find the *P*-value to assess the strength of the evidence against the null hypothesis, decide whether to reject the null hypothesis, and write a conclusion in context. This section will concentrate on calculating and interpreting *P*-values; the next section will concentrate on a decision-making approach to the testing process.

What Makes Strong Evidence?

Example 10.5

The four data sets plotted in Display 10.17 come from four different experiments. For each experiment, 50 people were paired based on similar pulse rates. Within each pair, one person was randomly assigned to stand and the other to sit. Then the difference in the two pulse rates, *standing – sitting*, was computed. The 25 differences for each experiment are plotted in Display 10.17.

a. To test whether pulse rates tend to be higher standing than sitting, what are the appropriate hypotheses?

b. For which experiment is the evidence strongest that pulse rates tend to be higher when people are standing than when they are sitting? For which experiment is the evidence weakest?

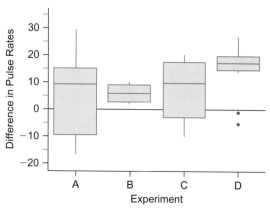

Display 10.17 Differences in pulse rates, *standing – sitting*.

Solution

a. The null hypothesis is $H_0: \mu = 0$, where μ is the true mean difference in pulse rates. The alternative hypothesis is $H_a: \mu > 0$.

b. Still being guided by shape, center, and spread, the strongest evidence for a positive difference is from Experiment B. The differences have a nearly symmetric distribution centered above 0 and with small variability. The weakest evidence is from Experiment A because, although the center is above 0, the distribution has large variability, compared with the others.

The hypotheses in this chapter will be written in terms of means rather than proportions, but the forms of the hypotheses are the same as in Chapter 8.

The Null Hypothesis and the Alternative Hypothesis

If you are doing a test of significance for a mean and wish to compare the results from a sample to a standard value, μ_0, begin by writing the **null hypothesis:**

H_0: The mean, μ, of the population is equal to the hypothesized, or standard, value, μ_0.

The **alternative hypothesis** has three forms, depending on what you want to establish:

H_a: $\mu \neq \mu_0$ The mean, μ, of the population is not equal to the hypothesized, or standard, value, μ_0.

H_a: $\mu > \mu_0$ The mean, μ, of the population is greater than the hypothesized, or standard, value, μ_0.

H_a: $\mu < \mu_0$ The mean, μ, of the population is less than the hypothesized, or standard, value, μ_0.

The first form defines a **two-sided test**. The second and third define **one-sided tests.**

The Test Statistic

In Section 8.3, you computed this test statistic to test whether a hypothesized population proportion p_0 is consistent with the sample proportion, $\hat{p}$:

$$z = \frac{\hat{p} - p_0}{\sqrt{\dfrac{p_0(1 - p_0)}{n}}} = \frac{statistic - parameter}{standard\ error}$$

The size of z tells you how many standard errors your sample proportion $\hat{p}$ lies from the hypothesized proportion, p_0. Note that the standard error need not be estimated using $\hat{p}$ because p_0 is stated in the null hypothesis (H_0: $p = p_0$).

What will change if you are testing a mean, H_0: $\mu = \mu_0$? It would be nice if you could use the z-statistic of Section 7.2 to measure how far a sample mean is from a hypothesized mean μ_0:

$$z = \frac{\bar{x} - \mu_0}{\sigma/\sqrt{n}}$$

The catch is that you don't know σ, the standard deviation of the population.

Perhaps you can guess what's coming next. You do the same thing you did in Section 10.1. If you don't know the value of σ, you substitute s for σ and t for z. The test statistic becomes

$$t = \frac{\bar{x} - \mu_0}{s/\sqrt{n}}$$

Mean Weight of Pennies	**Example 10.6**

The weights of newly minted U.S. pennies are approximately normally distributed and are targeted to have a mean weight of 3.11g. Samples of pennies are periodically selected from the production process and weighed to make sure the process average has not shifted from this target. The random sample of nine pennies shown in Display 10.18 was taken from a production line. Their mean weight is 3.16 g with standard deviation 0.065g. What is the value of the test statistic for a test to determine whether the mean has moved away from the target mean?

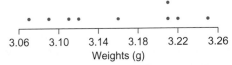

Display 10.18 Weights of newly minted pennies.

Solution

$$t = \frac{\bar{x} - \mu_0}{\dfrac{s}{\sqrt{n}}} = \frac{3.16 - 3.11}{\dfrac{0.065}{\sqrt{9}}} \approx 2.31$$

Constructing a *t*-Distribution

Before you can start interpreting the new test statistic t, you need to know how it behaves when the null hypothesis is true. If the value of t from your sample fits right into the middle of the distribution of t constructed by assuming the null hypothesis is true, you have no evidence against the null hypothesis. If the value of t from your sample is way out in the tail of the t-distribution, you have evidence against the null hypothesis. In other words, if the value of t from your sample would be a rare event

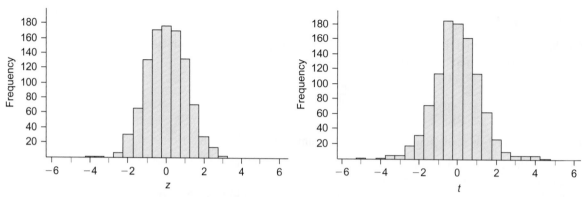

Display 10.19 *Left:* 1000 values of *z* computed from random samples taken from a normally distributed population with mean 100 and standard deviation 15. *Right:* 1000 values of *t* computed from random samples from the same population.

given the hypothesized mean, be suspicious of that mean. If the value of *t* from your sample is a reasonably likely event, the hypothesized mean is plausible.

The following simulation, based on a distribution of IQ scores, will show you how the distribution of *t* is different from that of *z*. Some IQ tests are constructed so that scores are normally distributed with mean $\mu = 100$ and standard deviation $\sigma = 15$. Suppose you repeatedly take random samples of size 9 from this distribution and compute the sample mean, $\bar{x}$. According to the explanation in Section 7.2, the distribution

$$z = \frac{\bar{x} - \mu}{\sigma/\sqrt{n}} = \frac{\bar{x} - 100}{15/\sqrt{9}}$$

will be normal even with this small sample size (because the population is normally distributed). The histogram on the left in Display 10.19 shows 1000 values of *z*, each computed from a random sample of size 9 from the population of IQs.

If you do not know σ (the usual situation), you have to use *s* to estimate it. You would then compute

$$z = \frac{\bar{x} - \mu}{s/\sqrt{n}} = \frac{\bar{x} - 100}{s/\sqrt{9}}$$

The histogram on the right in Display 10.19 shows 1000 values of *t*, each computed from a random sample of size 9 from the population of IQs.

Note that the values of *t* are more spread out than are the values of *z*. That makes sense. Not only does $\bar{x}$ vary from sample to sample, but *s* also varies from sample to sample. And, as you saw in Section 10.1, *s* tends to be smaller than σ more often than it tends to be larger than σ, making *t* larger in absolute value than the corresponding *z* more often than it is smaller. Thus, when you compute *t*, you tend to get more values in the tails of the distribution than when you compute *z*.

The box summarizes the most important facts about *t*-distributions. The values of *t* in Table B on page 761 that you used to construct confidence intervals in Section 10.1 come from these distributions.

The *t*-Distributions

Suppose you draw random samples of size *n* from a normally distributed population with mean μ and unknown standard deviation. The distribution of the values of

$$t = \frac{\bar{x} - \mu}{s/\sqrt{n}}$$

is called a **t-distribution**. There is a different *t*-distribution for each degree of freedom, $df = n - 1$, where *n* is the sample size.

A *t*-distribution is mound-shaped, with mean 0 and a spread that depends on the value of *df*. The greater the *df*, the smaller the spread. The spread of any *t*-distribution is greater than that of the standard normal distribution. Display 10.20 shows the *t*-distribution for $df = 4$ plotted on the same graph as the standard normal distribution.

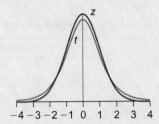

Display 10.20   The *t*-distribution for $df = 4$, and the standard normal distribution.

As the number of degrees of freedom gets larger, the *t*-distribution more closely approximates a normal distribution. In Display 10.21, the *t*-distribution with $df = 9$ is very much like the standard normal distribution.

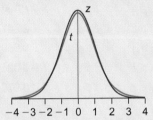

Display 10.21   The *t*-distribution for $df = 9$, and the standard normal distribution.

You can explore graphs of *t*-distributions using your calculator or a computer.

P-Values

Now that you know the distribution of *t* when the null hypothesis is true, the next step is to locate where the value of *t* computed from your sample lies on this distribution. The standard way to do that is to report a *P*-value, sometimes called an *observed significance level*, just as in Section 8.2. The next example will show you how to find the *P*-value and how to interpret it.

| **The *P*-Value for the Mean Weight of Pennies** | **Example 10.7** |

As you saw in Example 10.6 on page 485, the test statistic to see whether the mean weight of newly minted pennies has moved away from the target, 3.11g, is

$$t = \frac{\bar{x} - \mu_0}{\frac{s}{\sqrt{n}}} = \frac{3.16 - 3.11}{\frac{0.065}{\sqrt{9}}} \approx 2.31$$

Find and interpret the *P*-value that corresponds to this value of *t*.

Solution

This test requires a two-sided alternative because the goal of such tests is to see if the process has shifted from the target in other direction. The *P*-value for a two-sided test

Computation.

is the area under the *t*-distribution with *df* = 9−1, or 8, that lies above *t* = 2.31 and below *t* = −2.31, as shown in Display 10.22. It is approximately 2(0.0248) = 0.0496.

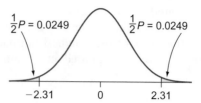

Display 10.22   *P*-value for *t* = 2.31 on a *t*-distribution with *df* = 8.

You can get the *P*-value from your calculator or software, as shown in Display 10.23.

JMP

Test Mean = value

Hypothesized Value	3.11
Actual Estimate	3.16
df	8
Std Dev	0.065

	t Test
Test Statistic	2.3077
Prob > |*t*|	0.0499
Prob > *t*	0.0249
Prob < *t*	0.9751

Minitab
One-Sample T: Mass
Test of mu = 3.11 vs not = 3.11

Variable	N	Mean	St Dev	SE Mean	95% CI	T	P
Mass	9	3.1600	0.0650	0.0217	(3.1100, 3.2100)	2.31	0.050

Display 10.23   *P*-values from JMP and Minitab.

Interpretation

If it is true that the mean has not slipped off the target, 3.11 g, there is only a 0.0499 chance of getting an absolute value of *t* as large as or even larger than the one from this sample (that is, *t* ≥ 2.31 or *t* ≤ −2.31). The small *P*-value, 0.0499, indicates that the sample is somewhat inconsistent with the null hypothesis. It looks as if the population mean has moved away from the target.

If you do not have access to technology that finds *P*-values, you can get an estimate of the *P*-value from Table B on page 761. The next example shows you how.

Example 10.8

Approximating the *P*-value from Table B

Suppose you have a two-sided test with *df* = 9 and *t* = 3.98 (see Display 10.24). Approximate the *P*-value for this test using Table B and using technology.

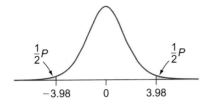

Display 10.24 A test statistic, *t*, of −3.98.

Solution

Go to Table B on page 761 and find the row with 9 degrees of freedom. Go across the row until you find the absolute value of your value of *t*, which probably will lie between two of the values in the table. The partial *t*-table here shows the two values of *t* that lie on each side of |*t*| = 3.98.

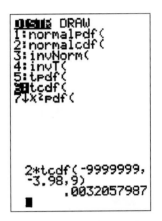

	Tail Probability *p*	
df	.0025	0.001
9	3.690	4.297

The "tail probability" gives the area that lies in each tail. Because you have a two-sided test, you will double the tail probability. If you must use Table B to find a *P*-value, all you can say is that the *P*-value is between 2(0.001), or 0.002, and 2(0.0025), or 0.005. That is, $0.002 < P\text{-value} < 0.005$. To find the *P*-value more precisely, use your calculator or computer to get approximately 0.0032.

As always, the ***P*-value** is a probability, computed under the assumption that the null hypothesis is true. It tells the chance of seeing a result from a sample that is as extreme as or even more extreme than the one computed from your data *when the hypothesized mean is correct*.

> The smaller the *P*-value, the stronger the evidence against the null hypothesis.

Using a *P*-Value in a Test of a Mean

The *P*-value is the probability of getting a random sample, from a distribution with the mean given in the null hypothesis, that has a value of *t* that is as extreme as or even more extreme than the value of *t* computed from the sample you observed.

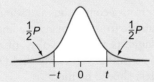

When the *P*-value is close to 0, you have convincing evidence against the null hypothesis. When the *P*-value is large, the result from the sample is consistent with the hypothesized mean and you don't have evidence against the null hypothesis.

One-Sided Tests

The penny weight example had a two-sided alternative hypothesis: The true mean, μ, is not equal to the standard, μ_0. But there are two other possible alternatives: μ might be less than the standard, or μ might be greater than the standard. In real applications,

> Sometimes it makes sense to consider alternatives that depart from the standard in only one direction.

you attempt to use the context to, first, decide if the alternative hypothesis should be one- or two-sided and, if one-sided, to rule out one of the two directions as meaningless, impossible, uninteresting, or irrelevant.

In *Martin v. Westvaco* (Chapter 1), a statistical analysis compared the ages of the workers who were laid off with the ages of workers in the population of employees working for Westvaco at the time of the layoff. An average age for the fired workers that was greater than the population mean would tend to support a claim of age discrimination. On the other hand, the opposite inequality is not relevant: An average age for the fired workers that was less than the overall average would not be evidence of age discrimination, because younger workers aren't protected under the law.

When the context tells you to use a one-sided alternative hypothesis, your *P*-value is the area on only one side of the *t*-distribution. Such *P*-values are called one-sided, or one-tailed, *P*-values, and the corresponding test of significance is called a one-sided or one-tailed test, just as in Chapter 8.

Example 10.9	**Blood Glucose**

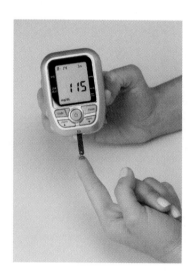

One of the authors is requested by the family physician to maintain morning blood glucose levels no higher than 110 mg/dL. Display 10.25 shows the last 25 blood glucose readings from this person's self-tests, taken about every three days. Is the request being met, on average?

119	104	113	114	106
120	108	105	117	109
118	106	109	115	106
114	93	119	123	117
109	105	113	115	101

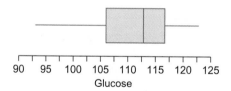

Display 10.25 Blood glucose readings.

Solution
The readings taken periodically from a continuous and somewhat random process (living) can be regarded as a random sample from the many readings that could have been taken over the 75 days. The plot suggests that normality of the population is not a bad assumption, because there are no outliers and only slight skewness. With 25 measurements, the conditions for a *t*-test should be met.

The request of keeping the mean at or below 110 (the null hypothesis) will be violated if the mean exceeds 110, so the alternative of importance is that the population mean might exceed 110. This results in a one-sided test with test statistic

$$t = \frac{\bar{x} - \mu_0}{\frac{s}{\sqrt{n}}} = \frac{111.12 - 110}{\frac{6.99}{\sqrt{25}}} \approx 0.8011$$

With 24 degrees of freedom, the *P*-value is between 0.20 and 0.25 from Table B, or about 0.22 from a computer. With this large a *P*-value, a value of 110 is a plausible population mean for the blood glucose levels. The physician should be pleased. Display 10.26 shows the region of the *P*-value under a *t*-distribution with 24 degrees of freedom.

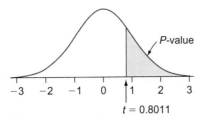

Display 10.26 *P*-value for the blood glucose example.

DISCUSSION
P-Values

D11. Study the formula for the test statistic when testing a hypothesis about a population mean, and think about the relationship of the test statistic to the *P*-value.

    a. What happens to the *P*-value if the sample standard deviation increases but everything else remains the same?

    b. What happens to the *P*-value if the sample size increases but everything else remains the same?

D12. For a given value of the test statistic and specified degrees of freedom, how does the *P*-value for a one-sided test compare to the *P*-value for a two-sided test? Argue to support the claim that it is always better to use a one-sided test whenever the question of interest allows it.

Summary 10.2: A Significance Test for a Mean: Interpreting a *P*-Value

A claim is made about the value of a population mean, μ, and you want to conduct a test of the significance of this claim by taking a random sample from the population. First, write the hypotheses:

- The null hypothesis gives value, μ_0, that you are hypothesizing is the true population mean, μ.

- The alternative (research) hypothesis, which can be one- or two-sided, conjectures how the true mean differs from the standard given in the null hypothesis.

You must check that you have a random sample from the population of interest and that it is reasonable to assume either that the sample came from a distribution that is approximately normal or that the sample size is large enough to invoke the Central Limit Theorem. If you have a random sample that is more than one-tenth the population size, then there are better formulas to use than those in this section.

Just as in constructing a confidence interval for a mean, when you use s to estimate σ, you must use t rather than z in doing a significance test for a mean. You can use the test statistic

$$t = \frac{\bar{x} - \mu_0}{s/\sqrt{n}}$$

The *P*-value gives the probability of seeing a test statistic t as extreme as or even more extreme than the one computed from the data, when the null hypothesis is true. It serves as a measure of the evidence against the null hypothesis. The smaller the *P*-value, the greater the evidence in the sample against the null hypothesis. You can get exact *P*-values from a calculator or statistical software. You can get approximate *P*-values from Table B on page 761.

Practice

The Test Statistic

P11. The thermostat in your lecture room is set at 72°F, but you think the thermostat isn't working well. On seven randomly selected days, you sit in the same seat near the thermostat and measure the temperature. Your measurements (in degrees Fahrenheit) are 71, 73, 69, 68, 69, 70, and 71. What is the test statistic for a significance test of whether the mean temperature at your seat is different from 72°F?

P12. In 2008, Pell grants and Minnesota State grants totaling $301,050,000 were awarded to 83,246 Minnesota resident undergraduates. (These grants do not have to be repaid.) You believe that the mean amount has changed since 2008. You take a random sample of 35 Minnesota college students who currently receive Pell grants and get a mean amount of $3862.14, with a standard deviation of $754.00. What is the test statistic for a test to determine whether the mean Pell grant amount for Minnesota college students has changed since 2008? [Source: *www.ohe.state.mn.us.*]

Constructing a *t*-Distribution

P13. One of the distributions in Display 10.27 is a *t*-distribution, and the other is the standard normal distribution. Which is which? Explain how you know.

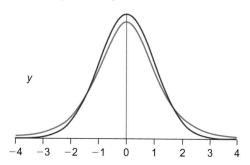

Display 10.27 A *t*-distribution and a standard normal distribution.

P14. You plan to use simulation to construct an approximate *t*-distribution with 6 degrees of freedom by taking random samples from a population of IQ scores that are normally distributed with mean, μ, 100 and standard deviation, σ, 15.

   a. Describe how you will do one run of this simulation.

   b. Produce three values of *t* using your simulation.

P-Values

P15. In P11, you computed the test statistic for testing whether the mean temperature at your seat is different from 72°F. Find and interpret the *P*-value.

P16. In P12, you computed the test statistic for testing whether the mean Pell grant amount for Minnesota college students has changed since 2008. Find and interpret the *P*-value.

One-Sided Tests

P17. The board of the Gainesville Home Builders' Association knows the selling price of all new houses sold in the city last month and believes that the mean selling price has gone up this month. To check this claim, the board obtains a random sample of the selling prices of houses sold this month.

   a. Should the alternative hypothesis be one-sided or two-sided?

   b. Tell what $\bar{x}$ and μ are for the problem.

   c. State the null and alternative hypotheses.

P18. The situation is the same as in P11. The thermostat in your classroom is set at 72°F, but this time, before you collect your data, you are convinced that the room tends to be colder. On seven randomly selected days, you measure the temperature at your seat. Your measurements are 71, 73, 69, 68, 69, 70, and 71. Test whether the mean temperature at your seat is lower than 72°F.

P19. Refer to the quantitative literacy data in E1 on page 479. To be considered at the "intermediate" level of quantitative literacy, a student must get a score of 290. Is there statistically significant evidence that the population mean for 2-year college students is larger than 290? Conduct the appropriate statistical test to answer this question, stating your conclusion in terms of the *P*-value.

P20. Refer to the quantitative literacy data in E1 on page 479. To be considered at the "proficient" level of quantitative literacy, a student must get a score of at least 350. Is there statistically significant evidence that the population mean for 4-year college students is less than at least 350? Conduct the appropriate statistical test to answer this question, stating your conclusion in terms of the *P*-value.

Exercises

E19. In E3 on page 479, Jack and Jill opened a water-bottling factory. The distribution of the number of ounces of water in the bottles is approximately normal. The mean, μ, is supposed to be 16 ounces, but the water-filling machine slips away from that amount occasionally and has to be readjusted. Jack and Jill take a random sample of bottles from today's production and weigh the water in each. The weights (in ounces) are

15.91	16.08	16.08	15.94	16.02
15.94	15.96	16.03	15.82	15.96

Should Jack and Jill readjust the machine? Conduct a test of significance, writing the conclusion in terms of the *P*-value.

E20. In E7 on page 479, a statistics class decided to check the weights of bags of small fries at a local McDonald's to see if, on average, they met the "target value" of 74 g. They bought 32 bags during two different time periods on two consecutive days and weighed the fries. The data are given in Display 10.13 on page 481. Is there evidence that this McDonald's wasn't meeting its target? Do all four steps in a test of significance, writing the conclusion in terms of the *P*-value.

E21. Fifteen students were given pieces of paper with five vertical line segments and asked to mark the midpoint of each segment. Then, using a ruler, each student measured how far each mark was from the real midpoint of each segment. The students then each averaged their five errors. If the average is positive, the student tended to place the midpoint too high. If the average is negative, the student tended to place the midpoint too low. The results are given in Display 10.28. Is there evidence of statistically significant measurement bias? That is, on average, do students tend to place the midpoints either too high or too low? Write the conclusion in terms of the *P*-value.

E22. Refer to E21. Fourteen different students were given the same instructions, but with pieces of paper with five *horizontal* line segments. The results are given in Display 10.29. Is there evidence of statistically significant measurement bias? That is, on average, do students tend to place the midpoints either too far to the right or too far to the left? Write the conclusion in terms of the *P*-value. Do the test twice, once with and once without the outlier. How do the results compare?

Average Error (cm)

1.6
1.2
1.1
1.1
1.0
0.8
0.4
0.4
0.4
0.4
0.2
0.2
0.0
0.0
−0.4

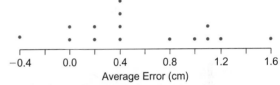

Display 10.28 Average error in estimating the location of the midpoint of vertical line segments. (A positive value indicates that the average estimate was too high.) [Source: Gretchen Davis, "Beware of Measurement Bias," *CMC ComMuniCator*, Vol. 22, 4 (June 1998), pp. 54–56.]

Average Error (cm)

−1.9
−0.4
−0.4
−0.2
−0.2
0.0
0.0
0.0
0.0
0.2
0.4
0.4
0.4
0.8

Display 10.29 Average error in estimating the location of the midpoints of horizontal line segments. (A positive value indicates that the average estimate was too far to the right.)

E23. Theory says that if you dissolve salt in water, the freezing point will be lower than it is for pure water (32°F). To test the theory, you plan to dissolve a teaspoon of salt in a bowl of water, put a thermometer into the mixture, and then put the bowl in your freezer, checking periodically so that you can observe the temperature at which the solution begins to freeze. You will repeat this ten times. (If you actually try this, you need to check often enough to be sure to remove the thermometer once ice begins to form, or you might end up with a broken thermometer.)

a. Should the alternative hypothesis be one-sided or two-sided?

b. Tell what $\bar{x}$ and μ are for the problem.

c. State the null and alternative hypotheses.

E24. At sea level, water boils at 212°F. Theory says that, at high altitude, water boils at a lower temperature than it does at sea level. You plan to test this theory by flying to the mile-high Denver airport and observing the boiling point of water there.

a. Should the alternative hypothesis be one-sided or two-sided?

b. Tell what $\bar{x}$ and μ are for the problem.

c. State the null and alternative hypotheses.

E25. Munchie's Potato Chip Company claims that the weight of the contents of a 10-oz bag of chips is normally distributed, with mean 10 oz. A consumer group, Snack Munchers for Truth (SMFT), says that the average weight is less than this; SMFT weighs the contents of 15 randomly selected bags of potato chips and gets the weights in Display 10.30. Test SMFT's claim, writing the conclusion in terms of the *P*-value.

E26. A random sample of 15 monthly rents for two-bedroom apartments was selected from a recent edition of the *Gainesville Sun*. The 15 values are symmetric, with no outliers. The mean of this sample is $636 with standard deviation $121.

Weights (oz)	Weights (oz)
9.8	9.6
9.9	9.7
10.2	10.1
9.9	10.1
9.9	9.9
9.9	9.8
10.0	9.3
9.7	

Display 10.30 Weights of bags of chips.

a. A group of students thinks the mean rent for two-bedroom apartments is at least $650. Test their claim by doing a test of significance, writing the conclusion in terms of the *P*-value.

b. A newspaper reports that the average monthly rent for two-bedroom apartments last year was $500. Is there statistical evidence of a change in mean rent from last year to this year? Test this claim by doing a test of significance, writing the conclusion in terms of the *P*-value.

E27. The heights of young women in the United States are approximately normally distributed, with mean 64.8 in. The heights of the 11 players on a recent roster of the WNBA Chicago Sky basketball team are (in inches) 72, 74, 76, 71, 76, 68, 69, 76, 71, 73, and 70. [Source: *www.wnba.com*.] Is there sufficient evidence to say that the mean height of these 11 players is so much larger than the population mean that the difference cannot reasonably be attributed to chance alone? (You might want to refer to "Inference for an Observational Study" in Section 9.3.)

E28. The heights of young men in the United States are approximately normally distributed, with mean 70.1 in. The heights of the young men in the Hamilton family are (in inches) 71, 72, 71, 72, and 70.

a. Is there sufficient evidence to say that the difference between the mean height of these family members and the population mean cannot reasonably be attributed to chance alone? (You might want to refer to "Inference for an Observational Study" in Section 9.3.)

b. Has the null hypothesis been rejected primarily because the sample size is so small, because the mean is so close to that of all young men, or because the variability is so small in the Hamilton family?

E29. Select the best way to complete the sentence: A *P*-value measures

A. the probability that the null hypothesis is true

B. the probability that the null hypothesis is false

C. the probability that an alternative hypothesis is true

D. the probability of seeing a value of t at least as extreme as the one observed, given that the null hypothesis is true

E30. Select the best way to complete the sentence: A t-distribution

A. is symmetric about zero and has less spread than a standard normal distribution

B. is symmetric about zero and has greater spread than the standard normal distribution

C. does not depend on the sample size as long as it is large

D. is named after a person famous for his involvement in the Boston Tea Party

10.3 ▶ Fixed-Level Tests

P-values are very informative about the strength of the evidence against the null hypothesis, but sometimes it is advantageous to know at what level you will reject the null hypothesis even before the data are collected. For what *P*-values will you decide to reset the penny machine? For what *P*-values will you decide to market the new drug? How much evidence will you require before replacing the thermostat? When a yes/no or stop/go decision is needed, a fixed-level test of significance, just as in Section 8.3, has three advantages:

- So that you are not influenced by the data, you decide before collecting data when you will reject the null hypothesis in favor of the alternative hypothesis.

- You can control the probability of rejecting a true null hypothesis (Type I error).

- Although not done in this book, you can calculate the probability of a Type II error for fixed alternative values of the parameter. Such power analyses depend on fixed-level testing.

Fixed-level tests usually state a **level of significance** (or **significance level** or simply level), α. If no level of significance is given, it is safe to assume that it is 5%. To decide whether to reject the null hypothesis in favor of the alternative hypothesis, you compare your *P*-value to α.

Fixed-Level Testing

- When you use fixed-level testing, you reject the null hypothesis in favor of the alternative hypothesis if your *P*-value is less than the level of significance, α.

- If your *P*-value is greater than or equal to the level of significance, you do not reject the null hypothesis.

- The significance level is equal to the probability of rejecting the null hypothesis when it is true (making a **Type I** error).

The smaller the significance level you choose, the stronger you are requiring the evidence to be in order to reject H_0. The stronger the evidence you require, the less likely you are to make a Type I error (to reject H_0 when it is true). However, the stronger the evidence you require, the more likely you are to make a **Type II** error (to fail to reject H_0 when it is false).

Fixed-Level Testing of the Mean Penny Weight

Example 10.10

State the null and alternative hypotheses for a test of significance to determine whether the mean penny weight has moved away from the target, 3.11 g (see Examples 10.6 and 10.7 on pages 485 and 487). Can you reject the null hypothesis at a level of significance, α, of 0.01? At $\alpha = 0.05$? At $\alpha = 0.10$?

Solution

The hypotheses are

$H_0: \mu = 3.11$, where μ denotes the mean weight of all pennies on this production line

$H_a: \mu \neq 3.11$

From Example 10.7, the *P*-value for this test is 0.0498.

Because the *P*-value is greater than $\alpha = 0.01$, you do not reject the null hypothesis at that level of significance. Because the *P*-value is less than $\alpha = 0.05$, you do reject the null hypothesis at that level of significance. Because the *P*-value is less than $\alpha = 0.10$, you do reject the null hypothesis at that level of significance.

Note that, in the previous example, if your calculator or software rounded the *P*-value to 0.05, you would not have rejected the null hypothesis at a significance level, α, of 0.05. That's the major drawback of fixed-level testing: It boils down all the data to the two choices *reject* or *don't reject*. To see what gets lost, think about a court trial where "not guilty" can mean anything from "This guy is so innocent he should never have been brought to trial" to "Everyone on the jury thinks the defendant did it, but the evidence presented wasn't quite strong enough so we had to let him go." In the same way, *do not reject* communicates only a small fraction of the information in the data. That's why reporting the *P*-value has become standard in modern statistical practice.

The *t*-Test

Tests of significance for means, called ***t*-tests**, have the same general structure as significance tests for proportions, although some details are a bit different.

Components of a Significance Test for a Mean

1. *Write a null and an alternative hypothesis.* The null hypothesis is that the population mean, μ, has a particular value μ_0. This is typically abbreviated

   $H_0: \mu = \mu_0$

   The alternative hypothesis can have any one of three forms:

   $H_a: \mu \neq \mu_0$ or $H_a: \mu < \mu_0$ or $H_a: \mu > \mu_0$

   Be sure to define μ in context.

2. *Check conditions.* For a test of significance for a mean, the methods of this section require that you check three conditions:

   - *Randomness.* In the case of a survey, the sample must have been randomly selected.

   - *Normality.* The sample must look like it's reasonable to assume that it came from a normally distributed population *or* the sample size must be large enough. There is no exact rule for determining whether a normally distributed population is a reasonable assumption or for what constitutes a large enough sample size, but for a moderately skewed population, the sample size should be at least 40.

   - *Sample/population size.* In the case of a sample survey, the population size should be at least ten times as large as the sample size.

3. *Compute the test statistic and a P-value.* The test statistic is the distance from the sample mean, $\bar{x}$, to the hypothesized value, μ_0, measured in standard errors:

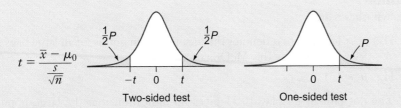

$$t = \frac{\bar{x} - \mu_0}{\frac{s}{\sqrt{n}}}$$

Two-sided test

One-sided test

The *P*-value given by your calculator or software (or estimated from Table B) is the probability of getting a value of *t* that is as extreme as or even more extreme than the one computed from the actual sample *if the null hypothesis is true.*

4. *Write your conclusion linked to your computations and in the context of the problem.* For fixed-level testing, reject the null hypothesis if your *P*-value is less than the level of significance, α. If the *P*-value is greater than or equal to α, do not reject the null hypothesis. (If you are not given a value of α, you can assume that α is 0.05.) Write a conclusion that relates to the situation and includes an interpretation of your *P*-value.

Example 10.11 demonstrates a fixed-level significance test with all four steps included.

Mass of a Large Bag of Fries

Example 10.11

The statistics class at the University of Wisconsin-Stevens Point (see E7 on page 481) also estimated the average mass of a large bag of french fries at McDonald's. They bought 30 bags during two different half-hour periods on two consecutive days at the same McDonald's and weighed the fries. The data are given in Display 10.31. McDonald's target value for the mass of a large order of fries was 171 g. Is there statistically significant evidence ($\alpha = 0.01$) that this McDonald's wasn't meeting that target?

Mass of Large Bag of Fries (g)	Mass of Large Bag of Fries (g)
125.7	155
139.9	162.7
132.9	140.2
131.6	145.7
152.1	156.6
151.3	154.3
145.2	143.7
138.6	115.9
142.7	162
132.8	175.4
127.5	139.6
141.1	153.8
152.4	133.6
145.8	142.2
143.8	137.9

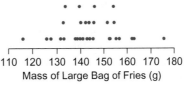

Mass of Large Bag of Fries (g)

Display 10.31 Mass of large bags of McDonald's fries. [Source: Nathan Wetzel, "McDonald's French Fries. Would You Like Small or Large Fries?" *STATS*, Vol. 43 (Spring 2005), pp. 12–14.]

Solution

The hypotheses are

Write a null and an alternative hypothesis.

H_0: $\mu \neq 171$, where μ denotes the mean mass, in grams, of all large bags of french fries produced by this McDonald's

H_a: $\mu \neq 171$

Check conditions.

The dot plot in Display 10.31 gives no indication that the large population of masses of large bags of french fries isn't approximately normally distributed. These certainly aren't a random sample of the bags of fries, because they were collected on two specific days during two half-hour time periods on each day. The mass of the fries might vary with the person bagging them, and the same person might bag the fries at those times while other people bag the fries on other days or at other times. With that in mind, we will proceed with a one-sample t-test for a mean.

Compute the test statistic and a P-value.

The statistics from the sample are $\bar{x} = 144.7$, $s = 12.28$, and $n = 30$. The test statistic is

$$t = \frac{\bar{x} - \mu}{s/\sqrt{n}} = \frac{144.07 - 171}{\frac{12.28}{\sqrt{30}}} \approx -12.01$$

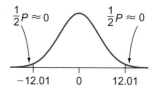

The P-value from a calculator is 8.85×10^{-13}, much less than $\alpha = 0.01$.

Write a conclusion linked to your computations and in the context of the situation.

A P-value this close to 0 is very strong evidence against the null hypothesis. A sample that gives a value of t of -12.01 —or one even more extreme—is almost never going to occur if the mean mass of the bags is 171 g. So you can reject the null hypothesis that the mean mass is 171 g. Even though this was not a true random sample, the sample of bags collected by these students might be typical of the bags produced during those times or on that day or by that shift of employees. So any generalizations should be made cautiously, with careful attention to describing the population to which the results of the analysis might apply.

In Example 10.12, you will learn a technique for dealing with outliers.

Example 10.12 | **Distance from the Sun**

"Earth is 93 million miles from the Sun." You have probably heard that assertion. But is it true? Display 10.32 shows 15 measurements made of the average distance between Earth and the Sun (called the astronomical unit, or AU), including Simon Newcomb's original measurement from 1895. Test whether the data are consistent with a true mean AU of 93 million mi.

Solution

Letting μ denote the true value of the astronomical unit, the hypotheses are

Write a null and alternative hypothesis.

H_0: $\mu = 93$ and H_a: $\mu \neq 93$

Check conditions.

This type of problem is commonly referred to as a *measurement error* problem. There is no population of measurements from which this sample was selected. The measurements were, however, independently determined by different scientists and can be thought of as a "random" sample taken from a conceptual population of all such measurements that could be made. We will assume that there is no systematic bias in the measurements.

Measurements of Astronomical Unit (millions of miles)
93.28
92.83
92.91
92.87
93.00
92.91
92.84
92.98
92.91
92.87
92.88
92.92
92.96
92.96
92.81

```
     Astronomical Unit (n = 15),
         Leaf Unit = 0.010
    ─────────────────────────────
    928 │ 1 3 4
      · │ 7 7 8
    929 │ 1 1 1 2
      · │ 6 6 8
    930 │ 0
      · │
    931 │
      · │
    932 │
      · │ 8
```

928 | 1 represents 92.81 million mi.

Display 10.32 Measurements of the astronomical unit (AU). [Source: W. J. Youden. *Experimentation and Measurement* (National Science Teachers Association, 1985), p. 94.]

The stemplot in Display 10.32 shows that one measurement is extremely large compared to the others. This sample does not look as if it could reasonably have been drawn from a normally distributed population. The large measurement, 93.28 million mi, is Newcomb's original measurement from 1895 and thus is different from the more modern measurements. There might be a good scientific reason to remove it from the data set. So the analysis will be done twice, once with Newcomb's measurement and once without it.

From the data, $\bar{x} = 92.93$ and $s = 0.112$. The test statistic then is given by

> Compute the test statistic and a *P*-value.

$$t = \frac{\bar{x} - \mu}{s/\sqrt{n}} = \frac{92.93 - 93.00}{\frac{0.112}{\sqrt{15}}} \approx -2.42$$

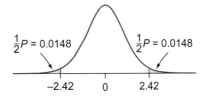

From a calculator or computer, the *P*-value for 14 degrees of freedom is 0.0297. From the table, the *P*-value is seen to be between 0.02 and 0.04.

If it is true that the AU is equal to 93 million mi, the chance of getting a mean from a random sample that is as far away as or even farther from 93 million mi as the mean from this sample is only 0.0297. Relative to an α of 0.05, this is convincing evidence against the null hypothesis. If there is no systematic bias in the measurements, you are pretty sure that the astronomical unit differs from 93 million mi.

> Write a conclusion linked to your computations and in the context of the situation.

What about the influence of the outlier? Removing Newcomb's original 1895 measurement, 93.28 million mi, gives a new observed test statistic of

$$t = \frac{\bar{x} - \mu_0}{\frac{s}{\sqrt{n}}} = \frac{92.93 - 93.00}{\frac{0.057}{\sqrt{14}}} \approx -6.56$$

The *P*-value is now 0.0000181. This value is even stronger evidence against the hypothesized value, 93 million mi. Whether the outlier is included or not, you come to the same conclusion: The AU is *not* 93 million mi.

Example 10.13 demonstrates a one-sided alternative.

Example 10.13 | Honda Civic Hybrid Gas Mileage

Honda Civic Hybrid.

Owners of hybrid cars have been known to exaggerate their gas mileage. For example, the EPA estimated that overall miles per gallon for the 2007 Honda Civic Hybrid is 42 miles per gallon (mpg). The first 40 owners to post their own miles per gallon on the EPA web site reported a mean of 43.3 mpg with a standard deviation of 5.06 mpg. Is this statistically significant evidence that all 2007 Honda Civic Hybrid owners would report, on average, higher gas mileage than the EPA estimate? Conduct the test at the 5% significance level. [Source: *www.fueleconomy.gov.*]

Solution

This situation requires a one-sided test, because you are interested only in the alternative (research) hypothesis that the population mean MPG reported is larger than 42.

Write a null and alternative hypothesis.

The hypotheses are

H_0: $\mu = 42$ mpg, where μ denotes the mean gas mileage that would be reported by all 2007 Honda Civic Hybrid owners

H_a: $\mu > 42$ mpg

Check conditions.

These are not a random sample of owners, but we might consider them representative of the owners who would report their gas mileage if they knew this site was available. The dot plot in Display 10.33 shows it is reasonable to assume that these miles per gallons came from a normally distributed population. There are more than $10(40) = 400$ owners of 2007 Honda Civic Hybrids.

Display 10.33 The reported overall miles per gallon for 40 Honda Civic Hybrids.

Compute the test statistic and a *P*-value.

The test statistic is given by

$$t = \frac{\bar{x} - \mu_0}{\frac{s}{\sqrt{n}}} = t = \frac{43.3 - 42}{\frac{5.06}{\sqrt{40}}} \approx 1.62$$

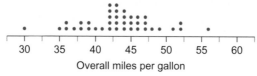

$P = 0.0561$

$t = 1.62$

The *P*-value is the area to the right of 1.62 under the *t*-distribution with $df = 40 - 1$, or 39. From a calculator or computer, the *P*-value is about 0.0561. From the table it is seen to be a little larger than 0.05.

Write a conclusion linked to computations and in the context of the situation.

A *P*-value of 0.0561 is borderline for a test at the 5% significance level but isn't very strong evidence against the null hypothesis that the mean miles per gallon reported by all owners would be 42 mpg. Thus, the data do not support the conjecture that 2007 Honda Civic Hybrid owners exaggerate their gas mileage.

The Meaning of Reject and Do Not Reject

Large *t* and Small *P*: Reject the Null Hypothesis

A test statistic that is large in absolute value tells you that there is a large discrepancy between your sample mean and the hypothesized mean, μ_0. There are three ways you can get a large value of *t* (and therefore a small *P*-value): The conditions are not met, a rare event occurred, or the null hypothesis is false.

- *Conditions not met.* The conditions for using *t*-procedures (that you have a random sample from a normally distributed population, a random sample with a large enough sample size, or random assignment of treatments) aren't consistent with the data.

- *Rare event.* The conditions are met and the null hypothesis is true, but just by chance an unlikely event occurred and gave you an unusually large value of the test statistic. In this situation, you will make a *Type I error*, rejecting a true null hypothesis.

A Type I error is rejecting a true H_0.

- *Reject is right.* The conditions are met, but the true population mean is not μ_0, so you are right to reject H_0.

Once you've checked the randomness of the sample and normality of the population as best you can or have a large random sample, you know you have a reasonable model, so either a rare event occurred or the null hypothesis is false. Standard practice is to reject the null hypothesis.

Small *t* and Large *P*: Do Not Reject the Null Hypothesis

What about a small absolute value of the test statistic? Assuming that your sample fits the conditions for your test, there are two ways you can get a small value of *t* (and therefore a large *P*-value): The hypothesized mean is right, or the hypothesized mean is wrong, but your test doesn't have enough power to allow you to reject it.

Power is the ability to reject H_0.

- *Do not reject is right.* The null hypothesis is true, and your sample is consistent with it. You don't reject the null hypothesis—the correct decision.

- *Not enough power.* When the sample size is small, there can be a lot of variability in $\bar{x}$ from sample to sample. Thus, your sample mean might end up being fairly close to the erroneous mean stated in the null hypothesis. You can't reject this false null hypothesis, and so you have made a *Type II error*. This occurs because a significance test's performance (its *power*) is measured by its chance of *rejecting* a hypothesized value. You want the power to be large for a hypothesized value that is not true. As you learned in Section 8.4, increasing your sample size is the best way to get more power in your test.

A Type II error is failing to reject a false H_0.

DISCUSSION
The Meaning of Reject and Do Not Reject

D15. Suppose you are doing a fixed-level test at the 1% significance level. Which of these questions can be answered?

A. What is the chance that the test will reject the null hypothesis?

B. If the null hypothesis is true, what is the probability that it will be rejected? That it will not be rejected?

C. If the null hypothesis is false, what is the probability that it will be rejected? That it will not be rejected?

D16. Discuss the relationship between a two-sided fixed-level test of a mean and a confidence interval estimate of the same mean.

Summary 10.3: Fixed-Level Tests

Instead of simply reporting a *P*-value to provide information about the strength of evidence against the null hypothesis, you can construct a fixed-level test with given significance level α, chosen before the data are analyzed. This procedure allows you to control the probability of a Type I error and allows the possibility of calculating the probability of a Type II error and the power for specified alternatives.

If you reject a true null hypothesis, you have made a Type I error. If you fail to reject a false null hypothesis, you have made a Type II error. The power of your test depends on its ability to reject the null hypothesis. The larger the sample size and the larger the level of significance, the greater the power of your test to reject a false null hypothesis.

The four steps of the hypothesis test are (1) write null and alternative hypotheses, (2) check the conditions for the statistical procedure you are to use, (3) calculate the test statistic and *P*-value, and (4) write a conclusion in the context of the original problem. The actual test statistic remains the same as in the previous section.

Practice

Fixed-Level Tests

P21. In P15 on page 492, you found the *P*-value for testing whether the mean temperature at your seat is different from 72°F.

a. State the hypotheses.

b. Would you reject the null hypothesis at the 10% level? The 5% level? The 1% level?

P22. In P16 on page 492, you found the *P*-value for testing whether the mean Pell grant in Minnesota has changed since 2008.

a. State the hypotheses.

b. Would you reject the null hypothesis at the 10% level? The 5% level? The 1% level?

P23. The provost states that the mean SAT score for students at State University is 1700. You are asked to validate his claim. You get the SAT scores of a random sample of nine students from State University and calculate a sample mean of 1535 with standard deviation 250.

a. State the hypotheses.

b. Would you reject the hypothesis that State University has a mean SAT score of 1700 at the 10% level? The 5% level? The 1% level?

The *t*-Test

P24. For the aldrin data of P1 on page 477, carry out a test of whether the true mean differs from 4 ng/L. Use $\alpha = 0.05$ and follow the four-step model for the test.

P25. The question about mean body temperature addressed with the data in Display 10.3 on page 471 could have been phrased differently: "Does the mean body temperature differ from 98.6°F?" Do these data provide convincing evidence that the population mean for males differs from 98.6°F? Answer the same question for the female population. Use a 10% significance level in the tests and cover all four steps of the testing procedure.

P26. In the manufacture of glass for airplane windows, samples from the process are taken periodically and their strengths measured. A sample of 31 such measurements and computer output is shown in Display 10.34.

a. Do the measurements look like they meet the conditions for testing the population mean?

b. Suppose the manufacturing operation has a standard of 32 ksi (1000 pounds per square inch) for the process mean. Is there statistically significant evidence that the process is not meeting that standard?

18.83	26.77	34.76
20.80	26.78	35.75
21.66	27.05	35.91
23.03	27.67	36.98
23.23	29.90	37.08
24.05	31.11	37.09
24.32	33.20	39.58
25.50	33.73	44.04
25.52	33.76	45.29
25.80	33.89	45.38
26.69		

Strength

Part b: One-Sample T: Strength

Test of mu = 32 vs < 32

Variable	N	Mean	St Dev	SE Mean	95% Upper Bound	T	P
Strength	31	30.81	7.25	1.30	33.02	−0.91	0.184

Part c: One-Sample T: Strength

Test of mu = 35 vs < 35

Variable	N	Mean	St Dev	SE Mean	95% Upper Bound	T	P
Strength	31	30.81	7.25	1.30	33.02	−3.22	0.002

Display 10.34 Strength measurements of glass in ksi. [Source: *www.itl.nist.gov/div898/ handbook/index.htm.*]

c. Suppose the standard for the mean is 35 ksi. Is there statistically significant evidence that the process is not meeting that standard?

Installing an airplane window.

The Meaning of Reject and Do Not Reject

P27. Return to the significance tests indicated and tell whether a Type I error, a Type II error, both errors, or neither error might have been made.

   a. Example 10.11 on page 497

   b. Example 10.12 on page 498

   c. Example 10.13 on page 500

P28. In each of the following situations, assume the null hypothesis is actually false. Tell which test will have more power, all else being equal.

   a. a test with $n = 12$; a test with $n = 16$

   b. a test with $\alpha = 0.05$; a test with $\alpha = 0.01$

Exercises

E31. Return to the significance test in E19. Rewrite the conclusion for a fixed-level test with α equal to 0.10.

E32. Return to the significance test in E20. Rewrite the conclusion for a fixed-level test with α equal to 0.04.

E33. Return to the significance test in E25.

   a. Rewrite the conclusion for a fixed-level test at the 0.05 level.

   b. In fact, these values were selected at random from a normal distribution with mean 9.9. Has an error been made? If so, which type?

   c. Could you have reached the conclusion in part a based only on a confidence interval? Explain.

E34. Return to the significance test in E26, part b.

   a. Rewrite the conclusion for a fixed-level test with α equal to 0.02.

   b. Could you have reached the conclusion in part a based only on a confidence interval? Explain.

E35. Select each correct description of the probability of a Type I Error.

   A. It is the same as the significance level of the test.

   B. It is the probability that the null hypothesis is false.

   C. It does not depend on the sample size.

   D. It does not depend on whether the test is one- or two-sided.

E36. Select each correct description of the power of a test.

   A. Power is the probability of rejecting the null hypothesis.

   B. Power is the probability of failing to reject the null hypothesis.

   C. To increase power, increase the sample size.

   D. To increase power, increase the level of significance.

E37. Golf is big business in Florida. A supplier of golf cart batteries to major courses around the state was suddenly deluged by complaints that the carts he was servicing were running out of power after about 14 holes of golf, making high-paying patrons more than a little unhappy. Calling in a statistician (in addition to an attorney) for advice, the supplier agreed to a plan that called for sampling batteries from a large new shipment from the manufacturer to check their weight. Weight is a key factor in performance because the plates inside the battery are made of lead, and a decrease in the weight of batteries is usually a sign of a decrease in the quality of these plates. Display 10.35 shows the weights (in pounds) of a random sample of 30 batteries, along with a plot of their distribution.

a. The manufacturer's standard for the weight of these batteries was 68 pounds. Should the supplier conclude that the mean weight of the new shipment is less than the standard?

b. Suppose the manufacturer's standard was 65 pounds. Should the supplier conclude that the mean weight of the new shipment is less than this standard?

61.5	64	63.8	63	66.5
63.5	65	63.5	62	67
63.5	64.2	64	63	66.5
63.8	64.5	64	63.5	65.5
63.8	66.5	63.2	64	66.5
64	63.5	66.5	63.5	66

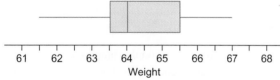

Display 10.35 Weights of a random sample of 30 golf cart batteries.

E38. Another key feature of golf cart batteries is the thickness of the lead plates inside the battery. The advertised standard for the thickness of plates in these batteries was 120 thousandths of an inch.

a. A sample of 12 plate thickness measurements from one battery had a mean of 117.3 and standard deviation of 1.23. Should the supplier conclude that the plates are inferior with regard to mean thickness? Justify your answer statistically.

b. A sample of 12 plate thickness measurements from a second battery had a mean of 119.5 and standard deviation of 1.30. Should the supplier conclude that the plates are inferior with regard to mean thickness? Justify your answer statistically.

E39. Suppose the null hypothesis is true and you are using a significance level of 5%. Does a one-sided test or a two-sided test give a larger chance of a Type 1 error, or do they give the same chance? Explain.

E40. If the null hypothesis is true, does using $\alpha = 0.05$ or $\alpha = 0.01$ give a larger chance of a Type I error, or do they give the same chance?

E41. Suppose that the null hypothesis is actually false. Which test will have more power, all other things being equal?

a. a test with $\alpha = 0.01$; a test with $\alpha = 0.10$

b. a test with $n = 45$; a test with $n = 29$

c. a one-sided test; a two-sided test

E42. Suppose $\mu > 0$ and you are using a significance level of 5%. Which of these three tests has the greatest power (probability of rejecting the false null hypothesis that $\mu = 0$)?

A. a one-sided test of H_0 versus H_a: $\mu > 0$

B. a two-sided test

C. a one-sided test of H_0 versus H_a: $\mu < 0$

E43. Histograms A and B in Display 10.36 were generated from 200 random samples of size 4, each selected from a normal distribution with $\mu = 100$ and $\sigma = 20$. One histogram shows the 200 z-values (one for each sample, using s) for testing the hypothesis that the population mean is, in fact, 100, and the other shows the 200 t-values (using s).

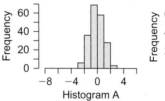 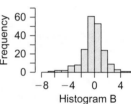

Display 10.36 Histograms of the t-values and z-values computed for 200 random samples of size 4 taken from a normal distribution with mean 100 and standard deviation 20.

a. Compare the shapes, centers, and spreads of these two distributions.

b. Choose which is the distribution of t-values, and give the reason for your choice.

E44. Degrees of freedom tell how much information in your sample is available for estimating the standard deviation of the population. In a t-test, you use s to estimate σ. The more deviations from the mean, $(x - \bar{x})$, you have available to use in the formula for s, the closer s should be to σ. But, as you will see in this exercise, not all of these deviations give you independent information.

a. For a sample of size n, how many deviations from the mean are there? What is their sum? (See Section 2.2.)

b. Suppose you have a random sample of size $n = 1$ from a completely unknown population. Call the sample value x_1. Does the value of x_1, by itself, give you information about the spread of the population?

c. Next suppose you have a sample of size $n = 2$, with values x_1 and x_2. If one deviation, $x_1 - \bar{x}$, is 3, what is the other deviation?

d. Now suppose you have a sample of size $n = 3$. Suppose you know two of the deviations, $x_1 - \bar{x} = 3$ and $x_2 - \bar{x} = -1$. What is the third deviation?

e. Finally, suppose you have a sample of size n. Show how to find the final deviation once you know all the others. With a sample of size n, how many deviations give you independent information about the size of σ and how many are redundant?

f. Explain what your answers to parts a–e have to do with degrees of freedom.

Chapter Summary ◂

To use the confidence intervals and significance tests of this chapter, you need a random sample from the population. When there is no randomness involved, you proceed with the test only after stating loudly and clearly the limitations of what you are doing. If you reject the null hypothesis in such a case, all you can conclude is that something happened that can't reasonably be attributed to chance.

You use a confidence interval if you want to find a range of plausible values for μ, the mean of your single population. The confidence interval has the form

$$statistic \pm (critical\ value) \cdot (estimated\ standard\ error) = \bar{x} \pm t^* \frac{s}{\sqrt{n}}$$

The significance test you studied has the following steps:

1. State the null hypothesis and the alternative hypothesis.

2. Justify your reasons for choosing this particular test. Discuss whether the conditions are met and, if they are not strictly met, decide whether it is okay to proceed.

3. Compute the test statistic which has the form

$$\frac{statistic - parameter}{estimated\ standard\ error} \quad \text{or} \quad t = \frac{\bar{x} - \mu_0}{\frac{s}{\sqrt{n}}}$$

and find the P-value. Draw a sketch of the situation.

4. Use the computations to decide whether to reject or not reject the null hypothesis by comparing the P-value to the level of significance, α. Then state your conclusion in terms of the context of the situation. ("Reject H_0" is not sufficient.) Mention any doubts you have about the validity of that conclusion.

There's only one way you can be confident of drawing correct conclusions from data: Use sound methods of data production, either random samples or randomized experiments. At the other extreme, there are many ways you can be confident that the conclusions might not be valid. For example, voluntary response samples are worthless when it comes to inference. In reality, many situations fall between the two extremes. What then? For example, what if you use your class as a sample instead of taking a random sample? There are no formal rules; the value of the inference methods is rarely all-or-nothing. The more reasonable it is to regard your data as coming from a random sample or a randomized experiment, the more reasonable it is to trust conclusions based on the inference methods. For many data sets, making a careful judgment about this issue is the hardest aspect of a statistician's job.

Review Exercises

E45. To find an estimate of the number of hours that highly trained athletes sleep each night, a researcher selects a random sample of 15 highly trained athletes and asks each how many hours of sleep he or she gets each night. The results are given in Display 10.37.

a. Are the conditions satisfied for computing a confidence interval for the mean?

b. Construct and interpret a 95% confidence interval for the mean of the population of athletes from which this sample was selected.

E46. A book about different colleges reports that the mean time students at a particular university study each week is 1015 minutes. A dean says she believes the mean is greater than 1015 minutes. To test her claim, she takes a

Stem-and-leaf of Number of Hours

N = 15

Leaf Unit = 0.10

```
3  7
4  29
5  149
6  15
7  115
8  58
9  39
```

Display 10.37 Hours of sleep for highly trained athletes.

random sample of 64 students and finds that the sample mean is 1050 minutes, with standard deviation 150 minutes. Is this strong evidence in favor of her claim?

E47. For the sleep data in E45, carry out a test of the claim that the population mean is 8 hours of sleep per night

　　a.　State the hypotheses, defining any symbols that you use.

　　b.　Calculate the value of the test statistic, showing the formula and the values substituted in.

　　c.　Find the *P*-value from your calculator and interpret it in the context of this situation.

E48. The distribution of annual incomes of the employees of a large firm is highly skewed toward the larger values because of the high salaries of the upper-level managers. A random sample of size *n* is to be selected for the purpose of testing the claim that the mean salary is over $50,000.

　　a.　How would you proceed if *n* = 10?

　　b.　How would you proceed if *n* = 50?

E49. Where are the best college values and what are the costs? According to a recent Kiplinger report, there are many good values for your education dollar at public colleges and universities. The data in Display 10.38 shows the typical costs (in dollars) for an in-state student and the typical debt at graduation for a random sample of ten out of the 100 best buys, according to Kiplinger.

　　a.　Estimate the mean in-state cost for the 100 best-buy colleges in a 95% confidence interval.

　　b.　Estimate the mean debt at graduation for the 100 best-buy colleges in a 90% confidence interval.

E50. Various sources report average college debt for students graduating from public colleges and universities, and a typical figure is around $20,000. Using the sample data in E49, does it look like the best-buy colleges are lowering the mean debt? Justify your answer with an appropriate statistical test.

E51. Suppose you have a sample of size 23 and the value of your test statistic is 1.645.

　　a.　Draw a sketch of a *t*-distribution and shade the area that corresponds to the *P*-value for a two-sided test.

　　b.　Identify the part of the shaded area that corresponds to evidence that $\mu > \mu_0$. What is the relationship

College	In-State Costs	Debt at Graduation
University of Colorado at Boulder	18,887	18,037
SUNY Oneonta	14,716	19,967
University of North Carolina at Wilmington	12,832	16,350
Georgia Institute of Technology	14,734	21,436
Georgia Southern University	12,848	18,618
University of Connecticut	19,438	20,658
University of Maryland, College Park	17,848	18,958
Western Washington University	13,344	15,280
Stony Brook University (SUNY)	15,841	16,096
University of California, Irvine	20,401	13,383

Display 10.38 Public college costs. [Source: *www.kiplinger.com/tools/colleges*.]

between the two-sided and one-sided *P*-values for a given value of the test statistic?

　　c.　If all else is equal and the alternative hypothesis is in the right direction, will the *P*-value be larger for a one-sided test or a two-sided test?

E52. For the aldrin data of P1 on page 477, find the *P*-value for each alternative hypothesis.

　　a.　$H_a: \mu \neq 4$

　　b.　$H_a: \mu < 4$

　　c.　$H_a: \mu > 4$

E53. Select the best answer. The confidence level measures

　　A.　the fraction of times the confidence interval will capture the parameter it is estimating in repeated use of the procedure on different samples

　　B.　the fraction of times the confidence interval will fail to capture the parameter it is estimating in repeated use of the procedure on different samples

　　C.　the fraction of times the confidence interval captures the sample statistic on which it is based

　　D.　None of the above is a good interpretation

E54. Select the best answer. The *P*-value measures

　　A.　the probability that the null hypothesis is true

　　B.　the probability that the confidence interval captures the true parameter

　　C.　the probability of getting a result from the sample as extreme as the one from the actual data

　　D.　None of the above is a good interpretation

C1. In measuring the angle formed by two intersecting laser beams in a physics lab, a student uses chalk dust to illuminate the beams and then uses a protractor to measure the angle between them. She takes ten measurements and then produces a 95% confidence interval estimate of the mean, but is unhappy with the large margin of error that she calculates. Which of the following is the worst plan for reducing the margin of error the next time she takes similar measurements?

Ⓐ Increase the number of measurements.

Ⓑ Use a more precise method of measuring angles.

Ⓒ Decrease the confidence level.

Ⓓ Check to be sure any outliers aren't mistakes.

Ⓔ Combine her results with those from other students.

C2. A statistics student constructs many confidence intervals for a population mean by computing $\bar{x} \pm 1.96\, s/\sqrt{n}$. For which of the following sample sizes will the proportion of her confidence intervals that contain the population mean be closest to 0.95?

Ⓐ $n = 30$

Ⓑ $n = 60$

Ⓒ $n = 95$

Ⓓ $n = 100$

Ⓔ The sample size does not matter, so the proportion is the same for all of these sample sizes.

C3. Which of the following describes a difference between the sampling distribution of $z = \dfrac{\bar{x} - \mu}{\frac{\sigma}{\sqrt{n}}}$ and the sampling distribution of $t = \dfrac{\bar{x} - \mu}{\frac{s}{\sqrt{n}}}$ in the case where $n = 5$?

Ⓐ The distribution of t is more symmetric.

Ⓑ The distribution of t is more skewed.

Ⓒ The distribution of t has a larger spread.

Ⓓ The distributions have different means.

Ⓔ The distribution of t is known, while the distribution of z is unknown because s is unknown.

C4. With a sample of size 15 and $t = -2.76$, should the null hypothesis be rejected in a two-sided significance test for a mean?

Ⓐ yes if $\alpha = 0.05$, and yes if $\alpha = 0.01$

Ⓑ yes if $\alpha = 0.05$, but no if $\alpha = 0.01$

Ⓒ no if $\alpha = 0.05$, but yes if $\alpha = 0.01$

Ⓓ no if $\alpha = 0.05$, and no if $\alpha = 0.01$

Ⓔ never, because t is negative

C5. To estimate the mean number of hours that students at a particular college sleep on a school night, a dean selects a random sample of 16 students and asks each how many hours they sleep on a school night. The number of hours are given in the following stem-and-leaf plot. The dean checks, and the outlier isn't a mistake. What is the best way for the dean to proceed?

Ⓐ Remove the outlier from the data set. This student is not at all typical.

Ⓑ Do the analysis with the outlier included.

Ⓒ Do two analyses, one with the outlier and one without and compare the results.

Ⓓ Because it is so easy to get this information, get a larger random sample.

Ⓔ Do not analyze these data by any inferential technique.

Stem-and-leaf of Number of Hours

$N = 16$

Leaf Unit = 0.10

5	9
6	1257
7	599
8	3589
9	126
10	
11	
12	9

C6. Several statistically trained knights seek to estimate the mean tail length of adult fire-breathing dragons. They collect a random sample of 10 dragons, observe that the distribution of tail lengths is symmetric with no outliers, and then compute a confidence interval. They publish a paper stating, "We are 95% confident that the mean tail length of adult fire-breathing dragons is between 8.2 feet and 11.3 feet." The knights later discover that in the population of all dragons, there are some with exceptionally short tails due to injuries, though none of these dragons were included in their sample. Which of the following is the best description of how the knights should revise the interpretation of their confidence interval?

Ⓐ They should lower the center.

Ⓑ They should decrease the width.

Ⓒ They should increase the width.

Ⓓ They should decrease the confidence level.

Ⓔ They should replace "adult" with "uninjured adult."

Investigation

C7. Another way to produce confidence interval estimates of population means is by use of a computer-intensive technique called *resampling*. As an example, take another look at the female temperature data of Display 10.3. The sample values are given here.

Female

97.8
98.0
98.2
98.2
98.2
98.6
98.8
98.8
99.2
99.4

The 95% confidence interval of plausible values for the population mean turned out to be (98.14, 98.90). The resampling method begins by assuming that these 10 values are all you know about the population, so resample from them. The resampling must be done *with replacement* as there are many values in the population that are repeats of these: that is, you could find many women with a body temperature of 98.2 if you searched a large population. The original sample size was 10; so to make a comparison with the preceding confidence interval, each resample should include 10 values. The resampling process involves the following steps:

a. Resample many samples of size 10 and calculate the sample mean for each resample. Display the distribution of results. Display 10.39 shows one resampling distribution with 500 resampling means. Describe the shape of this distribution. Is that surprising?

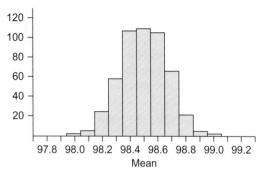

Display 10.39 Resampling means from the temperature data.

b. Because we are using the original sample as a model for the population, this distribution simulates the possible means that the population could have, with their relative frequency of occurrence. Thus, the middle 95% percent of these values represents the plausible values for the population mean.

Here are the smallest 15 and the largest 15 means from the simulated resampling distribution. Using them, find the set of plausible values for the population mean.

Lowest 15 Means	Highest 15 Means
98	98.8
98.06	98.8
98.06	98.8
98.1	98.82
98.14	98.82
98.14	98.84
98.16	98.84
98.18	98.84
98.18	98.84
98.18	98.86
98.18	98.86
98.18	98.88
98.18	98.88
98.2	98.88
98.2	99.02

c. Compare the resampling interval of plausible values with those produced by the standard confidence interval. What do you conclude?

d. The resampling method is quite general and can be used for estimating any population parameter. Try it out by finding a resampling interval of plausible values for the population median of female temperatures.

Chapter 11

Comparing Two Populations: Inference for the Difference of Two Means

Does it make a difference whether you take water samples at mid-depth or near the bottom when studying pesticide levels in a river? Inference for differences is fundamental to statistical applications because many surveys and all experiments are comparative in nature.

In Chapter 9, you learned to make inferences about the difference in the proportion of successes in two unknown populations. You did this by comparing the proportion of successes in two independent random samples taken from those populations. Now you will use similar ideas to compare the unknown means of two populations by making inferences from the sample means. What is the estimated difference between the mean number of yards men and women can walk within the sidelines of a football field when blindfolded? Can the improvement in mean SAT mathematics scores reasonably be attributed to chance alone? Are celebrities, on the average, more narcissistic than other people?

Means and proportions are the two most widely used summary statistics. In Chapter 9, you learned how to construct confidence intervals and perform significance tests for differences between proportions. This chapter will follow the same outline as Chapter 9 except that the methods are applied to means rather than proportions.

In this chapter, you will learn how to

▶ construct and interpret a confidence interval for estimating the difference between the means of two different populations

▶ conduct and interpret a significance test for the difference between the means of two different populations

▶ construct and interpret a confidence interval to estimate the mean of the differences from paired data

▶ conduct and interpret a significance test for the mean of the differences from paired data

11.1 ▶ A Confidence Interval for the Difference of Two Means

Most scientific studies involve comparisons. For example, for the purpose of studying pesticide levels in a river, does it make a difference whether you take water samples at mid-depth, near the bottom, or at the surface of the river? Do special exercises help babies learn to walk sooner? Inference about comparisons is more often used in scientific investigations than is inference about a single parameter. In fact, almost all experiments are comparative in nature. The study of inference for differences, then, is fundamental to statistical applications.

The Meaning of a Confidence Interval for the Difference of Two Means

Example 11.1 demonstrates how you can interpret a confidence interval for the difference of two means. Following this example, you will learn the details of constructing the confidence interval.

| Example 11.1 | A Confidence Interval for a Difference in Mean Aldrin Levels |

The Wolf River in Tennessee flows past an abandoned site once used by the pesticide industry for dumping wastes, including chlordane, aldrin, and dieldrin. These highly toxic organic compounds can cause various cancers and birth defects. The standard method to test whether these poisons are present in a river is to take samples at six-tenths depth, that is, six-tenths of the way from the surface to the bottom. Unfortunately, there are good reasons to worry that six-tenths is the wrong depth. The organic compounds in question don't have the same density as water, and their molecules tend to stick to particles of sediment. Both these facts suggest that you'd be likely to find higher concentrations near the bottom than near mid-depth.

Ten measurements of the aldrin concentration were taken at mid-depth in the Wolf River and ten measurements were taken at the bottom. The data and summary statistics are given in Display 11.1.

Two populations, two samples, two means.

Bottom		Mid-Depth
	2	
8	3	2 8
9 8	4	3 8 9
7 4 3	5	2 2
3	6	2 3 6
3	7	
8 1	8	
	9	
		2	8 = 2.8 ng/L

Bottom: $\bar{x}_{bottom} = 6.04$ $s_{bottom} = 1.579$

Mid-depth: $\bar{x}_{mid\text{-}depth} = 5.05$ $s_{mid\text{-}depth} = 1.104$

Display 11.1 Aldrin concentrations, in nanograms per liter. [Source: P. R. Jaffe, F. L. Parker, and D. J. Wilson "Distribution of Toxic Substances in Rivers," *Journal of the Environmental Engineering Division*, Vol. 108 (1982), pp. 639–649, *www.statsci.org/data/general/wolfrive.html.*]

A 95% confidence interval for the difference in the mean concentrations is 0.99 ± 1.2909.

a. Describe the populations from which these samples were taken. What parameter is being estimated?

b. What is the interpretation of 0.99 in the confidence interval? What name is given to the value 1.2909?

c. Write an interpretation of the confidence interval.

d. Is zero in the confidence interval? What can you conclude from your answer?

Solution

a. One population consists of all possible measurements taken from the bottom of the river, and the other population consists of all possible measurements taken from mid-depth. The parameter is the difference in the mean of all possible measurements taken from the bottom and the mean of all possible measurements taken from mid-depth: $\mu_{bottom} - \mu_{mid\text{-}depth}$.

b. The number 0.99 is the difference in the means of the two samples, $\bar{x}_{bottom} - \bar{x}_{mid\text{-}depth}$, or $6.04 - 5.05$. The number 1.2909 is the margin of error for the difference.

c. You are 95% confident that, if you had the population of all measurements at the bottom and the population of all measurements at mid-depth, the difference in their means, $\mu_{bottom} - \mu_{mid\text{-}depth}$, would fall in the interval 0.99 ± 1.2909.

d. Zero is in the confidence interval. Therefore, it is plausible that there is no difference in the means of the two populations. In other words, you don't have statistically significant evidence that, on average, it matters whether you take the measurement at the bottom of the river or at mid-depth.

DISCUSSION
The Meaning of a Confidence Interval for the Difference of Two Means

D1. When might you want a confidence interval for the difference of two means rather than a separate confidence interval for the mean of each population?

Constructing a Confidence Interval for the Difference of Two Means

A confidence interval for the difference between two means has the standard form you have seen before:

statistic ± (*critical value*) · (*estimated standard error*)

The main steps in the procedure for constructing a confidence interval, given in the following box, should also be familiar. Use this procedure when you want to estimate the size of the difference between the mean of one population and the mean of another population.

Confidence Interval for the Difference Between Two Means (Two-Sample *t*-Interval)

1. *Check conditions.* You must check three conditions:
   - The two samples were randomly and independently selected from two different populations. In the case of an experiment, check that the two treatments were randomly assigned to the available experimental units.
   - It is reasonable to assume that each of the two samples came from a normally distributed population *or* the sample sizes are sufficiently large (40 or more should be sufficient).
   - In the case of sample surveys, the population size should be at least 10 times larger than the sample size for both samples.

2. *Do computations.* A confidence interval for the difference between the means of two populations, $\mu_1 - \mu_2$, is given by

$$(\bar{x}_1 - \bar{x}_2) \pm t^* \sqrt{\frac{s_1^2}{n_1} + \frac{s_2^2}{n_2}}$$

Here, $\bar{x}_1$ and $\bar{x}_2$ are the means of the two samples, s_1 and s_2 are the respective standard deviations, and n_1 and n_2 are the sample sizes.

Subtract in the order that makes the difference positive.

It is best to use a calculator or statistics software to find this confidence interval because the value of t^* depends on a complicated calculation. If you are not using technology, you may use Table B with $(n_1 - 1) + (n_2 - 1)$ as an approximate number of degrees of freedom.

3. *Give interpretation in context and linked to computations.* For a 95% confidence interval, for example, you are 95% confident that if you knew the means of both populations, the difference between those means, $\mu_1 - \mu_2$, would lie in the confidence interval.

   For an experiment, the interpretation is like this: Imagine that all subjects had been assigned the first treatment and you compute the mean response. Now imagine that all subjects had been assigned the second treatment and you compute the mean response. Then you are 95% confident that the difference in these two means lies in the confidence interval.

   If zero is not in the confidence interval, then you have statistically significant evidence that there is a difference in the means of the two populations.

> If no confidence level is stated, it's safe to assume it is 95%.

At the end of this section, you will find some technical details about the procedure in the preceding box. Example 11.2 illustrates how to construct and interpret a confidence interval for the difference of two means estimated from a sample survey.

A Confidence Interval for the Three Stooges

Example 11.2

Larry and Moe appeared in all of the Three Stooges films. In the first 97 films, the third stooge was Curly. In the next 77 films, the third stooge was Shemp. (The next third stooge was Joe, and the fourth and final third stooge was Curly-Joe.)

Two professors at East Tennessee State University watched ten randomly selected films starring Larry, Moe, and Curly and ten randomly selected films starring Larry, Moe, and Shemp. They counted the number of times Moe acted violently (such as slapstick eye poking or slapping) against Curly and against Shemp. Their data and summary statistics are given in Display 11.2. Construct and interpret a 95% confidence interval for the difference between the mean number of violent acts against Curly and the mean number of violent acts against Shemp per film.

Solution

You have two random samples, taken independently from two different populations. However, the size of each sample is slightly more than 10% of the size of their respective populations. The confidence interval will be slightly wider than necessary, but not much wider because the proportions are not much larger than 10%.

The random sample for Shemp has no skewness or outliers, and it is reasonable to assume that it came from a normally distributed population. The random sample for Curly has two outliers, and it is questionable whether it is reasonable to assume that it came from a normal distribution. The conclusion will have to be stated with this in mind.

Using a calculator or computer (which generally does not give the value of t^*), you get a 95% confidence interval of

> Check randomness and normality.

> Do computations.

$$(\bar{x}_1 - \bar{x}_2) \pm t^* \sqrt{\frac{s_1^2}{n_1} + \frac{s_2^2}{n_2}} = (14.3 - 10.3) \pm t^* \sqrt{\frac{10.4568^2}{10} + \frac{6.2191^2}{10}}$$

$$= 4.0 \pm 8.2$$

Moe Against Curly		Moe Against Shemp	
Title	Number of Acts	Title	Number of Acts
Uncivil Warriors	27	*Shivering Sherlocks*	13
Whoops, I'm an Indian	13	*Punchy Cowpunchers*	3
Back to the Woods	12	*Love at First Bite*	20
Three Missing Links	9	*Three Arabian Nuts*	9
How High Is Up?	38	*Scrambled Brains*	11
Cookoo Cavaliers	14	*Corny Casanovas*	17
An Ache in Every Stake	6	*Cuckoo on a Choo Choo*	16
Sock-a-Bye Baby	10	*Knutzy Knights*	8
A Bird in the Head	11	*Shot in the Frontier*	2
Uncivil Warbirds	3	*Husbands Beware*	4

Curly: $\bar{x}_1 = 14.3$ $s_1 = 10.4568$ $n_1 = 10$

Shemp: $\bar{x}_2 = 10.3$ $s_2 = 6.2191$ $n_2 = 10$

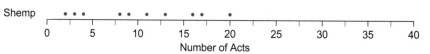

Display 11.2 Number of violent acts by Moe. [Source: Robert Gardner and Robert Davidson, "Hypothesis Testing Using the Films of the Three Stooges," *Teaching Statistics*, (2008).]

or $(-4.2, 12.2)$, where $df = 14.66$. If you are using Table B with $df \approx (n_1 - 1) + (n_2 - 1) = 18$, then $t^* = 2.101$ and the interval is 4.0 ± 8.1.

You are 95% confident that the true difference between the mean number of violent acts against Curly and Shemp, $\mu_{Curly} - \mu_{Shemp}$, is in the interval $(-4.2, 12.2)$. This interval overlaps 0. So there is insufficient evidence to say that if the professors had observed all Three Stooges films with Curly and Shemp, the mean of the number of violent acts against Curly would be different from the mean number of violent acts against Shemp.

Just as for any confidence interval, unless the conditions are satisfied, there is no automatic guarantee that your 95% confidence is warranted. However, if the two outliers for Curly are temporarily removed, 0 is still in the confidence interval; so the conclusion would not change.

> Give interpretation in context and linked to computations.

Example 11.3 illustrates how to construct and interpret a confidence interval for the difference of two means estimated from an experiment.

Example 11.3 | The Walking Babies Experiment

Some babies start walking well before they are a year old, while others still haven't taken their first unassisted steps at 18 months or even later. The data in Display 11.3 are from an experiment designed to see whether a program of special exercises for 12 minutes each day could speed up the process of learning to walk. In all, 23 male infants (and their parents) took part in the study and were randomly assigned either to the special exercise group or to one of three control groups.

To isolate the effects of interest, the scientists used three different control groups. In the exercise control group, parents were told to make sure their infant sons exercised at least 12 minutes per day but were given no special exercises and no other instructions about exercise. In the weekly report group, parents were given no instructions about exercise, but each week they were called to find out about their baby's progress. Parents in the final report group also were given no instructions about exercise, nor did they receive weekly check-in calls. Instead, they gave a report at the end of the study. The response is the age, in months, when the baby first walked without help.

Group	Age (months) at First Unaided Steps					
Special exercises	9	9.5	9.75	10	13	9.5
Exercise control	11	10	10	11.7	10.5	15
Weekly report	11	12	9	11.5	13.25	13
Final report control	13.2	11.5	12	13.5	11.5	

Display 11.3 Age (months) when 23 male infants first walked without support. [Source: Phillip R. Zelazo, Nancy Ann Zelazo, and Sarah Kolb, "Walking in the Newborn," *Science*, Vol. 176 (1972), pp. 314–315.]

Use these data to find a 95% confidence interval estimate of the difference between mean walking times if all 12 babies had been in the special exercises group and all 12 babies had been in the exercise control group.

Solution

Two treatments were randomly assigned to the babies.

Plot the ages from the two groups to see if there is any reason to doubt that they resemble samples from normally distributed populations. Display 11.4 shows that there is some doubt because each group has one relatively large value. Further, we were told that some babies are 18 months old before they walk, so the population of ages must be somewhat skewed right. However, the sampling distribution of the difference of the means tends to be more symmetric than that of either group mean by itself, so a confidence interval for the difference of two means should work for this amount of skewness.

Check conditions: randomness and normality.

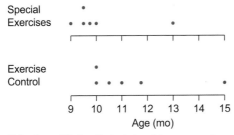

Display 11.4 Dot plots of two samples.

The summary statistics follow:

Do computations.

$$\text{Special exercises: } \bar{x}_1 = 10.125 \quad s_1 = 1.447 \quad n_1 = 6$$
$$\text{Exercise control: } \bar{x}_2 = 11.375 \quad s_2 = 1.896 \quad n_2 = 6$$

Display 11.5 shows the output from four commonly used statistical software packages. These give $df \approx 9.35$ and the confidence interval $(-3.44, 0.94)$.

You are 95% confident that the difference in the mean age at which these babies would have learned to walk had they all been in the special exercises group and all in

Give interpretation in context.

the exercise control group, is in the interval from −3.44 months to 0.94 month. Because 0 is in the confidence interval, you have no evidence that there would be any difference if you were able to give each treatment to all the babies. However, it is important to note that any conclusion is subject to doubt about the appropriateness of this procedure due to skewness of the population of potential measurements, so it would be good to confirm this finding with more data.

JMP

▼ **Means and Std Deviations**

Level	Number	Mean	Std Dev	Std Err Mean	Lower 95%	Upper 95%
EC	6	11.3750	1.89572	0.77392	9.3856	13.364
SE	6	10.1250	1.44698	0.59073	8.6065	11.644

▼ **t-Test**

SE–EC

Assuming unequal variances

Difference	−1.2500	t Ratio	−1.28388	
Std Err Dff.	0.9736	DF	9.349667	
Upper CL Dff	0.9400	Prob > |t|	0.2301	
Lower CL Dff	−3.4400	Prob > t	0.8850	
Confidence	0.95	Prob < t	0.1150	

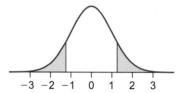

 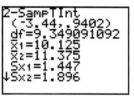

Fathom

Estimate of Walking Babies

First attribute (numeric): Special_Exercise
Second attribute (numeric or categorical): Exercise_Control

Interval estimate for the population mean of **Special_Exercise** minus that of **Exercise_Control**

	Special_Exercise	Exercise_Control
Count:	6	6
Mean:	10.125	11.375
Std dev:	1.44698	1.89572
Std error:	0.590727	0.773924

Confidence level: **95.0**
Using **unpooled variances**
Estimate: **-1.25 +/- 2.18997**
Range: **-3.43997 to 0.939967**

Minitab

TWO SAMPLE T FOR Special VS Exercise

	N	Mean	StDev	SEMean
Special	6	10.12	1.45	0.59
Exercise	6	11.38	1.90	0.77

95 PCT CI FOR MU Special − MU Exercise: (−3.45, 0.95)

TTEST MU Special − MU Exercise (VS NE): T = −1.28 P = 0.23 DF = 9

Display 11.5 Output for estimate of difference in mean walking age from four statistical software packages.

More on the Standard Error

The estimated standard error of the difference, $\sqrt{\frac{s_1^2}{n_1} + \frac{s_2^2}{n_2}}$, comes directly from the rules for random variables in the box in Section 6.2 on page 292. That is, if you have two independent random variables with estimated standard deviations $\frac{s_1}{\sqrt{n_1}}$ and $\frac{s_2}{\sqrt{n_2}}$, then the variance of the sampling distribution of their difference is found by adding the two estimated variances, $\frac{s_1^2}{n_1} + \frac{s_2^2}{n_2}$. Then take the square root to get the estimated standard error, $\sqrt{\frac{s_1^2}{n_1} + \frac{s_2^2}{n_2}}$.

> Variances of independent samples add.

> *Standard error* refers to the standard deviation of the sampling distribution of a statistic.

More on Sample Sizes

You may have been surprised that, in Example 11.3, constructing a confidence interval was allowed even with such small samples that appear to come from skewed distributions. The guideline of having a sample size of at least 40 for moderately skewed populations is a bit conservative for inference for the difference of two means. In other words, the minimum sample sizes for estimating differences are higher than they need to be. The reason is that, with skewed populations, the sampling distribution of the difference of two means tends to be more symmetric than either of the two separate sampling distributions of the sample mean. Subtracting two sample means brings in the tails. The net effect is that, for smaller sample sizes, the sampling distribution of the difference of two means will look more approximately normal than would be the case for the means themselves.

More on Degrees of Freedom

In Examples 11.2 and 11.3, the degrees of freedom, 14.66 and 9.35, aren't whole numbers. What's that all about? The matter is fairly complicated, as you might have guessed, and that's why it is better to let the computer or calculator find the confidence interval. The basic idea is that, unlike in the one-sample case, the sampling distribution of the statistic for the difference of two samples doesn't have a t-distribution. The exact distribution isn't even known. However, it is known that the distribution is reasonably close to a t-distribution if the right number of approximate degrees of freedom is used. If you are interested in the formula that technology uses to compute these degrees of freedom, see D2.

> Degrees of freedom is a fraction?

DISCUSSION
Constructing a Confidence Interval for the Difference of Two Means

D2. Your calculator or statistics software uses this formula to find df when doing a two-sample t-procedure. You might be curious as to why we need this complicated rule to calculate degrees of freedom. The simple answer is that using t in place of z does not provide quite the right adjustment in the two-sample case, as it does in the one-sample case, unless we make this additional adjustment to the degrees of freedom. (The whole theoretical story is complicated.)

$$\frac{\left(\frac{s_1^2}{n_1} + \frac{s_2^2}{n_2}\right)^2}{df} = \frac{\left(\frac{s_1^2}{n_1}\right)^2}{n_1 - 1} + \frac{\left(\frac{s_2^2}{n_2}\right)^2}{n_2 - 1}$$

  a. Verify the value of df given in Example 11.2 on page 513 and Example 11.3 on page 514.

  b. If $n_1 = n_2$, derive a simplified version of the formula for df.

  c. If $n_1 = n_2$ and, in addition, $s_1 = s_2$, derive an even simpler rule for df.

Summary 11.1: A Confidence Interval for the Difference of Two Means

In this section, you learned how to use the results from two independent random samples to construct a confidence interval for the difference of the means of two populations, $\mu_1 - \mu_2$. The conditions to check are familiar:

- Samples have been randomly and independently selected from two different populations (or two treatments were randomly assigned to the available experimental units).

- The two samples look as if they could reasonably have come from normally distributed populations *or* the sample sizes are sufficiently large. The sample size guideline of $n \geq 40$ being sufficient for single samples may be relaxed somewhat when comparing two means.

- In the case of sample surveys, the population size should be at least ten times larger than the sample size, for both samples.

A confidence interval for the difference between the means of two populations, $\mu_1 - \mu_2$, has the form

$$(\bar{x}_1 - \bar{x}_2) \pm t^* \sqrt{\frac{s_1^2}{n_1} + \frac{s_2^2}{n_2}}$$

Because the procedure for finding t^* is complicated, it is best to use a calculator or statistics software to compute this confidence interval.

If zero is not in the confidence interval, then you have statistically significant evidence that there is a difference in the means of the two populations. If zero is in the confidence interval, then you don't have statistically significant evidence that there is any difference in the means of the two populations.

Practice

The Meaning of a Confidence Interval for the Difference of Two Means

P1. Using a variety of methods, including random sampling of beggars, researchers in Brussels, Belgium, made estimates of the mean value of a "gift" made in coin to beggars. The beggars who didn't have children with them were classified into two groups. The first group consisted of beggars who were natives to Belgium or who were native French or Dutch speakers. These people often were homeless and had a history of drug addiction. The second group consisted of Roma (gypsy) beggars, who often had a precarious residential status. The mean amount given to native beggars was 0.79 euros and to Roma beggars was 0.76 euros. A 95% confidence interval for the difference in the mean "gift" in coin to the two groups (*native – Roma*), in euros (€), is 0.03 ± 0.24. [Source: Stefan Adriaenssens and Jef Hendrickx, "The Income of Informal Economic Activities: Estimating the Yield of Begging in Brussels," HUB (Hogeschool-Universiteit Brussel) Research Paper 2008/16, April 2008.]

a. Describe the populations from which these samples were taken. What parameter is being estimated?

b. What is the interpretation of 0.03 in the confidence interval? What name is given to the value 0.24?

c. Write an interpretation of the confidence interval.

d. The researchers used a two-sample *t*-procedure and concluded that there is not a statistically significant difference in the mean amounts. Do you agree? Explain.

Constructing a Confidence Interval for the Difference of Two Means

P2. Display 11.6 gives the heart rates of random samples of men and women under normal conditions. Is there sufficient evidence to say that the mean heart rates differ for the two groups? To answer this question, check conditions and then construct and interpret a 95% confidence interval.

Heart Rate (beats per minute)

Men	Women
74	75
80	66
75	57
69	87
58	89
76	65
78	69
78	79
86	85
84	59
71	65
80	80
75	74

Display 11.6 Heart rates. [Source: *Journal of Statistics Education*, April 15, 2002 data archive, *www.amstat.org*.]

P3. The data on the number of hours studied per week in Display 11.7 are from a survey taken in an introductory statistics class during the first week of a term (see P7 in Chapter 10 on page 478). This class may be considered a random sample taken from all students in this course.

   a. Are conditions met for inference about the difference of the two means?

   b. Estimate the difference in mean study hours per week for all the males and females taking this course, with confidence level 0.90.

   c. Interpret this interval.

	Gender	N	Mean	Median	StDev
Study Hours	F	46	10.93	10.00	6.22
	M	15	8.20	7.00	5.94

Stem-and-leaf of Study
Gender = Female N = 46
Leaf Unit = 1.0

```
0  22233
0  55555
0  666777
0  888
1  00000011
1  22233
1  555555
1
1  88
2  0000
2  3
2
2
2
3  0
```

Stem-and-leaf of Study
Gender = Male N = 15
Leaf Unit = 1.0

```
0  23
0  4445
0  677
0  8
1  00
1  3
1  5
1
1
2
2
2  5
```

Display 11.7 Two stemplots of study hours per week, classified by gender.

P4. The National Survey of America's College Students (NSACS) examines the literacy of U.S. college students. The survey sampled about 1000 students from four-year colleges and 800 students from two-year colleges. Among the four-year college students, the mean quantitative literacy (QL) score was 330 with a standard deviation of 111, while the mean was 310 with a standard deviation of 79 for the two-year college students. [Source: *www.air.org/news/documents/The%20 Literacy%20of%20Americans%20College%20Students_ final%20report.pdf.*]

   a. Assuming that the samples were randomly and independently selected, are conditions for inference met?

   b. Estimate the difference in mean QL scores for these two populations.

   c. Interpret the resulting interval in the context of the problem.

Exercises

E1. Kelly randomly assigned eight golden hamsters to be raised in long days or short days. She then measured the concentrations of an enzyme in their brains. (Refer to Section 4.3 on page 195 for more about Kelly's hamster experiment.) The resulting measurements of enzyme concentrations (in milligrams per 100 milliliters) for the eight hamsters are shown in Display 11.8.

   a. Are the conditions met for inference about the difference of two means?

   b. Regardless of your answer to part a, construct a 95% confidence interval.

Short days: 12.500 11.625 18.275 13.225
Long days: 6.625 10.375 9.900 8.800

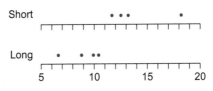

Display 11.8 A dot plot of Kelly's hamster data.

c. You are 95% confident that something is in the interval you constructed in part b. Describe exactly what that something is.

d. Does Kelly have statistically significant evidence to back her claim that the observed difference in enzyme concentrations between the two groups of hamsters is due to the difference in the hours of daylight?

E2. Suppose Kelly's means and the shapes of the distributions are the same as in E1 but the enzyme concentrations are more variable, as shown in Display 11.9.

Short days: 9.500   8.625   27.275   10.225

Long days: 4.625   12.375   11.900   6.800

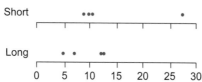

Display 11.9   Dot plots of altered hamster data.

a. With the altered data, are the conditions met for inference about the difference of two means?

b. Regardless of your answer to part a, construct and interpret a 95% confidence interval.

c. If the values had been this variable, would Kelly have had statistically significant evidence to back her claim that the observed difference in enzyme concentrations between the two groups of hamsters is due to the difference in the hours of daylight?

E3. The inflammation caused by osteoarthritis of the knee can be very painful and can inhibit movement. Leech saliva contains anti-inflammatory substances. To study the therapeutic effect of attaching four to six leeches to the knee for about 70 minutes, 51 volunteers were randomly assigned to receive either the leech treatment or a topical gel, diclofenac. [Source: Andreas Michalsen et al., "Effectiveness of Leech Therapy in Osteoarthritis of the Knee: A Randomized, Controlled Trial," *Annals of Internal Medicine*, Vol. 139, 9 (November 4, 2003), pp. 724–730.]

Medicinal leeches

a. This summary table gives the results of a pretreatment measure of the amount of pain reported by the two groups, before beginning therapy. A higher score means more pain. The researchers hoped that the randomization would result in two comparable groups with respect to this variable. Construct and interpret a 95% confidence interval for the difference in means. Is there statistically significant evidence that the randomization failed to yield groups with comparable means?

Leech: $\bar{x}_1 = 53.0$ $s_1 = 13.7$ $n_1 = 24$
Topical gel: $\bar{x}_2 = 51.5$ $s_2 = 16.8$ $n_2 = 27$

b. Because a high body mass can stress the knee, the researchers hoped that the randomization would result in two comparable groups with respect to this variable as well. This summary table gives the body mass index of the two treatment groups, before beginning therapy. Construct and interpret a 95% confidence interval for the difference in means. Is there statistically significant evidence that the randomization failed to yield groups with comparable means?

Leech: $\bar{x}_1 = 27.6$ $s_1 = 3.7$ $n_1 = 24$
Topical gel: $\bar{x}_2 = 27.1$ $s_2 = 3.7$ $n_2 = 27$

E4. Refer to E3. So far the researchers have been lucky, and the two treatment groups have comparable means. However, the researchers compared the two treatment groups on 12 initial variables. Another variable was an initial measure of stiffness.

a. This summary table gives summary statistics for the stiffness scores of the two groups, before beginning therapy. A higher score means more stiffness. Construct and interpret a 95% confidence interval for the difference in means. Is there statistically significant evidence that the randomization failed to yield groups with comparable means?

Leech: $\bar{x}_1 = 63.3$ $s_1 = 19.0$ $n_1 = 24$
Topical gel: $\bar{x}_2 = 48.6$ $s_2 = 22.2$ $n_2 = 27$

b. If you can consider the initial variables to be independent, how many of the 12 variables would you expect to show a statistically significant difference between two randomly assigned groups?

c. If you can consider the initial variables to be independent, what is the probability that none of the 12 variables shows a statistically significant difference between two randomly assigned groups?

d. At the end of the experiment, the leech treatment was shown to be significantly better than the topical gel. Does your result in part a tend to invalidate this conclusion?

E5. As you read in P6 in Chapter 10 on page 477, some students recruited 30 volunteers to attempt to walk the length of a football field while blindfolded. Each volunteer began at the middle of one goal line and was asked to walk to the opposite goal line, a distance of 100 yards. The dot plot and summary statistics in Display 11.10 show the distance at which the volunteer crossed a sideline of the field and whether the volunteer was left-handed or right-handed.

a. Assuming that these volunteers can be considered independent random samples from the populations

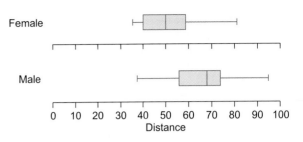

of left-handed volunteers and right-handed volunteers, are the conditions met for a confidence interval for the difference of two means?

b. Regardless of your answer to part a, construct a 95% confidence interval.

c. You are 95% confident that something is in the interval you constructed in part b. Describe exactly what that something is.

d. Do you have statistically significant evidence that left- and right-handed volunteers differ in the mean number of yards they can walk before crossing a sideline?

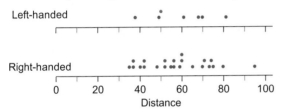

Male		Female
7	3	578
9	4	0228
62	5	02569
8100	6	5
54300	7	1
0	8	1
5	9	

Display 11.11 Boxplot and stemplot of yard-line data, categorized by gender.

Variable	Gender	N	Mean	Median	TrMean	StDev	SE Mean
Yard Line	F	15	51.40	50.00	50.38	13.40	3.46
	M	15	65.33	68.00	65.23	14.10	3.64

Variable	Gender	Min	Max	Q1	Q3
Yard Line	F	35.00	81.00	40.00	59.00
	M	37.00	95.00	56.00	74.00

Display 11.12 Summary statistics of yard-line data for males and females.

Descriptive Statistics

Variable	Handed	N	Mean	Median	TrMean	StDev	SEMean
Yard Line	L	7	59.57	61.00	59.57	14.77	5.58
	R	23	58.00	56.00	57.33	15.71	3.28

Variable	Handed	Min	Max	Q1	Q3
Yard Line	L	38.00	81.00	49.00	70.00
	R	35.00	95.00	42.00	71.00

Display 11.10 Yard-line data, categorized by handedness. [Source: Andrea Axtell, "When Direction Vanishes: Walking Straight or in Circles," *STATS*, Vol. 39 (2004), pp. 18–21.]

E6. Refer to E5, where students recruited 30 volunteers to attempt to walk the length of a football field while blindfolded. Each volunteer began at the middle of one goal line and was asked to walk to the opposite goal line, a distance of 100 yards. Display 11.11 shows the distance at which the volunteers crossed a sideline of the field and whether they were male or female.

a. Is this study more like an experiment or more like a survey?

b. Construct a 95% confidence interval for the difference in the mean number of yards walked for males and females. Use the summary statistics in the printout in Display 11.12. What can you conclude?

c. Display 11.13 shows the relationship between the height of the volunteer, in inches, and the number of yards he or she walked before crossing a sideline. What lurking variable can help explain your result in part b?

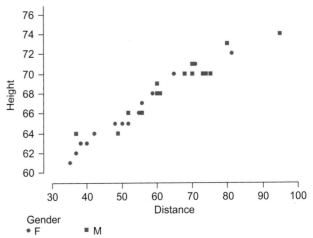

Display 11.13 Scatterplot of yard-line data, categorized by gender.

E7. A study to compare two insurance companies on length of stay (LOS) for pediatric asthma patients randomly sampled 393 cases from Insurer A. Summary statistics and a histogram for the data are shown in Display 11.14.

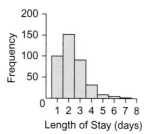

Summary of Insurer A

Length of Stay	
Count	393
Mean	2.32
SD	1.23

Display 11.14 Summary and data plot for lengths of hospital stays for Insurer A. [Source: R. Peck, L. Haugh, and A. Goodman, *Statistical Case Studies*, ASA-SIAM, 1998, pp. 45–64.]

An independent random sample of 396 cases from Insurer B gave the results on length of stay summarized in Display 11.15.

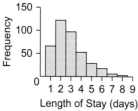

Summary of Insurer B:

Length of Stay	
Count	396
Mean	2.91
SD	1.58

Display 11.15 Summary and data plot for lengths of hospital stays for Insurer B.

a. Estimate the difference between the mean lengths of stay for the two insurance companies in a 95% confidence interval. Is there statistically significant evidence of a difference between the population means for the two companies?

b. Many other variables could contribute to the difference in mean length of stay for the two insurers. One is the number of full-time staff per bed. The sample means for this variable are 4.63 for Insurer A and 6.13 for Insurer B. The respective sample standard deviations are 1.70 and 2.40. Is

this difference in means statistically significant? If so, provide a practical explanation of how this difference might be related to the difference in mean length of stay.

c. Another contributing variable is the percentage of private hospitals versus public hospitals among the patients in each insurer's sample. For Insurer A, 93.6% of the sampled patients are in private hospitals; for Insurer B, that percentage is 73.0%. Is this a statistically significant difference? If so, how might this difference be related to the difference in mean length of stay?

E8. With regard to the speed of light measurements in E4 on page 480 in Chapter 10, Michelson actually took the measurements in five groups of 20, each successive group coming at a later time. Display 11.16 shows a plot and summary statistics for the first, fourth, and fifth groups of measurements. Assume that each group of 20 measurements was a random sample of the possible measurements Michelson could have taken at that time.

a. Using 95% confidence interval, estimate the difference in true velocities between groups 1 and 4.

b. Using 95% confidence interval, estimate the difference in true velocities between groups 4 and 5.

c. It is now known the Michelson's measurements were a little too high due to bias in his measurement process. Does he seem to be improving over time? Explain your reasoning.

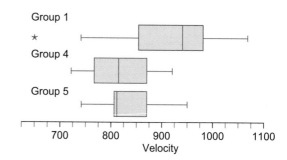

Group	Mean	Standard Deviation
1	909	105
4	820	60
5	832	54

Display 11.16 Velocity of light measurements in three groups.

E9. Heat flow (or heat transfer) is the passage of thermal energy from a hot to a cold body. This phenomenon is of particular interest to engineers, who attempt to understand and control the flow of heat through the use of thermal insulation and other devices. The data in Display 11.17 were taken from heat flow gauge readings for an industrial process over 50 equally spaced time intervals. A concern is that the process may be cooling down as time progresses. One way to check this statistically is to compare the measurements for the first 25 time periods (first half) to the measurements for the last 25 time periods (second half). The boxplots show that there may be a decrease in the mean, but the outliers make it difficult to confirm this through statistical analyses.

Time Period	Flow	Time Period	Flow
1	9.206	26	9.269
2	9.300	27	9.248
3	9.278	28	9.257
4	9.306	29	9.268
5	9.275	30	9.288
6	9.289	31	9.258
7	9.287	32	9.286
8	9.261	33	9.251
9	9.303	34	9.257
10	9.276	35	9.268
11	9.273	36	9.291
12	9.288	37	9.219
13	9.256	36	9.270
14	9.252	39	9.219
15	9.298	40	9.241
16	9.267	41	9.270
17	9.257	42	9.227
18	9.278	43	9.250
19	9.248	44	9.286
20	9.252	45	9.320
21	9.276	46	9.328
22	9.279	47	9.263
23	9.267	48	9.248
24	9.246	49	9.239
25	9.238	50	9.225

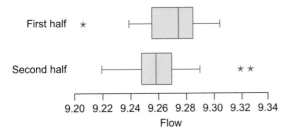

Display 11.17 Coded measures of heat transfer over 50 equally spaced time periods. [Source: *18www.illnist.gov/div898/hondbook/index.html.*]

a. Assume that these data can be regarded as random samples of all measurements that could have been collected during the first half and during the second half. Find a confidence interval estimate of the difference between population means for the two halves in the following output. Does this help confirm that the heat transfer is decreasing?

Two-Sample T-Test and CI: First Half, Second Half

Two-sample T for First Half vs. Second Half

	N	Mean	StDev	SE Mean
First Half	25	9.2702	0.0230	0.0046
Second Half	25	9.2622	0.0278	0.0056

Difference = mu (First Half) − mu (Second Half)
Estimate for difference: 0.00804
95% CI for difference: (−0.00646, 0.02254)
T-test of difference = 0 (vs. not =): T-value = 1.12
P-value = 0.270   DF = 46

b. Suppose that, on further checking, removal of the three outliers is justified. For example, the first outlier is the very first observation in the data set; so perhaps the system was not fully warmed up then. The following output gives the new confidence interval. Does this new interval help confirm that the heat transfer is decreasing?

Two-Sample T-Test and CI: First Half without Outlier, Second Half without Outliers

Two-sample T for First Half without Outlier vs. Second Half without Outliers

	N	Mean	StDev	SE Mean
First Half without Outlier	24	9.2729	0.0191	0.0039
Second Half without Outliers	23	9.2568	0.0215	0.0045

Difference = mu (First Half without Outlier) − mu (Second Half without Outliers)
Estimate for difference: 0.01609
95% CI for difference: (0.00413, 0.02805)
T-test of difference = 0 (vs. not =): T-value = 2.71
P-value = 0.010   DF = 43

E10. An important factor in the performance of a pharmaceutical product is how fast the product dissolves in vivo (in the body). This is measured by a dissolution test, which yields the percentage of the label strength (%LS) released after certain elapsed times. Laboratory tests of this type are conducted in vessels that simulate the action of the stomach. Display 11.18 shows %LS at certain time intervals for an analgesic (painkiller) tested in laboratories in New Jersey and Puerto Rico.

a. Check conditions and find a 90% confidence interval to estimate the difference in mean %LS at 40 minutes for New Jersey compared to Puerto Rico.

b. Check conditions and find a 90% confidence interval to estimate the difference in mean %LS at 60 minutes for New Jersey compared to Puerto Rico.

c. Good manufacturing practices call for equivalence limits of 15 percentage points for dissolution percentages below 90% and 7 percentage points for dissolution percentages above 90%. That is, if the 90%

confidence interval for the mean difference is within the equivalence limits (within an interval from –15% to +15% for percentages below 9), then the two sets of results are accepted as equivalent. Will the results in parts a and b be accepted as equivalent at 40 minutes? At 60 minutes?

New Jersey Elapsed Time (min)	Vessel No. (%LS)					
	V1	V2	V3	V4	V5	V6
0	0	0	0	0	0	0
20	5	10	2	7	6	0
40	72	79	81	70	72	73
60	96	99	93	95	96	99
120	99	99	96	100	98	100

Puerto Rico Elapsed Time (min)	Vessel No. (%LS)					
	V1	V2	V3	V4	V5	V6
0	0	0	0	0	0	0
20	10	12	7	3	5	14
40	65	66	71	70	74	69
60	95	99	98	94	90	92
120	100	102	98	99	97	100

Two-Sample T-Test and CI: NJ 40, PR 40

Two-sample T for NJ 40 vs. PR 40

	N	Mean	StDev	SE Mean
NJ 40	6	74.50	4.42	1.8
PR 40	6	69.17	3.31	1.4

Difference = mu (NJ 40) − mu (PR 40)
Estimate for difference: 5.33
90% CI for difference: (1.20, 9.46)
T-test of difference = 0 (vs. not =):
T-value = 2.37 P-value = 0.042 DF = 9

Two-Sample T-Test and CI: NJ 60, PR 60

Two-sample T for NJ 60 vs. PR 60

	N	Mean	StDev	SE Mean
NJ 60	6	96.33	2.34	0.95
PR 60	6	94.67	3.44	1.4

Difference = mu (NJ 60) − mu (PR 60)
Estimate for difference: 1.67
90% CI for difference: (−1.49, 4.83)
T-test of difference = 0 (vs. not =):
T-value = 0.98   P-value = 0.356   DF 8

Display 11.18 Percentage of label strength of analgesic for dissolution tests in two laboratories. [Source: R. Peck, L. Haugh, and A. Goodman, *Statistical Case Studies*, 1998, ASA-SIAM, pp. 37–44.]

11.2 ▶ A Significance Test for the Difference of Two Means

In the previous section, you learned to construct and interpret a confidence interval for the difference of two means. As you know, the other way to approach an inference problem is through a test of the significance, which is the subject of this section.

STATISTICS IN ACTION 11.1 ▶ How Strong Is the Evidence?

Before you begin to develop formal methods of a significance test for differences between means, a little practice can help you develop your intuition and your ability to make informal judgments based on comparing means.

Display 11.19 shows eight back-to-back stemplots. The first plot shows actual data. It compares concentrations of aldrin for 20 water samples taken from the Wolf River downstream from a dump site. (See Example 11.1 on page 511.) Ten of the samples were taken at mid-depth, and ten were taken at the bottom where the concentrations were expected to be higher. The other seven plots, numbered 1 through 7, show hypothetical data.

Actual Data		
Bottom		Mid-Depth
	2	
8	3	2 8
9 8	4	3 8 9
7 4 3	5	2 2
3	6	2 3 6
3	7	
8 1	8	
	9	

Hypothetical Data Set 1		
Bottom		Mid-Depth
	2	
	3	2 8
8	4	3 8 9
9 8	5	2 2
7 4 3	6	2 3 6
3	7	
3	8	
8 1	9	

Hypothetical Data Set 2		
Bottom		Mid-Depth
	2	
	3	8
9 8 8	4	2 3 8 9
7 4 3	5	2 2 6
3	6	2 3
8 3	7	
1	8	
	9	

Hypothetical Data Set 3		
Bottom		Mid-Depth
8	2	8
	3	2
9 8	4	3
7 4 3	5	8 9
3	6	2 2
3	7	2 6
8	8	3
1	9	

Hypothetical Data Set 4		
Bottom		Mid-Depth
	2	
8 8	3	2 2 8 8
9 9 8 8	4	3 3 8 8 9 9
7 7 4 4 3 3	5	2 2 2 2
3 3	6	2 2 3 3 6 6
3 3	7	
8 8 1 1	8	
	9	

Hypothetical Data Set 5		
Bottom		Mid-Depth
	2	
8	3	2
9	4	3 8
4 3	5	2
3	6	3 6
3	7	
8	8	
	9	

Hypothetical Data Set 6		
Bottom		Mid-Depth
	2	
	3	
8	4	2 8
9 8	5	3 8 9
7 4 3	6	2 2
3	7	2 3 6
3	8	
8 1	9	

Hypothetical Data Set 7		
Bottom		Mid-Depth
	2	
8	3	
9 8	4	2 8
7 4 3	5	3 8 9
3	6	2 2
3	7	2 3 6
8 1	8	
	9	

Display 11.19 Concentrations of aldrin (2|8 means 2.8 nanograms per liter) for samples from the Wolf River, actual data and seven hypothetical data sets.

1. Compare the plot of the actual data with the plot of each hypothetical data set, 1 through 7. Is the evidence that the mean concentration is higher at the bottom stronger in the actual data, stronger in the hypothetical data, or about the same for both? State your choice for each hypothetical data set and give the reason for your answer.

2. With everything else the same, is the evidence of a difference in concentration stronger when the difference in the means is smaller or larger? When the spread is smaller or larger? When the sample size is smaller or larger?

3. Here are the summary statistics and P-values for one-sided tests of significance of the difference in the means. Are the results consistent with the intuitive judgments you made about the strength of the evidence?

Actual Data: *P*-value = 0.0618

Bottom:	$\bar{x}_1 = 6.04$	$s_1 = 1.6$	$n_1 = 10$
Mid-depth:	$\bar{x}_2 = 5.05$	$s_2 = 1.1$	$n_2 = 10$

Hypothetical Data Set 1: *P*-value = 0.0026

Bottom:	$\bar{x}_1 = 7.04$	$s_1 = 1.6$	$n_1 = 10$
Mid-depth:	$\bar{x}_2 = 5.05$	$s_2 = 1.1$	$n_2 = 10$

Hypothetical Data Set 2: *P*-value = 0.0278

Bottom:	$\bar{x}_1 = 6.04$	$s_1 = 1.3$	$n_1 = 10$
Mid-depth:	$\bar{x}_2 = 5.05$	$s_2 = 0.8$	$n_2 = 10$

Hypothetical Data Set 3: *P*-value = 0.3663

Bottom:	$\bar{x}_1 = 6.04$	$s_1 = 1.9$	$n_1 = 10$
Mid-depth:	$\bar{x}_2 = 5.75$	$s_2 = 1.8$	$n_2 = 10$

Hypothetical Data Set 4: *P*-value = 0.0121

Bottom:	$\bar{x}_1 = 6.04$	$s_1 = 1.5$	$n_1 = 20$
Mid-depth:	$\bar{x}_2 = 5.05$	$s_2 = 1.1$	$n_2 = 20$

Hypothetical Data Set 5: *P*-value = 0.1449

Bottom:	$\bar{x}_1 = 5.97$	$s_1 = 1.7$	$n_1 = 7$
Mid-depth:	$\bar{x}_2 = 5.07$	$s_2 = 1.3$	$n_2 = 6$

Hypothetical Data Set 6: *P*-value = 0.0618

Bottom:	$\bar{x}_1 = 7.04$	$s_1 = 1.6$	$n_1 = 10$
Mid-depth:	$\bar{x}_2 = 6.05$	$s_2 = 1.1$	$n_2 = 10$

Hypothetical Data Set 7: *P*-value = 0.5064

Bottom:	$\bar{x}_1 = 6.04$	$s_1 = 1.6$	$n_1 = 10$
Mid-depth:	$\bar{x}_2 = 6.05$	$s_2 = 1.1$	$n_2 = 10$

A Significance Test for the Difference of Two Means

Because you learned the basic ideas of inference for the difference of two means in the previous section, you can get started right away with a significance test. Here's a summary of significance testing for the difference between two means.

Components of a Test for the Difference Between Two Means (the Two-Sample *t*-Test)

1. *Write a null and alternative hypothesis.* The null hypothesis is ordinarily that the two population means are equal. In symbols, $H_0: \mu_1 = \mu_2$ or, in terms of the difference between the means, $H_0: \mu_1 - \mu_2 = 0$. Here μ_1 is the mean of the first population and μ_2 is the mean of the second. There are several forms of the alternative or research hypothesis:

$$H_a: \mu_1 \neq \mu_2 \quad \text{or} \quad H_a: \mu_1 - \mu_2 \neq 0$$

$$H_a: \mu_1 < \mu_2 \quad \text{or} \quad H_a: \mu_1 - \mu_2 < 0$$

$$H_a: \mu_1 > \mu_2 \quad \text{or} \quad H_a: \mu_1 - \mu_2 > 0$$

For a randomized comparative experiment, the null hypothesis is that if all subjects had received the first treatment, the mean would be the same as if all subjects had received the second treatment.

2. *Check conditions.* The methods of this section require the same conditions as those for a confidence interval:

- The two samples were randomly and independently selected from two different populations. In the case of an experiment, check that the two treatments were randomly assigned to the available experimental units.

- It is reasonable to assume that each of the two samples came from normally distributed populations or the sample sizes are sufficiently large. As noted in the conditions for confidence intervals, you can get by with smaller sample sizes when taking a difference.

- In the case of sample surveys, the population size should be at least ten times larger than the sample size for both samples.

3. *Compute the test statistic, find the P-value, and draw a sketch.* Compute the difference between the sample means (because the hypothesized mean difference is zero), measured in estimated standard errors:

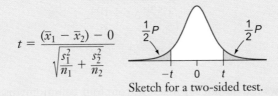

$$t = \frac{(\bar{x}_1 - \bar{x}_2) - 0}{\sqrt{\dfrac{s_1^2}{n_1} + \dfrac{s_2^2}{n_2}}}$$

Sketch for a two-sided test.

Use the two-sample *t*-test function of your calculator or statistics software to get the *P*-value.

> With Table B, use $df = (n_1 - 1) + (n_2 - 1)$.

4. *Write your conclusion, linked to your computations and in the context of the problem.* Write a conclusion that relates to the situation and includes an interpretation of your *P*-value as the strength of the evidence against the null hypothesis. If you are using fixed-level testing, reject the null hypothesis if your *P*-value is less than the level of significance, α. If the *P*-value is greater than or equal to α, do not reject the null hypothesis. (If you are not given a value of α, you can assume that α is 0.05.)

A Significance Test for the Aldrin Data

Example 11.4

You've been given the responsibility to analyze the Wolf River data in Display 11.1 on page 511 to test whether the true mean aldrin concentrations at the bottom and at mid-depth might differ. The confidence interval included zero, so you didn't have statistically significant evidence that the means differ. However, the information on page 511 says that aldrin should tend to settle on the bottom of the river. So, this actually is a one-sided situation, whereas the confidence interval from Example 11.1 on page 511 was two-sided. Conduct a one-sided test of significance for the difference between the mean aldrin concentration at the bottom and at mid-depth. Use $\alpha = 0.05$.

Solution
In terms of the difference between two means, your null hypothesis is

> Write a null and alternative hypothesis.

$$H_0: \mu_{\text{bottom}} = \mu_{\text{mid-depth}} \quad \text{or} \quad H_0: \mu_{\text{bottom}} - \mu_{\text{mid-depth}} = 0$$

where μ_{bottom} is the mean aldrin concentration at the bottom of the Wolf River and $\mu_{\text{mid-depth}}$ is the mean concentration at mid-depth. You are looking for a difference in favor of the bottom, so the alternative hypothesis is one-sided:

$$H_0: \mu_{\text{bottom}} > \mu_{\text{mid-depth}} \quad \text{or} \quad H_0: \mu_{\text{bottom}} - \mu_{\text{mid-depth}} > 0$$

Check conditions.

It is reasonable to assume that each sample in Display 11.1 was taken from a normally distributed population. But we have no information about randomness, so the conclusion will be subject to that concern.

Compute the test statistic, find the *P*-value, and draw a sketch.

Here are the summary statistics for the aldrin concentrations:

$$\text{Bottom:} \quad \bar{x}_1 = 6.04 \quad s_1 = 1.579 \quad n_1 = 10$$

$$\text{Mid-depth:} \quad \bar{x}_2 = 5.05 \quad s_2 = 1.104 \quad n_2 = 10$$

The value of the test statistic is

$$t = \frac{(\bar{x}_1 - \bar{x}_2) - 0}{\sqrt{\dfrac{s_1^2}{n_1} + \dfrac{s_2^2}{n_2}}}$$

$$= \frac{(6.04 - 5.05) - 0}{\sqrt{\dfrac{1.579^2}{10} + \dfrac{1.104^2}{10}}}$$

$$\approx 1.625$$

Display 11.20 shows the output for this test from four commonly used statistical software packages. Using a calculator or computer, you get an approximate *df* of 16.10 and a *P*-value of 0.0618.

Give the conclusion in context.

Because the *P*-value for a one-sided test is greater than $\alpha = 0.05$, you do not reject the null hypothesis. There is insufficient evidence to claim that the mean aldrin concentration at the bottom of the Wolf River is larger than the mean mid-depth concentration. In other words, although it appears from the stem-and-leaf plot that aldrin concentrations are greater near the bottom, with these sample sizes the difference is not large enough to rule out chance variation as a possible explanation.

JMP

Means and Std Deviations

Level	Number	Mean	Std Dev	Std Err Mean	Lower 95%	Upper 95%
Bottom	10	6.04000	1.57917	0.49938	4.9103	7.1697
Mid-Depth	10	5.05000	1.10378	0.34905	4.2604	5.8396

t-Test

Mid-Depth-Bottom

Assuming unequal variances

Difference	−0.9900	t Ratio	−1.62489
Std Err Dff.	0.6093	DF	16.0994
Upper CL Dff	0.3009	Prob > |t|	0.1236
Lower CL Dff	−2.2809	Prob > t	0.9382
Confidence	0.95	Prob < t	0.0613

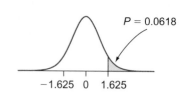

$P = 0.0618$

−1.625 0 1.625

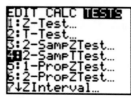  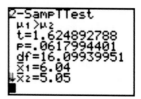

Fathom

Test of Test of Wolf River

First attribute (numeric): Bottom
Second attribute (numeric or categorical): Mid_Depth

Ho: Population mean of **Bottom** equals that of
Mid_Depth
Ha: Population mean of **Bottom is greater than** that of
Mid_Depth

	Bottom	**Mid_Depth**
Count:	**10**	**10**
Mean:	**6.04**	**5.05**
Std dev:	**1.57917**	**1.10378**
Std error:	**0.499377**	**0.349046**

Using **unpooled variances**
Student's t:	**1.625**
DF:	**16.0994**
P-value:	**0.062**

Minitab

TWOSAMPLE T FOR Bottom VS MidDepth

	N	MEAN	STDEV	SE MEAN
Bottom	10	6.04	1.58	0.50
MidDepth	10	5.05	1.10	0.35

TTEST MU Bottom $=$ MU MidDepth (VS GT): T $= 1.62$ P $= 0.062$ DF $= 16$

Display 11.20   Output for significance test from four statistical software packages.

For the aldrin data, it makes scientific sense to use a one-sided alternative. Aldrin tends to stick to particles of sediment and so should tend to end up on the bottom. Researchers wanted to know whether by taking samples at mid-depth they were underestimating the concentrations of aldrin (and might miss the presence of aldrin altogether). For Example 11.3 on page 514, a two-sided alternative makes more sense because, although the researchers were hoping that the special exercises help babies walk sooner, they should be open to the possibility that the 12 minutes of exercise selected by the parent might be superior.

Walking Babies Experiment, Revisited

Example 11.5

Refer to the data in Example 11.3 on page 514. Test the null hypothesis that the mean age at first unaided steps is the same for the special exercises and exercise control against the alternative hypothesis that the mean ages are different. Use $\alpha = 0.05$.

Solution
The null hypothesis is

$$H_0\text{: } \mu_{\text{special ex}} = \mu_{\text{control ex}}$$

or equivalently,

$$H_0\text{: } \mu_{\text{special ex}} - \mu_{\text{control ex}} = 0$$

where $\mu_{\text{special ex}}$ is the mean age that the babies in the experiment would first walk if they all could have been given the special exercises and $\mu_{\text{control ex}}$ is the mean age that the babies would first walk if all babies in the experiment could have received the exercise control treatment. (You can also use the symbols μ_1 and μ_2 as long as you define the symbols.)
For this two-sided test, the alternative hypothesis is

$$H_0\text{: } \mu_{\text{special ex}} \neq \mu_{\text{control ex}}$$

Write a null and alternative hypothesis.

or equivalently,

$$H_0: \mu_{\text{special ex}} - \mu_{\text{control ex}} \neq 0$$

The conditions were checked in Example 11.3 on page 514.

Check conditions.

Here are the summary statistics:

Special exercises: $\bar{x}_1 = 10.125$ $s_1 = 1.447$ $n_1 = 6$
Exercise control: $\bar{x}_2 = 11.375$ $s_2 = 1.896$ $n_2 = 6$

Compute the test statistic, find the *P*-value, and draw a sketch.

The test statistic is

$$t = \frac{(\bar{x}_1 - \bar{x}_2) - 0}{\sqrt{\dfrac{s_1^2}{n_1} + \dfrac{s_2^2}{n_2}}}$$

$$= \frac{(10.125 - 11.375) - 0}{\sqrt{\dfrac{1.447^2}{6} + \dfrac{1.896^2}{6}}}$$

$$\approx -1.284$$

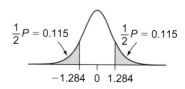

From a calculator or computer, *df* is 9.35 and the *P*-value is 0.230.

Give conclusion linked to computations and in context.

Because the *P*-value, 0.230, is greater than $\alpha = 0.05$, you do not reject the null hypothesis. The evidence isn't convincing that babies who are given the special exercises, on average, walk at a different age than babies who are given the control exercise. Even if each baby would have responded the same way no matter which treatment was assigned, it is reasonably likely to get a difference in means as large as or larger than the one from this experiment simply as a result of the particular random assignment of the treatments to these babies.

DISCUSSION
A Significance Test for the Difference of Two Means

D3. In the previous examples, the difference in the means was not statistically significant. Does this mean that the null hypothesis is true?

Increasing the Power of the Two-Sample *t*-Test

Power is the ability to reject the null hypothesis.

As always, the best way to get more power to reject a false null hypothesis is to increase the sample sizes. However, sometimes you have the resources to get, say, only 40 measurements total between the two samples. How should you divide them up—20 in each sample or some other way?

> ### Getting the Most Power Out of Your Two-Sample *t*-Test
>
> The best way to get more power is to have larger sample sizes.
>
> If you have reason to believe that the population standard deviations are about equal, make the sample sizes equal. If you have reason to believe that one population's standard deviation is larger than the other's, allocate your resources so that you take a larger sample from the population with the larger standard deviation. (Choose the sample sizes to be proportional to the estimated standard deviations.)
>
> Follow these same rules to get the smallest margin of error for a confidence interval, with a fixed sample size.

Why not two separate confidence intervals? Significance tests for the difference of two means take a little getting used to. It is natural to ask why you can't simply compute two separate confidence intervals, one for μ_1 and one for μ_2, and check to see if they overlap. This method will tell you if there are any values that are plausible means for both populations. If so, then you wouldn't reject the null hypothesis that the difference in the means is 0. The difficulty is that the method is too conservative, meaning that you won't reject a false null hypothesis often enough. In other words, you have sacrificed power.

However, you can use this rule: If you construct two separate confidence intervals for the means of two populations (at confidence level $1 - \alpha$) and they don't overlap, you are safe in rejecting the null hypothesis that the means are equal, at significance level α. If the intervals overlap, you cannot come to a conclusion.

A Special Case: Pooling When $\sigma_1 = \sigma_2$

Suppose you are doing a significance test for the purpose of comparing means and the two populations are actually identical. This happens, for example, when you randomly divide available experimental units into two groups for the purpose of comparing two treatments and the treatments have the same effect on the experimental units. Then, the true means do not differ (the usual null hypothesis), and the true variances should not differ either. In such cases, the two sample variances (that probably will differ) can be combined to estimate the single population variance. One way to combine them is simply to average the sample variances and use this average to estimate the population variance. (This average should be a weighted average if the sample sizes are not equal.) This process, called *pooling*, can give your *t*-test more power.

Your calculator and statistics software give you the choice of pooled or unpooled when doing two-sample *t*-procedures. Pooled should be used only when you have a good reason to believe that the population standard deviations are equal (which would almost never be the case). Even if you know this to be true, the two-sample (unpooled) procedure discussed in this chapter works almost as well as the pooled procedure, especially if the sample sizes are equal. The only situation in which the pooled procedure has a definite advantage is when the populations have equal standard deviations but your sample sizes are unequal. If you are in any doubt whether σ_1 and σ_2 are equal, choose the unpooled procedure.

> Almost always select the unpooled procedure.

Analysis of variance, ANOVA, which you will study in Chapter 14, is an extension of the two-sample procedures that allows you to compare more than two means. It is a generalization of the pooled procedure for two samples so it is good to know when pooling works and when it doesn't.

DISCUSSION
Increasing the Power of the Two-Sample *t*-Test

D4. *To pool or not to pool?* Here are three different situations for two independent samples.

$\quad$ I. $n_1 = 5$, $n_2 = 25$, $\sigma_1 = 10$, $\sigma_2 = 10$

$\quad$ II. $n_1 = 10$, $n_2 = 10$, $\sigma_1 = 10$, $\sigma_2 = 10$

$\quad$ III. $n_1 = 5$, $n_2 = 25$, $\sigma_1 = 10$, $\sigma_2 = 1$

For one of these situations, pooling is wrong. For a second situation, pooling, though not wrong, is not likely to offer an advantage over the unpooled approach. Finally, in the third situation, pooling not only is appropriate but also will likely give narrower intervals than the unpooled approach. Which is which?

Summary 11.2: A Significance Test for the Difference of Two Means

In this section, you learned how to perform a test of the significance of the difference between two sample means. You tested the null hypothesis that the means of the two populations from which the samples were taken are equal, $\mu_1 = \mu_2$. Whether you are doing a confidence interval or a significance test, the conditions to check are the same:

- Samples have been randomly and independently selected from two different populations (or two treatments were randomly assigned to the available experimental units).

- The two samples look as if they could have reasonably come from normally distributed populations or the sample sizes are sufficiently large. The sample size guideline of $n \geq 40$ that is sufficient for single samples may be relaxed somewhat when comparing two means.

- In the case of sample surveys, the population size should be at least 10 times larger than the sample size for both samples.

The test statistic for a test of significance is

$$t = \frac{(\bar{x}_1 - \bar{x}_2) - 0}{\sqrt{\dfrac{s_1^2}{n_1} + \dfrac{s_2^2}{n_2}}}$$

This statistic is called t, but it doesn't have exactly a t-distribution. Fortunately, it is reasonably accurate to proceed using the t-distribution with df approximated by a rather complicated rule. A calculator or statistics software is the most accurate method of finding a P-value. Always select the unpooled t-procedure unless you have good reason to do otherwise.

Practice

A Significance Test for the Difference Between Two Means

P5. In P3 on page 519, you saw the following data from a random sample of students taking introductory statistics. Each student reported the number of hours studied. You will decide whether there is a statistically significant difference in the mean number of hours studied by males and females.

a. Write the null and alternative hypotheses, defining your symbols.

b. What condition is not met?

c. Compute the value of t.

d. Find and interpret the P-value. What is your conclusion?

e. Remove the outlier for the males and recompute the P-value. Does your conclusion change?

	Gender	N	Mean	Median	StDev
Study Hours	F	46	10.93	10.00	6.22
	M	15	8.20	7.00	5.94

Stem-and-leaf of Study
Gender = Female N = 46
Leaf Unit = 1.0

```
0   22233
0   55555
0   666777
0   888
1   00000011
1   22233
1   555555
1
1   88
2   0000
2   3
2
2
2
3   0
```

Stem-and-leaf of Study
Gender = Male N = 15
Leaf Unit = 1.0

```
0   23
0   4445
0   677
0   8
1   00
1   3
1   5
1
1
2
2
2   5
```

P6. A researcher observed prices of typically purchased grocery products from stores in the richest (median family income of $46,594), middle (median family income of $33,765), and poorest (median family income of $16,371) neighborhoods in Buffalo, New York. The results are given in the following table. A relative price index of 1.05, for example, indicates that the item was priced 5% higher than the lowest price available in any of the other neighborhoods. You will decide whether there is statistically significant evidence that the mean relative price index is higher in the poorest neighborhoods than in the richest neighborhoods. While this wasn't a random sample of prices from the types of neighborhoods, you may assume that it was for the purposes of this problem.

Neighborhood	Number of Observed Prices	Price Index Mean	SD
Richest	1575	1.11	0.07
Middle	1485	1.14	0.05
Poorest	1912	1.22	0.03

[Source: Debabrata Talukdar, "Cost of Being Poor: Retail Price and Consumer Price Search Differences Across Inner-City and Suburban Neighborhoods," *Journal of Consumer Research*, Vol. 35 (October 2008), pp. 457–471.]

a. Is this a one-sided or a two-sided test? Write the null and alternative hypotheses, defining your symbols.
b. Because individual prices aren't given, you can't check the normality condition by making a plot. Does that matter in this situation?
c. Compute the value of *t*.
d. Find and interpret the *P*-value. Is the difference statistically significant?

P7. The fact that your food usually tastes good is no accident. Food manufacturers regularly check taste and texture by recruiting taste-test panels to measure palatability. A standard method is to form a panel with 50 persons—25 men and 25 women—to do the tasting. In one such experiment, simplified here, coarse versus fine texture was compared. Panel members were assigned randomly to the treatment groups as they were recruited. There were 16 panels of 50 consumers each. The variables in Display 11.21 are these:

Total palatability score for the panel of 50: A higher total score means the food was rated more palatable by the panel.
Texture: 0 (coarse), 1 (fine)

Is there a statistically significant difference in mean palatability score between the two texture levels? Show all four steps in a significance test when answering this question.

Score	Texture
35	0
39	0
77	0
16	0
104	1
129	1
97	1
84	1
24	0
21	0
39	0
60	0
65	1
94	1
86	1
64	1

Display 11.21 Food ratings. [Source: E. Street and M. G. Carroll, "Preliminary Evaluation of a Food Product," in *Statistics: A Guide to the Unknown*, ed. Judith M. Tanur, (San Francisco: Holden-Day, 1972), pp. 220–238.]

Increasing the Power of the Two-Sample *t*-Test

P8. The actual aldrin data appear in Display 11.22. Your boss sends you out to get ten more measurements. Would you suggest getting five additional measurements at each depth, or something different? Explain.

Bottom		Mid-Depth
	2	
8	3	28
98	4	389
743	5	22
3	6	236
3	7	
81	8	
	9	

Display 11.22 Aldrin concentrations taken from the river bottom and at mid-depth.

P9. Refer to Example 10.2 on page 470. From that example the 95% confidence intervals for the mean body temperature of males, (97.48, 98.28), and the mean body temperature of females, (98.14, 98.90), overlap. For example, if you look at the two genders separately, 98.2°F is a plausible mean temperature for both (but is near opposite ends of the confidence intervals).

a. Construct and interpret a 95% confidence interval for the difference in mean temperatures. You can assume that the conditions are met.

b. What can you conclude about whether males and females have the same mean body temperature?

c. When using the confidence intervals to test the hypothesis that there is no difference in population means, which procedure had more power: comparing two separate confidence intervals, or constructing one confidence interval for the difference?

Exercises

E11. In E21 and E22 of Chapter 10 on page 493, students were given pieces of paper with five vertical or horizontal line segments and asked to mark the midpoint of each segment. Then, using a ruler, each student measured how far each mark was from the real midpoint of each segment. He or she then averaged the five errors. (We aren't told if the assignments of the two treatments were random, but assume they were because this experiment was done in a statistics class.) The results are given in Display 11.23.

Average Error (cm) in Locating Midpoint of Segment

Vertical (positive means too high)	Horizontal (positive means too far right)
1.6	−0.9
1.2	−0.4
1.1	−0.4
1.1	−0.2
1.0	−0.2
0.8	0.0
0.4	0.0
0.4	0.0
0.4	0.0
0.4	0.2
0.2	0.4
0.2	0.4
0.0	0.4
0.0	0.8
−0.4	

Display 11.23 Average error in locating midpoints of horizontal and vertical line segments. [Source: Gretchen Davis, "Beware of Measurement Bias," *CMC ComMuni-Cator*, Vol. 22, 4 (June 1998), pp. 54–56.]

a. Before beginning this experiment, the instructor believed that students tend to make larger errors (in absolute value) when the line segments are vertical than when the line segments are horizontal. Make the necessary changes to the data in the table so that you can test the instructor's belief.

b. Conduct a significance test of the instructor's claim, showing all four steps.

E12. It has been speculated that the mean amount of calcium in the blood is higher in women than in men. A retrospective review of the medical charts of subjects tested in a certain city produced the summary statistics on calcium shown in Display 11.24.

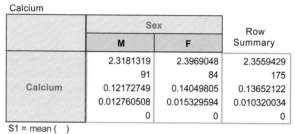

Calcium

	Sex		Row Summary
	M	**F**	
Calcium	2.3181319	2.3969048	2.3559429
	91	84	175
	0.12172749	0.14049805	0.13652122
	0.012760508	0.015329594	0.010320034
	0	0	0

S1 = mean ( )
S2 = count ( )
S3 = stdDev ( )
S4 = stdError ( )
S5 = count (missing ( ))

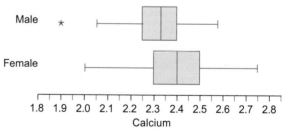

Display 11.24 Calcium in the blood of women and men, in millimoles per liter. [Source: JSE Data Archive, *www.amstat.org*.]

a. Discuss whether the conditions are met for doing a two-sample *t*-test of the difference in the means.

b. Find and interpret the *P*-value for this test.

E13. Aerobic exercise is good for the heart. A group of college students interested in the effect of stepping exercises on heart rate conducted an experiment in which subjects were randomly assigned to a stepping exercise on either a 5.75-in. step (coded 0) or an 11.50-in. step (coded 1). Each subject started with a resting heart rate and performed the exercise for 3 minutes, at which time his or her exercise heart rate was recorded. (This is a simplification of the actual design.) The data are shown in Display 11.25.

Height of Step (in.)	Resting Heart Rate	Exercise Heart Rate	Height of Step (in.)	Resting Heart Rate	Exercise Heart Rate
0	60	75	1	69	135
0	63	84	1	69	108
0	69	93	1	96	141
0	90	99	1	87	120
0	87	129	1	93	153
0	69	93	1	72	99
0	72	99	1	78	93
0	87	93	1	78	129
0	75	123	1	87	111
0	81	96	1	81	120
0	75	90	1	84	99
0	78	87	1	84	99
0	84	84	1	90	129
0	90	108	1	84	90
0	78	96	1	90	147

Display 11.25 Resting and exercise heart rates (in beats per minute), categorized by height of step. [Source: *lib.stat.cmu.edu*.]

a. You want to test whether the higher step resulted in a larger mean gain in heart rate than did the lower step. Are conditions met? Explain.

b. Test whether the higher step resulted in a larger mean increase in heart rate.

E14. Refer to E13. Is there statistically significant evidence that the initial random assignment of treatments to subjects did not do a satisfactory job of equalizing the mean resting heart rates between the two treatment groups?

E15. In warm and humid parts of the world, a constant battle is waged against termites. Scientists have discovered that certain tree resins are deadly to termites, and thus the trees producing these resins become a valuable crop. In one experiment typical of the type used to test the protective power of a resin, two doses of resin

(5 mg and 10 mg) were dissolved in a solvent and placed on filter paper. Eight dishes were prepared with filter paper at dose level 5 mg and eight with filter paper at dose level 10 mg. Twenty-five termites were then placed in each dish to feed on the filter paper. At the end of 15 days, the number of surviving termites was counted. The results are shown in Display 11.26, on the next page. In parts a–c, you'll determine if there is a statistically significant difference in the mean number of survivors for the two doses.

a. Are the conditions for a two-sample *t*-test met here?

b. Regardless of your answer to part a, carry out an appropriate test. Write your conclusion in the context of the study.

c. What are your main concerns about the conclusion you reached in part b? What advice would you give the experimenter?

Dose (mg)	Count of Survivors
5	11
5	11
5	12
5	12
5	5
5	9
5	6
5	10
10	16
10	13
10	1
10	0
10	0
10	0
10	0
10	3

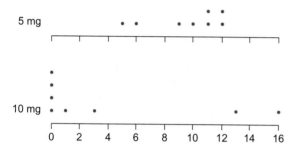

E16. The finishing stages in the manufacture of high-performance ceramic devices involve grinding the material for smooth finish and proper dimension. The ceramic material has a sort of grain, much like wood, and so the grinding can take place either with the grain (longitudinal) or across the grain (transverse). An experiment was conducted to compare the strength of the material after grinding in the two grinding directions. Two different batches of ceramic material were used in the study, resulting in an additional factor. The data and output, provided in Display 11.27, use numerical codes to indicate the categorical variables *direction* and *batch* and a coded measure of strength. In the experiment, the combinations of direction and batch were run in random order.

a. Is there a statistically significant difference between the means for direction of grinding?

b. Is there a statistically significant difference between the means for the two batches?

c. Regardless of your answer to part b, the means for the two batches do differ somewhat. Do you think this difference has any serious affect on the difference in means between the directions? Explain your reasoning.

E17. Refer to the walking babies data in Display 11.3 on page 515.

a. Test to see if there is a statistically significant difference in the mean walking age for the special exercises treatment and the weekly report treatment. Use $\alpha = 0.05$.

b. Test to see if there is a statistically significant difference in the mean walking age for the exercise control treatment and the weekly report treatment. Use $\alpha = 0.05$.

c. Compare your results to those in Example 11.3 on page 514. State a conclusion about comparisons among the three treatments.

Descriptive Statistics: Survivors

Variable	Dose	N	Mean	SE Mean	St Dev	Minimum	Q1	Median	Q3	Maximum
Survivors	5	8	9.500	0.945	2.673	5.000	6.750	10.500	11.750	12.000
	10	8	4.13	2.31	6.53	0.00	0.00	0.50	10.50	16.00

Two-Sample T-Test and CI: Survivors, Dose

Two-sample T for Survivors

Dose	N	Mean	St Dev	SE Mean
5	8	9.50	2.67	0.94
10	8	4.13	6.53	2.3

Difference = mu (5) − mu (10)
Estimate for difference: 5.38
95% CI for difference: (−0.27, 11.02)
T-Test of difference = 0 (vs. not =): T-value = 2.15   P-value = 0.060   DF = 9

Display 11.26 Number of termites, out of 25, surviving after being placed in a dish with 5 mg or 10 mg of a resin. [Source: *lib.stat.cmu.edu.*]

Direction −1 = longitudinal +1 = transverse	Batch	Strength
−1	−1	680.45
−1	−1	722.48
−1	−1	702.14
−1	−1	666.93
−1	−1	703.67
−1	−1	642.14
−1	−1	692.98
−1	−1	669.26
1	−1	491.58
1	−1	475.52
1	−1	478.76
1	−1	568.23
1	−1	444.72
1	−1	410.37
1	−1	428.51
1	1	491.47
−1	1	607.34
−1	1	620.80
−1	1	610.55
−1	1	638.04
−1	1	585.19
−1	1	586.17
−1	1	601.67
−1	1	608.31
1	1	442.90
1	1	434.41
1	1	417.66
1	1	510.84
1	1	392.11
1	1	343.22
1	1	385.52
1	1	446.73

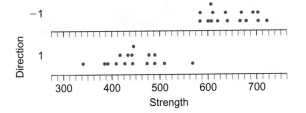

Two-Sample T-Test and CI: Strength, Direction

Two-sample T for Strength

Direction	N	Mean	St Dev	SE Mean
−1	16	646.1	45.3	11
1	16	447.7	54.7	14

Difference = mu (−1) −mu (1)
Estimate for difference: 198.5
95% CI for difference: (162.2, 234.8)
T-Test of difference = 0 (vs. not =): T-value = 11.18
    P-value = 0.000   DF = 29

Two-Sample T-Test and CI: Strength, Batch

Two-sample T for Strength

Batch	N	Mean	St Dev	SE Mean
−1	16	579	115	29
1	16	514	102	26

Difference = mu (−1) −mu (1)
Estimate for difference: 64.9
95% CI for difference: (−14.0, 143.8)
T-Test of difference = 0 (vs. not =): T-value = 1.68
    P-value = 0.103   DF = 29

Display 11.27 Strength of ceramic after grinding. [Source: *www.itl.nist.gov/div898/handbook/pri/section4/ pri471.htm.*]

E18. To study the effect of reducing television viewing time on children's obesity, researchers conducted a randomized experiment using as subjects 3rd- and 4th-grade students at two schools in the same city. The intervention being tested was to educate children about television watching and give them incentives to watch less TV. The outcome measures were body mass index and other variables related to obesity.

The randomization, however, was to schools rather than to students, with one school randomly assigned the intervention (for all of its participating students) and the other school being the control (no intervention for any student). This may or may not be a good method of randomization so the first question to study is "Do the treatment and control groups look alike before the intervention is applied, or are there important differences that cannot be explained by chance?" Display 11.28 gives the mean and standard deviation for a few of the items the researchers measured in that regard.

a. Conduct statistical tests to check if these differences in means can be relegated to chance.

b. For which of the three variables should you be most concerned about the lack of balance between the two groups?

	Intervention, $n = 92$	Control, $n = 100$
Body mass index (BMI)	18.38(3.67)	18.10(3.77)
TV watching (hours per week)	15.35(13.17)	15.46(15.02)
Physical activity (minutes per week)	396.8(367.8)	310.2(250.7)

Display 11.28. Initial variables with means (SD). [Source: *http://jama.ama-asm.org/cgi/reprint/282/16/1561.pdf.*]

c. Comment on whether you think the randomization by school was a good idea, based on these measures and other concerns you might have.

d. For the second and third variables in the table, the standard deviations are nearly as large as the means. What does this suggest about the shape of the distributions for these data sets? Should the shape be a major concern with these sample sizes?

e. After the intervention, the mean number of hours per week of TV watching was 8.80 ($s = 10.41$) for the intervention group and 14.46 ($s = 13.82$) for the control group. Assuming that you have independent random samples, is this a statistically significant difference?

f. For hours of TV watching, can you compare the before-intervention mean of 15.35 and the after-intervention mean of 8.80 by the methods of inference learned in this section? Why or why not?

E19. Refer to Kelly's hamster experiment in E1 on page 519.

a. What kind of error could Kelly be making in a significance test of the difference in mean enzyme concentration?

b. What should Kelly do in a similar experiment if she wants to reduce the probability of an error of this type?

c. What about this situation gave Kelly power to reject the null hypothesis even with such small treatment group sizes?

E20. Refer to the taste-test experiment in P7 on page 533.

a. What kind of error might investigators be making in this significance test?

b. What should the investigators do in a similar experiment if they want to reduce the probability of this type of error?

E21. *Narcissistic* is a term used to describe people who are excessively in love with themselves. To assess whether celebrities are more narcissistic than other people (duh!), the Narcissistic Personality Inventory (NPI) was given, confidentially, to 144 male and 56 female celebrities who appeared as guests on the radio show *Loveline*, which is hosted by Dr. Drew. The males averaged 17.27 with a standard deviation of 6.78 and the females averaged 19.26 with a standard deviation of 6.34. Higher scores indicate more narcissism. A mean score on this test of 15.3 for people from the United States was established through a very large sample. [Source: S. Mark Young and Drew Pinsky, "Narcissism

and Celebrity," *Journal of Research in Personality*, Vol. 40 (2006), pp. 463–471.]

a. Which test should you use to decide whether the difference between male celebrities and female celebrities is statistically significant? Perform this test and write a conclusion.

b. Which test should you use to decide whether male celebrities are more narcissistic, on average, than the general public? Perform this test using $\alpha = 0.05$ and write a conclusion.

c. The table in Display 11.29 gives the mean and standard deviation of the scores on the component scales of the NPI. For which component scales is the difference between males and females statistically significant? Use $\alpha = 0.01$.

d. In part c, you were asked to use $\alpha = 0.01$ because you are doing multiple tests. When doing multiple tests, you should use a smaller value of α. Which type of error does this practice help avoid?

E22. Eighty undergraduate students were randomly assigned to be either truth-tellers or liars. Here's how the researchers describe the setup for their experiment:

The 40 truth tellers participated in a staged event in which they played a game of Connect 4 with a confederate (who posed as another participant). During the game they were interrupted twice, first by another confederate who came in to wipe a blackboard and later by a third confederate who entered looking for his or her wallet. Upon finding the wallet, this latter confederate then claimed that money had gone missing from it. The participant was then told that she would be interviewed about the missing money.

The 40 liars did not participate in this staged event. Instead, they were asked to take the money from the wallet, but deny having taken it in a subsequent interview. They were told to tell the interviewer that they played a game of Connect 4 just as the truth tellers had.

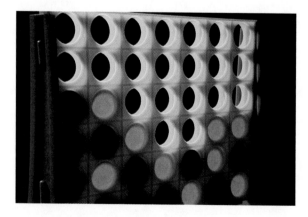

Participants were then asked by an interviewer, who was blind to the treatment they received, to describe in detail what had happened. Half of the truth-tellers and half of the liars were asked to tell the story in reverse order. Thus, there were four treatments, each with

	Authority		Exhibitionism		Superiority		Entitlement		Exploit-ativeness		Self-sufficiency		Vanity	
	Mean	SD	Mean	SD	Mean	SD	Mean	SD	Mean	SD	Mean	SD	Mean	SD
Male	4.93	2.03	2.35	1.93	1.67	1.20	2.01	1.58	1.97	1.35	3.33	1.40	1.01	1.03
Female	5.29	1.75	3.02	1.67	1.93	1.12	1.91	1.45	2.00	1.45	3.47	1.48	1.64	1.07

Display 11.29 Scores on component scales of the Narcissistic Personality Inventory for 144 male and 56 female celebrities.

20 subjects. The table in Display 11.30 lists some common behaviors that might distinguish truth-tellers from liars and the mean ratings on each behavior for each of the four treatment groups.

a. Consider only the control condition of describing the story in the order that it happened. For which behaviors is the difference between the mean for the truth-tellers and the mean for the liars statistically significant? Use $\alpha = 0.01$.

b. Now consider only the reverse order condition of describing what had happened in the reverse order.

For which behaviors is the difference between the mean for the truth-tellers and the mean for the liars statistically significant? Use $\alpha = 0.01$.

c. What can you conclude from parts a and b about how you might spot a liar?

d. In parts a and b, you were asked to use $\alpha = 0.01$ because you are doing multiple tests. Explain why doing multiple tests should make a difference in the level of significance you use.

	Reverse Order				Control Condition			
	Truth		Lie		Truth		Lie	
	Mean	SD	Mean	SD	Mean	SD	Mean	SD
Auditory details	9.00	4.6	5.35	3.6	6.25	4.2	8.85	4.2
Contextual embeddings	18.95	7.3	12.15	5.9	17.05	10.3	17.10	7.4
Speech hesitations	4.54	2.6	5.96	2.0	4.36	2.2	3.99	2.0
Speech rate	163.62	29.2	142.13	13.7	162.84	21.0	169.02	32.9
Hand/finger	11.78	9.8	15.41	13.1	14.55	8.7	7.33	6.4
Leg/foot	5.21	5.3	13.80	16.9	14.97	21.1	6.28	5.8
Cognitive operations	1.75	2.0	4.35	4.6	2.05	3.9	3.65	3.4
Speech errors	1.26	0.7	1.91	1.2	1.45	1.2	1.51	1.2
Eye blinks	17.38	9.7	27.07	12.7	19.56	9.1	24.75	12.1

Display 11.30 Behaviors for truth-tellers and liars under two conditions. [Source: Aldert Vrij et al., "Increasing Cognitive Load to Facilitate Lie Detection: The Benefit of Recalling an Event in Reverse Order," *Law and Human Behavior*, Vol. 32 (2008), pp. 253–265.]

11.3 ▶ Inference for Paired Comparisons

Now that you've seen the methods for comparing means, it's time to put them to work. In this section, you'll see confidence intervals and significance tests in action. Keep in mind that a *t*-test is no smarter than a chainsaw. Neither has any brains of its own. A chainsaw can't tell whether it's cutting an old dead tree into firewood or turning a valuable antique table into scrap wood. A *t*-test is every bit as oblivious. The difference between thoughtful and careless use is up to you, the operator. This final section looks at three issues on which you need to be clear in order to use your statistical tools with care.

• Do you really have two independent samples, or do you have only one sample of paired data?

• Is it meaningful to compare means?

• Does your inference have the chance it needs?

A Significance Test for Paired Observations

Sometimes your samples are independent and sometimes they contain paired observations. For example, suppose that you manufacture shoes and so need to know if, on average, left feet are shorter than right feet. One way to gather data would be to measure the left feet of a random sample of people and then measure the right feet of a different random sample of people. This design, with two independent random samples, may be analyzed using the techniques of Sections 11.1 and 11.2.

But you wouldn't gather your data this way. You would, instead, get one random sample of people and measure both of their feet. You should analyze these data by finding the differences, *right foot − left foot*, and then testing whether the mean difference is 0 using the procedure summarized in the following box. It turns out that this design, which seems natural and is less expensive to carry out, also has statistical advantages.

Components of a Test of the Mean Difference Based on Paired Observations

1. *Write a null and alternative hypothesis.* The null hypothesis is, ordinarily, that in the entire population the mean of the differences is 0. In symbols, $H_0: \mu_d = 0$.

   There are three forms of the alternative or research hypothesis:
   $$H_a: \mu_d \neq 0 \qquad H_a: \mu_d < 0 \qquad H_a: \mu_d > 0$$

2. *Check conditions.* For a test of significance of a mean difference based on paired observations, the method requires the same conditions as those for a confidence interval. You must check three conditions:

   - *Randomness.* In the case of a survey, you must have a random sample from one population, where a "unit" might consist of, say, a pair of twins or the same person's two feet. In the case of an experiment, the treatments must have been randomly assigned within each pair. If the same subject is given both treatments, the treatments must have been assigned in random order.

   - *Normality.* The differences must look like it's reasonable to assume that they came from a normally distributed population *or* the sample size must be sufficiently large.

   - *Sample size.* In the case of sample surveys, the size of the population of differences should be at least ten times as large as the number of differences in the sample.

3. *Compute the test statistic, find the P-value, and draw a sketch.* Compute the difference between the mean difference, $\bar{d}$, from the sample and the hypothesized difference μ_{d_0}, and then divide by the estimated standard error:

$$t = \frac{\bar{d} - \mu_{d_0}}{s_d/\sqrt{n}}$$

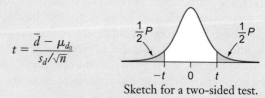

Sketch for a two-sided test.

Here, $\bar{d}$ is the mean of the differences between pairs of measurements, s_d is the sample standard deviation, and the degrees of freedom are $df = n-1$, where n is the number of differences.

4. *Write your conclusion linked to your computations and in the context of the problem.* Write a conclusion that relates to the situation and includes an interpretation of your *P*-value as the strength of the evidence against the null hypothesis. If you are using fixed-level testing, reject the null hypothesis if your *P*-value is less than the level of significance, α. If the *P*-value is greater than or equal to α, do not reject the null hypothesis. (If you are not given a value of α, you can assume that α is 0.05.)

Example 11.6 shows the steps in a test of the significance of the mean difference from paired data collected in a sample survey.

A Hard Grader

Example 11.6

Students in a large section of a business calculus class complained that their professor graded their midterms too hard, with not enough partial credit and too many points off for minor mistakes. The professor took their complaints seriously. She took a random sample of 12 of the 136 midterms and asked a colleague, who also teaches business calculus, to grade them. (She thought that 12 exams were about the most she could ask a colleague to grade.) The results are given in Display 11.31. Is there statistically significant evidence that the professor is grading harder than her colleague?

Midterm	Grade Given by Colleague	Grade Given by Professor	Difference, colleague – professor
A	97	98	−1
B	100	95	5
C	46	39	7
D	52	59	−7
E	99	87	12
F	99	89	10
G	93	89	4
H	65	65	0
I	72	67	5
J	70	61	9
K	63	52	11
L	85	87	−2

Display 11.31 Grades given by professor and by a colleague, and difference.

Solution
The grades on the 12 midterms are paired, one from the professor and one from her colleague: so you should use a one-sample *t*-test on the differences. This is a one-sided situation because the professor has been accused of grading harder than other instructors. The hypotheses are:

Write hypotheses.

H_0: $\mu_d = 0$, where μ_d is the mean of the differences, *grade given by colleague − grade given by professor*, for the population of all of the midterms given to that section of business calculus.

H_a: $\mu_d > 0$

Check conditions: randomness and normality.

The exams were randomly selected from a population that is more than ten times the sample size. The differences, *grade given by colleague − grade given by professor*, are shown in the boxplot. There are no outliers and little evidence of skewness. It's reasonable to assume that these differences are a random sample taken from a normal distribution.

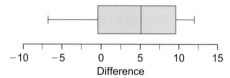

Do computations.

The mean of the differences is 4.41667 and the standard deviation is 5.88462. The test statistic is

$$t = \frac{\bar{x} - \mu_0}{\frac{s}{\sqrt{n}}} = \frac{4.41667 - 0}{\frac{5.88462}{\sqrt{12}}} \approx 2.60$$

With 11 degrees of freedom, the *P*-value is 0.012.

Give interpretation in context.

Reject the null hypothesis because the *P*-value is less than 0.05. The professor is busted. There is statistically significant evidence that she gives the same test a lower grade, on average, than her colleague. In terms of the *P*-value, suppose that the colleague also had graded all of the midterms. If the mean difference between the two sets of grades was 0, then there is only a 0.012 chance of getting a difference as large or larger than the mean difference of 4.41667 observed from this random sample. That is so unlikely, that we don't believe that the mean difference would be 0. A difference this small, only 4.4 points, may be of little practical importance, but it is statistically significant.

A Confidence Interval for Paired Observations

You can estimate the size of a mean difference by constructing a confidence interval. The next box summarizes how to construct a confidence interval for the mean difference.

> ### A Confidence Interval for the Mean Difference from Paired Observations
>
> 1. *Check conditions.* You must check three conditions:
>   - *Randomness.* In the case of a survey, you must have a random sample from one population, where a "unit" might consist of, say, a pair of twins or the same person's two feet. In the case of an experiment, the treatments must have been randomly assigned within each pair. If the same subject is assigned both treatments, the treatments must have been assigned in random order.
>   - *Normality.* The differences must look like it's reasonable to assume that they came from a normally distributed population *or* the sample size must be sufficiently large.
>   - *Sample size.* In the case of a sample survey, the size of the population of differences should be at least ten times as large as the sample size.
>
> 2. *Do computations.* A confidence interval for the mean difference, μ_d, is given by
>
> $$\bar{d} \pm t^* \frac{s_d}{\sqrt{n}}$$

where n is the sample size, $\bar{d}$ is the mean of the differences between pairs of measurements, s_d is the sample standard deviation, and t^* depends on the confidence level desired and the degrees of freedom, $df = n - 1$.

3. *Give interpretation in context.* An interpretation for a sample survey is of this form: "I am 95% confident that the mean of the population of differences, μ_d, is in this confidence interval."

   An interpretation for an experiment is of this form: "I am 95% confident that, if all subjects had received the first treatment and all subjects had received the second treatment, the mean of the differences would fall in this confidence interval."

Highway Versus City

Example 11.7

The table in Display 11.32 gives the highway and city gas mileage for a random sample of eight 2010 car models with automatic transmissions. Construct and interpret a 95% confidence interval for the mean difference in highway and city gas mileage.

Model	Highway mpg	City mpg
Buick LaCrosse/Allure	26	17
Chevrolet Aveo	34	25
Chevrolet Malibu	30	22
Honda Civic	36	24
Mercedes-Benz C300 4MATIC	24	17
Mitsubishi Eclipse	27	20
Toyota Camry	32	22
Volvo S40	31	20

Display 11.32 Highway and city gas mileage in miles per gallon for a random sample of 2010 car models. [Source: *Model Year 2010 Fuel Economy Guide*, U.S. Department of Energy, *www.fueleconomy.gov.*]

Solution

For these paired data, the eight differences, *highway − city*, are 9, 9, 8, 12, 7, 7, 10, and 11. Conditions are met because this is a random sample from a population of more than 8(10), or 80, car models and the following dot plot shows that it is reasonable to assume that the differences came from a normal distribution.

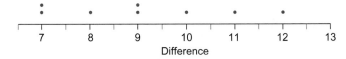

The differences have mean $\bar{d} = 9.125$ mpg and standard deviation $s_d \approx 1.808$ mpg. A 95% confidence interval for the mean difference is

$$\bar{d} \pm t^*_{n-1} \frac{s_d}{\sqrt{n}} = 9.125 \pm 2.365 \cdot \frac{1.808}{\sqrt{8}}$$

or (7.61, 10.64).

You are 95% confident that the true mean difference between highway and city gas mileage for all 2010 car models is between 7.61 mpg and 10.64 mpg. The confidence interval does not overlap 0, so the difference is statistically significant.

STATISTICS IN ACTION 11.2 ▶ Hand Spans

What You'll Need: a ruler marked in millimeters

Detective Sherlock Holmes amazed a man by relating "obvious facts" about him, such as that he had at some time done manual labor: "How did you know, for example, that I did manual labour? It is as true as gospel, for I began as a ship's carpenter." Sherlock replied, "Your hands, my dear sir. Your right hand is quite a size larger than your left. You have worked with it, and the muscles are more developed." [Source: Sir Arthur Conan Doyle, *The Adventures of Sherlock Holmes*, ed. Richard Lancelyn Green, (Oxford and New York: Oxford World Classics, 1988).] In fact, people's right hands tend to be bigger than their left, even if they are left-handed and even if they haven't done manual labor, but the difference is small. To detect it, you will have to design your study carefully.

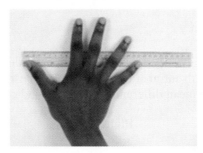

1. Measure your left and right hand spans, in millimeters. (An easy way to this is to spread your hand as wide as possible, place it directly on a ruler, and get the distance between the end of your little finger and the end of your thumb.) Record the data for each student in your class in a table with a column for the left hand span and another column for the right. There should be one row for each person.

2. For each row in the table, calculate the difference, *right − left*. Find the mean of the differences and the standard error of the mean difference using the formula $s_d = s/\sqrt{n}$.

3. Now make a new table, but this time randomize the order of the right hand spans so that people's left hand spans are no longer matched with their right. Then repeat step 2 with the new table.

4. Finally, treat the left hand spans and right hand spans as independent samples. Calculate the difference between the two sample means and the standard error of that difference using the formula

$$s_{x_R - x_L} = \sqrt{\frac{s_R^2}{n_R} + \frac{s_L^2}{n_L}}$$

5. Compare the standard errors from steps 2, 3, and 4. Which is smallest? Which two are most nearly the same size?

6. Suppose you make scatterplots of the data in your two tables from steps 1 and 3, with the data for the left hand spans on the horizontal axis. Which would you expect to have the higher correlation? Why? Make the scatterplots, and calculate the correlations to check your answer.

Two Independent Samples, or Paired Observations?

One of the recurring themes in statistics is how important it is to pay attention to data production. This theme was developed in Chapter 4, where the entire emphasis was on designing studies. Now the theme returns in the context of inference for means, where one of the key questions is, "Do I have paired data or independent samples?"

Paired observations can greatly reduce variation over independent samples and produce a much more powerful test and a more precise confidence interval estimate of the true mean difference. The reduction in variation is greatest when the underlying measurements vary greatly from pair to pair but the differences do not. (The measurements are *homogeneous* within pairs but *heterogeneous* between pairs.)

A class of 28 students conducted an experiment to compare sitting and standing pulse rates, using three different designs. You will now work through the analysis of each of the three sets of experimental results, beginning with the completely randomized design. As you will see, the type of design determines the type of analysis.

Sit and Stand—Completely Randomized Design (Two Independent Samples)	**Example 11.8**

The students first compared sitting and standing pulse rates using a completely randomized design. Half of the class was randomly assigned to sit and the other half to stand. The results are given in Display 11.33.

	Stand	Sit
	66	70
	82	88
	86	82
	102	88
	62	66
	70	70
	50	72
	62	86
	56	74
	104	86
	86	80
	86	46
	80	54
	96	86
Mean	77.71	74.86
Standard Deviation	17.04	13.00

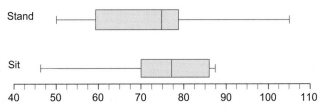

Display 11.33 Data for sitting versus standing pulse rates, in beats per minute, for a completely randomized design.

Is there statistically significant evidence that there is a difference in the mean pulse rates for standing and sitting?

> Here, you test that the difference of the means is 0.

Solution

As is typical in an experiment, the treatment groups were formed by randomly dividing the available subjects into two groups. The data can be analyzed using the two-sample t-test introduced in Section 11.2. The test statistic is

$$t = \frac{(\bar{x}_1 - \bar{x}_2) - 0}{\sqrt{\frac{s_1^2}{n_1} + \frac{s_2^2}{n_2}}} = \frac{77.71 - 74.86}{\sqrt{\frac{17.04^2}{14} + \frac{13.00^2}{14}}} \approx 0.4988$$

With $df \approx 24.31$, the P-value is 0.62.

The samples are not random samples from two larger populations, so the interpretation of the P-value should apply only to this group of subjects: Suppose that all 28 subjects were given the *stand* treatment and you compute their mean pulse rate. Suppose that all 28 subjects were given the *sit* treatment and you compute their mean pulse rate. Then, if the null hypothesis is true that there is no difference in these two means, there is a 0.62 chance of getting a difference in means as large or larger in absolute value than the difference from this particular random assignment of treatments to subjects.

With a P-value this large, you can't conclude that, with these subjects, one of these treatments produces a higher mean pulse rate than the other. The difference in the means is not statistically significant.

| **Example 11.9** | **Sit and Stand: Matched Pairs Design** |

In the matched pairs design, each student took his or her pulse rate while sitting. Then the students were matched (blocked), with the two students with the two highest pulse rates matched, the two students with the third and fourth highest pulse rates matched, and so on. Then, each pair of students flipped a coin to see who would sit and who would stand. Finally, each student counted his or her pulse rate using the treatment assigned. The results are given in Display 11.34. The last column shows the difference in pulse rates for each pair.

Is there statistically significant evidence that the mean difference in the pulse rates for standing and sitting is different from 0?

Solution

The data from this design should *not* be analyzed using the techniques for two independent samples. The matched pairs design has dependent observations within a pair, so the two-sample t-test is not a valid option. You can, however, look at the difference between the standing and sitting pulse rates for each pair and estimate the mean difference with a one-sample procedure. Observe in Display 11.34 that the difference between the sample means, $77.57 - 73.86$, is the same as the mean of the differences, 3.71. In general, the difference in the means is equal to the mean difference.

> The two sets of measurements become one set of differences.

The test statistic is

$$t = \frac{\bar{x} - \mu_0}{\frac{s}{\sqrt{n}}} = \frac{3.71 - 0}{\frac{12.38}{\sqrt{14}}} \approx 1.12$$

> Here, you test that the mean difference is 0.

With 13 degrees of freedom, the P-value is 0.28.

If the null hypothesis is true that the true mean value of the difference, *stand* − *sit*, is 0, there is a 0.28 chance of getting a mean difference as large or larger in absolute value than the mean difference from this particular random assignment of treatments to subjects.

While the P-value of 0.28 is smaller than the P-value of 0.62 for the completely randomized design, the mean difference is not statistically significant. You can't conclude that, with these subjects, one of these treatments produces a higher mean pulse rate than the other.

Stand	Sit	Difference (stand − sit)
68	62	6
78	74	4
80	82	−2
92	88	4
58	82	−24
96	66	30
72	64	8
100	84	16
82	72	10
76	82	−6
92	80	12
74	72	2
60	64	−4
58	62	−4
Mean 77.57	73.86	3.71
Standard Deviation 13.86	9.13	12.38

Display 11.34 Data for sitting versus standing pulse rates, in beats per minute, for a matched pairs design.

Sit and Stand: Repeated Measures Design

Example 11.10

In the matched pairs design, students were matched on a preliminary measure of pulse rate. Then, within each pair, one student was randomly assigned to sit and the other to stand. In the repeated measures design, each student did both sitting and standing, with a rest period in between and with the order of sitting and standing randomly assigned. This effectively doubles the sample size from 14 to 28, giving the test more power. But there is more going on here than just increased sample size. Check out Display 11.35. All of the differences are positive. Everyone's pulse rate was higher when standing. Pretty clearly, the mean difference now is going to be statistically significant.

Is there statistically significant evidence that the mean difference in the pulse rates for standing and sitting is different from 0?

Solution

Because the order in which each subject receives the two treatments is randomized, you can treat this as a random assignment of treatments to subjects. Again, you should

Stand	Sit	Difference (*stand − sit*)
64	60	4
72	70	2
76	72	4
82	78	4
92	80	12
98	84	14
68	60	8
64	62	2
70	66	4
86	72	14
100	82	18
80	74	6
58	50	8
54	52	2
66	64	2
76	70	6
86	76	10
88	70	18
96	88	8
86	80	6
56	54	2
82	68	14
96	86	10
74	68	6
80	68	12
58	48	10
72	64	8
94	74	20

	Stand	Sit	Difference
Mean	77.64	69.29	8.36
Standard Deviation	13.57	10.67	5.28

−30 −20 −10   0   10   20   30   40
Difference

Display 11.35 Data for sitting versus standing pulse rates, in beats per minute, for the repeated measures design.

> Here, you test that the mean difference is 0.

use a one-sample *t*-test because the measurements within each pair are dependent. The test statistic is

$$t = \frac{\bar{x} - \mu_0}{\frac{s}{\sqrt{n}}} = \frac{8.36 - 0}{\frac{5.28}{\sqrt{28}}} \approx 8.34$$

With 27 degrees of freedom, the *P*-value is less than 0.0001.

If the null hypothesis is true that the true mean value of the difference, *stand − sit*, is 0, there is less than a 0.0001 chance of getting a mean difference as large or larger in absolute value than the mean difference from this particular random assignment in the order of receiving treatments. In other words, the mean difference is greater than you typically would get if you randomly assigned one of the measurements within each pair to sitting and the other to standing and the treatment made no difference.

The mean difference is statistically significant. You can conclude that, with these subjects, standing produces a higher mean pulse rate than sitting.

Example 11.10 shows the value of the repeated measures design in increasing the power of your test. However, as you learned in Chapter 4, repeated measures designs often aren't possible when the effect of the first treatment doesn't wash out in a reasonable period of time.

DISCUSSION
Two Independent Samples, or Paired Observations?

D5. Based on what you recall from Chapter 4 and what you saw in the analysis of the pulse rate data in this section, discuss the relative merits of completely randomized, matched pairs, and repeated measures designs. Under what conditions will the analysis of differences between pairs pay bigger dividends than the analysis of differences of independent means?

D6. Display 11.36 shows the relationship between the sitting and standing pulse rates for pairing of values at random from the two samples in the completely randomized design (I) and from the pairings in the matched pairs design (II) and the repeated measures design (III). For which design is the correlation strongest? Weakest? Explain why this should be the case.

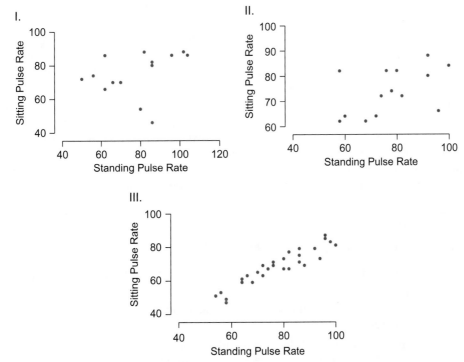

Display 11.36 Scatterplots of sitting versus standing pulse rates, in beats per minute, in a completely randomized design (with pairs formed at random), a matched pairs design, and a repeated measures design.

Is It Meaningful to Compare Means?

"When your only tool is a hammer, everything looks like a nail." By now you have quite a variety of statistical tools, but it's still worth reminding yourself to be thoughtful about when and how you use them. Before you start any computations, always ask yourself, "What is the question of interest?" Then ask whether it makes sense to try to answer that question by comparing means.

| Example 11.11 | **Comparing Aptitude Tests** |

Business firms often give aptitude tests to job applicants to see if they have the necessary skills and interests to adapt well to any training they might have to complete. Suppose a company desiring to use such a test has two versions it wants to compare, Test A and Test B. To see how well the exams predict aptitude for a job, the company decides to try both exams on a sample of recently hired employees who have completed the training (and whose aptitude they now know). Is the company interested in comparing the mean scores of the two groups?

Solution

Probably not. The two tests might be graded on different scales, so a comparison of means would be meaningless anyway. More important, the company really is interested in how well the tests select the applicants with the greatest aptitude, so they might want to compare, say, the top 10% of test takers on each exam to see if they really are the employees with the greatest aptitude. In general, the variability of the aptitude test scores is more important than the mean. Those with strong aptitude should score high. Those with weak aptitude should score low, with enough variability in scores so that the two groups are substantively different.

Does Your Inference Have the Chance It Needs?

In statistics, exploratory methods (like those you used in Chapters 2 and 3) look for patterns but make no assumptions about the process that created the data—with exploratory methods, what you see is what you get. Inference can deliver more than exploration, going beyond simply saying, "Here are some interesting patterns." For inference to be justified, however, you need the right kind of data. Provided your numbers come from random samples or randomized experiments, you can use probability theory and the predictable regularities of chance-like behavior to draw conclusions not only about the data you see but also about the unseen population from which they came or about the treatments that were assigned to the experimental units.

DISCUSSION
Is It Meaningful to Compare Means?

D7. Make a list of ways your data can fall short of the requirements for statistical inference on means.

D8. For which of the following scenarios would a comparison of means be of little interest?

    A. comparing crop yields for two brands of corn

    B. comparing the heights of two rivers while making a decision on which should receive a new dike system

    C. deciding on the smallest number of points that would result in getting the top score of a 5 on an AP exam so as to be consistent with previous years

> **D.** deciding which grocery store offers the lower prices for the foods you buy regularly

Summary 11.3: Inference for Paired Comparisons

Textbook illustrations of statistical procedures are always neater than their real-world applications. When confronted with a real problem involving two samples that have not been "sanitized" for textbook use, you should ask some key questions.

- "Are these paired observations, or two independent samples?" If the data are paired, they are likely to be correlated, and the two-sample procedures are not the correct method. The better method is to analyze the differences between pairs as a single random sample using the methods of this section.

- "How skewed is the population from which these differences came?" If you have paired differences, check to see if it's reasonable to assume that the differences are a random sample taken from a normally distributed population.

- "Is this a meaningful comparison?" Anyone armed with statistical software can compare the means of any two sets of data. But some might result in comparing the proverbial apples and oranges, and some might involve comparing samples that were constructed to have different means in the first place.

- "Is there any chance mechanism underlying the selection of these data?" Statistical inference is based on probability theory, and the procedures work only under the condition that the data come from random samples or randomized experiments. You can then draw conclusions not only about the data you see but also about the unseen population from which they came or about why the treatment groups differ.

When the data come paired, you should find the difference of the values within each pair and then use the one-sample t-test or t-confidence interval that you learned in Sections 10.1 and 10.2.

Practice

A Significance Test for Paired Observations

P10. A class conducted an experiment to compare how far they could launch a gummy bear using homemade catapults constructed to several different specifications. You would expect that teams launching gummy bears using the same catapult would get better with practice. Display 11.37 on the next page shows the results of the first and tenth launch for six teams. You will test whether there is statistically significant evidence that teams improved from launch 1 to launch 10. Use $\alpha = 0.01$

a. Should you use a two-sample test of the difference of two means or a one-sample test of the mean of the differences? Explain. Should this be a one-sided test or a two-sided test?

b. State hypotheses for your test, check conditions, do the computations, and write a conclusion in context.

c. Does the result of the test seem sensible? What type of error might you have made?

d. If you had used the other test, would the conditions have been satisfied?

P11. Refer to P10. This time you will test to see if there is statistically significant evidence that teams improved from launch 5 to launch 10. Use $\alpha = 0.01$. The data are given in Display 11.38.

a. Should you use a test of the difference of two means or a test of the mean of the differences? Explain. Should this be a one-sided test or a two-sided test?

Team	First Launch (in.)	Tenth Launch (in.)
1	15	74
2	44	120
3	18	92
4	13	33
5	10	37
6	125	174

Launch distance

Display 11.37 Bears-in-space data.

Team	Fifth Launch (in.)	Tenth Launch (in.)
1	19	74
2	37	120
3	88	92
4	41	33
5	65	37
6	125	174

Display 11.38 More bears-in-space data.

b. State hypotheses for your test, check conditions, do the computations, and write a conclusion in context.
c. What type of error might you have made?
d. Does the result of the test seem sensible compared with the result in P10?

A Confidence Interval for Paired Observations

P12. A class conducted an experiment to compare how far they could launch a gummy bear using homemade catapults constructed to several different specifications. Display 11.39 gives the average distance soared by the gummy bears launched by the six teams under two conditions: one book under the launching ramp and four books under the launching ramp. This was a randomized block design, with a team being the block. Whether the team did one book first or four books first was decided

by the flip of a coin. You want to estimate the mean difference of the distance soared (*four books − one book*) with 95% confidence.

a. Should you use a one-sample procedure or a two-sample procedure? Explain.
b. Are the conditions satisfied for constructing a confidence interval using your procedure?
c. Construct and interpret the 95% confidence interval.

Team	One Book (in.)	Four Books (in.)
1	87	246
2	43	67
3	87	244
4	81	38
5	49	103
6	44	64

Display 11.39 Gummy bear launch distances.

P13. The data in Display 11.40 show the typical costs (in dollars) for an in-state student and an out-of-state student for a random sample of ten out of the 100 best buys among public colleges, according to Kiplinger. Estimate the difference between out-of-state and in-state costs in a 90% confidence interval. To what population do these plausible means apply?

College	In-State Costs	Out-of-State Costs
University of Colorado at Boulder	18,887	32,781
SUNY Oneonta	14,716	20,976
University of NC at Wilmington	12,832	22,999
Georgia Institute of Technology	14,734	33,876
Georgia Southern University	12,848	22,430
University of Connecticut	19,438	34,150
University of Maryland, College Park	17,848	32,087
Western Washington University	13,344	24,418
Stony Brook University (SUNY)	15,841	22,101
University of California, Irvine	20,401	41,009

Display 11.40 Public college costs. [Source: *www.kiplinger.com/tools/colleges/.*]

Two Independent Samples, or Paired Observations?

P14. *Fish or fowl, which are smarter?* Display 11.41 lists brain weights of samples of birds and fish. When testing whether the mean brain weight is the same for birds and fish, are these independent samples or paired observations?

Birds	Brain Weight (g)	Fish	Brain Weight (g)
Canary	0.85	Barracuda	3.83
Crow	9.30	Brown trout	0.57
Flamingo	8.05	Catfish	1.84
Loon	6.12	Mackerel	0.64
Pheasant	3.29	Northern trout	1.23
Pigeon	2.69	Salmon	1.26
Vulture	19.60	Tuna	3.09

Display 11.41 Brain weights of samples of birds and fish.

P15. *Radioactive twins.* One feature of healthy lungs is that they are quick to get rid of any nasty stuff they breathe in. To investigate whether breathing city air or country air is healthier, researchers managed to find seven pairs of identical twins who satisfied two requirements: (1) One twin from each pair lived in the country and the other lived in a city, and (2) both twins in the pair were willing to inhale an aerosol of radioactive Teflon particles. The level of radioactivity was measured twice for each twin, right after inhaling and then again an hour later. The response was the percentage of original radioactivity still remaining 1 hour after inhaling. The data are shown in Display 11.42. This study was discussed in E65 in Chapter 4 on page 222.

Twin Pair	Environment	
	Rural	Urban
1	10.1	28.1
2	51.8	36.2
3	33.5	40.7
4	32.8	38.8
5	69.0	71.0
6	38.8	47.0
7	54.6	57.0

Display 11.42 Twin data on radioactivity, in percentage of radioactivity remaining. [Source: Per Camner and Klas Philipson, "Urban Factor and Tracheobronchial Clearance," *Archives of Environmental Health*, Vol. 27 (1973), p 82.]

a. Discuss whether a statistical test is appropriate.
b. Tell how to design a study that uses two independent samples (as an alternative to paired samples) to compare lung clearance rates for people living in urban and rural environments.

c. What are the advantages of using paired data from a single sample?
d. What are the advantages of using two independent samples?

Is It Meaningful to Compare Means?

P16. *Hens' eggs.* Harvard statistician Arthur Dempster went into his kitchen and measured the length and width of a dozen hens' eggs, which are given in Display 11.43. It would be possible to use these measurements to test the null hypothesis that the lengths and widths are equal. Do you think that would be a sensible comparison? Why or why not?

Length	Width	Length	Width
2.15	1.89	2.17	1.85
2.09	1.86	2.16	1.87
2.10	1.87	2.20	1.84
2.14	1.87	2.15	1.84
2.19	1.87	2.16	1.85
2.13	1.85	2.11	1.87

Display 11.43 Length and width measurements, in inches, of a dozen eggs. [Source: A. P. Dempster, *Elements of Continuous Multivariate Analysis* (Reading, MA: Addison-Wesley, 1969), p. 151.]

P17. An educational researcher is interested in determining whether calculus students who study with music playing perform differently from those who study with no music playing.

a. She asks for volunteers to participate in the study and finds a large number of students who study with music playing and a large number who study with no music playing. She then compares the mean scores on the next calculus exam, using a two-sample *t*-test. Is this an appropriate design and analysis for the problem at hand? Explain your reasoning.
b. Realizing that the abilities and backgrounds of the students might be important in this study, the researcher pairs students (one student from each group) from the volunteer groups on the basis of their current grade in the class. She then compares the mean scores on the next calculus exam, using a *t*-test. Is this an appropriate design and analysis for this problem? Is it better than the design in part a?
c. Explain how you would design a study to compare the effects of the two treatments on students' performance in calculus class.

Exercises

E23. Suppose the aldrin data (see Example 11.4, page 527) actually had been collected so that the data were paired, that is, if the bottom and mid-depth measurements in the same row in Display 11.44 (on the next page) were taken at the same time and in the same place.

a. Test whether the mean difference in the aldrin concentrations, *bottom − mid-depth*, is greater than 0. Do you have statistically significant evidence that the mean difference is not 0?

Bottom	Mid-Depth
3.8	3.2
4.8	3.8
4.9	4.3
5.3	4.8
5.4	4.9
5.7	5.2
6.3	5.2
7.3	6.2
8.1	6.3
8.8	6.6

Display 11.44 Paired aldrin data.

b. Do you reach the same conclusion as in Example 11.4?

c. Suppose the observations are paired and so are not independent and the null hypothesis is false. Which gives more power, a one-sample test of the mean difference or a two-sample test of the difference of the means?

E24. In the fitting of hearing aids to individuals, it is standard practice to test whether a particular hearing aid is right for a patient by playing a tape on which 25 words are pronounced clearly and then asking the patient to repeat the words as heard. Different lists are used for this purpose, and, in order to make accurate checks of hearing, the lists are supposed to be of equal difficulty with regard to understanding them correctly. The research question of interest is, "Are the test lists equally difficult to understand in the presence of background noise?" Display 11.45 shows the number of words identified correctly out of a list of 50 words by 24 people with normal hearing, in the presence of background noise. The two lists were presented to each person in random order.

List 1	List 2	List 1	List 2
28	20	32	28
24	16	40	38
32	38	28	36
30	20	48	28
34	34	34	34
30	30	28	16
36	30	40	34
32	28	18	22
48	42	20	20
32	36	26	30
32	32	36	20
38	36	40	44

Display 11.45 Number of correctly identified words in two lists. [Source: *lib.stat.cmu.edu*.]

a. First, suppose the data are not paired. For example, in the first row, the 28 words in list 1 and the 20 words in list 2 are not for the same person. Use the information in Display 11.46 to decide whether the conditions appear to be met for conducting a two-sample test to determine whether the mean number of words understood differs between the two lists.

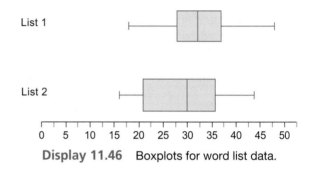

Display 11.46 Boxplots for word list data.

b. Do the computations for the two-sample *t*-test. Record the *P*-value.

c. Now use the fact that each row consists of the two measurements for the same person. Are the conditions met for conducting a test of significance for the mean difference?

d. Do the computations for the one-sample *t*-test. Record the *P*-value.

e. Which test is more powerful, the two-sample *t*-test or the one-sample *t*-test?

E25. *Sit and stand, revisited.* Display 11.47 shows a set of data on pulse rates, in beats per minute, from another class employing the same three experimental designs as in Examples 11.8, 11.9, and 11.10 on pages 551–555.

a. Display 11.48 shows the sitting pulse rate plotted against the standing pulse rate for each of the three designs. For which design is the relationship strongest? Weakest? Explain why that should be the case.

b. For the completely randomized design, perform a significance test that standing increases heart rate.

c. For the matched pairs design, perform a significance test that standing increases heart rate.

d. For the repeated measures design, perform a significance test that standing increases heart rate.

e. Compare the *P*-values for the three designs in parts b-d. Do they show the pattern you would expect?

Completely Randomized Design		Matched Pairs Design		Repeated Measures Design			
Sit	Stand	Sit	Stand	Sit	Stand	Sit	Stand
78	76	74	78	62	72	60	64
64	68	74	76	76	76	58	66
50	82	58	60	76	82	78	88
58	80	80	96	50	50	62	70
50	68	78	90	66	74	78	78
70	64	62	64	58	62	74	74
70	58	74	74	68	76	88	92
64	90	62	70	52	62	66	68
66	72	68	66	68	74	76	84
72	78	64	74	80	86	82	86
80	60	60	80	78	84	54	58
56	100	56	58	60	64	72	78
58	80	52	52	82	82	58	60
68	84	80	88	60	66	70	66

Display 11.47 Data tables and boxplots of pulse rate data from another class experiment.

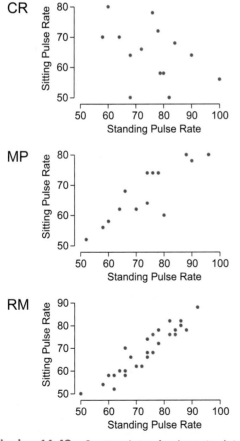

Display 11.48 Scatterplots of pulse rate data.

Summary Statistics

	CR		MP		RM	
	Sit	Stand	Sit	Stand	Sit	Stand
Mean	64.57	75.71	67.29	73.29	68.29	72.93
SD	9.33	11.68	9.37	12.71	10.24	10.38

E26. Did states improve their average SAT mathematics scores between 2005 and 2008? Display 11.49 (on the next page) shows the differences in mean scores, 2008 − 2005, and the summary statistics for the differences.

a. This is the entire population of states, so a significance test only helps answer the question of whether the observed mean difference could reasonably be due to chance alone. Even so, would a significance test, as presented in this section, be appropriate for these data? Explain why or why not.

b. If it is appropriate, test to see if the mean differs from zero.

E27. *M&Ms and IQ testing.* Twenty-two children, ages 5 to 7 and who liked candy, were matched into 11 pairs based on age, sex, and scores on Form L of the Stanford-Binet intelligence (IQ) test. Within each pair of children, one student was randomly chosen to receive the experimental treatment and the other child

State	Difference, 2008 – 2005	State	Difference, 2008 – 2005
Alabama	−2	Montana	8
Alaska	1	Nebraska	6
Arizona	−8	Nevada	−7
Arkansas	15	New Hampshire	−2
California	−7	New Jersey	−4
Colorado	10	New Mexico	1
Connecticut	−4	New York	−7
Delaware	−4	North Carolina	0
Florida	−1	North Dakota	−1
Georgia	−3	Ohio	1
Hawaii	−14	Oklahoma	9
Idaho	−2	Oregon	−1
Illinois	−5	Pennsylvania	−2
Indiana	0	Rhode Island	−7
Iowa	4	South Carolina	−2
Kansas	1	South Dakota	7
Kentucky	11	Tennessee	7
Louisiana	2	Texas	3
Maine	−39	Utah	0
Maryland	−13	Vermont	6
Massachusetts	−2	Virginia	−2
Michigan	19	Washington	−1
Minnesota	12	West Virginia	−10
Mississippi	2	Wisconsin	5
Missouri	9	Wyoming	31

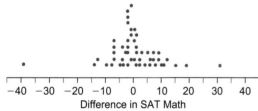

Mean of differences: 0.40
Standard deviation of differences: 9.8

Display 11.49 Differences in mean SAT mathematics scores 2008 − 2005, by state. [Source: College Board.]

to receive the control treatment. Seven weeks later, all 22 children were given Form M of the Stanford-Binet by the same school psychologist. The control group took the test according to the standard protocol. The experimental group took the test according to the standard protocol with the exception that the child was promised and given an M&M candy for each correct response. The results are given in Display 11.50. [Source: Calvin V. Edlund, "The Effect on the Behavior of Children, as Reflected in the IQ Scores, When Reinforced After Each Correct Response," *Journal of Applied Behavior Analysis*, Vol. 51, 3 (Fall 1972), pp. 317–319.]

	Experimental Group		Control Group	
Pair	Form L	Form M	Form L	Form M
A	75	85	71	76
B	71	83	75	82
C	78	97	79	76
D	66	79	61	68
E	100	119	111	96
F	94	106	91	85
G	87	107	81	86
H	61	72	74	73
I	83	90	79	90
J	85	95	84	85
K	107	107	98	97

Display 11.50 Scores of the 22 children on the Stanford-Binet used to form pairs, Form L, and the Stanford-Binet used as the response, Form M.

    a. Is this a completely randomized, matched pairs, or repeated measures design?

    b. What test should you use to determine if there is statistically significant evidence that the M&M treatment resulted in higher test scores on Form M, on average, than the control protocol? Perform this test and write a conclusion.

E28. Refer to E27.

    a. What test should be used to see how successful the randomization was in getting pairs of children of comparable IQ scores on Form L into the two treatment groups? Perform this test and write a conclusion.

    b. The two forms of the test were said to be of equal difficulty. How would you test whether that appears to be the case with these children? Perform this test and write a conclusion.

E29. Refer to E27.

    a. Is there statistically significant evidence that the children in the control group raised their scores in the 7 months between the test and retest? Is this what you would expect?

    b. Is there statistically significant evidence that the children in the experimental group raised their scores in the 7 months between the test and retest? Is this what you would expect?

E30. Refer to E27. One person analyzed the data this way: He found the difference in scores, *Form M – Form L*, for each student. He then subtracted the differences within each pair, *experimental group – control group*. Finally, he tested that the mean difference of the differences is 0. What was his conclusion? Is it the same as that from E27, part b? Which method do you like better? Why?

E31. An undesirable side effect of some antihistamines is drowsiness, which can be measured by the flicker frequency of patients (number of flicks of the eyelids per minute). Low flicker frequency is related to drowsiness because the eyes stay shut too long. One study reported data for nine subjects (Display 11.51), each given meclastine (A), a placebo (B), and promethazine (C), in random order. At the time of the study, A was a new drug and C was a standard drug known to cause drowsiness.

Patient	Drug A	Drug B	Drug C
1	31.25	33.12	31.25
2	26.63	26.00	25.87
3	24.87	26.13	23.75
4	28.75	29.63	29.87
5	28.63	28.37	24.50
6	30.63	31.25	29.37
7	24.00	25.50	23.87
8	30.12	28.50	27.87
9	25.13	27.00	24.63

Display 11.51 Flicker frequency of subjects, in eyelid flicks per minute. [Source: D. J. Hand et al., *A Handbook of Small Data Sets* (London: Chapman and Hall, 1994), p. 8.]

a. Are the conditions met for doing inference on the mean difference between drugs A and B? Between drugs B and C? Between drugs A and C?

b. Using a 98% confidence interval, estimate the mean of the differences between drugs A and B. Save your conclusion for part e.

c. Using a 98% confidence interval, estimate the mean of the differences between drugs B and C. Save your conclusion for part e.

d. Using a 98% confidence interval, estimate the mean of the differences between drugs A and C. Save your conclusion for part e.

e. Write a summary of what you learned about the three treatments in this analysis.

E32. Immediate release medications quickly liberate their drug content into the body, with the maximum concentration reached in a short time followed by a rapid decline in concentration. Sustained release medications, on the other hand, take longer to reach maximum concentration in the body but stay active for longer periods of time. A study of two such pain relief medications compared immediate release codeine (irc) with sustained release codeine (src). Thirteen healthy volunteers (nine males and four females) were randomly assigned to one of the two types of codeine and treated for 2.5 days. After a 7-day washout period, each volunteer was given the other type of codeine. Although the dosages were not exactly the same, since irc must be given in lower doses, the data were adjusted for this, and the measurements can be treated as if they came from equivalent doses.

Measurements in Display 11.52 include the maximum concentration (C) of codeine in the blood of the patient in nanograms per liter, the time (T) in hours it takes for the maximum concentration to be reached, and the total amount (A) of codeine absorbed by the patient's body over the life of the treatment (in ng · mL/hr).

a. Construct a 95% confidence interval estimate of the mean difference in concentrations for the two types of medications that would occur if all subjects could receive both treatments in both orders. Is there statistically significant evidence that the mean concentrations differ?

C_{irc}	C_{src}	T_{irc}	T_{src}	A_{irc}	A_{src}	Age	Gender
181.8	195.7	0.5	2.0	1091.3	1308.5	33	1
166.9	167.0	1.0	3.0	1064.5	1494.2	40	0
136.0	217.3	0.5	3.0	1281.1	1382.2	41	1
221.3	375.7	1.5	4.5	1921.4	1978.3	43	0
195.1	285.7	0.5	2.0	1649.9	2004.6	25	0
112.7	177.2	1.0	2.0	1423.6	*	30	1
84.2	220.3	2.0	1.5	1308.4	1211.1	24	0
78.5	243.5	1.0	3.0	1192.1	1002.4	44	0
85.9	141.6	1.5	1.5	766.2	866.6	42	1
85.3	127.2	2.0	4.5	978.6	1345.8	33	0
217.2	345.2	0.5	1.5	1618.9	979.2	38	0
49.7	112.1	1.5	1.0	582.9	576.3	39	0
190.0	223.4	0.5	1.0	972.1	999.1	43	0

Display 11.52 The time, in hours, it takes for the maximum concentration of a drug to be reached and the total amount of codeine absorbed by the patient's body over the life of the treatment [Source: Personal communication.]

b. Construct a 95% confidence interval estimate of the mean difference in time to maximum concentration for the two types of medications. Is there statistically significant evidence that the mean times differ?

c. Construct a 95% confidence interval estimate of the mean difference in the amount of codeine absorbed for the two types of medications. Is there statistically significant evidence that the mean amounts differ? Is there anything unusual about these data that you should consider in your analysis?

E33. *Bee stings*. Beekeepers sometimes use smoke from burning cardboard to reduce their risk of getting stung. It seems to work, but why?

One hypothesis is that smoke masks some other odor that induces bees to sting; in particular, smoke might mask some odor a bee leaves behind along with its stinger when it drills its target, an odor that tells other bees "Sting here." To test this hypothesis, J. B. Free suspended 16 cotton balls on threads from a square wooden board, in a four-by-four arrangement. Eight cotton balls had been freshly stung; the other eight were pristine and served as controls. Free jerked the square up and down over a hive that was open at the top and then counted the number of new stingers left behind in the treated and the control cotton balls. He repeated this whole procedure eight more times, each time starting with eight fresh and eight previously stung cotton balls and counting the number of new stingers left behind. For each of the nine occasions, he lumped together the eight balls of each kind and took as his response variable the total number of new stingers. His data are shown in Display 11.53.

a. The research hypothesis was that the previously stung cotton balls would receive more stings. State the null and alternative hypotheses for a one-sample test of significance of the mean difference.

Beekeeper using smoke to avoid being stung.

	Cotton Balls	
Occasion	Previously Stung	Fresh
1	27	33
2	9	9
3	33	21
4	33	15
5	4	6
6	22	16
7	21	19
8	33	15
9	70	10

Display 11.53 Number of new bee stingers left behind in cotton balls. [Source: J. B. Free, "The Stinging Response of Honeybees," *Animal Behavior*, Vol. 9 (1961), pp. 193–196.]

b. Are the conditions met for doing a test of the significance of the mean difference? If not, how do you suggest that you proceed?

c. Conduct your analysis and give your conclusion.

E34. The data in Display 11.54 are counts of leprosy bacilli colonies at specified sites on human subjects. In this experiment, the antiseptic group received a treatment where antiseptic was applied to the lesions, and the control group received a treatment that had no medical value. The columns show the counts on the same subjects before and after the treatment was applied. The order of the measurements in each pair cannot be randomized, but these subjects were randomly selected from a larger group, and which subjects were in the antiseptic group and which were in the control group was determined by random assignment. The research question is, "Is the antiseptic effective in reducing bacteria counts?"

a. Suppose that you want to perform a test of the mean difference based on paired observations for the antiseptic group. What concerns do you have about whether the conditions for inference are met?

b. Perform the test indicated in part a, regardless of your concerns.

c. Suppose that you want to perform a test of the mean difference based on paired observations for the placebo group. What concerns do you have about whether the conditions for inference are met?

d. Perform the test indicated in part c, regardless of your concerns.

Antiseptic		Placebo	
Before	After	Before	After
6	0	16	13
6	2	13	10
7	3	11	18
8	1	9	5
18	18	21	23
8	4	16	12
19	14	12	5
8	9	12	16
5	1	7	1
15	9	12	20

Display 11.54 Counts of leprosy bacilli colonies. [Source: G. Snedecor and W. Cochran, *Statistical Methods*, 6th ed. (Ames: Iowa State University Press, 1967), p. 422.]

E35. *Hospital carpets.* If you had to spend time in a hospital, would you want your room to have carpeting or a bare tile floor? Carpeting would keep down the noise level, which certainly would make the atmosphere more restful, but bare floors might be easier to keep free of germs. One way to measure the bacteria level in a room would be to pump air from the room over a growth medium, incubate the medium, and count the number of colonies of bacteria per cubic foot of air. This method was in fact used to compare the bacteria levels in 16 rooms at a Montana hospital. Eight randomly chosen rooms had carpet installed; the floors in the other eight rooms were left bare. The data are shown in Display 11.55.

a. Which of the two designs (two independent samples or paired data) is the more appropriate model? Why?

b. Tell how to run the experiment using the other design.

E36. *Hospital carpets.* Conduct the test that is appropriate for the design you selected in part a of E35.

E37. *Too tight?* You might have lived your entire life (until now) without once wondering what statistics has to do with the screw cap on a bottle of hair conditioner. (Brace yourself!) The machine that puts on the cap must apply just the right amount of turning force: too little, and conditioner can leak out; too much, and the cap might be damaged and be hard to get off. Imagine that you

Carpeted Floors		Bare Floors	
Room	Colonies/ft³	Room	Colonies/ft³
212	11.8	210	12.1
216	8.2	214	8.3
220	7.1	215	3.8
223	13.0	217	7.2
225	10.8	221	12.0
226	10.1	222	11.2
227	14.6	224	10.1
228	14.0	229	13.7

Display 11.55 Bacteria levels in hospital rooms. [Source: W. G. Walter and A. Stober, "Microbial Air Sampling in a Carpeted Hospitals," *Journal of Environment Health*, Vol. 30 (1968), p. 405.]

are in charge of quality control for a hair conditioner manufacturer. An engineer suggests that new settings on the capping machine will reduce the variability in the amount of force applied to the bottle caps so that fewer will be either too tight or too loose and more will be in the acceptable range. He offers to set up a comparison: Ten batches of bottles will be capped by each process, and the force needed to unscrew the cap will be measured. To make the comparison scientific, statistical methods will be used to test

$$H_0: \mu_{new} = \mu_{old} \quad \text{versus} \quad H_a: \mu_{new} > \mu_{old}$$

where μ in each case is the underlying true mean for the amount of force. What do you say to the engineer?

E38. One of the important measures of quality for a product that consists of a mixture of ingredients, such as cake mix, lawn fertilizer, or powdered medications, is how well the components are mixed. This is tested by taking small samples of the product from various places in the production process and measuring the proportions of various components in those samples. The sample data then will be a set of proportions for component A, a set of proportions for component B, and so on. Does a comparison of the means of these sets of proportions help measure the degree of mixing? If not, suggest another way to assess the degree of mixing.

E39. *Old Faithful.* The Old Faithful geyser, in Yellowstone National Park, got its name from the predictability of its eruptions, which used to occur about every 66 minutes. For decades, visitors throughout the warm months of the year have crowded the large circle of benches surrounding the geyser as each eruption time approached, waiting to see a 150-ft-high spout of superheated steam and water shoot into the air. In recent years, however, scientists have noticed that the times between eruptions have become somewhat more variable, and the distribution of times in fact appears to be bimodal, as seen in Display 11.56. Additional study suggests that the time until the next eruption might depend on the duration of the previous eruption. To test this hypothesis, a science class decides to classify eruptions as long or short and then compare the mean time until the next eruption for long eruptions and for short eruptions.

a. Is this a meaningful comparison? Why or why not?

b. Are the data on times between eruptions paired or unpaired?

E40. *Old Faithful revisited.* Suppose you have the data in Display 11.56 on times between eruptions but you don't have the data on the corresponding durations. "Aha!" you think. "We want to know whether the difference between the two modes is a 'real' pattern, that is, too big to be due just to chance variation. To test this, I'll just divide the histogram in half at the low point between the two modes and compare the means of the upper and lower halves of the data using a statistical test." Is this a reasonable application of hypothesis testing?

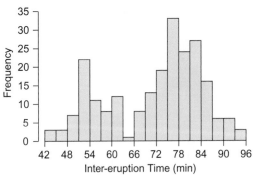

Display 11.56 Histogram based on a sample of 222 inter-eruption times taken during August 1978 and August 1979. [Source: S. Weisberg, *Applied Linear Regression*, 2nd ed. (Hoboken, NJ: John Wiley and Sons, 1985), pp. 231 and 234.]

Old Faithful.

E41. Auditors often are required to use sampling to check various aspects of accounts they are auditing. For example, suppose the accounts receivable held by a firm are being audited. Auditors typically will sample some of these accounts and compare the value the firm has on its books (*book value*) with what they agree is the correct amount (*audit value*), usually corroborated by talking with the customer. (Obtaining the audit values is a fair amount of work, hence the need for sampling.) Data of this sort are privileged information in most firms, but Display 11.57 shows what a typical data set might look like. Note that many of the book values and audit values agree, as they should. There is no question about the fact that these are paired data, so an estimate of the mean difference between the book values and audit values should proceed from the differences. The question is what to do with the differences that equal 0.

a. Construct a 95% confidence interval estimate using all 40 differences.

b. Construct a 95% confidence interval estimate using only the 15 nonzero differences.

c. How do the intervals in parts a and b compare? Which do you think provides the better answer? Why?

Book Value (B) ($)	Audit Value (A) ($)	B − A ($)
389	389	0
419	419	0
336	336	0
427	427	0
418	418	0
355	355	0
293	293	0
406	406	0
392	392	0
333	333	0
394	394	0
374	374	0
446	446	0
464	464	0
338	338	0
390	390	0
372	372	0
406	406	0
364	364	0
433	433	0
426	426	0
417	417	0
415	415	0
461	461	0
431	431	0
337	367	−30
429	392	37

Book Value (B) ($)	Audit Value (A) ($)	B − A ($)
433	403	30
412	379	33
383	341	42
484	462	22
471	428	43
400	430	−30
409	435	−26
386	361	25
370	375	−5
486	452	34
393	339	54
382	404	−22
480	432	48

Display 11.57 Book values versus audit values.

E42. In E9 of Chapter 10 on page 481, you read about the readability of cancer brochures. Can patients read the information they are given? Based on suspicions that the reading level of patient brochures is higher than the reading level of patients, a study was conducted to compare the two. A page was randomly selected from each of 30 brochures for cancer patients published by the American Cancer Society and the National Cancer Institute. The pages were judged for readability using standard readability tests. Then 63 patients were tested for their reading level. The results are summarized in Display 11.58.

a. Is there evidence that the materials are at a higher mean reading level than the patients' mean reading level? To answer this question, should you do a one-sample test or a two-sample test?

b. Perform the test you selected in part a and state your conclusion.

c. The question in part a isn't really the important one. Answer this more important question: "If a patient reads at the 10th-grade level and is given one of the booklets at random, what is the probability that the booklet will be at a higher reading level than that of the patient?"

d. If each of these patients is given one booklet at random, how many patients do you expect to get a booklet that is at too high a reading level for them?

E43. *The sign test.* Consider again the radioactive twin data of P15, which is shown in Display 11.59. The research (alternative) hypothesis is that people who live in a city have a larger percentage of radioactivity remaining than their twin in the country. For this exercise, assume that the treatments of country and city were randomly assigned within each independent pair of twins. Even with this assumption, standard analysis of these data using the procedures of this chapter would be problematic because of the extreme variation in the differences. The *sign test* allows you to ignore the numerical value of the differences and concentrate on the number of positive and negative differences.

Reading Level	Frequency Brochures	Frequency Patients
2	0	6
3	0	4
4	0	4
5	0	3
6	3	3
7	3	2
8	8	6
9	4	5
10	1	4
11	1	7
12	4	2
13	2	17
14	1	0
15	2	0
16	1	0

Display 11.58 Frequency of various reading levels, as measured for brochures and patients. [Source: Thomas H. Short et al., "Readability of Educational Materials for Patients with Cancer," *Journal of Statistics Education*, Vol. 3, 2 (July 1995).]

a. If there were no inherent difference in the percentage of radioactivity remaining between twins raised in the city and those raised in the country, what is the probability that any one difference would be positive? What is the probability that all of the seven observed differences would be positive? What is the probability that all but one of the seven observed differences would be positive? What is the probability that six or more of the seven observed differences would be positive?

b. Define the test statistic as the number of positive differences. What is the value of this test statistic?

Percentage of Radioactivity Remaining

Twin Pair	Urban	Rural	Difference, urban − rural
1	28.1	10.1	18.0
2	36.2	51.8	−15.6
3	40.7	33.5	7.2
4	38.8	32.8	6.0
5	71.0	69.0	2.0
6	47.0	38.8	8.2
7	57.0	54.6	2.4

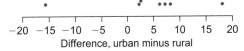

Difference, urban minus rural

Display 11.59 Setting up the sign test for the twin data.

Does a small value or a large value of the test statistic tend to support the alternative hypothesis?

c. If the null hypothesis is true that there is no difference between city and country residents in the percentage of radioactivity remaining, what is the P-value for this test (the probability of getting a test statistic as large as observed or even larger)?

d. Based on part c, what is your conclusion?

E44. Conduct a sign test (refer to the previous exercise) on the paired counts shown for the antiseptic group in E34 on page 558 (in the table on the left in Display 11.54) to determine if there is statistically significant evidence that the antiseptic is effective. (*Note:* You will have to ignore the fifth colony because there was a tie. Use the remaining nine colonies as your sample.) Do the same for the placebo group (the right-hand table) to see if there is a significant placebo effect. In each situation, state the hypotheses, give the value of the test statistic, find the P-value, and state a conclusion.

Chapter Summary ◀

To use the confidence intervals and significance tests of this chapter, you need

- two independent random samples from two distinct populations or, in the case of an experiment, two treatments randomly assigned to the available subjects

- a random sample from a population consisting of paired values or, in the case of an experiment, two treatments randomly assigned within paired units

When there is no randomness involved, you proceed with the test only after stating loudly and clearly the limitations of what you are doing. If you reject the null hypothesis in such a case, all you can conclude is that something happened that can't reasonably be attributed to chance. You use a confidence interval if you want to find a range of plausible values for

- $\mu_1 - \mu_2$, the difference between the means of your two populations

- μ_d, the mean of the differences in your population of pairs

Both of the confidence intervals you studied have the same form:

$$\text{statistic} \pm (\text{critical value}) \cdot (\text{estimated standard error})$$

The degrees of freedom for the two-sample *t*-procedure must be approximated from a rather cumbersome formula. For that reason, the two-sample *t*-procedures should be done with the aid of a calculator or statistics software.

All the significance tests you have studied involve the same steps:

1. Write the null hypothesis and the alternative hypothesis.

2. Justify your reasons for choosing this particular test. Discuss whether the conditions are met and, if they are not strictly met, decide whether it is okay to proceed.

3. Compute the test statistic, $\frac{statistic - parameter}{estimated\ standard\ error}$, and find the *P*-value. Draw a sketch of the situation.

4. Use the computations to decide whether to reject or not reject the null hypothesis by comparing the *P*-value to the level of significance, α. Then state your conclusion in terms of the context of the situation. ("Reject H_0" is not sufficient.) Mention any doubts you have about the validity of that conclusion.

Review Exercises

E45. *Old Faithful.* (Refer to E39 on page 559 for a description of the data.) Display 11.60 presents summary statistics and side-by-side boxplots comparing the distributions of the times until the next eruption following short eruptions (duration < 3 minutes) and long eruptions (duration > 3 minutes).

	Short Eruptions (min)($n = 67$)	Long Eruptions (min)($n = 155$)
Mean	54.46	78.16
Standard deviation	6.30	6.89

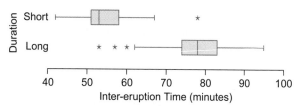

Display 11.60 Summary statistics and boxplots of times until the next eruption following short and long eruptions.

a. State appropriate null and alternative hypotheses in words. Then define notation and restate H_0 and H_a in symbols.

b. Tell whether the design of the study justifies the use of a probability model, and give reasons for your answer.

c. Based on the summary statistics and boxplots, tell whether the shapes of the distributions raise doubts about using a *t*-test.

d. Carry out the computations, find the *P*-value, and tell what your conclusion is if you take the results at face value.

e. Now tell what you think the results of the test really mean for this particular data set.

E46. The data in Display 11.61 show the life expectancies (in years) of males and females for a random sample of African countries. Is there evidence of a significant difference between the life expectancies of males and females in the countries of Africa? Give statistical justification for your answer and a careful explanation of your analysis.

E47. *Altitude and alcohol.* On every commercial passenger flight, there is an announcement that tells what to do if the cabin loses pressure. At high altitudes, air has less oxygen, and you can lose consciousness in a short time. To determine the effects of alcohol on the length of time people can stay conscious at high altitudes, ten subjects were put in an environment equivalent to an altitude of 25,000 feet and then monitored to see how long the subject could continue to perform a set of assigned tasks. As soon as performance deteriorated (the end of "useful consciousness"), the time was recorded and the

	Life Expectancy	
Country	Male	Female
Algeria	71.5	74.6
Gambia	51.9	55.6
Ghana	57.7	59.3
Kenya	48.9	47.1
Liberia	37.0	40.8
Libya	74.3	78.8
Mauritius	68.4	76.4
Morocco	68.4	73.1
Swaziland	32.5	34.0
Togo	55.0	59.1

Display 11.61 Life expectancies (in years) for a sample of African countries. [Source: *CIA Factbook*, 2006.]

environment was returned to normal. Three days later, each subject drank a dose of whiskey based on body weight—1 cc of 100-proof alcohol for every 2 lb-and then, after waiting 1 hour for the whiskey to take effect, was returned to the simulated altitude of 25,000 ft for another test. Display 11.62 gives the times (in seconds) until the end of useful consciousness under both conditions.

a. How can you tell from simply the description of the data, even before seeing the numbers, that this study gives you paired data from one sample rather than two independent samples?

Time Until End of Useful Consciousness (sec)

Subject	Control (C)	Alcohol (A)
1	261	185
2	565	375
3	900	310
4	630	240
5	280	215
6	365	420
7	400	405
8	735	205
9	430	255
10	900	900

Display 11.62 The effect of alcohol consumption on time of useful consciousness at high altitudes. [Source: Jay Devore and Roxy Peck, *Statistics: The Exploration and Analysis of Data* (Minneapolis, MN: West Publishing Co., 1986). Original source: "Effects of Alcohol on Hypoxia," *Journal of American Medical Association* (December 13, 1965), p. 135.]

b. Carry out an exploratory analysis. Include stemplots or boxplots of the recorded times under each of the two conditions and a stemplot or boxplot of the differences, as well as a scatterplot. Describe the patterns in words, and tell what questions, if any, the patterns raise about the validity of formal inference.

c. Find a 95% confidence interval for the true difference, μ_{C-A}. What does the "true" difference refer to in this context? What do you conclude about the effect of alcohol on the time of useful consciousness at high altitude?

E48. *Boosting your SATs?* Several commercial companies offer special courses designed to help you improve your score on the SAT. Suppose market research shows that students are willing to take such a course only if, on average, it raises SAT scores by more than 30 points. Here's a way to test a claim that a course is able to do that. You'll need some volunteers, the more representative the better. Randomly divide them into two groups. Those in the first group take the special course; those in the second group serve as controls. When the course is over, both groups take the SAT.

Let $\bar{x}_1$ and $\bar{x}_2$ be the sample means for the two groups, let s_1 and s_2 be the sample standard deviations, and let n_1 and n_2 be the sample sizes.

a. Define suitable notation, and state the null and alternative hypotheses twice, first in words and then in symbols.

b. If you construct a 95% confidence interval for $\mu_1 - \mu_2$, under what circumstances do you reject H₀?

c. Carry out the test using the *t*-statistic for $\bar{x}_1 = 1100$, $\bar{x}_2 = 1060$, $s_1 = 100$, $s_2 = 80$, and $n_1 = n_2 = 16$. Check your conclusion by constructing a confidence interval.

E49. *Little statisticians.* A series of research studies attempted to learn how babies are able to distinguish individual words from the flow of syllables in adult speech. One hypothesis is that babies are little statisticians. They distinguish the words in "hello baby" because they have discerned that the syllables "hel"/"lo" and the syllables "ba"/"by" occur in sequence more frequently than "lo"/"ba." In one experiment, twenty-four 8-month-old infants listened to 2 minutes of a continuous stream of speech, spoken in a monotone by a speech synthesizer, with no pauses or other clues between words. The "words" consisted of 4 three-syllable nonsense words, repeated in random order. Each infant then was tested on two sets of stimuli, presented in random order. One set of stimuli consisted of repetitions of some of the 4 three-syllable nonsense words previously heard. The other set of stimuli consisted of three-syllable nonwords made up from the same syllables heard before, but in a different order. Infants pay more attention to novel stimuli and so, if they paid attention longer to the nonwords, this was considered evidence that they had become familiar with the words. The results, as presented in the research report, are given in Display 11.63. [Source: Jenny R. Saffran, Richard N. Aslin, Elissa L. Newport, "Statistical Learning by 8-Month-Old Infants," *Science*, Vol. 274 (December 13, 1996), pp. 1926–1928.].

a. What was the design of this study: completely randomized, matched pairs, or repeated measures? Is the test a one-sample test or a two-sample test? Explain.

Mean Listening Times(sec)

Familiar Items	Novel Items	Matched-Pairs *t*-Test
7.97 (SE 0.41)	8.85 (SE 0.45)	$t(23) = 2.3, P < 0.04$

Display 11.63 Mean time spent listening to words and nonwords, with results of significance test.

b. In the table, how were the values of the SE computed? Compute the standard deviation of the times for the familiar items and for the novel items.

c. What is the meaning of the 23 in $t(23)$?

d. Using the information in the table, compute the standard error of the differences.

e. Is the result statistically significant? Write a conclusion for this study.

E50. In a second experiment in the research study described in E49, a different group of twenty-four 8-month-old infants also listened to 2 minutes of a continuous stream of speech, spoken in a monotone by a speech synthesizer, with no pauses or other clues between words. The "words" consisted of 4 three-syllable nonsense words, repeated in random order. This time each infant listened to words from the original speech and to two "part-words" made up by joining the final syllable of a word from the first two syllables of another word. For example, if two words were "bidaku-padoti," then the "part-word" would be "ku-pado." These two sets of stimuli were presented to each baby in random order. Again, listening longer was interpreted to mean that the babies recognized the stimulus as novel. The results, as presented in the research report, are given in Display 11.64.

Mean Listening Times (sec)

Familiar Items	Novel Items	Matched-Pairs *t*-Test
6.77 (SE 0.44)	7.60 (SE 0.42)	$t(23) = 2.4$, $P < 0.03$

Display 11.64 Mean time spent listening to "words" and "part words," with results of significance test, for second experiment.

a. What was the design of this study: completely randomized, matched pairs, or repeated measures? Is the test a one-sample test or a two-sample test? Explain.

b. In the table, how were the values of the SE computed? Compute the standard deviation of the times for the familiar items and for the novel items.

c. What is the meaning of the 23 in *t*(23)?

d. Using the information in the table, compute the standard error of the differences.

e. Is the result statistically significant? Write a conclusion for this study.

E51. *Schizophrenia and amphetamine.* A double-blind, placebo-controlled experiment was designed to determine the effect of dextroamphetamine on regional cerebral blood flow (rCBF) and performance on cognitive tasks by patients severely ill with schizophrenia. Ten institutionalized patients volunteered to participate. Each patient was given, in random order, a placebo and dextroamphetamine. After receiving the treatment, each patient was tested for rCBF and given several tasks, one of which was the Wisconsin Cart Sort Test (WCST). The WCST requires a person to manipulate abstract sets. The results on the WCST, as presented in the research report, appear in Display 11.65. [Source: David G. Daniel et al., "The Effect of Amphetamine on Regional Cerebral Blood Flow during Cognitive Activation in Schizophrenia," *Journal of Neuroscience*, Vol. 17, 7 (July 1991), pp. 1907–1917.]

a. What was the design of this study: completely randomized, matched pairs, or repeated measures? Is the test a one-sample test or a two-sample test? Explain.

b. How could this experiment have been made double-blind?

c. In the table, how were the values of the SE computed?

d. Verify the value of *t* and the *P*-value for *number of correct responses*.

e. For which measures is the result statistically significant? Write a conclusion for this part of the study.

Measure	Placebo	Amphetamine	*t*[a]	SE[b]	*P*	*N*
Number of correct responses	72	84	2.52	4.7	0.036	9
Percent conceptual level	44.7	54.6	2.51	3.9	0.036	9
Number of categories correctly completed	3.1	3.8	1.41	0.50	0.19	10
Failures to maintain set	2.33	2.22	0.36	0.31	0.73	9
Percentage perseverative error	19.6	18.7	0.20	4.3	0.85	9

[a] Matched-pairs *t* tests.
[b] Standard error of difference.

Display 11.65 Table giving results on the WCST by schizophrenic inpatients given a placebo and dextroamphetamine.

E52. In the study described in E51, regional cerebral blood flow (rCBF) at several sites in the brain was measured for each patient under both the placebo and amphetamine treatments. The effect of amphetamine on rCBF during the WCST task as presented in the research report, appear in Display 11.66.

Effect of Amphetamine on rCBF[c] During the WCST task

Region[a]	Slice[b]	Placebo	Amphetamine	*t*[d]	SE[*]	*P*
Left cortex	3	65.0	61.5	1.65	2.1	0.13
Right cortex	3	65.6	62.2	1.74	2.0	0.12
Left cortex	2	65.1	62.8	1.38	1.6	0.20
Right cortex	2	65.8	63.8	1.28	1.6	0.23
Left subcortex	2	64.2	62.7	0.77	1.9	0.46
Right subcortex	2	64.2	62.7	0.77	2.0	0.46

[a] Calculated by averaging the mean rCBF of all individual regions of interest contained within the region described.
[b] Slices 2 and 3 were centered at 40 and 60 mm above the canthomeatal line, respectively.
[c] Values are mean flow expressed in ml/100 g/min.
[d] Matched-pairs *t* tests; *n* = 10.
[*] Standard error of difference.

Display 11.66 Table giving measures of rCBF for schizophrenic inpatients given a placebo and dextroamphetamine.

a. Was this part of the study completely randomized, matched pairs, or repeated measures? Is the test a one-sample test or a two-sample test? Explain.

b. How could this part of the experiment been made double-blind?

c. In the table, how were the values of the SE computed?

d. Verify the value of t and the P-value for *left cortex*.

e. For which regions is the result statistically significant? Write a conclusion for this part of the study.

f. The rCBF was greater, on average, for each region of the brain for the placebo than for the amphetamine treatment. Propose and conduct a new test of significance using all six of these paired measurements.

Concept Test

C1. Which of the following are not required in estimating the difference between the means of two populations by use of the two-sample procedures of this chapter?

- **A** a random sample from each population, with the samples being independent
- **B** nearly normal populations or fairly large samples
- **C** equal population variances
- **D** equal sample sizes

C2. The main statistical purpose in pairing units in the design of experiments is to

- **A** simplify the testing procedure to a one-sided test
- **B** reduce the sample size
- **C** make data collection easier
- **D** reduce variation and improve the power of inference procedures

C3. In comparing the mean time to pain relief for two analgesics, L and M, a researcher conducted a randomized experiment and estimated, with 90% confidence, the L-mean minus the M-mean to be in the interval 6 ± 2 minutes. Which of the following is the most appropriate statistical conclusion?

- **A** The L-mean is larger than the M-mean, but a Type I error could have been made.
- **B** The L-mean is larger than the M-mean, but a Type II error could have been made.
- **C** We cannot conclude which mean is larger from a confidence interval estimate.
- **D** The interval is longer than the corresponding 95% confidence interval would be.

C4. In studies of the birth weights of newborn twins, what is sometimes reported is the mean weight of the larger and the mean weight of the smaller? Assuming that appropriate standard deviations could be found, does a test of the hypothesis that the mean difference in weights of twins is zero, based on these reported sample means, fit into any of the models studied in this chapter? Choose the best answer.

- **A** Yes, these are sample means from independent samples.
- **B** Yes, the difference of these means is the same as the mean of the differences, so a paired data analysis will work.
- **C** No, these means could not have come from a random sample of twins.
- **D** No, these means are not independent and are not the mean of the observed differences.

C5. Researchers wish to estimate, for the population of married couples, the average of the differences in the heart rate of each wife and her husband. They get a random sample of 100 married couples and measure the heart rate of each person. What is the best way to proceed?

- **A** Get a random sample of unmarried couples for comparison.

- **B** The independence assumption has been violated, so the researchers must start over and get a random sample of married women and independently get a random sample of married men.

- **C** Compute the mean heart rate of the women and the mean heart rate of the men. Compute a confidence interval for the difference between the mean heart rate of the women and the mean heart rate of the men.

- **D** Compute the mean heart rate of the women and the mean heart rate of the men. Compute separate confidence intervals for the mean heart rate of the women and for the mean heart rate of the men.

- **E** Subtract each husband's heart rate from his wife's heart rate, find the mean of these differences, and compute a single confidence interval for the mean difference.

C6. Scotland imposed a ban on smoking in bars on March 26, 2006. Before the ban, a researcher thought that the respiratory health of bar employees should improve after working in smokefree air. Before the ban went into effect, he scored the respiratory health of a random sample of Scottish bar employees. Two months after the ban, he obtained an independent random sample of bar employees and scored their respiratory health. The increase in the mean scores was statistically significant ($P = 0.049$). Which of the following is the best interpretation of this result?

- **A** An observed difference in sample means as large or larger than that in this sample is unlikely to occur if the mean score for all bar employees before the ban is equal to the mean after the ban.

- **B** The probability is only 0.049 that the mean score for all bar employees increased from before the ban to after the ban.

- **C** The mean score for all bar employees increased by more than 4.9%.

- **D** There is a 4.9% chance that the mean score of all bar employees after the ban is actually lower than before the ban, despite the increase observed in the samples.

- **E** Only 4.9% of bar employees had their scores drop while the other 95.1% had their scores increase.

Investigative Task

C7. The stereogram in Display 11.67 contains the embedded image of a diamond. [Source: W. S. Cleveland, *Visualizing Data* (1993). Original source: J. P. Frisby and J. L. Clatworthy, "Learning to See Complex Random-Dot Stereograms," *Perception*, Vol. 4 (1975), pp. 173–78, *lib.stat.cmu.edu*.] Look at a point between the two diagrams and unfocus your eyes until the two images merge into one and the diamond pops out.

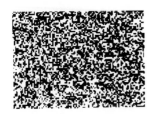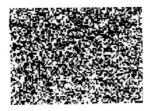

Display 11.67 Stereogram with embedded diamond.

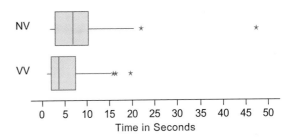

Stereograms

| | Group | | Row Summary |
	NV	V V	
	8.5604647	5.5514289	7.2102563
	43	35	78
Time_in_Seconds	8.0854116	4.8017389	6.9360082
	1.2330137	0.81164201	0.78534828
	0	0	0

S1 = mean ( )
S2 = count ( )
S3 = stdDev ( )
S4 = stdError ( )
S5 = count (missing ( ))

Display 11.68 Boxplot and summary statistics of time, in seconds, to see the diamond.

In an experiment, one group (NV) of subjects was either told nothing or told that a diamond was embedded. A second group (VV) of subjects was shown a drawing of the diamond. The times (in seconds) that it took the subjects to see the diamond are summarized in Display 11.68.

a. Check the conditions for doing a two-sample *t*-test of the difference in means.

b. The researchers did a two-sample *t*-test on the original data, with pooled variances ($\alpha = 0.05$). Do you agree with that decision? Why or why not?

c. Use your calculator or computer to replicate the test that the researchers did, and find the *P*-value. What conclusion do you come to if you take the results at face value?

d. What would the test decision have been if the variances weren't pooled?

e. When the standard deviations aren't comparable, what is the effect of using the pooled procedure rather than the unpooled procedure?

Chapter 12
Chi-Square Tests

Were the votes counted fairly in the 2009 presidential election in Iran? You can use a chi-square test to decide whether the reported vote counts are plausible.

You have now completed four chapters on statistical inference. Chapters 8 and 9 were about inference for proportions computed from categorical variables, and Chapters 10 and 11 were about inference for means computed from measurement (quantitative) variables. This chapter returns to categorical variables. You can think of it as a generalization of the results of Chapters 8 and 9.

In Chapters 8 and 9, there were only two categories—success or failure, heads or tails, and so on—and at most two populations. You used z-tests to determine statistical significance. In this chapter, you will use chi-square tests to answer questions about situations that may involve more than two categories and more than two populations:

- *Fair die?* You roll a die 60 times. You get 12 ones, 9 twos, 10 threes, 6 fours, 11 fives, and 12 sixes. Do you have evidence that the die is unfair?

- *Whose children select cigarettes?* Children aged 2 through 6 were told that some dolls were going to have a party and they should select products for them from a play supermarket. In the sample of 16 children whose parents went to high school or had job training, 56.3% purchased cigarettes. In the sample of 26 children who had one parent go to college, 38.5% purchased cigarettes. In the sample of 78 children where both parents went to college, 19.2% purchased cigarettes. Is this convincing evidence that the percentages of children who would purchase cigarettes vary with parent educational level? [Source: Madeline A. Dalton, et al., "Use of Cigarettes and Alcohol by Preschoolers While Role-Playing as Adults," *Archives of Pediatrics & Adolescent Medicine,* Vol. 159 (2005), pp. 854–859.]

You will use the chi-square technique to answer a new type of problem as well:

Independence of relationships? Each adult in a random sample of clients from a family counseling service is asked to categorize his or her family relationship as supportive, neutral, or not supportive and also to characterize it as happy, neutral, or unhappy. Do you have evidence of an association between level of supportiveness and level of happiness, or do these two variables appear to be independent?

In this chapter, you will learn to use a chi-square test of

▶ goodness of fit to decide whether it is plausible that the proportions that fall into each category in this population are the same as the hypothesized proportions

▶ homogeneity to decide whether it is plausible that the proportions that fall into each category are the same in different populations

▶ independence to decide whether it is plausible that two different categorical variables are independent in this population

The procedure for each type of chi-square test is almost the same as for the others. The difference is in the type of question asked, the method of sampling, and the conclusion that you can draw.

12.1 ▶ Testing a Probability Model: The Chi-Square Goodness-of-Fit Test

The chi-square test was developed by English statistician Karl Pearson in 1900. His was more than a small contribution to statistical methodology, as it transformed the way scientists view the world of data. *Science 84* (a publication of the American Association for the Advancement of Science) included the chi-square test as one of the top 20 discoveries that have changed our lives, along with the invention of the computer, the laser, plastic, and nuclear power. "The chi-square test was a tiny event in itself, but it was the signal for a sweeping transformation in the ways we interpret our numerical world. Now there could be a standard way in which ideas were presented to policy makers and to the general public.... For better or worse, statistical inference has provided an entirely new style of reasoning." [Source: *Science 84* (November 1984), pp. 69–70.]

Is the sample consistent with these proportions?

A **goodness-of-fit test** determines whether it is reasonable to assume that your sample came from a population in which, for each category, the proportion of the population that falls into the category is equal to some hypothesized proportion. If the result from the sample is very different from the expected results, then you have to conclude that the hypothesized proportions are wrong. To determine how different "very different" is, you will need a test statistic and its probability distribution.

A Test Statistic

To begin a chi-square goodness-of-fit test, first write hypotheses. Then make a table with the first column listing the possible outcomes. In the second column, give the frequency (or count) that each outcome was observed (O). In the third column, give the frequencies you would expect (E) if the hypothesized proportions are correct.

Example 12.1	Fair Die?

You want to test whether your die is fair, so you roll it 60 times and get 12 ones, 9 twos, 10 threes, 6 fours, 11 fives, and 12 sixes. Write your hypothesis. Then set up a table of observed and expected frequencies that you can use to see if you have statistically significant evidence that the die is unfair.

Solution
The hypotheses are

H_0: $p_1 = 1/6, p_2 = 1/6, p_3 = 1/6, p_4 = 1/6, p_5 = 1/6$, and $p_6 = 1/6$, where p_1 is the probability of rolling a 1, p_2 is the probability of rolling a 2, . . .

H_a: At least one of the probabilities in the null hypothesis is incorrect.

If the null hypothesis is true, you "expect" to get each face of the die on 1/6 of the 60 rolls, or 10 times each. Display 12.1 summarizes this information.

Outcome	Observed Frequency, O	Expected Frequency, E
1	12	$1/6 \cdot 60 = 10$
2	9	$1/6 \cdot 60 = 10$
3	10	$1/6 \cdot 60 = 10$
4	6	$1/6 \cdot 60 = 10$
5	11	$1/6 \cdot 60 = 10$
6	12	$1/6 \cdot 60 = 10$
Total	60	60

Display 12.1 Observed and expected frequencies for 60 rolls of a fair die.

You will conclude that the die is unfair if the observed frequencies are far from the expected frequencies. Are they? Here's where you need a test statistic to condense the information in the table into a single number that acts as an index of how far away the observed frequencies are from the expected frequencies.

> A test statistic is a measure of the distance between observation and model.

A Test Statistic

The test statistic for chi-square tests is

$$\chi^2 = \sum \frac{(O - E)^2}{E}$$

Here O is the observed frequency in the category, and E is the corresponding expected frequency. You sum over all categories.

> The symbol χ^2 is read "chi-square."

The Test Statistic for the Die

Example 12.2

Compute the value of χ^2 for the fair die problem in the previous example.

Solution
Organize your computations by extending the table of Display 12.1, as shown in Display 12.2.

Outcome	Observed Frequency, O	Expected Frequency, E	$(O - E)$	$\frac{(O - E)^2}{E}$
1	12	10	2	0.4
2	9	10	−1	0.1
3	10	10	0	0
4	6	10	−4	1.6
5	11	10	1	0.1
6	12	10	2	0.4
Total	60	60		2.6

Display 12.2 Table of χ^2 computations for the fair die problem.

The value of χ^2 is the sum of the values in the last column, $\sum \frac{(O - E)^2}{E}$, so $\chi^2 = 2.6$.

DISCUSSION
A Test Statistic

D1. Will you reject the hypothesis that the die is fair if χ^2 is relatively large, if it is relatively small, or both? How might you determine whether a χ^2 value of 2.6 is relatively large?

D2. Another test statistic that might be constructed is

$$\sum (O - E)$$

    a. Compute the value of this test statistic for the data in Display 12.1.

    b. Will your result in part a always occur? Prove your answer.

    c. Is this a good test statistic?

    d. You have seen two other situations in which the sum of the differences always turned out to be 0. What were those situations? What did you do in those situations?

D3. Still another test statistic that might be constructed is

$$\sum (O - E)^2$$

a. Display 12.3 shows the results from the rolls of two different dice. Die A was rolled 60 times and Die B was rolled 6000 times. For which die does the table give stronger evidence that the die is unfair?

b. Compute and compare the values of $\sum(O - E)^2$ for Die A and Die B. Does this appear to be a reasonable test statistic? Explain.

| | Die A | | Die B | |
Outcome	Observed Frequency	Expected Frequency	Observed Frequency	Expected Frequency
1	5	10	995	1000
2	16	10	1006	1000
3	18	10	1008	1000
4	4	10	994	1000
5	5	10	995	1000
6	12	10	1002	1000

Display 12.3 Results from 60 rolls of Die A and 6000 rolls of Die B.

c. What is χ^2 for Die A and for Die B in Display 12.3? Did the larger value of χ^2 correspond to the die that you thought seemed more unfair? What purpose does dividing by E serve in the formula for the test statistic?

D4. The χ^2 statistic involves a sum of squared differences. What other statistics have you seen that involve a sum of squared differences?

The Distribution of χ^2

From Display 12.2, χ^2 for the fair die problem is 2.6. If the observed and expected frequencies had been exactly equal, χ^2 would have been 0 and you would have had no reason to doubt that the die is fair. If the observed and expected frequencies had been much farther apart than they were, χ^2 would have been much larger than 2.6 and you would have had evidence that the die is not fair.

How can you assess whether a value of χ^2 of 2.6 is large enough to reject the null hypothesis that the die is fair? You need to see how much variation there is in the value of χ^2 when a die is fair.

| The effect of the sample size was accounted for when you divided by E. |

The distributions in Display 12.4 each show 2000 values of χ^2 computed by having software simulate the rolls of a fair die. In the histogram on the left, each value of χ^2 was computed from a sample of 1000 rolls of a fair die. The histogram on the right was constructed from repeated samples of only 60 rolls each.

As you can see, the χ^2 distribution does not change much with a change in sample size, as long as the sample size is large. Is there only one χ^2 distribution (like the z-statistic), or are there many (like the t-statistic)? Sample size doesn't change the shape, center, or spread, but there is one more variable to check—the number of categories.

Each histogram in Display 12.5 shows a distribution of 5000 values χ^2. Each of the 5000 values was computed from a sample of 60 rolls of a fair die. However, the histograms were made using dice with different numbers of sides. Notice how the distribution changes as the number of categories (sides) increases from 4 to 20.

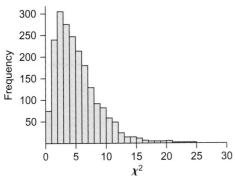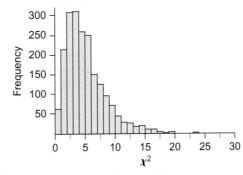

Display 12.4 Two histograms of 2000 values of χ^2, one for $n = 1000$ rolls of a fair die (left) and one with $n = 60$ rolls (right).

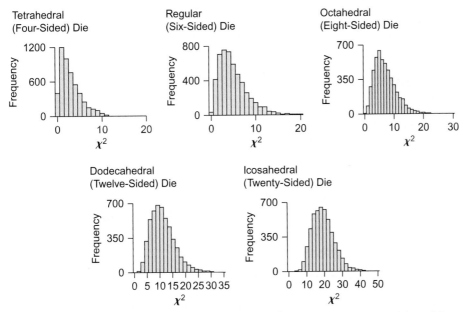

Display 12.5 Five histograms of 5000 values of χ^2, each value computed from 60 simulated rolls of a fair die.

STATISTICS IN ACTION 12.1 ▶ Generating a Chi-Square Distribution

What You'll Need: dice

1. Form five groups in your class. Your group should roll fair six-sided dice until your group has a total of 60 rolls. Compute χ^2 for your group's results. Get the values of χ^2 from the other four groups. From just these five results, does it appear that a χ^2 value of 2.6 is a reasonably likely outcome when a fair die is rolled?

2. Learn how to use a calculator or computer to simulate 60 rolls of a fair die.

3. Use a calculator or computer to simulate 60 rolls of a fair die. Compute χ^2 for your results.

4. Continue until your class has 200 values of χ^2, where each χ^2 value is computed from 60 rolls of a fair die. Display the values in a histogram and describe the shape of the histogram.

5. Should a one-sided test or a two-sided test be used to test if a die is fair? Using the results from your simulation, estimate the P-value for a χ^2 value of 2.6. Is this a reasonably likely outcome if the die is fair?

Chi-square probability density functions are continuous approximations to distributions like those in Display 12.5. In the output in Display 12.6, the numbers used in the software commands are *number of categories* -1 (the degrees of freedom), so the graphs show chi-square density functions for 4, 8, and 20 categories.

 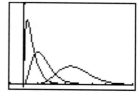

Display 12.6   Chi-square distributions for 4, 8, and 20 categories.

DISCUSSION
The Distribution of χ^2

D5. Refer to Display 12.5.

    a. Describe how the distribution of χ^2 changes as the number of categories increases. Discuss these changes in terms of shape, center, and spread.

    b. For which die are you most likely to get a value of χ^2 of 15 or more? For which die are you least likely to do so?

    c. For each die, make a rough approximation of the value of χ^2 that cuts off the upper 5% of the distribution.

Using the Chi-Square Table

df = *number of categories* -1.

To estimate the P-value for a chi-square test, you can use Table C in the Appendix on page 762. A partial table is shown in Display 12.7. In this table, *df* (*degrees of freedom*) is equal to the number of categories minus 1.

    The degrees of freedom concept is the same as it was in measuring variation from the mean. With n data values, the deviations from the mean sum to 0, so only $n - 1$ of the deviations are free to vary. Similarly, with k categories the deviations between observed and expected frequencies sum to 0, so only $k - 1$ of the deviations are free to vary.

df	.10	.05	...	Upper Tail Probability *p* .01	.005	...	.001
1	2.71	3.84	...	6.63	7.88	...	10.83
2	4.61	5.99	...	9.21	10.60	...	13.82
3	6.25	7.81	...	11.34	12.84	...	16.27
4	7.78	9.49	...	13.23	14.86	...	18.47
5	9.24	11.07	...	15.09	16.75	...	20.51
6	10.64	12.53	...	16.81	18.55	...	22.46
7	12.02	14.07	...	18.48	20.28	...	24.32
8	13.36	15.51	...	20.09	21.95	...	26.12
9	14.68	16.92	...	21.67	23.59	...	27.83
10	15.99	18.31	...	23.21	25.19	...	29.59
11	17.29	19.68	...	24.72	26.76	...	31.26
12	18.55	21.03	...	26.22	28.30	...	32.91

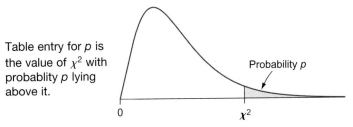

Display 12.7 A partial table of χ^2 and associated *P*-values from Table C.

Approximate *P*-Values | **Example 12.3**

In the fair die example, the observed value of the χ^2 statistic was 2.6. Would you reject the null hypothesis that the die is fair at the 5% level of significance? What is the approximate *P*-value?

Solution

Looking at the row of values for $6 - 1 = 5$ degrees of freedom in Display 12.7, you see that the smallest value is 9.24, which cuts off an upper tail area of 0.10. The computed value of χ^2 from the fair die example is only 2.6, which is well to the left of 9.24 under the χ^2 distribution (see Display 12.8). Therefore, all you can say about the *P*-value is that it is larger than 0.10. (Table C on page 762 tells you that it is also larger than 0.25.)

> Statistical software or a calculator can give you an exact *P*-value.

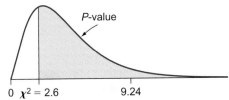

Display 12.8 Chi-square distribution with 5 degrees of freedom and $\chi^2 = 2.6$.

You have no evidence at the $\alpha = 0.05$ level that the die is unfair. As illustrated in Display 12.8, it is quite likely that you would get a value of 2.6 or greater with a fair die.

The Chi-Square Goodness-of-Fit Test

This section so far has used the fair die example to develop the ideas of a chi-square goodness-of-fit test. However, you can apply this test in important, real-life situations in which you wish to assess how well a given probability model fits your data. To perform a chi-square goodness-of-fit test, you should go through the same steps as for any test of significance.

Chi-Square Goodness-of-Fit Test

1. *State the hypotheses.*

   H₀: The proportions in the population are equal to the proportions in your model.

   (Continued)

H$_a$: At least one proportion in the population is not equal to the corresponding proportion in your model.

2. *Check conditions.*

- Each outcome in your population falls into exactly one of a fixed number of categories.
- You have a model that gives the hypothesized proportion of outcomes in the population that fall into each category.
- You have a random sample from your population.
- The expected frequency in each category is 5 or greater.

3. *Compute the value of the test statistic, approximate the P-value, and draw a sketch.* The test statistic is

$$\chi^2 = \sum \frac{(O - E)^2}{E}$$

Here, O is the observed frequency in the category and E is the corresponding expected frequency. Sum over all categories,

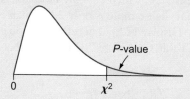

The *P*-value is the probability of getting a value of χ^2 as extreme as or even more extreme than the one from the sample, assuming the null hypothesis is true. Get a *P*-value from a calculator or computer, or approximate the *P*-value using Table C, where *df = number of categories* − 1.

4. *Write your conclusion linked to your computations and in the context of the problem.* If the *P*-value is smaller than the level of significance, α, then reject the null hypothesis. If not, then you don't have statistically significant evidence that the null hypothesis is false, and so you do not reject it.

Remember that you don't say you "accept" the null hypothesis. This is because you don't know that it is true; that is, you don't have any evidence that the hypothesized proportions are *exactly* the right ones, only evidence that your data are consistent with those proportions. Your data will be consistent with other proportions too.

Example 12.4	**Possible Election Fraud in Iran**

In 2009, Iran held an election for president, with incumbent Mahmood Ahmadinejad running against three other candidates. A victory for Ahmadinejad was speedily announced the morning after the election. The validity of the election was just as quickly challenged, and Iranians protested in the streets.

An independent poll of a random sample of voters was taken shortly before the election. The results of the poll and the (reported) election vote are given in Display 12.9. Are the results from the poll consistent with those of the election?

Candidate	Observed Number of Voters in Poll	Reported Votes in Election (%)
Ahmadinejad	338	63.29
Mousavi	136	34.10
Minor candidates	30	2.61
Total	504	100.00

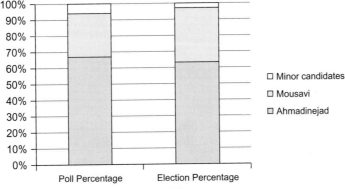

Display 12.9 Distribution of choices in pre-election poll of 1001 randomly selected voters, 504 of whom chose a candidate, and actual election results. [Source: "Poll Hint at Plausible Iran Vote," BBC News, *http://news. bbc.co.uk*.]

Solution

Suppose that the distribution of choices for the poll, if all voters had been asked, is the same as that in the election. Then, the expected number of voters who select each candidate in a poll of 504 voters is shown in Display 12.10. For example, Ahmadinejad got 63.29% of the votes in the election, so his expected frequency among 504 people polled would be $0.6329 \cdot 504 \approx 318.98$.

> Observed frequencies are always whole numbers. Expected frequencies typically aren't.

Candidate	Observed Number of Votes in Poll	Expected Number of Votes in Poll
Ahmadinejad	338	$0.6329 \cdot 504 \approx 318.98$
Mousavi	136	$0.3410 \cdot 504 \approx 171.86$
Minor candidates	30	$0.0261 \cdot 504 \approx 13.15$
Total	504	503.99

> It is not necessary that all of the expected frequencies be equal.

Display 12.10 Computation of expected number of votes in poll, based on election results.

The hypotheses are

> State the hypotheses.

H_0: The distribution of votes in the poll, if all voters had been polled, is the same as the distribution of votes in the election.

H_a: The distribution of votes in the poll, if all voters had been polled, is different from the distribution of votes in the election. That is, the proportion of votes for at least one candidate is different.

The conditions for a chi-square goodness-of-fit test are met in this situation. Each person polled could select only one candidate. The actual election results are the (hypothesized) model for how the poll would have turned out if every voter had been polled. You have a random sample of 504 voters. All of the expected frequencies are at least 5.

> Check conditions.

Compute the test statistic, approximate the *P*-value, and draw a sketch.	The test statistic is $$\chi^2 = \sum \frac{(O - E)^2}{E}$$ $$= \frac{(338 - 318.98)^2}{318.98} + \frac{(136 - 171.86)^2}{171.86} + \frac{(30 - 13.15)^2}{13.15}$$ $$= 1.13 + 7.48 + 21.59$$ $$= 30.20$$

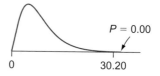

Write the conclusion in the context of the situation.	The value of χ^2 from the sample, 30.20, is quite far out in the tail of the chi-square distribution with $3 - 1$, or 2, degrees of freedom. The *P*-value is close to 0.

Reject the null hypothesis. You cannot attribute the difference in the expected and observed frequencies to variability due to sampling alone. A value of χ^2 this large is very unlikely to occur in random samples of this size if the distribution of votes at the time of the poll, had all voters been polled, was equal to the distribution of votes in the election. Conclude that the distributions are different.

This does not prove that there was vote fraud. People may have changed their minds between the poll and the election or may not have answered truthfully about how they intended to vote. Further, the large value of χ^2 is mainly due to the difference in the number of votes for the minor candidates. They got far more votes in the poll than would have been expected from the election results.

DISCUSSION
The Chi-Square Goodness-of-Fit Test

D6. In Example 12.4, does the value of the χ^2 statistic tell you where the lack of fit occurs (i.e., which candidates have polling numbers that match the worst to the election results)? If not, where can you look for such information?

D7. Why is a χ^2 test typically one-sided toward the large values?

Why Must Each Expected Frequency Be 5 or More?

The theoretical χ^2 distribution is a continuous distribution. For example, Display 12.11 shows a (continuous) χ^2 distribution for ten categories ($df = 9$) and another for six categories ($df = 5$).

The distribution of χ^2 computed from repeated sampling is discrete, however, because only a limited number of distinct values of χ^2 can be calculated for a given number of categories and a given sample size. Like the normal approximation to the binomial distribution, the χ^2 distribution is a continuous distribution that can be used to approximate this discrete distribution. The larger the expected frequencies, the closer the distribution of possible values of χ^2 is to a continuous distribution. In order to have a reasonable approximation, the expected frequency in each category should be 5 or greater. (This is a conservative rule, but it works well in most cases.)

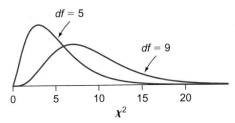

Display 12.11 Chi-square distributions for $df = 5$ and $df = 9$.

The Chi-Square Test Versus the *z*-Test

The chi-square test can be thought of as an extension of the *z*-test to more than two categories. When there are only two categories, you get the same *P*-value using a chi-square test as you do using a *z*-test. For example, according to the U.S. Census Bureau, 15% of the U.S. population is Hispanic and 85% is not Hispanic. You want to test whether it is plausible that these are the correct percentages in your county. You take a random sample of 1000 residents of your county and get 120 Hispanics. The χ^2 statistic is computed in Display 12.12. The value of χ^2 is about 7.059 and the *P*-value is 0.0079.

Category	Observed Frequency, O	Expected Frequency, E	$\frac{(O - E)^2}{E}$
Hispanic	120	150	6
Not Hispanic	880	850	1.059
Total	1000	1000	7.059

Display 12.12 Computation of χ^2 when testing the proportion of Hispanics.

If, instead, you use a two-sided *z*-test to test the null hypothesis that the proportion of Hispanics in your community is 0.15, the value of *z* is

$$z = \frac{\hat{p} - p_0}{\sqrt{\dfrac{p_0(1 - p_0)}{n}}} = \frac{0.12 - 0.15}{\sqrt{\dfrac{0.15(1 - 0.15)}{1000}}} \approx 2.6568$$

with a corresponding *P*-value of 0.0079. Not only are the *P*-values equal, but notice that $z^2 = (2.6568)^2 \approx 7.059 = \chi^2$. That is, the square of a *z*-statistic has a χ^2 distribution with 1 degree of freedom.

In summary, if there are only two types of outcomes (success and failure), you are back in the binomial situation and the *z*-test from Chapter 8 is equivalent to the chi-square test of this chapter. This implies, among other things, that the assumptions for the chi-square test are the same as those for the *z*-test: The sample must be random and large enough that the sample proportion has an approximately normal sampling distribution.

DISCUSSION
The Chi-Square Test Versus the *z*-Test

D8. Discuss the similarities and differences between a *z*-test using proportions and a chi-square test using frequencies.

Summary 12.1: Testing a Probability Model: The Chi-Square Goodness-of-Fit Test

The chi-square goodness-of-fit test is used when you want to know if it's plausible that the proportion of outcomes in the population that fall into each category is equal to the corresponding proportion in some hypothesized model.

To perform a chi-square goodness-of-fit test, you go through the same steps as for any test of significance (see page 575). The expected frequencies are calculated from the probabilities specified in the null hypothesis, and *df* = *number of categories* − 1.

If the results from your sample are very different from the results expected from the model, then χ^2 will be large and you will reject the hypothesis that the model is the correct one for your population. If the observed frequencies from your sample are close to the expected frequencies, then χ^2 will be relatively small and you don't have statistically significant evidence that the model is incorrect.

Practice

A Test Statistic

P1. Suppose you want to test whether a tetrahedral (four-sided) die is fair. You roll it 50 times and observe a one 14 times, a two 17 times, a three 9 times, and a four 10 times.

    a. What is the expected frequency in each category?

    b. Compute the value of χ^2.

P2. If each observed frequency equals the expected frequency, what is the value of χ^2?

The Distribution of χ^2

P3. Refer to Display 12.5 on page 573. Suppose you roll a 12-sided die 300 times to see if it is fair. You compute χ^2 and get 21.3. Use the appropriate histogram to approximate a *P*-value for this test. What is your conclusion if you are using $\alpha = 0.05$?

P4. Refer to Display 12.5. Suppose you roll an eight-sided die 100 times to see if it is fair, and get $\chi^2 = 21.3$. Use the appropriate histogram to approximate a *P*-value for this test. What is your conclusion if $\alpha = 0.05$?

Using the Chi-Square Table

P5. Give *df* for each situation, and then use Table C on page 762 to determine if the result is statistically significant ($\alpha = 0.05$).

    a. You roll a tetrahedral die 100 times and calculate a χ^2 value of 8.24.

    b. You roll a 20-sided die 500 times and calculate a χ^2 value of 8.24.

P6. Use Table C on page 762 to estimate the *P*-value for the case of

    a. rolling an 8-sided die, and $\chi^2 = 12$

    b. rolling a 12-sided die, and $\chi^2 = 21.3$

P7. In 1882, R. Wolf rolled a die 20,000 times. The results are recorded in Display 12.13.

    a. What is the expected number of the 20,000 rolls that would fall under each outcome if the die is fair?

    b. Is this statistically significant evidence that the die was unfair, or is it approximately what you would expect from a fair die?

Outcome	Frequency
1	3407
2	3631
3	3176
4	2916
5	3448
6	3422

Display 12.13 Results from 20,000 rolls of a die. [Source: D. J. Hand et al., *A Handbook of Small Data Sets* (London: Chapman & Hall, 1994), p. 29.]

The Chi-Square Goodness-of-Fit Test

P8. Display 12.14 gives the number of births for each month in a hospital in Switzerland. Only the 700 women who were having their first baby were included.

    a. What is the expected number of the 700 babies that would be born each month if births are spread evenly throughout the year? You may assume that each month has the same number of days.

    b. Assuming that these can be considered a random sample of births in Switzerland, is there statistically significant evidence that first births are not spread evenly throughout the months?

Month	Number of Births	Month	Number of Births	Month	Number of Births
January	66	May	64	September	54
February	63	June	74	October	51
March	64	July	70	November	45
April	48	August	59	December	42

Display 12.14 The number of births to first-time mothers per month in a hospital in Switzerland. [Source: D. J. Hand et al., *A Handbook of Small Data Sets* (London: Chapman & Hall, 1994), p. 77. Original source: P. Walser, "Untersuchung über die Verteilung der Geburtstermine bei der mehrebärenden Frau." *Helvetica Paediatrica Acta*, Suppl. XX, Vol. 24, 3 (1969), pp. 1–30.]

P9. It is sometimes said that older people are overrepresented on juries. Display 12.15 gives information about people on grand juries in Alameda County, California.

   a. What is the expected number of grand jurors in each age group, if 66 grand jurors are selected at random from the adults in the county?

   b. Assuming that these people can be considered a random sample of people on grand juries in Alameda County, is there statistically significant evidence that the distribution of ages of grand jurors is not the same as the distribution of ages in the county?

P10. Which of these statements describe properties of the chi-square goodness-of-fit test?

   A. If you switch the order of the categories, the value of the χ^2 statistic does not change.

   B. The observed frequencies are always whole numbers.

   C. The expected frequencies are always whole numbers.

   D. The number of degrees of freedom is 1 less than the sample size.

   E. A high value of χ^2 indicates a high level of agreement between the observed frequencies and the expected frequencies.

The Chi-Square Test Versus the z-Test

P11. You spin a coin 100 times and get tails 64 times. Do you have statistically significant evidence that this coin is unfair?

   a. Use inference for a proportion, as in Chapter 8.

   b. Use a chi-square goodness-of-fit test.

   c. Compare the two results in parts a and b.

P12. To demonstrate the differences between flipping and spinning, Professors A. Gelman and D. Nolan modified a checker by adding putty to the crown side, calling it heads. In 100 flips of the checker, they observed 54 heads. In 100 spins of the checker, they observed 23 heads. The results for the flips are consistent with the model that the probability of heads is 0.5, even with the putty. Are the results for the spins of checkers with putty consistent with the model that the probability of heads is 0.5? [Source: *stat-www.berkeley.edu.*]

   a. Answer the question by using inference for a proportion, as in Chapter 8.

   b. Answer the question by using a chi-square goodness-of-fit test.

   c. Compare the two results in parts a and b.

Age Group	Countywide Percentage in Age Group	Number of Grand Jurors
21–40	42	5
41–50	23	9
51–60	16	19
61 or older	19	33
Total	100	66

Display 12.15 Distribution of ages for jurors in Alameda County, California. [Source: David Freedman et al., *Statistics*, 2nd ed. (New York: Norton, 1991), p. 484. Original source: *UCLA Law Review.*]

Exercises

E1. The blood types of 254 black homicide victims in Philadelphia are shown in Display 12.16. Black blood donors in Philadelphia have a distribution with 49% Type 0, 27% Type A, 20% Type B, and 4% Type AB.

   a. What is the expected number of the 254 homicide victims who would have each blood type, if the homicide population has the same distribution of blood types as the donor population?

Blood Type	O	A	B	AB	Total
Frequency	134	70	45	5	254

Display 12.16 Frequency of blood types of 254 black homicide victims in Philadelphia. [Source: David Lester, "Distribution of Blood Types in a Sample of Homicide Victims," *Psychological Reports*, Vol. 53, 3 (1986), p. 802.]

   b. The article reported that $\chi^2 = 4.72$ with $df = 3$. Do you agree? Justify your answer.

   c. Do you have statistically significant evidence that black homicide victims do not have the same distribution of blood types as the donor population? Be sure to include all four steps of a test of significance.

E2. The blood types of 66 white homicide victims in Philadelphia are shown in Display 12.17. White blood donors in Philadelphia have a distribution with 45% Type O, 42% Type A, 9% Type B, and 4% Type AB.

   a. What is the expected number of the 66 homicide victims who would have each blood type, if the homicide population has the same distribution of blood types as the donor population?

Blood Type	O	A	B	AB	Total
Frequency	35	23	4	4	66

Display 12.17 Frequency of blood types of 66 white homicide victims in Philadelphia. [Source: David Lester, "Distribution of Blood Types in a Sample of Homicide Victims," *Psychological Reports*, Vol. 53, 3 (1986), p. 802.]

   b. The article reported that $\chi^2 = 3.15$ with $df = 3$. Do you agree? Justify your answer.

   c. Do you have statistically significant evidence that white homicide victims do not have the same distribution of blood types as the donor population? Be sure to include all four steps of a test of significance.

E3. Many people think that the full moon prompts strange behavior and occurrences. One commonly held belief is that epileptic seizures are more frequent during a full moon. To check this claim, a researcher observed that, among 470 epileptic seizures monitored at Tampa General Hospital, 103 occurred during the new moon, 121 during the first quarter, 94 during the full moon, and 152 during the last quarter. Each of these four quarters lasts the same number of days. Are these data consistent with a model of equal likelihood of seizures during the four periods? If not, where are the largest deviations from the model? [Source: *Los Angeles Times*, June 7, 2004, p. F2.]

E4. In another study to decide whether people behave differently under a full moon, researchers reviewed all records from an emergency unit over a 5-year period and found 859 patients admitted for seizures. Of these patients, 294 were admitted during the period when there was a full moon, 184 were admitted during the new moon, 193 were admitted during the first quarter of the lunar cycle, and 188 were admitted during the last quarter of the lunar cycle. Each of these four periods lasts the same number of days. Assuming that these patients can be considered a random sample of patients admitted for seizures, find appropriate expected numbers of admissions, and perform a chi-square goodness-of-fit test to test whether admission is related to the cycle of the moon. [Source: P. Polychronopoulos et al., "Lunar Phases and Seizure Occurrence: Just an Ancient Legend?" *Neurology*, Vol. 66 (May 9, 2006), pp. 1442–1443.]

E5. Gregor Mendel (1822–1884) performed experiments to try to validate the predictions of his genetic theory.

Mendel's experimental results match his predictions quite closely. In fact, statisticians think his results match too closely. In one experiment, Mendel predicted that he would get a 9:3:3:1 ratio among smooth yellow peas, wrinkled yellow peas, smooth green peas, and wrinkled green peas. So, for example, the proportion of peas that were smooth yellow would be 9/16. His experiment resulted in 315 smooth yellow peas, 101 wrinkled yellow peas, 108 smooth green peas, and 32 wrinkled green peas.

   a. What is the expected frequency for each type of pea, if Mendel's theory is correct?

   b. Where in the chi-square distribution will values of χ^2 lie when the reported numbers of peas of each type match the expected frequencies *too* closely?

   c. Are the observed counts suspiciously close to the counts predicted by Mendel's theory?

E6. The characteristics of the 529 offspring of the plants discussed in E5 are given in Display 12.18. According to Mendel's theory, these genotypes, starting from the top, should be in the ratio 1:1:1:1:2:2:2:2:4.

   a. What is the expected frequency for each genotype of pea, if Mendel's theory is correct?

   b. Are the data consistent with the theory?

   c. Do they support it better than you would expect?

Genotype	Frequency
AB	38
Ab	35
aB	28
ab	30
Abb	65
aBb	68
AaB	60
Aab	67
AaBb	138

A = round, a = wrinkled,
B = yellow, and b = green

Display 12.18 Frequencies of the genotypes of 529 pea plants. [Source: *www.mendelweb.org.*]

E7. One of the most famous U.S. elections of all time was the Truman–Dewey election of 1948. It became famous because Dewey was predicted to win by almost everyone and by almost every poll. Display 12.19 gives the final predictions of the three most important national polls of

	Crossley	Gallup	Roper	Actual Vote
Truman	45	44	38	50
Dewey	50	50	53	45
Thurmond	2	2	5	3
Wallace	3	4	4	2

Display 12.19 Poll predictions and actual vote (percentage of votes) for the 1948 U.S. presidential election. [Source: F. Mosteller, *The Pre-election Polls of 1948* (New York: Social Sciences Research Council, 1949).]

the day, along with the actual result. Suppose each poll sampled about 3000 voters. Do any of the three poll results demonstrate a reasonably good fit to the actual results?

E8. The 1948 presidential election polls of Crossley, Gallup, and Roper (see E7) were based on a method called "quota sampling," in which quotas such as so many males, so many females, so many retired, so many working, so many unemployed, and so on were to be filled by fieldworkers for the polling company. The state of Washington, however, tried out a fairly new method called "probability sampling" which made use of the random sampling ideas used today. Display 12.20 shows how that poll, based on about 1000 voters, came out. Perform a test of significance to determine if you can reject the hypothesis that the result from the probability sample is a good fit to the actual vote.

	Probability Sample	Actual Vote
Dewey	46.0	42.7
Truman	50.5	52.6
Wallace	2.9	3.5
Other	0.6	1.2

Display 12.20 Poll predictions and actual results (percentage of votes) for the 1948 U.S. presidential election for the state of Washington. [Source: F. Mosteller, *The Pre-election Polls of 1948* (New York; Social Sciences Research Council, 1949).]

E9. Does the number of people involved in traffic accidents increase after a Super Bowl telecast? To answer this question, investigators looked at the number of fatalities on public roadways in the United States for the first 27 Super Bowl Sundays. They compared the number of fatalities during the 4 hours after the telecast with the number during the same time period on the Sundays the week before and the week after the Super Bowl (54 control Sundays). If the telecast had no effect, there should be roughly twice the number of fatalities on the 54 control Sundays as on the 27 Super Bowl Sundays. [Source: D. A. Redelmeier and C. L. Stewart, "Do Fatal Crashes Increase Following a Super Bowl Telecast?" *Chance*, Vol. 18, 1 (2005), pp. 19–24.]

a. There were 662 fatalities in the 4 hours following Super Bowl telecasts and 936 in the corresponding hours on control Sundays. Compute the expected frequencies for each time period under the hypothesis that the telecast had no effect. Are these data consistent with the model that the telecast has no effect? What can you conclude?

b. Within the home state of the winning Super Bowl team, the number involved in traffic accidents (fatalities plus survivors) totaled 141 for Super Bowl Sundays and 265 for control Sundays. Compute the expected frequencies for each time period under the hypothesis that the telecast had no effect. Do these data fit the model that the telecast has no effect? What can you conclude?

E10. The International Mathematical Olympiad (IMO) is a contest in mathematical problem solving. Each of about 90 countries sends a team of six secondary school students. The U.S. team is chosen largely on the basis

of the American Mathematics Competitions, a series of increasingly difficult exams taken annually by almost half a million students. Display 12.21 gives the ethnicity of the 120 students who were on the U.S. IMO teams most recently and the approximate percentage of that ethnicity in the high school population. This is a situation where it is obvious that the distribution of IMO participants does not reflect the U.S. high school population.

	Number of Students	High School Population (approx)
Asian	42	4%
Jewish	26	2%
White (non-Jewish, non-Hispanic)	52	65%
Underrepresented Minority	0	29%

Display 12.21 Ethnicity of U.S. IMO participants, 1988–2007, and approximate percentage of the U.S. high school population. Some ethnicities estimated. [Source: Titu Andreescu et al., "Cross-Cultural Analysis of Students with Exceptional Talent in Mathematical Problem Solving," *Notices of the American Mathematical Society*, Vol. 55, 10 (November 2008), pp. 1248–1260; *Statistical Abstract of the United States*, 1998, Table 22.]

a. What is the expected number of students in the IMO for each ethnicity, if the distribution matched that of the U.S. high school population?

b. What is the value of χ^2 in a goodness-of-fit test? The *P*-value?

c. What category accounts for most of the lack of fit? Which category fits most closely?

E11. In a study reported in the prestigious journal *Science*, 18 college-educated volunteers were given five foods to taste: two expensive liver pâtés, liverwurst, Spam, and dog food. All were blended to the same consistency, garnished with parsley, and chilled. The people were asked to rate the foods and to identify the dog food.

a. The table in Display 12.22 gives the number of people who chose each food as the dog food. The researchers performed a chi-square test on these data to test whether the choices are compatible with random selection. Are conditions met for such a test?

b. Regardless of your answer to part a, compute the value of the chi-square test statistic and find a *P*-value. What is your conclusion?

c. The researchers got the following results:

Chi-Squared Test
$\chi^2 = 0.433$
2-tailed *P* value = 0.9797
* Not significantly different from random distribution

Can you figure out how they computed this (incorrect) value of chi-square? The table from the article, reproduced in Display 12.23, should provide a hint.

Food	Selected as Dog Food
Duck liver mousse	1
Spam	4
Dog food	3
Pork liver pâté	2
Liverwurst	8

Display 12.22 Number of taste testers who selected each food as the dog food. [Source: John Bohannon, "Gourmet Food, Served by Dogs," The Gonzo Scientist, *Science*, Vol. 323, 5917 (February 20, 2009), p. 100. Original source: John Bohannon, Robin Goldstein, and Alexis Herschkowitsch, "Can People Distinguish Pâté from Dog Food?" April 2009, American Association of Wine Economists Working Paper #36, *www.wine-economist.org.*]

Food Sample	Identified as Dog Food	Frequency	Expected If Random
Duck liver mousse	1	0.06	0.2
Spam	4	0.22	0.2
Dog food	3	0.17	0.2
Pork liver pâté	2	0.11	0.2
Liverwurst	8	0.44	0.2

Display 12.23 Table IV, Which sample is dog food? from the working paper. "Can People Distinguish Pâté from Dog Food?"

E12. A sign on a barrel of nuts in a supermarket says that it contains 30% cashews, 30% hazelnuts, and 40% peanuts by weight. You mix up the nuts and scoop out 20 pounds. When you weigh the nuts, you find that you have 6 pounds of cashews, 5 pounds of hazelnuts, and 9 pounds of peanuts.

   a. An introductory statistics book asks students to use a chi-square test to test whether there is evidence to doubt the supermarket's claim. Follow these instructions.

   b. Now convert the measurements to ounces and recalculate.

   c. What is your conclusion now?

   d. Do you see any problem with using a chi-square test on data like these?

E13. The *World Almanac and Book of Facts* lists 104 "Major Rivers in North America" with their lengths in miles. These data come from the U.S. Geological Survey. For example, the length of the Hudson River is given as 306 miles, the Columbia as 1243 miles, and the Missouri as 2315 miles. The final digits of the lengths of these three rivers are 6, 3, and 5, respectively.

   a. What distribution of the final digits of the lengths of the 104 major rivers would you expect if the U.S. Geological Survey measures length to the nearest mile?

   b. The distribution of the final digits is shown in Display 12.24. Test the hypothesis that the digits are equally likely. If they do not appear to be equally likely, what possible explanation can you offer?

Final Digit	Frequency
0	44
1	6
2	9
3	5
4	5
5	15
6	5
7	3
8	3
9	9

Display 12.24 Frequencies of the final digits of the lengths of the 104 major rivers of North America.

E14. If you look at the leading digits in a table of data, you might expect each digit to occur about 1/9 of the time. (There are only nine possibilities because the digit 0 is never a leading digit.) However, smaller digits tend to occur more frequently than larger digits. Benford's Law says that the digit k occurs with relative frequency

$$\log_{10}\left(\frac{k+1}{k}\right)$$

   a. Make a chart that shows the relative frequencies expected for the digits 1 through 9.

   b. Prove algebraically that the sum of the relative frequencies is 1.

   c. Display 12.25 shows the percentages of first digits seen in a sample of 100 numbers appearing on the first page of various newspapers, collected by Dr. Frank Benford himself. Do they obey his law?

Digit	1	2	3	4	5	6	7	8	9
Percent	30	18	12	10	8	6	6	5	5

Display 12.25 A sample of 100 first digits from newspapers. [Source: *mathworld.wolfram.com.*]

12.2 ▶ The Chi-Square Test of Homogeneity

A chi-square test of homogeneity tests whether it is reasonable to believe that when several different populations are sorted into the same categories, they will have the same proportion of units in each category. Such populations are called **homogeneous**. This sounds rather abstract, but the following situation will illustrate how such tests are used to make decisions that affect your life.

> Homogeneous populations have the same proportion of members in any given category.

Comparing Two or More Populations

In an effort to discourage drivers from running red lights, some communities have installed cameras at intersections. Drivers who run red lights are photographed and receive a ticket in the mail. Though the red-light cameras do cut down on crashes overall, decreasing the number of right-angle crashes, they produce somewhat more rear-end crashes as drivers brake suddenly to avoid a ticket. One concern is that, because the stops are so sudden, these rear-end crashes are more serious than rear-end crashes at intersections without red-light cameras. To address this concern, the Federal Highway Administration (FHWA) compared the distributions of injuries in rear-end crashes in two populations: at intersections before red-light cameras were installed and at intersections after cameras were installed. The two samples of data collected by the FHWA appear in Display 12.26. (Most rear-end crashes did not result in injury; these rear-end crashes without injuries were not included in the FHWA table.)

	Before Camera	After Camera
Death or disabling Injury	61	27
Evident Injury	210	136
Possible Injury	1659	845
Total	1930	1008

Display 12.26 Number of injuries of each type in rear-end accidents with injuries before and after red-light cameras installed in seven urban areas in the United States. [Source: *safety.fhwa.dot.gov/intersection/redlight/redl_faq.cfm*; Safety Evaluation of Red-Light Cameras, FHWA-HRT-05-048, FHWA, April 2005, Table 18.]

The table in Display 12.26 is called a two-way table because each rear-end injury accident is classified in two ways: according to whether it occurred before or after the installation of a red-light camera (the population it comes from) and the severity of the injury. Disregarding the labels and totals, it has three rows, two columns, and six cells. The data are called *categorical* because the only information recorded about each response is which category (or class) it falls into. A cell of the table consists of a frequency (or count).

The **grand total** is the sum of all entries in the cells of the table, 2938. The column totals, 1930 and 1008 are called the **marginal frequencies**. The summary that $1930/2938 \approx 0.66$, or 66%, of the crashes occurred before cameras were installed is called a **marginal relative frequency**. A total of 61 of the before-camera crashes resulted in death or disabling injury; this is called a **joint frequency** (joint between the column population, *before camera*, and the row category, *death or disabling injury*). Among the before-camera crashes, 61/1930, or 0.032, or 3.2% caused death or disabling injury; among the after-camera crashes, 27/1008, or 0.027, or 2.7%, caused death or disabling injury. These are called **conditional relative frequencies** for these columns.

Are the results from the FHWA's two samples consistent with the null hypothesis that the before-camera and after-camera distributions of severity of injury are identical?

A segmented bar chart stacks the categorical frequencies on top of each other.

As in any analysis, you should look first at a graphical display of the data. One possible plot is given in Display 12.27. This plot is called a **segmented**, or **stacked**, **bar chart**. For each population (before camera and after camera), the observed frequencies in each category are; stacked on top of each other. From this plot alone, it's clear that there isn't a major difference in the two distributions.

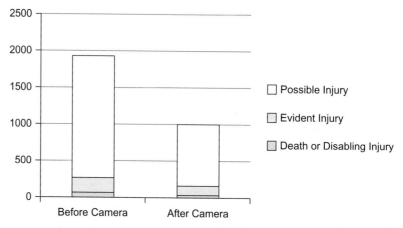

Display 12.27   A segmented bar chart for the red-light camera data.

DISCUSSION
Comparing Two or More Populations

D9. Suppose that it is the case that injuries tended to be more serious after the cameras were installed. What would the segmented bar chart look like?

Computing Expected Frequencies

What is the expected frequency in each cell if the null hypothesis is true that the distributions of severity of injury are identical for the population of all before-camera rear-end crashes and the population of all after-camera rear-end crashes? To determine this, begin by adding the marginal frequency to each row of the table, as in Display 12.28.

If the *before-camera* and *after-camera* population distributions are the same, then the best estimate of the proportion of crashes that fall into a given injury category is the marginal frequency for that injury category divided by the grand total. For example, the best estimate of the overall proportion of crashes with death or disabling injury is the marginal relative frequency:

This computation is like the one from Section 9.2, where you computed a pooled estimate, $\hat{p}$.

$$\frac{\text{total number with death or disabling injury}}{\text{grand total}} = \frac{88}{2938}$$

$$\approx 0.02995$$

	Before Camera	After Camera	Total
Death or Disabling Injury	61	27	88
Evident Injury	210	136	346
Possible Injury	1659	845	2504
Total	1930	1008	2938

Display 12.28   Number of injuries of each type in rear-end crashes with injuries before and after red-light cameras installed in seven urban areas in the United States with marginal frequencies.

If the null hypothesis is true, the proportion of all before-camera injury crashes with death or disabling injury would also be 0.02995. Then, the expected frequency of the number of before-camera crashes with death or disabling injury is

(proportion with death of disabling injury) *(number of before-camera crashes)*
$$= (0.02995)(1930) \approx 57.8$$

Note that the computations are equivalent to:

$$\frac{88 \cdot 1930}{2938} = \frac{row\ total \cdot column\ total}{grand\ total}$$

This pattern results in a formula for expected frequencies.

Formula for the Expected Frequency, E

$$E = \frac{row\ total}{grand\ total} \cdot (column\ total) = \frac{row\ total \cdot (column\ total)}{grand\ total}$$

Expected Frequencies for the Red-Light Camera Study | Example 12.5

Compute the expected frequencies for the red-light camera study in Display 12.29.

Solution
To find each frequency, E, use the formula

$$E = \frac{row\ total \cdot column\ total}{grand\ total}$$

The computations and expected frequencies appear in the table in Display 12.29.

Expected frequencies typically aren't whole numbers. Observed counts must be.

	Before Camera	After Camera	Total
Death or Disabling Injury	$\frac{88 \cdot 1930}{2938} \approx 57.8$	$\frac{88 \cdot 1008}{2938} \approx 30.2$	88
Evident Injury	$\frac{346 \cdot 1930}{2938} \approx 227.3$	$\frac{346 \cdot 1008}{2938} \approx 118.7$	346
Possible Injury	$\frac{2504 \cdot 1930}{2938} \approx 1644.9$	$\frac{2504 \cdot 1008}{2938} \approx 859.1$	2504
Total	1930	1008	2938

Display 12.29 Computation of expected frequencies for red-light camera data.

After you finish computing a table of expected frequencies, always check that the row and column totals are the same as in your table of observed frequencies (subject to round-off error).

DISCUSSION
Computing Expected Frequencies

D10. Review the reasoning for finding the expected frequencies in the red-light camera example. What does a segmented bar graph look like for the expected frequencies?

The Chi-Square Statistic, Degrees of Freedom, and *P*-value

The chi-square test of homogeneity is quite similar to a chi-square test for goodness of fit. The value of χ^2 is computed in the same way, using the observed and expected values for the cells of the table, but in a test of homogeneity the expected frequencies are estimated from the sample data. In the goodness-of-fit test, the expected frequencies were computed from the proportions specified in the null hypothesis.

| Example 12.6 | **The Test Statistic for the Red-Light Camera Study** |

Compute χ^2 for the red-light camera study, using the observed frequencies in Display 12.26 (page 585) and expected frequencies in Display 12.29.

Solution
You can organize your work in a table, like that in Display 12.30.

Outcome	Observed Frequency, O	Expected Frequency, E	$O - E$	$\frac{(O-E)^2}{E}$
Before camera, death or disabling injury	61	57.8	3.2	0.18
Before camera, evident injury	210	227.3	−17.3	1.32
Before camera, possible injury	1659	1644.9	14.1	0.12
After camera, death or disabling injury	27	30.2	−3.2	0.34
After camera, evident injury	136	118.7	17.3	2.52
After camera, possible injury	845	859.1	−14.1	0.23
Total	2938	2938	0	4.71

Display 12.30 Calculating χ^2 for red-light camera data.

The value of chi-square is the sum of the entries in the last column, so

$$\chi^2 = \sum \frac{(O - E)^2}{E} \approx 4.71$$

Before you can find the *P*-value for a chi-square test of homogeneity, you need to know the number of degrees of freedom, *df*.

For a chi-square test of homogeneity, the number of degrees of freedom is

$$df = (r - 1)(c - 1)$$

Here, *r* is the number of rows in the table of observed values and *c* is the number of columns (not counting the headings or totals in either case).

The Red-Light Camera Study, Coming to a Conclusion	**Example 12.7**

What is the number of degrees of freedom for the red-light camera data in Display 12.26? Find and interpret the *P*-value.

Solution

There are three rows and two columns. The number of degrees of freedom is

$$df = (r - 1)(c - 1) = (3 - 1)(2 - 1) = 2$$

Using a calculator or computer to determine the *P*-value for this test, where $\chi^2 = 4.71$, you get approximately 0.09, as shown in Display 12.31. Alternatively, if you look in Table C on page 762, you will find that this value of χ^2 is not significant at the 0.05 level. The *P*-value, 0.09, means that if the null hypothesis is true that the before-camera and after-camera distributions of severity of injury are identical, then it is reasonably likely to get observed frequencies as far from the expected frequencies as the FHWA did in their study. Consequently, you do not reject the hypothesis that the distributions are identical.

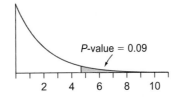

Display 12.31 χ^2 distribution with *df* = 2, χ^2 = 4.71.

You don't have statistically significant evidence that injuries are more or less severe after the red-light cameras were installed. The FHWA came to the same conclusion: "As can be seen, there is no apparent shift in the severity distribution for rear end injury crashes from before to after."

Procedure for a Chi-Square Test of Homogeneity

The steps in a chi-square test of homogeneity are much the same as the steps in the chi-square goodness-of-fit test.

Chi-Square Test of Homogeneity

1. *State the hypotheses.*

   H_0: The proportion that falls into each category is the same for every population.

(Continued)

H_a: For at least one category, it is not the case that each population has the same proportion in that category; that is, in some category, the proportion for at least one population is different from that for another population.

2. *Check conditions.*

- You should have independent random samples of fixed (but not necessarily equal) sizes taken from two or more large populations (or two or more treatments are randomly assigned to subjects).

- Each outcome falls into exactly one of several categories, with the categories being the same in all populations.

- The expected frequency in each cell is 5 or greater.

3. *Compute the value of the test statistic, approximate the P-value, and draw a sketch.* The test statistic is

$$\chi^2 = \sum \frac{(O - E)^2}{E}$$

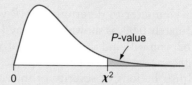

For each cell in a two-way frequency table, O is the observed frequency (count) and E is the expected frequency where

$$E = \frac{row\ total \cdot column\ total}{grand\ total}$$

The *P*-value is the probability of getting a value of χ^2 as extreme as or more extreme than the one from the sample, assuming the null hypothesis is true. Get the *P*-value from a calculator or computer, or approximate the *P*-value by comparing your value of χ^2 to the appropriate value of χ^2 in Table C on page 762 using $df = (r - 1)(c - 1)$ degrees of freedom. Here, r is the number of categories and c is the number of populations.

4. *Write your conclusion linked to your computations and in the context of the problem.* If the *P*-value is smaller than α, then reject the null hypothesis. If not, you don't have statistically significant evidence that the null hypothesis is false, so you do not reject it.

Example 12.8 | A Survey of Family Values

The Gallup Organization took a poll on family values in different countries. One question asked was, "For you personally, do you think it is necessary or not necessary to have a child at some point in your life in order to feel fulfilled?" Display 12.32 shows the results from five countries, for samples of 1000 adults.

From the plot, you can see that the percentages do not appear to be the same for each country. The difference between India and the United States seems too great. Test the hypothesis that the proportion of adults who would give each answer is the same for each country. Use $\alpha = 0.001$.

		U.S.	India	Mexico	Canada	Germany	Total
				Country			
Response	Necessary	460	930	610	590	490	3080
	Unnecessary	510	60	380	370	450	1770
	Undecided	30	10	10	40	60	150
	Total	1000	1000	1000	1000	1000	5000

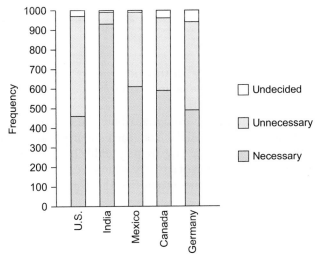

Display 12.32   Two-way table of observed frequencies and segmented bar chart for the Gallup poll results. [Source: *www.gallup.com*, May 2002.]

Solution

The hypotheses are

State the hypotheses.

H_0: If you could ask all adults, the distribution of answers would be the same for each country.

H_a: The distribution of answers would not be the same in each of the five countries. That is, in at least one country, the proportion of all adults who would give one of the answers is different from the proportion in another country.

Under the null hypothesis, the expected frequencies are given in the second row of each cell in the Minitab output in Display 12.33.

Check conditions.

Chi-Square Test

Expected counts are printed below observed counts

	U.S.	India	Mexico	Canada	Germany	Total
1	460	930	610	590	490	3080
	616.00	616.00	616.00	616.00	616.00	
2	510	60	380	370	450	1770
	354.00	354.00	354.00	354.00	354.00	
3	30	10	10	40	60	150
	30.00	30.00	30.00	30.00	30.00	
Total	1000	1000	1000	1000	1000	5000

ChiSq = 39.506 + 160.058 + 0.058 + 1.097 + 25.773 + 68.746 + 244.169 + 1.910 + 0.723 +
      26.034 + 0.000 + 13.333 + 13.333 + 3.333 + 30.000 = 628.075

df = 8, p = 0.000

Display 12.33   Minitab printout for the family values example.

Compute the test statistic, approximate the *P*-value, and draw a sketch.	

The conditions are met in this situation for a chi-square test of homogeneity. There are five large populations, and independent random samples of size 1000 were taken from each population. Each answer falls into exactly one of three categories. All of the expected frequencies are at least 5.

The test statistic shown in Display 12.33 was computed as follows:

$$\chi^2 = \frac{(460 - 616)^2}{616} + \frac{(930 - 616)^2}{616}$$
$$+ \cdots + \frac{(40 - 30)^2}{30} + \frac{(60 - 30)^2}{30}$$
$$= 628.075$$

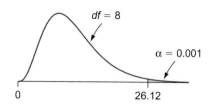

Write the conclusion in the context of the situation.	

Comparing the test statistic to the χ^2 distribution with $df = (3 - 1)(5 - 1)$, or 8, you can see that the value of χ^2 from the sample, 628.075, is extremely far out in the tail and the *P*-value is less than $\alpha = 0.001$.

Reject the null hypothesis. You cannot attribute the differences to the fact that you have only a random sample of adults from each country and not the entire adult population. A value of χ^2 this large is extremely unlikely to occur in five random samples of this size if the distribution of answers is the same in each country. Conclude that if you questioned all people in these countries, the distributions of answers would differ among some countries.

Example 12.9	**An Experiment About Treating Premature Infants**

Chronic lung disease is one of the primary long-term complications among premature infants. This condition generally impairs growth and can result in poor long-term cardiopulmonary function, an increased susceptibility to infection, and increased risk of abnormal neurological development.

A research hypothesis suggested that the use of inhaled nitric oxide would decrease the incidence of chronic lung disease and death in premature infants. To test this hypothesis, a randomized, double-blind, placebo-controlled study of inhaled nitric oxide versus a placebo in premature infants undergoing mechanical ventilation was designed and conducted. A total of 105 infants were randomly assigned to the nitric oxide treatment group, and 102 infants were assigned to the control group. Demographic and baseline clinical characteristics did not differ significantly between the control group and the group given inhaled nitric oxide. The results for the primary goals—reducing death and chronic lung disease—are shown in Display 12.34.

Do these data provide evidence that the inhaled nitric oxide and the placebo result in different distributions of outcomes?

Solution

State the hypotheses.	

The hypotheses are

H_0: If all 207 infants had received nitric oxide, the resulting distribution of outcomes would be the same as the distribution if all 207 infants had received the placebo.

H_a: If all 207 infants had received nitric oxide, the resulting distribution of outcomes would be different from the distribution if all 207 infants had received the placebo.

	Treatment	
	Nitric Oxide	Placebo
Death	16	23
Survival with Chronic Lung Disease	35	42
Survival without Chronic Lung Disease	54	37
Total	105	102

Outcome is indicated at the left spanning the three outcome rows.

Display 12.34 Table of observed frequencies for the premature infant study. [Source: Michael D. Schreiber et al., "Inhaled Nitric Oxide in Premature Infants with the Respiratory Distress Syndrome," *New England Journal of Medicine*, Vol. 349, 22 (November 27, 2003).]

Treatments were randomly assigned to subjects in large enough numbers so that all expected cell frequencies are at least 5 (see Display 12.35). The conditions are met.

Check conditions.

	Treatment	
	Nitric Oxide	Placebo
Death	19.78	19.22
Survival with Chronic Lung Disease	39.06	37.94
Survival without Chronic Lung Disease	46.16	44.84
Total	105	102

Display 12.35 Table of expected frequencies for the premature infant study.

These observed and expected frequencies result in a χ^2 value of

Compute the test statistic, approximate the P-value, and draw a sketch.

$$\chi^2 = \frac{(16 - 19.78)^2}{19.78} + \frac{(23 - 19.22)^2}{19.22} + \frac{(35 - 39.06)^2}{39.06} + \frac{(42 - 37.94)^2}{37.94}$$

$$+ \frac{(54 - 46.16)^2}{46.16} + \frac{(37 - 44.84)^2}{44.84}$$

$$= 5.026$$

Comparing this value to the tabled values of the χ^2 distribution with 2 degrees of freedom shows that the P-value is a little larger than 0.05 but smaller than 0.10. A calculator or computer gives a P-value of 0.08.

Write a conclusion linked to the computations and in the context of the problem.

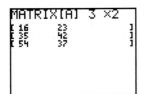

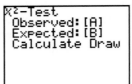
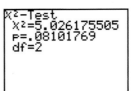

This is a borderline case. If you strictly interpret this as a fixed 5% level test, the decision is not to reject the null hypothesis and to say you don't have statistically significant evidence that the results are different when treating with nitric oxide and treating with a placebo. However, the P-value is fairly small and gives some evidence against the null hypothesis. As you will see in P19, the evidence is strengthened if you collapse the two undesirable outcomes *death* and *survival with chronic lung disease* into one category. The original article reporting this research ends with a positive recommendation for the nitric oxide treatment under certain conditions.

DISCUSSION
Procedure for a Chi-Square Test of Homogeneity

D11. Does it appear that the independent random sample condition is met in the red-light camera example? In what sense, then, is a chi-square test of homogeneity appropriate?

The Two-Sample *z*-Test Versus the Chi-Square Test of Homogeneity

In the red-light camera example, it is possible, but not a good idea, to use two-sample *z*-tests (see Section 9.2) to test three separate hypotheses about the difference in the proportion of injuries for the before-camera and after-camera populations.

- The proportion of rear-end injury crashes that result in a death or disabling injury is the same for the population of all before-camera crashes as for the population of all after-camera crashes.

- The proportion of rear-end injury crashes that result in evident injury is the same for the population of all before-camera crashes as for the population of all after-camera crashes.

- The proportion of rear-end injury crashes that result in a possible injury is the same for the population of all before-camera crashes as for the population of all after-camera crashes.

Never use several tests when one is available.

Why not use multiple *z*-tests rather than one chi-square test to determine whether the population proportions are homogeneous? This is a deep question, and the answer relies on complex statistical theory. The bottom line is that you have a greater chance of coming to an erroneous conclusion with multiple tests than you do with only one. A guiding statistical principle for tests of significance is never to use more tests than you absolutely need. The chi-square test is a clever way of combining many *z*-tests into one overall test. (However, it can be appropriate to use *z*-tests to compare proportions within a table once overall statistical significance has been established.)

If there are two populations and two categories, you have a choice of which test to use. Testing the null hypothesis that $p_1 = p_2$ in a two-sample *z*-test is equivalent to a chi-square test of homogeneity (with 1 degree of freedom). You will get exactly the same *P*-value. As in goodness-of-fit tests, $\chi^2 = z^2$.

DISCUSSION
The Two-Sample *z*-Test Versus the Chi-Square Test of Homogeneity

D12. Discuss the differences between a chi-square test of homogeneity and a chi-square goodness-of-fit test. Discuss the similarities and differences between a *z*-test of the equality of two proportions and the chi-square test of homogeneity.

Degrees of Freedom: Information About Error

Jordain: I've just been letting this "degrees of freedom" thing slide, but in this section they've gone too far.

Dr. C: What's the trouble?

Jordain: Well, just for a starter, back when we did the *t*-test, *df* was sample size minus 1. I was clueless why we subtracted 1, but never mind that now. Then, in Section 12.1, we had almost the same formula, number of categories minus 1. But why was it number of categories, not sample size? Now we get this hugely different formula, $df = (r - 1)(c - 1)$!

Dr. C: That's a good question. And there's going to be yet another formula for *df* in the next chapter, on regression.

Jordain: You keep saying I'm not supposed to just memorize stuff, but . . .

Dr. C: Okay, the entire statistical story is way out there, but I'll try to explain. It has to do with the amount of information in the error term, or *SE*. Do you remember the *SE* from the one-sample *t*-test?

Jordain: Sure, I remember. It's

$$\frac{s}{\sqrt{n}} \qquad \text{where} \qquad s = \sqrt{\frac{\Sigma(x - x)^2}{n - 1}}$$

Dr. C: Think about the deviations, $(x - x)$. There are *n* of them. Because the sum of the deviations from the mean is 0, if you know all the deviations but one, you can figure out the last one. So you need only $n - 1$ deviations to get all the information about the size of a typical deviation from the mean.

Jordain: Then the reason we didn't have to bother with *df* in a *z*-test for a proportion in Section 8.2 is that the standard error, $\sqrt{\frac{p_0(1 - p_n)}{n}}$, depended only on the known quantities *n* and p_0.

Dr. C: Right. You weren't estimating the error term from the data using a sum of squared deviations. So you didn't need to worry about *df*. Do you see what this has to do with *df* equaling the number of categories minus 1 for a chi-square goodness-of-fit test?

Jordain: Hmmm. The χ^2 statistic itself looks like one big error term. If you are testing whether a six-sided die is fair, there are six categories and six deviations, *O* minus *E*. I suppose if you know all but one of them, you can figure out the missing one.

Dr. C: Right again. The last deviation is determined by the others because the sum of the deviations from the "center" is always equal to 0:

$$\Sigma(O - E) = \Sigma O - \Sigma E = n - n = 0$$

Because the last deviation doesn't give you any new information about the size of the deviations from the center,

$$df = number\ of\ categories - 1$$

Jordain: But what about the formula in this section, $df = (r - 1)(c - 1)$?

Dr. C: Well, how many deviations, $O - E$, are you using in the formula?

Jordain: The number of rows times the number of columns.

Dr. C: Now all you have to do is figure out how many of these give you new information about the size of $O - E$ and how many are redundant.

Jordain: Well, in the red-light camera example, there were 3 times 2, or 6, values of $O - E$. They have to sum to 0 in each row and each column. So, for example, if you know just the deviations I've put in this table (Display 12.36), you can figure out the rest, meaning that they don't tell you anything new.

	Before Camera	After Camera	Total
Death or Disabling Injury	3.2	—?—	0
Evident Injury	−17.3	—?—	0
Possible Injury	—?—	—?—	0
Total	0	0	0

Display 12.36 Table of deviations with one row and one column missing.

Dr. C: That means you have only $(r - 1)$ times $(c - 1)$ independent deviations.

Jordain: Here's my rule: The concept of *df* applies when my test statistic involves a sum of squared deviations from a parameter or parameters and I must estimate the parameter or parameters from the data. I count the number of deviations that are free to vary. This is the number of degrees of freedom.

Dr. C: That covers it for everything you'll see in this class.

DISCUSSION
Degrees of Freedom: Information About Error

D13. Fill in the rest of the values of $O - E$ in Display 12.36. Use Display 12.33 to make a similar table of the values of $O - E$ for the family values example. How many rows and columns can you delete and still be able to complete the table?

Summary 12.2: The Chi-Square Test of Homogeneity

Use the chi-square test of homogeneity when

- you have independent samples from two or more populations
- you can classify the response from each member of the sample into exactly one of several categories
- you want to know if it's plausible that the proportion that falls into each category is the same for each population

This might sound familiar. In Section 9.2, you learned how to test whether the difference between two proportions is statistically significant. This test is equivalent to a chi-square test of homogeneity when there are two populations and two categories. When there are more than two populations or more than two categories, the chi-square test of homogeneity can test the differences among all the populations in one test.

In a chi-square test of homogeneity, the expected frequencies, E, are calculated from the sample data, using

$$E = \frac{row\ total \cdot column\ total}{grand\ total}$$

The test statistic is

$$\chi^2 = \sum \frac{(O - E)^2}{E}$$

The number of degrees of freedom for a two-way table with r rows and c columns is

$$df = (r - 1)(c - 1)$$

Practice

Comparing Two or More Populations

P13. For a project for her statistics class, Justine tested the dry strength of three brands of paper towels. Getting a random sample of towels from each brand is next to impossible on a student's budget, but she thought it was reasonable to assume that all rolls of towels of the same brand are pretty much identical. So Justine bought one roll of each of the three brands and used the first 25 towels from each. With the help of two friends, she stretched each towel tightly, dropped a golf ball on it from a height of 12 in., and recorded whether the golf ball went through the towel. Justine tested the 75 towels in random order because she realized that it would be impossible to hold her testing procedure completely constant. Display 12.37 shows her results.

| | | **Brand** | | | |
		A	B	C	Total
Towel Breaks?	Yes	18	7	5	30
	No	7	18	20	45
	Total	25	25	25	75

Display 12.37 Two-way table of observed frequencies for Justine's paper towels data.

a. Construct a segmented bar chart to help you see whether the brands are equally likely to break.

b. What is your conclusion?

P14. In the red-light camera study, the FHWA also looked at the severity of injuries in right-angle crashes at intersections. The data are given in Display 12.38.

a. Construct a segmented bar chart to help you see if it is plausible that the distributions are the same for before-camera crashes and after-camera crashes.

b. What is your conclusion?

A right-angle crash.

	Before Camera	After Camera
Death or Disabling Injury	152	59
Evident Injury	571	237
Possible Injury	1131	338
Total	1854	634

Display 12.38 Number of injuries of each type in right-angle crashes with injuries before and after red-light cameras installed in seven urban areas in the United States. [Source: *safety.fhwa.dot.gov/intersection/redlight/ redl_faq.cfm. Safety Evaluation of Red-Light Cameras*, FHWA-HRT-05-048, FHA, April 2005, Table 18.]

Computing Expected Frequencies

P15. Refer to P13. Assuming that the probability a towel will break is the same for each brand, construct a table of expected frequencies.

P16. Refer to P14. Assuming that the distributions are the same for the injuries in before-camera crashes and after-camera crashes, construct a table of expected frequencies.

The Chi-Square Statistic, Degrees of Freedom, and *P*-Value

P17. Refer to Justine's paper towel data in P13 and your expected frequencies in P15.

a. Compute χ^2.

b. What is the number of degrees of freedom?

c. What is the *P*-value? Is χ^2 statistically significant at the 5% level?

d. What do you conclude?

P18. Refer to the red-light camera data for right-angle crashes in P14 and your expected frequencies in P16.

a. Compute χ^2.

b. What is the number of degrees of freedom?

c. What is the *P*-value? Is χ^2 statistically significant at the 5% level?

d. What do you conclude?

Procedures for a Chi-Square Test of Homogeneity

P19. Refer to Example 12.9. Collapse the two undesirable outcomes, *death* and *survival with chronic lung disease*, in Display 12.34 into one category and analyze the resulting two-way table. Is there now a statistically significant difference between the two distributions?

P20. Another important aspect of the study on premature infants was the effect of the nitric oxide treatment on intraventricular hemorrhage (bleeding within the heart). Analyze the results reported in Display 12.39 to see if the two distributions differ significantly. If they do, explain the nature of the difference.

		Treatment	
		Nitric Oxide	Placebo
Degree of Hemorrhage	Severe	13	24
	Not Severe	27	10
	None	65	68
	Total	105	102

Display 12.39 Table of observed frequencies for the premature infants experiment.

The Two-Sample *z*-Test Versus the Chi-Square Test of Homogeneity

P21. Students in a statistics class surveyed random samples of 50 females and 50 males, asking each if they preferred a bath or a shower. Display 12.40 gives the results.

	Male	Female	Total
Bath	6	21	27
Shower	44	29	73
Total	50	50	100

Display 12.40 Table of observed preferences for bath versus shower.

a. What are the populations in this situation? Construct a segmented bar chart to help you see whether it's plausible that the choices for all males and all females have the same distribution. What is your conclusion?

b. In a two-sample *z*-test of the null hypothesis that males and females are equally likely to prefer baths, what is the value of *z*? What is the *P*-value?

c. In a chi-square test of homogeneity of the null hypothesis that the choices for all males and all females have the same distribution, what is the value of χ^2? What is the *P*-value?

d. Refer to your answers in parts b and c. Is it the case that $\chi^2 = z^2$? Are the *P*-values equal?

Exercises

E15. Since 1946, a committee of the American Bar Association (ABA) has rated the qualifications of presidential nominations to judgeships. If the committee agrees, it rates each candidate as well-qualified, qualified, or not qualified. If not, the rating can be split. For example, a candidate with a majority of not qualified votes and some qualified votes would be rated not qualified/qualified, while a candidate with a majority of qualified votes and some not-qualified votes would be rated qualified/not qualified. The table in Display 12.41 gives the ratings, categorized into three groups, for all 317 nominations that were rated for the last four presidents completing their terms.

a. What are the populations in this situation? Are conditions met for a chi-square test of homogeneity?

b. Is there statistically significant evidence that the distribution of ratings is different for the different presidents? (In light of your answer to part a, you will be deciding whether the difference reasonably can be attributed to chance alone.)

	Not Qualified or Split Between Qualified/Not Qualified	Qualified or Qualified/ Well-Qualified	Well-Qualified or Well-Qualified/ Qualified	Total
Reagan	13	15	30	58
G. H. W. Bush	4	18	31	53
Clinton	3	23	64	90
G. W. Bush	17	22	77	116
Total	37	78	202	317

Display 12.41 Distribution of ratings given to presidential nominations to circuit court judgeships by ABA committee, 1985–2008. [Source: Richard L. Vining, Jr., et al., "Bias and the Bar: Evaluating the ABA Ratings of Federal Judicial Nominees," paper prepared for presentation at the Annual Meeting of the Midwest Political Science Association, April 2–5, 2009, Chicago, Illinois.]

E16. Refer to the information and display in E15.

    a. The ABA ratings have sometimes been controversial, with Republicans complaining that the ABA committee exhibits a liberal bias. What test should you use to see if there is statistically significant evidence that the distribution of ratings of nominees of the Republican presidents is different from that of nominees of the Democratic president?

    b. Is there statistically significant evidence that the distribution of ratings of nominees of the Republican presidents is different from that of nominees of the Democratic president? Be sure to take into account that this is the population of all ratings when writing your conclusion.

    c. Suppose that your test in part b found evidence of a difference in the two distributions. Give another possible explanation other than liberal bias of the ABA committee.

E17. The Food and Drug Administration (FDA) of the United States government has the task of monitoring the country's food supply, whether domestically produced or imported. One of its duties is to check the amount of pesticide residue found on samples of various foods about to be sent to market. Foods that show residue are not necessarily in violation of a standard, as the standards allow small amounts of pesticide residue to enter our food supply. The food samples tested are selected by a rather complicated process involving the experience of experts, but you may regard the process as producing results similar to independent random samples. Display 12.42 shows the results of pesticide residue tests on samples of vegetables for a recent year.

    a. Construct a table of observed frequencies so that you can compare the counts with various degrees of residue for domestic and imported vegetables. What category must you add so that the total for domestic vegetables is 672 and for imported vegetables is 2447?

    b. Compute a table of expected frequencies, under the assumption that the proportions in each category are the same for domestic and imported vegetables.

    c. What are the populations in this situation? Are conditions met for inference using a chi-square test of homogeneity?

    d. Regardless of your answer to part c, compute the value of chi-square and find the *P*-value.

    e. What is your conclusion?

Pesticide Residue, Domestic Vegetables

Sample size	672
Percentage showing no residue (%)	73.8
Percentage showing residue in violation (%)	2.4

Pesticide Residue, Imported Vegetables

Sample size	2447
Percentage showing no residue (%)	60.4
Percentage showing residue in violation (%)	5.4

Display 12.42 Pesticide residue data for domestic and imported samples of vegetables. See Display 9.12 for source.

E18. Refer to the previous exercise. Display 12.43 shows the results of pesticide residue tests on samples of fruit for a recent year.

    a. Construct a table of observed frequencies so that you can compare the counts with various degrees of residue for domestic and imported fruits. What category must you add so that the total for domestic fruits is 344 and for imported fruits is 1136?

    b. Compute a table of expected frequencies, under the assumption that the proportions in each category are the same for domestic and imported fruits.

    c. What are the populations in this situation? Are conditions met for inference using a chi-square test of homogeneity?

    d. Regardless of your answer to part c, compute the value of chi-square and find the *P*-value.

    e. What is your conclusion?

Pesticide Residue, Domestic Fruits

Sample size	344
Percentage showing no residue (%)	44.2
Percentage showing residue in violation (%)	0.9

Pesticide Residue, Imported Fruits

Sample size	1136
Percentage showing no residue (%)	70.4
Percentage showing residue in violation (%)	3.6

Display 12.43 Pesticide residue data for domestic and imported samples of fruits.

E19. Children aged 2 through 6 were told that some dolls were going to have a party and they should select products for them from a play supermarket. In the sample of 16 children whose parents went to high school or had job training, 56.3% purchased cigarettes. In the sample of 26 children who had one parent go to college 38.5% purchased cigarettes. In the sample of 78 children where both parents went to college, 19.2% purchased cigarettes. [Source: Madeline A. Dalton et al., "Use of Cigarettes and Alcohol by Preschoolers While Role-Playing as Adults," *Archives of Pediatrics & Adolescent Medicine*, Vol. 159 (2005), pp. 854–859.]

    a. What are the populations in this situation?

    b. Make a table of observed counts.

    c. Construct a segmented bar chart to display the data.

    d. Perform a chi-square test of homogeneity to decide whether this is convincing evidence that the percentages

Grocery store setup used for role-playing.

of children this age who would purchase cigarettes vary with parental educational level. Use $\alpha = 0.05$.

e. Write a conclusion.

E20. A Gallup poll asked the same question it has asked every year for many years: "What do you think is the most important problem facing this country today?" Display 12.44 shows the percentage responses for three major concerns in 2003 and 2009. About 1000 people were surveyed each year in what you may consider is a random sample of U.S. adults. You will test whether there was a statistically significant change in the distribution of results from 2003 to 2009.

		Most Important Problem			
		War	Economy	Health Care	Other
Year	2003	19%	20%	9%	52%
	2009	9%	69%	9%	13%

Display 12.44 Results from Gallup polls: responses for three major concerns, 2003 and 2009. [Source: *www.gallup.com.*]

a. What are the populations in this situation?

b. Convert the table to one displaying observed frequencies.

c. Construct and interpret a segmented bar chart of these frequencies.

d. Perform a chi-square test of homogeneity. Use $\alpha = 0.05$.

e. What conclusion can you come to? Do you know a reason for the change?

E21. Joseph Lister (1827–1912), a surgeon at the Glasgow Royal Infirmary, was one of the first to believe in Pasteur's germ theory of infection. He experimented with using carbolic acid to disinfect operating rooms during amputations. Of 40 patients operated on using carbolic acid, 34 lived. Of 35 patients operated on without using carbolic acid, 19 lived. [Source: Richard Larson and Donna Stroup, *Statistics in the Real World: A Book of Examples* (New York: Macmillan, 1976), pp. 205–207. Original reference: Charles Winslow, *The Conquest of Epidemic Diseases* (Princeton, NJ: Princeton University Press, 1943), p. 303.]

a. Display these data in a two-way table and in a segmented bar chart.

b. Compute the value of the test statistic, z, for a test of the difference of two proportions. Find and interpret the *P*-value for this test.

c. Compute the value of χ^2 for a test of homogeneity. Find and interpret the *P*-value for this test.

d. Compare the two *P*-values from parts b and c. Compare the values of z^2 and χ^2.

E22. Obesity is associated with many long-term illnesses, such as diabetes and hypertension. In a Swedish study, a group of subjects who had undergone surgery for obesity were followed over a 10-year period and compared, over that time, to a matched sample of conventionally treated control subjects. The matching was done based

on 18 variables, including age, gender, weight, smoking habit, and many other health-related variables. Among 517 subjects who had surgery to reduce obesity, 7 developed diabetes. Among 539 subjects in the control group, 24 developed diabetes. [Source: Lars Sjöström et al., "Lifestyle, Diabetes, and Cardiovascular Risk Factors 10 Years After Bariatric Surgery," *New England Journal of Medicine*, Vol. 351, 26 (December 23, 2004).]

a. Display the data in a two-way table and in a segmented bar chart.

b. Compute the value of the test statistic, z, for a test of the difference of two proportions. Find and interpret the *P*-value for this test.

c. Compute the value of χ^2 for a test of homogeneity. Find and interpret the *P*-value for this test.

d Compare the two *P*-values in parts b and c. Compare the values of z^2 and χ^2.

E23. Experiments have been performed to determine the effects of a decorative bandage on the amount of pain a child perceives. [See, for example, C. C. Johnston et al., "The Effect of the Sight of Blood and Use of Decorative Adhesive Bandages on Pain Intensity Ratings by Preschool Children," *Journal of Pediatric Nursing*, Vol. 8 (June 1993), pp. 147–151.] You conduct an experiment to test the effect of bandages on the perceived pain of children at your local elementary school. Of the first 60 children who came to a school nurse for a skinned knee, 20 were selected at random to receive no bandage, 20 to receive a skin-colored bandage (matched to their skin color), and 20 to receive a brightly colored bandage. After 15 minutes, each child was asked, "Is the pain gone, almost gone, or still there?" Of the 20 children who got no bandage, 0 said the pain was gone, 5 said it was almost gone, and 15 said it was still there. Of the 20 children who got a skin-colored bandage, 9 said the pain was gone, 9 said it was almost gone, and 2 said it was still there. Of the 20 children who got a brightly colored bandage, 15 said the pain was gone, 2 said it was almost gone, and 3 said it was still there. Organize these data into a table, display them in a plot, and perform a chi-square test of homogeneity.

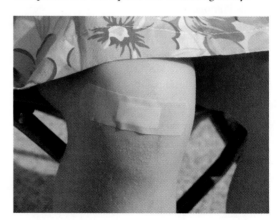

E24. The Department of Health Policy and Management of the Harvard School of Public Health conducted a sample survey of hospitals in the country to ask about the availability of electronic health record (EHR) systems.

A summary of their findings is found in Display 12.45. Comprehensive systems are those that cover the entire hospital, while basic systems are those that may be used in just one department of the hospital. The data are reported as percentages of the sample with the given type of EHR system.

a. Make a table of observed frequencies. Include a third category of medical record keeping, *No Electronic System*.

b. Make a table of expected frequencies.

c. Are conditions met for a chi-square test of homogeneity? (This is not a random sample, but all hospitals were surveyed and 2952 responded.)

e. Is there evidence of a difference in the population percentages of hospitals having basic EHR systems between the three types of hospital?

Type of Hospital	Sample Size	Comprehensive EHR System (%)	Basic EHR System (%)
For-profit	412	1.0	5.0
Private non-profit	1831	1.5	8.0
Public	709	2.7	6.9

Display 12.45 Percentages of hospitals having EHR systems. (Sample size approximated from Table 1 in the source.) [Source: Ashish K. Jha et al., "Use of Electronic Health Records in U.S. Hospitals," *New England Journal of Medicine*, Vol. 360 (2009), pp 1628–1638.]

E25. *Gallup Poll Monthly* reported on a survey built around the question, "What is your opinion regarding smoking in public places?" On the workplace part of the survey, respondents were asked to choose which of three policies on smoking in the workplace they favored. The percentages of the sample making each choice for two different years are shown in Display 12.46. Organize the data in a table of observed frequencies and display them in a segmented bar chart. Then conduct a chi-square test of homogeneity to test if the proportion of people who chose each response was the same in the 2 years. The sample size was approximately 1000 each year.

	Opinion		
Year	Designated Area	Ban Altogether	No Restrictions
2005	56%	41%	3%
1999	61%	34%	5%

Display 12.46 Results from Gallup polls on smoking in public places. [Source: *poll.gallup.com*.]

E26. Periodic Gallup polls beginning in 1997 have asked a randomly selected national sample of approximately 1000 adults, 18 years and older, "Do you think that global warming will pose a serious threat to you or

your way of life in your lifetime?" The responses for 1997 and 2009 are shown in Display 12.47. Organize the data in a table of observed frequencies and display them in a segmented bar chart. Then conduct a chi-square test of homogeneity to test whether it's plausible that the distribution of responses of all adults is the same for each year.

	Year	
Response	1997	2009
Yes	25	38
No	69	60
Unsure	6	2
Total	100%	100%

Display 12.47 Results from Gallup polls on global warming. [Source: *www.pollingreport.com*.]

E27. It is generally assumed that men are heavier drinkers of alcoholic beverages than are women. Display 12.48 gives the percentages of adults in independent surveys of about 500 men and 500 women from each of three countries who responded yes to the question of whether they had at least several alcoholic drinks per week.

	Great Britain	Canada	United States
Men	48%	37%	26%
Women	34%	17%	13%

Display 12.48 Percentages of people who drink regularly, out of 500 men and 500 women. [Source: *poll.gallup.com*, 2006.]

Describe the test you would use to answer each question. Show the test statistic (with all numbers substituted in) that you would compute.

a. Is the proportion of men who drink regularly significantly greater than the proportion of women who drink regularly in the United States?

b. Are there significant differences among the proportions of men who drink regularly across the three countries?

c. Is there a significant difference between the proportions of women who drink regularly in Canada and in the United States?

E28. Do active Christians behave differently from those with no active religious faith? Responses from independent random samples of 3011 active Christian adults and 1055 adults professing no active religious faith appear in Display 12.49. (This study did not control for differences in the demographic characteristics of the two groups.) Can the methods you have learned so far in this textbook be used to answer each of the following questions? If so, describe the test you would use to answer each question. If not, explain why not.

a. Is the proportion of all Christians who are registered to vote significantly higher than the proportion of all those professing no faith who are registered to vote?

	Christian	No Faith
Registered to vote	89%	78%
Active in the community	68%	41%
Suffer from personal addictions	12%	13%
Considered to be overweight	41%	26%
Feels stressed out much of the time	26%	37%

Display 12.49 Poll results for active Christian adults and adults professing no active religious faith. [Source: *www.barna.org*.]

b. Is the distribution of all Christians who fall into each category different from the distribution of all of those professing no faith who fall into each category?

E29. Gallup conducted a poll of about 750 Internet users in 2003 and a poll with a different sample of about 750 Internet users in 2005. Display 12.50 gives some of the findings. For example, 68% of those interviewed used the Internet for checking news and weather in 2003, and the percentage jumped to 72% in 2005.

a. Can this data table be used to conduct a chi-square test of homogeneity on these proportions across the two years? If so, find and interpret the *P*-value for this test. If not, explain why not.

b. Could you test for equality of proportions of those sending and reading email across the 2 years? If so, find and interpret the *P*-value for this test. If not, explain why not.

	2003	2005
Checking news and weather	68%	72%
Sending and reading email	84%	67%
Shopping	49%	52%

Display 12.50 Results of polls regarding Internet users' habits, for samples of 750 people in 2003 and 750 different people in 2005. [Source: *poll.gallup.com*, 2006.]

E30. Two questions were asked in random order of a random sample of about 1000 adults in a 2005 poll: "At the school your oldest child attends, do you think there is too much emphasis, the right amount, or too little emphasis on sports? Do you think there is too much emphasis, the right amount, or too little emphasis on preparing for standardized tests?" The results are shown in Display 12.51.

Can the methods of this section be used to test for homogeneity of proportions of responses about sports compared to responses about preparing for standardized tests? If so, find and interpret the *P*-value for this test. If not, explain why not.

	Too Much Emphasis	Right Amount	Too Little Emphasis	Not Applicable	No Opinion
Sports	18%	57%	20%	4%	1%
Preparing for Standardized Tests	18%	52%	24%	3%	3%

Display 12.51 Results of a poll of parents' opinions of the amount of emphasis schools place on sports and standardized test preparation, for a sample of 1000 adults. [Source: *poll.gallup.com*, 2005.]

12.3 ▶ The Chi-Square Test of Independence

Are these variables independent in this population?

Independent samples from different populations: test of homogeneity. One sample from one population: test of independence.

To protect the health of employees, California, like many states, has a law against smoking in bars and restaurants. Researchers visited a random sample of 121 stand-alone bars in San Francisco to see what characteristics of bars are associated with illegal smoking. One of the variables of interest was whether there were glamorized images of smoking in the bar, such as posters of movie stars holding lit cigarettes.

This situation is somewhat similar to those of the previous section. There are two categorical variables: *compliance with the nonsmoking law* (with categories *compliant* and *not compliant*) and *images of smoking* (with categories *glamorized images* and *no glamorized images*). For each of these two variables, each member of the population falls into exactly one category.

This situation is different from those of the previous section because the researchers observed *a single sample from one population*. When you observe a single sample from one population, such as the sample of 121 bars from San Francisco, you should use a chi-square test of independence to decide whether it is plausible that two different variables are independent in that population. In the previous section you used a chi-square test of homogeneity when you observed *independent samples from several populations* to decide whether it is plausible that the proportion of units that fall into each category is the same for all populations.

Terminology and Graphical Displays

Display 12.52 summarizes some of the data from the San Francisco study of compliance with the nonsmoking law. This is one sample from the population of stand-alone bars in San Francisco, with each bar categorized according to two variables: *compliance with the nonsmoking law* and *images of smoking*. The number of bars that were compliant and had glamorized images, 9, is a joint frequency, the number of bars with glamorized images, 24, is a row marginal frequency, and the number compliant bars, 61, is a column marginal frequency. The proportion of bars with glamorized images that are compliant, 9/24, is a conditional relative frequency; and the proportion of compliant bars, 61/121, is a marginal relative frequency.

	Compliant	Noncompliant	Total
Glamorized Images	9	15	24
No Glamorized Images	52	45	97
Total	61	60	121

Display 12.52   Randomly selected bars in San Francisco categorized by whether there were glamorized images of smoking and whether patrons complied with the nonsmoking law. [Source: Roland S. Moore et al., "Correlates of Persistent Smoking in Bars Subject to Smokefree Workplace Policy," *International Journal of Environmental Research and Public Health*, Vol. 6 (2009), pp. 1341–1357.]

The segmented bar chart was a natural graphic to use for studying homogeneity because, in that situation, you had samples from different populations that could be compared with side-by-side bars. In studying independence, you have one sample from one population categorized by two variables; so you could construct a segmented bar chart on either variable, as shown in Display 12.53. When two variables are independent, the bars are divided into segments according to the same proportions. That isn't exactly the case here, but it is close. In this sample, the noncompliant bars tend to be a bit more likely to have glamorized images of smoking than compliant bars, but it isn't clear if the difference is statistically significant.

> If categorical variables aren't independent, they are *dependent* or *associated*.

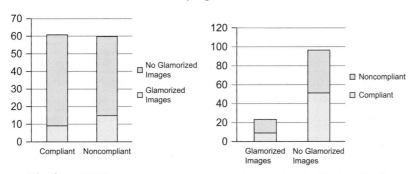

Display 12.53   Segmented bar charts for the data in Display 12.52.

DISCUSSION
Terminology and Graphical Displays

D14. Which pairs of variables do you believe are independent in the population of U.S. college students? Explain.

   a. hair color and eye color

   b. type of music preferred and ethnicity

   c. gender and color of shirt

   d. type of movie preferred and gender

   e. eye color and class year

   f. class year and whether taking statistics

Expected Frequencies in a Chi-Square Test of Independence

In a chi-square test of independence, the null hypothesis is that the variables are independent in the population. You learned a definition of independence in Chapter 5: Events A and B are independent if and only if $P(A$ and $B)$ equals $P(A) \cdot P(B)$. This definition enables you to find the expected frequencies for a chi-square test of independence.

Example 12.10	**Computing Expected Frequencies**

Solution

Display 12.54 gives the marginal frequencies from the San Francisco study of compliance with the nonsmoking law in Display 12.52. Under the hypothesis that compliance and glamorized images are independent variables, compute the expected frequencies for the cell in the first row and first column.

	Compliant	Noncompliant	Total
Glamorized Images	—?—	—?—	24
No Glamorized Images	—?—	—?—	97
Total	61	60	121

Display 12.54 Marginal frequencies for the San Francisco study of compliance with the nonsmoking law in Display 12.52.

Solution

Assuming that glamorized images and compliance are independent variables, the probability that a randomly selected bar has glamorized images of smoking and patrons who are compliant and is

$$P(\text{glamorized images and compliant}) = P(\text{glamorized images}) \cdot P(\text{compliant})$$

$$= \frac{24}{121} \cdot \frac{61}{121}$$

$$\approx 0.10$$

In a sample of 121 bars, then, the expected frequency that have glamorized images and patrons who are compliant is

$$P(\text{glamorized images and compliant}) \cdot (\text{grand total}) = 0.10(121)$$

$$= 12.1$$

In general, if two variables are independent, the probability that a randomly selected observation falls into the cell in column C and row R is

$$P(R \text{ and } C) = P(R) \cdot P(C) = \frac{\text{row } R \text{ total}}{\text{grand total}} \cdot \frac{\text{column } C \text{ total}}{\text{grand total}}$$

Thus, the expected number, E, of observations that fall into this cell is

$$E = P(R \text{ and } C) \cdot (\text{grand total}) = \frac{\text{row } R \text{ total}}{\text{grand total}} \cdot \frac{\text{column } C \text{ total}}{\text{grand total}} \cdot (\text{grand total})$$

$$= \frac{\text{row } R \text{ total} \cdot \text{column } C \text{ total}}{\text{grand total}}$$

This is the same formula as in Section 12.2.

Formula for the Expected Frequency, *E*

Under the hypothesis that the categories are independent, the expected frequency in a cell is

$$E = \frac{row\ total \cdot column\ total}{grand\ total}$$

Procedure for a Chi-Square Test of Independence

Now that you have reviewed the idea of independence, it's time to organize all the steps needed when you analyze a sample from a population to decide if it is plausible that two variables are independent in that population. The steps in a chi-square test of independence are much the same as the steps in the chi-square test of homogeneity.

Chi-Square Test of Independence

1. *State the hypotheses.*

    H_0: The two variables are independent in the population; that is, suppose one member is selected at random from the population. Then, for each cell, the probability that the member falls into both category *A* and category *B*, where *A* is a category from the first variable and *B* is a category from the second variable, is equal to $P(A) \cdot P(B)$.

    H_a: The two variables are not independent.

2. *Check conditions.*
    - A random sample is taken from one large population.
    - Each outcome can be classified into one cell according to its category on one variable and its category on a second variable.
    - The expected frequency in each cell is 5 or greater.

3. *Compute the value of the test statistic, approximate the P-value, and draw a sketch.* For each cell in the table, *O* is the observed frequency and *E* is the expected frequency where

    $$E = \frac{row\ total \cdot column\ total}{grand\ total}$$

    The test statistic is

    $$\chi^2 = \sum \frac{(O-E)^2}{E}$$

    The *P*-value is the probability of getting a value of χ^2 as extreme as or more extreme than the one from the sample, assuming the null hypothesis is true. Get the *P*-value from your calculator or approximate the P-value by comparing the value of χ^2 to the appropriate value of χ^2 in Table C on page 762. Use $df = (r - 1)(c - 1)$, where r is the number of categories for one variable and c is the number of categories for the other variable.

4. *Write your conclusion linked to your computation and in the context of the problem.* The smaller the P-value, the stronger the evidence against the null hypothesis. If the P-value is smaller than the level of significance, α, then reject the null hypothesis. If not, the evidence against the null hypothesis isn't statistically significant, so you do not reject it.

Example 12.11 | Scottish Children

Scottish children.

The classic data set in Display 12.55 shows the eye color and hair color of 5387 Scottish children. You might suspect that these two categorical variables are not independent because people with darker hair colors tend to have darker eye colors, whereas people with lighter hair colors tend to have lighter eye colors. Test to see if there is statistically significant evidence of an association between hair color and eye color.

		Hair Color				
		Fair	Red	Medium	Dark	Black
Eye Color	Blue	326	38	241	110	3
	Light	688	116	584	188	4
	Medium	343	34	909	412	26
	Dark	98	48	403	681	85

Display 12.55 Eye color and hair color data from a sample of Scottish children. [Source: R. A. Fisher, "The Precision of Discriminant Functions," *Annals of Eugenics*, Vol. 10 (1940), pp. 422–429.]

Solution

From the segmented bar chart in Display 12.56, you can see that it does indeed appear to be the case that children with darker hair colors tend to have darker eye colors, whereas children with lighter hair colors tend to have lighter eye colors. Compare, especially, the bars for fair hair and dark hair.

State the hypotheses.

The hypotheses are

H_0: Eye color and hair color are independent in the population of Scottish children.

H_a: Eye color and hair color are not independent.

Check conditions.

This situation satisfies the conditions for a chi-square test of independence if the children can be considered a random sample taken from one large population. Each child in the sample falls into one hair color category and one eye color category.

Compute the test statistic, calculate the P-value, and draw a sketch.

The second entry in each cell in the output in Display 12.57 shows the expected frequency under the assumption of independence. The expected frequency in each cell is 5 or greater. The third entry in each cell is the value of $\frac{(O - E)^2}{E}$.

The output gives a χ^2 value of about 1240, extremely far out in the tail, and a corresponding P-value of approximately 0.

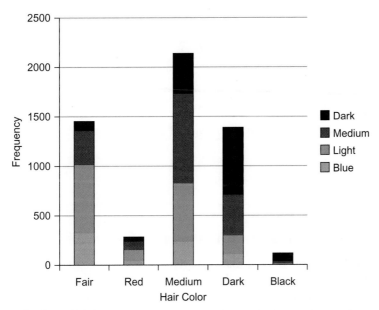

Display 12.56 Segmented bar chart for hair color and eye color.

Chi-Square Test

Expected counts are printed below observed counts
Chi-Square contributions are printed below expected counts

	Fair	Red	Medium	Dark	Black	Total
1	326	38	241	110	3	718
	193.93	38.12	284.83	185.40	15.73	
	89.946	0.000	6.744	30.663	10.300	
2	688	116	584	188	4	1580
	426.75	83.88	626.78	407.98	34.61	
	159.934	12.297	2.920	118.610	27.072	
3	343	84	909	412	26	1774
	479.15	94.18	703.74	458.07	38.86	
	38.686	1.101	59.869	4.634	4.255	
4	98	48	403	681	85	1315
	355.17	69.81	521.65	339.55	28.80	
	186.215	6.816	26.989	343.355	109.633	
Total	1455	286	2137	1391	118	5387

Chi-Sq = 1240.039, DF = 12, P-Value = 0.000

Display 12.57 Minitab output of a chi-square test of independence for the Scottish children example.

A sketch of the distribution is shown in Display 12.58, on the next page.

Reject the null hypothesis. These are not results you would expect from a random sample from a population in which there is no association between eye color and hair color. As you can see from Display 12.58, a χ^2 value of 1240 is much larger than the value of 32.91 (given in Table C) that cuts off an upper tail of 0.001. Thus, a value of χ^2 of 1240 or larger is extremely unlikely to occur in a sample of this size if hair color and eye color are independent. Examining the table and the segmented bar chart, you conclude that darker eye colors tend to go with darker hair colors and lighter eye colors tend to go with lighter hair colors.

> Write the conclusion in the context of the situation.

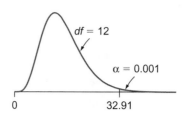

Display 12.58 Chi-square distribution with 12 degrees of freedom.

DISCUSSION
Procedure for a Chi-Square Test of Independence

D15. In Chapter 5, you showed that two events *A* and *B* are independent by verifying that one of these statements is true (and if one statement is true, all three are true):

$$P(A) = P(A|B) \qquad P(B) = P(B|A) \qquad P(A \text{ and } B) = P(A) \cdot P(B)$$

Why do you need a test of independence here when you already have this test from Chapter 5?

STATISTICS IN ACTION 12.2 ▶ Independent or Not?

1. In step 2, you will count the number of females and males in your class. You will also ask if the last digit of their phone number is even or odd. Do you think that the categorical variables *gender* and *even/odd phone number* are independent? Explain.

2. Collect the data described in step 1 and organize the data in a two-way table.

3. Using the definition of independence in Chapter 5, determine if *female* and *odd phone number* are independent events in the selection of a student at random from your class. Are you surprised by this result?

4. Determine your eye dominance by holding your hands together in front of you at arm's length. Look through a small space between your hands, with both eyes open, at a small object that is far away. Close your right eye. Can you still see the object? If so, you are left-eye dominant. If not, you are right-eye dominant. Make a two-way table for your class in which one categorical variable is *eye dominance* (*left* or *right*) and one categorical variable is *handedness* (*left* or *right*).

5. For a randomly selected student from your class, are the events *right-eye dominant* and *right-hand dominant* independent according to the definition of independence in Chapter 5?

6. Construct two different segmented bar charts to display the data your class collected on the variables *gender* and *even/odd phone number*. Interpret your plots.

7. Steps 1 through 6 (should have) illustrated the need for a statistical test of independence as a substitute for using the strict mathematical definition of independent events.

   a. Assuming your class is a random sample of students, test the hypothesis that gender and even or odd phone number are independent characteristics among students.

b. Assuming your class is a random sample of students, test the hypothesis that hand dominance and eye dominance are independent characteristics among students.

c. Are the results of the tests in parts a and b consistent with your expectations?

Homogeneity Versus Independence

Dr. C: Did you notice that a chi-square test of homogeneity and a chi-square test of independence are performed in exactly the same way, except for the wording of the hypotheses and the conclusion? The two null hypotheses say exactly the same thing about the table: The columns and rows are proportional.

Jordain: So how do we know which test to use?

Dr. C: You can't tell from the table itself. You have to find out how the data were collected. Did they take one sample of a fixed size from one population and then classify each person according to two categorical variables? If so, it's a test of independence. Or did they sample separately from two or more populations and then classify each person according to one categorical variable? If so, it's a test of homogeneity.

Jordain: Can you give me an example?

Dr. C: Suppose you want to find out whether there is an association between age and wearing blue jeans. You take a random sample of 50 people under age 40. Then you take a random sample of 50 people age 40 or older. Your table might look like Display 12.59.

| | | Age | |
		Under 40	40 or Older
Wears Blue Jeans?	Now	21	20
	Not Now, but Sometimes	23	25
	Never	6	5
	Total	50	50

Display 12.59 Sample table for blue jeans data, collected from two populations.

You use a test of homogeneity because you sampled separately from two populations: people under 40 and older people. There are two populations and one variable about jeans.

Jordain: Now I see what you mean by the columns being proportional. For people under 40, 42% are wearing jeans now, 46% are not wearing them now, and 12% never wear them. That's about the same distribution as for people 40 or older, so I won't be able to reject the hypothesis that these populations are homogeneous with respect to wearing jeans. Jeans-wearing behavior is about the same for both age groups.

Dr. C: Right. In the homogeneity case, you have one variable but several populations. The column—or sometimes the row—proportions represent the separate distributions for those populations. Now suppose you go out and take a random sample of 100 people, classify them, and get the data in Display 12.60. This time you sampled from just one population.

		Age		
		Under 40	40 or Older	Total
Wears Blue Jeans?	Now	23	17	40
	Not Now, but Sometimes	28	22	50
	Never	6	4	10
	Total	57	43	100

Display 12.60 Sample table for blue jeans data, collected from one population.

Jordain: I use a test of independence because there is only one sample, from the population of all people. Each person in the sample is categorized according to the two variables, *age* and *jeans-wearing behavior*. The columns are roughly proportional, so I won't be able to reject the hypothesis that the two variables are independent. Jeans-wearing behavior and age aren't associated. But this time we didn't predetermine how many people would be in each age group.

Dr. C: Right again, Jordain. Also, it now makes sense to talk about conditional distributions for the rows. You can estimate that 23/40, or about 58%, of people now wearing jeans are under age 40 and about 42% are age 40 or older. That statement wouldn't make sense in the homogeneity case.

Jordain: Then why wouldn't we always design our study as a test of independence? Then we can look at the table both ways, and as a bonus we get an estimate of the percentage of people in each age group!

Dr. C: It depends on what you want to find out. Suppose your research hypothesis is that the distribution of jeans-wearing behavior is different for the two age groups. You would design the study as a test of homogeneity, taking an equal sample size from each age group. That gives you the best chance of rejecting a false null hypothesis. In short, if you want to compare populations, take independent random samples from each and conduct a test of homogeneity. If you want to describe association between two variables in a single population, take a single random sample and conduct a test of independence.

DISCUSSION
Homogeneity Versus Independence

D16. You are working for an advertising agency and want to pick a model for a commercial for a hair dryer. You wonder whether the model should have straight or curly hair and so want to find out whether people with straight hair and people with curly hair are equally likely to use hair dyers. You are considering two designs for your survey:

I. You take a random sample of 50 people with straight hair and ask them whether they use a hair dryer. You take a random sample of 50 people with curly hair and ask them whether they use a hair dryer.

II. You take a random sample of 100 people, note if they have curly hair or straight hair, and ask them whether they use a hair dryer.

a. For which of the two designs should you use a test of homogeneity, and for which should you use a test of independence?

b. How are the data from the samples likely to be different?

c. Is there any reason why one design would be better than the other?

d. Which of the two designs could also be correctly analyzed by using a z-test for the difference between two proportions?

e. Just as a guess, do you think that type of hair and use of hair dryers are independent?

Strength of Association and Sample Size

A chi-square test of independence tells you if there is statistically significant evidence of an association between two categorical variables. It does not tell you the *strength of the association* or whether the association is of any practical importance. As an illustration, Display 12.61 shows three separate samples of responses on the same two variables.

Associated = not independent.

An association can be weak but still statistically significant.

	Sample 1		Sample 2		Sample 3	
	A	B	A	B	A	B
Yes	48	52	96	104	960	1040
No	52	48	104	96	1040	960
Total	100	100	200	200	2000	2000

	Sample 1	Sample 2	Sample 3
Difference between proportions of A and B saying yes	0.04	0.04	0.04
χ^2	0.32	0.64	6.40
P-value	0.57	0.42	0.01

Display 12.61 Tables showing how sample size can affect the *P*-value. For each sample, *df* = 1.

In each sample, the columns are almost proportional—the difference between the proportions in the first and second columns is only 0.04. This is not a very strong association. In most applications, a difference of 0.04 between two proportions would be of little practical significance. However, if you do a chi-square test of independence on each of the three samples, you will see that as the sample size increases, the test statistic gets larger and the *P*-value gets smaller. (For fixed conditional distributions, the numerical value of the χ^2 statistic is directly proportional to the sample size.) Even though the strength of the association is no greater than for the smaller samples, with a sample size of 4000 the association between the variables becomes highly statistically significant.

Be wary when using a chi-square test (or any test) with very large sample sizes, and always compare proportions to see if the statistical significance has any practical value.

Statistical significance is not the same as practical significance.

DISCUSSION
Strength of Association and Sample Size

D17. As in all tests of significance, the *P*-value measures the strength of the evidence against the null hypothesis provided by the data. If the *P*-value is very small, there is much evidence in favor of the alternative hypothesis.

a. In the context of a chi-square test of independence, discuss the difference between the statements "There is strong evidence in favor of an association" and "There is evidence of a strong association."

b. Suppose you have 3 two-way tables of frequencies for two categorical variables and a chi-square test of independence gives these results:

Table A: very small *P*-value and a small difference between conditional proportions

Table B: very small *P*-value and a large difference between conditional proportions

Table C: very large *P*-value and a small difference between conditional proportions

For Table A, is there much or little evidence of association? Is there evidence of strong association or evidence of weak association? Answer these questions for Table B and Table C.

D18. One of the basic principles you have learned in this textbook is that, all else being equal, it is better to have a larger sample than a smaller one.

a. Explain why a larger sample is better.

b. As long as each expected frequency is at least 5, is there any advantage to having a larger sample in a chi-square goodness-of-fit test?

Summary 12.3: The Chi-Square Test of Independence

Even when two variables in a population are independent, it would be very unusual to get a random sample in which the variables satisfy the definition of independence from Chapter 5. Requiring $P(A \text{ and } B)$ to be exactly equal to $P(A) \cdot P(B)$ for every cell of the table is too rigid a position to take when dealing with random samples. A chi-square test of independence is much more useful: Is it reasonable to assume that the random sample came from a population in which the two variables are independent? Use the chi-square test of independence when

- a random sample of a fixed size is taken from one large population

- each outcome can be classified into one cell according to its category on one variable and its category on a second variable

- you want to know if it's plausible that this sample came from a population in which these two categorical variables are independent

In a chi-square test of independence, the expected frequencies are calculated from the sample data:

$$E = \frac{row\ total \cdot column\ total}{grand\ total}$$

Each value of *E* should be 5 or greater. Again, this rule is conservative and you sometimes can proceed when the expected frequency is less than 5 in a small proportion of cells.

The number of degrees of freedom for a two-way table with *r* rows and *c* columns is $df = (r - 1)(c - 1)$ and the test statistic is

$$\chi^2 = \sum \frac{(O - E)^2}{E}$$

If the result from your sample is very different from the result expected under independence, then χ^2 will be large and you will reject the hypothesis that the variables are independent. However, even if the test tells you there is evidence of an association (dependence), it does not tell you anything about the strength of the association or whether it is of practical importance.

Practice

Terminology and Graphical Displays

P22. Refer to the data in Display 12.62. Give the meaning of each fraction, and tell whether each is a marginal relative frequency, a conditional relative frequency, or neither.

a. $\frac{27}{100}$

b. $\frac{21}{27}$

c. $\frac{50}{100}$

d. $\frac{6}{100}$

	Male	Female	Total
Bath	6	21	27
Shower	44	29	73
Total	50	50	100

Display 12.62 Table of observed preferences for bath versus shower.

Expected Frequencies in a Chi-Square Test of Independence

P23. Display 12.63 shows the marginal totals for a sample of 2237 residents of the United States classified according to gender and handedness.

a. On a copy of the table, fill in the expected frequencies for each cell under the hypothesis that gender and handedness are independent variables in the population of U.S. residents.

b. Do you think these variables really are independent?

	Men	Women	Total
Right-Handed			2004
Left-Handed			205
Ambidextrous			28
Total	1067	1170	2237

Display 12.63 Table of expected frequencies for a test of independence between gender and handedness.

Procedure for a Chi-Square Test of Independence

P24. Refer to Display 12.52.

a. Compute the value of χ^2.

b. Find the *P*-value. Is there statistically significant evidence that the two variables are associated?

c. Write a conclusion.

P25. Refer to Display 12.63. Are handedness and gender independent? A good set of data to help you answer this question comes from the government's 5-year Health and Nutrition Survey (HANES), which recorded the gender and handedness of a random sample of 2237 individuals from across the country. The observed frequencies for men and women in each of three categories are shown in Display 12.64. Use an appropriate statistical test to answer the question posed. If the patterns differ significantly, explain where.

	Men	Women	Total
Right-Handed	934	1070	2004
Left-Handed	113	92	205
Ambidextrous	20	8	28
Total	1067	1170	2237

Display 12.64 Table of observed frequencies of gender and handedness. [Source: D. Freedman, R. Pisani, and R. Purves, *Statistics*, 3rd ed. (New York: Norton, 1998) p. 537.]

P26. Each year, the United States Patent and Trademark Office examines about 450,000 applications from people who want to patent inventions. To expedite the process, applicants are encouraged to file an Information Disclosure Statement (IDS) that details the most pertinent references about similar inventions known to the applicant. A blogger (and professor at the University of Missouri Law School) created a random sample of about 3000 applications, which were sorted into the categories in the table in Display 12.65. [Source: Dennis Crouch, Prelim: The Association Between Grant Rate and Pre-Filing Searches, *www.patentlyo.com*, June 29, 2009.]

a. Make a table of expected counts, under the hypothesis that there is no association between IDS filing and status of an application.

b. Conduct a chi-square test of the independence of the two variables.

c. What is your conclusion?

	Application Abandoned	Invention Patented	Application Still Pending	Total
No IDS filed	300	255	195	750
IDS filed with application	203	359	218	780
IDS filed after application filing date	353	588	529	1470
Total	856	1202	942	3000

Display 12.65 Status of each application (abandoned, patented, or pending) as of June 26, 2009—about 4½ years after filing.

Homogeneity Versus Independence

P27. Many studies have found a relationship between eating breakfast and academic performance in children. You want to test this using a random sample of the students at a nearby middle school. You asked each child in your sample, "Did you eat breakfast this morning?" and you observed whether the child was (mostly) paying attention during the math lesson that day. Your data are shown in Display 12.66.

a. Should this be a test of homogeneity or independence? How would you obtain your information in order to do a test of the other type?

b. Conduct the appropriate chi-square test.

c. Does your result from part b mean that eating or not eating breakfast affects a child's ability to concentrate on a math lesson?

		Paying Attention?	
		Yes	No
Breakfast This Morning?	Yes	22	19
	No	5	15

Display 12.66 Table of responses from middle school students.

Strength of Association and Sample Size

P28. The American Community Survey is a nationwide sample survey sent to households. In 2007, about 1,968,362 households received and returned the survey. The table in Display 12.67 shows (approximately) how many of these households had complete kitchen facilities and how many did not, by region of the country.

a. The value of χ^2 turns out to be 1067.404. Is there a statistically significant association between region of the country and kitchen facilities? If so, is this evidence of a strong association, or strong evidence of a weak association?

b. Now suppose that the sample size had been about 1/1000 as big—1969 rather than 1,968,362—but that the proportion of responses falling into each cell remained the same. The table would have looked like that in Display 12.68. What is the value of χ^2 now?

c. In large sample surveys like the American Community Survey, almost any small difference can show up as being statistically significant but not necessarily of practical importance. Would you call the difference seen in the actual survey practically important? Why or why not?

	Northeast	Midwest	South	West	Total
Complete kitchen facilities	392,897	389,997	670,958	494,092	1,947,945
Lacking complete kitchen facilities	3,779	3,832	5,641	7,165	20,417
Total	396,676	393,829	676,599	501,258	1,968,362

Display 12.67 Data from American Community Survey showing kitchen facilities by region of the country. [Source: U.S. Census Bureau, 2007 American Community Survey.]

	Northeast	Midwest	South	West	Total
Complete kitchen facilities	393	390	671	494	1,948
Lacking complete kitchen facilities	4	4	6	7	21
Total	397	394	677	501	1,969

Display 12.68 Hypothetical data from a survey of only 1969 households from American Community Survey.

Exercises

E31. *Characters in action movies.* From a stratified random sample of 42 top-grossing action films, 159 male characters were analyzed. In terms of physical appearance, 76.1% were coded as muscular. One research hypothesis was that the muscular characters would be more likely to be in a romantic relationship than nonmuscular characters. The coders found that 37% of the muscular characters were in a romantic relationship and 63% were not. Of the nonmuscular characters, 16% were in a romantic relationship and 84% were not. The authors reported the following: "A significant association was observed, $\chi^2 (1, N = 159) = 6.08, p = .01.$" [Source: Todd G. Morrison and Marie Halton, "Buff, Tough, and Rough: Representations of Muscularity in Action Motion Pictures," *Journal of Men's Studies*, Vol. 17, 1 (Winter 2009), pp. 57–74.]

a. Overall, what proportion of the 159 characters were in a romantic relationship?

b. Write the hypotheses for the chi-square test.

c. Verify the value of chi-square given by the authors of the study.

d. Verify the degrees of freedom and the *P*-value. Is the *P*-value for this chi-square test one-sided or two-sided? Does it matter in this case that the research hypothesis was one-sided? Explain.

E32. Refer to E31, where 76.1 % of 159 male characters in action films were coded as muscular. Another research hypothesis was that the muscular characters would be more likely to experience a positive outcome than nonmuscular characters. The coders found that 58% of the muscular characters experienced a positive outcome but only 21% of the nonmuscular characters experienced a positive outcome. The authors reported the following: "Results indicated a statistically significant association, $\chi^2 (1, N = 159) = 15.67, p < .01.$"

a. Overall, what proportion of the 159 characters experienced a positive outcome?

b. Write the hypotheses for the chi-square test.

c. Verify the value of chi-square given by the authors of the study.

d. Verify the degrees of freedom and the *P*-value. Is the *P*-value for this chi-square test one-sided or two-sided? Does it matter in this case that the research hypothesis was one-sided? Explain.

E33. According to a National Center for Education Statistics report, 1,524,092 bachelor's degrees were granted in 2006–2007. About 22% were in business, 18% in the humanities, 17% in the social and behavioral sciences, and 43% in other fields (no field is more than 9.3%). About 43% of the degrees went to men and 57% to women. [Source: *nces.ed.gov*, *Digest of Education Statistics, 2008*, Tables 268 and 274.]

a. Construct a two-way table showing the proportion of degrees that fall into each cell under the assumption of independence.

b. Construct a two-way table showing the number of degrees that fall into each cell under the assumption of independence.

c. Do you think that the variables *field of degree* and *gender* are associated rather than independent? Explain.

E34. The first U.S. National Health and Nutrition Examination Survey in the 1980s reported on the age of a mother at the birth of her first child and whether the mother eventually developed breast cancer. Of the 6168 mothers in the sample, 26.39% had their first child at age 25 or older and 98.44% had not developed breast cancer. [Source: Jessica Utts, *Seeing Through Statistics*, 2nd ed. (Pacific Grove, CA: Duxbury, 1999), p. 209.]

a. Construct a two-way table showing the proportion of mothers who fall into each cell under the assumption of independence.

b. Construct a two-way table showing the number of mothers who fall into each cell under the assumption of independence.

c. The *P*-value for a test of independence is 0.186. Write a conclusion for this study.

E35. A CBS/*New York Times* poll of 1112 adults nationwide got the results in Display 12.69 regarding whether men and women think that in our society, there are more advantages in being a man, there are more advantages in being a woman, or there are no more advantages in being one than the other.

	Men	Women
Advantages Being a Man	32%	48%
Advantages Being a Woman	9%	4%
No Advantages	58%	46%
Unsure	1%	2%

Display 12.69   Results of a CBS/*New York Times* poll of 1112 adults taken January, 11-15, 2009. [Source: *www.pollingreport.com.*]

a. Change the percentages into observed frequencies and perform a chi-square test to determine if opinion and gender are independent. (Assume the 1112 respondents are equally split between men and women.)

b. Could you have made a Type I error?

c. Describe how the survey should have been designed in order to test for homogeneity of male and female populations with regard to opinion on this topic.

E.36. A student surveyed a random sample of 300 students in her large college and collected the data in Display 12.70, on the next page, on the variables *class year* and *favorite team sport*.

a. Perform a chi-square test to determine if the variables *class year* and *favorite team sport* are independent.

b. Could you have made a Type I error?

c. Describe how the survey should have been designed to test for homogeneous proportions among the

		Favorite Team Sport			
	Basketball	Soccer	Baseball/Softball	Other	Total
Freshman	12	40	10	1	63
Sophomore	12	44	16	8	80
Junior	9	43	11	11	74
Senior	10	49	18	6	83
Total	43	176	55	26	300

Class Year (row label grouping Freshman–Senior)

Display 12.70 Results of a survey of 300 college students.

populations of freshmen, sophomores, juniors, and seniors with respect to favorite team sports.

E37. The 44 adults and 45 children of the Donner party were trapped in the Sierra Nevada mountain range over the winter of 1846–1847. Their fate is shown in Display 12.71.

Donner Pass today.

		Adult	Child
Survived?	Yes	17	31
	No	27	14

Display 12.71 Donner party survival data. [Source: *www.utahcrossroads.org*, January 2010.]

a. Construct a plot that displays these data to see whether the variables *age* and *survival status* appear to be independent.

b. Test to see if the association between the variables *age* and *survival status* can reasonably be attributed to chance or if you should look for some other explanation. These data cannot reasonably be considered a random sample from any well-defined population, so state your conclusion accordingly.

E38. The fates of the 55 males and 34 females of the Donner party are shown in Display 12.72.

a. Construct a plot that displays these data to see whether the variables *gender* and *survival status* appear to be independent.

b. Test to see if the association between the variables *gender* and *survival status* can reasonably be attributed to chance or if you should look for some other explanation. These data cannot reasonably be considered a random sample from any well-defined population, so state your conclusion accordingly.

		Gender	
		Male	Female
Survived?	Yes	23	25
	No	32	9

Display 12.72 Donner party survival data. [Source: *www.utahcrossroads.org*, January, 2010.]

E39. Are people involved in alcohol-related accidents on public roadways in the United States just as likely to survive on Super Bowl Sunday as on other Sundays? To help answer this question, investigators looked at data for alcohol-related crashes during the 4 hours after the telecast of the first 27 Super Bowls. They compared these data with the data for alcohol-related crashes during the same time period on the Sundays the week before and immediately after the Super Bowl Sundays. Display 12.73 gives the number of people who were killed in and who survived alcohol-related crashes for the 27 Super Bowl Sundays and the 54 control Sundays.

	Super Bowl Sundays	Control Sundays	Total
Killed	284	163	447
Survived	48	218	266
Total	332	381	713

Display 12.73 Alcohol-related traffic fatalities on Super Bowl Sundays and control Sundays. [Source: D. A. Redelmeier and C. L. Stewart, "Do Fatal Crashes Increase Following a Super Bowl Telecast?" *Chance*, Vol. 18, 1 (2005), pp. 19–24.]

a. Is this an experiment, a sample survey, or an observational study?

b. Does the comparison of the survival rates on the two types of Sundays call for a test of homogeneity or a test of independence? Defend your choice.

c. Is there a statistically significant association between the type of Sunday and whether a person survived an alcohol-related accident? State your conclusion accordingly, in light of parts a and b.

E40. Display 12.74 gives the smoking behavior by occupation type for a sample of white males in 1976. Assuming this can be considered a random sample of all white males in the United States in 1976, perform a chi-square test of independence to see if it is reasonable to assume that the variables *occupation type* and *smoking behavior* are independent.

	Occupation			
Smoker?	Blue Collar	Professional	Other	Total
Yes	43.2%	6.3%	50.5%	13,112
Former	30.9%	11.0%	58.1%	8,509
Never	30.5%	11.8%	58.7%	9,694

Display 12.74 Smoking behavior sorted by type of occupation. [Source: D. J. Hand et al, *A Handbook of Small Data Sets* (London: Chapman & Hall, 1994), p. 284. Original source: T. D. Sterling and J. J. Weinkam, "Smoking Characteristics by Type of Employment," *Journal of Occupational Medicine*, Vol. 18 (1976), pp.743–753.]

E41. When a person has life-threatening irregularities to the normal heart rhythm, an electronic device called a defibrillator can be used to give an electric shock to the heart. The shock can reestablish normal rhythm. Defibrillation should be begun as soon as possible after such an event, preferably in less than 2 minutes. Researchers identified 6789 already hospitalized patients who had cardiac arrest due to abnormal heart rhythm. Display 12.75 presents some of their results.

a. The researchers concluded, "Delayed defibrillation is common and is associated with lower rates of survival after in-hospital cardiac arrest." What test would you use to verify their claim?

b. Conduct your test. What is your conclusion?

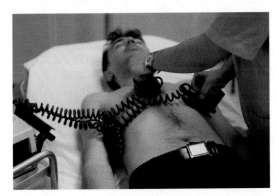

Patient receiving defibrillation.

	Defibrillation Begun ≤ 2 min	Defibrillation Begun > 2 min	Total
Survived to discharge	1863	455	2318
Did not survive to discharge	2881	1590	4471
Total	4744	2045	6789

Display 12.75 Whether patient survived to be discharged from, the hospital versus time to defibrillation. [Source: Paul S. Chan et. al., "Delayed Time to Defibrillation After In-Hospital Cardiac Arrest," *New England Journal of Medicine*, Vol. 358, (January 3, 2008), pp. 9–17, Table 3.]

E42. Refer to E41. Researchers also checked for functional outcomes among those who survived to discharge. The results are given in Display 12.76.

	Defibrillation Begun ≤ 2 min	Defibrillation Begun > 2 min	Total
No major disability	533	100	633
Major disability	1,009	281	1,290
Total	1,542	381	1,923

Display 12.76 Whether surviving patient survived had functional disability versus time to defibrillation.

a. How can you tell that data were not available for all of the patients who survived to discharge? What might be a reason why data were not available for everyone who survived?

b. The researchers concluded that among those surviving to discharge, delayed defibrillation was associated with a significantly lower likelihood of having no major disabilities in functional status. What test would you use to verify their claim?

c. Conduct your test. What is your conclusion?

E43. Many years ago, Smith College, a residential college, switched to an unusual academic schedule that made it fairly easy for students to take most or all of their classes on the first three days of the week. (Smith has long since abandoned this experiment.) At the time, the infirmary staff wanted to know about after-hours use of the infirmary under this schedule. They gathered data for an entire academic year, recording the time and day of the week of each after-hours visit, along with this nature of the problem, which they later classified as belonging to one of four categories: A visit to the infirmary was unnecessary for the problem, the problem required a nurse's attention, the problem required a doctor's attention, the problem required admission to the infirmary or to the local hospital. The results appear in Display 12.77.

	Mon.	Tues.	Wed.	Thurs.	Fri.
Visit	77	67	77	78	70
Nurse needed	80	66	53	73	62
Doctor needed	90	71	76	95	75
Admitted	61	49	42	52	28

Display 12.77 After-hours use of the infirmary. [Source: George W. Cobb, *Design and Analysis of Experiments* (New York: Springer, 1998), pp. 239–40.]

a. Construct a plot to display these data.

b. Are there any interesting trends in the table? What might explain them?

c. Could you use a test of independence on these data? A test of homogeneity? Does the design of the study fit the conditions of these tests?

d. Perform the test you think is best, giving any necessary cautions about your conclusion.

E44. Refer to E43.

a. How might you group the data to see whether the pattern for the weekdays when students attend class differs from the pattern for the weekdays when students don't attend class?

b. Perform a chi-square test on your regrouped data.

Chapter Summary

In this chapter, you have learned about three chi-square tests: a test of goodness of fit, a test of homogeneity, and a test of independence. Although the computations are similar, the questions the tests answer are different.

In a chi-square goodness-of-fit test, you ask, "Does this look like a random sample from a population in which the proportions that fall into these categories are the same as the proportions hypothesized?" This test is an extension of the test of a single proportion developed for the binomial case.

In a chi-square test of homogeneity, you ask, "Do these samples from different populations look like random samples from populations in which the proportions that fall into these categories are equal?" This test is an extension of the test for the equality of two binomial proportions.

In a chi-square test of independence, you ask, "Does this sample look like a random sample from a population in which these two categorical variables are independent (not associated)?" This test is not equivalent to any test developed earlier in this book. It is, however, related to the concept of independent events and the computation of $P(A \text{ and } B)$.

Review Exercises

E45. Professors Andrew Gelman and Deborah Nolan asked their students to roll a die that had the corners on the 1 side slightly rounded. The results of 120 rolls are shown in Display 12.78.

Outcome	Frequency
1	41
2	18
3	28
4	10
5	22
6	1

Display 12.78 Frequencies of outcomes of 120 rolls of a shaved die. [Source: *www.berkeley.edu.*]

Clearly, there are more 1's than you would expect and fewer 6's, the outcome on the side opposite the 1. Name and then conduct a test to determine if there is statistically significant evidence that the other four outcomes are not equally likely.

E46. Milk Chocolate M&M's are 24% blue, 13% brown, 16% green, 20% orange, 13% red, and 14% yellow. A random sample of 200 Peanut Butter M&M's yielded the distribution of colors shown in Display 12.79. Do you have statistically significant evidence that the distribution of colors of Peanut Butter M&M's is different from the distribution in Milk Chocolate M&M's?

E47. A great deal of research has been done on the consequences of being left-handed in a world where most people are right-handed. The marginal frequencies from one such study are given in Display 12.80.

a. Under the hypothesis that handedness and marital status are independent variables, compute the expected frequencies for each cell of the table.

Color	Observed Number of M&M's
Blue	40
Brown	19
Green	45
Orange	34
Red	25
Yellow	37
Total	200

Display 12.79 Distribution of colors in a random sample of peanut butter M&M's.

	Handedness		
	Right	Left	Total
Married	—?—	—?—	1678
Not married	—?—	—?—	799
Total	2150	327	2477

Display 12.80 Two-way table for the variables *handedness* and *marital status* for a sample of 2477 men. [Source: Kevin Denny and Vincent O' Sullivan, "The Economic Consequences of Being Left-Handed: Some Sinister Results," *Journal of Human Resources*, Vol. 42, 2 (2007), pp. 353–374.]

b. Sixty-eight percent of the right-handed men were married and 66% of the left-handed men were married. Make a table of observed counts.

c. Is there statistically significant evidence that handedness and marital status are associated?

E48. Foresters often are interested in the patterns of trees in a forest. For example, they gauge the integration or segregation of two species of trees (say, I and II) by looking at a sample of one species and then observing the species of its nearest neighbor. Data from such a study can be arrayed in a two-way table such as the one in Display 12.81.

		Tree Species of Its Nearest Neighbor	
		I	II
Tree	I	A	B
Species	II	C	D

Display 12.81 Two-way table for organizing nearest neighbor data.

a. Which cell counts, among A, B, C, and D, would be large if segregation of species is high? Would this lead to a large or small value of the χ^2 statistic in a test of independence?

b. Which cell counts, among A, B, C, and D, would be large if segregation of species is low? Would this lead to a large or small value of the χ^2 statistic in a test of independence?

E49. Vehicles can turn right, turn left, or continue straight ahead at a given intersection. It is hypothesized that vehicles entering the intersection from the south will continue straight ahead 50% of the time and that a vehicle not continuing straight ahead is just as likely to turn left as to turn right. A sample of 50 vehicles gave

the counts shown in Display 12.82. Are the data consistent with the hypothesis?

	Straight	Left Turn	Right Turn
Number of vehicles	28	12	10

Display 12.82 Vehicle turning data.

E50. In 1988, there were 540 spouse-murder cases in the 75 largest counties in the United States. Display 12.83 gives the outcomes.

		Defendant	
		Husband	Wife
Results	Not prosecuted	35	35
	Pleaded guilty	146	87
	Convicted at trial	130	69
	Acquitted at trial	7	31

Display 12.83 Outcomes for spouse-murder defendants in large urban counties. [Source: Bureau of Justice Statistics, *Spouse Murder Defendants in Large Urban Counties, Executive Summary*, NCJ-156831 (September 1995).]

a. What are the populations?

b. Construct a suitable plot to display the data. Describe what the plot shows.

c. Write suitable hypotheses and perform a chi-square test.

E51. The data in Display 12.84 show the *Titanic* passengers sorted by two variables: *class of travel* and *survival status*. These data cannot reasonably be considered a random sample. They are the population itself. Test to see if the apparent lack of independence between the variables *class of travel* and *survival status* can reasonably be attributed to chance or if you should look for some other explanation. If the latter, what is that explanation?

		Class of Travel			
		First	Second	Third	Total
Survived?	Yes	203	118	178	499
	No	122	167	528	817
	Total	325	285	706	1316

Display 12.84 *Titanic* passengers sorted by class of travel and survival status.

E52. In 2009, Iran held an election for president, with incumbent Mahmoud Ahmadinejad running against three other candidates. Many Iranians did not believe the result reported that Ahmadinejad won.

Because of restricted access to the voting itself, some people attempted to make a case for election fraud by looking at the numbers released by the Iranian Ministry of the Interior. One method was to see if the last digits of the vote counts in various localities looked like a random sample from a distribution where the digits were equally likely. (The last digits of made-up numbers don't tend to look random.)

The vote counts for Ahmadinejad and Mousavi, the two major candidates, appear in Display 12.85. Is there statistically significant evidence that the last digits of the counts don't occur equally often?

Region	Ahmadinejad	Mousavi
East Azerbaijan	1,131,111	837,858
West Azerbaijan	623,946	656,508
Ardabil	325,911	302,825
Isfahan	1,799,255	746,697
Ilam	199,654	96,826
Bushehr	299,357	177,268
Tehran	3,819,495	3,371,523
Chahar Mahaal and Bakhtiari	359,578	106,099
South Khorasan	285,984	90,363
Khorasan Razavi	2,214,801	884,570
North Khorasan	341,104	113,218
Khuzestan	1,303,129	552,636
Zanjan	444,480	126,561
Semnan	295,177	77,754
Sistan and Baluchestan	450,269	507,946
Fars	1,758,026	706,764
Qazvin	468,061	177,542
Qom	422,457	148,467
Kurdistan	315,689	261,772
Kerman	1,160,446	318,250
Kermanshah	573,568	374,188
Kohgiluyeh and Boyer-Ahmad	253,962	98,937
Golestan	515,211	325,806
Gilan	998,573	453,806
Lorestan	677,829	219,156
Mazandaran	1,289,257	585,373
Markazi	572,988	190,343
Hormozgan	482,990	241,988
Hamadan	765,723	218,481
Yazd	337,178	255,799

Display 12.85 Reported Iranian election results for the two major candidates. [Source: *www.guardian.co.uk/news/datablog/2009/jun/15/iran_1.*]

E53. In testing a random-digit generator, one criterion is to be sure that each digit occurs 1/10 of the time.

a. Generate 100 random digits on your calculator, and use a chi-square test to see if it's reasonable to assume that this criterion is met.

b. This criterion isn't sufficient. Give an example of a sequence of digits that clearly is not random but in which each digit occurs 1/10 of the time. How might you test against this possibility?

E54. "How does the amount of crime this year compare with the amount last year?" This question was asked of a random sample of 1010 people in Great Britain and of an independent random sample of 1012 people in the United States. In Great Britain, 43% said more, 25% said less, 21% said the same, and the rest had no opinion. In the United States, 47% said more, 33% said less, 18% said the same, and the rest had no opinion. Is there evidence of a statistically significant difference in the response pattern between the two countries? Organize these data in a table, display them in a plot, and perform a chi-square test of homogeneity. [Source: *poll.gallup.com*, 2005.]

E55. A statistics teacher noted, in Display 12.86, which kind of writing instrument her students used on the final exam and whether they passed the exam with a grade of 70% or greater. Assume that these students can be considered a random sample of all of her students. Is there statistically significant evidence that whether a student passes the final is associated with the writing instrument used?

		Writing Instrument			
		Pencil	Ballpoint Pen	Felt-Tip or Other Pen	Total
Passed Final?	Yes	18	7	5	30
	No	14	11	20	45
	Total	32	18	25	75

Display 12.86 Two-way table of observed frequencies for writing instrument use on final exam.

E56. Display 12.87 contains the response percentages from recent surveys that asked this question of randomly selected adults in three countries: "How often, if ever, do you drink alcoholic beverages such as liquor, wine, or beer—every day, a few times a week, about once a week, less than once a week, only on special occasions such as New Year's and holidays, or never?" These categories have a natural order. Does a chi-square test of whether the three countries differ with respect to drinking patterns take the order into account?

	Every Day	A Few Times a Week	About Once a Week	Less Than Once a Week	Only on Special Occasions	Never	Total Sample Size
United States	5%	14%	11%	10%	29%	31%	1012
Canada	7%	20%	18%	13%	31%	11%	1003
Great Britain	10%	31%	17%	9%	19%	14%	1010

Display 12.87 Results of surveys on drinking patterns. [Source: *poll.gallup.com*, 2006.]

C1. This partially completed table of expected frequencies is for a test of the independence of a father's handedness and his oldest child's handedness. What is the expected frequency in the cell marked "–?–"?

		Father		
		Left	Right	Total
Oldest Child	Left	64	72	136
	Right	–?–		455
	Total			591

A 64

B $455 \cdot \dfrac{64}{136}$

C $591 \cdot \dfrac{64}{136}$

D $455 \cdot \dfrac{64}{72}$

E Cannot be determined from the information given.

C2. In a pre-election poll, 13,660 potential voters were categorized based on their annual income (into one of eight categories) and their intended vote (two categories, Republican or Democratic candidate). The resulting chi-square test of independence of these variables gave a test statistic of 270. How many degrees of freedom are there?

A 2 **B** 7 **C** 8 **D** 16 **E** 13,659

C3. Researchers interested in determining whether there is a relationship between a mother's birthday and the birthday of her oldest child took a random sample of 200 mothers. The mother's birthday was categorized as within a week, more than a week but less than a month, or more than a month from her oldest child's birthday. Assuming that there are 365 days in a year and the same number of people are born on each day, what is the expected frequency of mothers in the *within at most a week* category?

A 8 **B** 8.219 **C** 66 **D** 66.67 **E** 67

C4. In a study of whether two acne medications are equally effective, researchers got eight volunteers and randomly chose one side of each person's face to receive each medication. After one month of use, they counted the number of blemishes on each side of each person's face. Which test is most appropriate in this situation?

A matched-pairs *t*-test

B two-sample *t*-test

C chi-square test for goodness of fit

D chi-square test for homogeneity of proportions

E chi-square test of independence

C5. A random sample of 1000 adults from each of the 50 states is taken, and the people are categorized as to whether they have ever been convicted of a felony, generating a 50-by-2 table. Which is the most appropriate test for determining whether the felony rates differ among the states?

A Compare each count to the expected frequency of 500, and use a chi-square goodness-of-fit test.

B Use a chi-square test of homogeneity of proportions.

C Use a chi-square test of independence.

D Use a one-proportion *z*-test.

E Use a *t*-test for the difference of means.

C6. For a project, a student plans to roll a six-sided die 1000 times. The professor becomes suspicious when the student reports getting 168 ones, 165 twos, 170 threes, 167 fours, 164 fives, and 166 sixes. The professor performs a chi-square test with the alternative hypothesis that the student's reported observed counts are closer to the expected frequencies than can reasonably be attributed to chance. What is the test statistic and *P*-value for this test?

A $\chi^2 = 0.14$; *P*-value = 0.0004

B $\chi^2 = 0.14$; *P*-value = 0.9996

C $\chi^2 = 1000$; *P*-value = 0

D $\chi^2 = 23.33$; *P*-value = 0.0004

E $\chi^2 = 23.33$; *P*-value = 0.9996

C7. A researcher wanted to determine whether Barbarians and Vandals have similar food preferences. Independent random samples were taken from each tribe, and each person was categorized as to whether he or she prefers to eat gruel, turnips, or raw meat. The value of χ^2 was statistically significant. What conclusion can be drawn?

A The proportion of all Barbarians who prefer gruel is different from the proportion of all Vandals who prefer gruel.

B The proportion of all Barbarians who prefer gruel is equal to the proportion of all Vandals who prefer gruel.

C The population proportions are different for at least one of the three choices.

D The population proportions are different for all three of the choices.

E The population proportions are equal for all three of the choices.

C8. Random samples are taken from two populations (Barbarians and Vandals), and each person is categorized as a pillager or a burner. Which of the following, when done appropriately, is equivalent to a chi-square test of homogeneity for this situation?

A one-sample *z*-test

B two-sample *z*-test

C one-sample *t*-test

D two-sample *t*-test

E chi-square goodness-of-fit test

Investigations

C9. To study the effectiveness of seat belts and air bags, researchers took a careful look at data that were collected on accidents between 1997 and 2002. The data came from a division of the National Highway Traffic Safety Administration that collects detailed data on a random sample of accidents "in which there is a harmful event and from which at least one vehicle is towed." A total of 22,804 front-seat occupants were involved in the accidents studied. [Source: Mary C. Meyer and Tremika Finney, "Who Wants Airbags?" *Chance*, Vol. 18, 2 (2005), pp. 3–16. The data here are approximations created from scaling down the population projections in the article.]

a. Do the data in the table show a statistically significant association between seat belt use and surviving an accident?

	Seat Belts Used	Seat Belts Not Used
Killed	35	67
Survived	16,694	6,008

b. Do the data in the table show a statistically significant association between air bag use and surviving an accident?

	Air Bags	No Air Bags
Killed	42	60
Survived	12,315	10,387

c. From the two analyses in parts a and b, which safety feature appears to have the stronger association with survival?

C10. The analysis in C9 does not account for seat belts and air bags interacting with each other. This table splits the air bag data according to whether seat belts were also in use.

a. What do the data suggest as to the effect of air bag use when seat belts are also used? Is there a significant association?

b. What do the data suggest about the effect of air bag use when seat belts are not used? Is there a significant association?

	Seat Belts Used		Seat Belts Not Used	
	Air Bags	No Air Bags	Air Bags	No Air Bags
Killed	19	16	23	44
Survived	10,464	6,230	1,851	4,157
Total	10,483	6,246	1,874	4,201

Chapter 13
Inference for Regression

On Mars, do rocks with more sulfur tend to be redder? Data from the
Pathfinder mission to the red planet were used to explore questions like this.

Who are Shark, Barnacle Bill, Half Dome, Wedge, and Yogi?

Answer: Mars rocks, found at the landing site of Mars *Pathfinder*.

Trivia Question 2: Mars is often called the red planet. What makes it red?

Answer: Sulfur?

Display 13.1 shows how the redness of the five Mars rocks is related to their sulfur content. The response variable, *redness,* is the ratio of red to blue in a spectral analysis: the higher the value, the redder the rock. The explanatory variable is the percentage, by weight, of sulfate in the rock.

The *Sojourner* robot rover rolls out of Mars *Pathfinder* to meet Barnacle Bill and Yogi.

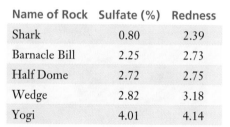

Name of Rock	Sulfate (%)	Redness
Shark	0.80	2.39
Barnacle Bill	2.25	2.73
Half Dome	2.72	2.75
Wedge	2.82	3.18
Yogi	4.01	4.14

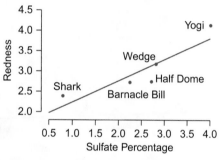

Display 13.1 Scatterplot and data table for five Mars rocks. [Source: Harry Y. McSween, Jr., and Scott L. Murchie, "Rocks at the Mars *Pathfinder* Landing Site," *American Scientist*, Vol. 87 (January–February 1999), pp. 36–45. Data taken from a graph.]

Shark: Have you figured out how change in the sulfate content is reflected in change in redness?

Yogi: According to my calculations, the slope of the regression line is 0.525. Rocks that differ by 1 percentage point in their sulfate content will differ by about 0.525, on average, in their measure of redness. I learned how to do this in Chapter 3.

Shark: That's the story if you take the numbers at face value.

Yogi: Why shouldn't I?

Shark: Well, after all, we aren't the only rocks on the planet. The slope, 0.525, is an estimate based on a sample—us! If that cute little rolling *Sojourner* robot had nuzzled up to different rocks, you would have a different estimate. So the goal of this chapter is to show how to construct confidence intervals and test the significance of the slope of a regression line.

Yogi: You need a whole chapter just for that? I know from Chapter 3 how to estimate the slope. So, if you tell me how to find the standard error for the slope, I can do the rest. To get a confidence interval, I just do the usual: $0.525 \pm t^* \cdot SE$. To test a null hypothesis, I simply compute

$$t = \frac{0.525 - \textit{hypothesized slope}}{SE}$$

and use the t-table to find the P-value.

Shark: It sounds familiar, all right. Maybe this can be a short chapter!

This chapter addresses how to make inferences about the unknown true relationship between two quantitative variables. The methods and logic you will learn in this chapter apply to a broad range of such questions.

In this chapter, you will learn

▷ what affects the variability of the slope of a regression line

▷ how to estimate the standard error of the slope

▷ how to construct and interpret a confidence interval for the slope

▷ how to test whether the true slope is different from a hypothesized value

▷ how to know when to trust confidence intervals and tests

13.1 ▶ Variation in the Slope from Sample to Sample

As you know by now, statistical thinking, though powerful, is never as easy or automatic as simply plugging numbers into formulas. To use statistical methods appropriately, you need to understand their logic, not just the computing rules. The logic in this chapter is designed for bivariate populations that you can think of as modeled by "linear fit plus random deviation."

Bivariate: paired variables, usually called x and y.

Linear Models

In Chapter 3, you learned to summarize a linear relationship with a least squares regression equation

$$\hat{y} = b_0 + b_1 x$$

That equation is a complete description if you have the entire population. But if you have only a random sample from the population, $\hat{y} = b_0 + b_1 x$ should be considered an estimate of the "true" linear relationship, just as you use a sample mean, $\bar{x}$, as an estimate of a population mean, μ.

Linear Model: True Regression Equation Plus Error

The model of a linear relationship between x and y has the form

$$\textit{response} = \textit{prediction from true regression equation} + \textit{random deviation}$$

$$y = (\beta_0 + \beta_1 x) + \varepsilon$$

b_0 and b_1 are estimates of β_0 and β_1.

ε is the observed value minus the value predicted by the true regression line.

Here, β_0 and β_1 refer, respectively, to the intercept and slope of a line that you don't ordinarily get to see—the **true regression line** you would get if you had data for the whole population instead of only a sample. The letter ε indicates the size of the *random deviation*—how far a point falls above or below the true regression line. The true regression equation, sometimes called the **line of means** or the **line of averages**, can be written

$$\mu_y = \beta_0 + \beta_1 x$$

where μ_y stands for the mean value of y at the given value of x.

Because such linear models are often used to predict unknown values of y from known values of x, or to explain how x influences the variation in y, y is called the *response variable* and x is called the *predictor variable* or the *explanatory variable*.

The following example will help you understand the roles of these regression equations and the relationship between them.

Example 13.1	**A Model of the Growth of Children**

The heights of children ages 8 to 13 increase about 2 in. per year, on the average, with 8-year-olds having a mean height of about 51 in. At each of these ages, heights of children are approximately normally distributed with a standard deviation of about 2.1 in. What is the equation of the true regression line for predicting the height of a child given his or her age? Simulate taking a random sample of ten children at each age from 8 to 13 and find an estimate of the true regression equation.

Solution

The true regression equation, or line of means, is the linear equation relating the mean heights to age (x), ignoring the variation in heights at each age. As shown in Display 13.2, the true regression line has slope 2 and goes through the points (8, 51), (9, 53), (10, 55), etc. Thus, its equation is

$$\mu_y = 35 + 2x$$

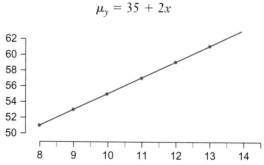

Display 13.2 True regression line for the growth model of children ages 8 to 13.

To simulate taking a random sample of ten 8-year-old children, use technology to randomly select ten values from a normal distribution with mean 51 in. and standard deviation 2.1 in. To get a random sample of 9-year-olds, sample from a normal distribution with mean 53 in. and standard deviation of 2.1 in., and so on. Sample results are shown in Display 13.3. The equation of the regression line, $\hat{y} = 33.6 + 2.13x$, is a reasonably good approximation to the true regression equation, $\mu_y = 35 + 2x$.

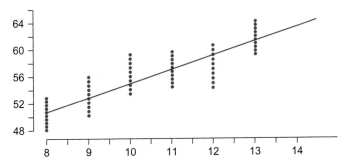

Display 13.3 Scatterplot and regression line for simulated random sample of children's heights.

Real data do not get created by the construction of an ideal model and then sampling from it, as was done for the children's heights in Example 13.1. However, that model (mean as a linear function plus random deviation from that mean) does serve as a good approximation to many data sets seen in practice. In addition, the model provides a basis for statistical inference, as you will soon see. The validity of the inference depends on properties of the conditional distributions of y given x.

Conditional Distribution of *y* Given *x*

The conditional distribution of y given x refers to all the values of y for a fixed value of x.

Conditional Distributions in the Children's Growth Model

Example 13.2

Refer to Example 13.1, which describes the distribution of the heights of children ages 8 through 13. Describe the conditional distribution of y given x when x is 9 years.

Solution
If you were to make a scatterplot of (*age, height*) for a sample of hundreds of children, the vertical column of points for all the 9-year-old children would approximate the conditional distribution of the height, y, given that the age, x, is 9, much as shown in Display 13.3 but with many more dots. This distribution is normal, with mean $51 + 2$, or 53 in., and standard deviation 2.1 in.

A linear model is appropriate for modeling the relationship between y and x if the means of the conditional distributions fall near a line. For the methods of inference in this chapter, the conditional distributions must be approximately normal, and the variability in y should be the same for each conditional distribution. That is, if you picked a value of x and computed the standard deviation of all the associated values of y in the population, you'd get the same number, σ, as you would if you picked any other x, as shown in Display 13.4. This implies that σ also measures the variability of *all* values of y about the true regression line. You can estimate σ using the sum of the squared residuals.

You can think of the true regression line as the graph of the averages of the y-values for each value of x.

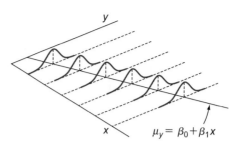

Display 13.4 The variability of all values of *y* about the true regression line is the same as the variability of the *y*-values at each fixed value of *x*.

Variability in *y*

> *σ* measures the variability of the residuals.

The common variability of *y* at each *x* is denoted σ. It is estimated by *s*, which can be thought of as the standard deviation of the residuals. You compute *s* using all *n* values of *y* in the sample:

$$\sigma \approx s = \sqrt{\frac{\sum(y_2 - \hat{y}_2)^2}{n-2}} = \sqrt{\frac{SSE}{n-2}}$$

Divide by $n-2$ because $\hat{y}$ is computed using *two* estimates, b_0 and b_1.

DISCUSSION
Linear Models

D1. Refer to the information about children's heights in Example 13.1 on page 626 . Describe the conditional distribution of the heights given that the age is 10. Given that the age is 12.

D2. Explain the difference between the linear model $\mu_y = \beta_0 + \beta_1 x + \varepsilon$ and the fitted equation $\hat{y} = b_0 + b_1 x$. In particular, what is the difference between β_1 and b_1? What is the difference between a random deviation, ε, from the model and an observed residual, $y - \hat{y}$?

D3. If samples were to be selected from populations represented by the rectangles in Display 13.5, would you have any concerns about using a straight line to model the relationship between *x* and *y*? Why or why not?

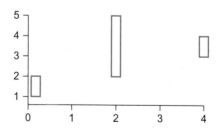

Display 13.5 A "scatterplot" for D3.

The Variability of b_1 from Sample to Sample

To estimate the equation of the true regression line, $\mu_y = \beta_0 + \beta_1 x$, you take a random sample from the population, compute the least squares regression equation $\hat{y} = b_0 + b_1 x$, and use the intercept, b_0, as an estimate of β_0 and use the slope, b_1, as an estimate of β_1. Typically, you will be most interested in estimating β_1.

To make inferences about the slope, β_1, of the true regression line, you will use formulas for the test statistic and confidence interval that are of a familiar form:

$$t = \frac{b_1 - hypothesized\ slope}{estimated\ standard\ error\ of\ the\ slope}$$

$$b_1 \pm t^* \cdot (estimated\ standard\ error\ of\ the\ slope)$$

To move toward finding a formula for the estimated standard error of b_1, think about the situations when b_1 would tend to be close to β_1, and when it would tend to be farther away. As you might have guessed, with a larger sample size, b_1 tends to be closer to β_1.

> Larger n, less variability in b_1.

Display 13.6 can help you see what other factors affect the variability in b_1. For each plot, four regression equations were computed by taking a sample of the heights of eight children from a population like that in Display 13.3. Plots A and B each have regression lines constructed using random samples of four 8-year-olds and four 13-year-olds. In Plot A, the standard deviation, σ, was 2.1 in. In Plot B, however, the standard deviation, σ, was assumed to be 4.2 in. It is no surprise that the population with more variability in values of y at each fixed value of x generates regression lines with more variability in the slope.

> Less variability in y, less variability in b_1.

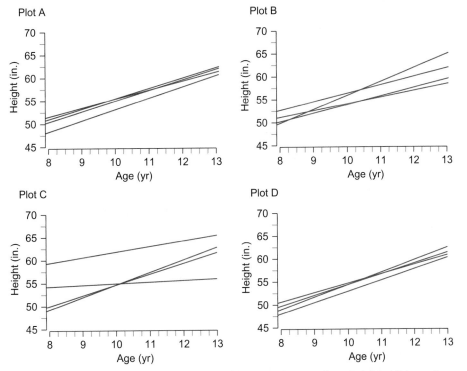

Plot A   Plot B

Plot C   Plot D

> Less spread in x, more variability in b_1.

Display 13.6   Regression lines computed from random samples of eight children, demonstrating how variability in b_1 is related to variation in y (Plots A and B) and spread in x (Plots C and D).

In constructing Plots C and D, the only difference was in how the ages were selected. In Plot C, four of the children were age 10 and four were age 11. In Plot D, four children were age 8 and four were age 13. Note how the slopes of the regression lines are far less variable when the ages are farther apart.

As you have seen, the spread of the values of x is important in determining the variability of the slope. This spread is quantified as shown in the next box.

Spread in the Values of x

The spread in the values of x is measured by $\sqrt{\Sigma(x - \bar{x})^2}$, where $\bar{x}$ is computed using all n values of x in the sample.

The tightrope walker's hands are held far apart for stability.

STATISTICS IN ACTION 13.1 ▶ **Variability in the Estimate of the Growth Rate of Children**

As you read in Example 13.1 on page 626, on average, kids from the ages of 8 to 13 grow taller at the rate of 2 in. per year. Heights of 8-year-olds average about 51 in. At each age, the distribution of heights is approximately normal, with a standard deviation of roughly 2.1 in.

1. Use this information to fill in the "Average Height" column of a copy of this table.

Age, x	Average Height from Model, μ_y	Random Deviation, ε	Observed Height, y, of Your Child
8	51	—?—	—?—
9	—?—	—?—	—?—
10	—?—	—?—	—?—
11	—?—	—?—	—?—
12	—?—	—?—	—?—
13	—?—	—?—	—?—

On TI-calculators, use randNorm (0, 2.1) from the MATH PRB menu.

2. Now simulate taking a randomly selected child of each age. To find how much the height of your 8-year-old child deviates from the average height of 51 in., use a calculator or computer to randomly select a deviation, ε, from a normal distribution with mean 0 and SD 2.1. Record ε in the third column of the table, and then add ε to the average height for an 8-year-old. Record the sum in the "Observed Height" column. Repeat, with a new deviation, for each age.

3. Fit a regression equation to the pairs (x, y) where x is the age and y is the observed height. Record your estimated slope, b_1.

4. Collect the value of b_1 from each member of your class and plot these values. Discuss the shape, center, and spread of the distribution of estimated slopes. How well does b_1 do as an estimator of β_1?

DISCUSSION
The Variability of b_1 from Sample to Sample

D4. In the "scatterplots" in Display 13.7, imagine taking a sample of responses from within each rectangle. That is, the rectangles define the regions in which the responses lie for each value of x. Suppose you take a sample and calculate a regression line. Then you repeat the process many times. Which of the three "plots" should produce regression lines with the smallest variation in slope? Which should produce regression lines with the largest variation in slope?

The Standard Error of the Slope

You now know what affects how much the slope, b_1, of the regression line varies from sample to sample: the sample size, n; the variability, σ, of y at each fixed value of x (estimated by s); and the spread in the values of x. This box gives the formula for estimating the standard error, s_{b_1}, of the sampling distribution of the slope.

I.

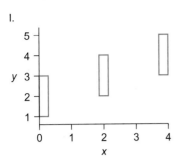

II.

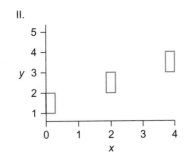

III.

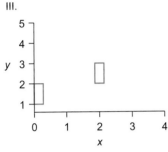

Display 13.7   Three "scatterplots" for D4.

Formula for Estimating the Standard Error of the Slope

$$s_{b_1} = \frac{s}{\sqrt{\sum(x - \bar{x})^2}} = \frac{\sqrt{\dfrac{\sum(y - \hat{y})^2}{n - 2}}}{\sqrt{\sum(x - \bar{x})^2}}$$

The formula does what you would expect: The slope varies less from sample to sample when the sample size is larger, when the values of y tend to be closer to the regression line, and when the values of x are more spread out.

Do Easy Professors Get Higher Ratings?

Example 13.3

A web site allows students anonymously to rate various characteristics of their professors. Mean ratings for a random sample of seven mathematics instructors from California State University, Northridge appear in Display 13.8. One of the variables is ease, a judgment of how difficult the professor is and how much work is required. The second variable is an overall rating of the professor (which is computed without taking the ease rating into consideration).

   a. The regression equation is $\widehat{rating} = 1.31 + 0.79\ ease$. Interpret the slope.

   b. Does "line plus random deviation" appear to be an appropriate model?

   c. Calculate the estimated standard error of the slope and then locate it on the output in Display 13.9. Interpret this value. Then locate and interpret s.

   d. For the population of 92 instructors who had at least 10 ratings, the true regression equation is $rating = 0.53 + 0.96\ ease$ (with $r = 0.8$). How many estimated standard errors is the estimate of the slope from the slope of the true regression line?

Earase	Overall Rating
2.9	3.5
3.6	3.9
1.7	3.0
4.3	4.8
1.9	2.6
3.1	4.2
2.5	2.9

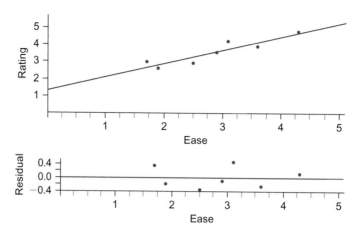

Display 13.8 Mean ratings, scatterplot with regression line, and residual plot for a random sample of seven Cal State Northridge mathematics instructors. *Ease* is rated from 1 (hard) to 5 (easy A). Overall rating is 1 (poor) to 5 (good). [Source: *www. ratemyprofessors.com.*]

The regression equation is
Overall Rating = 1.31 + 0.787 Ease

Predictor	Coef	SE Coef	T	P
Constant	1.3078	0.4546	2.88	0.035
Ease	0.7873	0.1525	5.16	0.004

s = 0.343588   R-Sq = 84.2%   R-Sq(adj) = 81.0%

Display 13.9 Minitab regression output of *overall rating* versus *ease*.

Solution

a. If one instructor's mean rating is one point higher than another's on the Ease scale, then the first instructor's mean overall rating tends to be 0.79 points higher.

b. A linear model appears to be appropriate because the points on the residual plot are randomly scattered with no apparent pattern.

c. The slope of the least squares regression line has an estimated standard error of

$$s_{b_1} = \frac{s}{\sqrt{\sum(x - \bar{x})^2}} = \frac{\sqrt{\frac{\sum(y - \hat{y})^2}{n - 2}}}{\sqrt{\sum(x - \bar{x})^2}} \approx \frac{\sqrt{\frac{0.590264}{7 - 2}}}{\sqrt{5.07714}} \approx \frac{0.343588}{2.25325} \approx 0.1525$$

If you were to take repeated random samples of seven instructors and compute the slope of the regression line for predicting *overall rating* from *ease*, the standard deviation of the distribution of these slopes is estimated to be 0.1525. From the computation or the output, *s* is 0.343588. This is the estimate of the common variability in *overall rating* at each fixed value of *ease*. Both s_{b_1} and *s* are relatively small because the points are clustered tightly about the regression line, so there should be little variability in the slope from sample to sample.

d. The estimated slope of 0.79 from the random sample of seven instructors is 1.11 estimated standard errors lower than the true slope of 0.96:

$$\frac{0.79 - 0.96}{0.1525} \approx -1.11$$

D5. You plan to collect data to be able to predict the actual temperature of your oven from the temperature shown on its thermostat. Suppose this relationship is linear. How will you design this study? What have you learned in this section that will help you?

D6. Write s_{b_1} in terms of s, the standard deviation of the residuals; s_x, the standard deviation of the x's; and n, the sample size. Use that formulation to argue that the standard deviation of the slope decreases with increasing sample size so long as the variation in the residuals and in the x's does not change appreciably.

Summary 13.1: Variation in the Slope from Sample to Sample

Suppose you want to estimate an underlying linear relationship, $y = \beta_0 + \beta_1 x + \varepsilon$ where $\mu_y = \beta_0 + \beta_1 x$ is the equation of the true regression line. The intercept, b_0, and slope, b_1, in the least squares regression equation $\hat{y} = b_0 + b_1 x$ serve as your estimates of the population parameters, β_0 and β_1. Typically, you will be most interested in how closely b_1 estimates β_1. This formula can be used to estimate the standard error of the slope:

$$s_{b_1} = \frac{s}{\sqrt{\sum (x - \bar{x})^2}} = \frac{\sqrt{\dfrac{\sum (y - \hat{y})^2}{n - 2}}}{\sqrt{\sum (x_i - \bar{x})^2}}$$

The slope, b_1, of the regression line varies less from sample to sample when

- the sample size is larger
- the random deviations between the points and the line are smaller
- the values of x are farther apart

In Section 13.2, you will learn to compute a confidence interval and test statistic for the slope, but your conclusions won't be valid unless line plus random deviations is the appropriate model for your data. So always plot your data first to judge whether a line is a reasonable model.

Practice

Linear Models

P1. The table and scatterplot in Display 13.10 give the number of grams of fat and the number of calories in different kinds of taco salad available at Taco Bell. Fat contains 9 cal/g.

    a. What number would you predict for the slope of the true regression line for such data?

    b. Approximate the y-intercept of the line. What does the y-intercept tell you?

    c. Find the least squares regression equation, $\hat{y} = b_0 + b_1 x$, for these data. Interpret the slope.

Compare the estimated slope to your predicted slope from part a. Are they close?

    d. What are some reasons why not all of the points fall exactly on a line with the slope in part a?

    e. Suppose that the true regression equation is $\mu_y = 200 + 9x$. Calculate the random deviation, ε, for the Express Taco Salad.

    f. Calculate the residual from the estimated regression line for the Express Taco Salad.

    g. Compute s, the estimate of the common variability of y at each x.

Taco Salad	Fat (gms)	Calories
Chicken Ranch Fully Loaded Taco Salad	57	960
Chipotle Steak Fully Loaded Taco Salad	59	950
Fiesta Taco Salad	43	820
Fiesta Taco Salad Without Shell	22	460
Express Taco Salad	30	600

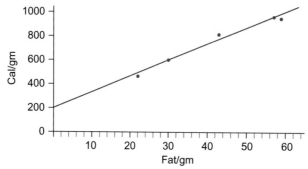

Display 13.10 Fat content and calories in Taco Bell's taco salads. [Source: *www. tacobell.com/nutrition/information/*, July 2009.]

P2. Refer to the table in Display 13.1 on page 624, which gives the sulfate percentage, x, and redness, y, of the Mars rocks. Compute s, the estimate of the common variability of y at each x.

The Variability of b_1 from Sample to Sample

P3. Refer to the plots in Display 13.11. Each plot shows five regression lines. Each line was computed from a random sample of size 8, with four points at each value of x at the two ends of the line segments. The populations from which the samples were drawn have $\sigma = 3$ or $\sigma = 5$.

   a. Which population, for Plot I or for Plot III, has $\sigma = 3$ and which has $\sigma = 5$? How do you know?

   b. Which population, for Plot II or for Plot IV, has $\sigma = 3$ and which has $\sigma = 5$? How do you know?

   c. Is there more variability in the slopes with wider spread in x or with smaller spread in x?

   d. For which plot is the variability in the slope the greatest? What accounts for that fact?

P4. Five quantities follow, along with two possible values of each. For each quantity, decide which value will give you the larger variability in b_1, or state that there is no difference in the variability, and give a reason why. (Assume that all other things stay the same.)

   a. the standard deviation, σ, of the individual response values of y at each value of x: 3 or 5

   b. the spread of the x-values: 3 or 10

   c. the number of observations, n: 10 or 20

   d. the true slope, β_1: 1 or 3

   e. the true intercept, β_0: 1 or 7

Plot I

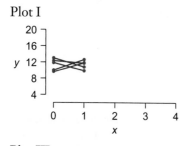

Plot II

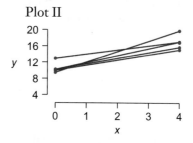

Plot III

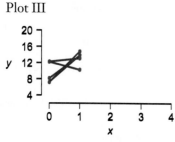

Plot IV

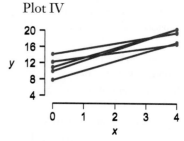

Display 13.11 Regression lines relating variability in b_1 to variation in y and to spread in x.

The Standard Error of the Slope

P5. Look again at the Mars rocks data in Display 13.1 on page 624.

    a. Compute the estimate of the standard error of the slope, s_{b_1}.

    b. On the computer output for the regression in Display 13.12, locate the standard error of the slope. Then locate your estimate from P2 of the variation in y about the line. What is the equation of the regression line?

The regression equation is
Redness = 1.72 + 0.525 Sulfate

Predictor	Coefficient	Stdev	t-ratio	P
Constant	1.7153	0.4010	4.28	0.023
Sulfate	0.5249	0.1472	3.57	0.038

s = 0.3414   R-sq = 80.9%   R-sq(adj) = 74.6%

Display 13.12 Regression analysis for the Mars rocks data.

    c. If the values in the "Sulfate" row were missing in Display 13.12, explain how you would compute "Coefficient" and "Stdev."

P6. From normal distributions with $\sigma = 3$ and $\mu_y = 10$ at $x = 0$ and $\mu_y = 18$ at $x = 4$, ten observations of y were randomly selected at each value of x. The results are given in Display 13.13.

    a. Find both the true regression equation, $\mu_y = \beta_0 + \beta_1 x$, and the least squares regression equation, $\hat{y} = b_0 + b_1 x$, and compare them.

    b. Compute s and compare it to $\sigma = 3$.

    c. Compute s_{b_1}, the estimate of the standard error of b_1, and compare it to the true value of the standard error. (*Note:* The true value of the SE for b_1 is
$$\frac{\sigma}{\sqrt{\Sigma(x - \bar{x})^2}}.)$$

    d. Does the least squares regression line go through the mean of the responses at both $x = 0$ and $x = 4$?

Observations of y	
At $x = 0$	At $x = 4$
10.1	17.2
9.5	16.3
8.0	20.7
10.6	19.5
13.6	14.1
11.9	15.9
11.9	21.9
2.6	16.1
14.7	20.9
6.2	17.7

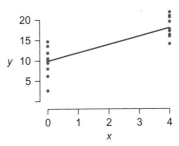

Display 13.13 Data and scatterplot for P6.

P7. Each of the lists in I–V gives the values of x used to compute a regression line. Assuming $s = 1$ in all cases, order the lists from the one with the largest estimated standard error of the slope to the one with the smallest. (You should be able to rank these without doing much, if any, computation.)

    I. $x = 1, 2, 3, 4, 5, 6, 7, 8, 9, 10, 11, 12$

    II. $x = 1, 2, 3, 4, 5, 6, 7, 8, 9, 10$

    III. $x = 1, 1, 1, 1, 1, 1, 12, 12, 12, 12, 12, 12$

    IV. $x = 1, 1, 1, 4, 4, 4, 9, 9, 9, 12, 12, 12$

    V. $x = 1, 2, 3, 4, 5, 6, 7$

Exercises

E1. Every spring, visitors eagerly await the opening of the spectacular Going-to-the- Sun Road in Glacier National Park, Montana. A typical range of yearly snowfall for the area is from 30 to 70 in. The amount of snow is measured at Flattop Mountain, near the top of the road, on the first Monday in April, and is given in swe (snow water equivalent: the water content obtained from melting). From analysis of past data, when the amount of snow was 30 in. of swe, the road opened, on average, on the 150th day of the year. Every additional 0.57 in. of swe measured at Flattop Mountain meant another day on average until the road opened. [Source: "Spring Opening of the Going-to-the-Sun Road and Flattop Mountain SNOTEL Data," Northern Rocky Mountain Science Center, U.S. Geological Survey, December 2001, *www.nrmsc.usgs.gov.*]

Going-to-the-Sun Road.

a. If you write an equation summarizing the information given, should you use the form $y = \beta_0 + \beta_1 x$ or $\hat{y} = b_0 + b_1 x$? Explain.

b. Write an equation that predicts when the road will be open, given the swe. In this situation, what is the response variable, y? The predictor variable, x?

c. In 2005, the Flattop Mountain station recorded 31.0 in. of swe. What date would you predict the road opened?

d. Do you think the random deviation, ε, in this situation tends to be relatively large or small ? Make a guess as to what it might be on average.

E2. According to Leonardo da Vinci, a person's arm span and height are about equal. Display 13.14 gives height and arm span measurements for a sample of 15 students.

Arm Span (cm)	Height (cm)	Arm Span (cm)	Height (cm)
168.0	170.5	129.0	132.5
172.0	170.0	169.0	165.0
101.0	107.0	175.0	179.0
161.0	159.0	154.0	149.0
166.0	166.0	142.0	143.0
174.0	175.0	156.5	158.0
153.5	158.0	164.0	161.0
95.0	95.5		

Display 13.14 *Height* versus *arm span* for 15 students.

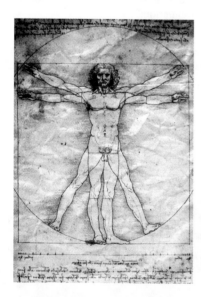

Drawing by Leonard o Da Vinci.

a. What is the regression equation, $\mu_y = \beta_0 + \beta_1 x$, that Leonardo is proposing for this situation? Use *arm span* as the explanatory variable.

b. Find the least squares regression equation, $\hat{y} = b_0 + b_1 x$, for these data. Interpret the slope. Compare the

estimated slope and intercept to Leonardo's proposed slope and intercept from part a. Are they close?

c. Calculate the random deviation, ε, for each student, using Leonardo's proposed regression line. Plot these random deviations against the arm spans and comment on the pattern.

d. For each student, calculate the residual from the least squares regression equation. Plot the residuals against the arm spans and comment on the pattern. Is the pattern similar to that for the random deviations?

E3. Greenhouse gases that have been released into the atmosphere since the start of industrialization (around 1750) are thought to be the fundamental cause of global warming. As part of the worldwide effort to control the emission of these gases, scientists and manufacturers in the United States are now measuring the greenhouse gas emissions (mostly carbon dioxide, nitrous oxide, and methane) of automobiles. In this quest, it is important to know how the gas emissions are related to other variables that can be controlled by the manufacturer, at least to some degree, and by the consumer. Display 13.15 provides data for a sample of 15 randomly selected car models on greenhouse gas emissions (*GGE*) in estimated tons per year, expected highway gas mileage (*HGM*) in miles per gallon, engine size (*ES*) in liters, number of cylinders (*CY*), and passenger compartment size (*PCS*) in cubic feet. The scatterplots in the display show how each of the possible explanatory variables is related to *GGE*.

a. Refer to the scatterplot for each pair of variables and tell whether you think a line gives a suitable summary of their relationship.

b. Compute the slope of the regression line for $x = HGM$ and $y = GGE$. Interpret this slope in context. Compute the estimated standard error of the slope.

c. Find your values from part b on the computer output in Display 13.16. Also find and interpret s, the estimate of σ.

E4. How does energy consumption in a residence relate to the outdoor temperature? Some interesting patterns can be detected through study of available monthly records on natural gas and electricity usage for a gas-heated single-family residence (with no air-conditioning) in the Boston area, along with monthly data on outdoor temperature and heating degree days. Heating degrees for a day with mean temperature less than 65°F are defined as $(65 - mean\ daily\ temperature)$. The data are given on page 644, with these variables:

mean temp	mean monthly temperature in Boston, in degrees Fahrenheit
mean gas	mean natural gas usage per day for the month, in therms
mean kWh	mean electricity usage per day for the month, in kilowatt-hours
heat DD	total heating degree days for the month

Model	GGE	HGM	ES	CY	PCS
Acura TSX	7.3	30	2.4	4	95
Buick Lucerne	10.2	22	4.6	8	108
Cadillac CTS	13.1	18	6.2	8	100
Chevrolet Malibu	7.1	33	2.4	4	95
Honda Civic	6.3	34	1.8	4	91
Dodge Challenger	11.4	22	6.1	8	91
Ford Fusion	9.6	24	3.5	6	101
Kia Forte	6.1	36	2.0	4	97
Lincoln MKS	9.6	23	3.5	6	106
Mazda 3	6.6	33	2.0	4	94
Saturn Aura	9.2	26	3.6	6	98
Scion tC	7.7	29	2.4	4	85
Toyota Camry	7.1	32	2.5	4	101
Volkswagen Eos	7.3	31	2.0	4	77
Volvo S80	10.2	22	4.4	8	98

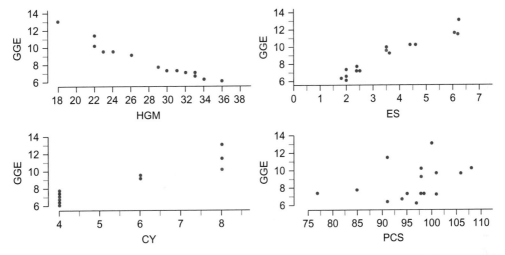

Display 13.15 Data for a random sample of 2010 car models. [Source: *www.fueleconomy.gov.*]

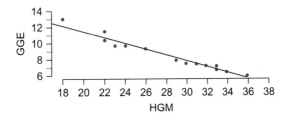

The regression equation is
GGE = 18.6 − 0.362 HGM

Predictor	Coef	Stdev	t-ratio	P
Constant	18.5907	0.6941	26.79	0.000
HGM	−0.36159	0.02463	−14.68	0.000

s = 0.5115   R-sq = 94.3%   R-sq(adj) = 93.9%

Display 13.16 Regression analysis of GGE versus HGM for 15 car models.

(These data are from the 1990s, but the patterns among the variables do not change much over the years.)

a. Refer to the scatterplots in Display 13.17. For each pair of variables, tell whether you think a line gives a suitable summary of the relationship.

Month	Mean Temp	Mean Gas	Mean kWh	Heat DD
June 95	69	1.3	11.9	30
August 95	73	0.7	15.6	2
October 95	59	1.5	18	214
November 95	46	4.5	14.9	685
December 95	29	8.9	18.1	1023
January 96	30	11.6	18.1	1074
February 96	31	10.7	18.8	981
March 96	37	11.6	37.8	875
April 96	48	7.5	17.6	510
May 96	57	3.5	17.9	264
June 96	68	1.5	13	20
August 96	71	0.8	17.4	6
October 96	53	1.9	17.7	358
November 96	40	5	22.3	739
December 96	39	7.3	20.7	792
January 97	29	9.3	32.8	1104
February 97	36	9.7	30.5	806
March 97	37	7.9	24.6	868
April 97	46	5.8	17.2	551
May 97	56	3.2	26.2	269

b. Compute the slope of the least squares regression line for the relationship between y = *mean gas* and x = *mean temp*. Interpret the slope in context. Compute the estimated standard error of the slope.

c. Find your values from part b on the computer printout in Display 13.18. Also find and interpret s, the estimate of σ.

Predictor	Coef	Stdev	t-ratio	P
Constant	16.988	1.196	14.20	0.000
MeanTemp	−0.23643	0.02401	−9.85	0.000

$s = 1.548$ R-sq = 84.3%   R-sq(adj) = 83.5%

Display 13.18 Regression analysis for *mean gas* versus *mean temp*.

E5. Suppose you collect a dozen cups, glasses, round bowls, and plates of various sizes, and you then plot the distance around the rim versus the distance across, both measured in centimeters.

a. What would you predict for the slope of the true regression line for this situation? Interpret this slope in context.

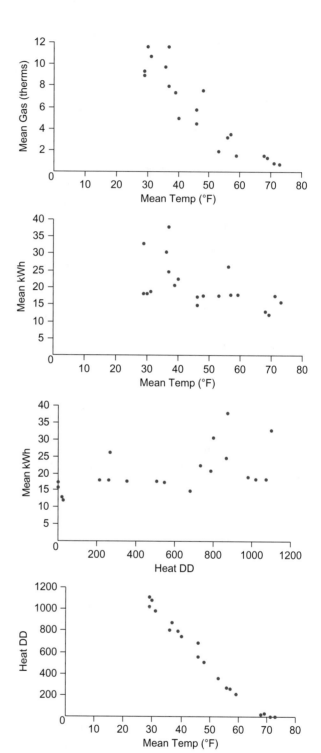

Display 13.17 Scatterplots of household energy data. [Source: R. Carver, "What Does It Take to Heat a New Room? Estimating Utility Demand in a Home," *Journal of Statistics Education*, Vol. 6, 1 (1998).]

b. What are some reasons why not all of the points fall exactly on a line with the slope in part a?

E6. Suppose you have a collection of equilateral triangles of different sizes, as in the structural steel of a bridge support. You have both the measured areas and the measured side lengths of these triangles, and you plot the area inside the triangle against the square of the length of a side.

a. What would you predict for the slope of the true regression line for this situation? Interpret the slope in this context.

b. What are some reasons why some of the points may not fall exactly on a line with the slope in part a?

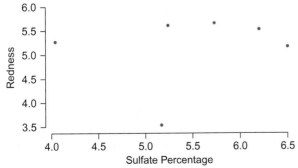

E7. Display 13.19 gives the sulfate content and redness of six soil samples from Mars.

a. Compare the plot in Display 13.19 to the one for the Mars rocks in Display 13.1 on page 624. Which value of s_{b_1} do you expect to be larger—the one for the five rocks, or the one for the six soil samples? Why?

b. Compute s_{b_1} for the soil samples, and check your conjecture from part a. (You computed s_{b_1} for the rocks in P5.)

Soil Sample	Sulfate (%)	Redness
1	4.04	5.27
2	5.17	3.94
3	5.24	5.61
4	5.73	5.66
5	6.20	5.53
6	6.50	5.17

Display 13.19 Sulfate percentage and redness of six Mars soil samples.

E8. In the Mars soil sample data of Display 13.19, one of the points appears to be an outlier. Suppose that point is removed.

a. Suppose a line is fit to the remaining five points. Do you expect the estimated *SE* of the slope to be larger for the original data or for the revised data?

b. Compute s_{b_1} using the remaining five points, and check your conjecture from part a. (You computed s_{b_1} for all six points in E7.)

E9. Example 13.1 on page 626 gives a model for the growth of children ages 8 to 13. The response variable is *height*, the predictor variable is *age*. The assumptions were that the mean height of 8-year-olds is 51 in.; that children grow, on average, 2 in. per year; and that, at any one age, the distribution of heights is approximately normal with the same standard deviation. The scatterplots of Display 13.20, on the next page, show the sort of results you would expect to get by simulating taking random samples of children of various ages. For each plot, the standard deviation used was different and for one plot it differed for different ages. Match the standard deviations described in A, B, and C with the appropriate scatterplot I, II, or III.

A. $\sigma = 3$ in.

B. $\sigma = 1$ in.

C. A different value of σ is used for different ages: age 8, $\sigma = 1$ in.; age 9, $\sigma = 2$ in.; age 10, $\sigma = 3$ in.; age 13, $\sigma = 6$ in.

E10. Refer to E9. The scatterplots in Display 13.21 (on the next page) show the sort of data you would expect to get when simulating taking random samples of children of various ages from the new theoretical models in A and B. Match A and B to the appropriate scatterplot I or II.

A. With $\sigma = 3$, assume that children grow faster and faster as they get older so that the average height at age x is given by $50 + (x - 8)^2$.

B. Keep everything the same as in part a, except this time use the values of σ from E9, description c.

E11. Each of the lists in I–V gives the values of *x* used to compute a regression equation. Order the lists from the one with the largest estimated standard error of the slope to the one with the smallest. (No computation should be necessary.)

I. $s = 5; x = 1, 1, 1, 1, 3, 3, 3, 3$

II. $s = 2; x = 1, 1, 1, 1, 2, 2, 2, 2, 3, 3, 3, 3$

III. $s = 3; x = 1, 1, 2, 2, 3, 3, 4, 4, 5, 5, 6, 6$

IV. $s = 2; x = 1, 1, 2, 2, 3, 3, 4, 4, 5, 5, 6, 6$

V. $s = 7; x = 1, 2, 2, 3$

E12. Each of the lists in I–V gives the values of *x* used to compute a regression equation. Order the lists from the one with the largest estimated standard error of the slope to the one with the smallest. (For these, you might need to do some computation.)

I. $s = 2; x = 1, 1, 1, 1, 3, 3, 3, 3$

II. $s = 1; x = 1, 1, 1, 1, 2, 2, 2, 2, 3, 3, 3, 3$

III. $s = 3; x = 1, 1, 2, 2, 3, 3, 4, 4, 5, 5, 6, 6$

IV. $s = 1; x = 1, 1, 2, 2, 3, 3, 4, 4, 5, 5, 6, 6$

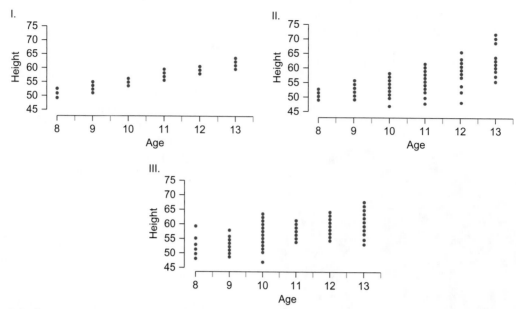

Display 13.20 Scatterplots of simulated height data for E9. The sample size is 20 for each age, so many dots are hidden behind others.

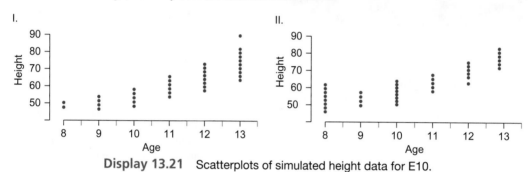

Display 13.21 Scatterplots of simulated height data for E10.

E13. You want to create a model for the Mars rocks data in Display 13.1 on page 624 using the same approach as in modeling the heights of children in Example 13.1 on page 626.

a. What is the best estimate of the true regression equation?

b. What is the best estimate of σ?

c. Write out directions for a simulation telling how to use the model to find the redness of a randomly selected rock that has the same sulfur content as Half Dome. Do you need to make any additional assumptions?

d. In what ways do you consider your model a reasonable one for this situation? In what ways is it not so reasonable?

E14. Refer to E4. You want to create a model for predicting mean gas consumption from mean temperature.

a. What is the best estimate of the true regression equation?

b. What is the best estimate of σ?

c. Write out directions for a simulation telling how to use the model to find the mean gas consumption for a randomly selected month when the mean monthly temperature is 50°F. Do you need to make any additional assumptions?

d. In what ways do you consider your model a reasonable one for the situation? In what ways is it not so reasonable?

13.2 ▶ Making Inferences About Slopes

The first part of this section presents and illustrates a significance test for a slope, and the second part gives a confidence interval. The logic here is important, but you've seen it before, and the computations follow a familiar pattern. Pay special attention to checking how well the model fits the data. This is the part of inference that separates humans from computers. Any old box of microchips can compute a P-value, but it takes experience and judgment to figure out whether the P-value really tells you anything.

The Test Statistic for a Slope

Often there is no positive or negative linear relationship between two variables. For example, the number of letters in a student's full name should not be related to the day of the year (numbered 1 to 365) on which he or she was born. Checking this out in a class of 25 students gave the data and plot shown in Display 13.22. The equation of the regression line is $\hat{y} = 16.4 + 0.0084x$. The slope of the regression line is slightly positive, but is this enough to indicate a statistically significant linear relationship?

Day of Year Born (x)	Number of Letters in Full Birth Name (y)	Day of Year Born (x)	Number of Letters in Full Birth Name (y)
196	16	108	15
78	24	231	20
48	15	77	19
315	14	160	23
8	16	70	14
189	29	259	17
131	25	311	20
310	10	114	16
354	24	210	15
55	18	131	10
178	22	10	10
323	24	252	15
323	17		

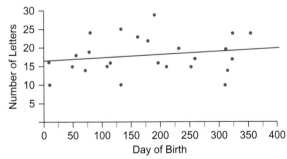

Display 13.22 Letters in name versus day of birth.

Although the scatterplot in Display 13.22 shows little association, the slope, b_1, isn't exactly equal to 0. Even when the true slope, β_1, is 0, the estimate, b_1, will usually turn out to be different from 0. In such cases, the estimated slope differs from 0 simply because cases were picked at random. If another class collected information like that in Display 13.22, the value of b_1 probably would not be 0 either and probably would be different from the value of b_1 for the class in Display 13.22.

A significance test for the slope of a regression line asks, "Is that trend real, or is it plausible that the numbers came out the way they did by chance?" The test statistic is based on how far b_1 is from 0 (or some other hypothesized value of β_1) in terms of the standard error.

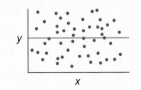

No linear relationship between x and y:

> ### Test Statistic for Testing the Significance of the Slope
>
> The **test statistic for the slope** is the difference between the slope, b_1, estimated from the sample, and the hypothesized slope, β_{1_0}, measured in standard errors:
>
> $$t = \frac{b_1 - \beta_{1_0}}{s_{b_1}}$$
>
> If a linear model is correct and the null hypothesis is true, then the test statistic has a t-distribution with $n - 2$ degrees of freedom.

Some calculators and statistical software do not give you the value of s_{b_1}. However, when testing that $\beta_1 = 0$, they will give you t and b_1, so you can compute s_{b_1} from

$$t = \frac{b_1}{s_{b1}} \quad \text{or} \quad s_{b_1} = \frac{b_1}{t}$$

Example 13.4 — Day of Birth and Length of Name

Refer to the data on day of birth and length of names in Display 13.22. You may consider these as having come from a random sample of students. Compute the test statistic for a test of whether there is a statistically significant linear relationship between the day of birth and the number of letters in a student's full name. Then give the degrees of freedom, find the P-value, and write a conclusion.

Solution
To test whether there is a statistically significant linear relationship between the day of birth and the number of letters in a student's full name, you use $\beta_{1_0} = 0$. The slope of the regression line is $b_1 = 0.0084$. From the formula for s_{b_1} on page 631 (or technology), the estimated standard error of the slope is 0.00954. Thus, the test statistic is

$$t = \frac{b_1 - \beta_{1_0}}{s_{b1}} = \frac{0.0084 - 0}{0.00954} \approx 0.88$$

With $n - 2 = 25 - 2 = 23$ degrees of freedom, the P-value from technology is 0.39 (Table B gives P-value > 0.30). It is plausible that $\beta_1 = 0$. You do not have statistically significant evidence of a linear relationship between the day of birth and the number of letters in a student's full name in the population of all students.

DISCUSSION
The Test Statistic for a Slope

D7. Using the output for the Mars rocks data in Display 13.12 on page 635, verify the value of the t-statistic. Then verify the value of df, and check that the P-value in the computer output is consistent with that found from Table B on page 761.

D8. Display 13.23 shows the combined data for the five Mars rocks and six soil samples. In relation to the t-statistics for the rocks and for the soil samples, where do you expect the value of the t-statistic for the combined data to fall? Why?

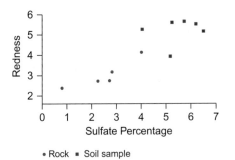

Jimmy Tran works on Rocky, an early prototype for the Mars *Rover*.

Display 13.23 *Redness* versus *sulfate percentage* for the five Mars rocks and six soil samples together.

Significance Test for a Slope

The next box describes the steps for testing the significance of a slope. Generally, you will use this test when you have bivariate data from a sample that appear to have a positive (or negative) linear association and you want to establish that this association is "real." That is, you want evidence that the nonzero slope you see didn't happen just by chance—that there actually is a true linear relationship with a nonzero slope—so knowing the value of x is helpful in predicting the value of y.

The "Check conditions" step in the box lists two methods of getting a sample. For example, suppose you want to model the heights of children from ages 8 through 13 and can afford a sample of size 600. There are two basic ways you might get your sample. The first is to put all of the children into one group and take a random sample of 600 of them. The second is to divide the children into the six age groups 8, 9, 10, 11, 12, and 13 and then take a random sample of size 100 from each age group. This is called *sampling with fixed values of x*. One advantage of using fixed values of x is that it guarantees a wide spread in the values of x. The methods of this section were developed for sampling with fixed values of x, but will work well for simple random sampling as long as the values of x in the sample cover the range of interest.

Sample with or without fixed values of *x*.

Components of a Significance Test for a Slope

1. *State the hypotheses.* The null and alternative hypotheses usually will be $H_0: \beta_1 = 0$ and $H_a: \beta_1 \neq 0$, where β_1 is the slope of the true regression line. However, the test may be one sided, and the hypothesized value, β_{1_0}, may be some constant other than 0.

2. *Check conditions.* For the test to be trustworthy, the conditional distributions of y for fixed values of x must be approximately normal, with means that lie on a line and standard deviations that are constant across all values of x. Of course, you can't check the population for this; you have to use the sample:

   - Verify that you have one of these situations:

     i. a single random sample from a bivariate population

     ii. a set of independent random samples, each taken at a fixed value of the explanatory variable, x

     iii. an experiment with a random assignment of treatments to units

   - Make a scatterplot and residual plot to check for a linear relationship.

   - Use the residual plot to verify that the residuals have approximately the same variability at all values of x.

Randomness

Linearity

Residuals of equal variability

Normality

- Make a univariate plot (dot plot, stemplot, or boxplot) of the residuals to see if it's reasonable to assume that they came from a normal distribution.

3. *Compute the value of the test statistic, find the P-value, and draw a sketch.* The test statistic is

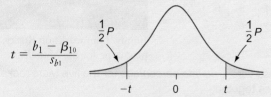

$$t = \frac{b_1 - \beta_{1_0}}{s_{b_1}}$$

Here, b_1 and s_{b_1} are computed from your sample and β_{1_0} is the hypothesized value of the slope. To find the P-value, use a calculator's or computer's t-distribution with $n - 2$ degrees of freedom, where n is the number of ordered pairs in your sample. The P-value illustrated is for a two-sided test.

4. *Write your conclusion linked to your computations and in context.* The smaller the P-value, the stronger the evidence against the null hypothesis. Reject H_0 if the P-value is less than the given value of α, typically 0.05.

Yogi: *Four* conditions! What happened to good old "line plus random variation"?

Shark: It's still there. But "variation" takes in a lot of territory. The variation about the line has to be both random and equal. *Equal* here means that the vertical spread is the same as you go from left to right across your scatterplot and that the distribution of points in each vertical slice is roughly normal.

Yogi: This is starting to sound complicated.

Shark: Not really. Sometimes a violation of the conditions will be obvious from the scatterplot. To be safe, you should look at a residual plot as well as a dot plot or boxplot of the residuals.

Yogi: That still leaves one more condition. Surely you're not going to tell me I can check randomness by looking at a plot?

Shark: No. For that condition, you need to check how the data were collected. The observations should have been selected randomly, which means either by taking a single random sample from the entire population or by taking separate random samples from the conditional distribution for selected values of *x*.

Yogi: [*Loud sigh*]

Shark: Just read the next example, and you'll see how easy it is.

Example 13.5	***Greenhouse Gas Emissions* Versus *Highway Gas Mileage***

In E3 on page 636, you investigated *greenhouse gas emissions (GGE)* versus *highway gas mileage (HMG)* for a random sample of car models. Display 13.24 shows the data and a scatterplot. The equation of the least squares regression line through these points is $\hat{y} = 18.59 - 0.362x$.

On the face of it, the relationship looks strong enough for you to conclude that the pattern is not simply the result of random sampling: In the population as a whole, there really must be a relationship between *greenhouse gas emissions* and *highway gas mileage*. You would expect a formal test to lead to the same conclusion, and it does. Carry out a significance test for the slope of the true regression line for *greenhouse gas emissions* versus *highway gas mileage*.

Model	GGE	HGM
Acura TSX	7.3	30
Buick Lucerne	10.2	22
Cadillac CTS	13.1	18
Chevrolet Malibu	7.1	33
Honda Civic	6.3	34
Dodge Challenger	11.4	22
Ford Fusion	9.6	24
Kia Forte	6.1	36
Lincoln MKS	9.6	23
Mazda 3	6.6	33
Saturn Aura	9.2	26
Scion tC	7.7	29
Toyota Camry	7.1	32
Volkswagen Eos	7.3	31
Volvo S80	10.2	22

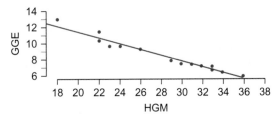

Display 13.24 *Greenhouse gas emissions* versus *highway gas mileage* for a random sample of 15 car models.

Solution

The hypotheses are $H_0: \beta_1 = 0$ and $H_a: \beta_1 \neq 0$, where β_1 is the slope of the true regression line that relates greenhouse gas emissions to highway gas mileage of a car model. The four conditions are met:

Hypotheses

- You have a single random sample from a bivariate population.
- The relationship looks reasonably linear in the scatterplot in Display 13.24.

Four conditions: random, linear, equal variance, and normal.

- If you examine the residual plot in Display 13.25, you can see that, while the relationship between *y* and *x* appears to be generally linear, the variation from the regression line appears to exhibit some curvature. That is, the values of *y* tend to show somewhat of a curved pattern as the value of *x* increases, but this is not serious enough to invalidate the regression analysis.

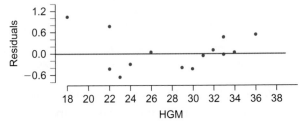

Display 13.25 Residual plot from the regression analysis of *greenhouse gas emissions* versus *highway gas mileage*.

- The residuals, plotted in Display 13.26, show a fairly symmetric distribution. It is reasonable to assume that they could have come from a normal distribution.

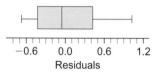

Display 13.26   Boxplot of residuals.

Using technology, the t-statistic is approximately -14.7, and the P-value is approximately 0.

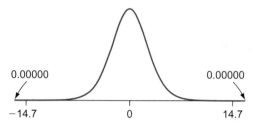

$t : 0.$

$$t = \frac{b_1 - \beta_{10}}{s_{b_1}}$$

$$= \frac{-0.362 - 0}{0.0246} = -14.7$$

With a P-value near zero, there is strong evidence against the null hypothesis. Conclude that the slope of the true regression line for *greenhouse gas emissions* versus *highway gas mileage* is different from 0. A linear model with 0 for the true slope probably would not have produced these data.

The results in the previous example may be found in using statistical software, which gives output similar to that in Display 13.27. Note that s, the standard deviation of the residuals, is equal to the square root of the mean square for residuals. The regression coefficients are given in the column labeled "Coefficient." The estimated standard error of b_1 is shown in the HGM row in the column labeled "s.e. of Coeff." The t-ratio and its corresponding P-value (prob) finish out this row.

Dependent variable is: GGE
No Selector
R squared = 94.3%   R squared (adjusted) = 93.9%
s = 0.5115 with 15 − 2 = 13 degrees of freedom

Source	Sum of Squares	df	Mean Square	F-ratio
Regression	56.3963	1	56.3963	216
Residual	3.40104	13	0.261618	

Variable	Coefficient	s.e. of Coeff	t-ratio	prob
Constant	18.5907	0.6941	26.8	≤ 0.0001
HGM	−0.361592	0.0246	−14.7	≤ 0.0001

Display 13.27   Data Desk summary of the regression of *greenhouse gas emissions* versus *highway gas mileage* for a random sample of cars.

Yogi: *F*-ratio! Aaargh! See, there is another distribution!

Shark: True, but you can ignore the *F*-ratio until the next chapter. Here, it is equivalent to the *t*-test.

As for all statistical tests, there are three ways to get a tiny P-value like the one in Display 13.27: The conditions were not met, the sample could be truly unusual, or the null hypothesis could be false.

- *Conditions not met.* The relationship looks linear, and the sample is random, But we are a bit concerned about the fact that the relationship between y and x may be curved.

- *Unusual sample?* The P-value tells just how unusual the sample would be if the null hypothesis were true. Here, less than one sample in 10,000 would give such a large value of the test statistic.

- *False H_0?* By a process of elimination, this is the most reasonable explanation. What's the bottom line? In the population as a whole, there's a positive linear relationship between *greenhouse gas emissions* versus *highway gas mileage*.

DISCUSSION
Significance Test for a Slope

D9. Refer to Example 13.4 on page 642. Enter the data into a calculator or statistical software, and learn to use it to find the test statistic and P-value.

D10. Refer to Display 13.27.
   a. What is being tested in the row that includes a t-statistic for the constant?
   b. What does the P-value for this test represent? What is the conclusion if the P-value is very small, say, 0.001?

Confidence Interval Estimation

Now that you know that the greenhouse gas emissions decrease as gas mileage increases, your next question might be "By how much?" Whenever you reject a null hypothesis that a slope is 0, it is good practice to construct a confidence interval for the slope. If the interval is extremely wide, due to large variation in the residuals and small sample size, you know that the estimate b_1 is practically useless. For example, an increase in annual income estimated to be between $50 and $10,000 for every additional year of work experience doesn't tell you much about what an individual employee's salary increase might be. It also might happen that the slope is so small as to be practically meaningless, even though a large sample size makes it "statistically significant." An exercise program that lowers blood pressure by 1 to 1.5 points for every additional 5 hours per week spent exercising will not attract many takers, even if a huge study proves that the decrease is real. Always temper statistical significance with practical significance.

> Statistical significance doesn't always mean the results are useful.

As in past chapters, a confidence interval takes the form

$$\text{statistic} \pm (\text{critical value}) \cdot (\text{estimated standard error})$$

Constructing a Confidence Interval for a Slope

1. *Check conditions.* To get a capture rate equal to the advertised rate, the conditional distributions of y for fixed values of x must be approximately normal, with means that lie on a line and standard deviations that are constant across all values of x. You need to do several things to check conditions:
   - Verify that you have one of these situations:
     i. a single random sample from a bivariate population

> Randomness

ii. a set of independent random samples, each taken at a fixed value of the explanatory variable, x

iii. an experiment with a random assignment of treatments to units

Linearity

- Make a scatterplot and residual plot to check for a linear relationship.

Residuals of equal variability

- Use the residual plot to verify that the residuals have the same variability at all values of x.

Normality

- Make a univariate plot (dot plot, stemplot, or boxplot) of the residuals to see if it's reasonable to assume that they came from a normal distribution.

2. *Do computations.* The confidence interval is

$$b_1 \pm t^* \cdot s_{b_1}$$

The value of t^* depends on the confidence level and the number of degrees of freedom, *df*, which is $n - 2$.

3. *Give interpretation in context.* For a 95% confidence interval, you are 95% confident that the slope of the true (population) regression line lies in the interval. By 95% confidence, you mean that out of every 100 such confidence intervals you construct from random samples, you expect the true value, β_1, to be in 95 of them.

| **Example 13.6** | **A Confidence Interval for the Slope in Example 13.5** |

Refer to Example 13.5 on page 646. Construct a 95% confidence interval for the slope of the line that relates greenhouse gas emissions to highway gas mileage.

Solution

Check conditions: random, linear, equal variability, and normal.

The conditions are the same as for a test of significance and were checked on page 646.

From Example 13.5, the slope is $b_1 = -0.362$ and s_{b_1} is 0.0246. with have $15 - 2$, or 13, degrees of freedom, the value of t^* from Table B on page 761 is 2.160. Thus, a 95% confidence interval estimate of the true slope is given by

Do computations.

$$b_1 \pm t^* \cdot s_{b_1} \approx -0.362 \pm 2.160(0.0246) = -0.362 \pm 0.053$$

or $(-0.415, -0.309)$.

Typically, you will use a calculator or computer to find this interval.

Give interpretation in context.

You are 95% confident that the slope of the true linear relationship between greenhouse gas emissions and highway gas mileage is between -0.415 and -0.309. That is, you are reasonably confident that the reduction in greenhouse gas emissions achieved by moving to a car that produces 1 mile per gallon greater gas mileage would be between 0.309 and 0.415 tons per year.

This result means that any true slope β_1 between -0.415 and -0.309 could have produced data with a slope like that observed in this sample as a reasonably likely outcome. The observed slope would not be a reasonably likely outcome for any value of β_1 outside the confidence interval.

DISCUSSION
Confidence Interval Estimation

D11. Construct 95% confidence intervals for the slopes of these regression lines. State which interval is narrowest and which is widest. Explain why the interval widths differ.

    a. for predicting *redness* from *sulfate percentage* for the Mars rocks (use the information in Display 13.12 on page 635)

    b. for predicting *redness* from *sulfate percentage* for the Mars soil samples (use the information in Display 13.19 on page 639)

    c. for predicting *redness* from *sulfate percentage* for the rocks and soil samples taken together (the fitted slope is 0.6268 and the estimated standard error is 0.11)

D12. Suppose a two-sided test of the null hypothesis that the slope of the true regression line is 0 produces a *P*-value of 0.02. What do you know about the 95% confidence interval estimate of the slope?

Checking Whether the Model Is Appropriate

More than any other part of inference, model checking is what makes the difference between statistical thinking and mindless number crunching. Model checking is also the hardest part of inference because often the answers are not clear-cut: A plot shows a hint of curvature at one end; a point is an outlier, but just barely; a sample is not strictly random, but there are other reasons to think that it is representative. If you are the sort of person who likes definite answers, you might have to push yourself to do justice to model checking.

In some instances, a condition will be clearly violated or clearly satisfied. In others, patterns in the data might merely raise questions without providing clear answers. In still others, there might be so little evidence that there's not much you can say. In general, you can be more confident about your judgment when you have a larger number of points than when you have fewer points, but there is no simple rule.

You should not use the techniques of this section for *time-series* data, that is, for cases that correspond to consecutive values over a period of time. In these situations, the individual observations typically aren't selected at random and so are highly dependent. Today's temperature depends on yesterday's. The unemployment rate next quarter is unlikely to be very far from the rate this quarter. If your cases have a natural order in time, chances are good that you should use special inference methods for analysis of time series rather than the methods of this chapter.

> Use other techniques for time-series data.

DISCUSSION
Checking Whether the Model Is Appropriate

D13. Four scatterplots, I–IV, along with their regression lines, residual plots, and dot plots and boxplots of the residuals, are shown in Display 13.28.

    a. What features of a plot suggest that the variation in the response is not constant but depends on *x*? Which plots show this?

    b. What features of a plot suggest that the conditional distributions of *y* given *x* aren't normal? Which plots show this?

    c. Which plot(s) contains an influential point? Imagine fitting a least squares regression line to this plot using all the data and then removing the influential point and refitting the line. How will the two lines differ?

    d. All in all, which plots raise questions about whether the data are suitable for inference about the slope?

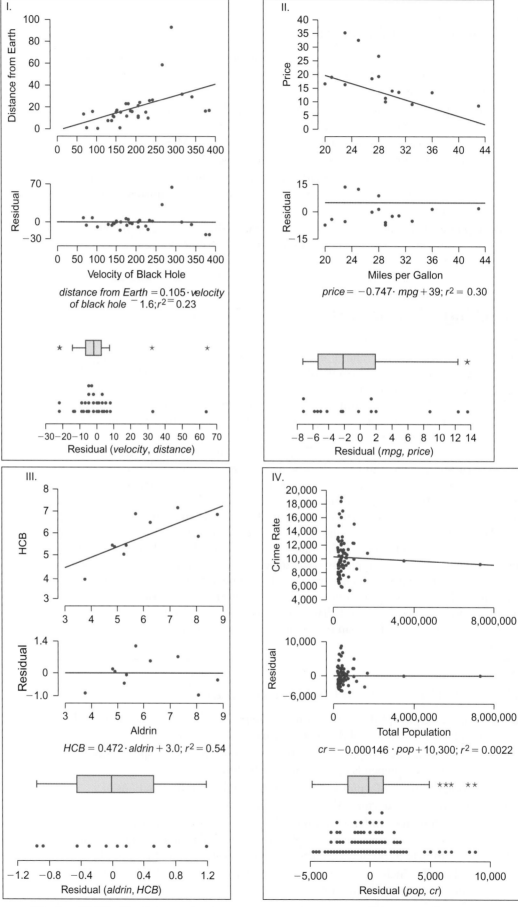

Display 13.28 Four scatterplots with their regression line, residual plot, and dot plot and boxplot of the residuals.

Summary 13.2: Making Inferences About Slopes

When you compute a regression line from a random sample, the slope is very unlikely to be exactly zero, even if there is no positive or negative linear relationship in the population. A test of significance of the slope or a confidence interval for the slope tells you whether the nonzero slope you see from the sample can reasonably be considered evidence of a nonzero slope in the population.

For inference about the slope of a regression line, there are four conditions to check:

1. You have a *random sample* taken either from the population or from the conditional distributions at fixed values of x or a *random assignment of treatments to subjects*. (Find out how the data were collected.)

2. The relationship is *linear*. (Check the scatterplot and residual plot. Sometimes nonlinearity is obvious from the original scatterplot and it is unnecessary to make a residual plot.)

3. Residuals have *equal standard deviations* across values of x. (Check the residual plot. Again, sometimes it is unnecessary to make a residual plot because it's obvious from the original scatterplot that the residuals tend to grow or shrink as x increases.)

4. Residuals are approximately *normal* at each fixed x. (Make a dot plot or boxplot of only the residuals.)

All is not lost if the last three conditions are not met, because more advanced techniques, such as transforming the variables, often can be used. But if the observations weren't selected at random from the population (or, in an experiment, if treatments weren't randomly assigned), then if you proceed with inference, you must state your conclusions very, very cautiously.

To test H_0: $\beta_1 = 0$, compute the test statistic

$$t = \frac{b_1 - 0}{s_{b_1}}$$

To find the *P*-value compare the value of the test statistic with a *t*-distribution with $n - 2$ degrees of freedom. If you cannot reject the null hypotheses, then you don't have statistically significant evidence of a linear relationship between x and y.

For a confidence interval for the slope of the true regression line, compute $b_1 \pm t^* \cdot s_{b_1}$. Again, use $n - 2$ degrees of freedom.

As a rule, first do a test to answer the question, "Is there an effect?" Then, if you reject H_0, construct a confidence interval to answer the question, "How big is the effect?"

Don't confuse statistical significance with practical importance. *Significant* means big enough to be detected with the data available. *Important* means big enough to care about.

Practice

The Test Statistic for a Slope

P8. The regression analysis in Display 13.29 is for the six soil samples from Mars (see E7 on page 639). The value of the test statistic and corresponding *P*-value for "Sulfate" are missing. Use only the information in the rest of the printout and your calculator or Table B to find these values. What is your conclusion?

Dependent variable is: Redness

R squared = 3.3%   R squared (adjusted) = − 20.9%

s = 0.7095 with 6 − 2 = 4 degrees of freedom

Source	Sum of Squares	df	Mean Square	F-ratio
Regression	0.068487	1	0.068487	0.136
Residual	2.01345	4	0.503362	

Variable	Coefficient	s.e. of Coeff	t-ratio	prob
Constant	4.46564	2.003	2.23	0.0897
Sulfate	0.133399	0.3617	—?—	—?—

Display 13.29   Regression analysis for the Mars soil samples.

This image of the terrain on Mars was taken by *Spirit*, the NASA exploration rover.

P9. Display 13.30 shows data on name length and the sum of the last four digits of the phone number for a sample of 25 people, along with the scatterplot 657 and regression line. Will the *t*-statistic for testing that the slope is not significantly different from zero be large or small in absolute value? Explain your reasoning.

Sum of Digits	Letters in Name	Sum of Digits	Letters in Name
11	12	14	10
20	14	17	12
12	16	27	12
21	12	19	8
12	9	14	13
17	13	17	8
24	10	21	10
10	7	16	14
8	13	17	12
20	14	21	10
22	13	18	13
11	9	20	15
10	11		

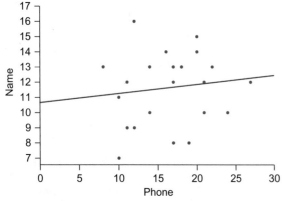

Display 13.30 Sum of phone number digits and number of letters in name.

Significance Test for a Slope

P10. For the Mars rocks and soil samples taken together, the fitted slope is 0.6268, with an estimated standard error of 0.11. Compute the *t*-statistic and find the *P*-value for testing the hypothesis that the true slope is 0. State your conclusion.

P11. *Chirping.* Display 13.31 shows the number of times a cricket chirped per second and the air temperature at the time of the chirping. Some people claim that they can tell the temperature by counting cricket chirps.

   a. Refer to the Minitab output in Display 13.32. What is the equation of the least squares regression line that could be used to predict the temperature from the chirp rate? Interpret the slope.

   b. Make a residual plot of (*chirp rate*, *residual*) and a dot plot of the residuals to check the conditions.

   c. Using the output in Display 13.32, test whether there is a statistically significant linear relationship between the chirp rate and the temperature.

Chirps per Second	Temperature (°F)
15.4	69.4
14.7	69.7
16.0	71.6
15.5	75.2
14.4	76.3
15.0	79.6
17.1	80.6
16.0	80.6
17.1	82.0
17.2	82.6
16.2	83.3
17.0	83.5
18.4	84.3
20.0	88.6
19.8	93.3

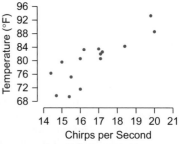

Display 13.31 Temperature and number of chirps per second by a *Nemobius fasciatus fasciatus cricket*, as measured electronically. [Source: George W. Pierce, *The Songs of Insects* (Cambridge, MA: Harvard University Press, 1949), pp. 12–21.]

The regression equation is
Temp = 25.2 + 3.29 Chirps

Predictor	Coef	Stdev	t-ratio	p
Constant	25.23	10.06	2.51	0.026
Chirps	3.2911	0.6012	5.47	0.000

s = 3.829   R-sq = 69.7%   R-sq(adj) = 67.4%

Display 13.32   Minitab regression analysis for the chirping problem.

P12. Whether head size is related to intelligence is a very controversial topic. Display 13.33 shows IQ and head circumference (in centimeters) for a sample of 20 people. (These data were explored in D21 in Chapter 3). With *head circumference* as the predictor and *IQ* as the response, is there statistically significant evidence of a linear trend?

Head Circumference (cm)	IQ
54.7	96
54.2	89
53.0	87
52.9	87
57.8	101
56.9	103
56.6	103
55.3	96
53.1	127
54.8	126
57.2	101
57.2	96
57.2	93
57.2	88
55.8	94
57.2	85
57.2	97
56.5	114
59.2	113
58.5	124

Display 13.33   IQ and head circumference. [Source: M. J. Tramo et al., "Brain Size, Head Size, and IQ in Monozygotic Twins," *Neurology*, Vol. 50 (1998), pp. 1246–1252.]

Confidence Interval Estimation

P13. Mars rocks contain a high proportion of silicon dioxide (the predominant compound in sand and glass) and much smaller amounts of titanium dioxide (similar to silicon dioxide but with titanium replacing the silicon). Display 13.34 shows computer output for the regression of *titanium dioxide percentage* versus *silicon dioxide percentage* for the five Mars rocks.

Dependent variable is: %TiO2

5 total cases

R squared = 96.1%   R squared (adjusted) = 94.8%
s = 0.0257 with 5 − 2 = 3 degrees of freedom

Variable	Coefficient	s.e. of Coeff	t-ratio	prob
Constant	2.76962	0.2217	12.5	0.0011
%SiO2	−0.033718	0.0039	−8.61	0.0033

Display 13.34   Regression of *titanium dioxide percentage* versus *silicon dioxide percentage* for the five Mars rocks.

   a. Construct and interpret a 90% confidence interval for the slope.

   b. What does the value $s = 0.0257$ tell you?

P14. *Soil samples.* Display 13.35 shows information for the six soil samples that is similar to that in P13 for the Mars rocks.

   a. Construct and interpret a 90% confidence interval for the slope.

   b. Based on your confidence intervals for the rocks in P13 and the soil samples in part a, would you say the relationship between silicon dioxide content and titanium dioxide content is the same or different for rocks and soil?

Variable	Coefficient	s.e. of Coeff	t-ratio	prob
Constant	−0.177459	2.289	−0.078	0.9419
%SiO2	0.026856	0.0461	0.582	0.5916

Display 13.35   Excerpt from regression of *titanium dioxide percentage* versus *silicon dioxide percentage* for the six soil samples.

Checking the Fit of the Model

P15. A good method of measuring the thickness of a sheet of paper is to press a large number of sheets together and measure the stack. A student who wanted to estimate the thickness of a textbook page collected the data in Display 13.36.

   a. Does it appear that the conditions for inference are met here?

   b. Estimate the slope of the regression line in a 95% confidence interval. Interpret the interval in the context of this problem.

Sheets	Thickness (mm)
50	6.0
100	11.0
150	12.5
200	17.0
250	21.0

Display 13.36   Thickness of textbook sheets.

P16. *Creature features.* The typical human's heart beats 2.21 billion times, more than any other creature listed in Display 13.37, and we also live a long time. In this problem, you will explore whether lifetime heartbeats are a good predictor of longevity or whether weight or heart rate is better.

Creature	Weight (g)	Heart Rate bpm	Longevity (yr)	Lifetime Heartbeats (billions)
Human	90,000	60	70	2.21
Cat	2,000	150	15	1.18
Small dog	2,000	100	10	0.53
Medium dog	5,000	90	15	0.71
Large dogs	8,000	75	17	0.67
Hamster	60	450	3	0.71
Chicken	1,500	275	15	2.17
Monkey	5,000	190	15	1.50
Horse	1,200,000	44	40	0.93
Cow	800,000	65	22	0.75
Pig	150,000	70	25	0.92
Rabbit	1,000	205	9	0.97
Elephant	5,000,000	30	70	1.10
Giraffe	900,000	65	20	0.68
Large whale	120,000,000	20	80	0.84

Display 13.37 Characteristics of 15 creatures. [Source: *www.sjsu.edu/faculty/watkins/longevity.htm.*]

a. Interpret the *P*-value in a significance test of the slope when using lifetime heartbeats as a predictor of longevity. Now check the conditions. Do you trust the result of the significance test? Explain why or why not.

b. Interpret the *P*-value in a significance test of the slope when using weight as a predictor of longevitys?

Now check the conditions. Do you trust the result of the significance test? Explain why or why not.

c. What is the *P*-value in a significance test of the slope when using heart rate as a predictor of longevity? Now check the conditions. Do you trust the result of the significance test? Explain why or why not.

Exercises

E15. Display 13.38 shows the *greenhouse gas emissions (GGE)* and *the passenger compartment sizes (PCS)* for the random sample of car models in E3. The scatterplot and output for the regression of GGE on PCS are shown in Display 13.39.

Model	GGE	PCS
Acura TSX	7.3	95
Buick Lucerne	10.2	108
Cadillac CTS	13.1	100
Chevrolet Malibu	7.1	95
Honda Civic	6.3	91
Dodge Challenger	11.4	91
Ford Fusion	9.6	101
Kia Forte	6.1	97
Lincoln MKS	9.6	106
Mazda 3	6.6	94
Saturn Aura	9.2	98
Scion tC	7.7	85
Toyota Camry	7.1	101
Volkswagen Eos	7.3	77
Volvo S80	10.2	98

Display 13.38 *Greenhouse gas emissions (GGE) and the passenger compartment sizes (PCS).*

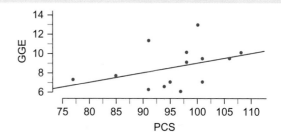

The regression equation is
GGE = −0.97 + 0.0998 PCS

Predictor	Coef	Stdev	t-ratio	p
Constant	−0.973	6.521	−0.15	0.884
PCS	0.09979	0.06786	1.47	0.165

s = 1.986   R-sq = 14.3%   R-sq(adj) = 7.7%

Display 13.39 Regression analysis for *greenhouse gas emissions* versus *passenger compartment sizes*.

a. Locate the estimated standard error of b_1.

b. Is there statistically significant evidence that the slope of the true regression line is different from 0? Use $\alpha = 0.05$.

c. Find a 90% confidence interval estimate of the slope of the true regression line. Interpret the result in the context of the variables.

E16. The scatterplot and output in Display 13.40 show the regression of *mean monthly gas usage* (in therms) versus *mean monthly temperature* for a single-family residence over a sample of months. (See the table on page 638 for the complete set of data.)

   a. Locate the estimated standard error of the slope.

   b. Is there sufficient evidence to say that the slope of the true regression line is different from 0? Use $\alpha = 0.05$.

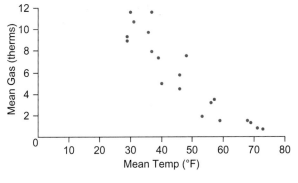

Predictor	Coef	Stdev	t-ratio	p
Constant	16.988	1.196	14.20	0.000
Mean Temp	−0.23643	0.02401	−9.85	0.000

s = 1.548   R-sq = 84.3%   R-sq(adj) = 83.5%

Display 13.40 Scatterplot and printout for *mean monthly gas usage* versus *mean monthly temperature*.

   c. Find a 90% confidence interval estimate of the slope of the true regression line. Interpret the result in the context of the variables.

E17. Display 13.41 shows the *greenhouse gas emissions (GGE)* and *number of cylinders (CY)* for the random sample of car models in E3. Is there statistically significant evidence that the mean *GGE* is a linear function of the *CY*? Complete all four steps of the test of significance.

Model	GGE	CY
Acura TSX	7.3	4
Buick Lucerne	10.2	8
Cadillac CTS	13.1	8
Chevrolet Malibu	7.1	4
Honda Civic	6.3	4
Dodge Challenger	11.4	8
Ford Fusion	9.6	6
Kia Forte	6.1	4
Lincoln MKS	9.6	6
Mazda 3	6.6	4
Saturn Aura	9.2	6
Scion tC	7.7	4
Toyota Camry	7.1	4
Volkswagen Eos	7.3	4
Volvo S80	10.2	8

Display 13.41 *Greenhouse gas emissions (GGE) and number of cylinders (CY).*

E18. Refer to the data in the table on page 638 on mean monthly electricity usage (in kWh) and mean monthly temperature (in degrees Fahrenheit) for a single-family residence over a sample of months. Is there statistically significant evidence that the mean monthly electricity usage is a linear function of the mean monthly temperature?

   a. Do all four steps of the test of significance.

   b. Would you say that this is strong evidence of a linear relationship or evidence of a strong linear relationship? Explain.

E19. Refer to the data on cricket chirp rate in P11.

   a. Construct and interpret a 95% confidence interval for the slope of the true regression line

      i. for predicting the temperature from the chirp rate.

      ii. for predicting the chirp rate from the temperature.

   b. Explain why the interval widths in part a are not the same.

E20. Arsenic is an odorless and tasteless strong poison that is sometimes found in groundwater, either from natural sources or from agricultural and industrial runoff. The Environmental Protection Agency has set a standard for drinking water of 0.010 parts per million (ppm). [Source: *www.epa.gov/safewater/arsenic/index.html*.] Display 13.42 shows the arsenic concentrations in the toenails of a sample of 21 people who used water from their private wells and the arsenic concentration in their well water.

Arsenic

   a. Find and interpret a 95% confidence interval estimate of the slope of the true regression line

      i. for predicting arsenic in toenails from arsenic in well water.

      ii. for predicting arsenic in well water from arsenic in toenails.

   b. Explain why the interval widths in part a are not the same.

   c. Choose the data point that you think is exerting the strongest influence on the results. What happens to the answers in part a if this data point is removed?

Age	Arsenic in Water (ppm)	Arsenic in Toenails (ppm)
44	0.00087	0.119
45	0.00021	0.118
44	0	0.099
66	0.00115	0.118
37	0	0.277
45	0	0.358
47	0.00013	0.08
38	0.00069	0.158
41	0.00039	0.31
49	0	0.105
72	0	0.073
45	0.046	0.832
53	0.0194	0.517
86	0.137	2.252
8	0.0214	0.851
32	0.0175	0.269
44	0.0764	0.433
63	0	0.141
42	0.0165	0.275
62	0.00012	0.135
36	0.0041	0.175

Display 13.42 Arsenic in drinking water and toenails (ppm). [Source: M. R. Karagas et. al., "Toenail Samples as an Indicator of Drinking Water Arsenic Exposure," *Cancer Epidemiology, Biomarkers and Prevention*, Vol. 5 (1996), pp. 849–852.]

E21. Refer to the *height* versus *arm span* data in Display 13.14 on page 636. Is there statistically significant evidence in the data to refute Leonardo's claim that, on average, height is equal to arm span?

E22. When traveling by air, would you rather arrive on time or be relatively sure that your bags would arrive at the same time you do? Or perhaps there is no relationship between these two variables. Use the data in Display 13.43 to test whether there is evidence of a linear relationship that cannot reasonably be explained by chance between the percentage of on-time arrivals and the rate of mishandled bags for these major airlines.

E23. Oak trees continue to grow in girth until they die. They are said to add the same amount of cross-sectional area each year. [Source: *www.forestry.gov.uk/PDF/fcin12.pdf/$FILE/fcin12.pdf.*] Display 13.44 gives the cross-sectional areas (in square inches at chest height) of a sample of oak trees, along with the ages of the trees (in years).

a. Which conditions for inference might be violated a bit by these data?

Airline	Mishandled Bag Reports per 1000 Passengers	Percent On-Time Arrivals
AirTran Airways	1.42	77.1
Hawaiian Airlines	1.89	75.3
Northwest Airlines	2.12	82.8
JetBlue Airways	2.27	71.7
Frontier Airlines	2.48	78.4
US Airways	2.88	79.9
Continental Airlines	3.06	70.7
Southwest Airlines	3.24	84.5
Alaska Airlines	3.24	83.8
United Airlines	3.35	81.2
Mesa Airlines	3.92	81.1
Express Jet Airlines	4.28	68.4
American Airlines	4.65	74.7
Pinnacle Airlines	5.04	85.5
Delta Airlines	5.05	76.2
Comair	5.07	68.2
SkyWest Airlines	5.28	85.7
American Eagle Airlines	8.53	75.8
Atlantic Southeast Airlines	9.17	67.8

Display 13.43 Number of mishandled bags and percentage of on-time arrivals. [Source: U.S. Department of Transportation, *Air Travel Consumer Report, 2009, http://airconsumer.dot.gov/reports/atcr09.htm.*]

Age (yrs)	Area (sq in.)	Age (yrs)	Area (sq in.)
4	0.503	23	17.349
5	0.503	25	33.183
8	0.785	28	28.274
8	3.142	29	15.904
8	7.069	30	28.274
10	9.621	30	38.485
12	18.857	33	50.265
13	9.621	34	33.183
14	4.909	35	38.485
16	15.904	38	19.635
18	16.619	38	38.485
20	23.758	40	44.179
22	26.421	42	44.179

Display 13.44 Oak yree data. [Source: Herman H. Chapman and Dwight B. Demeritt, *Elements of Forest Mensuration*, 2nd ed. (J.B. Lyon, 1936).]

b. Estimate the slope of the regression line used to predict area from age in a 90% confidence interval. Interpret your answer in the context of the growth of an oak tree.

E24. The longer you stay in school the more you eventually earn! True? Display 13.45 provides data on educational attainment, median weekly earnings and unemployment rates for the respective educational groups.

Education Attained	Years of Schooling (approximate)	Median Weekly Earnings ($)	Unemploy-ment Rate (%)
Professional degree	20	1,474	1.1
Doctoral degree	20	1,441	1.4
Master's degree	18	1,140	1.7
Bachelor's degree	16	962	2.3
Associate degree	14	721	3.0
Some college, no degree	13	674	3.9
High school graduate	12	595	4.3
Less than a high school diploma	10	419	6.8

Display 13.45 Educational attainment, earnings, and employment. Data are 2006 annual averages for persons age 25 and over. Earnings are for full-time wage and salary workers. [Source: Bureau of Labor Statistics.]

a. Test to see if there is a statistically significant positive linear trend in the relationship between years of schooling and earnings.

b. Test to see if there is a statistically significant negative linear trend between years of schooling and unemployment.

c. Do any of the conditions for inference appear to be suspect?

E25. To test the potential effectiveness of laetisaric acid in controlling fungal diseases in crop plants, various concentrations of the acid were applied to Petri dishes containing the fungus *Pythium ultimum*. After 24 hours, the average radius of the fungus colonies at each concentration was calculated. The data are shown in Display 13.46. Find the linear regression equation and interpret the slope. Then determine whether the slope is statistically significant.

E26. Display 13.47 shows the selling price, the area, the number of bedrooms, and the number of bathrooms for a random sample of 20 previously owned homes resold in Boulder, Colorado. Is the slope of the least squares line for predicting *price* from *area* statistically significant? If so, interpret the slope.

Pythium ultimum damage to corn seedling.

Acid Concentration (μg/ml)	Average Colony Radius (mm)
0	33.3
0	31.0
3	29.8
3	27.8
6	28.0
6	29.0
10	25.5
10	23.8
20	18.3
20	15.5
30	11.7
30	10.0

Display 13.46 Acid concentration and average radius of fungus colonies. [Source: Myra L. Samuels and Jeffrey A. Witmer, *Statistics for the Life Sciences*, 3rd ed. (Upper Saddle River, N: Prentice Hall, 2003), p. 538. Original source: W. S. Bowers, H. C. Hoch, P. H. Evans, and M. Katayama, "Thallophytic Allelopathy: Isolation and Identification of Laetisaric Acid," *Science*, Vol. 232 (1986), pp. 105–106.]

E27. Refer to E26 and Display 13.47. Fit a least squares regression equation to *price* as a function of *number of bedrooms*. Interpret the slope of the line. Do you see any weaknesses in this analysis?

E28. Refer to E26 and Display 13.47. Fit a least squares regression equation to *price* as a function of *number of bathrooms*. Interpret the slope of the line. Do you see any weaknesses in this analysis?

Sale Price ($)	Area (sq ft)	Number of Bedrooms	Number of Bathrooms
167,000	771	2	1
160,000	956	1	1
260,000	1050	3	1
160,000	1057	2	1
229,000	1327	3	1
309,000	1402	2	1
260,000	1404	2	1
270,000	1418	2	1
336,000	1610	3	1
347,000	1729	3	1
487,000	1861	3	1
525,000	2080	4	2
450,000	2197	3	1
260,000	2269	2	1
486,000	2506	3	1
305,000	2812	4	1
700,000	4412	4	3
753,000	5266	4	2
1,550,000	5511	4	2
1,650,000	6194	4	3

Display 13.47 House selling prices, square footage, and number of bedrooms and bathrooms, late 2008 and early 2009. [Source: *www.homes.com.*]

▶ Chapter Summary

When two variables are both quantitative, you can display their relationship in a scatterplot, and you might be able to summarize that relationship by fitting a line to the data. Although many relationships are more complicated than this, linear relationships occur often enough that many questions can be recast as questions about the slope of a fitted line.

That is old news. What is new in this chapter is that you can use the *t*-statistic to test whether the slope of the true regression line is 0. If you can reject that possibility, your regression line will be useful in predicting *y* given *x*. If you can't, then it's plausible that the positive or negative trend you see is due solely to the chance variation that always results when you have only a sample.

You also have learned to find a confidence interval for the slope of the true regression line, which helps you decide not only whether the linear relationship is statistically significant but whether it has any practical significance. One way to put the methods of this chapter into a larger picture is to remind yourself of the inference methods you've learned so far: for one proportion and for the difference of two proportions (Chapters 8 and 9), for one mean and for the difference of two means (Chapters 10 and 11), for the relationship between two categorical variables (Chapter 12), and now for the relationship between two quantitative variables.

Review Exercises

E29. Study the scatterplots in Display 13.48.

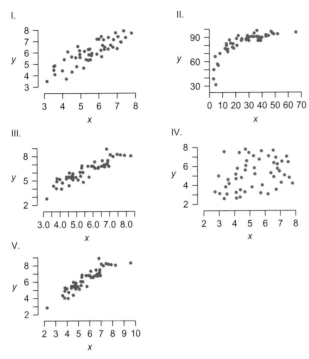

Display 13.48 Five scatterplots.

a. In which scatterplots is it reasonable to model the relationship between *y* and *x* with a straight line?

b. If you fit a line through each scatterplot by the method of least squares, which plot will give a line with slope closest to 0?

c. Which plot shows a correlation coefficient closest to 1?

E30. A class of 25 students performed the following activity:

1. Put at least 50 regular-sized dice in a bag.

2. Reach into the bag with your dominant hand and grab as many dice as you can. Turn your hand over (in the bag) so that any you haven't totally grabbed will fall back into the bag. Count the number of dice that you could grab.

3. Measure the width of your dominant hand with fingers spread out as far as you can by placing your hand on a ruler. Measure to the nearest tenth of a centimeter.

Their data appear in Display 13.49. Is there evidence of a statistically significant linear relationship that allows the prediction of number of dice in a handful from the width of the hand?

Hand Width (cm)	Number of Dice	Hand Width (cm)	Number of Dice
22.5	28	23.4	33
23.0	33	22.0	16
24.1	31	21.0	21
18.0	18	21.6	36
19.5	25	19.5	23
17.0	11	23.0	21
24.0	24	19.0	17
21.0	25	21.0	27
19.0	18	21.5	23
21.4	25	20.0	18
23.7	32	19.0	19
22.1	23	23.0	23
22.6	26		

Display 13.49 Hand width and number of dice in a handful.

E31. *More on pesticides in the Wolf River.* Refer to Example 11.1 on page 511. Display 13.50 shows four scatterplots of HCB concentration versus aldrin concentration. HCB (hexachlorobenzene) is a toxic pesticide and fungicide. The four scatterplots are for the measurements taken on the bottom, at mid-depth, at the surface, and for all three locations together. Based on the plots, which depth do you expect to give the narrowest confidence interval for the slope? The widest? Give your reasoning.

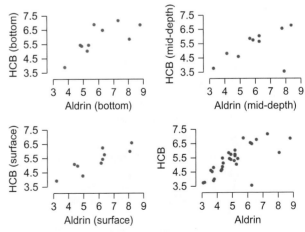

Display 13.50 Four scatterplots of HCB concentration versus aldrin concentration.

E32. "If I change to a brand of pizza with lower fat, will I also reduce the number of calories?" Display 13.51 shows the output for a significance test of calories versus fat per 5-oz serving for 17 popular brands of pizza.

a. Use Display 13.51 to estimate, in a 95% confidence interval, the reduction in calories you can expect per 1-gram decrease in the fat content of the pizza.

b. Use your answer to part a to estimate, in a 95% confidence interval, the reduction in calories you can expect per 5-gram decrease in the fat content of the pizza.

The regression equation is

Calories = 241 + 7.26 Fat

Predictor	Coef	Stdev	t-ratio	P
Constant	240.55	14.88	16.17	0.000
Fat	7.263	1.064	6.82	0.000

s = 14.36   R-Sq = 75.6%   R-Sq(adj) = 74.0%

Display 13.51 Regression analysis of *calories* on *fat* for 17 pizza brands.

E33. Espionage played a critical role in the American Civil War. One of the more interesting examples concerns the role of detective Allan Pinkerton in using spying methods to estimate the number of troop regiments in Virginia from the various states. Display 13.52 shows his estimates for September 1861, along with the actual number as discovered after the war.

a. Plot the data, with Pinkerton's estimate as the explanatory variable. Does the plot have a definite linear trend? If so, fit a least squares regression equation to the data.

b. Although these data cover all states with Confederate regiments, inference might still be meaningful in deciding if a slope of this size could (reasonably) have happened merely by chance. Conduct a test to see. What is your conclusion?

Allan Pinkerton, President Lincoln, and Major General McClernand, Antietam, Maryland, 1862.

State	Actual Number of Regiments	Pinkerton's Estimate
Alabama	11	18
Arkansas	2	2
Florida	1	2
Georgia	22	18
Kentucky	1	4
Louisiana	11	18
Maryland	1	4
Mississippi	10	12
North Carolina	15	14
South Carolina	9	18
Tennessee	7	8
Texas	3	3
Virginia	50	45

Display 13.52 Actual and estimated numbers of Confederate regiments. [Source: Chris Olsen, "Was Pinkerton Right?" *STATS* (Winter 2000), p. 24.]

c. Interpret the slope of the least squares line in part a. Estimate plausible values for the "true" slope in a 95% confidence interval. If Pinkerton was on target with his estimates, what value would you get for the slope of the line? Is this value in the confidence interval?

d. Virginia contributed by far the most troops to the Confederate war effort. What influence does Virginia have on the analysis in parts a–c?

E34. How are film ratings associated with film lengths, and are either of these variables associated with the year of the film's release? The data in Display 13.53 are a random sample of 20 films (TV films excluded) taken from Netflix. Each movie is rated from one to five stars by its viewers, with five being excellent. The reported rating is a composite score of all viewer ratings.

Because data of this type are highly variable and you are going to be looking for general trends, use a 10% level of significance on all statistical tests.

a. Is there a statistically significant linear relationship between year and length? Remember to do all parts of the analysis, including the construction of appropriate plots.

b. Is there a statistically significant relationship between length and rating? Again, do all parts of the analysis.

c. Given the results from parts a and b, what do you think the relationship between year and rating will be? Is this, in fact, the case? Make a plot and conduct a significance test of a slope as part of your answer.

d. Display 13.54 shows a plot of ratings by year, separated by short versus long movies, with *short* being defined as a running length of 90 minutes or less. (In the plot, a square represents "short.") How does the trend in ratings across the years for the short films compare to the trend for long films? Use this information to develop a partial explanation for the result in part c.

Title	Year	Length (min)	Rating (stars)
Car Wash	1976	97	3.4
Jason's Lyric	1994	120	3.6
Mad Love	1995	96	3.1
It Came from Outer Space	1953	81	3.4
And the Ship Sails On	1984	127	3.5
Action in the North Atlantic	1943	126	3.7
A Ticklish Affair	1963	89	3.4
Four Jills in a Jeep	1944	89	3.1
Hitler—Dead or Alive	1943	70	2.1
City Lights	1931	87	4.1
Galileo	1973	138	2.7
Heaven's Prisoners	1996	132	3.2
Raising the Heights	1998	88	2.5
Dancer in the Dark	2000	141	3.7
Waking Up in Reno	2002	90	3.0
Harry Potter and the Order of the Phoenix	2007	138	4.2
What Happens in Vegas	2008	101	3.6
The Queen	2006	103	3.7
Two Lovers	2008	108	2.9
Push	2009	111	3.4

Display 13.53 Film lengths and ratings.

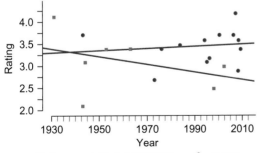

• Rating = −1.18 + 0.00232 Year; r^2 = 0.013
■ Rating = 20.06 − 0.00865 Year; r^2 = 0.14

Short ■   Long •

Display 13.54 Ratings by year, with regression lines for short and long movies.

E35. Display 13.55 shows the data on the number of police officers from a random sample of the 50 U.S. states. The explanatory variables are the amount of money it takes to keep the given number of police officers on the job for one year and population of the state.

a. Fit a regression equation to predict the number of officers from the yearly expenditure. How many new police officers would you expect could be added for an increase of $1 million in expenditure?

b. Does removal of the most influential point have much of an effect on your answer to part a?

c. Fit a regression equation to predict number of officers from the population. If one state has 1 million more people than another state, how many more police officers would the first state be expected to have?

State	Officers (thousands)	Yearly Expenditure ($ millions)	Population (millions)
California	96.9	7663	34
Colorado	12	753	4.3
Florida	55.2	3371	16
Illinois	44.1	2718	12.4
Iowa	7.3	346	2.9
Louisiana	16.1	635	4.5
Maine	3.1	118	1.3
Mississippi	8.6	337	2.8
New Jersey	33.4	1829	8.4
Tennessee	18.1	828	5.7
Texas	58.9	2866	20.9
Virginia	18.8	965	7.1
Washington	14.1	854	5.9

Display 13.55 Number of police officers, expenditures, and population for a sample of U.S. states. [Source: U.S. Census Bureau, *Statistical Abstract of the United States, 2004–2005.*]

d. Does removing the most influential point have much of an effect on your answer to part c?

E36. How do the violent crime rates relate to the property crime rates in U.S. cities?

a. Find a good-fitting model relating the violent crime rate to the property crime rate for the random sample of cities in Display 13.56.

b. Construct and interpret a 95% confidence interval estimate of the slope of the true regression line.

E37. In a bivariate population, what is the relationship between the standard deviation of the responses and the standard deviation of the residuals?

City	Violent Crime Rate	Property Crime Rate
New York, NY	734	2183
Philadelphia, PA	1378	4175
San Antonio, TX	598	6844
Honolulu, HI	288	5336
Columbus, OH	856	7758
Memphis, TN	1577	8494
Boston, MA	1216	4726
Denver, CO	624	5136
Washington, DC	1569	5606
Oklahoma City, OK	890	9123
Albuquerque, NM	947	6249
Sacramento, CA	778	6427
Atlanta, GA	1970	8869
Miami, FL	1875	6909
Minneapolis, MN	1193	5390
Wichita, KS	610	5382

Display 13.56 Violent crimes and property crimes (per 100,000 population) in a random sample of cities. [Source: U.S. Census Bureau, *Statistical Abstract of the United States, 2006,* Table 296.]

C1. A statistics exam has two parts, free response and multiple choice. A regression equation for predicting the score, f, on the free response part, from the score, m, on the multiple choice part, is $f = 50 + 0.25m$. This equation was based on the scores of 17 students. The standard deviation of their multiple choice scores was 30, the standard deviation of their free response scores was 16, and the sum of the squared residuals was 2940. What is the estimated standard error of the slope?

(A) 0.117   **(B)** 0.133   **(C)** 0.219

(D) 0.467   **(E)** 3.5

C2. Which of the following is not an important condition to check before constructing a confidence interval for the slope of the true regression line?

(A) You have a random sample.

(B) The points fall in an elliptical cloud.

(C) The residuals for small values of x have about the same variability as the residuals for large values of x.

(D) The sum of the squared residuals is small.

(E) The residuals are approximately normally distributed.

C3. Data on shoe size and height were gathered for 82 randomly selected women. Part of a regression analysis follows. Which of the following is the best interpretation of the P-value for shoe size?

Predictor	Coefficient	Standard Error	t-Statistic	P-value
Constant	55.4174	1.1681	47.443	
Shoe size	1.2116	0.1438	8.425	0.0000

$s = 1.87862$

(A) An error has been made in the data collection or analysis.

(B) There is statistically significant evidence of a nonzero slope in the true linear relationship between shoe size and height.

(C) There isn't statistically significant evidence of a nonzero slope in the true linear relationship between shoe size and height.

(D) A confidence interval for the slope would have a width of 0.

(E) A confidence interval for the slope would include 0.

C4. Refer to C3. What is the best, explanation of what is measured by $s = 1.87862$?

(A) the variability in the slope from sample to sample

(B) the variability in the y-intercept from sample to sample

(C) the variability in the shoe sizes

(D) the variability in the heights

(E) the variability in the residuals

C5. Refer to C3. What is the best explanation of what is measured by standard error = 0.1438?

(A) the variability in the slope from sample to sample

(B) the variability in the y-intercept from sample to sample

(C) the variability in the shoe sizes

(D) the variability in the heights

(E) the variability in the residuals

C6. Refer to C3. Which of the following is the appropriate computation for a 95% confidence interval for the slope?

(A) $1.2116 \pm 1.990 \cdot 1.87862$

(B) $1.2116 \pm 1.990 \cdot 0.1438$

(C) $1.2116 \pm 1.96 \cdot \frac{0.1438}{\sqrt{80}}$

(D) $1.2116 \pm 1.990 \cdot \frac{0.1438}{\sqrt{80}}$

(E) $1.2116 \pm 11.990 \cdot \frac{1.87862}{\sqrt{80}}$

C7. In an attempt to predict adult heights, researchers randomly selected men and collected their heights at age 2 and their adult heights, and then computed a least squares regression equation, *adult height* = $1.9 \cdot$ *height at age 2* + 3.5. A 95% confidence interval for the slope was given as (1.8, 2.0). Which of the following is the best interpretation of this confidence interval?

(A) If the researchers took 100 more random samples, they would expect 95 of the regression equations to have a slope between 1.8 and 2.0.

(B) 95% of 2-year-old boys have heights between 1.8 and 2.0.

(C) The researchers are pretty sure that if they studied all men, the slope of the true regression line would be between 1.8 and 2.0.

(D) If a 2-year-old boy's height is known, there's a 95% chance that his adult height will be between 1.8 and 2.0 times his current height.

(E) In predicting the height of an adult man using his height at age 2, 95% of the errors will be between 1.8 and 2.0 in.

C8. A friend is doing a regression analysis to predict the mass of a small bag of french fries given the number of fries in the bag. He finds a linear regression equation and then makes the residual plot in Display 13.57. He asks your advice about whether to proceed with a test of significance of the slope. Which is the best advice you could give?

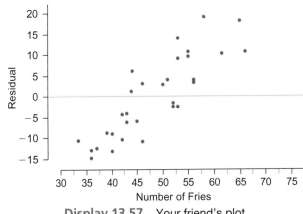

Display 13.57 Your friend's plot.

663

Ⓐ Something's wrong here. This can't be a residual plot.

Ⓑ The relationship is linear. It's okay to proceed with the significance test.

Ⓒ The relationship doesn't have constant variability across all values of *x*. It's not okay to proceed.

Ⓓ Try a transformation before proceeding.

Ⓔ It's clear even without a significance test that the true slope isn't 0.

Investigation

C9. Do animals make optimal decisions? Professor Tim Penning of Hope College studied the strategy his dog, Elvis, uses in retrieving a ball thrown into the water of Lake Michigan. Looking at the diagram in Display 13.58, suppose Tim and Elvis stand on the edge of the lake at *A* and Tim throws the ball to *B* in the lake. Elvis could jump into the water immediately at *A* and swim to the ball, but he seems to know that he can run faster than he can swim. He could run all the way to *C* and then swim the perpendicular distance to the ball, but that is not the most time-efficient strategy either. What Elvis actually does is run to a point *D* and then swim diagonally to the ball at *B*. But does he determine *D* so as to minimize the time it takes him to get to the ball?

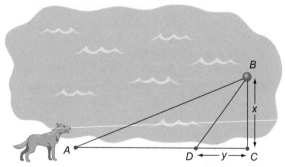

Display 13.58 Diagram of paths Elvis might follow to get from *A* to *B*.

Methods of calculus can be used to show that, for Elvis's running speed of about 6.40 m/s and swimming speed of about 0.91 m/s, the time is minimized when *y* is related to *x* by the formula

$$y = 0.144x$$

Display 13.59 shows the measurements *x* and *y* (in meters) for 35 ball-tossing experiences with Elvis.

a. Make a scatterplot and describe the pattern in the scatterplot.

b. Assuming this to be a random sample of Elvis's work, construct a 90% confidence interval estimate of the slope of the true regression line relating

distances *x* and *y*. Does it look like Elvis is making an optimal decision?

x	y	x	y	x	y	x	y
10.5	2.0	6.6	1.0	13.5	1.8	12.5	1.5
7.2	1.0	14.0	2.6	14.2	1.9	15.3	2.3
10.3	1.8	13.4	1.5	14.2	2.5	11.8	2.2
11.7	1.5	6.5	1.0	10.9	2.2	7.5	1.4
12.2	2.3	11.8	2.4	11.2	1.3	11.5	2.1
19.2	4.2	4.7	0.9	15.0	3.8	12.7	2.3
11.4	1.3	11.6	2.2	14.5	1.9	6.6	0.8
17.0	2.1	11.5	1.8	6.0	0.9	15.3	3.3
15.6	3.9	9.2	1.7	14.5	2.0		

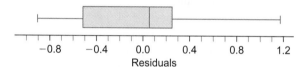

Residuals

Display 13.59 Data from Elvis. Measurements are in meters. [Source: *College Mathematics Journal* 34 (May 2003).]

c. Assuming this to be a random sample of Elvis's work, do a test of the hypothesis, at the 0.05 significance level, that the slope of the true regression line relating distances *x* and *y* is the optimal 0.144. Now does it look like Elvis is making an optimal decision?

d. Choose the point that you think is most influential on the slope of the regression line and the correlation. Describe what will happen to the slope of the regression line and the correlation if it is removed and the data reanalyzed.

e. The regression analysis in Display 13.60 shows that a regression model fit to the data in Display 13.59 does not have a *y*-intercept of 0, as the optimal model does. Can the hypothesis that $\beta_0 = 0$ be rejected? Explain what will happen to the slope of the regression line shown in the plot if you force the line to go through the origin.

The regression equation is

$y = -0.328 + 0.196x$

Predictor	Coef	Stdev	t-ratio	P
Constant	-0.3277	0.3207	-1.02	0.314
x	0.19647	0.02625	7.49	0.000

$s = 0.5162$ R-Sq = 62.9%   R-Sq(adj) = 61.8%

Display 13.60 Analysis of Elvis's data.

Chapter 14
One-Way Analysis of Variance

To see better in bright sunlight, does it matter if an athlete uses eye-black grease, antiglare stickers, or petroleum jelly on the cheekbones? Analysis of variance (ANOVA) can help you decide if the difference in the means is statistically significant.

In Chapter 11, you learned how to do a significance test to compare the means of two groups. However, several of the situations you saw actually had more than two groups. For example, in the walking babies experiment on page 520, there were four treatment groups. The Michelson measurements on the speed of light in E8 on page 528 were taken in five different groups. In this chapter, the two-sample *t*-test will be generalized to situations like these in which the means of more than two groups are to be compared.

The comparison of two means was handled quite efficiently by looking at the difference between the sample means and by comparing that difference to its standard error, but three or more means cannot be compared in this simple way. To make comparisons among multiple means, a new statistic will have to be developed. It is based on the *t*-statistic and so the fundamental concepts will remain the same. The basic question for surveys remains, "Is the result a reasonably likely outcome from random sampling if there is no difference in the population means?" The basic question for experiments remains, "Is the result a reasonably likely outcome from random assignment of treatments if there is no difference in the effect of the treatments?"

In this chapter you will learn

▷ the terminology of analysis of variance

▷ to perform and interpret a one-way analysis of variance for comparing more than two means acquired from independent random samples or from an experiment with a completely randomized design

14.1 ▷ A New Look at the Two-Sample *t*-Test

Your goal in this chapter is to learn how to test whether it is plausible that several samples all come from populations with the same mean. The statistical method you will learn, called **analysis of variance**, or **ANOVA** for short, comes from extending the logic of the two-sample *t*-test to apply to more than two samples. In this first section, you'll review the two-sample *t*-test and learn about the changes that make it possible to extend the *t*-test to more than two samples.

Here is a quick comparison of the *t*-tests you have already seen and the new method, analysis of variance, which is the topic of this chapter.

Purpose

t-test: Compare the means from two samples to see whether they are so far apart that the observed difference cannot reasonably be attributed to sampling variability alone ($H_0: \mu_1 = \mu_2$).

ANOVA: Compare the means from two *or more* samples to see whether they are so far apart that the observed differences cannot all reasonably be attributed to sampling variability alone ($H_0: \mu_1 = \mu_2 = \cdots = \mu_k$).

Method

t-test: Compute a ratio.

$$t = \frac{\bar{x}_1 - \bar{x}_2}{SE_{(\bar{x}_1 - \bar{x}_2)}}$$

Large values of this ratio lead to small *P*-values. If the *P*-value is small enough, reject the null hypothesis and conclude that the population means are not equal.

ANOVA: Compute a ratio.

$$F = \frac{\textit{variability between the sample means}}{\textit{variability within the samples}}$$

Large values of this ratio lead to small *P*-values. If the *P*-value is small enough, reject the null hypothesis and conclude that the population means are not all equal.

As you can see, the *t*-test and ANOVA are a lot alike. The most important difference is that the *t*-test works for only two samples at a time while ANOVA can handle more than two samples. Another difference is that the *t*-test compares means *directly*, by subtracting one sample mean from the other, while ANOVA compares means *indirectly*, by comparing the sample means to an overall mean. Finally, on the *t*-test can be one-sided.

From the *t*-Test to ANOVA in the Two-Sample Case

Pooling Variances

When you have two samples, the *t*-test and ANOVA are equivalent, but only if you use a *pooled variance* in the denominator of the test statistic.

Pooled Variance for Two Samples

The **pooled variance**, s_p^2, is an average of the sample variances, s_1^2, and s_2^2, weighted by the sample sizes, n_1 and n_2:

$$s_p^2 = \frac{(n_1 - 1)s_1^2 + (n_2 - 1)s_2^2}{(n_1 + n_2 - 2)}$$

"To pool" means to combine into one statistic.

When sample sizes are equal, the pooled variance is simply the average of the sample variances.

You can pool variances only if you reasonably can assume that the population variances are equal or nearly so. *A rough guideline is that the larger sample standard deviation should be no more than three times the smaller.* Because of this change—pooling—the method of ANOVA should be reserved for data sets where the samples have standard deviations that are all roughly the same.

A *t*-Test with Pooled Variance for Aldrin in the Wolf River | **Example 14.1**

In Section 11.2, you used a two-sample *t*-test to compare the aldrin concentrations at the bottom and at mid-depth in the Wolf River. (The data are in Display 11.1 on page 517, and the *t*-test is on page 533.) Display 14.1 shows a dot plot of the concentrations for the two samples:

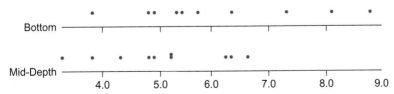

Display 14.1 Aldrin concentrations in the Wolf River for samples taken at the bottom and at mid-depth.

The hypotheses are

$$H_0: \mu_{\text{bottom}} = \mu_{\text{mid-depth}} \quad \text{and} \quad H_a: \mu_{\text{bottom}} \neq \mu_{\text{mid-depth}}$$

The summary statistics are

$$\text{Bottom:} \quad \bar{x}_1 = 6.04 \quad s_1 = 1.579 \quad n_1 = 10$$
$$\text{Mid-depth:} \quad \bar{x}_2 = 5.05 \quad s_2 = 1.104 \quad n_2 = 10$$

Because the two-sample standard deviations are not very different, pooling the variance is a reasonable strategy for the aldrin data. The pooled variance is

$$s_p^2 = (n_1 - 1)s_1^2 + (n_2 - 1)s_2^2/(n_1 + n_2 - 2)$$

$$= \frac{(10 - 1)1.579^2 + (10 - 1)1.104^2}{(10 + 10 - 2)}$$

$$\approx 1.856$$

The value of the test statistic is then

$$t = \frac{(\bar{x}_1 - \bar{x}_2)}{\sqrt{\frac{s_p^2}{n_1} + \frac{s_p^2}{n_2}}} = \frac{6.04 - 5.05}{\sqrt{\frac{1.856}{10} + \frac{1.856}{10}}} \approx 1.625$$

With $n_1 + n_2 - 2 = 10 + 10 - 2 = 18$ degrees of freedom, the P-value, for a a two-sided test this time, is 0.12 and there is insufficient evidence to claim that the mean aldrin concentration at the bottom of the Wolf River is different from the mid-depth concentration. In other words, with these sample sizes the difference in sample means is not large enough to rule out chance variation as a plausible explanation.

Pooling Means

In ANOVA, each sample mean is compared to the overall mean, $\bar{x}$, which is computed by pooling all samples into one big sample.

> ### The Overall Mean
>
> The **overall mean**, $\bar{x}$, is the mean of all the sample data pooled together. It can be computed by weighting the sample means, $\bar{x}_1$ and $\bar{x}_2$, by their respective sample sizes, n_1 and n_2:
>
> $$\bar{x} = \frac{n_1\bar{x}_1 + n_2\bar{x}_2}{n_1 + n_2}$$

The overall mean is sometimes called the *grand mean*.

A New Version of the *t*-Statistic

When the two-sample *t*-test is generalize to ANOVA, the square root in the denominator of the *t*-statistic turns out to be a bit of a nuisance; so the *t*-statistic is squared. (That's a minor change.) After squaring and a *lot* of algebra, you get

$$t^2 = \left| \frac{\bar{x}_1 - \bar{x}_2}{\sqrt{\frac{s_p^2}{n_1} + \frac{s_p^2}{n_2}}} \right|^2 = \cdots = \frac{n_1(\bar{x}_1 - \bar{x})^2 + n_2(\bar{x}_2 - \bar{x})^2}{s_p^2}$$

What has been accomplished? You now have a new way of looking at the *t*-statistic that hints at how it can be generalized to the case of more than two samples. The summary statistics for a third sample can be added into the formula for t^2 quite easily.

Calculations for the Aldrin Measurements	**Example 14.2**

Compute t^2 to compare the bottom and mid-depth concentrations of aldrin in the Wolf River.

Solution
The overall mean is

$$\bar{x} = \frac{n_1 \bar{x}_1 + n_2 \bar{x}_2}{n_1 + n_2} = \frac{10 \cdot 6.04 + 10 \cdot 5.05}{10 + 10} \approx 5.545$$

The pooled variance, $s_p^2 = 1.856$, was computed in Example 14.1.
The test statistic, t^2, is

$$t^2 = \frac{n_1(\bar{x}_1 - \bar{x})^2 + n_2(\bar{x}_2 - \bar{x})^2}{s_p^2}$$

$$= \frac{10(6.04 - 5.545)^2 + 10(5.05 - 5.545)^2}{1.856}$$

$$\approx 2.642$$

> These computations easily can be extended to three or more samples.

Taking square roots gives $t \approx 1.625$, as in Example 14.1

DISCUSSION
From the *t*-Test to ANOVA

D1. How might you extend the t^2-statistic to the case where you want to compare three samples?

Between-Sample Variation and Within-Sample Variation

The t^2-statistic has a sum of squares in both the numerator and denominator.

> **Between-Sample Variation and Within-Sample Variation**

The numerator of t^2 is a sum of squared differences, where each term measures the variation *between* a sample mean and the overall mean, $\bar{x}$. This sum of squares (*SS*) is called **between-sample variation:**

$$SS(\textit{between}) = n_1(\bar{x}_1 - \bar{x})^2 + n_2(\bar{x}_2 - \bar{x})^2$$

If $SS(between)$ is big compared to $SS(within)$, there is a statistically significant difference in the means.

The denominator of t^2 also has a sum of squares, where each term measures the variation *within* each sample using the sample variances, s_1^2 and s_2^2. This sum of squares is called **within-sample variation**:

$$SS(within) = (n_1 - 1)s_1^2 + (n_2 - 1)s_2^2$$

If the between-sample variation is large compared to the within-sample variation, the t^2-statistic will be large and you will have strong evidence against the null hypothesis of equal population means. If, on the other hand, the within-sample variation swamps the between-sample variation, the t^2-statistic will be small and there will not be sufficient evidence to reject the null hypothesis of equal population means. This approach to comparing means by looking at between-sample variation and within-sample variation is the basis of analysis of variance.

To get a practical feel for why the ratio of between- to within-sample variation is a good measure of the evidence against equal population means, think of taking a random sample of patrons at the movie theater in each of two different small towns. A difference of, say, 3 years between the mean ages probably would not seem statistically significant because there is a great amount of variation in ages within each population. In other words, the between-sample variation is small compared to the within-sample variation. Now, suppose the two theaters are on two college campuses. In this scenario a difference of 3 years between the mean ages is of much higher statistical significance because it is large compared to the variability within the ages of college students. The large between-sample variation compared to the relatively small within-sample variation might well lead you to conclude that there is a real difference in the mean ages of college students frequenting the two theaters.

Look again at the dot plot of the aldrin data in Display 14.1. The variability in each group of dots is considerable. The within-sample variation obscures any real difference that might exist between the population means.

STATISTICS IN ACTION 14.1 ▶ Dividing Up Your Class

In this activity you will split the members of your class into two groups based on height. If you have an even number of students, make the groups the same size. If not, you'll have one extra person in one of the groups. First, have everyone stand up who wants to participate.

1. Have one person divide your class into two groups so that the between-group variability in height is as large as possible.

2. Have another person exchange two people to try to get even larger between-group variability.

3. Which division does your class think has the larger between-group variability? What computation could you do to decide which division is better?

4. Have another person divide your class into two groups so that the within-group variability in height is as large as possible. What happened to the between-group variability compared to the division in steps 1 and 2?

5. Have another person exchange two people to try to get even larger within-group variability.

6. Which division does your class think has the larger within-group variability? What computation could you do to decide which division is better?

DISCUSSION
Between-Sample and Within-Sample Variation

D2. Why do the terms *between-* and *within-*sample variation make good sense? Sometimes the within-sample variation is referred to as the error. Why?

D3. Nine small data sets, each of which has two samples, and each sample is of size 2, are shown in Display 14.2. They are designed to help you practice comparing variability between and within groups.

Data Set	Sample 1	Sample 2
A	0 2	0 2
B	0 2	4 6
C	0 2	8 10
D	0 4	0 4
E	0 4	4 8
F	0 4	8 12
G	0 8	0 8
H	0 8	4 12
I	0 8	8 16

Display 14.2 Nine pairs of small data sets.

a. Without doing any computing, divide the pairs of data sets into those with "small," "medium," and "large" between-group variability.

b. Without doing any computing, divide the pairs of data sets into those with "small," "medium," and "large" within-group variability.

c. Use your answers to parts a and b to classify each of the nine data sets by putting its letter in a table like the following one. (For example, if you think that Data Set A has small within-group variability and small between-group variability, you would put the letter A in the upper left cell of the table.)

	Between-Group Variation		
Within-Group Variation	**Small**	**Medium**	**Large**
Small			
Medium			
Large			

d. Now make a table that starts like the listing of the data sets at the beginning of the problem but that has additional columns for $\bar{x}_1$, $\bar{x}_2$, s_1^2, s_2^2, the between sum of squares, the within sum of squares, and t^2. Carry out the computations you need to complete the table. (All the numbers should be simple.)

e. Go back to your table in part c, and write the value of t^2 next to each data set. Then write a sentence or two describing the pattern in the table: How is the value of t^2 related to the sizes of the two kinds of variability?

Summary 14.1: A New Look at the Two-Sample *t*-Test

The goal of this chapter is to learn to analyze data in which more than two means must be compared. The technique is called analysis of variance (ANOVA), and it is a generalization of the (two-sided) two-sample *t*-test (with pooled variances).

How the t-statistic generalizes to more than two samples is most easily seen when t^2 is written in the following form:

$$t^2 = \frac{n_1(\bar{x}_1 - \bar{x})^2 + n_2(\bar{x}_2 - \bar{x})^2}{s_p^2}$$

Here, s_p^2 is the pooled variance

$$s_p^2 = \frac{(n_1 - 1)s_1^2 + (n_2 - 1)s_2^2}{(n_1 + n_2 - 2)}$$

and $\bar{x}$ is the overall (pooled) mean

$$\bar{x} = \frac{n_1\bar{x}_1 + n_2\bar{x}_2}{n_1 + n_2}$$

You use $df = n_1 + n_2 - 2$.

Just like the two-sample t-statistic, ANOVA is based on the ratio of the variability between the means of the samples (what you are interested in determining for the population) and the variability within the samples themselves (the "noise" that you wish would go away).

- If the ratio is large (the differences in the sample means are quite large compared to the variability within the samples), you will reject the null hypothesis of equal population means ($H_0: \mu_1 = \mu_2$).
- If the ratio is small (there is little difference between samples compared to the differences within the samples), conclude that it is plausible that $\mu_1 = \mu_2$ and the variability between sample means is due to chance.

In the next section, you will learn to use the F-statistic, which is a generalization of the t-statistic used in the two-sample t-test, with pooled variance, to ANOVA. It turns out that, in the two-sample case, $t^2 = F$.

Practice

From the t-Test to ANOVA in the Two-Sample Case

P1. Data about the effects of special exercises to speed up infant walking were given in Display 11.3 on page 521. Display 14.3 shows the data and summary statistics for two of the treatments:

Group	Age (months) at First Unaided Steps					
Special exercises	9	9.5	9.75	10	13	9.5
Exercise control	11	10	10	11.7	10.5	15
Special exercises	$\bar{x}_1 = 10.125$	$s_1 = 1.447$	$n_1 = 6$			
Exercise control	$\bar{x}_2 = 11.375$	$s_2 = 1.896$	$n_2 = 6$			

Display 14.3 Walking baby data.

a. Compute the overall mean, $\bar{x}$.
b. Compute the pooled variance, s_p^2.
c. Compute t^2 to test the hypothesis that the treatment makes no difference in mean age at first unaided steps.
d. Using the square root of t^2 and $df = 6 + 6 - 2 = 10$, find the P-value. What is your conclusion?

P2. Display 14.4 shows two sets of hypothetical data. In each pair, the sample means are the same as in the actual

aldrin samples (6.04 and 5.05). However, the standard deviations are different from those for the actual samples.

I.

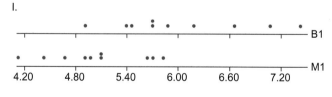

III.

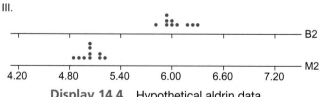

Display 14.4 Hypothetical aldrin data.

a. Does plot I or plot II show more variation within the samples, or is the within-sample variability the same?
b. Does plot I or plot II show more variation between the sample means, or is the between-sample variability the same?
c. Which set of data—plot I or plot II—will have the larger value of t^2? Which will have the smaller? Why?

Between-Sample and Within-Sample Variation

P3. Refer to the data in Display 14.3 in P1.

    a. Compute *SS(between)* and *SS(within)*.

    b. Is *SS(between)* large compared to *SS(within)*? What do you conclude?

P4. Refer to Display 14.5, which shows the final exam scores in a freshman level business calculus course, by class year.

    a. Did any of the class years clearly do better than the rest? Worse?

    b. Is *SS(between)* large compared to *SS(within)*? What do you conclude?

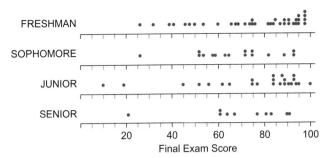

Display 14.5 Scores on final exam in business calculus, by class year.

Exercises

E1. *Pulse rates.* Display 14.6 gives the pulse rates for independent random samples of men and women (as presented in Display 11.6 of Section 11.1).

Men	Women
74	75
80	66
75	57
69	87
58	89
76	65
78	69
78	79
86	85
84	59
71	65
80	80
75	74

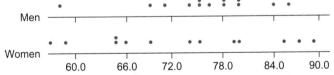

Display 14.6 Pulse rates (beats per minute) for random samples of men and women.

    a. A goal of the study was to see if there is evidence of a difference between the mean pulse rates of men and women. What are the null and alternative hypotheses for this study?

    b. Compare the distributions of men's and women's pulse rates. Do you think that the null hypothesis will be rejected?

    c. Calculate the t^2-statistic. Is there strong evidence against the null hypothesis?

E2. *Sleep and reaction time.* In a study of the effects of the amount of sleep on school-age children, children were randomly assigned to go to sleep 1 hour earlier or 1 hour later than their regular bedtime. After three nights, several tests were given to the two groups. One of them was a simple reaction-time test where students were to press a button as quickly as possible when given a certain signal. A goal of the study was to determine whether the amount of sleep has an effect on mean reaction time. The results are summarized here (measurements are in thousandths of a second and shorter reaction times are better). [Source: Avi Sadeh, Reul Gruber, and Amiram Raviv, "The Effects of Sleep Restriction and Extension on School-Age Children: What a Difference an Hour Makes," *Child Development*, Vol. 74 (March–April 2003), pp. 444–455.]

Sleep-restricted:	$\bar{x}_1 = 458.2$	$s_1 = 77.1$	$n_1 = 28$
Sleep-extended:	$\bar{x}_2 = 418.7$	$s_2 = 58.4$	$n_2 = 21$

    a. What are the null and alternative hypotheses for this experiment? Do you think there is strong evidence against the null hypothesis?

    b. Calculate the t^2-statistic. Should you reject the null hypothesis?

E3. The boxplots in Display 14.7 show the season batting averages of the players on the 16 National League baseball teams who had 100 or more at bats in 2008.

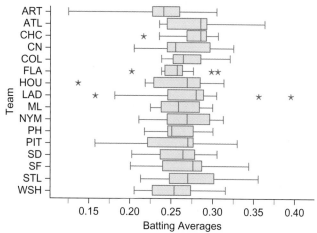

Display 14.7 Batting averages of players with 100 or more at bats, by National League Team. [Source: *http://mlb.mlb.com/*.]

a. Which team shows the most within-team variation? Which team shows the least?

b. Which pair of teams shows the most between-team variation?

c. Does it look like there might be any problem with the guideline that the largest standard deviation should be no more than three times the smallest?

d. Does it look to you as though there is a statistically significant difference between the mean batting averages?

E4. As explained in E4 of Chapter 10, in 1879, A. A. Michelson made 100 determinations of the velocity of light in air. These measurements, in kilometers per second but coded by subtracting 299,000 from each, were actually made in five groups of 20 measurements each, perhaps with some adjustments to the methodology between groups. Display 14.8 provides boxplots of the five groups of 20 measurements.

A. A. Michelson (1852–1931).

a. Which group shows the most within-group variation? Which group shows the least?

b. Would you judge the between-group variation to be large or small, as compared to the within-group variation?

c. Does it look like there might be a problem with the guideline that the largest standard deviation should be no more than three times the smallest?

d. Does it look to you like there are statistically significant differences among the groups overall?

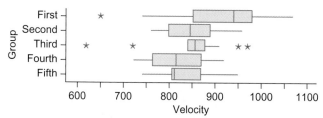

Display 14.8 Michelson's measurements of the speed of light.

E5. The table and dot plots of Display 14.9 give the data from Kelly's hamster experiment described on page 201.

Enzyme Concentration	Length of Day
6.625	L
10.375	L
9.900	L
8.800	L
12.500	S
11.625	S
18.275	S
13.225	S

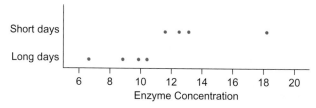

Display 14.9 Hamster data.

a. From just looking at the numbers in the table or at the dot plots, is there more within-sample variability for long days or for short days?

b. Combine all eight measurements into one group and compute the sum of squared differences from the overall mean. Then compute *SS(within)* and *SS(between)*. What do you notice about these three sums of squares?

c. Calculate the value of t^2. Is there a statistically significant difference in the means?

d. Does it look like there might be a problem with the guideline that the largest standard deviation should be no more than three times the smallest?

E6. In E15 of Chapter 11, you read about an experiment to test the protective power of a resin. Two doses of resin (5 mg and 10 mg) were dissolved in a solvent and placed on filter paper. Eight dishes were prepared with filter paper at dose level 5 mg and eight with filter paper at dose level 10 mg. Twenty-five termites were then placed in each dish to feed on the filter paper. At the end of 15 days, the surviving termites were counted. The results are shown in Display 14.10.

Dose (mg)	Count of Survivors
5	11
5	11
5	12
5	12
5	5
5	9
5	6
5	10
10	16
10	13
10	1
10	0
10	0
10	0
10	0
10	3

Display 14.10 Number of termites, out of 25, surviving after being placed in a dish with 5 mg or 10 mg of a resin. [Source: *lib.stat.cmu.edu.*]

a. From just looking at the numbers in the table or at the dot plots, is there more within-sample variability for a dose of 5 mg or a dose of 10 mg?

b. Combine all 16 measurements into one group and compute the sum of squared differences from the overall mean. Then compute *SS(within)* and *SS(between)*. What do you notice about these three sums of squares?

c. Calculate the value of t^2. Is there a statistically significant difference in the means?

d. Does it look like there might be a problem with the guideline that the largest standard deviation should be no more than three times the smallest?

E7. Suppose you begin with two identical samples so that each has mean $\bar{x}$ and standard deviation s.

a. What value do you get for s_p? For t^2?

b. Now suppose you increase the standard deviation of one or both samples while leaving the means unchanged. What will happen to s_p? To t^2?

c. If you decrease the standard deviations instead, what will happen to s_p? To t^2?

d. Suppose, instead, that you leave the standard deviations as they were originally, but add 5 to each value in one of the samples. What will happen to s_p? To t^2?

E8. Display 14.11 shows the hypothetical aldrin data used in P2 (Display 14.4). Check your answer to part c of P2 by calculating t^2 for both pairs. What do you observe about the two numerators? Why should this be the case?

I.	B1	M1	II.	B2	M2
	4.92	4.125		5.816	4.865
	5.42	4.425		5.916	4.925
	5.47	4.675		5.926	4.975
	5.67	4.925		5.966	5.025
	5.72	4.975		5.976	5.035
	5.87	5.125		6.006	5.065
	6.17	5.125		6.066	5.065
	6.67	5.625		6.166	5.165
	7.07	5.675		6.246	5.175
	7.42	5.825		6.316	5.205

Display 14.11 Hypothetical aldrin data.

14.2 ► One-Way ANOVA: When There Are More Than Two Groups

The researchers who studied aldrin pollution in the Wolf River took measurements on the surface of the water, as well as on the bottom and at mid-depth. So there really are three independent samples. Display 14.12 includes the complete set of measurements, along with summary statistics and a dot plot of each sample.

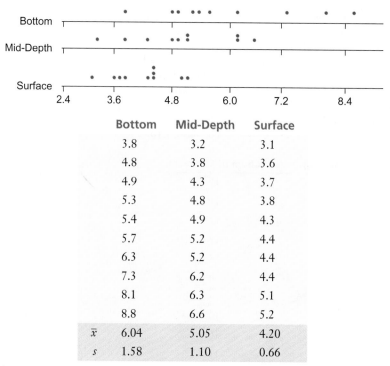

	Bottom	Mid-Depth	Surface
	3.8	3.2	3.1
	4.8	3.8	3.6
	4.9	4.3	3.7
	5.3	4.8	3.8
	5.4	4.9	4.3
	5.7	5.2	4.4
	6.3	5.2	4.4
	7.3	6.2	4.4
	8.1	6.3	5.1
	8.8	6.6	5.2
$\bar{x}$	6.04	5.05	4.20
s	1.58	1.10	0.66

Display 14.12 Aldrin concentration (nanograms per liter) at three levels of depth.

These samples suggest that the surface measurements tend to be a little smaller than the others. But could this just be due to the variability you always get by taking samples? To test whether the differences in the sample means reflect a real difference in the population means, you will use the same ideas as you used for a *t*-test. The null hypothesis will be that all three population means are equal, and the alternative will be that at least two population means differ from each other

H_0: $\mu_1 = \mu_2 = \mu_3$, where μ_1, μ_2, μ_3 are the mean aldrin concentrations at the bottom, mid-depth, and surface, respectively.

H_a: At least one of the following is true:

$$\mu_1 \neq \mu_2, \quad \mu_1 \neq \mu_3, \quad \mu_2 \neq \mu_3$$

Now that there are three means to compare, you will need a new test statistic.

The Test Statistic for ANOVA

The test statistic for ANOVA, called *F*, in honor of R. A. Fisher, is the ratio given in the following box. The numerator is the variability *between* the means of the samples (what you really are interested in). The denominator is based on the standard deviations of the samples and so measures how much individual observations vary *within* the samples (the noise or error).

In 1921, R. A. Fisher (1890–1962) was the first to use ANOVA.

One-way means that there is only one factor (for an experiment) or you are comparing just one set of means (for a sample survey).

The *F*-Statistic for Comparing *k* Means

The test statistic, *F*, for a one-way analysis of variance to test the equality of *k* population means is

$$F = \frac{\dfrac{SS(between)}{k-1}}{\dfrac{SS(within)}{n-k}}$$

$$= \frac{\dfrac{n_1(\bar{x}_1 - \bar{x})^2 + n_2(\bar{x}_2 - \bar{x})^2 + \cdots + n_k(\bar{x}_k - \bar{x})^2}{(k - 1)}}{\dfrac{(n_1 - 1)s_1^2 + (n_2 - 1)s_2^2 + \cdots + (n_k - 1)s_k^2}{(n - k)}}$$

$$= \frac{\sum n_i(\bar{x}_i - \bar{x})^2/(k - 1)}{s_p^2}$$

Here, n is the total number of observations in all k samples, $\bar{x}$ is the overall mean of all observations in the k samples, and $\bar{x}_i$, s_i, and n_i are the mean, standard deviation, and number of observations for the ith sample.

A new wrinkle in the formula for F is that the numerator needs a divisor (its degrees of freedom, $k - 1$) to keep it from getting too large simply because of the larger number of samples. (The divisor was not needed in the formula for t^2 in the two-sample case because $df = k - 1 = 2 - 1 = 1$.)

Sometimes F is written using this notation.

$df = k - 1$ because there are $k - 1$ independent terms in $SS(between)$.

$$F = \frac{MS(between)}{MS(within)}$$

Here, MS stands for **mean square**, a sum of squares divided by its degrees of freedom. The following example shows the steps for computing F.

Computing F for the Three Aldrin Samples

Example 14.3

Refer to Display 14.12. Compute the F-statistic for comparing the mean aldrin level at the bottom, mid-depth, and surface.

Solution
The overall mean for all three aldrin samples put together is the average of all 30 observations. It can be computed by adding all 30 measurements and dividing by 30 or by the generalizing formula:

$$\bar{x} = \frac{n_1\bar{x}_1 + n_2\bar{x}_2 + n_3\bar{x}_3}{n_1 + n_2 + n_3} = \frac{10 \cdot 6.04 + 10 \cdot 5.05 + 10 \cdot 4.20}{10 + 10 + 10} \approx 5.10$$

(Because the sample sizes are equal, you also could find the overall mean simply by averaging the individual sample means: $\frac{6.04 + 5.05 + 4.20}{3} \approx 5.10$.)

The two mean sum of squares are

$$MS(between) = \frac{n_1(\bar{x}_1 - \bar{x})^2 + n_2(\bar{x}_2 - \bar{x})^2 + n_3(\bar{x}_3 - \bar{x})^2}{(k - 1)}$$

$$= \frac{10(6.04 - 5.10)^2 + 10(5.05 - 5.10)^2 + 10(4.20 - 5.10)^2}{(3 - 1)}$$

$$\approx \frac{16.96}{2}$$

$$\approx 8.48$$

$$MS(within) = \frac{(n_1 - 1)s_1^2 + (n_2 - 1)s_2^2 + (n_3 - 1)s_3^2}{(n - k)}$$

$$= \frac{(10 - 1)1.58^2 + (10 - 1)1.10^2 + (10 - 1)0.66^2}{(10 + 10 + 10 - 3)}$$

$$\approx \frac{37.33}{27}$$

$$\approx 1.38$$

Dividing the two mean sums of squares gives F:

$$F = \frac{MS(between)}{MS(within)} = \frac{8.48}{1.38} \approx 6.13$$

This value of F is certainly larger than the test statistic of $t^2 = F = 2.642$ for the difference between the mean concentration at the bottom and at mid-depth in Example 14.2, on page 674. But when you add another sample, you should expect more variability. Is this value of F, 6.13, large enough to reject the null hypothesis that the three population means are equal? That is the next part of the story.

DISCUSSION
The *F*-Statistic

D4.   The numerator of the F-statistic has $k - 1$ degrees of freedom. Verify that $k - 1$ is the right number of degrees of freedom by showing that you can compute the kth squared deviation between a treatment mean and the overall mean if you know the overall mean, the other $k - 1$ squared deviations, and the sample sizes.

The Sampling Distribution of the *F*-Statistic

The final step in completing the analysis of variance for the aldrin data is to estimate the P-value. The P-value is defined in the same way as for all significance tests.

The *P*-Value for ANOVA

The P-value is the probability of getting an F-statistic as large as or larger than that computed from your samples if the null hypothesis is correct that there is no difference in the population means.

One way to determine if a sample summary statistic is reasonably likely to occur, under the null hypothesis, is to construct a sampling distribution using simulation.

If the depth at which you take the measurement in the river makes no difference, you can estimate the values of F you might get from random sampling by dumping all 30 aldrin measurements together and randomly dividing them into three groups of size 10 each. Here are the steps in that simulation:

1.   Randomly divide the 30 measurements into three groups of equal size.
2.   Compute the F-statistic for this random allocation.
3.   Repeat this process many times, say, 200.
4.   Make a histogram or other plot of the distribution of the 200 values of F.

Display 14.13 shows a simulated sampling distribution of the F-statistic constructed using these instructions. The value of this statistic observed from the aldrin samples ($F = 6.13$) was exceeded only once in 200 attempts, so that observed value would be very unusual if the 30 measurements had been assigned to the depths by a chance device. This is very strong evidence against the null hypothesis, and you conclude that at least one pair of mean concentrations must differ. Depth does matter, at least at this location and at this point in time.

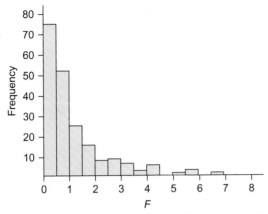

Display 14.13 Simulated distribution of the F-statistic for the aldrin data.

The distribution of the F-statistic from repeated sampling when the null hypothesis is true looks much like a chi-square distribution, with a shape that is skewed toward the larger values.

A simulation helps you understand the sampling distribution of the statistic, but generally you will use statistical software to find P-values. From statistical software, the P-value for the aldrin example is 0.006, which is close to the estimate of $\frac{1}{200} = 0.005$ from the simulation in Display 14.13.

Get the P-value from software or Table D.

If you aren't using software, refer to Table D on page 762. Find the entry for $k-1$ degrees of freedom for the numerator and $n-k$ degrees of freedom for the denominator. If your computed value of F is larger than the entry in the table (called the *critical value*), you reject the null hypothesis at $\alpha = 0.05$.

Large F, reject H_0.

DISCUSSION
The Sampling Distribution of the F-Statistic

D5. Display 14.14 shows the Michelson data on speed of light for the five trials pictured in Display 14.8 on page 680. Describe how to construct the sampling distribution for the value of F needed to test the hypothesis that the true mean speeds are equally for the processes used in the five trials.

Trial 1	Trial 2	Trial 3	Trial 4	Trial 5
850	960	880	890	890
740	940	880	810	840
900	960	880	810	780
1070	940	860	820	810
930	880	720	800	760
850	800	720	770	810
950	850	620	760	790
980	880	860	740	810
980	900	970	750	820
880	840	950	760	850
1000	830	880	910	870
980	790	910	920	870

(Continued)

930	810	850	890	810
650	880	870	860	740
760	880	840	880	810
810	830	840	720	940
1000	800	850	840	950
1000	790	840	850	800
960	760	840	850	810
960	800	840	780	870

Display 14.14 Michelson's speed of light measurements (coded).

Reading ANOVA Tables

The computations for an analysis of variance typically are organized in a table formatted like the one in Display 14.15.

Source	SS	df	MS	F
Between	SS(between)	$k - 1$	SS(between)/$(k - 1)$	MS(between)/MS(within)
Within	SS(within)	$n - k$	SS(within)/$(n - k)$	
Total	SS(total)	$n - 1$		

Display 14.15 The format of an ANOVA table. Here, *k* is the number of samples and *n* is the total number of observations in the *k* samples.

The summary table in Display 14.15 involves a third sum of squared differences, SS(*total*), called the total sum of squares.

The Total Sum of Squares

The third sum of squared differences, the SS(*total*) or total sum of squares, is the sum of the squared deviations of the individual observations from the overall mean $\bar{x}$:

$$SS(total) = \sum (x - \bar{x})^2$$

Here, you sum over all of the individual observations, x, in all samples.

The sums of squares are related by this equation:

$$SS(total) = SS(between) + SS(within)$$

Example 14.4 **The Total Sum of Squares for the Three Aldrin Samples**

Compute the total sum of squares for the three aldrin samples in Display 14.12 and verify that $SS(total) = SS(between) + SS(within)$.

Solution

As previously computed, the overall mean for the three aldrin samples is

$$\bar{x} = \frac{n_1\bar{x}_1 + n_2\bar{x}_2 + n_3\bar{x}_3}{n_1 + n_2 + n_3} = \frac{10 \cdot 6.04 + 10 \cdot 5.05 + 10 \cdot 4.20}{10 + 10 + 10} \approx 5.10$$

Because there are 30 measurements in the three samples, there are 30 terms in the computation of $SS(total)$:

$$SS(total) = \sum(x - \bar{x})^2$$
$$= (3.8 - 5.10)^2 + (4.8 - 5.10)^2 + (4.9 - 5.10)^2 + \cdots + (4.4 - 5.10)^2$$
$$+ (5.1 - 5.10)^2 + (5.2 - 5.10)^2$$
$$= 54.29$$

From Example 14.3, $SS(between) = 16.96$ and $SS(within) = 37.33$, for a total of 54.29.

The output for the aldrin analysis from Minitab, JMP, Excel, and the TI-84 Plus appear in Display 14.16.

Minitab

ANALYSIS OF VARIANCE ON Aldrin

SOURCE	DF	SS	MS	F	P
Depths	2	16.96	8.48	6.13	0.006
ERROR	27	37.33	1.38		
TOTAL	29	54.29			

JMP

Oneway Analysis of Aldrin By Depth

Oneway Anova

Summary of Fit

Rsquare	0.312411
Adj Rsquare	0.261478
Root Mean Square Error	1.175821
Mean of Response	5.096667
Observations	30

Analysis of Variance

Source	DF	Sum of Squares	Mean Square	F Ratio	Prob > F
Depth	2	16.960667	8.48033	6.1338	0.0064
Error	27	37.329000	1.38256		
C. Total	29	54.289667			

Excel

Anova: Single Factor

SUMMARY

Groups	Count	Sum	Average	Variance
Column 1	10	60.4	6.04	2.49377778
Column 2	10	50.5	5.05	1.21833333
Column 3	10	42	4.2	0.43555556

ANOVA

Source of Variation	SS	df	MS	F	P-value	F crit
Between Groups	16.9606667	2	8.48033333	6.13381017	0.00636713	3.3541312
Within Groups	37.329	27	1.38255556			
Total	54.2896667	29				

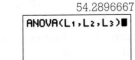

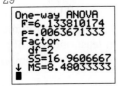

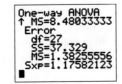

Display 14.16 Analysis of variance output from three software packages and a calculator.

The outputs are very similar except that they vary in their language. Both Minitab and JMP substitute the name of the variable (*depth*, in this case) for "between" and the calculator uses "Factor." Minitab, JMP, and the calculator use "error" instead of "within." The table in Display 14.17 lists some terms used by different statistical software.

Sum of Squares Formula	Terms Used	Terms for *SS/df*
Between: $\sum_{i=1}^{k} n_i(\bar{x}_i - \bar{x})^2$	SS(*between*) SS(*group*) SS(*between groups*) SS(*treatment*) SS(*factor*) SSB	*mean square for groups* *mean square for treatments* *mean square for factor* MSB
Within: $\sum_{i=1}^{k} (n_i - 1)s_i^2$	SS(*within*) SS(*within groups*) SS(*error*) SSE	*mean square error* MSE

Display 14.17 Terminology used in ANOVA tables.

As you learned in Chapter 4, two different types of studies employ randomization, sample surveys and experiments. Like the *t*-test, ANOVA can be used to compare means in either type of study. The aldrin example was a sampling situation. Example 14.5 on page 690, is an experiment. Because testing for equality of means is more often used in experiments and experiments will be the focus of the rest of the chapter, the terms SS(*treatment*) and SS(*error*) generally will be used from now on.

DISCUSSION
Reading ANOVA Tables

D6. The following ANOVA compares the batting averages of the 16 teams in the National League from E3 on page 680.

Source	DF	SS	MS	F	P
Teams	15	0.0214	0.0014	1.110	0.3490
Error	217	0.2795	0.0013		
Total	232	0.3009			

a. Explain how each of the values in the table was computed.

b. What conclusion should you draw about the batting averages across the 16 teams? Your conclusion should reflect the fact that this is an observational study rather than a randomized study.

Steps in a Complete One-Way ANOVA

The following is a summary of the steps in a one-way analysis of variance. (An analysis of variance is called one-way when there is only one set of means to be compared.)

Components of the One-Way Analysis of Variance

1. *State your hypotheses.* The null hypothesis in standard analysis of variance problems is that all k population means are equal. The alternative hypothesis is that in at least one pair of population means, the two population means are different. In symbols,

$$H_0: \mu_1 = \mu_2 = \cdots = \mu_k$$

where $\mu_1, \mu_2, \ldots, \mu_k$ are the means of the k populations (or the means for each treatment if all subjects had been given that treatment).

$$H_a: \mu_i \neq \mu_j \text{ for at least one pair from the } k \text{ populations}$$

2. *Check conditions.* For testing the significance of differences among k (two or more) sample means, the analysis of variance requires conditions similar to those of the pooled t-test.

   - Check that the samples were randomly and independently selected from the k populations or that the k treatments were randomly assigned to the available experimental units.

   | Randomness. |

   - Plot each sample to check that it looks like it reasonably could have come from a population that is approximately normal. (Alternatively, the sample sizes must be large enough for the k sample means to have approximately normal sampling distributions.)

   | Normality. |

   - Check that the standard deviations of the k samples are approximately equal (**homogeneity of variance**). You can use this guideline: The largest standard deviation should be no more than three times the smallest.

   | Equal variances. |

3. *Compute the test statistic.* The test statistic is

$$F = \frac{MS(treatmemt)}{MS(error)} = \frac{\sum_{i=1}^{k} n_i(\bar{x}_i - \bar{x})^2/(k-1)}{\sum_{i=1}^{k}(n_i - 1)s_i^2/(n-k)} = \frac{\sum_{i=1}^{k} n_i(\bar{x}_i - \bar{x})^2/(k-1)}{s_p^2}$$

   Here, $\bar{x}$ is the mean of all n observations in the k samples, n_i is the number of observations in the ith sample, $n = n_1 + n_2 + \cdots + n_k$, $\bar{x}_i$ is the mean of the ith sample, and s_i is the standard deviation of the ith sample. The degrees of freedom are $k-1$ for the numerator and $n-k$ for the denominator.

   You usually will get this test statistic from a graphing calculator or computer, which also will give the P-value.

4. *Write a conclusion that is linked to your computations and is stated in context.* Report the P-value and interpret it as the amount of evidence against the null hypothesis of equal means. The smaller the P-value, the stronger the evidence. If the evidence against the hypothesis of equal means is strong, the alternative tells you that at least two means differ, but it does not specify which means differ.

The Conditions Are More What You Would Call Guidelines Than Actual Rules

As with many statistical tests, the theoretical assumptions used in the mathematical justification of ANOVA are stronger than necessary in statistical practice.

- *Randomization.* Like the t-test, ANOVA may be used for comparing two means, whether the means come from random assignment of treatments to subjects in an experiment or from random samples in a sample survey. ANOVA also may be used with observational studies (no randomization) to help answer the question, "Is it reasonable to conclude that the observed difference in means could have happened by chance alone?" In such studies there can be no cause-and-effect conclusion or generalization to a population.

> Robust: The technique tends to give correct results even when the assumptions for the test aren't met exactly.

- *Normal population.* As with *t*-procedures, the ANOVA *F*-test is quite *robust* against lack of normality. That is, it is fairly insensitive to departures from normality with moderate to large sample sizes. Small samples that might come from highly skewed populations will require a transformation (or another statistical method).

- *Homogeneity of variance.* The guideline on homogeneity of variance given in the box should be applied regardless of sample size, but the lack of homogeneity is more serious when the sample sizes are large. Thus, for large samples the guideline that the largest standard deviation should be no more than three times the smallest should be interpreted rather strictly.

One-Way ANOVA and Completely Randomized Designs

> Review completely randomized designs on page 210 of Chapter 4.

The aldrin measurements came from sampling from the population of water in the Wolf River. The next example describes an experiment in which three treatments were randomly assigned to 30 available subjects so that 10 subjects received each treatment. If *k* treatments are randomly assigned to *n* experimental units, with one observation per unit (i.e., no repeated measures and no blocking), the design is referred to as being *completely randomized*. This is an ideal situation for analysis of variance because, under the null hypothesis of no difference in the mean response resulting from the treatment, the means and standard deviations of the *k* treatment groups should vary only because of the random assignment. That makes the equal variance assumption quite tenable under the null hypothesis (although it still should be checked with the data, if possible).

| Example 14.5 | **Caffeine and Finger Tapping** |

> A small, 8-oz, cup of coffee contains about 100 mg of caffeine.

To test the stimulus effect of caffeine, a double-blind experiment was conducted. Thirty male college students were randomly assigned to one of three groups of ten students each. Each group was given one of three doses of caffeine (0, 100, or 200 mg) and 2 hours later the students were given a finger tapping exercise. The response is the number of taps per minute, as shown in Display 14.18. Does caffeine affect performance on this task? If so, how?

0 mg Caffeine	100 mg Caffeine	200 mg Caffeine
242	248	246
245	246	248
244	245	250
248	247	252
247	248	248
248	250	250
242	247	246
244	246	248
246	243	245
242	244	250
$\bar{x}$: 244.8	246.4	248.3
s: 2.394	2.066	2.214

Display 14.18 Finger taps per minute after different d of caffeine. [Source: Norman Draper and Harry Smith, *Applied Regression Analysis*, 2nd ed. (Wiley, 1981).]

Solution

1. *State hypotheses.*

   H_0: If each subject had received 0 mg of caffeine, the mean number of taps would have been equal to the mean number of taps if each subject had received 100 mg of caffeine, which would have been equal to the mean number of taps if each subject received 200 mg of caffeine.

   H_a: At least one of the means would have been different from another.

2. *Conditions are met.*
   - One of three treatments was randomly assigned to each available subject.
   - The boxplots in Display 14.19 indicate that it is reasonable to assume that the samples come from populations that are approximately normal.
   - Display 14.18 also shows sample standard deviations that are approximately equal.

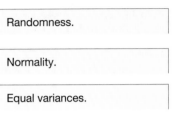

Randomness.

Normality.

Equal variances.

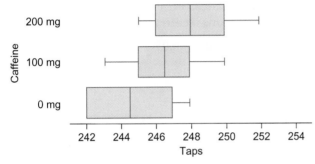

Display 14.19 Boxplots of finger taps per minute.

3. *Do computations.* The analysis of variance is summarized in the following output.

Conditions are met.

ANOVA

Source of Variation	SS	df	MS	F	P-value
Between Groups	61.4	2	30.7	6.1812	0.0062
Within Groups	134.1	27	4.9667		
Total	195.5	29			

4. *Write a conclusion in context.* The probability is only 0.0062 of getting a value of F as large or larger than 6.1812 from a random assignment of treatments to subjects if caffeine has no effect on finger tapping. With a *P*-value of only 0.0062, you have strong evidence that caffeine had an effect on the number of finger taps per minute.

DISCUSSION
Steps in a Complete ANOVA

D7. Do the aldrin measurements meet the conditions for ANOVA?

D8. Do the speed of light data from D5 on page 685 meet the conditions for ANOVA?

D9. Use a calculator and the analysis of variance formulas to confirm the calculations in the analysis of the finger tapping data.

Multiple Comparisons: But Which Means Differ?

If an ANOVA results in rejection of the null hypothesis of equal means, then a natural second step is determining *which* means differ.

The caffeine experiment involves three treatments, and three comparisons can be made: 0 mg versus 100 mg, 0 mg versus 200 mg, and 100 mg versus 200 mg. If you construct three 95% confidence intervals for these three differences, the probability is larger than 0.05 that at least one of the confidence intervals doesn't capture the true difference. Similarly, if you do multiple significance tests, each at, say, $\alpha = 0.05$, you end up with an overall error rate that is larger than 0.05.

> The probability that at least one of three intervals doesn't capture the population mean is $1 - (0.95)^3 \approx 0.143$.

One way to guard against the overall error rate being exorbitantly high is to make the individual decisions at a lower error rate. A good rule to follow is to divide the desired overall error rate by the number of individual decisions (or estimates) to be made within the context of the same analysis. This rule is based on the fact that the probability that at least one of a set of events occurs can be no greater than the sum of the probabilities of the separate events. Thus, if on each test you have a probability of 0.05/3 of making an incorrect decision when the null hypothesis is true, then the probability of at least one incorrect decision is no greater than

$$\frac{0.05}{3} + \frac{0.05}{3} + \frac{0.05}{3} = 0.05$$

Making Multiple Comparisons

If you construct c confidence intervals within the same analysis, each estimating the difference between two treatment means, and you want 95% confidence that *all* of the intervals capture the true parameter, use the formula

$$(\bar{x}_i - \bar{x}_j) \pm t^* s_p \sqrt{\frac{1}{n_i} + \frac{1}{n_j}}$$

Compute the pooled standard deviation, s_p, from all treatment groups, not just the two you are comparing at the time:

$$s_p^2 = \frac{(n_1 - 1)s_1^2 + (n_2 - 1)s_2^2 + \cdots + (n_k - 1)s_k^2}{(n_1 + n_2 + \cdots + n_k - k)}$$

> This is known as the Bonferroni correction for multiple comparisons.

Construct each interval at the $1 - 0.05/c$ level of confidence using $n - k$ degrees of freedom when finding t^*.

| Example 14.6 | **Confidence Intervals for Finger Tapping Differences** |

Construct confidence intervals for the differences between the means of the treatments in Example 14.5 on page 690 so that you have an overall level of confidence of 95%.

Solution
For each individual confidence interval, use the level of confidence

$$1 - \frac{0.05}{3} \approx 1 - 0.02 = 0.98$$

The pooled standard deviation (calculated using all three treatment groups) is $s_p = \sqrt{4.967} \approx 2.23$ with $(n - 3) = 27$ degrees of freedom, and $t^* = 2.473$. The confidence interval is

$$(\bar{x}_1 - \bar{x}_2) \pm t^* s_p \sqrt{\frac{1}{n_i} + \frac{1}{n_j}}$$

This gives the three 98% confidence interval estimates

$$\mu_0 - \mu_{100}: (-4.07, 0.87)$$
$$\mu_0 - \mu_{200}: (-5.97, -1.03)$$
$$\mu_{100} - \mu_{200}: (-4.37, 0.57)$$

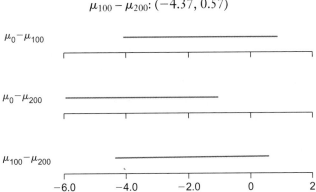

You are 95% confident that all three intervals have captured the difference of the true means. It is now clear that the significant difference is between the 0-mg caffeine treatment and the 200-mg caffeine treatment, because this is the only confidence interval that does not include 0. The others are close to being statistically significant, however, suggesting that these differences might be checked with a larger number of subjects in the future.

DISCUSSION
Multiple Comparisons: But Which Means Differ?

D10. Suppose the F-test for equality of treatment means does not indicate any significant differences among treatment means. Would you go on to look for differences by constructing confidence intervals anyway?

D11. Suppose that you do 10 independent significance tests, each with $\alpha = 0.05$, and the null hypothesis is true in each case.

    a. What is the probability that you falsely reject the null hypothesis on the first of these tests? On the fifth? On the tenth?

    b. What is the probability that you falsely reject the null hypothesis on at least one of these tests?

Summary 14.2 One-Way ANOVA: When There Are More Than Two Groups

One-way ANOVA can be used to test whether treatment means differ in an experiment with one set of treatments laid out in a completely randomized design. It can also be used to test whether population means differ when you take random samples from k populations. To perform an ANOVA, follow the steps on pages 688–689.

The computations are usually arranged in an ANOVA table:

Source	Sum of Squares (SS)	df	MS = SS/df	F	P-value
Treatment	$\sum_{i=1}^{k} n_i(\bar{x}_i - \bar{x})^2$	$k-1$	SS(treat)/(k − 1)	$\dfrac{MS(treat)}{MS(error)}$	
Error	$\sum_{i=1}^{k} (n_i - 1)s_i^2$	$n-k$	SS(error)/(n − k)		
Total	$\sum_{i=1}^{n} (x_i - \bar{x})^2$	$n-1$	SS(total)/(n − 1)		

The sums of squares are related by SS(total) = SS(treatment) + SS(error).

The ANOVA F-test is a simultaneous test of the equality of several treatment means. If there is insufficient evidence to reject the null hypothesis of equal means, then you generally do not perform individual t-tests or construct confidence intervals for the purpose of looking for significant differences. In other words, if the F-test says there are no significant differences among the observed treatment means, do not go looking for them.

Practice

The Test Statistic for ANOVA

P5. The table in Display 14.20 gives the summary statistics for an experiment in which 30 subjects were randomly assigned to four treatments. Treatments 1 and 2 were expected to result in more variability, so twice as many subjects were assigned to them as to treatments 3 and 4.

   a. Find the overall mean and pooled standard deviation.

   b. Compute SS(between) and SS(within).

   c. What are the degrees of freedom for SS(between)? For SS(within)?

   d. Compute the value of the F-statistic.

Treatment	n	Mean	Standard Deviation
1	10	3.26	0.16
2	10	3.12	0.13
3	5	3.06	0.09
4	5	2.65	0.06

Display 14.20 Data for experiment in P5.

The Sampling Distribution for F

P6. Display 14.21 gives the simulated sampling distribution of the F-statistic for the experiment in P5. It was constructed assuming that the null hypothesis is true that the treatments have the same mean effect on the available subjects.

   a. Use the simulated sampling distribution to approximate the probability of getting a value of F as large or larger than the one you got in P5, assuming the null hypothesis is true.

   b. What is your conclusion?

   c. Approximate the P-value if you had gotten F = 3. What would your conclusion be now?

   d. Approximate the P-value if you had gotten F = 0.5. What would your conclusion be now?

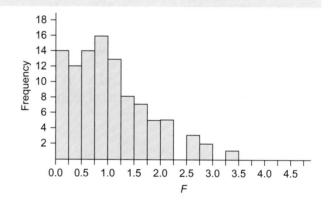

Display 14.21 Simulated distribution of the F-statistic, 100 repetitions.

Reading ANOVA Tables

P7. Fill in the missing numbers in this ANOVA table. Is the result statistically significant at $\alpha = 0.05$?

Analysis of Variance

Source	df	SS	MS	F	P
Treatment	2	0.985	-?-	-?-	0.085
Error	32	5.907	-?-		
Total	-?-	-?-			

P8. The table in Display 14.22 gives some simple numbers so that you can practice building an ANOVA table from scratch.

	Sample 1	Sample 2	Sample 3
	1	5	2
	2	6	4
	3	7	6
Sample Mean	2	6	4
Overall Mean		4	

Display 14.22 Hypothetical data for P8.

a. Compute *SS(between)*, *SS(within)*, and *SS(total)*.

b. Compute the degrees of freedom for *SS(between)*, *SS(within)*, and *SS(total)*.

c. Compute *MS(between)* and *MS(within)*.

d. Find *F* and summarize the results in a table like that in Display 14.23.

Source	df	SS	MS	F	P
Between					
Within					
Total					

Display 14.23 An empty ANOVA table for P8.

P9. Refer to the aldrin data analyzed in Display 14.16. There are two sources of variability in the individual measurements: the depth at which the measurements were taken and the random error (or unexplained variability in measurements taken at the same depth). Which of these is reflected in *MS(between)*? Which in *MS(within)*?

P10. The readerships of three magazines *Scientific American* (1), *People* (2), and *True Confessions* (3) were rated as having high, medium, and low educational levels, respectively. In a study of readability of advertisements, six advertisements were randomly selected from each of the three magazines. Among the variables measured was the number of sentences in each advertisement. The data and dot plots are provided in Display 14.24. [Source: *lib.stat.emu.edu/ DASL/Datafiles/magadsdat.html.* The original investigation covers more magazines and additional variables.]

a. Do you think the *F*-statistic will be relatively close to 1 or much larger than 1 (say, 5 or more)?

b. Compute *SS(between)*, *SS(within)*, and *SS(total)*.

c. Complete an ANOVA table. Was your estimate in part a confirmed?

Magazine	Number of Sentences
1	9
1	20
1	18
1	16
1	9
1	16
2	9
2	9
2	17
2	13
2	11
2	7
3	18
3	12
3	12
3	17
3	11
3	20

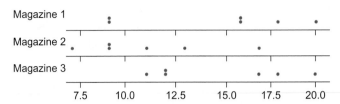

Display 14.24 Number of sentences in ads for three magazines: *Scientific American* (1), *People* (2), and *True Confessions* (3).

Steps in a Complete One-Way ANOVA

P11. Students conducted an experiment in which six teams (the treatments) launched gummy bears from Popsicle stick launchers. Display 14.25 provides dot plots of the distances (in inches) traveled by the bears in ten launches for each team, as well as the summary statistics.

a. Which team has the most within-team variation? The least? Is the within-team variation associated with the mean launch length?

b. Between which two teams is there the most between-team variation? The least?

c. Write the hypotheses to test whether team matters and compute *F*.

d. Write a conclusion.

e. Do you have any concerns about pooling to get a common estimate of the variance, s_p^2

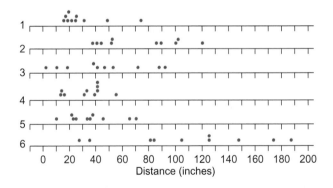

	Team	Mean	Standard Deviation
1	1	29.3	18.3609
2	2	72.1	30.4538
3	3	46	31.1091
4	4	32.4	13.9936
5	5	36.2	19.2342
6	6	109	53.6346

Display 14.25 Summary statistics of bear launch distances for six teams.

P12. Athletes often have trouble distinguishing objects when there is glare from sunlight or stadium lights. An experiment was conducted to measure the effect of applying eye-black grease, antiglare stickers, or petroleum jelly to the cheekbones on the ability of students to distinguish objects from background in bright sunlight. The subjects were 46 students between the ages of 18 and 30 who don't wear eyeglasses. They were randomly assigned to one of the three treatment groups. Each subject had his or her treatment applied, faced into the sun, and was asked to read a chart. [Source: Brian M. DeBroff and Patricia J. Pahk, "The Ability of Periorbitally Applied Antiglare Products to Improve Contrast Sensitivity in Conditions of Sunlight Exposure," *Archives of Ophthalmology*, Vol. 121 (2003), pp. 997–1001.]

a. Are the conditions for ANOVA met? (You can't check the normality condition because you don't have the individual observations.)

The results are given in Display 14.26. Larger scores are better.

Treatment	Mean	SD	n
Eye-black grease	1.87	0.09	16
Petroleum jelly	1.78	0.11	16
Antiglare sticker	1.75	0.14	14

Display 14.26 Resistance to glare for three treatments below the eye. Larger scores are better.

b. Complete an ANOVA table.

c. What is your conclusion?

Multiple Comparisons: But Which Means Differ?

P13. What level of confidence would you use in each individual confidence interval for the following situations?

a. An experiment has five treatments and you want to construct confidence intervals for four pairwise differences with an overall confidence level of 95%.

b. An experiment has three treatments and you want to construct confidence intervals for all possible differences with an overall confidence level of 98%.

P14. Return to the antiglare data in P12. Construct confidence intervals for the differences between the means of the treatments so that you have an overall level of confidence of 95%. Which treatments differ from the others?

Exercises

E9. Three different fertilizers were randomly assigned to 24 fields, with eight fields receiving each treatment. The response was the yield per field in bushels of tomatoes.

a. Fill in the missing values in this ANOVA table.

Source	df	SS	MS	F	P
1	-?-	745.36	-?-	-?-	-?-
2	-?-	-?-	-?-		
Total	-?-	1203.56			

Pooled standard deviation -?-

b. Is there a statistically significant difference in the mean yields?

E10. In an experiment, there were seven treatments and ten observations on each.

a. Fill in the missing values in this ANOVA table.

b. Is there a statistically significant difference in the means?

Source	df	SS	MS	F	P
Treatment	-?-	-?-	-?-	16.50	-?-
Error	-?-	-?-	40.1		
Total	-?-	-?-			

E11. The example on page 520 of Chapter 11 provides details and data (Display 11.3) on the age (in months) at which first steps occur for babies assigned to four different treatments. The data and ANOVA are given in Display 14.27. Test the hypothesis that all four methods have equal mean ages for first steps.

E12. *Peaches.* The yields (in bushels) of four different varieties of peach tree were compared by randomly selecting six trees of each variety grown under the same conditions in a greenhouse. The ANOVA table follows. Write a conclusion for the grower.

Source	df	SS	MS	F	P
Yield	3	0.019	0.006	0.05	0.985
Error	20	2.572	0.129		
Total	23	2.591			

E13. Participants in a summer workshop on statistics enhanced their study of experimental design by carrying out a series of experiments to measure and compare characteristics of various brands of paper towels. In one experiment a tablespoon of water was placed in the center of a towel stretched over an embroidery hoop. After waiting 30 seconds, a 50-g weight was added

Special Exercise	Exercise Control	Weekly Report	Final Report
9	11	11	13.25
9.5	10	12	11.5
9.75	10	9	12
10	11.75	1.5	13.5
13	10.5	13.25	11.5
9.5	15	13	

ANOVA: Single Factor

Summary

Groups	Count	Sum	Average	Variance
Column 1	6	60.75	10.125	2.09375
Column 2	6	68.25	11.375	3.59375
Column 3	6	69.75	11.625	2.39375
Column 4	5	61.75	12.350	0.925

ANOVA

Source of Variation	SS	df	MS	F	P
Between groups	14.4481	3	4.8160	2.0746	0.1375
Within groups	44.1062	19	2.3214		
Total	58.5543	22			

Display 14.27 Age (months) baby took first steps.

to the center of the towel, and then a series of 50-g weights were added every 3 seconds until the towel broke. The response was the total weight on the towel immediately before it broke. The data are provided in Display 14.28.

a. Beginning with one roll of each brand, how would you randomize the treatments to the towels to reduce various types of bias?

b. Is there evidence of statistically significant differences among the mean strengths of the towels?

Scott	So-Dri	Bounty	Brawny	Sparkle
800	450	950	450	300
700	500	1050	500	300
750	450	1000	450	300
700	450	1050	450	300

Display 14.28 Weight (g) when damp paper towel broke. [Source: NCSSM Statistics Leadership Institute Notes, *www.ncssm.edu/math/resources.php.*]

E14. Refer to E13. In a second experiment on paper towels, the participants measured absorption times by cutting three 1-in. by 6-in. strips from each brand and drawing marks 1- in. from each end of each strip. The bottom inch of a strip was then placed in a cup of water, and the

recorded response (Display 14.29) was the time it took the water to reach the top mark.

a. Beginning with one roll of each brand, how would you randomize the treatments to the towels to reduce various types of bias?

b. Is there evidence of statistically significant differences among the mean absorption times for the towels?

School	Brawny	Viva	Scott	Bounty	7th Gen	So-Dri	Sparkle
0.4	1.5	1.1	0.9	1.4	1.3	1.1	1.6
0.4	1.5	1.2	1.0	1.4	1.5	1.2	1.6
0.4	1.7	1.2	0.9	1.4	1.3	1.2	1.6

Display 14.29 Time (minutes) for water to travel up 4 in. [Source: NCSSM Statistics Leadership Institute Notes, *www.ncssm.edu/math/resources.php.*]

E15. Coaches of professional sports get paid well but generally have short tenures on the job. Is this lack of longevity true of all the major sports in the United States, or are there differences among the sports? Display 14.30 shows the data on coaching longevity (in seasons) as of December 2008 for four professional sports: the National Football League (NFL), the National Basketball Association (NBA), the National Hockey League (NHL), and Major League Baseball (MLB).

a. This is an observational study, as there was no random assignment of coaches to sports. Plot the data to see if the other conditions for analysis of variance appear to be met reasonably well.

b. Carry out an analysis to see if the mean longevities differ among the sports more than it is reasonable to attribute to chance alone.

c. If the outliers are removed, does the result found in part b persist?

Mike Singletary, head coach of the San Francisco 49ers.

E16. Refer to the Michelson data on the speed of light given in Display 14.14 on page 679.

a. Complete the analysis of variance and write a conclusion.

b. The first trial might be considered a learning experience for the scientist as he adjusted new laboratory equipment. If the first trial is removed, are there statistically significant differences among the means of the other four trials?

E17. *Mums.* Because plants can produce and use only so much energy, within species a plant with a long stem is likely to have a smaller flower than a plant with a short stem. Researchers experimented with growth inhibitors to see which were most effective in reducing the length of stems, anticipating that reduced stem growth would enhance the quality of the flower. Individual chrysanthemum plants of the same age were grown under nearly identical conditions, except for the growth-inhibitor treatment. Each of four treatments was randomly assigned to ten plants, whose heights (in centimeters) were measured at the outset of the experiment and after a period of 10 weeks. The difference between the initial and final heights will be called the "growth." The data are provided in Display 14.31. A goal of the study is to see if there are statistically significant differences in mean growth among the four treatments.

a. Do the design and the data appear to fit the conditions for analysis of variance?

NFL	NBA	NHL	MLB
1.9	4.3	3.4	4.0
0.9	4.3	0.4	18.6
0.9	0.3	1.4	1.6
2.9	0.3	10.4	9.0
6.9	3.3	1.4	2.0
4.9	0.3	0.2	5.0
5.9	3.8	0.4	1.0
3.9	0.3	0.4	6.0
1.9	2.3	1.0	6.9
13.9	1.3	6.4	3.0
2.9	1.3	3.4	2.0
2.9	5.3	7.4	1.2
2.9	3.3	2.4	1.0
6.9	1.3	10.4	9.0
5.9	0.3	1.4	1.0
2.9	0.3	0.4	0
0.9	0.1	3.6	7.0
2.9	4.8	0.4	1.0
8.9	4.3	3.2	1.0
2.9	0.3	3.4	2.0
4.9	0.2	3.0	4.0
2.9	1.3	0.4	1.0
0.6	0	1.0	2.0
9.9	0.3	0.2	2.0
1.9	3.3	0.4	0
2.9	0	2.4	13.0
0.4	12.0	1.1	5.0
9.9	0.1	0.4	2.0
0.6	20.0	0.4	0.5
13.9	0.1	7.4	2.0
0.9			
6.9			

Display 14.30 Longevities of head coaches or managers in four professional sports. [Source: *www.nytimes.com/interactive/2008/12/18/ sports/20081218-COACHES.html.*]

Chrysanthemum plant.

Treatment	Initial Height (cm)	Final Height (cm)	Growth (cm)
1	6	48	42
1	3	32.5	29.5
1	4	20	16
1	3	32	29
1	1	37	36
1	2	29	27
1	2.5	34	31.5
1	4.5	46	41.5
1	3.5	33	29.5
1	3.5	30	26.5
2	2	32	30
2	2	37	35
2	1.5	15.5	14
2	4	41	37
2	2	28	26
2	2	31	29
2	2	25	23
2	3	23.5	20.5
2	3	35.5	32.5
2	5	42.5	37.5
3	3	48	45
3	3.5	43.5	40
3	3	40.5	37.5
3	4	38	34
3	3	43	40
3	4	44	40
3	1	26.5	25.5
3	2	23	21
3	4	47	43
3	4	37	33
4	1.5	44.5	43
4	0	39	39
4	2.5	45.5	43
4	3	42.5	39.5
4	1	43	42
4	2	45	43
4	1.5	38	36.5
4	1	46	45
4	1.5	38.5	37
4	2.5	32	29.5

Display 14.31 Mum growth for four treatments. [Source: University of Florida Institute for Food and Agricultural Sciences.]

b. Are there statistically significant differences among the growth means?

c. Use multiple comparison procedures to sort out where the major differences in mean growth occur.

d. Which of the four treatments would you recommend?

E18. *Mercury in lakes.* Data on mercury content in water samples from three types of lakes are presented in Display 14.32. Type 1 lakes are balanced between decaying vegetation and living organisms, type 3 lakes have high decay rates and little oxygen, and type 2 lakes are between the other two types.

a. Check on the homogeneity of variance assumption. Does it appear to be met?

b. Use analysis of variance to decide if the mean mercury levels are the same for the three types of lakes. If the outliers are removed, would your conclusion remain the same?

Mercury	Type	Mercury	Type
1.050	3	0.450	3
0.230	3	1.120	2
0.100	2	0.320	3
0.770	3	0.370	2
0.910	3	0.540	2
0.250	3	0.860	2
0.130	1	0.770	3
0.290	3	0.670	2
0.410	2	0.600	2
0.210	2	0.680	3
0.940	3	0.220	2
0.360	1	0.470	2
1.220	3	0.370	2
0.240	1	0.290	3
0.900	2	0.430	3
2.500	3	0.160	1
0.340	2	0.490	2
0.400	2		

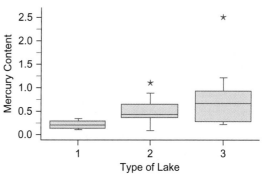

Display 14.32 Mercury content (parts per million) of lakes in Maine classified by type. [Source: R. Peck, L. Haugh, and A. Goodman, *Statistical Case Studies*, 1998, ASA-SIAM, pp. 1–14.]

c. If you decide in part b that the means differ, sort out where the differences occur.

E19. Refer to the walking babies experiment in E11.

   a. Describe how to perform a simulation to approximate the distribution of the F-statistic for this situation, constructed assuming that the null hypothesis is true that the treatments have the same mean effect on the available babies.

   b. The histogram of Display 14.33 displays the result of 200 repetitions of such a simulation. Using your F-statistic from E11, approximate a P-value from this distribution.

   c. What can you conclude?

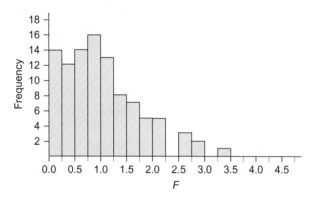

Display 14.33 Two hundred runs of a simulation to approximate the distribution of an F-statistic.

E20. Display 14.34 gives the simulated distribution of the F-statistic for the gummy bear launch data in P11. It was constructed assuming that the null hypothesis is true that the teams average equal launching distances.

   a. Use the simulation to approximate the probability of getting a value of F as large as or larger than the one you got in P11, assuming the null hypothesis is true.

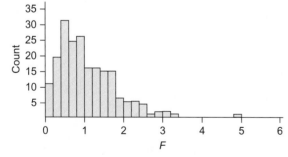

Display 14.34 Simulated distribution of the F-statistic, with 200 repetitions.

b. What is your conclusion?

c. Approximate the P-value if you had gotten $F = 3$. What would your conclusion be now?

d. Approximate the P-value if you had gotten $F = 0.14$. What would your conclusion be now?

E21. Return to Example 14.5 on page 684.

   a. Perform a different analysis on these data, a test of the slope of a regression line, using the number of milligrams of caffeine as x and the number of taps as y, as in the scatterplot of Display 14.35.

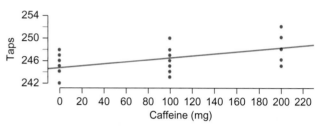

Display 14.35 Regression of finger taps on caffeine consumption.

   b. If you compare the P-value you got for this test with the one from the ANOVA on page 685, you will find that the P-value for the regression is smaller. Why should that be the case?

   c. What are the differences in the ANOVA and regression analyses?

E22. Return to E18 on page 693 about mercury in three types of lakes.

   a. Perform a different analysis on these data, a test of the slope of a regression, using the type of lake as x and the amount of mercury as y, as in the scatterplot of Display 14.36.

   b. If you compare the P-value you got for this test with the one from the ANOVA in E18, you will find that the P-value for the regression is smaller. Why should that be the case?

   c. What are the differences in the ANOVA and regression analyses?

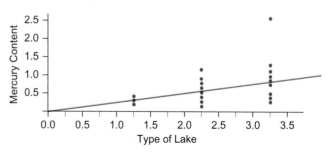

Display 14.36 Regression of mercury content on type of lake.

Chapter Summary ◀

Most real problems of practical and/or scientific interest involve more than two categories or treatments. In studying attitudes and habits of college students, you might want to make comparisons among freshmen, sophomores, juniors, and seniors. In assessing a new treatment for diabetes, you might want to compare the new treatment with the standard treatment as well as with a placebo. The methods of this chapter, called analysis of variance (ANOVA), extend the notion of testing the equality of two means with a t-test to testing for equality of k means with an F-test. The conditions necessary to insure that the test works well are basically the same as those for the t-test extended to more groups.

ANOVA is so named because the F-test is based on comparing the variation among "treatment" means (the between-treatment variation) with the pooled variation within the data for each "treatment" (the within-treatment variation). Suppose, for example, that monthly salaries within a sample of female employees of a firm vary by hundreds of dollars and salaries within a sample of male employees also vary by hundreds of dollars. If the mean of the sampled male salaries is only $50 greater than the mean of the sampled female salaries, then there is little reason to sound an alarm. If, however, the variability within the male and female salaries is only $10, a difference of $50 between their means justifies an alarm. The between-group variation becomes alarming only when it is large compared to the within-group variation.

Analysis of variance is a very widely used statistical procedure, perhaps too widely used in situations that do not come close to meeting the conditions under which the procedure was designed to operate. As in the case of all powerful tools, make sure you use it wisely!

Review Exercises

E23. *Porosity of metal.* Porosity of metal, an important determinant of strength, can be determined by looking at a cross section of the metal under a microscope. If a grid is laid on the microscopic field, the number of times the grid lines and the pore boundaries intersect is proportional to the length of the boundary per unit area and so is proportional to the pore surface area per unit volume. Such counts are given in Display 14.37 for samples of antimony, Linde copper, and electrolytic copper.
   a. What is the design of this study?
   b. Is the difference between the mean counts for the three metals statistically significant?
   c. Is there a statistically significant difference between the mean count for antimony and the mean count for Linde copper?

Antimony	Linde Copper	Electrolytic Copper	Antimony	Linde Copper	Electrolytic Copper
10	14	42	14	17	49
11	17	34	8	12	43
7	15	46	9	14	39
10	13	41	8	13	36
8	10	34	7	12	50
9	12	46	9	10	28
14	16	42	7	17	39
6	11	42	10	10	50
12	20	37	11	14	37
9	13	38	8	6	32
9	15	44	8	16	34
11	11	40	10	13	42
9	16	43			

Display 14.37 Counts of grid lines intersections and pores. Higher numbers mean higher porosity. [Source: Student experiment at the University of Florida.]

E24. Whether large doses of vitamin C help cancer patients survive longer still has not been established. In one early study, patients with terminal cancer of the stomach, colon, and breast were treated with ascorbate (vitamin C). The number of days the patients survived is given in Display 14.38.

Stomach	Colon	Breast
124	248	1235
42	377	24
25	189	1581
45	1843	1166
412	180	40
51	537	727
1112	519	3808
46	455	791
103	406	1804
876	365	3460
146	942	719
340	776	
396	372	
	163	
	101	
	20	
	283	

Display 14.38 Days survived by terminal cancer patients after treatment with ascorbate. [Source: Ewan Cameron and Linus Pauling. "Supplemental Ascorbate in the Supportive Treatment of Cancer: Reevaluation of Prolongation of Survival Times in Terminal Human Cancer," *Proceedings of the National Academy of Science*, Vol. 75, (1978), pp. 4538–4542.]

a. Make an appropriate plot to determine if there appears to be a difference in survival times for the three sites.

b. State the null and alternative hypothesis for an ANOVA.

c. Are the conditions for an analysis of variance met?

d. Take the square root of each survival time. With these transformed data, are conditions better met?

e. Perform an analysis of variance and write a conclusion.

E25. *Bears in space.* Display 14.39 shows the results from one class for a gummy bear launch experiment in which the launchers were set at two different angles, one with the rear of the launch pad one book high and another with the rear of the launch pad four books high. Six teams of students were formed, with three randomly selected teams assigned to each launch angle. The data are distances traveled by the bears from the front of the launcher.

a. What is the design of this experiment? How is it different from others you have analyzed?

b. What method might you use to analyze the results to determine if launch angle matters?

c. Does team matter?

d. Does launch angle matter?

Flatter Launch: One Book			Steeper Launch: Four Books		
Team 1	Team 2	Team 3	Team 4	Team 5	Team 6
15	44	18	13	10	125
24	40	2	39	35	147
48	51	10	16	24	35
25	100	41	41	22	81
19	37	88	41	65	125
17	52	46	31	45	27
19	102	72	41	70	187
21	86	53	55	21	84
31	89	38	14	33	105
74	120	92	33	37	174

Display 14.39 Gummy bear launch distances (in.).

E26. *Presses for pellets.* An experiment was conducted to study the effect of darkness on the learning behavior of rats. Twenty rats were taught to press a bar to get pellets of food. After the rat had learned that it got a pellet each time it pressed the bar, the number of times it pressed the bar per minute was counted. Those results are listed in the "Before" column in Display 14.40 [as 100 log (# number of bar presses per minute)]. Then, ten randomly selected rats were assigned to total darkness and the other ten rats were kept in normal lighting. At this stage, they learned that they no longer got rewarded with pellets when they pressed the bar. Later, all 20 rats were tested a second time. The numbers in the "After" column give the number of bar presses [as 100 log (# number of bar presses per minute)] during this second test period.

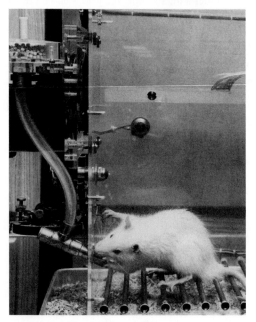

Lab rat presses a bar to release food.

a. What is the design of this study?

b. What kind of analysis should be done here?

c. Does the treatment of dark versus light seem to matter?

	Light			Dark	
Rat	Before	After	Rat	Before	After
1	271	201	11	275	157
2	231	203	12	150	123
3	176	92	13	201	176
4	249	161	14	284	135
5	240	239	15	234	74
6	290	253	16	225	138
7	175	123	17	244	153
8	236	131	18	227	88
9	256	210	19	271	155
10	239	170	20	264	160

Display 14.40 Rat learning data. [Source: George Cobb, *Introduction to Design and Analysis of Experiments* (Key College, 1998), pp. xxii–xxiii.]

d. Is there a statistically significant difference in the "before" rates of the rats assigned to the two treatment groups?

E27. The existence or nonexistence of the Mozart effect, that cognitive ability increases after listening briefly to Mozart's music, continues to be controversial. One of the leading critics conducted an experiment in which students were randomly assigned to listen to Mozart, sit in silence, or listen to Philip Glass (a modern composer). The students took a test consisting of 16 paper-folding and-cutting visualization tasks before receiving the assigned treatment. After the treatment, the students were given a different but similar set of 16 tasks. The response was the number of items answered correctly. The results are given in Display 14.41.

	Sample Size	Pretest		Posttreatment	
		Mean	SE	Mean	SE
Mozart	44	9.66	0.56	11.77	0.48
Silence	42	9.88	0.47	11.60	0.43
Glass	39	9.90	0.70	12.15	0.62

Display 14.41. Mean number of tasks answered correctly, before and after three randomly assigned treatments. [Source: K. M. Steele, K. E. Bass, and M. D. Crook "The Mystery of the Mozart Effect: Failure to Replicate." *Psychological Science*, Vol. 10 (1999), pp. 366–369.]

a. The researchers claim, "The pretest results indicate that random assignment was successful in creating groups not significantly different in initial task performance, $F(2, 122) = 0.05, P = .95$." Verify this statement.

b. Are conditions met for a test of equality of the post-treatment means?

c. Is there a statistically significant effect of treatment as shown by the post-treatment means? What is your overall conclusion?

E28. An observational study enlisted 24 adults age 50 or older who participated in a yoga class and 27 adults age 50 or older who participated in mall walking groups. Some of the variables measured by the researchers are listed in Display 14.42.

a. The researchers report that "the results of the ANOVA revealed that level of depression, $F(1, 49) = 5.38, p < 0.05$, was significantly greater, and satisfaction with life was significantly lower, $F(1, 49) = 6.60, p < 0.05$, in the yoga group." Verify this statement.

b. Write a conclusion that takes into account the design of the study.

	Yoga (n = 24)		Walking (n = 27)	
Variable	M	SD	Mean	SD
Degree of exercise	33.6	16.4	34.6	19.6
Quality of life	25.9	5.3	29.9	5.8
Level of depression	10.0	6.9	6.3	4.5

Display 14.42 For degree of exercise, higher numbers indicate higher levels of physical activity; for level of depression, higher numbers indicate more depression; for satisfaction, higher scores indicate more satisfaction with quality of life. [Source: Jennifer M. Kraemer and David X. Marquez, "Psychosocial Correlates and Outcomes of Yoga or Walking Among Older Adults," *Journal of Psychology*, Vol. 143 (2009), pp. 390–404.]

E29. Researchers asked 721 first-time pregnant women to report how much they ate of 133 individual foods in the year before conception. They found that only the amount of breakfast cereal was associated with sex of the baby. Women who ate less than one bowl per week had a smaller proportion of boys than women who ate two to six bowls a week, who had a smaller proportion of boys than women who ate seven or more bowls a week. [Source: Fiona Mathews, Paul J. Johnson, and Andrew Neil, "You Are What Your Mother Eats: Evidence for Maternal Preconception Diet Influencing Foetal Sex in Humans," *Proceedings of the Royal Society B: Biological Sciences*, Vol. 275 (2008), pp. 661–668.]

a. If the researchers had used a significance level of 0.05 and, in fact, none of the foods are associated with sex of the baby in the population of all pregnant women, what is the expected number of foods that would reach statistical significance?

b. The researchers reported that "Given the multiplicity of testing, we interpreted p-values conservatively for individual nutrient items." Explain what they mean by this. What level of significance should the researchers use to keep the overall probability of at least one Type I error in the 133 tests to at most 0.05?

c. The P-value associated with breakfast cereal was reported as 0.004. Would you consider this statistically significant?

E30. Researchers assessed eight characteristics of the sperm of 361 men consulting a fertility clinic. Based on their cell phone use, the men were divided into four groups: no use, less than 2 hours per day, 2 to 4 hours per day, and more than 4 hours per day. For four of eight variables, the difference in means was statistically significant, with sperm count, motility, viability and morphology being lower in the men who used cell phones more. Summary statistics for sperm count are given in Display 14.43.

	No Cell Phone Use	Less Than 2 Hours	2 to 4 Hours	More Than 4 Hours
Mean	85.9	69.0	58.9	50.3
SD	35.6	40.3	51.9	41.9
Sample size	40	107	100	114

Display 14.43 Cell phone use and sperm count (millions/ml). [Source: Ashok Agarwal, et al., "Effect of Cell Phone Usage on Semen Analysis in Men Attending Infertility Clinic: An Observational Study," *Fertility and Sterility*, Vol. 89 (2008), pp. 124–128.]

a. Write null and alternative hypotheses for testing the equivalence of the means.

b. Does it appear that conditions are met for a one-way analysis of variance?

c. Compute the value of F for a one-way analysis of variance. What are the degrees of freedom? Is the difference in means statistically significant?

d. For the sperm counts, the researchers also constructed all possible pairwise confidence intervals for the difference in means using the Bonferroni correction. How many confidence intervals was this? What level of confidence should they use with each if they want to be 95% confident that each interval contains the true difference?

e. Name a lurking variable and explain how it might account for any statistically significant differences.

C1. In a one-way ANOVA with four groups, the *P*-value is 0.034. What is the best interpretation of this result?

(A) The four population means all are different from each other.

(B) The four sample means all are different from each other.

(C) It's plausible that all four population means are equal.

(D) It's plausible that all four sample means are equal.

(E) At least one population mean is different from another population mean.

(F) At least one sample mean is different from another sample mean.

C2. The Clippers and the Lakers are basketball teams in Los Angeles. On each team, the shortest player is about 6 feet tall and the tallest player is about 7 feet tall. Consider the heights of the following pairs of groups of people:

    I. players for the Clippers
       adults in Los Angeles

    II. adults in Los Angeles
        adults in San Francisco

    III. players for the Clippers
         players for the Lakers

a. Which pair (I, II, or III) will have the largest between-sample variation in height?

b. Which pair (I, II, or III) will have the smallest within-sample variation in height?

C3. The boxplots show the mean life expectancy in years for men in random samples of about 15 countries from different regions of the world. Conditions are met for analysis of variance. Circle the best conclusion for a test that the mean life expectancy, by country, is the same in each region.

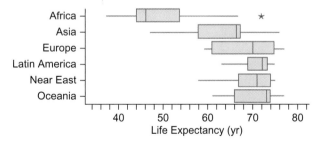

(A) The null hypothesis will be rejected because the boxplot for Africa barely overlaps the one for Latin America.

(B) The null hypothesis won't be rejected because the boxplot for Africa barely overlaps the one for Latin America.

(C) The null hypothesis will be rejected because the two boxplots for the Near East and Oceania pretty much overlap.

(D) The null hypothesis won't be rejected because the two boxplots for the Near East and Oceania pretty much overlap.

(E) The null hypothesis will be rejected because there is so much variability in Africa, Europe, and Asia.

(F) The null hypothesis won't be rejected because there is so much variability in Africa, Europe, and Asia.

C4. Refer to the boxplots in C3. Which of these regions has the smallest within-sample variability?

(A) Africa

(B) Asia

(C) Europe

(D) Latin America

(E) Near East

C5. Within an analysis comparing the regions in C3, you construct four 95% confidence intervals, each estimating the difference between the mean life expectancy for men in two different regions. What is the probability that all of the intervals capture the true difference?

(A) $(0.95)^4$

(B) $(1 - 0.95)^4$

(C) $(0.95) + (0.95) + (0.95) + (0.95)$

(D) 0.95

(E) $1 - (0.05)^4$

C6. Which of the following is *not* a reasonable description of *between-sample variation*?

(A) It is a sum of squares.

(B) It measures the variability in the sample means.

(C) It is a measure of the total distance of the sample means from the overall mean.

(D) It is a weighted average of the sample means.

(E) It appears in the numerator of the *F*-statistic.

C7. Suppose that conditions are met for ANOVA and you compute a pooled variance, s_p^2. Which of the following is *not* a reasonable interpretation of this quantity?

(A) It is an estimate of the variance in the largest population.

(B) If it is large, the *P*-value will be larger than if it is small, all other things being equal.

(C) It is the ratio of the variances of the samples.

(D) It is a weighted average of the sample variances.

(E) It is an estimate of the common variance in all of the populations.

Investigation

C8. As you saw in Exercise 31 in Chapter 11, an undesirable side effect of some antihistamines is drowsiness, which can be measured by the flicker frequency of patients (number of flicks of the eyelids per minute). Low flicker frequency is related to drowsiness because the eyes stay shut too long. The following table gives the flicker frequency for nine subjects, each given meclastine (A), a placebo (B), and promethazine (C), in random order. At the time of

699

the study, A was a new drug and C was a standard drug known to cause drowsiness.

Patient	Drug A	Drug B	Drug C
1	31.25	33.12	31.25
2	26.63	26.00	25.87
3	24.87	26.13	23.75
4	28.75	29.63	29.87
5	28.63	28.37	24.50
6	30.63	31.25	29.37
7	24.00	25.50	23.87
8	30.12	28.50	27.87
9	25.13	27.00	24.63

a. What type of experimental design was used?

b. Why isn't a one-way analysis of variance appropriate here?

c. Regardless of your answer, find the P-value using a one-way analysis of variance, as if this were a completely randomized design with 27 different patients. Is the result statistically significant?

This situation calls for a *two-way* analysis of variance, which you may study later. The null hypothesis is that the true treatment means are equal, just as in the com-

pletely randomized design. In other words, the only reason that the treatment means differ is the particular random assignment of the order of the treatments. The alternative hypothesis is that at least one pair of treatment means are different from each other.

d. The following analysis of variance table is for the two-way ANOVA. Now there are two P-values, one for the differences in treatment means and one for the differences in the means for the different patients. Compare the P-value for the treatments with your P-value from part c. What is the advantage of blocking?

Analysis of Variance for Flicker Frequency

Source	df	SS	MS	F	P
Treat	2	11.945	5.972	6.71	0.008
Block	8	163.093	20.387	22.90	0.000
Error	16	14.241	0.890		
Total	26	189.278			

e. Refer to the "Block" line in the ANOVA table in part d. Is there statistically significant evidence that the subjects vary in their mean response to the three drugs. Is this what you would expect?

Chapter 15
Multiple Regression

What characteristics of a car contribute to greenhouse gas emission? Multiple regression allows you to test many characteristics at once.

Cristiano Ronaldo

In Chapter 13, you studied how to model and make appropriate statistical inferences about the relationship between a response variable and a single explanatory variable. But most response variables are related to more than one explanatory variable, and concentrating on bivariate relationships alone limits the scope of potentially interesting discoveries you might make.

This chapter is about multiple regression, which allows you to expand your models to include more than one explanatory variable. For example, was the great soccer player Cristiano Ronaldo worth the $131.6 million "transfer fee" that Real Madrid paid Manchester United for him? To find out, Frontier Economics collected information on the the transfer fee for many soccer players, along with their age, experience, and quality such as number of goals scored. The response variable was the transfer fee. The other variables were used as the explanatory variables. The technique used to model the relationship between transfer fee and the explanatory variables was *multiple regression*. (The researchers concluded that the transfer fee was about right.) [Source: *wwwfrontier-economics.com*.]

In this chapter, you will learn

▷ how to fit a linear regression equation when there is more than one explanatory variable

▷ how to interpret the regression coefficients in a multiple regression model

▷ how to test whether the true regression coefficients are different from zero

▷ how to build categorical variables into a regression model

15.1 ▷ From One to Two Explanatory Variables

The generalization from a *simple linear regression model* (which has one explanatory variable) to a model with two explanatory variables is straightforward. As you know, a regression model is of the form

response = prediction from true regression line + random deviation

> The "true" regression equation is the one computed from the entire population.

In simple linear regression, the *prediction from true regression line* was a linear function of one explanatory variable. Now it will be a linear function of two explanatory variables, so the response will be modeled as

$$y = (\beta_0 + \beta_1 x_1 + \beta_2 x_2) + \varepsilon$$

Estimates of the parameters β_0, β_1, and β_2 will be produced by the same method as before, least squares, and the resulting equation will be of the form

$$\hat{y} = b_0 + b_1 x_1 + b_2 x_2$$

As on page 106, this least squares equation still will be the equation that minimizes the sum of the squared errors (squared residuals), SSE, where

$$\text{SSE} = \sum(y - \hat{y})^2$$

The variability, σ, in the random deviations will be estimated by

$$\sigma \approx s = \sqrt{\frac{\text{SSE}}{n - 3}}$$

The denominator is $n - 3$ because now three coefficients, β_0, β_1, and β_2 are being estimated.

As on page 129, R^2 still will be the ratio that compares SSE to SST:

$$R^2 = \frac{\text{SST} - \text{SSE}}{\text{SST}} = \frac{\Sigma(y - \bar{y})^2 - \Sigma(y - \hat{y})^2}{\Sigma(y - \bar{y})^2}$$

> An upper case R is generally used in multiple regression.

Thus, R^2 still measures the proportion of the total variation in the response that can be attributed to its linear relationship with the explanatory variables in the model.

Predicting Greenhouse Gas Emissions | Example 15.1

Suppose you want a linear model that can be used to predict approximate yearly greenhouse gas emissions (*GGE*) from a car's highway gas mileage (*HGM*) and engine size (*ES*). Each car model in the random sample in Display 15.1 (from E3 page 636) can be represented by a point (*HGM*, *ES*, *GGE*) in three-dimensional space. The least squares regression equation turns out to be

$$\widehat{GGE} = 11.649 - 0.194HGM + 0.702ES$$

Model	GGE (tons/yr)	HGM (mpg)	ES (L)
Acura TSX	7.30	30	2.4
Buick Lucerne	10.20	22	4.6
Cadillac CTS	13.10	18	6.2
Chevrolet Malibu	7.10	33	2.4
Honda Civic	6.30	34	1.8
Dodge Challenger	11.40	22	6.1
Ford Fusion	9.60	24	3.5
Kia Forte	6.10	36	2.0
Lincoln MKS	9.60	23	3.5
Mazda 3	6.60	33	2.0
Saturn Aura	9.20	26	3.6
Scion tC	7.70	29	2.4
Toyota Camry	7.10	32	2.5
Volkswagen Eos	7.30	31	2.0
Volvo S80	10.20	22	4.4

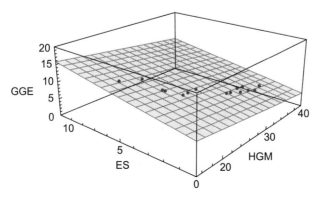

Display 15.1 Points representing (*HGM*, *ES*, *GGE*) for a random sample of car models, along with the best fitting plane. The origin is at the front of the box.

This equation represents the plane shown in Display 15.1. The points cluster fairly closely about this plane. Note that the plane slants downward with slope -0.194 as *HGM* increases and upward with slope 0.702 as *ES* increases, indicating that greenhouse gas emissions tend to go down as gas mileage increases and to go up as engine size increases. That should make practical sense as well as statistical sense.

Find the residual for each car model in Display 15.1 and then compute SSE, *s*, SST, and R^2. Interpret the value of *s* and the value of R^2.

Solution

For the Acura TSX, for example, the predicted *GGE* is

$$\widehat{GGE} = 11.649 - 0.194(30) + 0.702(2.4)$$

$$\approx 7.51$$

The residual for this car model is

$$GGE - \widehat{GGE} = 7.30 - 7.51 = -0.21$$

For the Acura, the prediction of *GGE* from the regression equation is 0.21 tons per year too low. From similar computations, all residuals and squared residuals are given in the table in Display 15.2.

Model	GGE	$\widehat{GGE}$	Residual	Squared Residual
Acura TSX	7.30	7.51	−0.21	0.0441
Buick Lucerne	10.20	10.6	−0.4	0.1600
Cadillac CTS	13.10	12.51	0.59	0.3481
Chevrolet Malibu	7.10	6.92	0.18	0.0324
Honda Civic	6.30	6.31	−0.01	0.0001
Dodge Challenger	11.40	11.66	−0.26	0.0676
Ford Fusion	9.60	9.44	0.16	0.0256
Kia Forte	6.10	6.06	0.04	0.0016
Lincoln MKS	9.60	9.64	−0.04	0.0016
Mazda 3	6.60	6.64	−0.04	0.0016
Saturn Aura	9.20	9.13	0.07	0.0049
Scion tC	7.70	7.7	0	0
Toyota Camry	7.10	7.19	−0.09	0.0081
Volkswagen Eos	7.30	7.03	0.27	0.0729
Volvo S80	10.20	10.46	−0.26	0.0676
		Total	0	0.8362

Display 15.2 Residuals and squared residuals for the car models of Display 15.1.

The SSE is the sum of the last column, 0.8362. No other equation will yield a smaller SSE.

The estimated value of σ is

$$\sigma \approx s = \sqrt{\frac{SSE}{n-3}} = \sqrt{\frac{0.8362}{12}} \approx 0.26$$

To find the SST, first compute the mean *GGE*, 8.587. Subtract this mean from each *GGE* value, square, and add the results:

$$SST = \Sigma(GGE - 8.587)^2 = (7.30 - 8.587)^2 + \cdots + (10.20 - 8.587)^2 \approx 59.797$$

Finally

$$R^2 = \frac{\text{SST} - \text{SSE}}{\text{SST}} = \frac{59.797 - 0.8362}{59.797} \approx 0.986$$

The value of s, 0.26, is an estimate of how much the responses in the population vary from the predictions given by the true regression equation. Specifically, it is an estimate of the standard deviation of all of the random deviations in the population.

The value of R^2 means that about 98.6% of the variation in the greenhouse gas emissions among different models of cars can be attributed to their differences in gas mileage and engine size. That leaves only about 1.4% of the variability in greenhouse gas emissions unaccounted for.

> Think of s as the standard deviation of the residuals.

The next example will show you how to tell whether two explanatory variables are better than just one.

Does Adding *ES* in Addition to *HGM* Help Predict *GGE*?

Example 15.2

Engine size and gas mileage are highly correlated with each other. Thus, it may seem that nothing much can be gained by having both variables in the prediction model. You'll see if this checks out. Using the random sample of 15 car models from Display 15.1, determine if adding *ES* as an explanatory variable helps predict greenhouse gas emissions (*GGE*) appreciably better than using *HGM* alone.

Solution
Display 15.3 contains the linear regression analysis using *HGM* as the single explanatory variable. As *HGM* increases, emissions decrease and this single explanatory variable accounts for 94.3% of the variation in *GGE*.

The regression equation is
GGE = 18.6 − 0.362 HGM

Predictor	Coef	Stdev	t-ratio	P
Constant	18.5907	0.6941	26.79	0.000
HGM	−0.36159	0.02463	−14.68	0.000

s = 0.5115   R-sq = 94.3%   R-sq (adj) = 93.9%

Analysis of Variance

Source	DF	SS	MS	F	P
Regression	1	56.396	56.396	215.57	0.000
Error	13	3.401	0.262		
Total	14	59.797			

Display 15.3 Linear regression of *GGE* on *HGM* from Minitab.

Now, add *ES* to the model and check the results in Display 15.4.

The two explanatory variables together account for 98.6% of the variation in *GGE*. Because there was not much variation left after accounting for gas mileage, this represents a modest but important increase over that for *HGM* alone. Note also that s, the standard deviation of the residuals, dropped from about 0.51 to 0.26, indicating that the predictions of *GGE* will have much smaller errors using the second model.

The regression equation is
GGE = 11.6 − 0.194 HGM + 0.702 ES

Predictor	Coef	Stdev	t-ratio	P
Constant	11.649	1.205	9.67	0.000
HGM	−0.19426	0.03052	−6.37	0.000
ES	0.7019	0.1163	6.04	0.000

s = 0.2650   R-sq = 98.6%   R-sq (adj) = 98.4%

Analysis of Variance

Source	DF	SS	MS	F	P
Regression	2	58.955	29.477	419.77	0.000
Error	12	0.843	0.070		
Total	14	59.797			

Display 15.4 Linear regression of *GGE* on *HGM* and *ES*.

Multiple regression equations are found using statistical software because the computations can be rather involved.

DISCUSSION
From One to Two Explanatory Variables

D1. How can you estimate the residual for the Acura, for example, from the plot in Display 15.1?

D2. Assuming that you have the regression equation, does moving from simple linear regression to regression with two explanatory variables significantly complicate the computation and interpretation of R^2?

Statistical Inference in Multiple Regression

When doing simple linear regression, you were able to ignore the analysis of variance (ANOVA) table in the computer output because it supplied much the same information as the *t*-test for the statistical significance of the slope. For example, in Display 15.3, the analysis of variance *F*-statistic is the square of the *t*-statistic for testing for significance of the slope. So these two tests are equivalent. However, in the multiple regression of Display 15.4, the analysis of variance *F*-statistic is much greater than the square of the *t*-statistic for testing either slope. Thus, the tests are not equivalent and will have to be interpreted separately.

To determine statistical significance for a multiple regression, then, you need to do three things:

1. Be sure conditions are met (more on that later).
2. Check the *P*-value in the analysis of variance table. The *F*-test tells you whether, overall, the regression is statistically significant.
3. If the regression is statistically significant overall (*F*-test), check the *P*-values of the explanatory variables (*t*-tests) to determine which of them contribute significantly to the overall statistical significance.

ANOVA: Overall, Is the Regression Statistically Significant?
The analysis of variance tests the null hypothesis

H_0: $\beta_1 = 0$ and $\beta_2 = 0$

versus the alternative

H_a: At least one of β_1, β_2 differs from 0.

The test statistic commonly used for this test is based on sums of squares. The regression sum of squares (which we will label SSR) is the difference between the total sum of squares, SST, and the sum of squared errors (residuals), SSE. That is, SSR = SST − SSE or

$$SST = SSE + SSR$$

You can think of SSR as the variability in the response that is accounted for by its linear relationship with the explanatory variables. Thus, the closer the points are to the plane, the larger the SSR will be, the smaller the SSE will be, and so the larger their ratio SSR/SSE will be. This ratio is the basis of the F-statistic, which is used to test the significance of a regression:

| F is always positive. |

$$F = \frac{\dfrac{SSR}{k}}{\dfrac{SSE}{n-k-1}} = \frac{MSR}{MSE}$$

Here, k is the number of explanatory variables and n is the sample size. The denominators are the degrees of freedom of the SSR and SSE. It is always the case that $df_{SST} = df_{SSR} + df_{SSE} = n - 1$. The "mean square" values, MSR and MSE, are the SSR and SSE divided by their degrees of freedom. You can find them in the "MS" column of an ANOVA table.

| Use technology to compute F and the P-value. |

For the same sample size and the same number of explanatory variables, the larger F is, the smaller the P-value will be.

The ANOVA Table for Predicting *GGE*

Example 15.3

Write the null and alternative hypotheses, defining any symbols you use, for the *GGE* example of Display 15.4. Then, use Display 15.4 to determine if the overall regression is statistically significant and write a conclusion.

Solution
The hypotheses are

| The true regression equation would be computed using *all* car models rather than a random sample. |

H₀: $\beta_1 = 0$ and $\beta_2 = 0$, where β_1 is the coefficient of *HGM* in the true regression equation and β_2 is the coefficient of *ES* in the true regression equation.

Hₐ: At least one of β_1, β_2 differs from 0.

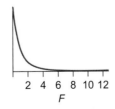

The P-value in the ANOVA table is 0.000. Thus, reject the null hypothesis that $\beta_1 = 0$ and $\beta_2 = 0$ in favor of the alternative that at least one of those coefficients is not zero. There is strong evidence that at least one of the two true regression slopes (for *HGM* or *ES*) must differ from zero, and so gas mileage is linearly related to at least one of highway gas mileage and engine size.

| The value of F, 419.77, is far out in the right tail of this F-distribution. |

The *t*-Test: Is the Contribution of This Explanatory Variable Statistically Significant?

Once you have established the overall statistical significance of the multiple regression model, you can check the individual explanatory variables to see which ones contribute significantly to the regression. These tests are not independent, and the errors (chances of making a wrong decision) can compound so that the overall error rate will exceed 5% if each individual test is made at the 5% significance level. By first conducting the ANOVA test of both parameters simultaneously, you have limited your chance of making a Type I error.

Tests of significance of the parameters in a multiple regression proceed along the same lines as in Section 13.2. The *t*-test for β_1, for example, tests the null hypothesis

$$H_0: \quad \beta_1 = 0, \text{ given that } \beta_2 \text{ is in the model}$$

versus the alternative

$$H_a: \quad \beta_1 \neq 0, \text{ given that } \beta_2 \text{ is in the model}$$

As the hypotheses suggest, the *t*-test is conditional on the other variable(s) being in the model. Thus, it is not equivalent to a separate test of the explanatory variable as if it were the only explanatory variable. For example, the *t*-statistic for *HGM* is not the same in the multiple regression in Display 15.4 as in the regression in Display 15.3 where *HGM* in the only explanatory variable. (The *P*-values aren't exactly the same either, but both round to 0.000.) Testing the statistical significance of a regression parameter in a simple linear regression can give a different picture from testing it as part of a multiple regression.

> Type I error: rejecting a true H_0.

Example 15.4

Statistical Significance of the Explanatory Variables in Predicting *GGE*

Write appropriate hypotheses and conclusions for the *t*-tests of the explanatory variables in the *GGE* regression in Display 15.4.

Solution

For *HGM*, the hypotheses are

$$H_0: \quad \beta_1 = 0, \text{ given that } ES \text{ is in the model, where } \beta_1 \text{ is the coefficient of } HGM \text{ in the true regression equation}$$

$$H_a: \quad \beta_1 \neq 0, \text{ given that } ES \text{ is in the model}$$

For *ES*, the hypotheses are

$$H_0: \quad \beta_2 = 0, \text{ given that } HGM \text{ is in the model, where } \beta_2 \text{ is the coefficient of } ES \text{ in the true regression equation}$$

$$H_a: \quad \beta_2 \neq 0, \text{ given that } HGM \text{ is in the model}$$

For the *GGE* multiple regression, both *t*-statistics are large in absolute value (-6.37 for *HGM* and 6.04 for *ES*) and both *P*-values are close to 0. Both null hypotheses should be rejected. This shows that both *HGM* and *ES* have a statistically significant effect on *GGE* *when the other is included in the regression*, even though the two explanatory variables are highly related to each other.

> The same conclusion as in Example 15.2.

DISCUSSION
Statistical Inference in Multiple Regression

D3. Describe what a Type I error and a Type II error would be in an analysis of variance F-test.

D4. Describe what a Type I error and a Type II error would be in a t-test of a coefficient.

Checking Conditions in Multiple Regression

Plotting data is essential to the modeling process. Multiple regression certainly is no exception even though you will be unable to look at the entire data set in one plot unless you have three-dimensional graphing capabilities on your software.

The conditions you must check and the methods for checking them are similar to those for simple linear regression on page 644.

> ### Checking Conditions for Inference for Multiple Regression
>
> - Verify that you have one of these situations:
>   **i.** a single random sample from a multivariate population
>   **ii.** a set of independent random samples, one for each combination of the explanatory variables
>   **iii.** an experiment with a random assignment of treatments to units
> - Check for a linear relationship by plotting the response variable against each explanatory variable. Each plot should look linear (form an elliptical cloud). In these plots, look for curvature and outliers, especially potentially influential ones.
> - As a further check on linearity and to be sure that the residuals are about the same size across all values of the explanatory variables, make a residual plot with the residuals plotted against the predicted values. Also, make a residual plot for each explanatory variable, with the explanatory variable plotted on the horizontal axis. All residual plots should look patternless.
> - Make a univariate plot (dot, stem, or box) of the residuals to see if it's reasonable to assume that they came from a normal distribution.

Randomness.

Linearity.

Equal variability in residuals.

Normality.

When multiple regression is used on an entire population or for an observational study, the randomness condition is not met. In those situations, you can proceed with the regression analysis, but your conclusion should state that you are modeling only the data you have. You are not making an inference to a larger population or establishing cause and effect. If you have statistical significance, you can say that the linear relationship is unlikely to have occurred just by chance. Refer to Section 9.3 for some examples from the univariate situation.

What if there is no randomness?

If a plot of the response variable plotted against an explanatory variable, say x_1, has a curved pattern, there are two things you can try. Applying a transformation to x_1 or to the response may straighten the plot. Common transformations are log and square root. Alternatively, you can add another term to the model, such as x_1^2. (See E7 and E8 for examples of the latter.)

What if there is curvature?

| **Example 15.5** | **Checking the Conditions for Predicting *GGE*** |

Check the car data in Display 15.1 to be sure that conditions are satisfied for proceeding with inference for a multiple regression linear model that can be used to predict greenhouse gas emissions (*GGE*) from a car's gas mileage (*HGM*) and engine size (*ES*).

Solution

- The sample is a random sample from all car models.
- The scatterplots of *GGE* versus *HGM* and *GGE* versus *ES* in Display 15.5 show a strong negative linear relationship between *GGE* and *HGM* and a strong positive linear relationship between *GGE* and *ES*.

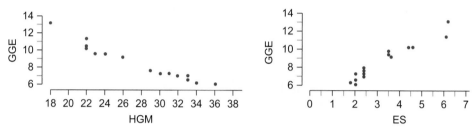

Display 15.5   Scatterplots for *GGE* versus *HGM* and *ES*.

- The three residual plots and the dot plot of the residuals from the multivariate equation are shown in Display 15.6. None of the residual plots show

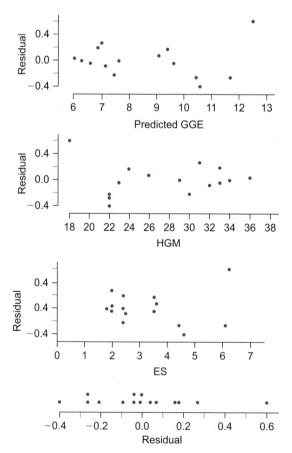

Display 15.6   Residual plots and dot plot of the residuals for *GGE*, as predicted from *HGM* and *ES*.

much of a pattern, but the Cadillac shows up as the most extreme data point in all situations.

- The residuals in the last plot in Display 15.6 look like they could have come from a normal distribution, even with some skewness induced by the Cadillac.

So it looks like the linear model with *HGM* and *ES* as explanatory variables fits the data well.

As a further check, try deleting the potential influential point, the Cadillac, from the regression. The coefficients in the regression in Display 15.7, done without the Cadillac, are not much different from those in Display 15.4. All coefficients remain statistically significant, and the value of R^2 has increased a bit due to the removal of a large residual.

The regression equation is
GGE = 12.1 − 0.199 HGM + 0.595 ES
14 cases used

Predictor	Coef	Stdev	t-ratio	P
Constant	12.0727	0.7501	16.09	0.000
HGM	−0.19907	0.01888	−10.54	0.000
ES	0.59519	0.07561	7.87	0.000

s = 0.1637   R-sq = 99.2%   R-sq (adj) = 99.1%

Analysis of Variance

Source	DF	SS	MS	F	P
Regression	2	37.677	18.839	703.12	0.000
Error	11	0.295	0.027		
Total	13	37.972			

Display 15.7   Regression analysis without the Cadillac.

To conclude, all conditions are met for ensuring the validity of the ANOVA test for the overall significance of the regression and *t*-tests of the statistical significance of the explanatory variables, *HGM* and *ES*.

DISCUSSION
Checking Conditions in Multiple Regression

D5.   If conditions are met for inference concerning a multiple regression with two explanatory variables, what does the cloud of points look like in a three-dimensional plot like that of Display 15.1?

Summary 15.1: From One to Two Explanatory Variables

Multiple regression with two explanatory variables is a generalization of the simple linear regression model

$$y = \beta_0 + \beta_1 x_1 + \varepsilon$$

to

$$y = \beta_0 + \beta_1 x_1 + \beta_2 x_2 + \varepsilon$$

Interpretations of the SSE (sum of squared residuals), s (estimate of the standard deviation of the random deviations), and R^2 (proportion of the variability in the response that is accounted for by its linear relationship with the explanatory variables) follow directly from those for simple linear regression.

With multiple regression, you first examine the analysis of variance (ANOVA) to determine the overall statistical significance of the regression. ANOVA uses the F-statistic to test the significance of a regression:

$$F = \frac{\dfrac{\text{SSR}}{k}}{\dfrac{\text{SSE}}{n - k - 1}}$$

Here, SSR is the sum of squares due to regression (SST − SSE), k is the number of explanatory variables, and n is the sample size. A small P-value for this test lets you reject the null hypothesis that *each* of the two coefficients is equal to 0 ($\beta_1 = \beta_2 = 0$) in favor of the alternative that at least one coefficient is nonzero.

Once you have established the overall statistical significance of a multiple regression, a second part of the regression analysis allows you to use the familiar t-test to check each explanatory variable to see if it contributes significantly to the regression. A small P-value for a t-test for β_1, for example, is evidence against the null hypothesis and so allows you to reject null hypothesis that $\beta_1 = 0$, given that β_2 is in the model, in favor of the alternative that β_2 differs from 0, given that β_2 is in the model.

Practice

From One to Two Explanatory Variables

P1. Match each description of three variables to a graph of the regression plane.

  A. When x_1 and x_2 both increase, y tends to increase.

  B. When x_1 and x_2 both increase, y tends to decrease.

  C. When x_1 increases and x_2 decreases, y tends to increase.

  D. When x_1 increases and x_2 decreases, y tends to decrease.

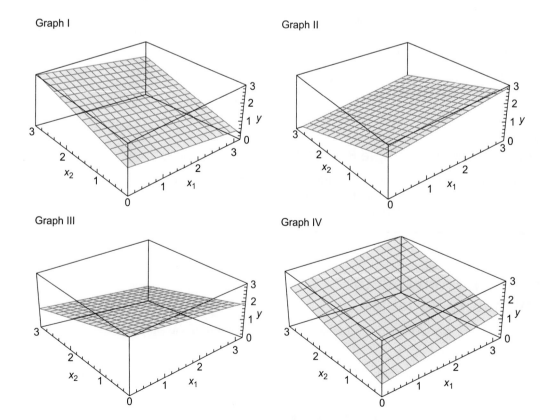

P2. The table in Display 15.8 gives information about a random sample of eight earthquakes taken from all 1043 earthquakes with a magnitude of 2.5 or greater on the Earth over a 40-day period. The least squares multiple regression equation for predicting magnitude from latitude and depth is

$$\widehat{magnitude} = 4.6 - 0.02 \; latitude - 0.002 \; depth$$

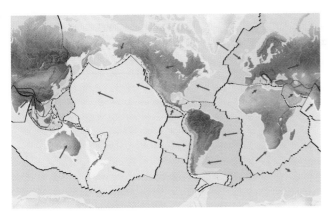

Direction of movement of Earth's tectonic plates

a. Find the residuals and then compute SSE, SST, s, and R^2.

b. Interpret the value of R^2.

Magnitude	Latitude (deg)	Depth (km)
3.9	−7	422
4.4	−34	35
4.1	28	6
5.3	−4	35
3.1	3	24
3.1	60	44
5.7	−23	38
3.8	53	0

Display 15.8 A random sample of eight earthquakes from those with a magnitude of 2.5 or greater over a 40-day period. [Source: United States Geological Survey, Earthquake Hazards Program, *http://neic.usgs.gov.*]

P3. Refer to P2 and then study the two regression outputs in Display 15.9. Does adding *depth* as an explanatory variable help predict the magnitude of an earthquake appreciably better than using *latitude* alone?

The regression equation is
Magnitude = 4.64 − 0.0219 Latitude − 0.00216 Depth

Predictor	Coef	Stdev	t-ratio	P
Constant	4.6417	0.3057	15.18	0.000
Latitude	−0.021905	0.007366	−2.97	0.031
Depth	−0.002156	0.001859	−1.16	0.299

s = 0.6651   R-sq = 64.2%   R-sq(adj) = 49.9%

Analysis of Variance

Source	DF	SS	MS	F	P
Regression	2	3.9633	1.9816	4.48	0.077
Error	5	2.2117	0.4423		
Total	7	6.1750			

The regression equation is
Magnitude = 4.45 − 0.0195 Latitude

Predictor	Coef	Stdev	t-ratio	P
Constant	4.4457	0.2620	16.97	0.000
Latitude	−0.019513	0.007271	−2.68	0.036

s = 0.6839   R-sq = 54.5%   R-sq(adj) = 47.0%

Analysis of Variance

Source	DF	SS	MS	F	P
Regression	1	3.3684	3.3684	7.20	0.036
Error	6	2.8066	0.4678		
Total	7	6.1750			

Display 15.9 Two regression analyses for the earthquake magnitudes.

Statistical Inference in Multiple Regression

P4. Verify that the entries in the output of Display 15.4 satisfy the following equations.

$$SST = SSE + SSR$$

$$F = \frac{\dfrac{SSR}{k}}{\dfrac{SSE}{n - k - 1}} = \frac{MSR}{MSE} = 419.77$$

$$df_{SST} = df_{SSR} + df_{SSE}$$

P5. After looking at the car data in Display 15.1, someone suggests that the physical size of the car should affect the greenhouse gas emissions. Thus, a model for predicting *GGE* with both gas mileage (*HGM*) and size of the passenger compartment (*PCS*), in cubic feet, as explanatory variables should work well. Do you agree? The data and analysis are given in Display 15.10 on the next page.

P6. Refer to P3 and the first regression analysis in Display 15.9.

a. Write hypotheses for the ANOVA *F*-test and the *t*-tests.

b. Write a conclusion for this analysis.

Checking Conditions in Multiple Regression

P7. The table in Display 15.11 (on the next page) describes all of the earthquakes in 2006 that had a magnitude of 7.0 or greater. You want to model the relationship between magnitude of the earthquake and the explanatory variables of time of day and latitude.

a. What type of study is this? Is the randomization condition met?

b. Check the plots to determine if the other conditions are met for statistical inference on the multiple regression. Explain any concerns you might have.

c. Write an appropriate conclusion for the regression analysis in Display 15.11.

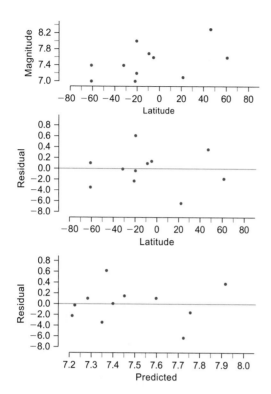

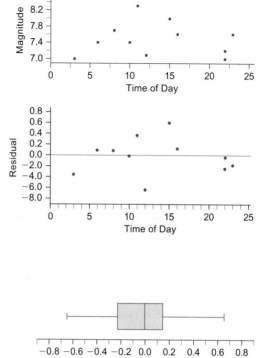

Model	GGE	HGM	PCS
Acura TSX	7.30	30	95
Buick Luarne	10.20	22	108
Cadillac CTS	13.10	18	100
Chevrolet Malibu	7.10	33	95
Honda Civic	6.30	34	91
Dodge Challenger	11.40	22	91
Ford Fusion	9.60	24	101
Kia Forte	6.10	36	97
Lincoln MKS	9.60	23	106
Mazda 3	6.60	33	94
Saturn Aura	9.20	26	98
Scion tC	7.70	29	85
Toyota Camry	7.10	32	101
Volkswagen Eos	7.30	31	77
Volvo S80	10.20	22	98

Display 15.10 Linear regression of *GGE* on *HGM* and *PCS*.

Region	Magnitude	Latitude	Time of Day (24-hr clock)
East of the South Sandwich Islands	7.4	−60.957	6
Fiji	7.2	−19.926	22
Banda Sea	7.6	−5.473	16
Mozambique	7.0	−21.324	22
Koryakia, Russia	7.6	60.949	23
Tonga	8.0	−20.187	15
Kermadec Islands	8.0	−20.187	15
South of Java, Indonesia	7.4	−31.81	10
Scotia Sea	7.7	−9.284	8
Kunril Islands	7.0	−61.029	3
	8.3	46.592	11
Taiwan	7.1	21.799	12

Display 15.11 Earthquakes in 2006 that had a magnitude of 7.0 or greater. [Source: United States Geological Survey, Earthquake Hazards Program, *http://neic.usgs.gov/neis/ eq_depot/2006/2006_stats. html.*]

The regression equation is
GGE = 19.7 − 0.368 HGM − 0.0101 PCS

Predictor	Coef	Stdev	t-ratio	P
Constant	19.723	2.339	8.43	0.000
HGM	−0.36758	0.02796	−13.15	0.000
PCS	−0.01009	0.01985	−0.51	0.620

s = 0.5267   R-sq = 94.4%   R-sq(adj) = 93.5%

Analysis of Variance

Source	DF	SS	MS	F	P
Regression	2	56.468	28.234	101.76	0.000
Error	12	3.329	0.277		
Total	14	59.797			

The regression equation is
Magnitude = 7.83 + 0.00683 Latitude − 0.0213 Time

Predictor	Coef	SE Coef	T	P
Constant	7.8313	0.3136	24.97	0.000
Latitude	0.006831	0.003544	1.93	0.090
Time	−0.02133	0.02040	−1.05	0.326

s = 0.382514   R-sq = 31.8%   R-sq(adj) = 14.8%

Analysis of Variance

Source	DF	SS	MS	F	P
Regression	2	0.5458	0.2729	1.87	0.216
Residual Error	8	1.1705	0.1463		
Total	10	1.7164			

Exercises

E1. You want to model the relationship between a country's life expectancy at birth and the explanatory variables of the proportion of the country that is forested and its military expenditure as a percentage of the gross domestic product (GDP). The table in Display 15.12 includes all Latin American countries for which the data were available, plus a few Caribbean island nations.

a. What kind of study is this? Will inference to a larger population be possible?

Country	Life Expectancy at Birth	Military Expenditure (% of GDP)	Proportion of Country That Is Forested
Argentina	74.8	0.97	0.12
Bahamas	71.0	0.65	0.37
Bolivia	64.8	1.87	0.53
Brazil	71.2	1.42	0.56
Chile	78.2	3.67	0.21
Colombia	72.8	3.71	0.53
Ecuador	74.7	2.41	0.38
El Salvador	71.3	0.62	0.14
Guatemala	67.9	0.41	0.36
Honduras	68.6	0.64	0.41
Jamaica	70.9	0.71	0.31
Mexico	75.4	0.39	0.33
Nicaragua	70.4	0.69	0.40
Paraguay	71.4	0.75	0.45
Peru	70.7	1.25	0.53
Uruguay	75.6	1.38	0.09
Venezuela	74.2	1.11	0.52

Display 15.12 Life expectancy data for Latin American countries for which the data were available. [Source: World Bank Database, *web.worldbank.org/data*.]

b. Judging from the plots in Display 15.13, are the linearity, equal variability in the residuals, and normality conditions met?

c. The regression analysis is given in Display 15.14. Is the regression statistically significant overall? Which explanatory variables are statistically significant?

d. Write an appropriate conclusion for this situation.

The regression equation is
Life = 74.4 − 11.5 Forested + 1.39 Military

Predictor	Coef	SE Coef	T	P
Constant	74.368	1.858	40.02	0.000
Forested	−11.526	4.417	−2.61	0.021
Military	1.3939	0.6489	2.15	0.050

s = 2.66325   R-sq = 42.2%   R-sq(adj) = 33.9%

Analysis of Variance

Source	DF	SS	MS	F	P
Regression	2	72.429	36.214	5.11	0.022
Residual Error	14	99.301	7.093		
Total	16	171.729			

Unusual Observations

Obs	Forested	Life	Fit	SE Fit	Residual	St Resid
3	0.530	64.800	70.865	0.998	−6.065	−2.46 R

R denotes an observation with a large standardized residual.

Display 15.14 Linear regression analysis of *life expectancy at birth* on *military expenditure as a percentage of the GDP* and *proportion of the country that is forested*.

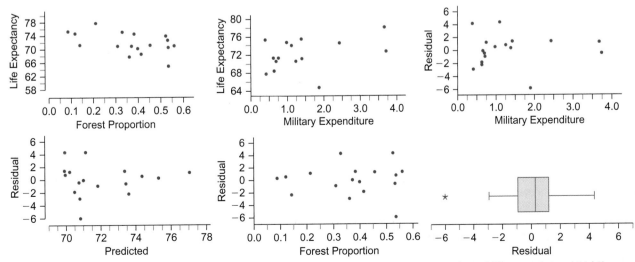

Display 15.13 Scatterplot and residual plots and plot of residuals for the linear regression of *life expectancy at birth* on *military expenditure as a percentage of the GDP* and *proportion of the country that is forested*.

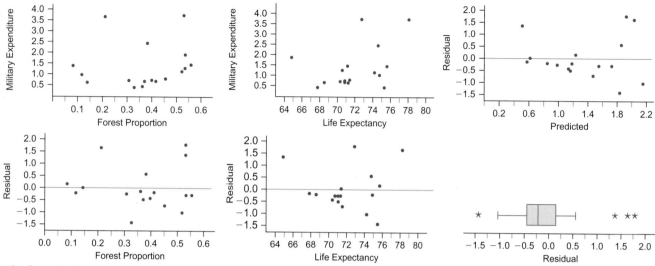

Display 15.15 Scatterplot, residual plots, and plot of the residuals for the linear regression of *military expenditure as a proportion of the GDP* on *life expectancy at birth* and *proportion of the country that is forested* for Latin American countries.

E2. Refer to the table in Display 15.12. Suppose that this time you want to model the relationship between military expenditure as the response variable and the explanatory variables of proportion of the country that is forested and the life expectancy at birth.

a. What kind of study is this? Will inference to a larger population be possible?

b. Judging from the plots in Display 15.15, are the linearity, equal variability in the residuals, and normality conditions met?

c. The regression analysis is given in Display 15.16. Is the regression statistically significant overall? If so, which explanatory variables are statistically significant?

The regression equation is
Military = −12.5 + 0.178 Life + 2.67 Forested

Predictor	Coef	SE Coef	T	P
Constant	−12.451	6.306	−1.97	0.068
Life	0.17783	0.08279	2.15	0.050
Forested	2.672	1.786	1.50	0.157

s = 0.951256   R-sq = 25.9%   R-sq(adj) = 15.3%

Analysis of Variance

Source	DF	SS	MS	F	P
Regression	2	4.4275	2.2137	2.45	0.123
Residual Error	14	12.6684	0.9049		
Total	16	17.0959			

Display 15.16 Linear regression analysis of *military expenditure as a proportion of the GDP* on *life expectancy at birth* and *proportion of the country that is forested* for Latin American countries.

d. Write an appropriate conclusion for this situation, performing any further analysis that you might need.

E3. You want to find a good model to predict the yearly change in the urban population for all countries of the world using two explanatory variables. You obtain the information in Display 15.17 about a random sample of 15 world countries.

a. Use your software to experiment with different models. Which do you think is the best? Be sure to check conditions for inference.

b. For the model you have selected, interpret the value of R^2.

c. For the model you have selected, interpret the result from the analysis of variance.

d. For the model you have selected, interpret the results from the *t*-tests.

e. Write a summary of your findings.

E4. You want to find a good model to predict a student's height from two other measurements. You obtain the information in Display 15.18 from a random sample of 15 students.

a. Use your software to experiment with different models. Which do you think is the best? Be sure to check conditions for inference.

b. For the model you have selected, interpret the value of R^2.

c. For the model you have selected, interpret the result from the analysis of variance.

d. For the model you have selected, interpret the results from the *t*-tests.

e. Write a summary of your findings.

Country	Yearly Change in Urban Population (%)	Yearly Change in Rural Population (%)	Infant Mortality Rate (per 1000 live births)	Men's Life Expectancy at Birth (yr)	Women's Life Expectancy at Birth (yr)
Bangladesh	3.5	1.4	52	63	65
French Polynesia	1.4	2	8	72	77
Bhutan	5.1	1.9	45	64	67
Ghana	3.8	0.7	57	60	60
China	3.1	−0.9	23	71	75
Lebanon	1.2	0.1	22	70	74
Brunei	2.9	0.6	6	75	80
Thailand	1.4	−2.1	11	66	75
Moldova	−0.1	−0.5	16	65	72
Sweden	0.4	0.1	3	79	83
Liberia	2.7	−0.4	133	45	47
Isle of Man	0	−0.1	2	74	80
Burundi	6.1	2.7	99	48	51
Laos	4.1	1.9	51	63	66
Canada	1.2	0.3	5	78	83

Display 15.17 Yearly change in the urban population for a random sample of 15 world countries. [Source: Official national life tables compiled by the United Nations Demographic Yearbook system, published by the Secretariat of the Pacific Community. Data refer to census year.]

Student	Height (cm)	Arm Span (cm)	Kneeling Height (cm)	Hand Length (cm)
1	170.5	168	126	18
2	170	172	129.5	18
3	107	101	79.5	10
4	159	161	116	16
5	166	166	122	18
6	175	174	125	19.5
7	158	153.5	116	16
8	95.5	95	71.5	10
9	132.5	129	95	11.5
10	165	169	124	17
11	179	175	131	20
12	149	154	109.5	15.5
13	143	142	111.5	16
14	158	156.5	119	17.5
15	161	164	121	16.5

Display 15.18 Physical measurements of a random sample of 15 students.

Player	Total Player Rating	Runs Scored	Number of Times Caught Stealing
Willie Upshaw	−5.8	596	59
Jerry Hairston	−3.3	216	5
Shanty Hogan	3.5	288	2
Joe Kuhel	−16.5	1236	90
Rick Reichardt	0	391	41
Stan Lopata	6.6	375	11
Heinie Manush	5.2	1287	58
Mike Ryan	−9.3	146	4
Dave Duncan	−6.9	274	13
Ron Hansen	14.5	446	14
Tim McCarver	6.8	590	49
John Shelby	−15.3	389	40
Jim Rice	30.9	1249	34
Lenny Green	−13.7	461	41
Eddie Robinson	−1.3	546	12
Dave Nelson	−5.9	340	73
Gene Clines	−6.1	314	40
Bob Aspromonte	−16.9	386	24

Display 15.19 Random sample of players eligible for the Major League Baseball Hall of Fame in 2000. [Source: James J. Cochran, "Career Records for All Modern Position Players Eligible for the Major League Baseball Hall of Fame," *Journal of Statistics Education*, Vol. 8, 2 (2000), *www.amstat.org/publications/jse/secure/v8n2/datasets.cochran.new.cfm.*]

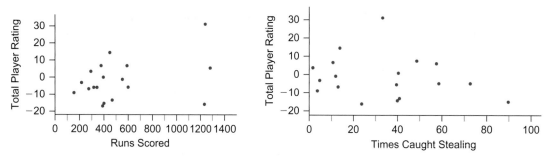

Display 15.20 Scatterplots for the linear regression of *total player rating* on *runs scored* and *times caught stealing*.

E5. *Too few explanatory variables.* The players listed in Display 15.19 are a random sample from the 1,340 baseball players who were eligible for the Major League Baseball Hall of Fame (were retired and had played at least 10 seasons). *Total player rating* is a combination of various statistics, adjusted for the player's position.

a. Do a simple linear regression of *total player rating* (response) on *runs scored* and a simple linear regression of *total player rating* (response) on *times caught stealing*. Is there a statistically significant linear relationship between *total player rating* and *runs scored*? Is there a statistically significant linear relationship between *total player rating* and *time caught stealing*? Does your impression from the two scatterplots in Display 15.20 agree with the computations?

b. Perform a multiple regression of *total player rating* on *runs scored* and *times caught stealing*. What conclusion can you draw from the analysis of variance? Are either of the explanatory variables statistically

significant, given that the other variable is in the model?

c. Are all conditions met for the regression from part b?

d. What seems paradoxical about the results from parts a and b? Can you resolve this paradox?

E6. *Too few explanatory variables.* The scatterplot in Display 15.21 shows the crime rate plotted against the high school graduation rate for a random sample of Florida counties.

a. Do a simple linear regression of *crime rate* (response) on *graduation rate*. Is the relationship between *graduation rate* and *crime rate* statistically significant? Is the direction of the relationship what you would expect?

b. What pattern in the scatterplot might account for the surprising relationship?

c. The table in Display 15.22 includes a third variable, percentage of the population that lives in urban areas. Do a multiple regression analysis adding *percentage urban* as a second explanatory variable. Describe your findings.

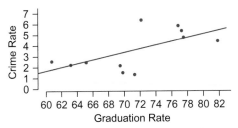

— Crime = 169 Grad Rate − 8.5e + 03; $r^2 = 0.39$

Display 15.21 Scatterplot of *crime rate* versus *high school graduation rate* for a random sample of Florida counties. [Source: Florida Department of Education, Florida Bureau of Law Enforcement.]

County	Crimes per 100,000 Residents	High School Graduation Rate (%)	Percentage Urban
Franklin	2542.8	60.7	26.2
Hamilton	2266.3	63.2	19.2
Washington	1521.3	69.8	17.2
Holmes	1348.6	71.3	21.1
Orange	6347.2	72.2	96.6
Manatee	5831.7	76.9	91.8
Hillsborough	5336.7	77.3	94.4
Volusia	4337.5	81.9	90.5
Suwannee	2507.8	65.1	18.5
Bradford	2178.6	69.5	33.7
Bay	4703.5	77.5	89.2

Display 15.22 Data for linear regression of *crime rate* on *high school graduation rate* and *percentage urban* for a random sample of Florida counties. [Source: Florida Department of Education, Florida Bureau of Law Enforcement.]

E7. *From straight lines to curves.* As you have observed in studying bivariate data, straight lines are not always adequate to capture the relationship. This exercise will show you how to fit a quadratic equation using multiple regression.

Galileo performed an experiment to help him understand the motion of falling bodies. The end of a ramp was set at various heights above the floor. He rolled a ball down the ramp and allowed the ball to drop to the floor. The response was the horizontal distance the ball traveled from the end of the ramp to the point where it hit the floor. Display 15.23 provides the data, scatterplot, and residual plot, which show that a linear equation is not a good fit. Galileo conjectured that a quadratic equation such as *distance* = $b_0 + b_1(height) + b_2(height^2)$ might be a better fit.

a. Square each height. Then fit a multiple regression equation to the response variable of *distance* using the explanatory variables of *height* and *height²*. What equation do you get?

b. Graph this quadratic equation on a copy of the scatterplot. How well does it fit the points?

c. Is your conclusion from part b consistent with the value of R^2 in the regression analysis?

Galileo Galilei (1564–1642).

Distance	Height
573	1000
534	800
495	600
451	450
395	300
337	200
253	100

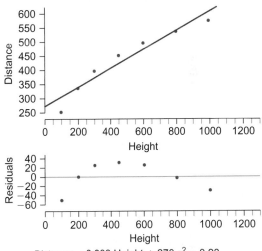

Distance = 0.333 Height + 270 $r^2 = 0.93$

Diplay 15.23 Data and scatterplot for Galileo's height and distance experiment. [Source: D. A. Dickey and T. Arnold, "Teaching Statistics with Data of Historic Significance: Galileo's Gravity and Motion Experiments," *Journal of Statistics Education*, Vol. 3, l (1995).]

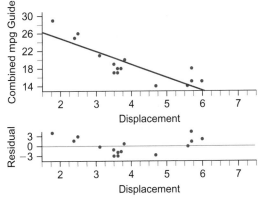

— COMB MPD GUIDE = −2.87 DISPLACEMENT + 30.4: $r^2 = 0.71$

Manufacturer	Model	Displacement	Combined MPG
Dodge	Charger AWD	5.7	18
Mitsubishi	Raider Pickup 2WD	3.7	18
Nissan	Armada 4WD	5.6	14
Infiniti	FX35 AWD	3.5	17
Audi	A6	3.1	21
Porsche	Carrera 2 S Cabriolet	3.8	20
Jeep	Grand Cherokee 4WD	5.7	15
Cadillac	SRX 2WD	3.6	17
Dodge	RAM 1500 Pickup 4WD	4.7	14
BMW	760Li	6.0	15
Nissan	Altima Coupe	2.5	26
Nissan	Quest	3.5	19
Kia	Optima	2.4	25
Honda	Civic	1.8	29
Saturn	Outlook AWD	3.6	18

Display 15.24 Data and scatterplot of *combined mpg* versus *engine displacement* for a random sample of 2008 car models. [Source: *www.fueleconomy.gov/feg/download.shtml*.]

E8. *From straight lines to curves.* Refer to E7 for a discussion of including a quadratic term in a regression. To find a model that predicts a car's combined miles per gallon (mpg) from its engine displacement, a random sample of 15 car models was examined. The results are given in Display 15.24. The scatterplot, and residual plot show that a linear equation is not a good fit.

a. Square each displacement. Then fit a multiple regression equation to the response variable of *mpg* using the explanatory variables of *displacement* and *displacement*^{2}. What equation do you get?

b. Graph this quadratic equation on a copy of the scatterplot. How well does it fit the points?

c. Is your conclusion from part b consistent with the value of R^2 in the regression analysis?

E9. *The interpretation of $R^2(adj)$.* In a multiple regression, R^2(adj) is an adjustment of the value of R^2 to account for the fact that when any variable is added to a multiple regression model, statistically significant or not, the value of R^2 increases. R^2(adj) is the better value to look at when judging the proportion of the variability in the response that is accounted for by its linear relationship with the explanatory variables. You can interpret R^2(adj) the same way that you interpret R^2.

a. Refer to the two regression analyses in P3. Is the contribution of *depth* to the regression statistically significant?

b. How much does the value of R^2 change when *depth* is removed from the regression? How much does the value of R^2 (adj) change? Which appears to give

a better indication of the importance of *depth* to the regression?

c. Verify in both regressions that

$$R^2(adj) = 1 - \frac{\dfrac{SSE}{n - k - 1}}{\dfrac{SST}{n - 1}}$$

E10. *The interpretation of $R^2(adj)$.* Refer to the explanation of R^2(adj) in the previous exercise and to the regressions in Display 15.3 and Display 15.4.

a. Is the contribution of *ES* to the regression in Display 15.4 statistically significant?

b. How much does the value of R^2 change when *ES* is removed from the regression? How much does the value of R^2(adj) change? Does this reflect the contribution of *ES*?

c. Verify in both regressions that the formula IN E9, part c, holds.

E11. *The interpretation of SST.* Refer to the two regression analyses in P3.

a. Compare the value of SST in the first regression to that in the second.

b. Explain your finding in part a.

E12. *The interpretation of SST.* Refer to the two regression analyses in Display 15.3 and Display 15.4.

a. Compare the value of SST in the first regression to that in the second.

b. Explain your finding in part a.

15.2 ► From Two to More Explanatory Variables, Including Categorical Variables

Response variables generally are affected by many more than two possible explanatory variables, some numerical and some categorical. The multiple regression methodology remains the same for any number of explanatory variables: Fit a linear equation by minimizing the SSE, check the F-test in the analysis of variance for overall significance, check the t-tests for significance of individual predictors, and plot the data to make sure you are not missing anything important. In the following examples, four explanatory variables are evaluated.

Modeling Greenhouse Gas Emissions from Four Explanatory Variables

Example 15.6

In Display 15.25, two additional variables, number of cylinders (CY) and passenger compartment size (PCS), are added to the greenhouse gas emissions (GGE) data from the last section. Certainly large cars would emit more greenhouse gases than small cars. The two new variables are an indication of the size of the car and so might help predict greenhouse gas emissions. Compare the regression analysis in Display 15.26 (on the next page), which includes the two new explanatory variables of CY and PCS, with the original analysis in Display 15.27, which includes only HGM and ES. Which explanatory variables do you recommend stay in the model?

Model	GGE (tons/year)	HGM (mpg)	ES (L)	CY	PCS (cu ft)
Acura TSX	7.30	30	2.4	4	95
Buick Luceme	10.20	22	4.6	8	108
Cadillac CTS	13.10	18	6.2	8	100
Chevrolet Malibu	7.10	33	2.4	4	95
Honda Civic	6.30	34	1.8	4	91
Dodge Challenger	11.40	22	6.1	8	91
Ford Fusion	9.60	24	3.5	6	101
Kia Forte	6.10	36	2.0	4	97
Lincoln MKS	9.60	23	3.5	6	106
Mazda 3	6.60	33	2.0	4	94
Satum Aura	9.20	26	3.6	6	98
Scion tC	7.70	29	2.4	4	85
Toyota Camry	7.10	32	2.5	4	101
Volkswagen Eos	7.30	31	2.0	4	77
Volvo S80	10.20	22	4.4	8	98

Display 15.25 Data related to automobile greenhouse gas emissions (GGE): highway gas mileage (HGM), engine size (ES), number of cylinders (CY), and passenger compartment size (PCS).

Solution
The CY coefficient could be considered significantly different from zero at the 10% level, but not at the 5% level of significance. The PCS is nowhere near significance in the presence of the other explanatory variables. Adding CY and PCS does very little to improve the ability of the model to predict GGE (both R^2 and s change very little). In the interest of keeping the model as simple as possible, it is reasonable to recommend that they both should be dropped from the analysis.

> Use as few explanatory variables as will be sufficient.

Regression Statistics

R	0.9950
R square	0.9899
Adjusted R Square	0.9859
Standard Error	0.2453
Number of Cases	15

ANOVA

	df	*SS*	*MS*	*F*	*P-level*
Regression	4	59.1958	14.7989	246.0030	0.0000
Residual	10	0.6016	0.0602		
Total	14	59.7973			

	Coefficients	*Standard Error*	*t-Stat*	*P*
Intercept	13.5901	1.6165	8.4070	0.0000
HGM	−0.2246	0.0321	−6.9927	0.0000
ES	0.8659	0.1439	6.0187	0.0000
CY	−0.2282	0.1316	−1.7343	0.0719
PCS	−0.0041	0.0097	−0.4236	0.9439

Display 15.26   StatPlus regression analysis of the car emission data in Display 15.25.

The reason C don't add much to the regression is that the number of cylinders and passenger compartment size both are closely related to the size of the engine, and so the "car size" piece of the variability has already been accounted for without them.

Regression Statistics

R	0.9929
R Square	0.9859
Adjusted R Square	0.9836
Standard Error	0.2650
Number of Cases	15

ANOVA

	df	*SS*	*MS*	*F*	*p-level*
Regression	2	58.9547	29.4773	419.7717	0.0000
Residual	12	0.8427	0.0702		
Total	14	59.7973			

	Coefficients	*Standard Error*	*t-Stat*	*p-level*
Intercept	11.6495	1.2049	9.6685	0.0000
HGM	−0.1943	0.0305	−6.3651	0.0000
ES	0.7019	0.1163	6.0359	0.0000

Display 15.27   Regression analysis of the car emission data without *CY* and *PCS*.

Example 15.7 — Modeling Prices of Houses

What variables help to explain the price of a single-family house? Some are obvious, like the area of the living space, but others are not so obvious. How about number of bedrooms or bathrooms? Are both important? To find out, variables on 93 houses sold in one month in Gainesville, Florida, were analyzed. Display 15.28 contains the first six cases. Find a good fitting regression equation for modeling the relationship of *price* with the given explanatory variables.

Price ($1000s)	Area (1000 sq. ft.)	Bedrooms	Bathrooms	Age (Old = 0; New = 1)
48.6	1.10	3	1	0
55.0	1.01	3	2	0
68.0	1.45	3	2	0
137.0	2.40	3	3	0
309.4	3.30	4	3	1
17.5	0.40	1	1	0

Display 15.28 House price and related variables for six cases.

Solution

There are four possible explanatory variables offered and all seem reasonable predictors of the price of a house, so all four are included in the regression analysis in Display 15.29.

Response: Price

Summary of Fit

RSquare	0.868863
RSquare Adj	0.862902
Root Mean Square Error	16.35994
Mean of Response	99.53333
Observations	93

Analysis of Variance

Source	DF	Sum of Squares	Mean Square	F-Ratio	Prob > F
Model	4	156052.88	39013.2	145.7634	< .0001
Error	88	23552.98	267.6		
Total	92	179605.87			

Parameter Estimates

Term	Estimate	Std Error	t-Ratio	Prob > |t|
Intercept	−41.7945	12.10408	−3.45	0.0009
Area	64.760606	5.629568	11.50	< .0001
Bedrooms	−2.765564	3.959807	−0.70	0.4868
Bathrooms	19.203105	5.649951	3.40	0.0010
Age	18.984237	3.872689	4.90	< .0001

Display 15.29 Regression analysis of house prices with four explanatory variables, produced by JMP software.

The huge F-ratio (145.7) in the ANOVA indicates that at least some of the explanatory variables are important. The individual t-statistics indicate that all but "bedrooms" may have a statistically significant affect on price. Dropping "bedrooms" from the model results in the analysis in Display 15.30 (on the next page). The three remaining explanatory variables all remain statistically significant, and the value of R^2 decreases very little by simplifying the model in this way.

As you learned earlier, it is important to examine residuals when exploring the fit of a model. Display 15.31 (on the next page) shows the plot of residuals against the predicted values of price for the model of Display 15.30. The fifth house on the list, selling for over $309,000 is way out of the pattern set by the rest of the data. It could be highly influential in the analysis, but an analysis with this data point deleted shows that not to be the case.

Response: Price

Summary of Fit

RSquare	0.868136
RSquare Adj	0.863691
Root Mean Square Error	16.31279
Mean of Response	99.53333
Observations	93

Analysis of Variance

Source	DF	Sum of Squares	Mean Square	F-Ratio	Prob > *F*
Model	3	155922.33	51974.1	195.3127	< .0001
Error	89	23683.53	266.1		
Total	92	179605.87			

Parameter Estimates

Term	Estimate	Std Error	t-Ratio	Prob > |*t*|
Intercept	−47.99179	8.208585	−5.85	< .0001
Area	62.262676	4.33488	14.36	< .0001
Bathrooms	20.072031	5.495382	3.65	0.0004
Age	18.370963	3.760955	4.88	< .0001

Display 15.30   Regression analysis of house prices with three explanatory variables.

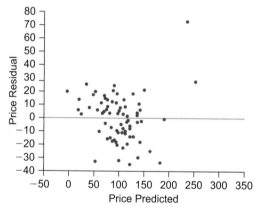

Display 15.31   Residual plot from regression analysis with three explanatory variables.

DISCUSSION
From Two to More Explanatory Variables, Including Categorical Variables

D6. The output in Display 15.32 gives the simple linear regression of *house price* on *bedrooms* for the complete data set of Display 15.28. Compare the statistical significance of the coefficient of *bedrooms* in this regression and the one in Display 15.29. Provide a plausible reason as to why these are so different.

The regression equation is

Price = −37.2 + 43.0 Bedrooms

Predictor	Coef	Stdev	t-ratio	p
Constant	−37.23	19.96	−1.87	0.065
Bedrooms	42.969	6.160	6.98	0.000

s = 35.86   R-sq = 34.8%   R-sq (adj) = 34.1%

Analysis of Variance

Source	DF	SS	MS	F	p
Regression	1	62578	62578	48.66	0.000
Error	91	117028	1286		
Total	92	179606			

Display 15.32   Minitab regression of price on number of bedrooms.

Categorical Explanatory Variables

You may have noticed that one of the explanatory variables used in modeling the price of houses was of a different nature from the others. The age variable was categorical, old or new, with an arbitrary assignment of the numerical value 0 to indicate "old" and 1 to indicate "new." The regression coefficient of age, 18.37 in Display 15.30, gets added to the predicted value of price when age = 1 (new house) and does not get added when age = 0 (old house). In other words, a new house tends to cost $18,370 more than an old house with the same square footage and same number of bathrooms.

Categorical explanatory variables occur often in multiple regression analyses. Suppose, for example, you want to model income as a function of years of formal education for a sample of adults from your area of the country. Incomes for females generally are not as high as incomes for males; so a second explanatory variable should be gender. But gender is categorical rather than numerical. No problem—use a numerical code for gender and proceed with regression analysis as usual. The next example, from an experiment, will explore categorical explanatory variables a little further.

For convenience of analysis, categorical outcomes can be coded numerically.

Flowering Meadowfoam	**Example 15.8**

Meadowfoam grows in the Pacific Northwest and is valued for the nongreasy and highly stable oil that comes from its seeds. Researchers interested in improving the flower production of this plant conducted an investigation involving the intensity and starting time of the lighting used in the greenhouse growing meadowfoam seedlings. The numerical variable of *light intensity* was set at one of six levels. The lighting was started either when flowering was induced (late) or 24 days prior to flowering inducement (early), so there were two levels of the categorical variable *starting time*.

The resulting 12 treatments were randomly assigned to 24 groups of seedlings, so that each treatment was measured on two groups. The response for each group was the average number of flowers per plant. The results are shown in Display 15.33. The categorical variable *starting time* is arbitrarily coded as 0 and 1 for the two categories, but these codes cannot be interpreted as measures.

Flowering meadowfoam.

Mean Number of Flowers	Light Intensity (μmol/m²/sec)	Starting Time of Light (0 = late, 1 = early)
62.30	150	0
77.40	150	0
55.30	300	0
54.20	300	0
49.60	450	0
61.90	450	0
39.40	600	0
45.70	600	0
31.30	750	0
44.90	750	0
36.80	900	0
41.90	900	0
77.80	150	1
75.60	150	1
69.10	300	1
78.00	300	1
57.00	450	1
71.10	450	1
62.90	600	1
52.20	600	1
60.30	750	1
45.60	750	1
52.60	900	1
44.40	900	1

Display 15.33 Flowering meadowfoam data. [Source: M. Seddigh and G. D. Jolliff, "Light Intensity Effects on Meadowfoam Growth and Flowering," *Crop Science*, Vol. 34 (1994), pp. 497–503.]

Find an appropriate regression model for predicting flower production from a combination of light intensity and starting time of the lighting.

Solution

First, look at the plot of the data in Display 15.34. The responses from the early starting time are the squares and the ones from the late starting time are the circles. Two things stand out: (1) a strong decreasing trend in mean number of flowers as the light intensity increases and (2) a tendency for the responses for the early starting times to lie above those for the late starting times. Intensity and timing both matter.

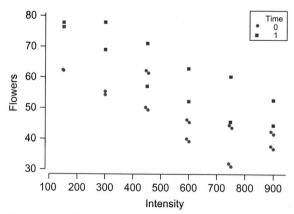

Display 15.34 Scatterplot of the meadowfoam data. Early starting times of light are represented by squares and late starting times by circles.

Even though *starting time* is categorical, its numerically coded values can be used as an explanatory variable in a regression model. The results of modeling the mean number of flowers as a linear function of *intensity* and *starting time* are shown in Display 15.35.

The regression equation is
Flowers = 71.3 − 0.0405 Intensity + 12.2 Time

Predictor	Coef	SE Coef	T	P
Constant	71.306	3.274	21.78	0.000
Intensity	−0.040471	0.005132	−7.89	0.000
Time	12.158	2.630	4.62	0.000

S = 6.44107   R-Sq = 79.9%   R-Sq(adj) = 78.0%

Analysis of Variance

Source	DF	SS	MS	F	P
Regression	2	3466.7	1733.4	41.78	0.000
Residual Error	21	871.2	41.5		
Total	23	4337.9			

Display 15.35 Regression of mean number of flowers on light intensity and starting time of the light.

The *F*-ratio of 41.78 is highly significant, yielding a *P*-value close to 0. So one or both of the explanatory variables must be making a significant contribution. The *t*-tests on the regression coefficients verify that both the intensity of the light and the starting time are playing a statistically significant role.

The equation of the prediction model is

$$\widehat{flowers} = 71.3 - 0.04 \, intensity + 12.2 \, time$$

Regardless of the timing, the number of flowers tends to decrease by about 4 for each additional 100 units of light intensity. But the mean number of flowers produced changes with the timing. At *starting time* = 0, the prediction equation is

$$\widehat{flowers} = 71.3 - 0.04 \, intensity + 12.2 \cdot 0 = 71.3 - 0.04 \, intensity$$

At *starting time* = 1, the equation is

$$\widehat{flowers} = 71.3 - 0.04 \, intensity + 12.2 \cdot 1 = 83.5 - 0.04 \, intensity$$

The plot of these two equations in Display 15.36 shows two parallel regression lines. These two lines capture the important features of the data much better than would a single regression line running through the center of the cloud of points.

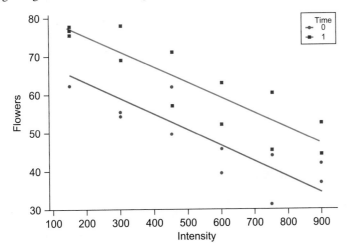

Display 15.36 Scatterplot of the flower data with regression lines for the two levels of *time*.

Interaction

The modeling for the meadowfoam data could be taken a step further. Perhaps the two regression lines, one for each starting time, would fit the data better if they were allowed to have different slopes. That implies that the regression coefficient for *intensity* would have to have a component that could change with *time*, which can be accomplished by adding another variable, the product of *intensity* and *time*. (Such terms are often called *interactions*.) The regression analysis that includes the variable (*intensity*) · (*time*) is shown in Display 15.37.

The regression equation is
Flowers = 71.6 − 0.0411 Intensity + 11.5 Time + 0.0012 Intensity x Time

Predictor	Coef	SE Coef	T	P
Constant	71.623	4.343	16.49	0.000
Intensity	−0.041076	0.007435	−5.52	0.000
Time	11.523	6.142	1.88	0.075
Intensity x Time	0.00121	0.01051	0.12	0.910

S = 6.59795   R-Sq = 79.9%   R-Sq(adj) = 76.9%

Analysis of Variance

Source	DF	SS	MS	F	P
Regression	3	3467.3	1155.8	26.55	0.000
Residual Error	20	870.7	43.5		
Total	23	4337.9			

Display 15.37 Regression analysis of meadowfoam data with intensity by time interaction.

The *intensity by time* interaction is not even close to being statistically significant; so you would not add an interaction term to the model. But, if it had been statistically significant, you would use the following two equations.
At *starting time* = 0, the equation is

$$\widehat{flowers} = 71.6 − 0.0411 \cdot intensity + 11.5\ time + 0.0012(intensity)(time)$$
$$= 71.6 − 0.0411 \cdot intensity + 11.5 \cdot 0 + 0.0012(intensity)(0)$$
$$= 71.6 − 0.0411 \cdot intensity$$

At *starting time* = 1, the equation is

$$\widehat{flowers} = 71.6 − 0.0411 \cdot intensity + 11.5\ time + 0.0012(intensity)(time)$$
$$= 71.6 − 0.0411 \cdot intensity + 11.5 \cdot 1 + 0.0012(intensity)(1)$$
$$= 83.1 − 0.0399 \cdot intensity$$

Adding the interaction term would add only 0.0012 units to the slope as you moved from late to early starting time. The second line would not look much different from that in Display 15.36.
You will see examples of statistically significant interaction in E19 and E20.

DISCUSSION
Categorical Variables

D7. Will the two regression lines from a multivariable regression (such as those from the meadowfoam example in Display 15.35) where there is one categorical explanatory variable and one quantitative explanatory variable always be parallel?

D8. In general, will doing separate regressions as if the other variable didn't exist (such as *flowers* on *intensity* and *flowers* on *starting time*) give the same regression coefficients as those from a multivariable analysis?

Summary 15.2: From Two to More Explanatory Variables, Including Categorical Variables

Fitting a multiple regression model to data with more than two explanatory variables proceeds along the same lines as those used in Section 15.1, except that you can no longer draw a picture of the regression surface in three dimensions. In checking conditions, it is important to plot the residuals against the predicted values to make sure that a few data points are not exerting undue influence on the results.

Categorical explanatory variables occur quite commonly in multiple regression analyses. Numerical coding of the categories allows the use of such variables in fitting a regression model, but the resulting regression coefficients require a different interpretation. For a categorical variable, the coefficient measures the change in the intercept as you move from one category to another. If you have an interaction term, the slope also changes as you move from one category to another.

Practice

From Two to More Explanatory Variables

P8. Determining the gestation period (length of pregnancy) of mammals that live in the wild can be difficult. Perhaps the gestation period can be predicted fairly accurately from other variables that can be more easily observed. The output in Display 15.38 shows the output of a multiple regression analysis with *gestation period* (days) as the response and *female maturity* (days to maturity), *male maturity* (days to maturity), *weaning* (typical number of days baby is nursed), *litter size* (typical), and *birth weight* (grams). The cases were a random sample of 117 mammals from all mammals for which the data are available.

Pregnant gorilla.

a. As a group, are these variables helpful in predicting the gestation period for mammals? How can you tell from the output?

b. Which variables are statistically significant?

Regression Statistics

R	0.8205
R Square	0.6732
Adjusted R Square	0.6585
Standard Error	75.1731
Number Of Cases	117

ANOVA

	df	SS	MS	F	p-level
Regression	5	1292433.0	258486.500	45.7418	0.0000
Residual	111	627259.7	5650.988		
Total	116	1919692.0			

	Coefficients	Standard Error	t-Stat	p-level
Intercept	110.8647	16.2809	6.8095	0.0000
Female maturity	0.0354	0.0359	0.9875	0.4363
Male maturity	0.0217	0.0312	0.6952	0.7881
Weaning	0.1378	0.0521	2.6459	0.0000
Litter size	−17.0889	3.8631	−4.4236	0.0000
Birth weight	0.0001	0.0000	2.0330	0.0018

Display 15.38 StatPlus multiple regression with *gestation period* as the response variable.

c. What proportion of the variability in the gestation period from mammal to mammal is accounted for by this regression?

d. The standard error is about 75 days. Does this indicate that the error in your estimate of a gestation period is likely to be relatively small or relatively large?

e. The coefficient on *litter size* is negative. What does this mean in the context of the situation?

P9. In this problem, you will consider various relationships between the number of police officers in a state and expenditures for keeping those officers on the job, population of the state, and its violent crime rate. Using the random sample of states given in Display 15.39, find a good fitting equation to predict the number of officers as a function of the other variables, and justify your decision.

State	Police (thousands)	Expenditures ($ millions)	Population (millions)	Violent Crime Rate (per 100,000)
Colorado	12	753	4.3	334
Florida	55.2	3371	16	812
Illinois	44.1	2718	12.4	657
Iowa	7.3	346	2.9	266
Louisiana	16.1	635	4.5	681
Maine	3.1	118	1.3	110
Mississippi	8.6	337	2.8	361
New Jersey	33.4	1829	8.4	384
Tennessee	18.1	828	5.7	707
Texas	58.9	2866	20.9	545
Virginia	18.8	965	7.1	282
Washington	14.1	854	5.9	370

Display 15.39 Number of police officers for a random sample of states. [Source: U.S. Census Bureau, *Statistical Abstract of the United States, 2004–2005.*]

Categorical Explanatory Variables

P10. Refer to the meadowfoam data in Display 15.33 and regression output in Display 15.35.

   a Is the randomization condition for statistical inference met for this experiment?

   b. Check the plots in Display 15.40 to determine if the other conditions are met for statistical inference on the multiple regression.

   c What plots were not given? Why can they be ignored?

P11. Refer to the regression analysis for the meadowfoam experiment, in Display 15.35.

   a. Write hypotheses for the *t*-tests of the statistical significance of the explanatory variables.

   b. Write a conclusion for each *t*-test.

P12. An observational study of a sample of 1498 French middle-aged workers (day shift) was conducted to explore the association between sleep duration and daily coffee consumption (for people who drink the equivalent of less caffeine than is contained in 8 cups of coffee). A multiple regression with total sleep time as the response variable and various independent variables resulted in the estimates in Display 15.41.

   a. Which of these were categorical variables? From the information in the table, how do you think each variable was coded?

   b. Which of the variables are statistically significant at the 5% level?

   c. How can you tell that you should write different modeling equations for each gender? Write these two equations. (The intercept was not given in the report, so write it as b_0.) Are the slopes, intercepts, or both different in your two equations?

   d. How can you tell that you should not write two different modeling equations for alcohol use?

Independent Variables	Slope	P-Value
Number of cups of coffee/day	−0.007	0.78
Gender (female)	0.064	<0.05
Smoking status (smokers)	0.042	0.12
Alcohol use (heavy drinkers)	−0.026	0.35
Use of hypnotics (regular use)	0.020	0.44
Age (years)	−0.054	0.06

Display 15.41 Multiple regression analysis of total sleep time (minutes) on various independent variables for a sample of 1498 French middle-aged workers. [Source: Montserrat Sanchez-Ortuno et al., "Sleep Duration and Caffeine Consumption in a French Middle-Aged Working Population," *Sleep Medicine*, Vol. 6 (2005), pp. 247–251.]

   e. The slopes of three variables are negative. Explain what each means in the context of this situation.

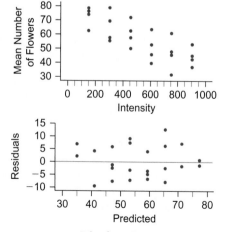
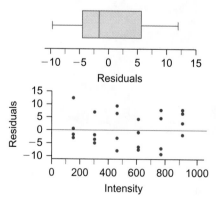

Display 15.40 Plots of meadowfoam data and residuals.

f. The researchers concluded, "Multiple linear regression analysis did not find a significant relationship between total sleep time and daily caffeine intake less than 8 cups of coffee per day, after controlling for age, gender, alcohol intake, smoking status, and use of hypnotics." Do you agree?

P13. Refer to P12. The researchers reported that they did not find a statistically significant interaction between gender and caffeine intake.

a. Sketch a plot , similar to Display 15.36, of the relationship among total sleep time, gender, and caffeine consumption.

b. Sketch a plot of what the relationship between total sleep time, gender, and caffeine would look like if there had been interaction. Assume that females who drink a lot of caffeine sleep longer than those who do not and that the opposite is true of males.

Exercises

E13. You want to be able to estimate the speed of mammals. You suspect that maximum life span and whether the mammal is a predator are strongly related to speed. You collect this information for a random sample of mammals, using the code 1 for predator and 0 for nonpredator, and get the regression analysis of Display 15.42. All conditions for inference for regression are met.

The regression equation is
Speed = 35.9 + 12.5 Predator − 0.148 MaxLife

Predictor	Coef	Stdev	t–ratio	P
Constant	35.873	7.308	4.91	0.0000
Predator	12.492	5.593	2.23	0.037
MaxLife	−0.1481	0.1768	−0.84	0.411

s = 12.06   R–sq = 29.5%   R–sq(adj) = 22.7%

Analysis of Variance

Source	DF	SS	MS	F	P
Regression	2	1275.8	637.9	4.39	0.026
Error	21	3054.7	145.5		
Total	23	4330.5			

Display 15.42 Linear regression analysis for *speed of mammals* on *predator* and *maximum life span*.

a. How many mammals are in your sample?

b. Write a conclusion based on the analysis of variance table.

c. What can you conclude from the *t*-tests for the explanatory variables?

d. How can you tell that you should write separate modeling equations for the predators and for the nonpredators? Write these two equations. Are the slopes, intercepts, or both different in your two equations? What is the effect of being a predator?

E14. You want to be able to estimate the maximum life span of various mammals. You suspect that gestation period and whether the mammal is a predator are strongly related to maximum life span. You collect this information for a random sample of mammals, using the code 1 for predator and 0 for nonpredator, and get the regression analysis in Display 15.43. All conditions for inference for regression are met.

a. How many mammals are in your sample?

b. Write a conclusion based on the analysis of variance table.

c. What can you conclude from the *t*-tests for the explanatory variables?

d. Should you write separate modeling equations for the predators and for the nonpredators? If so, do this. If not, explain why not.

The regression equation is
MaxLife = 15.8 + 0.824 Gestation − 1.37 Predator

Predictor	Coef	Stdev	t–ratio	P
Constant	15.777	3.592	4.39	0.000
Gestation	0.08239	0.01291	6.38	0.000
Predator	−1.370	4.179	−0.33	0.745

s = 10.92   R–sq = 58.0%   R–sq(adj) = 55.6%

Analysis of Variance

Source	DF	SS	MS	F	P
Regression	2	5778.5	2889.3	24.21	0.000
Error	35	4176.4	119.3		
Total	37	9954.9			

Display 15.43 Linear regression analysis for *maximum life span* on *predator* and *gestation period*.

E15. *Too many explanatory variables*. Refer to E1. In Display 15.44, a new variable for the Central American countries, annual population growth rate, has been added. You want to model the relationship between life expectancy at birth and the explanatory variables of proportion of country that is forested, military expenditures as a percentage of the gross domestic product, and the annual population growth rate.

a. What type of study is this? Is the randomization condition met?

b. Check the plots in Display 15.45 to determine if the other conditions are met for statistical inference on the multiple regression.

c. The regression is given in Display 15.46. Write an appropriate conclusion. Compare your conclusion to your conclusion for E1.

Country	Annual Population Growth Rate (%)	Life Expectancy at Birth	Military Expenditures (% of GDP)	Proportion of Country That Is Forested
Argentina	0.974145	74.8	0.97	0.12
Bahamas	1.34026	71.0	0.65	0.37
Bolivia	1.90176	64.8	1.87	0.53
Brazil	1.3461	71.2	1.42	0.56
Chile	1.05672	78.2	3.67	0.21
Colombia	1.41806	72.8	3.71	0.53
Ecuador	1.43474	74.7	2.41	0.38
El Salvador	1.73732	71.3	0.62	0.14
Guatemala	2.44462	67.9	0.41	0.36
Honduras	2.19465	68.6	0.64	0.41
Jamaica	0.473126	70.9	0.71	0.31
Mexico	1.01335	75.4	0.39	0.33
Nicaragua	0.516704	70.4	0.69	0.40
Paraguay	1.91019	71.4	0.75	0.45
Peru	1.46175	70.7	1.25	0.53
Uruguay	0.120876	75.6	1.38	0.09
Venezuela	1.70769	74.2	1.11	0.52

Display 15.44 Annual population growth rate added to the life expectancy data for Latin American countries. [Source: World Bank Database, *web.worldbank.org/data.*]

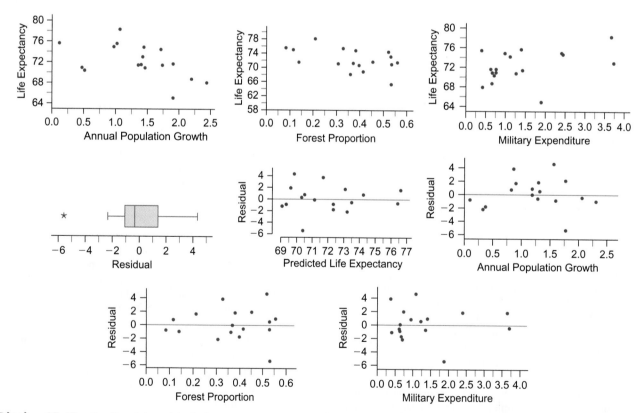

Display 15.45 Scatterplot, residual plots, and plot of residuals for the linear regression of *life expectancy at birth* on *military expenditure as a percentage of the GDP, annual population growth rate,* and *proportion of the country that is forested.*

Response: Life Exp

Summary of Fit

R Square	0.500694
R Square Adj	0.38547
Root Mean Square Error	2.56823
Mean of Response	71.99412
Observations	17

Analysis of Variance

Source	DF	Sum of Squares	Mean Square	F-Ratio
Model	3	85.98392	28.6613	4.3454
Error	13	85.74549	6.5958	**Prob > F**
Total	16	171.72941		0.0250

Parameter Estimates

Term	Estimate	Std-Error	t-Ratio	Prob > |t|
Intercept	75.673717	2.01023	37.64	<.0001
Growth Rate	−1.684836	1.175267	−1.43	0.1753
Military Exp.	1.2440826	0.63444	1.96	0.0717
Forest	−8.316139	4.812347	−1.73	0.1076

Display 15.46 JMP linear regression of *life expectancy at birth* on *military expenditure as a percentage of the GDP, annual population growth rate,* and *proportion of the country that is forested* for Latin American countries.

E16. *Too many explanatory variables.* Refer to P7 on page 713. In Display 15.47, for all of the earthquakes in 2006 that had a magnitude of 7.0 or greater, a new variable, depth of the earthquake, has been added. You want to model the relationship between magnitude of the earthquake and the explanatory variables of depth beneath the surface of the Earth, time of day, and latitude.

Magnitude	Latitude	Depth (km)	Time of Day
7.4	−60.957	13	6
7.2	−19.926	583	22
7.6	−5.473	397	16
7.0	−21.324	11	22
7.6	60.949	22	23
8.0	−20.187	55	15
7.4	−31.81	152	10
7.7	−9.284	20	8
7.0	−61.029	13	3
8.3	46.592	10	11
7.1	21.799	10	12

Display 15.47 Data for all 2006 earthquakes with magnitude ≥ 7. [Source: U.S. Geological Survey, Earthquake Hazards Program, *http://neic. usgs.gov/neis/eq_depot/2006/2006_stats.html.*]

a. What type of study is this? Is the randomization condition met?

b. Check the plots in Display 15.48 to determine if the other conditions are met for statistical inference on the multiple regression.

c. The regression is given in Display 15.49. Write an appropriate conclusion. Compare your conclusion to your conclusion for P7.

E17. The chemical process of oxidizing ammonia to form nitric acid allows some of the ammonia to escape through the exhaust stacks, this amount being referred to as the *stack loss.* The key variables in the production process are the *air flow* into the processor (which controls the rate at which the process operates), the *temperature* of the cooling water (which helps control the temperature at which the process operates), and the *concentration* of nitric acid in the ammonia being processed. It is desirable to keep stack loss to a minimum, but accomplishing that requires knowledge of how stack loss is related to the key variables that govern the process. Display 15.50 provides data on stack loss for 17 runs of this process at various levels of the three controllable explanatory variables.

a. Find a good fitting model that allows the prediction of stack loss from some or all of the process variables. Explain why you settled on this model.

b. For the model chosen in part *a,* describe what each of the regression coefficients is measuring in the context of the problem.

E18. Refer to E17 and the data of Display 15.50. The first data point has a very high value for *x,* and an outlier *y.* Delete this case and find what you believe to be the "best" fitting model for predicting *y* based on some or all of the three explanatory variables. Does your result differ markedly from what you found in part a?

E19. *Adding an interaction term.* This small example shows what you can do when you have a categorical variable for which the regression lines clearly should not be parallel. Six chimpanzees—three youngsters and three older chimps—were taught to touch, in numerical order, numbers that appeared in random position on a screen. After learning this, the task was changed so that once the first number was touched, the others were replaced with blank squares. Thus to complete touching the numbers in order, the chimp had to remember where each number was. The table in Display 15.51 gives the age of the chimp, the time to touch the first numeral after the numerals appeared on the screen, and the percentage of trials in which the chimp managed to touch all squares in order. The multiple regression equation is

$$\overline{percentage\ correct} = 107.9 - 13.9\ age - 0.03\ time$$

a. What is the equation for the young chimps (*age* = 0)? For the older chimps (*age* − 1)? Are the slopes of these two equations equal?

b. In the scatterplot in Display 15.52, circles represent the young chimps and squares the older chimps. Does it appear that the slope of the equation for the young chimps should be equal to that for the older chimps?

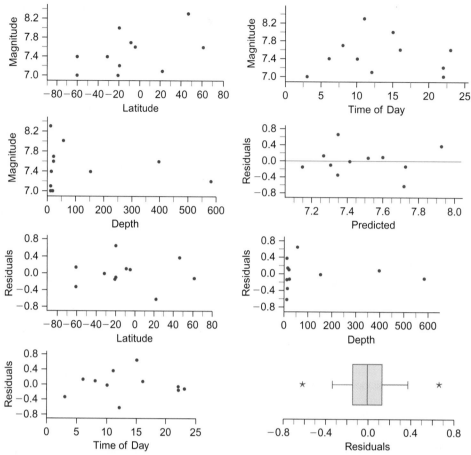

Display 15.48 Scatterplot, residual plots, and plot of residuals for the linear regression of *magnitude* on *depth, time of day*, and *latitude*.

The regression equation is
Magnitude = 7.87 + 0.000260 Depth − 0.0259 Time + 0.00735 Latitude

Predictor	Coef	SE Coef	T	P
Constant	7.8669	0.3498	22.49	0.000
Depth	0.0002602	0.0007910	0.33	0.752
Time	−0.02588	0.02569	−1.01	0.347
Latitude	0.007351	0.004079	1.80	0.114

s = 0.405799   R−sq = 32.8%   R−sq(adj) = 4.1%

Analysis of Variance

Source	DF	SS	MS	F	P
Regression	3	0.5637	0.1879	1.14	0.397
Residual Error	7	1.1527	0.1647		
Total	10	1.7164			

Display 15.49 Regression analysis of linear regression of *magnitude* on *depth, time of day*, and *latitude*.

c. Add a third explanatory variable to the analysis by adding an interaction term. Multiply the response time by the coded age: *time • age*. Add this variable to your model.

d. What is the multiple regression equation now? What is the equation for the young chimps? For the older chimps? Are the two slopes about what you would want?

E20. *Adding an intersection term.* The table in Display 15.53 gives the hand length and height of a sample of 11 male students and 19 female students. A multiple regression using *height* as the response and *hand length* and *gender* as the explanatory variables gives the equation

$$\widehat{height} = 48.8 + 2.39\ hand\ length + 3.59\ gender$$

Air Flow (coded)	Temperature of the Cooling Water (°C)	Concentration of Acid 10(actual % −50)	Stack Loss 10(actual % loss)
80	27	88	37
62	22	87	18
62	23	87	18
62	24	93	19
62	24	93	20
58	23	87	15
58	18	80	14
58	18	89	14
58	17	88	13
58	18	82	11
58	19	93	12
50	18	89	8
50	18	86	7
50	19	72	8
50	19	79	8
50	20	80	9
56	20	82	15

Display 15.50 Stack loss in oxidizing ammonia. [Source: S. B. Vardeman, *Statistics for Engineering Problem Solving*, (PWS, 1994), p. 128.]

Chimp	Age (young = 0; older = 1)	Time to Begin (ms)	Percentage of Correct Trials
Ayumu	0	885	81
Cleo	0	672	83
Pal	0	929	83
Ai	1	635	84
Chloe	1	659	67
Pan	1	944	62

Display 15.51 Chimpanzee data. [Source: Sana Inoue and Tetsuro Matsuzawa, "Working Memory of Numerals in Chimpanzees," *Current Biology*, Vol. 17 (December 4, 2007), pp. R1004–R1005 and supplements.]

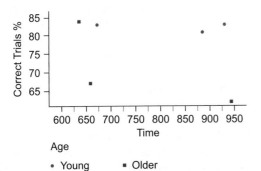

Display 15.52 Scatterplot of the percentage of correct trials versus time where the circles represent the young chimps and the squares represent the older chimps.

The numbers are replaced by white squares and Ayumu correctly touches the other squares in the correct order.

Hand Length (in.)	Height (in.)	Gender (0 = female, 1 = male)
7.25	70.5	0
7	63	0
7.25	66	0
7	69	0
6.5	61.5	0
7.5	65	0
8	71	1
7.25	68	0
7.63	71	1
6.5	61.5	0
8	62	0
7.75	73	1
6	62	0
6.7	70	0
6.75	64	1
6.625	62	0
7.8	73	1
7	64	0
5.8	62	0
8.5	71.5	1
7.5	70.5	1
8	71.5	1
6	66	0
6.5	66	0
7.5	68.5	1
8.5	76	1
7	65	0
6.5	68	0
7.25	66	0
8	72	1

Display 15.53 Height and hand length by gender for a sample of students.

a. What is the equation for the females (*gender* = 0)? For the males (*gender* = 1)? Are the slopes of these two equations equal?

b. In the scatterplot in Display 15.54, the regression lines were computed separately for the females and for males. Does it appear reasonable to assume that the slopes are equal?

c. Add an interaction term as in E19 to construct a multivariate regression model that produces modeling lines with different slopes for males and females.

d. What is the equation for the females? For the males? Are the two slopes about what you would expect?

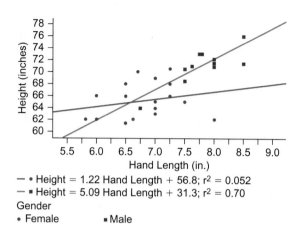

— • Height = 1.22 Hand Length + 56.8; r^2 = 0.052
— ▪ Height = 5.09 Hand Length + 31.3; r^2 = 0.70
Gender
 • Female   ▪ Male

Display 15.54 Scatterplot for *height* versus *hand length* where the circles represent the females and squares represent the males.

Chapter Summary

Multiple regression is a generalization of the simple linear regression model

$$y = \beta_0 + \beta_1 x_1 + \varepsilon$$

to one with more than one explanatory variable

$$y = \beta_0 + \beta_1 x_1 + \beta_2 x_2 + \cdots + \beta_k x_k + \varepsilon$$

In multiple regression, some of the explanatory variables can be categorical variables, where the categories typically are coded with integer values.

Interpretations of the SSE (sum of squared residuals), s (estimate of the standard deviation of the random deviations), and R^2 (proportion of the variability in the response that is accounted for by its linear relationship with the explanatory variables) follow directly from those for simple linear regression.

With multiple regression you first conduct an analysis of variance (ANOVA) of the overall significance of the regression. ANOVA uses the F-statistic to test the significance of a regression:

$$F = \frac{\dfrac{\text{SSR}}{k}}{\dfrac{\text{SSE}}{n - k - 1}}$$

Here, SSR is the sum of squares due to regression (SST−SSE), k is the number of explanatory variables, and n is the sample size. A small P-value for this test lets you reject the null hypothesis that *each* coefficient is equal to 0: $\beta_1 = \beta_2 = \beta_3 = \cdots = \beta_k = 0$ in favor of the alternative that at least one coefficient is nonzero.

Once you have established the overall statistical significance of a multiple regression, a second part of the regression analysis allows you to use the familiar t-test to check each explanatory variable to see if it contributes significantly to the regression. A small P-value for a t-test for β_1, for example, is evidence against the null hypothesis and so allows you to reject the null hypothesis that $\beta_1 = 0$, given that $\beta_2, \beta_3, \ldots, \beta_k$ are in the model, in favor of the alternative that β_1 differs from 0, given that $\beta_2, \beta_3, \ldots, \beta_k$ are in the model.

If your study does not include random selection of units from a larger population or random assignment of treatments to available units, you can still use multiple regression. However, your regression is modeling only the cases you have. You cannot generalize to a larger population or make a cause-and-effect conclusion. If the result is statistically significant, you can say that the linear relationship cannot be reasonably attributed to chance alone.

Multiple regression is an incredibly useful tool, especially in the analysis of experiments. While this section has provided you with the basic concepts, there is a lot more you could learn. Entire books and courses are devoted to this subject.

Review Exercises

E21. A mother was concerned about the sizing of girls' shoes because she noticed that shoes for boys tend to be wider than shoes for girls. Sales representatives told her that boys have wider feet than girls. To test this claim, she measured the feet of two classes of 4th graders in a single school. These measurements, and

Length of Longer Foot (cm)	Width of Longer Foot (cm)	Relative Age (months: 1 = youngest)	Sex	Foot Measured	Handedness
23.2	9.8	6	B	L	R
23.1	8.9	7	B	L	R
24.1	9.6	7	B	L	R
23.7	9.7	8	B	R	R
22.8	8.9	8	B	L	R
25	9.8	8	B	L	R
21.6	9	9	B	R	L
22.7	8.6	9	B	R	L
22.4	8.4	10	B	L	R
22	9.2	11	B	R	R
22.4	8.6	12	B	L	R
25.1	9.4	12	B	L	R
24.1	9.1	1	B	L	R
21.9	9.3	13	B	R	L
25.5	9.8	2	B	R	R
20.9	8.8	14	B	R	L
23.4	8.8	3	B	L	L
22.2	8.9	16	B	L	R
22.5	9.7	5	B	R	R
23.5	9.5	5	B	R	R
19.6	7.9	6	G	R	R
24.1	9.5	7	G	L	R
22.7	8.8	8	G	R	R
23.5	9.5	8	G	R	R
22.5	8.6	8	G	L	R
20.9	8.5	8	G	L	R
24	9	8	G	L	R
21.7	7.9	9	G	R	R
21.6	9.3	10	G	R	R
21	8.8	11	G	L	R
24.7	9	11	G	L	L
22	9.3	11	G	L	R
20.5	8.6	12	G	R	R
22	8.3	14	G	R	L
22.5	9	14	G	L	R
22.6	8.8	14	G	L	R
22.2	8.1	15	G	L	R
24	9.3	4	G	L	R
22	8.7	5	G	R	L

Display 15.55 Data on two classes of 4th graders in a single school. [Source: Mary C. Meyer, "Wider Shoes for Wider Feet?" *Journal of Statistics Education*, Vol. 14, 1 (March 2006), *www.amstat. org/publications/jse/v14n1/datasets.meyer.html.*]

some additional data about the children, are given in Display 15.55.

a. What type of sample is this? Regardless of your answer, assume for the rest of this exercise that the children may be considered the equivalent of a random sample of all 4th graders.

b. Perform an appropriate procedure to test whether the width of 4th-grade boys feet are wider than the width of 4th-grade girls feet, on average.

c. The 4th-grade children themselves understood that boys might simply tend to be bigger than girls and would tend to have longer feet as well. Is there evidence that this is true?

d. Conduct (and interpret) an appropriate analysis of the widths of boys' and girls' feet that takes into account any difference in the lengths of their feet.

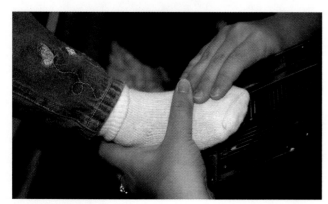

Measuring a child's foot.

E22. Refer to the data in E21. The multiple regression with *width* as the response variable and *length, sex,* and *age* is given in Display 15.56.

Response attribute (numeric): width cm

Predictor	Coefficient	Std Error	t	P	ΔR²
Constant	4.4430	1.3049	3.405	0.0017	
Length	0.2095	0.0536	3.909	0.0004	0.2335
Boy0 Girl1	−0.2264	0.1309	−1.730	0.0924	0.0457
Age	−0.0114	0.0188	−0.605	0.5488	0.0056

Source	DF	SS	MS	F	P	ΔR²
Regression	3	4.5899	1.5300	10.146	0.0001	0.4651
Residual	35	5.2778	0.1508			
Total	38	9.8677				

R²:	0.465143
Adjusted R²:	0.419299
Std dev error:	0.388322

Display 15.56 Fathom linear regression analysis of *width* on *length, sex,* and *age*.

a. Find the residual for the first child in the table.

b. Find and interpret *s*.

c. Interpret the value of R^2.

d. Write hypotheses and a conclusion as to whether the regression is statistically significant overall.

e. Which explanatory variables are statistically significant? Write hypotheses and a conclusion for the *t*-test of the statistical significance of *age*.

E23. The table in Display 15.57, first seen in Chapter 3, gives information collected over a representative set of races at 21 NASCAR racetracks. The variables are the angle at which the turns are banked, the circumference of the raceway in miles, and the top average speed over the set of races (TRAS).

a. In a simple linear regression, is banking a statistically significant explanatory variable for TRAS?

b. In a simple linear regression, is circumference a statistically significant explanatory variable for predicting TRAS?

c. In a multiple regression, are banking and circumference both statistically significant explanatory variables for predicting TRAS?

d. How do you explain any seeming inconsistencies among the three answers in parts a–c?

Speedway	Banking	Circumference	TRAS
Atlanta Motor Speedway	24	1.54	163.633
Bristol Motor Speedway	36	0.53	101.074
California Speedway	14	2	155.012
Chicagoland Speedway	18	1.5	121.2
Darlington Raceway	24	1.37	139.958
Daytona International Speedway	31	2.5	177.602
Dover International Speedway	24	1	132.719
Homestead-Miami Speedway	6	1.5	140.335
Indianapolis Motor Speedway	12	2.5	155.912
Kansas Speedway	15	1.5	110.576
Las Vegas Motor Speedway	12	1.5	146.53
Lowes Motor Speedway	24	1.5	160.306
Martinsville Speedway	12	0.53	82.223
Michigan International Speedway	18	2	173.997
New Hampshire Speedway	12	1.06	117.134
North Carolina Speedway	23	1.02	131.103
Phoenix International Raceway	10	1	118.132
Pocono Raceway	9	2.5	144.892
Richmond International Raceway	14	0.75	108.707
Talladega Superspeedway	33	2.66	188.354
Texas Motor Speedway	24	1.5	144.276

Display 15.57 NASCAR speedway data.

E24. *Indicator variables.* During the construction of the Alaskan pipeline, defects in the pipe were a serious concern. In the field, the depth of a defect was measured using ultrasound. To ensure accuracy of the in-field measurements, they were checked by bringing samples of the defective pipes into the laboratory and measuring the same defects on much more accurate laboratory equipment. Your goal is to determine how well the in-field measurements predict the laboratory measurements.

Trans-Alaska Pipeline transports oil the length of Alaska.

The data in the table at right are a sample of the defects measured both ways. A model you might use for these data would be

$$\hat{y} = b_0 + b_1 L$$

Here, L gives the depth measured in the field and $\hat{y}$ is the depth measured in the lab. However, these pipes come from three different batches, so the possible batch effect must be taken into account in the analysis. To account for three batches, the regression model must have two *indicator variables*. As set up in the table, $x_2 = 1$ if the measurement is for a pipe from batch 2 and is 0 otherwise. Similarly, $x_3 = 1$ if the measurement is for a pipe from batch 3 and is 0 otherwise. Both of these variables are 0 if the measurement is for a pipe from batch 1.

The following model uses interaction terms to fit three regression lines simultaneously, which are shown in the scatterplot in Display 15.58 (on the next page). This allows for differences among both slopes and intercepts for the three batches:

$$\hat{y} = b_0 + b_1 L + b_2 x_2 + b_3 x_3 + b_4 (L \cdot x_2) + b_5 (L \cdot x_3)$$

Field Measurement (0.001 in.)	Lab Measurement (0.001 in.)	x_2 (1 = batch 2)	x_3 (1 = batch 3)
15	19	0	0
37	55.5	0	0
15	12.3	0	0
18	18.4	0	0
11	11.5	0	0
35	38	0	0
20	18.5	0	0
40	38	0	0
50	55.3	0	0
36	38.7	0	0
50	54.5	0	0
38	38	0	0
10	12	0	0
75	81.7	0	0
10	11.5	0	0
85	80	0	0
13	18.3	0	0
50	55.3	0	0
58	80.2	0	0
58	80.7	0	0
21	15.5	1	0
19	23.7	1	0
10	9.8	1	0
33	40.8	1	0
16	17.5	1	0
5	4.3	1	0
32	36.5	1	0
23	26.3	1	0
30	30.4	1	0
45	50.2	1	0
33	30.1	1	0
25	25.5	1	0
12	13.8	1	0
53	58.9	1	0
36	40	1	0
5	6	1	0
63	72.5	1	0
43	38.8	1	0
25	19.4	1	0
73	81.5	1	0
45	77.4	1	0
52	54.6	0	1
9	6.8	0	1
30	32.6	0	1
22	19.8	0	1
56	58.8	0	1
15	12.9	0	1
45	49	0	1

a. Fit this complete model to these data.

b. What equation would you use to predict the laboratory measurement when the pipe is from batch 1? From batch 2? From batch 3?

c. Does any batch differ from the others? If so, which one?

d. In light of this analysis, refine the complete model to one that you think best represents these data.

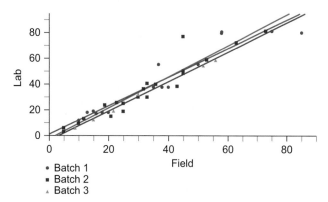

- • Batch 1
- ■ Batch 2
- ▲ Batch 3

Display 15.58 Measurements and scatterplot of pipeline defects in thousandths of an inch for three batches of pipe. [Source: *www.itl.nist.gov/div898/handbook/pmd/section6/pmd621.htm.*]

E25. Refer to the data in Display 15.33 on page 726. The scatterplot in Display 15.59 shows, for each group, the mean number of flowers plotted against the starting time of the light.

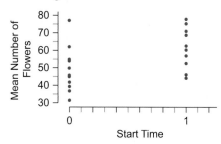

Display 15.59 Scatterplot of mean number of flowers per plant plotted against starting time of light (0 = late; 1 = early).

a. Do a simple linear regression of *flowers* on *start time*. Describe what the *P*-value of this analysis tells you.

b. What other test could you use to determine whether there is a statistically significantly realtionship between the mean number of flowers and the starting time of the lighting? Do this test and give your conclusion.

c. Compare the results from the two tests.

E26. Calories! Almost everyone, it seems, thinks about how to reduce them, increase them, or at least manage them in the daily diet. It might help to manage a healthy diet if there was a good way to predict calories from the other common dietary measures attached to most foods, fat, carbohydrates, and protein. The data in Display 15.60 show the amount of each of these measures, as well as calories, for eight popular snack foods.

a. Find a good fitting regression model to predict calories from some function of the other three dietary measures.

b. In looking at the relationship between fat and calories, it can be seen that there is a definite curvature to the pattern of the scatterplot. Square each of the *fat* data values to form a *fat-squared* explanatory variable. Refit the model you found in part a by adding this new explanatory variable. Does that variable add significant improvement to the model?

c. Fit a model that predicts calories from a quadratic model using *fat* and *fat-squared* alone. How does it compare to the best fitting model you found in part a?

Food	Fat	Carbohydrates	Protein	Calories
M&M's	10.0	34	2	240
Cheez It	16.0	31	7	290
Cracker Jack	1.5	21	2	100
Oreo	10.0	36	2	240
Twix	14.0	37	3	280
Nutri-Grain	3.0	27	2	140
Trail mix	20.0	23	11	295
Planters peanuts	25.0	9	13	300

Display 15.60 Dietary data for snack foods (grams per typical serving). [Source: S. D. Szydik, *Mathematics Teacher*, Vol. 103 (August 2009), p. 21.]

E27. You are working your way through school at the medical center on campus. You have noticed that, in an emergency, physicians occasionally must quickly administer a drug to an unconscious patient. The optimal dosage of the drug depends on the weight of the person. But the weight of an unconscious person is hard to estimate, and you wonder if the weight can be predicted well from measurements that are easier and faster to get. You find [at *www.statsci.org/data/oz/pysical.html*] that a student has randomly selected 22 males between the ages of 16 and 30 and measured them. Display 15.61 gives the results of a multiple regression analysis for predicting weight in terms of the other measurements.

a. As a group, are these variables helpful in predicting the weight of a male between the ages of 16 and 30? How can you tell from the output?

b. Which variables are statistically significant?

c. What proportion of the variability in the weight from male to male is accounted for by this regression?

d. The standard error is about 2.7 kg. Does this indicate that the error in your estimate of a young male's weight is likely to be relatively small or relatively large?

e. A second regression analysis, using *forearm*, *waist*, *height*, and *thigh* is given in Display 15.62. What is surprising about the *t*-test for *thigh*? How can you explain this?

f. Did dropping most the variables make much difference in how well you can predict weight?

Regression Analysis: Weight Versus Forearm, Bicep, ...

The regression equation is
Weight = −69.5 + 1.78 Forearm + 0.155 Bicep −
 0.189 Chest − 0.482 Neck − 0.029 Shoulder +
 0.661 Waist + 0.318 Height + 0.446 Calf +
 0.297 Thigh − 0.920 Head

Predictor	Coef	SE Coef	T	P
Constant	−69.52	29.04	−2.39	0.036
Forearm	1.7818	0.8547	2.08	0.061
Bicep	0.1551	0.4853	0.32	0.755
Chest	0.1891	0.2258	0.84	0.420
Neck	−0.4818	0.7207	−0.67	0.518
Shoulder	−0.0293	0.2394	−0.12	0.905
Waist	0.6614	0.1165	5.68	0.000
Height	0.3178	0.1304	2.44	0.033
Calf	0.4459	0.4125	1.08	0.303
Thigh	0.2972	0.3051	0.97	0.351
Head	−0.9196	0.5201	−1.77	0.105

s = 2.28679   R-sq = 97.7%   R-sq(adj) = 95.6%

Analysis of Variance

Source	DF	SS	MS	F	P
Regression	10	2466.62	246.66	47.17	0.000
Residual Error	11	57.52	5.23		
Total	21	2524.15			

Display 15.61 Regression analysis of weight (kg) on other measurements (cm) for a random sample of 16- to 30-year-old males.

Regression Analysis: Weight Versus Forearm, Waist, Height, Thigh

The regression equation is
Weight = −113 + 2.04 Forearm + 0.647 Waist +
 0.272 Height + 0.540 Thigh

Predictor	Coef	SE Coef	T	P
Constant	−113.31	14.64	−7.74	0.000
Forearm	2.0356	0.4624	4.40	0.000
Waist	0.6469	0.1043	6.20	0.000
Height	0.27175	0.08548	3.18	0.005
Thigh	0.5401	0.2374	2.27	0.036

s = 2.24885   R-sq = 96.6%   R-sq(adj) = 95.8%

Analysis of Variance

Source	DF	SS	MS	F	P
Regression	4	2438.17	609.54	120.53	0.000
Residual Error	17	85.97	5.06		
Total	21	2524.15			

Display 15.62 Regression analysis of weight (kg) on fewer measurements (cm) for a random sample of 16- to 30-year-old males.

C1. Fitting a multiple regression line to data, differs from fitting a bivariate regression line only in that (choose the best response)

- **Ⓐ** minimizing the SSE is no longer the criterion for fitting the model.
- **Ⓑ** the square root of R^2 no longer measures a correlation between the response and one of the explanatory variables.
- **Ⓒ** regression coefficients no longer estimate an intercept and slopes.
- **Ⓓ** inference for the regression parameters no longer requires random sampling or assignment.
- **Ⓔ** t-tests for the significance of regression parameters are no longer used.

C2. Which *two* of the following are not important to check before doing inference in a multiple regression analysis?

- **Ⓐ** The residuals plotted against the predicted values show a plot that is relatively free of pattern.
- **Ⓑ** It is plausible that the residuals could have come from a normal distribution.
- **Ⓒ** The explanatory variables are independent of one another.
- **Ⓓ** The data come from a random sample or a random assignment of treatments.
- **Ⓔ** The explanatory variables all have the same units of measurement.

Data collected from a sample of 33 commercial airplanes was used to model *cost* (dollars per hour of operation) as a function of typical flight *length* (miles), *fuel* consumption (gallons per hour) and number of *seats* on the plane. For C3-C5, refer to the following analysis.

The regression equation is
Cost = 266 − 0.713 Length + 2.93 Fuel + 3.74 Seats

Predictor	Coef	Stdev	t-Ratio	p
Constant	266.1	307.0	0.87	0.393
Length	−0.7132	0.2495	−2.86	
Fuel	2.9347	0.4046	7.25	0.000
Seats	3.743	3.882	0.96	0.343

s = 762.1   R-sq = 90.2%   R-sq(adj) = 89.1%

Analysis of Variance

Source	DF	SS	MS	F	P
Regression		154178208	51392736	88.48	0.000
Error		16843800	580821		
Total		171022016			

C3. The degrees of freedom for regression, error and total, respectively, are

- **Ⓐ** 2, 31, 33
- **Ⓑ** 2, 30, 32
- **Ⓒ** 30, 2, 32
- **Ⓓ** 29, 3, 32
- **Ⓔ** 3, 29, 32

C4. The P-value for testing the significance of the coefficient of length is

- **Ⓐ** close to 1.0
- **Ⓑ** close to 0.5
- **Ⓒ** close to 0.01
- **Ⓓ** close to 0.10
- **Ⓔ** unknown.

C5. Choose the best interpretation of the regression analysis among the following options.

- **Ⓐ** All three explanatory variables contribute significantly to predicting cost.
- **Ⓑ** Among these three explanatory variables, only length and fuel contribute significantly to cost.
- **Ⓒ** Among these three explanatory variables, only length and seats contribute significantly to cost.
- **Ⓓ** None of these three explanatory variables contribute significantly to predicting cost.
- **Ⓔ** There must be a mistake in the analysis because longer flights should cost more than shorter flights, not less.

Investigation

C6. What variables appear to be most strongly associated with high school graduation rates? The table in Display 15.63 shows the high school graduation rates (*GRate*, percentage of ninth graders graduating within 5 years) for each state, along with four other education-related variables: average teacher salary (*Sal*), education spending per student (*Exp*), student-to-teacher ratio (*STRatio*), and percentage of the state's total taxable resources spent on education (*Tax*).

- **a.** Is there a significant simple linear relationship between *GRate* and *Exp*?
- **b.** Is there a significant simple linear relationship between *GRate* and *Tax*?
- **c.** When a model is fit using both *Exp* and *Tax* as explanatory variables, do both turn out to be statistically significant? Explain how this can happen, given the results in parts a and b.
- **d.** Does adding *STRatio* to the model of part c improve the fit of the model appreciably?
- **e.** Based on these data, what is the best fitting model for predicting *GRate* from any combination of these possible explanatory variables?

State	GRate	Sal	Exp	STRatio	Tax
Alabama	61	38.3	8.8	12.6	3.5
Alaska	66	51.1	12.1	17.2	3.3
Arizona	69	42.3	7.1	21.3	3.3
Arkansas	72	39.2	9.8	14.7	4.2
California	67	56.4	7.6	21.1	3.3
Colorado	73	43.3	8.5	16.9	3.0
Connecticut	79	56.5	11.9	13.6	4.0
Delaware	66	51.1	11.4	15.2	2.4
Florida	57	40.6	8.4	13.8	3.1
Georgia	56	45.8	8.8	17.9	3.9
Hawaii	64	45.5	10.4	15.7	4.2
Idaho	77	40.1	8.0	16.5	3.5
Illinois	74	53.8	8.8	17.9	3.5
Indiana	73	45.8	10.1	16.5	4.4
Iowa	81	38.4	10.0	16.9	3.5
Kansas	75	38.6	10.2	13.8	4.1
Kentucky	72	39.8	8.7	14.4	3.6
Louisiana	62	37.1	9.8	16.1	2.7
Maine	76	39.9	13	14.4	4.6
Maryland	74	50.3	10.1	11.5	3.9
Massachusetts	76	53.3	11.5	15.8	3.9
Michigan	70	54.5	9.8	13.6	4.6
Minnesota	79	45.0	9.5	18.1	3.6
Mississippi	61	36.2	8.6	16.3	3.9
Missouri	74	38.2	9.1	15.1	3.7
Montana	76	37.2	11.7	13.9	3.7
Nebraska	79	39.6	11	14.4	3.6
Nevada	47	43.2	7.2	13.6	2.8
New Hampshire	77	42.7	11.2	19.0	4.0
New Jersey	82	53.7	13.2	13.7	4.9
New Mexico	56	38.5	9.5	12.7	3.7
New York	68	55.2	13.1	15.0	4.2
North Carolina	63	43.2	7.8	13.3	2.6
North Dakota	79	35.4	10.9	15.1	3.1
Ohio	74	47.8	10.1	12.7	4.3
Oklahoma	71	35.1	8.3	15.2	3.2
Oregon	75	47.8	9.5	16.0	3.2
Pennsylvania	78	52.6	11.3	20.6	4.1
Rhode Island	73	54.8	12.5	15.2	4.1
South Carolina	66	41.2	9.0	13.4	4.1
South Dakota	77	33.2	10.2	15.3	2.9
Tennessee	69	40.3	7.6	13.6	2.8
Texas	65	40.5	7.6	15.7	3.3
Utah	72	39.0	6.0	15.0	3.2
Vermont	79	43.0	15.1	22.4	5.3
Virginia	69	43.9	8.7	11.3	3.3
Washington	62	45.4	7.7	13.2	3.0
West Virginia	72	38.5	11.2	19.3	4.6
Wisconsin	82	41.7	10.5	14.0	4.1
Wyoming	73	39.5	14.1	15.1	3.5

Display 15.63 Education data for the states. [Source: *www.edweek.org/rc/2007/06/07/edcounts.html.*]

Chapter 16
Martin v. Westvaco Revisited: Testing for Possible Discrimination in the Workplace

In Chapter 1, you first read about Bob Martin, who was laid off from his job at the Westvaco Corporation when he was 54. Martin made the claim that he had been terminated because of his age, which, if true, would be against the law. Of necessity, his case was based largely on statistics. No one said to him, "Bob, we're going to let you go because you are too old."

The statistical analysis presented during the lawsuit was quite a bit more involved than the simplified version you saw back in Chapter 1. With what you have learned, you can now carry out a much more thorough analysis.

Display 16.1 shows the data used in the lawsuit. Each row of the table represents one of the 50 people who worked in the engineering department of the Envelope Division of Westvaco when layoffs began.

Analysis Using Age Categories

By federal law, all employees 40 or older belong to what is called a "protected class": To discriminate against them on the basis of their age is against the law. At the time of the layoffs at Westvaco, 36 of the 50 people working in the engineering department were 40 or older. A total of 28 employees were terminated; 21 of them were 40 or older.

| **Comparing Termination Rates of Younger and Older Employees** | **Example 16.1** |

The two-way table in Display 16.2 categorizes each of the 50 employees by age group and whether he or she was terminated or retained.

a. Should you use a two-sample z-test or a chi-square test to determine if there is statistically significant evidence that older employees were more likely to be chosen for termination?

b. The Supreme Court has set the criterion for statistical significance in discrimination cases as P-value < 0.025 for a one-sided test. Compute a P-value for the test you think is most appropriate. What is your conclusion if you take the test at face value (that is, you don't worry about the conditions being met)?

Solution

a. A two-sample z-test for the difference of two proportions and a chi-square test of homogeneity are equivalent for a two-sided alternative hypothesis. However, a one-sided alternative makes the most sense here because there will be evidence of possible discrimination only if the proportion of older employees who were terminated is greater than the proportion of younger employees who were terminated. The chi-square test is inherently two-sided, so the two-sample z-test for the difference of two proportions is the better choice.

b. *State your hypotheses.* The null hypothesis is, "A difference in proportions this large can reasonably be attributed to chance alone." In other words, if all of the employees' names had been put into a hat and Westvaco drew out 28 to be terminated, it would be reasonably likely to get 21 or more age 40 or older. The alternative hypothesis is that the difference is larger than you would reasonably expect if workers had been selected at random, and so Westvaco has some explaining to do.

Compute the test statistic and draw a sketch. For the employees age 40 or older, $n_1 = 36$ and $\hat{p}_1 = 21/36 \approx 0.5833$, where $\hat{p}_1$ is the proportion of

Row	Job Title	Hourly (H) or Salaried (S)	Seniority (yr)	Round Terminated (6 = retained)	Age (yr)
1	Engineering Clerk	H	1.5	6	25
2	Engineering Tech II	H	12.4	6	38
3	Engineering Tech II	H	25.5	6	56
4	Secretary to Engin Manag	H	24.3	6	48
5	Engineering Tech II	H	16.3	1	53
6	Engineering Tech II	H	30.8	1	55
7	Engineering Tech II	H	27.9	1	59
8	Parts Crib Attendant	H	1.2	1	22
9	Engineering Tech II	H	13.8	2	55
10	Engineering Tech II	H	39.1	2	64
11	Technical Secretary	H	17.2	2	55
12	Engineering Tech II	H	28.8	3	55
13	Engineering Tech II	H	14.2	4	33
14	Engineering Tech II	H	13.7	4	35
15	Customer Serv Engineer	S	24.3	6	61
16	Customer Serv Engr Assoc	S	2.7	6	29
17	Design Engineer	S	23.3	6	48
18	Design Engineer	S	16.6	6	54
19	Design Engineer	S	12.9	6	55
20	Design Engineer	S	23.8	6	60
21	Engineering Assistant	S	4.5	6	31
22	Engineering Associate	S	5.8	6	34
23	Engineering Manager	S	27.2	6	59
24	Machine Designer	S	0.8	6	32
25	Packaging Engineer	S	7.2	6	53
26	Prod Spec-Printing	S	16.2	6	47
27	Proj Engineer-Elec	S	19.8	6	48
28	Project Engineer	S	17.3	6	42
29	Project Engineer	S	26.8	6	48
30	Project Engineer	S	9.4	6	57
31	Supv Engineering Serv	S	18.6	6	37
32	Supv Machine Shop	S	26.8	6	54
33	Chemist	S	36.8	1	69
34	Design Engineer	S	3.1	1	53
35	Engineering Associate	S	5.3	1	30
36	Machine Designer	S	5.8	1	52
37	Machine Parts Cont-Supv	S	37.4	1	63
38	Prod Specialist	S	47.2	1	64
39	Project Engineer	S	31.3	1	66
40	Chemist	S	38.2	2	61
41	Design Engineer	S	1.7	2	31
42	Electrical Engineer	S	4.8	2	42
43	Machine Designer	S	22.1	2	56
44	Machine Parts Cont Coor	S	23.2	2	54
45	VH Prod Specialist	S	35.3	2	56
46	Printing Coordinator	S	29.0	3	50
47	Prod Dev Engineer	S	5.2	3	32
48	Prod Specialist	S	36.0	4	59
49	VH Prod Specialist	S	28.8	4	49
50	Engineering Associate	S	1.7	5	23

Display 16.1   The Westvaco data. [Source: CA No. 92-03121-F.]

	Terminated	Retained	Total
Under age 40	7	7	14
40 or older	21	15	36
Total	28	22	50

Display 16.2   Two-way table categorizing employees by age group and whether terminated or retained.

these employees terminated. For the employees under age 40, $n_2 = 14$ and $\hat{p}_2 = 7/14 = 0.5$. The pooled value of p is

$$\hat{p} = \frac{\text{total number of terminations in both groups}}{n_1 + n_2} = \frac{21 + 7}{36 + 14} = 0.56$$

The value of z for a one-sided two-sample test of the difference of two proportions is

$$z = \frac{(\hat{p}_1 - \hat{p}_2) - 0}{\sqrt{\hat{p}(1 - \hat{p})\left(\frac{1}{n_1} + \frac{1}{n_2}\right)}} = \frac{(0.5833 - 0.5) - 0}{\sqrt{(0.56)(0.44)\left(\frac{1}{36} + \frac{1}{14}\right)}} \approx 0.5330$$

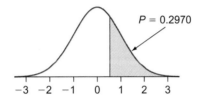

$P = 0.2970$

Write a conclusion in context. The P-value for this one-sided test is 0.2970. The evidence of possible discrimination isn't statistically significant. Although the direction of the difference, $0.5833 - 0.5 = 0.0833$, is consistent with what you would expect to see if discrimination were present, the P-value is not small enough to satisfy the criterion ($P < 0.025$) set by the Supreme Court. You must conclude that it is reasonably likely, just by chance, to get a difference of 0.0833 or larger in the proportion of employees age 40 or older who were terminated and the proportion of employees under age 40 who were terminated.

Conditions Rarely Match Reality

The differences between reality and the statistical model are often the basis for heated arguments between opposing lawyers in discrimination cases. The two-sample z-test for the difference of two proportions, used in Example 16.1, is based on serveral assumptions:

- *Independent random samples.* A random sample is taken from a binomial population. Independently, a second random sample is taken from a second binomial population.

- *Large samples from even larger populations.* When finding the P-value, you use the normal distribution to approximate the sampling distribution for $\hat{p}_1 - \hat{p}_2$. This approximation is reasonable provided that all of

$$n_1\hat{p}_1, \ n_1(1 - \hat{p}_1), \ n_2\hat{p}_2, \qquad n_2(1 - \hat{p}_2)$$

are at least 5 and each population is at least 10 times the sample size.

On the surface, the mismatch between the assumptions and the reality of the Westvaco situation is striking:

z-Test for the Difference of Two Proportions (assumptions)	The *Westvaco* Case (reality)
Two populations that are	One population that is
• both very large	• small (50 employees)
Two samples that are	Two groups (age 40 or older and under age 40) that are
• large enough	• large enough
• randomly selected	• not randomly selected
• from two different populations	• from the same population
• independent	• as dependent as can be, because if you know one groups you automatically know the other

The sample sizes are large enough, but these are not independent random samples. This does not make the test invalid, however. You can still use the z-test to answer this question: "If the process had been random, how likely would it have been, just by chance, to get a difference in proportions as big as Westvaco got?" As long as it is made clear that this is the question being answered and that the sample sizes are large enough to use the normal approximation, the two-sample z-test is quite valid and can be very informative.

So you may proceed with a significance test, but you must make the limitations of what you have done very clear. If you reject the null hypothesis, all you can conclude is that something happened that can't reasonably be attributed to chance alone.

Analysis Using Actual Ages

The test in Example 16.1 was based on dividing employees into two groups, "under 40" and "40 or older." A quantitative variable (*age*) was replaced with a categorical variable (*age group*). As well as involving substantial loss of information, the cutoff age that defines who is classified as an older employee can be arbitrary (as long as it is at least 40) and, as you will see in E5, can change the results of the analysis. Looking at the mean ages of those terminated and those retained can avoid such an arbitrary decision.

Example 16.2	Comparing the Mean Ages of Employees Terminated and Retained

Display 16.3 shows the means and standard deviations of the ages of the terminated and retained employees. What is the most appropriate test to use to see if the greater mean age of the terminated employees compared to the retained employees' mean age is statistically significant? Compute a *P*-value for the test you think is most appropriate. What is your conclusion if you take the test at face value (that is, you don't worry about the conditions being met)?

	n	Mean Age	Standard Deviation
Terminated	28	49.86	13.40
Retained	22	46.18	11.00
All employees	50	48.24	12.42

Display 16.3 Summary statistics of ages of terminated and retained employees.

Solution
Use a two-sample *t*-test.

State your hypotheses. The null hypothesis is that the difference in the mean age of employees terminated and the mean age of employees retained is no larger than you would reasonably expect if Westvaco were picking the 28 people at random for termination. This should be a one-sided test, so the alternative hypothesis is that the difference in the mean age of employees terminated and the mean age of those retained is larger than you would reasonably expect if people were selected at random for termination.

Compute the test statistic and draw a sketch. The test statistic is

$$t = \frac{\bar{x}_1 - \bar{x}_2}{\sqrt{\dfrac{s_1^2}{n_1} + \dfrac{s_2^2}{n_2}}} = \frac{49.86 - 46.18}{\sqrt{\dfrac{13.40^2}{28} + \dfrac{11.00^2}{22}}} \approx 1.066$$

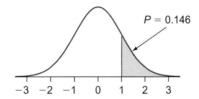

The approximate *df* is 47.89, which gives a (one-sided) *P*-value of 0.146.

Write a conclusion in context. Do not reject the null hypothesis. If 28 employees were selected totally at random to be terminated, then it is reasonably likely to get a difference of 49.86 − 46.18 = 3.68 years or more in the average age of employees terminated minus the average age of the employees retained. Taking the results of the test at face value, the conclusion again is that, although the pattern is consistent with what you would expect to see if discrimination were present, the *P*-value is not small enough to satisfy the criterion ($P < 0.025$) set by the Supreme Court.

Note that the *P*-value of 0.146 in Example 16.2, where actual ages were used, is less than half of that in Example 16.1, where age groupings were used. Though neither reaches the level of statistical significance, using the actual ages gives Martin a stronger argument.

Again, Conditions Don't Match Reality
The *t*-test for the difference of two means is based on independent random samples from two large populations, which must be approximately normally distributed if the sample sizes are small. In Example 16.2, there is only one population, and this population gets sorted into two groups. Again, the mismatch between the conditions needed for a *t*-test and the reality of the Westvaco situation is striking:

On the surface, there's no apparent reason to think the *t*-test is appropriate. Remarkably, though, extensive simulations have shown that, despite the mismatch

t-Test for the Difference of Means (assumptions)	The *Westvaco* Case (reality)
Two populations that are	One population that is
• both very large • both normal	• small • not normal (ages skewed left)
Two samples that are	Two groups that are
• randomly selected • from two different populations • independent	• not randomly selected • from the same population • as dependent as can be, because if you know one groups—those terminated—you automatically know the other

between the two sets of conditions, the *t*-test nevertheless tends to give a good approximation of the *P*-value that you would get using an approach consistent with the reality of situations like the *Westvaco* case.

Looking for a Better Approach: A Randomization Test

What is the approach that is consistent with the real conditions of the *Westvaco* case? The null hypotheses of Examples 16.1 and 16.2 suggest a way to proceed. In these examples, the null hypothesis states that the process used by Westvaco was equivalent to randomly choosing the 28 employees for termination from the total population of 50 employees. Repeated random selection of 28 employees will allow you to test whether the difference in the observed mean ages of those terminated and those retained could reasonably have been produced by random selection alone.

Example 16.3

Using a Randomization Test to Compare Mean Ages of Those Terminated and Retained

Display 16.4 shows the results of 1000 random selections of 28 employees to be terminated. Each value plotted is the mean age of the 28 employees randomly selected for termination minus the mean age of the 22 employees retained.

a. Locate the actual mean difference on this distribution, estimate and interpret a *P*-value, and give your conclusion.

b. Compare the *P*-value to that in Example 16.2. Are they similar or different?

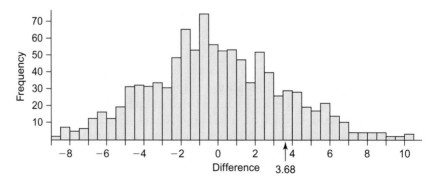

Display 16.4   Randomization distribution for the mean age of 28 employees selected at random for termination minus the mean age of the 22 employees retained. The arrow shows the actual difference in the mean ages, 3.68.

Solution

a. For the actual data, the difference, *mean age of those terminated − mean age of those retained*, is $49.86 − 46.18 = 3.68$. This value is located on the histogram by the arrow. A difference as large or larger than 3.68 occurred in 149 of the 1000 repetitions, for an estimated *P*-value of 0.149. That is, if employees were selected at random for termination, the probability of getting a difference in mean ages, *mean age of those terminated − mean age of those retained*, of 3.68 years or larger is about 0.149. Because such a difference is reasonably likely to occur just by chance, you cannot reasonably attribute the difference to any other cause, such as discrimination.

b. The *P*-value in Example 16.2 is 0.146, which is almost identical to the *P*-value of 0.149 from the randomization test. This illustrates that the two-sample *t*-test can give results that are consistent with a randomization test, even when the samples aren't selected independently from larger populations. (The *t*-test generally works well when the randomization distribution is approximately normal.)

In the previous example, failure to reject the null hypothesis of "randomly choosing the employees to terminate" does not imply that the company actually did select them randomly, but it does provide evidence that there is no obvious nonrandom pattern in the ages of those laid off and those retained.

DISCUSSION
Looking for a Better Approach: A Randomization Test

D1. In what ways does the two-sample *t*-test in Example 16.2 involve approximations to get the *P*-value? In what ways does the randomization test in Example 16.3 involve approximations to get the *P*-value? Which seems like the more "exact" way to get a *P*-value?

D2. Agree or disagree, and tell why: "A hypothesis test is based on a probability model. Like all probability models, it assumes certain outcomes are random. But in the *Westvaco* case, the decisions about which people to terminate weren't random. There's no probability model, and so a statistical test is invalid."

The End of the Story

About all you can ever show by using statistical methods in discrimination cases is that the process doesn't look like random selection. Statistical analysis alerts you to such questionable situations, but it cannot reconstruct the intent of the people who did the terminations. Knowing the intent is crucial because it may be perfectly legal: Perhaps employees in obsolete jobs were the ones picked for termination, and it just happens that the obsolete jobs were held by employees in a protected class.

The *Westvaco* case never got as far as a jury. After a lot of statistical analysis and some arguments about the statistics, and just before the case was about to go to trial, the two sides agreed on a settlement. Details of such settlements are not public informations. So, as with many problems based on statistics, this case has no final verdict.

Exercises

In these exercises, you will investigate other approaches to the Westvaco data. Some are methods of statistical inference that you have learned in this course, such as chi-square tests and inference for regression. Others are more advanced techniques, such as logistic regression and Fisher's exact test. While you haven't learned the mechanics of these tests, you know enough statistics to understand their hypotheses, *P*-value, and conclusion.

Analysis Using Age Categories

E1. Bob Martin was a salaried employee. Perhaps looking only at salaried employees would turn up more evidence in his favor. Display 16.5 summarizes the information for the 36 salaried employees.

	Terminated	Retained	Total
Under age 40	4	5	9
40 or older	14	13	27
Total	18	18	36

Display 16.5 Two-way table categorizing salaried employees by age group and whether the employee was terminated or retained.

a. This table presents a new problem, the normal approximation will not be satisfactory if you conduct a two-sample *z*-test. Why not?

b. Nevertheless, find the *P*-value for such a (one-sided) test and save it for use in E3.

c. Taking your *P*-value at face value, is there statistically significant evidence that older salaried employees were more likely to be laid off than younger ones?

E2. Martin might also be interested in whether hourly and salaried workers were treated differently. The summary table is given in Display 16.6.

	Terminated	Retained	Total
Hourly	10	4	14
Salaried	18	18	36
Total	28	22	50

Display 16.6 Two-way table categorizing employees by whether hourly or salaried and whether terminated or retained.

a. Will the normal approximation be satisfactory if you conduct a two-sample *z*-test?

b. Nevertheless, find the *P*-value for such a test and save it for use in E4.

c. Taking your *P*-value at face value, is there statistically significant evidence that hourly and salaried employees were not equally likely to be terminated?

E3. *Fisher's exact test* does not depend on any minimum count in each cell of the table and so is appropriate for the situation in E1.

a. To see how this test works, begin by filling in the cells of the table in Display 16.7. Do this in such a way that the row and column totals remain the same as with the actual data in Display 16.5, but the new cell counts make Bob Martin's case better than the table of actual data.

	Terminated	Retained	Total
Under age 40			9
40 or older			27
Total	18	18	36

Display 16.7 Table with empty cells and the same row and column totals as Display 16.5.

b. If Martin has a case, do you expect observations to cluster in the upper right and lower left cells, or do you expect the observations to cluster in the upper left and lower right cells?

c. The *P*-value in Fisher's exact test gives the probability of getting a table that would be as extreme or more extreme in making a case for Bob Martin than the actual table of data in Display 16.5, assuming that employees are selected randomly for termination and the row and column totals remain fixed. The computations are extensive; so it is best to use one of the numerous web sites that will calculate exact *P*-values. One such web site gives the results in Display 16.8. Use "2-Tail" for a two-sided test. Use "Left" when you expect observations to cluster in the upper right and lower left cells. Use "Right" when you expect the observations to cluster in the upper left and lower right cells. Which *P*-value is the appropriate one to use in this situation?

Fisher's Exact Test

Table = [4, 5, 14, 13]

Left: *P*-value = 0.500000000000006

Right: *P*-value = 0.7784912529072688

2-Tail: *P*-value = 1

Display 16.8 *P*-values for Fisher's exact test
[Source: *www.langsrud.com/fisher.htm.*]

d. Is the difference in the proportion of employees under age 40 and age 40 or older who were terminated statistically significant?

e. Compare the *P*-value to the one from E1. Are they close or quite different?

E4. As you read in E3, Fisher's exact test does not depend on any minimum count in each cell of the table and so is appropriate for the situation in E2.

a. Does the situation in E2 call for a one-sided or a two-sided test?

b. The *P*-value in Fisher's exact test gives the probability of getting a table that would be as extreme or more extreme than the actual table of data in Display 16.6, assuming that employees are selected randomly for termination and the row and column totals remain fixed. One web site gives the results in Display 16.9. Use "2-Tail" for a two-sided test. Use "Left" when you expect observations to cluster in the upper right and lower left cells. Use "Right" when you expect the observations to cluster in the upper left and lower right cells. Which *P*-value is the appropriate one to use in this situation?

Fisher's Exact Test

Table = [10, 4, 18, 18]

Left: *P*-value = 0.9563238295663672

Right: *P*-value = 0.14603368417286103

2-Tail: *P*-value = 0.215017518522793

Display 16.9 *P*-value for Fisher's exact test. [Source: *www.langsrud.com/fisher.htm*.]

c. Is the difference in the proportion of salaried and hourly employees who were terminated statistically significant?

d. Compare the *P*-value to the one from E2. Are they close or quite different?

E5. In employment law, 40 is a special age because only those 40 or older belong to what is called the "protected class," the group covered by federal law against age discrimination. Thus, only people over age 40 can sue for age discrimination. Having age 40 or older as a protected class does not mean that all analyses must be done by splitting employees into the two groups of those age 40 or older and those under age 40. Martin can make his case by, for example, showing that people age 50 or older were disproportionately laid off. That is reasonable because age discrimination doesn't necessarily kick in until employees are quite a bit older than 40.

Display 16.10 shows the employees divided at age 50 and whether they were terminated or not.

a. Use an appropriate test to find the *P*-value for this situation.

	Terminated	Retained	Total
Under age 50	9	13	22
50 or older	19	9	28
Total	28	22	50

Display 16.10 Two-way table classifying employees by age group and whether the employee was terminated or retained.

b. Compare your *P*-value to the one in Example 16.1. To make the best case for age discrimination, should Martin split the groups at age 40 or age 50?

E6. Refer to Display 16.10 in E5 that classifies employees by age group (under age 50 or age 50 or older) and whether the employee was terminated or retained.

a. Fill in the cells in Display 16.11 so that it presents the strongest possible case for discrimination against those 50 and over. Keep the marginal totals the same as in Display 16.10.

	Terminated	Retained	Total
Under age 50			22
50 or older			28
Total	28	22	50

Display 16.11 Table with empty cells but the same margin totals as Display 16.10.

b. Make another copy of Display 16.11, but this time, fill in the cells to make the variables *terminated* and *age* as close to independent as you can. If you select an employee at random from those represented in your table, are *terminated* and *age* independent events according to the definition in Chapter 5?

c. Use the results of Fisher's exact test from a web site to get a *P*-value for the situation in E5.

d. Compare the *P*-value from part c to the one from E5, part a. Is there much difference? Why should this be or not be the case?

E7. According to the U.S. Supreme Court, if you do a statistical test in a discrimination case, you should reject the null hypothesis if your *P*-value from a one-sided test is less than or equal to 2.5% or your *P*-value from a two-sided test is less than or equal to 5%.

a. According to this standard, should the null hypothesis be rejected in any of the tests so far?

b. Which of these statements correctly completes the phrase: If you use a *P*-value of 2.5% to determine "guilt," then

   I. 2.5% of not-guilty companies would be declared guilty by the test.

   II. 2.5% of guilty companies would be declared not-guilty by the test.

c. Is 2.5% the probability of a Type I or Type II error?

d. Describe a Type II error in a lawsuit alleging discrimination.

E8. Consider two companies: Seniors, Inc., with almost all of its 50 employees over 40, and Youth Enterprises, with roughly half of its 50 employees under 40. Suppose both companies discriminate against older employees in a layoff. If you use a significance test for the difference of two proportions, will you be more likely to detect the discrimination at Seniors, Inc. or Youth Enterprises? Explain your reasoning.

E9. Refer to Display 16.1 to get the ages of the Westvaco employees and whether each employee was terminated or retained.

a. Divide the employees into three age groups, under age 40, age 40 through 54, and age 55 or older. Make a table that classifies each employee according to age group and whether the employee was terminated or retained.

b. What is an appropriate test to use to decide whether there is statistically significant evidence that age group and termination/retention are independent?

c. Are conditions met for this test? Make a chart like that on page 748 that compares this situation to the conditions needed for the test.

d. Perform the test and write a conclusion.

E10. Refer to your division of the employees into three age groups in E9.

a. Compute the proportion of employees in each age group. Twenty-eight employees were terminated. How many would you expect to fall into each age group if employees were selected randomly for termination?

b. How many employees actually did get terminated in each age group?

c. What test can you use to determine whether the difference in the counts in parts a and b is statistically significant?

d. Perform this test, including a check of conditions.

e. Is this test or the one in E9 better to use? Explain.

E11. The planning that led up to the layoffs at Westvaco took place in several stages. In the first stage, the head of the engineering department drew up a list of 11 employees to lay off. His boss reviewed the list and decided it was too short: they needed to reduce the size of their work force even further. The department head added a second group of people to the list and checked again with his boss: still too few. He added a third group of names, then a fourth, and finally, one more person was added in the fifth round of planning. Display 16.1 shows this information in the column headed "Round Terminated." An entry of a 1 means "chosen for termination in Round 1 of the planning" and similarly for 2, 3, 4, and 5. An entry of a 6 "not chosen for termination."

Older employees fared much worse in the earlier rounds of the planning than in the later rounds, as you can see in Display 16.12.

a. Judging from the table in Display 16.12, do you think that the evidence will be statistically significant that, in earlier rounds, older employees were more likely to be terminated than younger employees?

| | Rounds 1–2 | | | Percentage |
	Terminated	Retained	Total	Terminated
Under age 50	4	18	22	18.18%
50 or older	16	12	28	57.14%
Total	20	30	50	

| | Rounds 3–5 | | | Percentage |
	Terminated	Retained	Total	Terminated
Under age 50	5	13	18	27.78%
50 or older	3	9	12	25.00%
Total	8	22	30	

Display 16.12 Tables of downsizing by round and age group.

b. Conduct an appropriate test on the table for rounds 1 and 2 to confirm your opinion.

c. Time plots give another way to see whether older employees were more likely to be laid off in the earlier rounds. The two lines in Display 16.13 show the numbers of employees remaining at Westvaco after each round of the planning, for the two age groups, under 50 and 50 or older. Is there evidence here to support Martin's case? Explain.

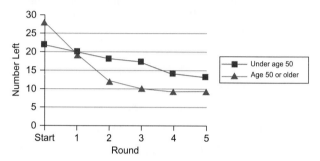

Display 16.13 A plot of the number of employees left in each age group after each round.

E12. Refer to E11.

a. Judging from the table in Display 16.12, do you think that the evidence will be statistically significant that, in later rounds, older employees were more likely to be terminated than younger employees?

b. Conduct an appropriate test using the table for rounds 3–5 to confirm your opinion.

c. What is your overall conclusion from the results here and in E11?

Analysis Using Actual Ages

E13. In E1, you looked only at the salaried employees and performed a two-sample z-test comparing termination rates for two age groups.

a. Conduct a two-sample t-test of the difference of the mean ages for the salaried employees terminated and retained.

b. Compare the P-value with the one from E1. Which is smaller? Why?

E14. Refer to E11.

a. Conduct a two-sample t-test of the difference of the mean ages for the employees terminated and retained in rounds 1 and 2.

b. Compare the P-value with the one from E11. Which is smaller? Why?

E15. Refer to E13.

a. Conduct a one-way analysis of variance to see if there is a statistically significant difference in the mean age of the salaried employees terminated and the salaried employees retained.

b. Should the P-value from part a be the same as or different from that in E13? Explain.

E16. Refer to Example 16.2.

    a. Conduct a one-way analysis of variance to see if there is a statistically significant difference in the mean age of those terminated and those retained.

    b. Should the *P*-value from part a be the same as or different from that in Example 16.2? Explain.

Analysis Using Regression Techniques

E17. The scatterplot and output in Display 16.14 show a regression analysis of age of the employee versus the round when he or she was terminated. If the employee was not terminated, the round was coded as 6.

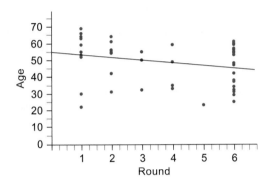

The regression equation is
Age = 54.2 − 1.57 Round

Predictor	Coef	Stdev	t-ratio	p
Constant	54.237	3.515	15.43	0.000
Round	−1.5700	0.8044	−1.95	0.057

s = 12.08   R-sq = 7.4%   R-sq(adj) = 5.4%

Analysis of Variance

Source	DF	SS	MS	F	p
Regression	1	555.5	555.5	3.81	0.057
Error	48	6999.6	145.8		
Total	49	7555.1			

Unusual Observations

Obs.	Round	Age	Fit	Stdev. Fit	Residual	St. Resid
8	1.00	22.00	52.67	2.84	−30.67	−2.61R

R denotes an obs. with a large st. resid.

Display 16.14   Scatterplot and regression analysis of age versus round terminated (not terminated = 6).

    a. Interpret the slope of the regression line.

    b. Is the slope of the regression line statistically significant at the 0.05 level? At the 0.10 level?

    c. What makes the unusual observation unusual?

E18. Display 16.15 shows the number of employees terminated by round and the number and percentage

terminated who were age 40 or older. Was there a tendency to terminate older employees in the earlier rounds?

Round	Number Terminated	Number Terminated Age 40 or Older	Percentage Terminated Age 40 or Older
1	11	9	82
2	9	8	89
3	3	2	67
4	4	2	50
5	1	0	0

Display 16.15   Table showing layoffs by round with number age 40 or older.

    a. Make a plot, with round on the horizontal axis and the number you think most appropriate on the vertical axis.

    b. Is the slope of the least squares regression line statistically significant?

    c. What, if anything, can you conclude from this test?

E19. The scatterplot in Display 16.16 shows the ages of the 36 salaried employees and whether they were terminated (1) or retained (0). The response variable (*terminated or retained*) is binary. One technique that can be used to model such data is *logistic regression*. The resulting logistic curve, graphed on the plot, can be interpreted as modeling the proportion of salaried employees of a specified age who would be terminated.

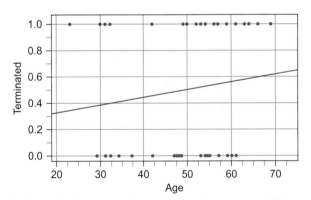

Display 16.16   Scatterplot of whether terminated (1) or retained (0) versus age of salaried employee. The curve is the result of a logistic regression.

The logistic curve on the plot in Display 16.16 looks linear, but it is actually *S*-shaped, as can be seen from the plot in Display 16.17 that has a wider range on the horizontal scale.

    a. Use the logistic curve in Display 16.16 to estimate the proportion of 40-year-old salaried employees who would be terminated. Use the logistic curve to estimate the proportion of 60-year-old salaried employees who would be terminated.

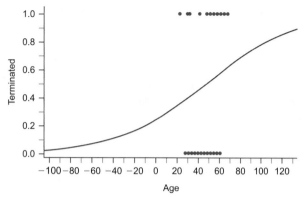

Display 16.17 Scatterplot and logistic curve showing whether terminated (1) or retained (0) versus age of salaried employee, with rescaled horizontal axis.

b. A test of the statistical significance of the logistic curve is called the Hosmer-Lemeshow test. The null hypothesis is that the predictions from the model are consistent with the proportions terminated at each age. The *P*-value for this test is 0.107. Can you conclude that the model is an acceptable predictor of the proportion of salaried employees of a specified age who would be terminated?

E20. The scatterplot in Display 16.18 shows the ages of all 50 employees and whether they were terminated (1) or retained (0). The response variable (terminated or retained) is binary. As in E19, one technique that can be used to model such data is *logistic regression*. The resulting logistic curve, graphed on the plot, can be interpreted as modeling the proportion of employees of a specified age who would be terminated.

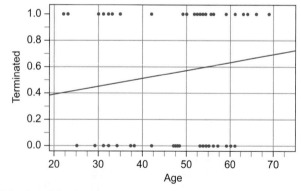

Display 16.18 Scatterplot of whether terminated (1) or retained (0) versus age of employee. The curve is the result of a logistic regression.

The logistic curve on the plot in Display 16.18 looks linear, but it is actually S-shaped, as can be seen from the plot in Display 16.19 that has a wider range on the horizontal scale.

a. Use the logistic curve in Display 16.18 to estimate the proportion of 40-year-old employees who would be terminated. Use the logistic curve to estimate the proportion of 60-year-old employees who would be terminated.

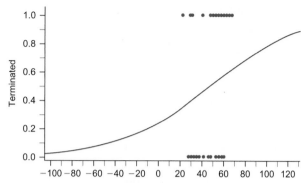

Display 16.19 Scatterplot and logistic curve showing whether terminated (1) or retained (0) versus age of employee, with rescaled horizontal axis.

b. A test of the statistical significance of the logistic curve is called the Hosmer-Lemeshow test. The null hypothesis is that the predictions from the model are consistent with the proportions terminated at each age. The *P*-value for this test is 0.156. Can you conclude that the model is an acceptable predictor of the proportion of employees of a specified age who would be terminated?

Looking for a Better Approach: A Randomization Test

E21. Can the difference between the mean age of the 18 salaried employees terminated and the mean age of the 18 salaried employees retained reasonably be attributed to chance variation? Display 16.20 shows the results of 1000 random selections of 18 of the 36 salaried employees to be terminated. Each value plotted is the mean age of the 18 salaried employees randomly selected for termination minus the mean age of the 18 salaried employees retained.

a. What is the actual difference in the mean ages (see your work for E13)? Locate this difference on this distribution.

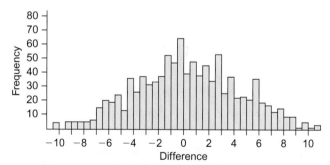

Display 16.20 Randomization distribution for the mean age of 18 salaried employees selected at random for termination minus the mean age of the 18 salaried employees retained.

b. Estimate the proportion of repetitions that had a mean difference as large as or larger than the actual difference from part a.

c. Interpret your estimated *P*-value.

d. Write a conclusion.

E22. Can the difference between the mean age of the 10 hourly employees terminated and the mean age of the 4 hourly workers retained reasonably be attributed to chance variation? The table in Display 16.21 gives the data.

Ages

Terminated	22	33	35	53	55	55	55	55	59	64
Retained		25	38	48	56					

Display 16.21 Ages of the 14 hourly employees terminated and retained.

a. Compute the actual difference in the mean ages.
b. Display 16.22 shows the results of 1000 random selections of 10 of the 14 hourly employees to be terminated. Each value plotted is the mean age of the 10 hourly employees randomly selected for termination minus the mean age of the 4 hourly employees retained. Estimate the *P*-value. What is your conclusion?

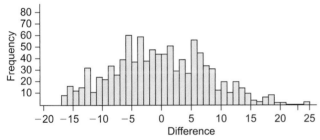

Display 16.22 Randomization distribution for the mean age of 10 hourly employees selected at random for termination minus the mean age of the 4 hourly employees retained.

E23. Compare the *P*-value for the difference in mean ages of the salaried employees terminated and salaried employees retained from the two-sample *t*-test in E13 with the *P*-value from the randomization test in E21. Are they similar or different? Why? Do the conclusions differ?

E24. Refer to E22.

a. Use the data in Display 16.21 to conduct a two-sample *t*-test of the difference of the mean ages of the hourly employees terminated and those retained. Does the condition that the samples must look like they reasonably could have come from normal distributions appear to be met for such a test?
b. Compare the *P*-value from the two-sample *t*-test in part a with the *P*-value from the randomization test in E22. Are they similar or different? Why? Do the conclusions differ?

E25. Randomization tests also can be used with categorical data. Refer to the situation in E1 and the data in Display 16.5.

a. What is the difference in the proportion of employees age 40 or older who were terminated and the proportion of employees under age 40 who were terminated?
b. To construct a randomization distribution, make 36 cards to represent the 36 salaried employees. Refer

to Display 16.1 and on each card, write the age of the employee. Shuffle the cards and deal them into two piles of 18 cards each. The pile on the left will represent employees who are terminated, and the pile on the right will represent employees who were retained. In your first randomization, what is the difference in the proportion of employees age 40 or older who were terminated and the proportion of employees under age 40 who were terminated?

c. Describe what the random dealing of cards into piles in part b would represent in the actual situation.
d. A randomization distribution of 1000 differences, made following the instructions in part b, appears in Display 16.23. The values plotted are *proportion of employees age 40 or older who were terminated − proportion of employees under age 40 who were terminated*. Use it and your answer to part a to estimate the *P*-value for the situation in E1.
e. Compare the *P*-value in part d to the *P*-value from Fisher's exact test in E3.

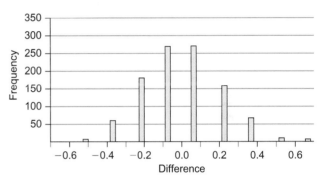

Display 16.23 Randomization distribution for the situation in E1.

E26. Randomization tests also can be used with categorical data. Refer to the situation in E2 and the data in Display 16.6.

a. What is the difference in the proportion of hourly employees who were terminated and the proportion of salaried employees who were terminated?
b. To construct a randomization distribution, make 50 cards to represent the 50 employees. Because 14 employees were hourly, write "hourly" on 14 cards. Write "salaried" on the other 36 cards. Shuffle the cards and deal them into one pile of 28 cards to represent the 28 terminated employees and another pile to represent the 22 employees who were retained. In your first randomization, what is the difference in the proportion of hourly employees who were terminated and the proportion of salaried employees who were terminated?
c. Describe what the random dealing of cards into piles in part b would represent in the actual situation.
d. A randomization distribution of 1000 differences, made following the instructions in part b, appears in Display 16.24. The values plotted are *proportion of hourly employees who were terminated − proportion of salaried employees who were terminated*. Use it and your answer to part a to estimate the *P*-value for the situation in E2.

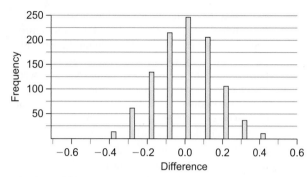

Display 16.24 Randomization distribution for the situation in E1.

e. Compare the *P*-value in part d to the *P*-value from Fisher's exact test in E4.

E27. Review the work you have done in this chapter. As Martin's lawyer, how would you present his case to a jury that is knowledgeable about statistics?

E28. Martin's lawyer has just finished presenting his case that there was age discrimination in the terminations at Westvaco. As Westvaco's lawyer, how would you present its rebuttal to a jury that is knowledgeable about statistics?

Appendix: Statistical Tables

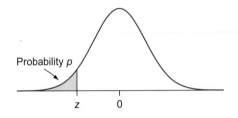

Table entry for z is the
probability lying below z.

Probability p

TABLE A Standard Normal Probabilities

z	.00	.01	.02	.03	.04	.05	.06	.07	.08	.09
−3.8	.0001	.0001	.0001	.0001	.0001	.0001	.0001	.0001	.0001	.0001
−3.7	.0001	.0001	.0001	.0001	.0001	.0001	.0001	.0001	.0001	.0001
−3.6	.0002	.0002	.0001	.0001	.0001	.0001	.0001	.0001	.0001	.0001
−3.5	.0002	.0002	.0002	.0002	.0002	.0002	.0002	.0002	.0002	.0002
−3.4	.0003	.0003	.0003	.0003	.0003	.0003	.0003	.0003	.0003	.0002
−3.3	.0005	.0005	.0005	.0004	.0004	.0004	.0004	.0004	.0004	.0003
−3.2	.0007	.0007	.0006	.0006	.0006	.0006	.0006	.0005	.0005	.0005
−3.1	.0010	.0009	.0009	.0009	.0008	.0008	.0008	.0008	.0007	.0007
−3.0	.0013	.0013	.0013	.0012	.0012	.0011	.0011	.0011	.0010	.0010
−2.9	.0019	.0018	.0018	.0017	.0016	.0016	.0015	.0015	.0014	.0014
−2.8	.0026	.0025	.0024	.0023	.0023	.0022	.0021	.0021	.0020	.0019
−2.7	.0035	.0034	.0033	.0032	.0031	.0030	.0029	.0028	.0027	.0026
−2.6	.0047	.0045	.0044	.0043	.0041	.0040	.0039	.0038	.0037	.0036
−2.5	.0062	.0060	.0059	.0057	.0055	.0054	.0052	.0051	.0049	.0048
−2.4	.0082	.0080	.0078	.0075	.0073	.0071	.0069	.0068	.0066	.0064
−2.3	.0107	.0104	.0102	.0099	.0096	.0094	.0091	.0089	.0087	.0084
−2.2	.0139	.0136	.0132	.0129	.0125	.0122	.0119	.0116	.0113	.0110
−2.1	.0179	.0174	.0170	.0166	.0162	.0158	.0154	.0150	.0146	.0143
−2.0	.0228	.0222	.0217	.0212	.0207	.0202	.0197	.0192	.0188	.0183
−1.9	.0287	.0281	.0274	.0268	.0262	.0256	.0250	.0244	.0239	.0233
−1.8	.0359	.0351	.0344	.0336	.0329	.0322	.0314	.0307	.0301	.0294
−1.7	.0446	.0436	.0427	.0418	.0409	.0401	.0392	.0384	.0375	.0367
−1.6	.0548	.0537	.0526	.0516	.0505	.0495	.0485	.0475	.0465	.0455
−1.5	.0668	.0655	.0643	.0630	.0618	.0606	.0594	.0582	.0571	.0559
−1.4	.0808	.0793	.0778	.0764	.0749	.0735	.0721	.0708	.0694	.0681
−1.3	.0968	.0951	.0934	.0918	.0901	.0885	.0869	.0853	.0838	.0823
−1.2	.1151	.1131	.1112	.1093	.1075	.1056	.1038	.1020	.1003	.0985
−1.1	.1357	.1335	.1314	.1292	.1271	.1251	.1230	.1210	.1190	.1170
−1.0	.1587	.1562	.1539	.1515	.1492	.1469	.1446	.1423	.1401	.1379
−0.9	.1841	.1814	.1788	.1762	.1736	.1711	.1685	.1660	.1635	.1611
−0.8	.2119	.2090	.2061	.2033	.2005	.1977	.1949	.1922	.1894	.1867
−0.7	.2420	.2389	.2358	.2327	.2296	.2266	.2236	.2206	.2177	.2148
−0.6	.2743	.2709	.2676	.2643	.2611	.2578	.2546	.2514	.2483	.2451
−0.5	.3085	.3050	.3015	.2981	.2946	.2912	.2877	.2843	.2810	.2776
−0.4	.3446	.3409	.3372	.3336	.3300	.3264	.3228	.3192	.3156	.3121
−0.3	.3821	.3783	.3745	.3707	.3669	.3632	.3594	.3557	.3520	.3483
−0.2	.4207	.4168	.4129	.4090	.4052	.4013	.3974	.3936	.3897	.3859
−0.1	.4602	.4562	.4522	.4483	.4443	.4404	.4364	.4325	.4286	.4247
−0.0	.5000	.4960	.4920	.4880	.4840	.4801	.4761	.4721	.4681	.4641

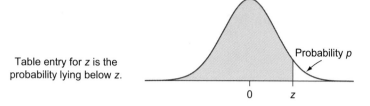

Table entry for z is the
probability lying below z.

Probability p

0   z

TABLE A   Standard Normal Probabilities (continued)

z	.00	.01	.02	.03	.04	.05	.06	.07	.08	.09
0.0	.5000	.5040	.5080	.5120	.5160	.5199	.5239	.5279	.5319	.5359
0.1	.5398	.5438	.5478	.5517	.5557	.5596	.5636	.5675	.5714	.5753
0.2	.5793	.5832	.5871	.5910	.5948	.5987	.6026	.6064	.6103	.6141
0.3	.6179	.6217	.6255	.6293	.6331	.6368	.6406	.6443	.6480	.6517
0.4	.6554	.6591	.6628	.6664	.6700	.6736	.6772	.6808	.6844	.6879
0.5	.6915	.6950	.6985	.7019	.7054	.7088	.7123	.7157	.7190	.7224
0.6	.7257	.7291	.7324	.7357	.7389	.7422	.7454	.7486	.7517	.7549
0.7	.7580	.7611	.7642	.7673	.7704	.7734	.7764	.7794	.7823	.7852
0.8	.7881	.7910	.7939	.7967	.7995	.8023	.8051	.8078	.8106	.8133
0.9	.8159	.8186	.8212	.8238	.8264	.8289	.8315	.8340	.8365	.8389
1.0	.8413	.8438	.8461	.8485	.8508	.8531	.8554	.8577	.8599	.8621
1.1	.8643	.8665	.8686	.8708	.8729	.8749	.8770	.8790	.8810	.8830
1.2	.8849	.8869	.8888	.8907	.8925	.8944	.8962	.8980	.8997	.9015
1.3	.9032	.9049	.9066	.9082	.9099	.9115	.9131	.9147	.9162	.9177
1.4	.9192	.9207	.9222	.9236	.9251	.9265	.9279	.9292	.9306	.9319
1.5	.9332	.9345	.9357	.9370	.9382	.9394	.9406	.9418	.9429	.9441
1.6	.9452	.9463	.9474	.9484	.9495	.9505	.9515	.9525	.9535	.9545
1.7	.9554	.9564	.9573	.9582	.9591	.9599	.9608	.9616	.9625	.9633
1.8	.9641	.9649	.9656	.9664	.9671	.9678	.9686	.9693	.9699	.9706
1.9	.9713	.9719	.9726	.9732	.9738	.9744	.9750	.9756	.9761	.9767
2.0	.9772	.9778	.9783	.9788	.9793	.9798	.9803	.9808	.9812	.9817
2.1	.9821	.9826	.9830	.9834	.9838	.9842	.9846	.9850	.9854	.9857
2.2	.9861	.9864	.9868	.9871	.9875	.9878	.9881	.9884	.9887	.9890
2.3	.9893	.9896	.9898	.9901	.9904	.9906	.9909	.9911	.9913	.9916
2.4	.9918	.9920	.9922	.9925	.9927	.9929	.9931	.9932	.9934	.9936
2.5	.9938	.9940	.9941	.9943	.9945	.9946	.9948	.9949	.9951	.9952
2.6	.9953	.9955	.9956	.9957	.9959	.9960	.9961	.9962	.9963	.9964
2.7	.9965	.9966	.9967	.9968	.9969	.9970	.9971	.9972	.9973	.9974
2.8	.9974	.9975	.9976	.9977	.9977	.9978	.9979	.9979	.9980	.9981
2.9	.9981	.9982	.9982	.9983	.9984	.9984	.9985	.9985	.9986	.9986
3.0	.9987	.9987	.9987	.9988	.9988	.9989	.9989	.9989	.9990	.9990
3.1	.9990	.9991	.9991	.9991	.9992	.9992	.9992	.9992	.9993	.9993
3.2	.9993	.9993	.9994	.9994	.9994	.9994	.9994	.9995	.9995	.9995
3.3	.9995	.9995	.9995	.9996	.9996	.9996	.9996	.9996	.9996	.9997
3.4	.9997	.9997	.9997	.9997	.9997	.9997	.9997	.9997	.9997	.9998
3.5	.9998	.9998	.9998	.9998	.9998	.9998	.9998	.9998	.9998	.9998
3.6	.9998	.9998	.9999	.9999	.9999	.9999	.9999	.9999	.9999	.9999
3.7	.9999	.9999	.9999	.9999	.9999	.9999	.9999	.9999	.9999	.9999
3.8	.9999	.9999	.9999	.9999	.9999	.9999	.9999	.9999	.9999	.9999

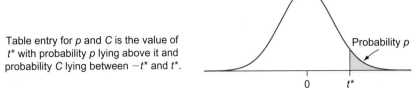

Table entry for p and C is the value of
t* with probability p lying above it and
probability C lying between −t* and t*.

TABLE B   *t*-Distribution Critical Values

Tail Probability p

df	.25	.20	.15	.10	.05	.025	.02	.01	.005	.0025	.001	.0005
1	1.000	1.376	1.963	3.078	6.314	12.71	15.89	31.82	63.66	127.3	318.3	636.6
2	.816	1.061	3.386	1.886	2.920	4.303	4.849	6.965	9.925	14.09	22.33	31.60
3	.765	.978	1.250	1.638	2.353	3.182	3.482	4.541	5.841	7.453	10.21	12.92
4	.741	.941	1.190	1.533	2.132	2.776	2.999	3.747	4.604	5.598	7.173	8.610
5	.727	.920	1.156	1.476	2.015	2.571	2.757	3.365	4.032	4.773	5.893	6.869
6	.718	.906	1.134	1.440	1.943	2.447	2.612	3.143	3.707	4.317	5.208	5.959
7	.711	.896	1.119	1.415	1.895	2.365	2.517	2.998	3.499	4.029	4.785	5.408
8	.706	.889	1.108	1.397	1.860	2.306	2.449	2.896	3.355	3.833	4.501	5.041
9	.703	.883	1 100	1.383	1.833	2.262	2.398	2.821	3.250	3.690	4.297	4.781
10	.700	.879	1.093	1.372	1.812	2.228	2.359	2.764	3.169	3.581	4.144	4.587
11	.697	.876	1.088	1.363	1.796	2.201	2.328	2.718	3.106	3.497	4.025	4.437
12	.695	.873	1.083	1.356	1.782	2.179	2.303	2.681	3.055	3.428	3.930	4.318
13	.694	.870	1.079	1.350	1.771	2.160	2.282	2.650	3.012	3.372	3.852	4.221
14	.692	.868	1.076	1.345	1.761	2.145	2.264	2.624	2.977	3.326	3.787	4.140
15	.691	.866	1.074	1.341	1.753	2.131	2.249	2.602	2.947	3.286	3.733	4.073
16	.690	.865	1.071	1.337	1.746	2.120	2.235	2.583	2.921	3.252	3.686	4.015
17	.689	.863	1.069	1.333	1.740	2.110	2.224	2.567	2.898	3.222	3.646	3.965
18	.688	.862	1.067	1.330	1.734	2.101	2.214	2.552	2.878	3.197	3.611	3.922
19	.688	.861	1.066	1.328	1.729	2.093	2.205	2.539	2.861	3.174	3.579	3.883
20	.687	.860	1.064	1.325	1.725	2.086	2.197	2.528	2.845	3.153	3.552	3.850
21	.686	.859	1.063	1.323	1.721	2.080	2.189	2.518	2.831	3.135	3.527	3.819
22	.686	.858	1.061	1.321	1.717	2.074	2.183	2.508	2.819	3.119	3.505	3.792
23	.685	.858	1.060	1.319	1.714	2.069	2.177	2.500	2.807	3.104	3.485	3.768
24	.685	.857	1.059	1.318	1.711	2.064	2.172	2.492	2.797	3.091	3.467	3.745
25	.684	.856	1.058	1.316	1.708	2.060	2.167	2.485	2.787	3.078	3.450	3.725
26	.684	.856	1.058	1.315	1.706	2.056	2.162	2.479	2.779	3.067	3.435	3.707
27	.684	.855	1.057	1.314	1.703	2.052	2.158	2.473	2.771	3.057	3.421	3.690
28	.683	.855	1.056	1.313	1.701	2.048	2.154	2.467	2.763	3.047	3.408	3.674
29	.683	.854	1.055	1.311	1.699	2.045	2.150	2.462	2.756	3.038	3.396	3.659
30	.683	.854	1.055	1.310	1.697	2.042	2.147	2.457	2.750	3.030	3.385	3.646
40	.681	.851	1.050	1.303	1.684	2.021	2.123	2.423	2.704	2.971	3.307	3.551
50	.679	.849	1.047	1.299	1.676	2.009	2.109	2.403	2.678	2.937	3.261	3.496
60	.679	.848	1.045	1.296	1.671	2.000	2.099	2.390	2.660	2.915	3.232	3.460
80	.678	.846	1.043	1.292	1.664	1.990	2.088	2.374	2.639	2.887	3.195	3.416
100	.677	.845	1.042	1.290	1.660	1.984	2.081	2.364	2.626	2.871	3.174	3.390
1000	.675	.842	1.037	1.282	1.646	1.962	2.056	2.330	2.581	2.813	3.098	3.300
∞	.674	.841	1.036	1.282	1.645	1.960	2.054	2.326	2.576	2.807	3.091	3.291
	50%	60%	70%	80%	90%	95%	96%	98%	99%	99.5%	99.8%	99.99%

Confidence level C

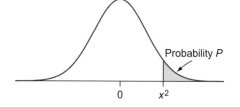

Table entry for p is the value of x^2 with probability p lying above it.

TABLE C χ^2 Critical Values

df	\multicolumn{11}{c}{Tail Probability p}
	.25	.20	.15	.10	.05	.025	.02	.01	.005	.0025	.001
1	1.32	1.64	2.07	2.71	3.84	5.02	5.41	6.63	7.88	9.14	10.83
2	2.77	3.22	3.79	4.61	5.99	7.38	7.82	9.21	10.60	11.98	13.82
3	4.11	4.64	5.32	6.25	7.81	9.35	9.84	11.34	12.84	14.32	16.27
4	5.39	5.59	6.74	7.78	9.49	11.14	11.67	13.23	14.86	16.42	18.47
5	6.63	7.29	8.12	9.24	11.07	12.83	13.33	15.09	16.75	18.39	20.51
6	7.84	8.56	9.45	10.64	12.53	14.45	15.03	16.81	18.55	20.25	22.46
7	9.04	9.80	10.75	12.02	14.07	16.01	16.62	18.48	20.28	22.04	24.32
8	10.22	11.03	12.03	13.36	15.51	17.53	18.17	20.09	21.95	23.77	26.12
9	11.39	12.24	13.29	14.68	16.92	19.02	19.63	21.67	23.59	25.46	27.83
10	12.55	13.44	14.53	15.99	18.31	20.48	21.16	23.21	25.19	27.11	29.59
11	13.70	14.63	15.77	17.29	19.68	21.92	22.62	24.72	26.76	28.73	31.26
12	14.85	15.81	16.99	18.55	21.03	23.34	24.05	26.22	28.30	30.32	32.91
13	15.98	16.98	18.20	19.81	22.36	24.74	25.47	27.69	29.82	31.88	34.53
14	17.12	18.15	19.41	21.06	23.68	26.12	26.87	29.14	31.32	33.43	36.12
15	18.25	19.31	20.60	22.31	25.00	27.49	28.26	30.58	32.80	34.95	37.70
16	19.37	20.47	21.79	23.54	26.30	28.85	29.63	32.00	34.27	36.46	39.25
17	20.49	21.61	22.98	24.77	27.59	30.19	31.00	33.41	35.72	37.95	40.79
18	21.60	22.76	24.16	25.99	28.87	31.53	32.35	34.81	37.16	39.42	42.31
19	22.72	23.90	25.33	27.20	30.14	32.85	33.69	36.19	38.58	40.88	43.82
20	23.83	25.04	26.50	28.41	31.41	34.17	35.02	37.57	40.00	42.34	45.31
21	24.93	26.17	27.66	29.62	32.67	35.48	36.34	38.93	41.40	43.78	46.80
22	26.04	27.30	28.82	30.81	33.92	36.78	37.66	40.29	42.80	45.20	48.27
23	27.14	28.43	29.98	32.01	35.17	38.08	38.97	41.64	44.18	46.62	49.73
24	28.24	29.55	31.13	33.20	36.42	39.36	40.27	42.98	45.56	48.03	51.18
25	29.34	30.68	32.28	34.38	37.65	40.65	41.57	44.31	46.93	49.44	52.62
26	30.43	31.79	33.43	35.56	38.89	41.92	42.86	45.64	48.29	50.83	54.05
27	31.53	32.91	34.57	36.74	40.11	43.19	44.14	46.96	49.64	52.22	55.48
28	32.62	34.03	35.71	37.92	41.34	44.46	45.42	48.28	50.99	53.59	56.89
29	33.71	35.14	36.85	39.09	42.56	45.72	46.69	49.59	52.34	54.97	58.30
30	34.80	36.25	37.99	40.26	43.77	46.98	47.96	50.89	53.67	56.33	59.70
40	45.62	47.27	49.24	51.81	55.76	59.34	60.44	63.69	66.77	69.70	73.40
50	56.33	53.16	60.35	63.17	67.50	71.42	72.61	76.15	79.49	82.66	86.66
60	66.98	68.97	71.34	74.40	79.08	83.30	84.58	88.38	91.95	95.34	99.61
80	88.13	90.41	93.11	96.58	101.9	106.6	108.1	112.3	116.3	120.1	124.8
100	109.1	111.7	114.7	118.5	124.3	129.6	131.1	135.8	140.2	144.3	149.4

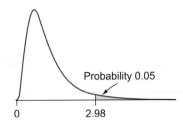

Table entry is the critical value F^* with probability 0.05 lying above it.

Probability 0.05

0   2.98

TABLE D   *F*-Distribution Critical Values for $\alpha = 0.05$

	\multicolumn{9}{c	}{Degrees of freedom for the numerator}
	1	2	3	4	5	6	7	8	9
1	161.4	199.5	215.7	224.6	230.2	234.0	236.8	238.9	240.5
2	18.51	19.00	19.16	19.25	19.30	19.33	19.35	19.37	19.38
3	10.13	9.55	9.28	9.12	9.01	8.94	8.89	8.85	8.81
4	7.71	6.94	6.59	6.39	6.26	6.16	6.09	6.04	6.00
5	6.61	5.79	5.41	5.19	5.05	4.95	4.88	4.82	4.77
6	5.99	5.14	4.76	4.53	4.39	4.28	4.21	4.15	4.10
7	5.59	4.74	4.35	4.12	3.97	3.87	3.79	3.73	3.68
8	5.32	4.46	4.07	3.84	3.69	3.58	3.50	3.44	3.39
9	5.12	4.26	3.86	3.63	3.48	3.37	3.29	3.23	3.48
10	4.96	4.10	3.71	3.48	3.33	3.22	3.14	3.07	3.02
11	4.48	3.98	3.59	3.36	3.20	3.09	3.01	2.95	2.90
12	4.75	3.89	3.49	3.26	3.11	3.00	2.91	2.85	2.80
13	4.67	3.81	3.41	3.18	3.03	2.92	2.83	2.77	2.71
14	4.60	3.74	3.34	3.11	2.96	2.85	2.76	2.70	2.65
15	4.54	3.68	3.29	3.06	2.90	2.79	2.71	2.64	2.59
16	4.49	3.63	3.24	3.01	2.85	2.74	2.66	2.59	2.54
17	4.45	3.59	3.20	2.96	2.81	2.70	2.61	2.55	2.49
18	4.41	3.55	3.16	2.93	2.77	2.66	2.58	2.51	2.46
19	4.38	3.52	3.13	2.90	2.74	2.63	2.54	2.48	2.42
20	4.35	3.49	3.10	2.87	2.71	2.60	2.51	2.45	2.39
21	4.32	3.47	3.07	2.84	2.68	2.57	2.49	2.42	2.37
22	4.30	3.44	3.05	2.82	2.66	2.55	2.46	2.40	2.34
23	4.28	3.42	3.03	2.80	2.64	2.53	2.44	2.37	2.32
24	4.26	3.40	3.01	2.78	2.62	2.51	2.42	2.36	2.30
25	4.24	3.39	2.99	2.76	2.60	2.49	2.40	2.34	2.28
26	4.23	3.37	2.98	2.74	2.59	2.47	2.39	2.32	2.27
27	4.21	3.35	2.96	2.73	2.57	2.46	2.37	2.31	2.25
28	4.20	3.34	2.95	2.71	2.56	2.45	2.36	2.29	2.24
29	4.18	3.33	2.93	2.70	2.55	2.43	2.35	2.28	2.22
30	4.17	3.32	2.92	2.69	2.53	2.42	2.33	2.27	2.21
40	4.08	3.23	2.84	2.61	2.45	2.34	2.25	2.18	2.12
60	4.00	3.15	2.76	2.53	2.37	2.25	2.17	2.10	2.04
120	3.92	3.07	2.68	2.45	2.29	2.17	2.09	2.02	1.96
∞	3.84	3.00	2.60	2.37	2.21	2.10	2.01	1.94	1.88

Degrees of freedom for the denominator

TABLE D   *F*-Distribution Critical Values for $\alpha = 0.05$ (continued)

$F_{.05}$

Degrees of freedom for the numerator

10	12	15	20	24	30	40	60	120	∞
241.9	243.9	245.9	248.0	249.1	250.1	251.1	252.2	253.3	254.3
19.40	19.41	19.43	19.45	19.45	19.46	19.47	19.48	19.49	19.50
8.79	8.74	8.70	8.66	8.64	8.62	8.59	8.57	8.55	8.53
5.96	5.91	5.86	5.80	5.77	5.75	5.72	5.69	5.66	5.63
4.74	4.68	4.62	4.56	4.53	4.50	4.46	4.43	4.40	4.36
4.06	4.00	3.94	3.87	3.84	3.81	3.77	3.74	3.70	3.67
3.64	3.57	3.51	3.44	3.41	3.38	3.34	3.30	3.27	3.23
3.35	3.28	3.22	3.15	3.12	3.08	3.04	3.01	2.97	2.93
3.14	3.07	3.01	2.94	2.90	2.86	2.83	2.79	2.75	2.71
2.98	2.91	2.85	2.77	2.74	2.70	2.66	2.62	2.58	2.54
2.85	2.79	2.72	2.65	2.61	2.57	2.53	2.49	2.45	2.40
2.75	2.69	2.62	2.54	2.51	2.47	2.43	2.38	2.34	2.30
2.67	2.60	2.53	2.46	2.42	2.38	2.34	2.30	2.25	2.21
2.60	2.53	2.46	2.39	2.35	2.31	2.27	2.22	2.18	2.13
2.54	2.48	2.40	2.33	2.29	2.25	2.20	2.16	2.11	2.07
2.49	2.42	2.35	2.28	2.24	2.19	2.15	2.11	2.06	2.01
2.45	2.38	2.31	2.23	2.19	2.15	2.10	2.06	2.01	1.96
2.41	2.34	2.27	2.19	2.15	2.11	2.06	2.02	1.97	1.92
2.38	2.31	2.23	2.16	2.11	2.07	2.03	1.98	1.93	1.88
2.35	2.28	2.20	2.12	2.08	2.04	1.99	1.95	1.90	1.84
2.32	2.25	2.18	2.10	2.05	2.01	1.96	1.92	1.87	1.81
2.30	2.23	2.15	2.07	2.03	1.98	1.94	1.89	1.84	1.78
2.27	2.20	2.13	2.05	2.01	1.96	1.91	1.86	1.81	1.76
2.25	2.18	2.11	2.03	1.98	1.94	1.89	1.84	1.79	1.73
2.24	2.16	2.09	2.01	1.96	1.92	1.87	1.82	1.77	1.71
2.22	2.15	2.07	1.99	1.95	1.90	1.85	1.80	1.75	1.69
2.20	2.13	2.06	1.97	1.93	1.88	1.84	1.79	1.73	1.67
2.19	2.12	2.04	1.96	1.91	1.87	1.82	1.77	1.71	1.65
2.18	2.10	2.03	1.94	1.90	1.85	1.81	1.75	1.70	1.64
2.16	2.09	2.01	1.93	1.89	1.84	1.79	1.74	1.68	1.62
2.08	2.00	1.92	1.84	1.79	1.74	1.69	1.64	1.58	1.51
1.99	1.92	1.84	1.75	1.70	1.65	1.59	1.53	1.47	1.39
1.91	1.83	1.75	1.66	1.61	1.55	1.55	1.43	1.35	1.25
1.83	1.75	1.67	1.57	1.52	1.46	1.39	1.32	1.22	1.00

TABLE E   Random Digits

Row										
1	10097	32533	76520	13586	34673	54876	80959	09117	39292	74945
2	37542	04805	64894	74296	24805	24037	20636	10402	00822	91665
3	08422	68953	19645	09303	23209	02560	15953	34764	35080	33606
4	99019	02529	09376	70715	38311	31165	88676	74397	04436	27659
5	12807	99970	80157	36147	64032	36653	98951	16877	12171	76833
6	66065	74717	34072	76850	36697	36170	65813	39885	11199	29170
7	31 060	10805	45571	82406	35303	42614	86799	07439	23403	09732
8	85269	77602	02051	65692	68665	74818	73053	85247	18623	88579
9	63573	32135	05325	47048	90553	57548	28468	28709	83491	25624
10	73796	45753	03529	64778	35808	34282	60935	20344	35273	88435
11	98520	17767	14905	68607	22109	40558	60970	93433	50500	73998
12	11805	05431	39808	27732	50725	68248	29405	24201	52775	67851
13	83452	99634	06288	98083	13746	70078	18475	40610	68711	77817
14	88685	40200	86507	58401	36766	67951	90364	76493	29609	11062
15	99594	67348	87517	64969	91826	08928	93785	61368	23478	34113
16	65481	17674	17468	50950	58047	76974	73039	57186	40218	16544
17	80124	35635	17727	08015	45318	22374	21115	78253	14385	53763
18	74350	99817	77402	77214	43236	00210	45521	64237	96286	02655
19	69916	26803	66252	29148	36936	87203	76621	13990	94400	56418
20	09893	20505	14225	68514	46427	56788	96297	78822	54382	14598
21	91499	14523	68479	27686	46162	83554	94750	89923	37089	20048
22	80336	94598	26940	36858	70297	34135	53140	33340	42050	82341
23	44104	81949	85157	47954	32979	26575	57600	40881	22222	06413
24	12550	73742	11100	02040	12860	74697	96644	89439	28707	25815
25	63606	49329	16505	34484	40219	52563	43651	77082	07207	31790
26	61196	90446	26457	47774	51924	33729	65394	59593	42582	60527
27	15474	45266	95270	79953	59367	83848	82396	10118	33211	59466
28	94557	28573	67897	54387	54622	44431	91190	42592	92927	45973
29	42481	16213	97344	08721	16868	48767	03071	12059	25701	46670
30	23523	78317	73208	89837	68935	91416	26252	29663	05522	82562
31	04493	52494	75246	33824	45862	51025	61962	79335	65337	12472
32	00549	97654	64051	88159	96119	63896	54692	82393	23287	29529
33	35963	15307	26898	09354	33351	35462	77974	50024	90103	39333
34	59808	08391	45427	26842	83609	49700	13021	24892	78565	20106
35	46058	85236	01390	92286	77281	44077	93910	83647	70617	42941
36	32179	00597	87379	25241	05567	07007	86743	17157	85394	11838
37	69234	61406	20117	45204	15956	60000	18743	92423	97118	96338
38	19565	41430	01758	75379	40419	21585	66674	36806	84962	85207
39	45155	14938	19476	07246	43667	94543	59047	90033	20826	69541
40	94864	31994	36168	10851	34888	81553	01540	35456	05014	51176
41	98086	24826	45240	28404	44999	08896	39094	73407	35441	31880
42	33185	16232	41941	50949	89435	48581	88695	41994	37548	73043
43	80951	00406	96382	70774	20151	23387	25016	25298	94624	61171
44	79752	49140	71961	28296	69861	02591	74852	20539	00387	59579
45	18633	32537	98145	06571	31010	24674	05455	61427	77938	91936
46	74029	43902	77557	32270	97790	17119	52527	58021	80814	51748
47	54178	45611	80993	37143	05335	12969	56127	19255	36040	90324
48	11664	49883	52079	84827	59381	71539	09973	33440	88461	23356
49	48324	77928	31249	64710	02295	36870	32307	57546	15020	09994
50	69074	94138	87637	91976	35584	04401	10518	21615	01848	76938

Glossary

Symbols

b_0	y-intercept of the sample regression line
b_1	slope of the sample regression line
E	margin of error
$E(X)$	expected value (mean) of the random variable X
H_0	null hypothesis
H_a	alternative hypothesis (sometimes H_1)
IQR	interquartile range
n	sample size or the number of trials
N	population size
p	population proportion or the probability of a success on any one trial
p_0	hypothesized value of the population proportion
$\hat{p}$	sample proportion
$\tilde{p}$	estimate of the common value of a population proportion found by combining two samples
$P(A)$	probability that event A happens
$P(A \text{ or } B)$	probability that event A happens or event B happens or both [also $P(A \cup B)$]
$P(A \text{ and } B)$	probability that event A and event B both happen [also $P(A \cap B)$]
$P(B \mid A)$	conditional probability that event B happens given that event A happens
$P(x)$	probability that the random variable X takes on the value x [also $P(X = x)$]
Q_1	first or lower quartile
Q_3	third or upper quartile
r	sample correlation
s	sample standard deviation; in regression, the estimate from the sample of the common variability of y at each value x
s^2	sample variance
s_{b_1}	estimated standard error of the slope of the regression line
SD	standard deviation
SE	standard error
SRS	simple random sample
SSE	sum of squared errors
t	a test statistic using a t-distribution
$Var(X)$	variance of the random variable X
$\bar{x}$	sample mean
x	observed value of a variable
X	a random variable
y	observed value of a variable
$\hat{y}$	predicted value of a variable
z	standardized value (z-score)
z^*	critical value

Greek Letters

α	the significance level; probability of a Type I error
β	probability of a Type II error
β_0	y-intercept of the population regression line
β_1	slope of the population regression line
ε	difference of the value of y for a point and the value predicted by the population regression line
μ	population mean
$\mu_{\bar{x}}$	mean of the sampling distribution of the sample mean
$\mu_{x \mid y}$	mean of the conditional distribution of y given x
σ	population standard deviation; in regression, the common variability of y at each value of x
$\sigma_{\bar{x}}$	standard error of the mean
σ^2	population variance
σ_x^2	variance of the random variable X
χ^2	test statistic for chi-square tests

addition rule   For any two events A and B, $P(A \text{ or } B) = P(A) + P(B) - P(A \text{ and } B)$. If events A and B are disjoint, then $P(A \text{ or } B) = P(A) + P(B)$.

adjacent values   On a modified boxplot, the largest and smallest non-outliers.

alternative hypothesis   The set of values, as compared to that of the null hypothesis, that an investigator believes might contain the plausible values of a population parameter in a test of significance. Sometimes called the research hypothesis.

average   See **mean**.

balanced design   An experimental design in which each treatment is assigned to the same number of units.

bar chart (or bar graph)   A plot that shows frequencies for categorical data as heights or lengths of bars, with one bar for each category.

bias   The difference between the actual value of a parameter being estimated and the average value, over repeated sampling, of an estimator of that parameter.

bias due to sampling   See **sample selection bias**.

bimodal   Describes a distribution with two well-defined peaks.

binomial distribution   A distribution of the random variable X where X represents the number of successes in n independent trials, with the probability of a success the same on each trial.

bins The intervals on the real number line that determine the width of the bars of a histogram.

bivariate data Data that involve two variables per case. For quantitative variables, often displayed on a scatterplot. For categorical variables, often displayed in a two-way table.

blind Describes an experiment in which the subjects do not know which treatment they received.

blocking The process of setting up an experiment by dividing the units into groups (blocks) of similar units and then assigning the treatments at random within each block.

blocks In an experiment, groups of similar units, with treatments randomly assigned within these groups.

box-and-whiskers plot See **boxplot**.

boxplot (or box-and-whiskers plot) A graphical display of the five-number summary. The "box" extends from the lower quartile to the upper quartile, with a line across it at the median. The "whiskers" run from the quartiles to the minimum and maximum.

capture rate The proportion of confidence intervals produced by a particular method that capture the population parameter. See also **confidence level**.

case The subject (or unit) on which a measurement is made.

categorical variable A variable that can be grouped into categories, such as "yes" and "no." Categories sometimes can be ordered, such as "small," "medium," and "large."

census A collection of measurements on all units in the population of interest.

Central Limit Theorem The shape of the sampling distribution of the sample mean becomes more normal as n increases.

chance model See **probability model**.

chi-square test of homogeneity A chi-square test used to determine whether it is reasonable to believe that when several different populations are broken down into the same categories, they have the same proportion of units in each category.

chi-square test of independence A chi-square test of the hypothesis that two categorical variables measured on the same units are independent of each other in the population.

clinical trial A randomized experiment comparing the effects of medical treatments on human subjects.

cluster(s) On a plot, a group of data "clustering" close to the same value, away from other groups. In sampling, non-overlapping and exhaustive groupings of the units in a population.

cluster sampling Selecting a simple random sample of clusters of units (such as classrooms of students) rather than individual units (students)

coefficient of determination The square of the correlation r. Tells the proportion of the total variation in y that can be explained by the relationship with x.

column chart A three-dimensional plot of frequencies taken from a two-way table, which depicts those frequencies as heights of columns.

column marginal frequency The total of all frequencies across row categories for a particular column of a two-way table of frequencies for categorical variables.

comparison group In an experiment, a group that receives one of the treatments, often the standard treatment.

complement In probability, the outcomes in the sample space that lie outside an event of interest.

completely randomized design An experimental design in which treatments are randomly assigned to units without restriction.

conditional distribution of y given x With bivariate data, the distribution of the values of y for a fixed value of x.

conditional probability The notion that a probability can change if you are given additional information. The conditional probability that event A happens given that event B happens is given by $P(A \mid B) = \frac{P(A \text{ and } B)}{P(B)}$ as long as $P(B) > 0$.

conditional relative frequency The joint frequency in a column divided by the marginal frequency for that column, or the joint frequency in a row divided by the marginal frequency for that row.

confidence interval A set of plausible values for a population parameter, any one of which could be used to define a population for which the observed sample statistic would be a reasonably likely outcome.

confidence level The probability that the method used will give a confidence interval that captures the parameter.

confounding variables (or **confounding**) Two variables in an observational study whose effects on the response are impossible to separate.

continuous variable A quantitative variable that can take on any value in an interval of real numbers.

control group In an experiment, a group that provides a standard for comparison to evaluate the effectiveness of a treatment; often given a placebo.

convenience sample A sample in which the units chosen from the population are the units that are easy (convenient) to include, rather than being selected randomly.

correlation   A numerical value between −1 and 1, inclusive, that measures the strength and direction of a linear relationship between two variables.

critical value   The value to which a test statistic is compared in order to decide whether to reject the null hypothesis. Or, the multiplier used in computing the margin of error for a confidence interval.

cumulative percentage plot   See **cumulative relative frequency plot**.

cumulative relative frequency plot (or **cumulative percentage plot**) A plot of ordered pairs in which each value x in the distribution and its cumulative relative frequency, that is, the proportion of all values less than or equal to x, are plotted.

data   A set of numbers or observations with a context and drawn from a real-life sample or population.

data analysis   See **statistics**.

degrees of freedom   The number of freely varying pieces of information on which an estimator is based. For example, when using a sample to estimate the variability in the population, the number of independent deviations from the estimate of center.

dependent events   Events that are not independent.

deviation   The difference from the mean, $x - \bar{x}$, or from some other measure of center.

disjoint events (or **mutually exclusive events**)   Events that cannot occur on the same opportunity. If event A and event B are disjoint, $P(A \text{ and } B) = 0$.

distribution, data   The set of values that a variable takes on in a sample or population, together with how frequently each value occurs.

distribution, probability   The set of values that a random variable takes on, together with a means of determining the probability of each value (or interval of values in the case of a continuous distribution).

dot plot   A graphical display that shows the values of a variable along a number line.

double-blind   Describes an experiment in which neither the subjects nor the researcher making the measurements knows which treatment the subjects received.

event   Any subset of a sample space.

expected value, μ_X or $E(X)$ The mean of the probability distribution for the random variable X.

experimental units   In an experiment, the subjects or objects to which treatments are assigned.

explanatory variable (or **predictor**)   A variable used to predict (or explain) the value of the response variable. Placed on the x-axis in a regression analysis.

exploratory analysis (or **data exploration**)   An investigation to find patterns in data, using tools such as tables, statistical graphics, and summary statistics to display and summarize distributions.

exponential relationship   A relationship between two variables in which the response variable, y, is multiplied by a constant for each unit of increase in the explanatory variable, x. Mathematically, $y = ab^x$ where a and b are constants.

extrapolation   Making a prediction when the value of the explanatory variable, x, falls outside the range of the observed data.

factor   An explanatory variable, usually categorical, in a randomized experiment or an observational study.

first quartile, Q_1   See **lower quartile**.

fitted value   See **predicted value**.

five-number summary   A data summary that lists the minimum and maximum values, the median, and the lower and upper quartiles for a data set.

fixed-level test   A test in which the null hypothesis is rejected or not rejected based on comparison of the test statistic with the critical value for some predetermined level of significance.

frame   See **sampling frame**.

frequency (or **count**)   The number of times a value occurs in a distribution. With categorical data, the number of units that fall into a specific category.

frequency table   A table that gives data values and their frequencies.

Fundamental Principle of Counting   If there are k stages in a process, with n_i possible outcomes for stage i, then the number of possible outcomes for all k stages taken together is $n_1 n_2 n_3 \cdots n_k$.

gap   On a plot, the space that separates clusters of data.

goodness-of-fit test   A chi-square test used to determine whether it is reasonable to assume that a sample came from a population in which, for each category, the proportion of outcomes in the population that fall into that category is equal to some hypothesized proportion.

heteroscedasticity   The tendency of points on a scatterplot to fan out at one end, indicating that the relationship varies in strength.

histogram   A plot of a quantitative variable that groups cases into rectangles or bars. The height of the bar shows the frequency of measurements within the interval (or bin) covered by the bar.

homogeneous populations   Two or more populations that have nearly equal proportions of units in each category of study.

hypothesis test   See **test of significance**.

incorrect response bias   A bias resulting from responses that are systematically wrong, such as from intentional lying, inaccurate measurement devices, faulty memories, or misinterpretation of questions.

independent events   Events A and B for which the probability of event A happening doesn't depend on whether event B happens. Events A and B are independent if and only if $P(A \mid B) = P(A)$ or, equivalently, $P(B \mid A) = P(B)$ or, equivalently, $P(A \text{ and } B) = P(A) \cdot P(B)$.

inference (or inferential statistics) Using results from a random sample to draw conclusions about a population or using results from a randomized experiment to compare treatments.

influential point On a scatterplot, a point that strongly influences the regression equation and correlation. To judge a point's influence, you compare the regression equation and correlation computed first with and then without the point.

interpolation Making a prediction when the value of the explanatory variable, x, falls inside the range of the observed data.

interquartile range, *IQR* A measure of spread equal to the distance between the upper and lower quartiles; $IQR = Q_3 - Q_1$.

joint frequency The frequency within a particular cell of a two-way table of frequencies for categorical variables.

judgment sample A sample selected using the judgment of an expert to choose units that he or she considers representative of a population.

Law of Large Numbers A theorem that guarantees that the proportion of successes in a random sample will converge to the population proportion of successes as the sample size increases. In other words, the difference between a sample proportion and a population proportion must get smaller (except in rare instances) as the sample size gets larger, if the sample is randomly selected from that population.

least squares line See **regression line**.

level One of the values or categories making up a factor.

level of significance, α The maximum P-value for which the null hypothesis will be rejected.

line of averages See **line of means**.

line of means (or **line of averages**) Another term for the regression line, if points form an elliptical cloud. In theory, the population regression line contains the means (expected values) of the conditional distribution of y at each value of x.

linear shape The characteristic of an elliptical cloud of points where the means of the conditional distributions of y given x tend to fall along a line.

lower quartile (or **first quartile, Q_1**) In a distribution, the value that separates the lower quarter of values from the upper three-quarters of values. The median of the lower half of all the values.

lurking variable A variable other than those being plotted that possibly can cause or help explain the behavior of the pattern on a scatterplot. More generally, a variable that is not included in the analysis but, once

identified, could help explain the relationship between the other variables.

margin of error, E Half the length of a confidence interval; $E = $ (critical value) $\cdot$ (standard error).

marginal frequency The total of the joint frequencies across row categories for a given column or across column categories for a given row of a two-way table of frequencies for categorical variables.

marginal relative frequency The marginal frequency of a two-way table of categorical data divided by the total frequency (number of units represented in the table).

matched pairs design See **randomized paired comparison design**.

maximum The largest value in a data set.

mean, $\bar{x}$ A measure of center, often called the average, computed by adding all the values of x and dividing by the number of values, n. On a plot, the place where you would put a pencil point below the horizontal axis in order to balance the distribution.

measure of center A single-number summary that measures the "center" of a distribution; usually the mean (or average). Median, midrange, mode, and trimmed mean are other measures of center.

measure of spread (or **measure of variability**) A single-number summary that measures the variability of a distribution. Range, IQR, standard deviation, and variance are measures of spread.

median A measure of center that is the value that divides an ordered set of values into two equal halves. To find it, you list all the values in order and select the middle one or, if the number of values is even, the average of the two middle ones. If there are n values, the median is at position $(n + 1)/2$. On a plot of a distribution, the median is the value that divides the area between the distribution curve and the x-axis in half.

method of least squares A general approach to fitting functions to data by minimizing the sum of the squared residuals (or errors).

midrange The midpoint between the minimum and maximum values in a data set, or $(max + min)/2$.

minimum The smallest value in a data set.

mode A measure of center that is the value with the highest frequency in a distribution. On a plot of a distribution, it occurs at the highest (maximum) peak.

modified boxplot A graphical display like the basic boxplot except that the whiskers extend only as far as the largest and smallest non-outliers (sometimes called adjacent values) and any outliers appear as individual dots or other symbols.

Multiplication Rule For any two events A and B, $P(A \text{ and } B) = P(A) \cdot P(B \mid A) = P(B) \cdot P(A \mid B)$. If events A and B are independent, then $P(A \text{ and } B) = P(A) \cdot P(B)$.

mutually exclusive events See **disjoint events**.

negative trend The tendency of a cloud of points to slope downward as you go from left to right, or the tendency of the value of y to get smaller as the value of x gets larger.

nonresponse bias A bias that can occur when people selected for the sample do not respond to the survey.

normal distribution A useful probability distribution that has a symmetric bell or mound shape and tails extending infinitely far in both directions.

null hypothesis The standard or status quo value of a parameter that is assumed to be true in a test of significance until possibly refuted by the data in favor of an alternative hypothesis.

observational study A study in which the conditions of interest are already built into the units being studied and are not randomly assigned.

one-sided (one-tailed) test of significance A test in which the P-value is computed from one tail of the sampling distribution. Used when the investigator has an indication of which way any deviation from the standard should go, as reflected in the alternative hypothesis.

outlier A value that stands apart from the bulk of the data.

parameter A summary number describing a population or a probability distribution.

percentile The quantity associated with any specific value in a univariate distribution that gives the percentage of values in the distribution that are equal to or below that specific value. The median is the 50th percentile.

placebo A nontreatment that mimics the treatment(s) being studied in all essential ways except that it does not involve the crucial component.

placebo effect The phenomenon that when people believe they are receiving the special treatment, they tend to do better even if they are receiving the placebo.

plot of distribution (or graphical display or statistical graphic) A graphical display of the distribution of a variable that provides a sense of the distribution's shape, center, and spread.

point estimator A statistic from a sample that provides a single point (number) as a plausible value of a population parameter.

point of averages The point $(\bar{x}, \bar{y})$, where $\bar{x}$ is the mean of the explanatory variable and $\bar{y}$ is the mean of the response variable. This point falls on the regression line.

pooled estimate The weighted average of two statistics estimating the same parameter, with the weights usually determined by the sample sizes or degrees of freedom.

population The entire set of people or things (units) that you want to know about.

population regression line See **true regression line**.

population size The number of units in the population.

population standard deviation, σ See **standard deviation of a population**.

positive trend The tendency of a cloud of points to slope upward as you go from left to right, or the tendency of the value of y to get larger as the value of x gets larger.

power of a test The probability of rejecting the null hypothesis.

power relationship A relationship between two variables in which the response variable, y, is proportional to the explanatory variable, x, raised to a power. Mathematically, $y = ax^b$, where a and b are constants.

predicted (or **fitted**) **value** An estimated value of the response variable calculated from the known value of the explanatory variable, x, often by using a regression equation.

prediction error The difference between the actual value of y and the value of y predicted from a regression line. Usually unknown except for the points used to construct the regression line, whose prediction errors are called residuals.

predictor See **explanatory variable**.

probability A number between 0 and 1, inclusive (or between 0% and 100%), that measures how likely it is for a chance event to happen. At one extreme, events that can't happen have probability 0. At the other extreme, events that are certain to happen have probability 1.

probability density A probability distribution, such as the normal or x^2 distribution, where x is a continuous variable and probabilities are identified as areas under a curve.

probability distribution See **distribution, probability**.

probability model (or **chance model**) A description that approximates—or simulates—the random behavior of a real situation, often by giving a description of all possible outcomes with an assignment of probabilities.

probability sample A sample in which each unit in the population has a known probability of ending up in the sample.

protocol A written statement telling exactly how an experiment is to be designed and conducted.

P-value For a test, the probability of seeing a result from a random sample that is as extreme as or more extreme than the one computed from the random sample, if the null hypothesis is true. (Sometimes called the observed significance level.)

quantitative variable (or **numerical variable**) A variable that takes on numerical values.

quartiles   Three numbers that divide an ordered set of data values into four groups of equal size.

questionnaire bias   Bias that arises from how the interviewer asks and words the survey questions.

random sample   A sample in which individuals are selected by some chance process. Sometimes used synonymously with **simple random sample**.

random variable   A variable that takes on numerical values determined by a chance process.

randomization (or random assignment)   Assigning subjects to different treatment groups using a random procedure.

randomized block design   An experimental design in which similar units are grouped into blocks and treatments are then randomly assigned to units within each block.

randomized comparative experiment   An experiment in which two or more treatments are randomly assigned to experimental units for the purpose of making comparisons among treatments.

randomized paired comparison (matched pairs)   An experimental design in which two different treatments are randomly assigned within pairs of similar units.

randomized paired comparison (repeated measures)   An experimental design in which each treatment is assigned (in random order) to each unit.

range   A measure of spread equal to the difference between the maximum and minimum values in a data set.

rare events   Values or outcomes that lie in the outer 5% of a distribution or in the upper 2.5% and lower 2.5% of a distribution. Compare **reasonably likely**.

reasonably likely   Describes values or outcomes that lie in the middle 95% of a distribution. Compare **rare events**.

recentering   Adding the same number c to all the values in a distribution. This procedure doesn't change the shape or spread but slides the entire distribution by the amount c, adding c to the measures of center.

rectangular distribution   A distribution in which all values occur equally often.

regression   The statistical study of the relationship between two (or more) quantitative variables, such as fitting a line to bivariate data. (Can be extended to categorical variables.)

regression effect (or regression toward the mean)   On a scatterplot, the difference between the regression line and the major axis of the elliptical cloud.

regression line (or least squares line or least squares regression line)   The line for which the sum of squared errors (residuals), SSE, is as small as possible.

regression toward the mean   See **regression effect**.

relative frequency   A proportion computed by dividing a frequency by the number of values in the data set.

relative frequency histogram   A histogram in which the length of each bar shows proportions (or relative frequencies) instead of frequencies.

repeated measures design   See **randomized paired comparison (repeated measures)**.

replication   Repetition of the same treatment on different units.

rescaling   Multiplying all the values in a distribution by the same nonzero number d. This process doesn't change the basic shape but instead stretches or shrinks the distribution, multiplying the IQR and standard deviation by $|d|$ and multiplying the measures of center by d.

residual (or error)   For points used to construct the regression line, the difference between the observed value of y and the predicted value of y, that is, $y - \hat{y}$.

residual plot   A scatterplot of residuals, $y - \hat{y}$, versus predictor values, x, or versus predicted values, $\hat{y}$. A diagnostic plot used to uncover nonlinear trends in a relationship between two variables.

resistant to outliers   Describes a summary statistic that does not change very much when an outlier is removed from the data set.

response variable   The outcome variable used to compare results of different treatments in an experiment or the outcome variable that is predicted by the explanatory variable or variables in regression analysis. Placed on the y-axis in a regression analysis.

robustness   The comparative insensitivity of a statistical procedure to departure from the assumptions on which the procedure is based.

row marginal frequency   The total of all frequencies across column categories for a particular row of a two-way table of frequencies for categorical variables.

sample   The set of units selected for study from the population.

sample selection bias (or sampling bias or bias due to sampling)   The extent to which a sampling procedure produces samples for which the estimate from the sample is larger or smaller, on average, than the population parameter being estimated.

sample space   A complete list or description of disjoint (mutually exclusive) outcomes of a chance process.

sampling bias   See **sample selection bias**.

sampling distribution   The distribution of a sample statistic under some prescribed method of probability sampling.

sampling distribution of a sample proportion, $\hat{p}$ The theoretical distribution of the sample proportion in repeated random sampling.

sampling distribution of the sample mean, $\bar{x}$ The theoretical distribution of the sample mean in repeated random sampling.

sampling frame (or frame) The listing of units from which the sample is actually selected.

sampling with replacement In sequential sampling of units from a population, a procedure in which each sampled unit is placed back into the population before the next unit is selected.

sampling without replacement In sequential sampling of units from a population, a procedure in which each sampled unit is not placed back into the population before the next unit is selected.

scatterplot A plot that shows the relationship between two quantitative variables, usually with each case represented by a dot.

segmented bar graph (or stacked bar graph) A plot in which categorical frequencies are stacked on top of one another.

sensitive to outliers Describes a summary statistic that changes considerably when an outlier is removed from the data set.

shape One of the characteristics, along with center and spread, that is used to describe distributions. Univariate distributions sometimes have a standard shape such as normal, uniform, or skewed. Bivariate distributions may form an elliptical cloud. Descriptions of shape should consider possible outliers, clusters, and gaps.

simple random sample, SRS A sample generated from a sampling procedure in which all possible samples of a given fixed size are equally likely.

simulation A procedure that uses a probability model to imitate a real situation. Often used to compare an actual result with the results that are reasonable to expect from random behavior.

size bias A type of sample selection bias that gives units with a larger value of the variable a higher chance of being selected.

skewed Describes distributions that show bunching at one end and a long tail stretching out in the other direction. Often happens because the values "bump up against a wall" and hit either a minimum that values can't go below or a maximum that values can't go above.

skewed left A skewed distribution with a tail that stretches left, toward the smaller values.

skewed right A skewed distribution with a tail that stretches right, toward the larger values.

slope For linear relationships, the change in y (rise) per unit change in x (run).

split stem A stem-and-leaf plot in which the leaves for each stem are split onto two or more lines. For example, if the second digit is 0, 1, 2, 3, or 4, it is placed on the first line for that stem. If the second digit is 5, 6, 7, 8, or 9, it is placed on the second line for that stem.

spread See **variability**.

stacked bar graph See **segmented bar graph**.

standard deviation of a population, σ A measure of spread equal to the square root of the sum of the squared deviations divided by n. For a probability distribution, it is the square root of the expected squared deviation from the mean.

standard deviation of a sample, s A measure of spread equal to the square root of the sum of the squared deviations divided by $n - 1$.

standard error The standard deviation of a sampling distribution.

standard error of the mean, $\sigma_{\bar{x}}$ The standard deviation of the sampling distribution of $\bar{x}$, or $\sigma/\sqrt{n}$.

standard error of the mean (estimated) The estimated standard deviation of the sampling distribution of $\bar{x}$, or $s/\sqrt{n}$.

standard normal distribution A normal distribution with mean 0 and standard deviation 1. The variable along the horizontal axis is called a z-score.

standard units, z The number of standard deviations a given value lies above or below the mean:

$$z = \frac{value - mean}{standard\ deviation}$$

standardizing Converting to standard units; the two-step process of recentering and rescaling that turns any normal distribution into a standard normal distribution.

statistic A summary number calculated from a sample taken from a population. For example, the sample mean, $\bar{x}$, and standard deviation, s, are statistics.

statistically significant Describes the situation when the difference between the estimate from the sample and the hypothesized parameter is too big to reasonably be attributed to chance variation.

statistics (or data analysis) The study of the production, summarization, and analysis of data, along with the processes for drawing conclusions from the data.

stem-and-leaf plot (or stemplot) A graphical display with "stems" showing the leftmost digit of the values separated from "leaves" showing the next digit or set of digits.

stemplot See **stem-and-leaf plot**.

strata (singular, stratum) Subgroups of the population, usually selected for homogeneity or sampling convenience, that cover the entire population. See also **stratified random sampling**.

stratification A classification of the units in a population into homogeneous subgroups, known as strata, prior to sampling.

stratified random sampling Stratifying the population and then taking a simple random sample from within each stratum.

strength In the context of regression analysis, two variables are said to have a strong relationship if there is little variation around the regression line. If there is a lot of variation around the regression line, the relationship is weak.

sum of squared errors, SSE The sum of the squared residuals: $\sum(y - \hat{y})^2$.

summary statistic See **statistic**.

systematic sampling with random start A sample selected by taking every nth member of the population, starting at a random spot—for example, having people count off and then picking one of the numbers at random.

table of random digits A string of digits constructed in such a way that each digit, 0 through 9, has probability $\frac{1}{10}$ of being selected and each digit is selected independently of the previous digits.

t-distribution The distribution, for example, of the statistic below, when the data are a random sample from a normally distributed population:

$$t = \frac{\bar{x} - \mu}{s\sqrt{n}}$$

test of significance (or **hypothesis test**) A procedure that compares the results from a sample to some predetermined standard in order to decide whether the standard should be rejected.

test statistic Typically, in significance testing, the distance between the estimate from the sample and the hypothesized parameter, measured in standard errors.

third quartile, Q_3 See **upper quartile**.

treatment group In an experiment, a group that receives an actual treatment being studied. Compare with **control group**.

treatments Conditions assigned to different groups of subjects to determine whether subjects respond differently to different conditions.

tree diagram A diagram used to calculate probabilities for sequential events.

trend On a scatterplot, the path of the means of the vertical strips (conditional distributions of y given x) as you move from left to right. More simply, the path taken by a line through the center of the data in a scatterplot.

true regression line (or **population regression line**) The regression line that would be computed if you had the entire population. Theoretically, the line through the means of the conditional distributions of y given x. See also **line of means**.

t-test A test of significance of a population mean (or comparison of means) using the t-distribution. See also **test of significance** and **t-distribution**.

two-sided (two-tailed) test of significance A test in which the P-value is computed from both tails of the

sampling distribution. Used if the investigator is interested in detecting a change from the standard in either direction.

two-stage sampling A sampling procedure that involves two steps. For example, taking a random sample of clusters and then taking a random sample from each of those clusters.

two-way table A table of frequencies that lists outcomes in the cells formed by the cross-classification of two categorical variables measured on the same units.

Type I error The error made when the null hypothesis is true and you reject it.

Type II error The error made when the null hypothesis is false and you fail to reject it.

unbiased Describes an estimator (statistic) that has an average value in repeated sampling (expected value) equal to the parameter it is estimating.

uniform distribution A distribution whose frequencies are constant across the possible values. Its plot is rectangular.

unimodal Describes a distribution of univariate data with only one well-defined peak.

units Individuals that make up the population from which samples may be selected or to which treatments may be applied.

univariate data Data that involve a single variable per case. A quantitative variable often is displayed on a histogram. A categorical variable often is displayed on a bar chart.

upper quartile (or **third quartile, Q_3**) In a distribution, the value that separates the lower three-quarters of values from the upper quarter of values. The median of the upper half of all the values.

variability (or **spread**) The degree to which values in a distribution differ. Measures of variability for quantitative variables include the standard deviation, variance, interquartile range, and range.

variability due to sampling (or variation in sampling) A description of how an estimate varies from sample to sample.

variable A characteristic that differs from case to case and defines what is to be measured or classified.

variance A measure of spread equal to the square of the standard deviation.

voluntary response bias The situation in which statistics from samples are not fair estimates of population parameters because the sample data came from volunteers rather than from randomly selected respondents.

voluntary response sample A sample made up of people who volunteer to be in it.

z-score See **standard normal distribution** and **standard units**.

Brief Answers to Practice and Odd-Numbered Exercises

The answers below are not complete solutions but are meant to help you judge whether you are on the right track. If you round computations in intermediate steps or use tables rather than a calculator, your numerical answers might not exactly match those given.

Chapter 1

Section 1.1
P1. Older hourly workers were far more likely to be laid off in Rounds 1–3 than were younger hourly workers.

P2. B

P3. hourly, although the patterns are similar

E1. a. $\frac{3}{6}$; $\frac{3}{6}$ b. $\frac{3}{10}$; $\frac{7}{10}$
- c. A higher proportion of those age 50 and older were laid off than those under age 50 (0.875 versus 0.50).
- d. Hourly workers. The difference in the proportions for salaried workers (0.60 versus 0.375) is smaller than in part c. But note the small number of hourly workers.

E2. a. $\frac{14}{27}$; $\frac{14}{18}$ b. $\frac{4}{9}$; $\frac{5}{9}$
- c. A higher proportion of those age 40 and older were laid off than those under age 40 (0.52 versus 0.44).
- d. age 50, because the difference in proportions is greater

E3. a. The hourly workers who kept their jobs tended to be younger than the salaried workers who kept their jobs.
- b. Not unless you can compare them with dot plots of the ages of the hourly and salaried workers before layoffs began.

E5. b. The percentage of those laid off who were age 40 or older, by round, were 82%, 89%, 67%, 50%, and 0%. Most layoffs came early, and older workers were hit harder in earlier rounds.

E7. a. 0.293, not 293
- b. $-8\frac{5}{8}$; -6.30; 170; $47\frac{1}{2}$; -15.79; -12; -6.45

Section 1.2
P4. a. about 37 out of 200, or 0.185

- b. An average ages this high would be fairly easy to get just by chance, so there is no evidence of age discrimination

P5. a. 48.6
- b. Write 14 ages on cards. Draw 10 at random and find the average age. Repeat many times and find where 48.6 falls in the distribution.
- c. about 45 out of 200, or 0.225   d. no

E9. b. 24 out of 50, or 0.48
- c. The proportion can reasonably be attributed to chance.

E11. C

E13. a. 4   b. $\frac{4}{45}$ c. no

Chapter Summary
E15. b. There is no reason to look for an explanation, because there isn't much difference in the two distributions.

E17. B

E19. a. II   d. 13%; no

E21. a. 1001   b. 5, 6, 7, 8, or 9
- c. i. 360   ii. 90   iii. 5   d. $\frac{455}{1001}$

E23. a. $\frac{7}{1000}$ c. yes   d. $\frac{6}{1225}$

Chapter 2

Section 2.1
P1. a. collected by a statistics class; a penny; age of the penny
- b. The shape is strongly skewed right, with a wall at 0. The median is about 8 years, and the spread is quite large, with the middle 50% of ages falling between 3 and 15 years; however, it is not unusual to see a penny that is more than 30 years old.
- c. The skew results because about the same number of pennies is produced each year and a penny has about the same chance of going out of circulation each year.

P2. a. w: skewed to the left; x: mound-shaped or approximately normal; y: two mounds, possibly representing two groups of objects
- b. x

c. mean around 52 and standard deviation around 5

P3. **a.** A typical SAT math score is roughly 500, give or take about 100 or so.

b. A typical ACT score is about 20, give or take 5 or so.

c. A typical college-age woman is about 65 in. tall, give or take 2.5 in. or so.

d. A typical baseball player had a batting average of about .260 or .270, give or take about .040 or so.

P4. median = 19; lower quartile = 17; upper quartile = 25; distribution is skewed toward the higher ages

P5. The distribution is skewed right, with no obvious gaps or clusters and a wall at 0. The elephant is the only possible outlier. About half the mammals have gestation periods of more than 160 days, and half less. The middle 50% have gestation periods between 63 and 284 days. Large mammals have longer gestation periods.

P6. **a.** Approximately normal with mean around 46 and standard deviation around 6

c. .25 to .30

P7. **a.** about 0.15

b. about 33, because $223(0.15) \approx 33$

c. skewed left , with median between 70 and 75 and the middle 50% between about 65 and 75–80

P8. **b.** The average longevity distribution is skewed right, with two possible outliers, while the distribution of maximum longevity is more uniform but has a peak at 20–30 years and a possible outlier. The center and spread of the distribution of maximum longevity are larger.

c. maxima are larger than averages by definition; maxima tend to spread out more because of the possibility of extremely large values, not constrained by averaging

P9. See P8 for a discussion of shapes. For the average longevity, width 4 works well; for the maximum longevity, width 8 works well because of the wider spread of the data.

P10. The number of deaths per month is fairly uniform, with about 190,000–220,000 per month. Summer months have the smallest numbers of deaths, and winter months the largest.

P11. Students are cases; age, hair color, number of siblings, gender and miles are variables; age, number of siblings, and miles are quantitative; hair color and gender are categorical.

P12. The distribution of digits is approximately uniform over the range 0 to 9.

P13. **a.** The cases are the individual males in the labor force age 25 and older. The variable is their educational attainment.

b. The proportion increases through the first three levels with a huge jump at the high school graduation level. Then it decreases and is especially low at the associate degree level.

c. The distributions for males and females are similar in shape.

d. Relative frequency bar charts account for the different numbers of males and females.

E1. I. D II. A III. C IV. B

E3. **a.** each of the approximately 92 officers; the age at which the officer became a colonel

b. This distribution is skewed left, with no outliers, gaps, or clusters. The middle 50% of the ages are between 50 and 53, with half above 52 and half below.

c. mandatory retirement; age discrimination; an "up or out" rule according to which if you haven't been promoted beyond colonel by your 55th birthday you must retire

E5. **a.** skewed left **b.** skewed right

c. approximately normal **d.** skewed right

E9. **b.** about 71 in.; about 3 in.

c. about 0.92

d. about 0.17

e. The distribution isn't smooth and ranges only from about 63 to 79 in.

E11. **a.** The western states tend to be large, the eastern states small.

b. Yes; note that western states spread out more.

E13. **a.** The distribution has wide spread, with two clusters (low GDP and medium GDP) and one extremely large value.

b. Norway and Switzerland; only Norway appears to be an outlier.

c. The higher GDPs belong to Western Europe and North America; the lower GDPs belong to Eastern Europe and Asia.

d. No, countries in the same region tend to have similar economies, producing a clustering effect.

E15. **a.** 1

b. 0.5, 1, and 1.5

c. 0.5 and 1.5

d. 15%

e. 0.05 and 1.95

E17. **c.** Both distributions are slightly skewed right, with possible outliers on the high end. The median of both distributions is 12, but the spread of the distribution of values is larger for wild mammals.

E19. Most layoffs occurred in rounds 1 and 2.

E21. a. the number of nonpredators that fall into the categories Domesticated and Wild, and the total number of nonpredators

b. wild, because the second bar is taller than the first

c. predator, because the second bar is a larger fraction of the third bar for predators than it is for nonpredators

E23. a. 12

b. two-tenths of 18, or 3.6

c. The speeds for the wild mammals must be estimates.

Section 2.2

P14. a. 2.5; 2.5 b. 3; 3 c. 3.5; 3.5 d. 49.5; 49.5 e. 50; 50

P15. about 4 ft 4 in.; about 4 ft

P16. a. Africa: 55; Europe: 81

b. For Africa, the median is smaller than the mean because of the right skew. For Europe, the mean is slightly lower than the median because of the left skew.

P17. a. 2 and 5; 3 b. 2 and 6; 4

P18. a. predators: 12, 9.5 and 15.5; nonpredators: 12, 8, and 15

b. The distributions are centered at exactly the same place and have about the same spread, but the distribution for nonpredators has two outliers on the high side. Both are essentially mound-shaped.

P19. Both distributions are somewhat symmetric, except for the outlier in Europe. Percentages for Africa center at about 38, with quartiles around 30 and 56. Percentages for Europe center around 68, with lower quartile about even with Africa's upper quartile and upper quartile around 73. Europe has much higher percentages with less spread in its distribution.

P20. a. The median coincides with one of the quartiles. More variability in speeds of wild mammals.

P21. a. min: 1; $Q1$: 8; median: 12; $Q3$: 15; max: 41

b. 7

c. −2.5; no outliers on the low end

d. Outliers are elephant at 35 years and hippopotamus at 41 years; grizzly bear at 25 years is the largest non-outlier.

P22. The *SD* is 3.21.

P23. a. i b. iii c. iv d. vii e. ii f. v g. vi

P24. b. 1967: mean = 3.63, *SD* = 1.6

2007: mean = 3.09, *SD* = 1.3; family sizes got smaller and less variable

P25. a. The median is 3 for each year.

b. 1967: $Q_1=2, Q_2=3, Q_3=5, IQR = 3$
2007: $Q_1=2, Q_2=3, Q_3=4, IQR = 2$

d. Family size has decreased slightly over the 40 year period from 1967 to 2007. The median size is 3 for both years, but the mean size has decreased from 3.63 to 3.09. In addition, there was less variability among family sizes in 2007 than there was in 1967.

E25. 10

E27. b. The speeds of wild mammals tend to include some faster speeds than the speeds of domesticated mammals because some wild animals depend on speed for food and survival.

E29. a. A – mound-shaped symmetric; B – skewed right; C – skewed left; D - uniform

b. A – III; B- IV; C – II; D –I

E31. b-c. Distributions for all regions are skewed right, with at least one outlying country in each (United States, Canada and Uruguay; Kuwait and United Arab Emirates; Denmark). Central Asia and the Americas center around 2.5, with Europe centering around 5. Central Asia has the largest spread, followed closely by the Americas.

d. About 68%

E33. No; the minimum score on the second plot is lower than the minimum on the combined plot (and other similar reasons).

E35. a. II; III   b. II and III

E37. heights of basketball players; heights of all athletes

E39. a. 39.29; 30.425; yes, because it is computed using only the maximum and minimum

b. 10.54

E41. a. 3.11 g   b. 0.043 g   c. yes

E43. a. 72.9

b. Finding the median of the combined groups requires having the ordered values.

E45. $\sum(x - \bar{x}) = \sum x - \sum \overline{x} = \sum x - n\bar{x} = \sum x - \sum x = 0$

Section 2.3

P26. a. Values tend to be skewed right, because some houses cost a lot more than most houses in a community while very few houses cost a lot less.

b. $43,964,124.05

c. $4,508.68

P27. a. The distribution of car ages is strongly skewed right, so the mean may seem unduly large.

b. vehicles proving more durable; people unwilling or unable to buy new cars

P28. a. 4 ft ; 3.75 ft ; 0.2 ft ; 0.25 ft

b. 50 in.; 47 in.; 2.4 in.; 3 in.

c. 4 1/3 ft ; 4 1/12ft ; 0.2 ft ; 0.25 ft

P29. a. 1; b. 10; c. 5; d 100

P30. a. outliers occur above $-30 + 1.5(21)$, or 1.5, so Hawaii, at 12, is an outlier.

b. count 49; mean -41.5; median about -40; SD about 16; min -80; max between -5 and 0; range between 75 and 80; Q_1 about -51; Q_3 about -30

P31. a. 18

b. middle 90% between about 325 and 730; middle 95% between about 275 and 750

c. about 513

P32. 90%; percentiles 2.5 and 97.5

P33. $Q_1 \approx 450$, $Q_3 \approx 640$, median ≈ 550, $IQR \approx 229$

E47. a. It depends on the purpose of computing a measure of center. Real estate agents usually report the median because it is lower and tells people that half the houses cost more and half less. The tax collector wants the mean price because the mean times the tax rate times the number of houses gives the total amount of taxes collected.

b. If the reason is to find the total crop in Iowa, use the mean. An individual farmer might use the median to see whether his or her yield was typical.

c. Survival times usually are strongly skewed right. Telling a patient only the mean survival time would give perhaps too optimistic a picture. The smaller median would inform the person that half the people survive for a longer time and half shorter. On the other hand, the mean would help a physician estimate the total number of hours he or she will spend caring for patients with this disease.

E49. a. The temperatures at the tick marks go from 36.67 to 58.89, but the shape remains the same.

b.

Variable	N	Mean	Median	StDev
HighTemp	50	45.61	45.56	3.72

Variable	Min	Max	Q1	Q3
HighTemp	37.78	56.67	43.33	47.78

c. yes

E51. Show that the mean of $(x_1 + c) + (x_2 + c) (x_3 + c) + (x_4 + c) + (x_5 + c)$ is equal to the original mean, $\bar{x}$, plus c.

E53. a. 23¢ or 24¢

b. about 10 and about 70; 60

c. skewed right

E55. about 32 mi/h or, if rounded down, 30 mi/h

E57. A

Section 2.4
P34. a. 1.29% b. 4.75% c. 34.46% d. 78.81%

P35. a. -0.47 b. -0.23 c. 1.13 d. 1.555

P36. a. 85.58% b. 99.74% (calculator: 0.9973)

P37. a. -1.645 to 1.645 b. -1.96 to 1.96

P38. a. cancer, because it is 1.43 SDs below the mean, compared to 0.85 SD for heart disease

b. cancer, because it is 1.13 SDs above the mean, compared to 0.89 SDs for heart disease

c. The death rate for heart disease in Colorado (1.83 SDs below the mean) is more extreme than the death rate for cancer in Georgia (1.20 SDs below the mean).

P39. a. about 29.7% or 29.8% (using the table)

b. about 64.1 in.

P34. a. about 143 to 295 b. about 173 to 265

P41. a. outside both b. not outside either

c. not outside either d. not outside either

E59. a. 84.13%; 99.43%

b. 15.87%; 0.57%

c. 93.32%

d. 68.27% (68.26% using Table A)

E61. a. 2

b. 1

c. 1.5

d. 3

e. -1

f. -2.5

E63. a. i. 0.6340 (calculator: 0.6319)

ii. 0.0392 (calculator: 0.0395)

iii. 0.3085 (calculator: 0.3101)

b. about 287 to 723

E65. 70%; 88%

E67. a. 0.1587

b. 8.16 (calculator: 8.17)

c. 10.02

d. 8 (or 7.89)

E69. 68%; 95%; 16%; 84%; 2.5%; 97.5%

E71. a. 0.2119

b. 19,000,000(0.0827) = 1,571,300 (calculator: 1,582,700)

c. 74.24 in.

E73. a. skewed right

b. 0.1151 (calculator: 0.1148)

c. If all values in the distribution must be positive and if two standard deviations below the mean is less than 0, the distribution isn't approximately normal.

Chapter Summary

E75. b. min: 0; Q_1: 2; median: 10; Q_3: 35; max: 232

c. Florida and Texas

e. Both show the strong skewness and outliers. In the stemplot, you can see that half the states have fewer than ten tornadoes. The cleanness of the boxplot makes it clear how much of an outlier Texas actually is. However, you can't see from the boxplot that many states have at most one tornado. Because the stemplot is easy to read while showing the values, it is reasonable to select it as the more informative plot.

f. The distribution is strongly skewed right, with two outliers and a wall at 0. The median number of tornadoes is 10, with the middle 50% of states having between 2 and 35 tornadoes.

E77. a. scores below 457.5 or above 797.5

b. probably skewed right

E79. a. Region 1 is Africa; Region 2 is the Middle East; Region 3 is Europe.

b. Region 1 is C; Region 2 is B; Region 3 is A.

E81. No; for example, values {2, 2, 4, 6, 8, 8} are symmetric with mean and median 5. Only two of the six values, or about 33%, are within one *SD*, 2.76, of the mean, 5.

E83. a. Half the cities have fewer than 13 pedestrian deaths per year, and half have more.

b. The outliers—New York, Los Angeles, and Phoenix—are among the most populous cities in the United States.

c. A stemplot reveals the three outliers and the right skew and retains the original values.

d. New York 1.91; Miami 6.68. The rate is adjusted for population size and gives a more accurate picture of pedestrian safety.

E85. 2.98; 1.33

E87. An example is {1, 1, 1, 1, 1, 2, 2, 10}.

E89. a. i. 6.325 vs. 6.667

ii. 2.000 vs. 2.010

iii. 0.632 vs. 0.633

b. No; the ratio s/α is close to 1 for large n.

E91. a. The state with the lowest per capita income in 1980 had an average income per person of $6,573.

b. There is at least one outlier on the high end for both 2000 and 2007, but none for 1980. There are no outliers on the low end for any of the years.

c. There was a noticeable improvement in position as the z-scores went from −1.30 in 1980 to −0.86 in 2007.

d. As the years progress, the histograms become more skewed toward the higher values and both the centers and spreads increase.

e. The z-score would work best for the 1980 data, as it is more mound-shaped and symmetric then the others.

E93. 126.65 mg/dL and 225.35 mg/dL.

E95. a. about .260; about .040.

b. Both distributions are approximately normal in shape. The distribution for the American League has a higher mean (by about .010) and less spread.

c. about .300

Index

Photo Credits

Chapter 1
1: iStockphoto. **4:** iStockphoto. **8:** ©AP/Wide World Photos. **18:** ©Stu Porter /Alamy. **20:** Masterfile.

Chapter 2
21: Kevin Steele/Getty Images, Inc. **24:** ©Cheryl Fenton Photography. **25:** Photo from *The Journal of Heredity*. **28:** FoxTrot ©2008 Bill Amend/Reprinted with permission of UNIVERSAL PRESS SYNDICATE. All rights reserved. **28:** iStockphoto. **33:** Robert van der Hilst /Getty Images, Inc. **42** (left): David McNew/Staff/Getty Images, Inc. **42** (right): Paul Gilham/Staff/Getty Images, Inc. **51:** iStockphoto. **54:** ©Cheryl Fenton Photography. **69:** ©Stockbyte/Alamy. **74:** iStockphoto. **80:** Image Source/Getty Images, Inc. **87:** ©Scott T. Smith/©Corbis.

Chapter 3
93: Comstock/Getty Images, Inc. **98:** Ian Wilson/iStockphoto. **109:** ©Mary Watkins. **101:** Michael Krinke/iStockphoto. **114:** Dane Wirtzfeld/iStockphoto. **120:** Mel Yates/Alamy. **124:** Sheila Terry/Photo Researchers, Inc. **137:** Agencyby/iStockphoto. **138:** Bob Jacobson/Alamy. **144:** Petr Podzemny/iStockphoto. **148:** ©Laura Murray. **156:** Universal/The Kobal Collection, Ltd. **161:** Science Source/Photo Researchers, Inc. **168:** David Madison/Getty Images Inc.

Chapter 4
169: Heidi and Hans-Jürgen Koch /Minden Pictures /NG Image Collection. **177:** ©Robert Harding Picture Library, Ltd./Alamy. **180:** John McIntire/Alamy. **193:** iStockphoto. **197:** Fritz Goro/Time & Life Pictures/Getty Images, Inc. **200:** ©Scott Adams/ Reprinted with permission of Universal Press Syndicate. All rights reserved. **207:** ©Matt Perry. **208:** iStockphoto. **209:** James Kay/ Alamy. **213:** iStockphoto. **217:** ©Cheryl Fenton Photography. **219** (left): PhotoDisc, Inc./Getty Images, Inc. **219** (right): Kenneth H. Thomas/Photo Researchers, Inc. **225:** iStockphoto.

Chapter 5
226: ©Ralph White/©Corbis. **228:** iStockphoto. **232:** ©Cheryl Fenton Photography. **237** (left): ©StockShot/Alamy. **237** (right): ©Cheryl Fenton Photography. **241:** © Ilene MacDonald/Alamy. **245:** ©Corbis/SuperStock. **247:** Courtesy of the Library of Congress. **256:** ©Image Asset Management Ltd./SuperStock. **256** (top): Jose Luis Pelaez, Inc./Getty Images, Inc. **256** (bottom): ©Image Asset Management, Ltd./SuperStock. **265:** Anton Robert/Getty Images, Inc. **269:** PlusTwentySeven/Getty Images Inc. **271:** Courtesy of the National Human Genome Research Institute.

Chapter 6
273: Tek Image/Photo Researchers, Inc. **284** (top): ©James Andrew/Alamy. **284** (bottom): ©Cheryl Fenton Photography. **285:** iStockphoto. **295:** Courtesy of Scientific Games, International. **297:** ©C.C. Lockwood/Animals Animals/Earth Scenes. **299:** iStockphoto. **302:** ©Cheryl Fenton Photography. **303:** iStockphoto. **305:** Photo reprinted with permission from *Proceedings of the National Academy of Sciences*. **306:** iStockphoto. **307:** Dean Krakel II/Photo Researchers, Inc. **308:** iStockphoto. **309:** iStockphoto.

Chapter 7
312: Joseph Van Os/Stone/Getty Images, Inc. **316:** iStockphoto. **319:** M81 Spiral Galaxy/NASA. **321:** Courtesy of NASA. **324:** iStockphoto. **336:** MLB Photos/Getty Images, Inc. **343:** Jose Luis Pelaez, Inc./©Blend Images/Alamy. **348:** Saul Loeb/Getty Images, Inc. **349:** © Ian Cruickshank /Alamy. **353:** Barry Lewis/Alamy. **355:** iStockphoto.

Chapter 8
359: Aldo Murillo/iStockphoto. **363:** ©José Elias - Abstract series/Alamy. **374:** Alexandra Milgram. **381** (bottom): iStockphoto. **384:** ©Scott Adams/Reprinted by permission of United Features Syndicate, Inc. **385:** ©Cheryl Fenton Photography. **389:** Victoria Horner and Andrew Whiten, *Animal Cognition*. **396:** iStockphoto. **400:** iStockphoto. **404:** iStockphoto. **416:** Illustration from *Journal of the American Medical Association*.

Chapter 9
420: Tony Anderson/Getty Images, Inc. **429** (left): Hulton Archive/Stringer/Getty Images, Inc. **429** (right): ©Corbis Premium RF/ Alamy. **439:** ©Cheryl Fenton Photography. **440:** ©Francesco Ridolfi/Alamy. **443:** Paul Grebliunas/Getty Images, Inc. **452:** Photo by James Moulin/Courtesy of Professor David Strayer. **456** (left): Joe McNally/Getty Images, Inc. **456** (right): Time & Life Pictures/Getty Images, Inc. **458** (top): Ronald C. Modra/Getty Images, Inc. **458** (bottom): iStockphoto.

Chapter 10

465: Age fotostock/SuperStock. **467:** Stockbyte/SuperStock. **477:** Willard Clay/Photolibrary Group Limited. **482:** ©Photodisc/ Alamy. **490:** FreezeFrameStudio/iStockphoto. **493:** ©Mary Watkins. **494:** Daniel H. Bailey/Alamy. **497:** Jim West/Alamy. **498:** Photodisc/SuperStock. **500:** Mark Renders/Stringer/Getty Images, Inc. **503:** Jim West/Alamy. **504:** Robert Michael/Alamy.

Chapter 11

509: ©Paul Glendell/Alamy. **515:** iStockphoto. **518:** ©David Grossman/Alamy. **520:** Martin Dohrn/Photo Researchers, Inc. **521:** NewsCom. **527:** iStockphoto. **535** (left): Stockbyte /Getty Images, Inc. **535** (right): ©age fotostock/SuperStock. **538:** Glow Images/Getty Images, Inc. **544:** ©Cheryl Fenton Photography. **551:** ©Cheryl Fenton Photography. **554:** iStockphoto. **558:** National Geographic/Getty Images, Inc. **559:** iStockphoto.

Chapter 12

568: AFP/Getty Images, Inc. **572:** ©Andres Liivamagi/Alamy. **584:** Copyright John Wiley & Sons, Inc. **585:** ©Ilene MacDonald/ Alamy. **597** (left): ©Cheryl Fenton Photography **597** (right): iStockphoto. **599:** Photo used with permission from Archives of Pediatrics & Adolescent Medicine. **600:** iStockphoto. **602:** iStockphoto. **606:** ©Steve Vidler/SuperStock. **608:** Pando Hall/Getty Images, Inc. **614:** ©Diana Ninov/Alamy. **616:** ©SCPhotos/Alamy. **617:** iStockphoto.

Chapter 13

623: Courtesy of NASA. **624:** Courtesy of NASA. **629:** iStockphoto. **635:** Courtesy of the National Park Service. **636:** iStockphoto. **639:** iStockphoto. **643:** ©Peter Menzel/menzelphoto.com. **652:** Courtesy of NASA. **655:** iStockphoto. **657:** Nigel Cattlin/Photo Researchers, Inc. **659:** ©Mary Watkins. **660:** Courtesy of the Library of Congress.

Chapter 14

665: Blend Images/SuperStock. **673:** iStockphoto. **674:** Courtesy of Smithsonian Institution Libraries, Washington D.C., **676:** A. Barrington Brown/Photo Researchers, Inc. **692** (left): Getty Images, Inc. **692** (right): iStockphoto. **696:** Omikron/Photo Researchers, Inc.

Chapter 15

701: Getty Images, Inc. **702:** Getty Images, Inc. **708:** iStockphoto. **713:** Gary Hincks/Photo Researchers, Inc. **719:** ©The London Art Archive/Alamy. **725:** iStockphoto. **729:** iStockphoto. **735:** Photo reprinted with permission from *Current Biology*. **738:** iStockphoto. **739:** Altrendo Images/Getty Images, Inc.

Chapter 16

744: iStockphoto. **751:** B.A.E. Inc./Alamy.